STOCHASTIC MODELS IN OPERATIONS RESEARCH

Volume II

Stochastic Optimization

STOCHASTIC MODELS IN OPERATIONS RESEARCH

Volume II

Stochastic Optimization

Daniel P. Heyman
Lincroft, New Jersey

Matthew J. Sobel
Weatherhead School of Management
Case School of Engineering
Case Western Reserve University

DOVER PUBLICATIONS, INC.
Mineola, New York

Copyright

Bibliographical Note

This Dover edition, first published in 2004, is an unabridged republication of the work originally published in the "McGraw-Hill Series in Quantitative Methods for Management" by McGraw-Hill Book Company, New York, in 1984. This Dover edition has been published by special arrangement with Lucent Technologies, Inc., 600 Mountain Avenue, Murray Hill, N.J. 07974.

There is a list of corrections to *Stochastic Models in Operations Research, Volumes I and II* at http://weatherhead.cwru.edu/orom/sobel/index.htm. This website also has information that instructors will find useful such as solutions to many problems and comments on each chapter.

International Standard Book Number: 0-486-43260-2

Manufactured in the United States of America
Dover Publications, Inc., 31 East 2nd Street, Mineola, N.Y. 11501

To our wives
Anita and Susan

CONTENTS

* Starred sections can be skipped without interfering with the understanding of subsequent unstarred sections.

PREFACE

This volume covers the optimization of stochastic models in which time plays an essential role. The emphasis is on Markov decision processes (also known as probabilistic dynamic programming) and specially structured models. The models are prompted by many phenomena, such as inventories, cash balances, fisheries, advertising, maintenance, capacity expansion, congestion, and reservoirs. The book is intended primarily as a text at the graduate school level for students in operations research, management science, computer science, all branches of engineering, applied mathematics, statistics, business, public administration, and economics. The comprehensive coverage should make the book useful for practitioners, teachers, and researchers who want an up-to-date review of stochastic optimization and its applications. Volume I is devoted to stochastic processes and operating characteristics of stochastic systems.

The principal objectives of this volume are:

1. To give the central facts and ideas of stochastic optimization and show how they are used in various models which arise in applied and theoretical investigations
2. To demonstrate the interdependence of three areas of study that usually appear separately: stochastic processes, operating characteristics of stochastic systems, and stochastic optimization
3. To show the importance of structured models by formulating and analyzing models in many contexts
4. To provide a comprehensive treatment that emphasizes the practical importance, intellectual stimulation, and mathematical elegance of stochastic models

ORGANIZATION AND COVERAGE

This volume is organized on the basis of generic properties of many models rather than on the types of applications which prompt models. That choice fosters the intellectual unity of the three areas that have usually been treated disparately. However, we frequently employ inventory and production models to illustrate general principles. Such models figure prominently in applications of operations research and in the development of stochastic optimization. Stocks and flows of stocks frequently occur in models of other phenomena, so Chapter 2 begins with the elements of inventory and production models. The remainder of the chapter justifies the use of expected present value to make trade-offs in probabilistic dynamic (i.e., time-varying) models.

Dynamic models are usually more difficult to optimize than static models. However, some dynamic models can be optimized by solving static models. Chapter 3 presents models with this simplifying feature which often yields an optimal policy that depends on only a single parameter.

Chapters 4, 5, and 6 concern Markov decision processes. The basic properties of the discrete-time model are contained in Chapter 4, generalizations to continuous-time models make up Chapter 5, and computational details are explored in Chapter 6.

Chapters 7 and 8 examine classes of models with specially structured optimal policies. The relatively few parameters which characterize optimal policies in these chapters facilitate computation and implementation. Production and inventory models arise frequently in these chapters.

Models with two or more decision makers have long been used in the social sciences and in military applications of operations research. If such models are dynamic, then they are sequential games. Sequential game models have recently been advocated for other kinds of applications, and we present some of the basic properties of sequential games in Chapter 9. The sequential game model and its analysis and properties are natural generalizations of Chapters 3 through 6.

Two appendixes are included so that prerequisite facts can be looked up as needed. Appendix A reviews probability theory, the exponential distribution, and mathematical analysis. Appendix B reviews convex sets and convex functions.

As mentioned, Chapters 4, 5, and 6 cover Markov decision processes, and Chapters 7 and 8 cover structured stochastic models. This grouping achieves intellectual unity at the expense of evenness in the level of difficulty. The unstarred sections in Chapters 7 and 9 are easier to read than most of Chapters 5, 6, and 8.

MATHEMATICAL PREREQUISITES

The reader ought to have the following mathematical prerequisites:

1. Elementary matrix algebra (among unstarred sections, needed only for Section 4-6)
2. An elementary knowledge of linear programming (among unstarred sections, needed only for Sections 4-5 and 4-6)

3. A first course in probability theory with a calculus prerequisite
4. Mathematical maturity

"Mathematical maturity" is a state of grace which is usually attained through course work in mathematics or rigorous applied mathematics. Most students will not have completed items 1 through 4 before the senior year of their undergraduate programs. We have found that students can satisfy items 1 and 2 at the same time as beginning this volume.

It is unnecessary to study stochastic processes, such as Part A of Volume I, before reading unstarred sections of this volume. However, readers who have done so will appreciate some nuances they would otherwise miss. The sole exception to this general rule is Section 4-6, which depends on elementary properties of Markov chains.

COURSE OUTLINES

We have designed Volume II for several courses:

1. Stochastic optimization: one quarter, one semester, or two quarters
2. Stochastic dynamic programming (Markov decision processes): one quarter, one semester, or two quarters
3. Inventory and other structured stochastic models: one quarter or one semester
4. Advanced Markov decision processes: one quarter or one semester

For the first option, we find that in one semester we can review probability theory and cover part of Chapter 2, most of Chapters 3 and 4, and part of Chapter 7. In a one-quarter course, we omit most of Chapter 2 and all of Chapter 7.

For the second option, we cover most of Chapter 4 and parts of Chapters 5 and 6 in one semester. This presentation assumes prerequisite introductory knowledge of Markov chains (for Section 4-6). If the prerequisite material must be incorporated in the course, then there is hardly any time for Chapters 5 and 6.

For the third option, in one semester we cover all of Chapters 3 and 7 and sample parts of Chapters 8 and 9 according to the background and interests of the instructor and students.

For the fourth option, we select topics from Chapters 5, 6, 8, and 9 according to the preferences of the instructor and students.

ADVICE TO INSTRUCTORS

We find it useful to review basic probability (Section A-1) at the start of a first course in stochastic optimization. A review of the exponential distribution (Section A-2) usually is needed before covering Chapter 5. Facts from mathematical analysis (Section A-3) should be covered as they come up. A concentrated dose at the start will give the wrong idea of the thrust of the course. A *selective* review of convex sets and convex functions (sampling from Appendix B) usually is needed before covering Chapter 8.

When we use the book as a text, we find that (particularly in second courses) we skip around to accommodate students with different preparations and interests. We do not recommend that you try to cover everything in each chapter or section. We do not. The book contains material which we think is important and interesting, but the usual academic calendars make it impossible to cover it all.

We frequently leave proofs, or parts of proofs, as exercises. An entire proof deferred to an exercise parallels a previous proof or is easy, or else hints are given. A portion of a proof is left as an exercise to avoid breaking the thread of an argument with tedious manipulations. We recommend that you assign these exercises.

We occasionally refer to mathematical ideas that are beyond the scope of the book. This is done for the benefit of those readers who know such things and to point out further directions of study. The bibliographic guides at the end of the chapters provide numerous references for further reading.

INCLUSIONS AND OMISSIONS

Stochastic Models in Operations Research is too large a subject even for two volumes. Many readers will no doubt wish we had made different choices concerning what material to include in these volumes. For example, Chapter 2 in this volume is nonstandard, but we think that graduate students in operations research should know some of the foundations of stochastic optimization.

The omissions of simulation and statistics were easy decisions because these areas are covered by numerous texts and are not usually taught together with the material that *is* contained in both volumes. Other omissions were sometimes made with regrets. We would have liked to include material on martingales in Volume I, but book size constraints had already severely truncated the material we wished to include on approximations of queueing systems. Also, martingale theory has yet to attain the significance for operations research models that it has reached in some other fields.

In this volume, we had hoped to include material on the optimization of adaptive and partially observable processes, deterministic dynamic programming, and the optimization of deterministic inventory and production models. Alas, book size constraints and our limited stamina led to the omission of these topics (some of which are included in the chapter-ending bibliographic guides).

In spite of our own interest in "stochastic programming," we had not planned to include it in these volumes. It is usually regarded as a branch of mathematical programming, but recent research in the subject is less distinct from Markov decision processes than used to be the case. We chose also not to cover the optimal control of diffusion processes. The prerequisite material in mathematics and probability is well beyond the level of the rest of this volume. See pages 242 and 364 for references to the literatures on stochastic programming and controlled diffusions.

ACKNOWLEDGMENTS

During the years we spent writing both volumes, we received valuable advice from many people. We may not always have recognized the wisdom of their counsel, so any flaws in the book should not be attributed to them. It is a pleasure to thank for their comments on preliminary drafts Drs. David Y. Burman, Shlomo Halfin, Richard Serfozo, Diane D. Sheng, Donald R. Smith, Donald M. Topkis, Ward Whitt, and Eric Wolman of the Bell Telephone Laboratories; Glenn Cordingley and Bruce J. Linskey and Profs. Robert G. Jeroslow, George E. Monahan, Loren K. Platzman, Robert P. Kertz, and Jonathan E. Spingarn of the Georgia Institute of Technology; Prof. Ray Rischel of the University of Kentucky; Prof. Mordecai Henig of the University of Tel Aviv; and Dr. Roy Mendelssohn of the National Marine Fisheries Service. The following people read the entire manuscript of at least one volume, and their suggestions greatly improved the content and presentation of both volumes: Jagadeesh Chandramohan and Prof. Ralph L. Disney of the Virginia Polytechnic Institute and State University; Prof. Sudhakar D. Deshmukh of Northwestern University; Profs. Edward J. Ignall and Linda Green of Columbia University; Prof. William S. Jewell of the University of California at Berkeley; Prof. Edward P. C. Kao of the University of Houston; Profs. Steven Lippman and Bruce Miller of the University of California, Los Angeles; and Prof. Haim Mendelson of the University of Rochester.

The preparation of this book was supported by the Bell Telephone Laboratories, the University of Arizona, the Georgia Institute of Technology, the National Science Foundation, and Yale University. We are especially grateful to Bell Laboratories for offering us the opportunity to collaborate on this book.

The seemingly endless task of typing the manuscript (and its many revisions) was accomplished by Judy Bone, Lisa Giummo, Janet Greene, and the Word Processing Center supervised by Viola Trenta.

Daniel P. Heyman
Matthew J. Sobel

STOCHASTIC MODELS IN OPERATIONS RESEARCH

Volume II

Stochastic Optimization

CHAPTER
ONE
INTRODUCTION

Uncertainties, hence risks, are important characteristics in many settings where operations research is applied. In a warehouse, daily demands for an item may fluctuate from day to day. In a telephone company, requests for assistance from a team of operators may arrive in an irregular manner. A computer center often receives programs that require different amounts of processing time, which causes congestion. A power company's turbine sometimes needs emergency repair. In these settings, the exact times and magnitudes of the events cannot be predicted long in advance. *Uncertainty is pervasive.*

Applications of operations research focus on decision making. Stochastic models in operations research are used to compare alternative decisions when uncertainty is too important to be ignored. For the warehouse, we would compare alternative rules for ordering from the manufacturer. For the telephone company, we would compare the effects of alternative sizes for the team of operators. For the computer center, we would compare the congestion caused by alternative rules for operating the computer system. For the power company, we would compare alternative preventive-maintenance rules. *Uncertainty is manageable.*

The inclusion of uncertainty in a model usually makes the model more complex. It is tempting sometimes to simplify a stochastic model by replacing all its random variables by their expected values. Suppose a stochastic model is replaced by a deterministic model, and decisions based on an analysis of the deterministic model are implemented. In contexts where uncertainty is a peripheral consideration, this replacement may cause negligible deterioration of system performance. In other contexts, including the ones discussed in this book, such a replacement may lead to serious degradation of performance. *Uncertainty is important.*

During the twentieth century, there has been a continual interplay between the formulation of stochastic models and mathematical developments that permit us to analyze the models more completely. The models and their attendant mathematics now comprise an enormous literature. The motivating contexts are extraordinarily varied, and the models and their analysis often exhibit ingenuity and elegance. *Uncertainty is intellectually stimulating.*

Operations research is not, of course, the only applied science in which stochastic models play a major role. Other areas include engineering, particularly industrial engineering, computer science, most of the social sciences, particularly economics, and most of the biological sciences. Some of the phenomena ordinarily not associated with operations research but where stochastic models play a central role include reservoir management, animal population dynamics, and stock price fluctuations.

Methods in this book have been applied successfully to these phenomena and many others ordinarily not associated with operations research.

1-1 OVERVIEW OF BOTH VOLUMES

The modeling phase of operations research can be divided into four parts: Understanding the operation, choosing a model, validating the appropriateness of the model, and drawing conclusions from the model. Volumes I and II concentrate on descriptions of settings for which stochastic models have been built and on derivations of the properties of the models.

All branches of applied mathematics consist of formal mathematical theorems, intuitive results, and a collection of applications. Intuitive explanations of the theorems are essential for successful applications and for extensions of the theory. We emphasize both the intuitive basis of the theorems and the formal proofs.

We present fundamental ideas and facts that provide a background for future learning and are apt to be useful in new applications. The examples have been chosen to illustrate the large number of potential applications and the power of general methods of analysis.

These volumes, like ancient Gaul, are divided into three parts. Volume I contains Parts A and B. Part A describes the basic stochastic processes and descriptive models that are used in operations research and allied fields. The stochastic processes are used in Part B to obtain the operating characteristics of a wide variaety of models; most of them concern congestion and storage. This volume contains Part C. It is devoted to the characterization and computation of optimal policies for controlling a stochastic process.

1-2 USING THIS VOLUME

Neither volume assumes prior knowledge of stochastic models other than the topics usually covered in a calculus-based probability course. Neither volume assumes prior knowledge of operations research methods except for Sections 4-5 and 4-6 in this volume, which make some use of linear programming.

With the exception of Appendix A, the material in this volume builds sequentially. Sections marked with an asterisk treat special topics which may be skipped without obstructing the understanding of subsequent unstarred sections.

This volume is independent of Volume I in the following sense. Statements and proofs of the major results in unstarred sections do not depend on Volume I. However, discussions of the results sporadically refer to stochastic process concepts in Volume I. In that sense, the reader is not obliged to have read Volume I beforehand but is better off for having done so. The sole exception to this assertion is Section 4-6, which is unstarred but uses elementary notions concerning Markov chains.

The examples and exercises are essential for appreciating the content of the theorems and for developing intuition. Some important results that are used in subsequent sections are given in the exercises. Occasional questions have been placed in the text for readers to check their understanding of earlier material.

Although we provide rigorous proofs, we do not believe the main content of a proof is its maneuvering of epsilons and deltas and its justifications of limiting operations. These are details (albeit important ones) and must be mastered. It is the conception, or what some call the "physics," of the proof that should be understood first.

Definitions, theorems, propositions, examples, exercises, equations, and figures are numbered sequentially within each chapter. For example, the fourth theorem in Chapter 2 is numbered Theorem 2-4. The ninth proposition in Appendix A is numbered Proposition A-9.

We often find it necessary to refer to results presented earlier in the book. For example, Theorem 2-4 is presented in Section 2-5 but is cited in Section 2-7. It is easy to find the page on which such an item first occurs. The inside covers and facing pages at the front and rear of the book list all such items with their pages. Thus, the rear inside cover shows that Theorem 2-4 is stated on page 47.

1-3 NOTATION

We try not to use the same symbol to denote different quantities. Since there are only 26 Latin letters, and some of them (such as E, P, and O) have special uses, occasionally we use the same symbol in two ways. In those instances, we warn the reader.

One example of dual notation concerns **boldface** type. In most of the volume, boldface letters denote vectors; the same letters without boldface denote elements of the vectors. However, we depart from this convention in several sections where boldface type denotes random variables and ordinary typeface denotes their realizations. We warn the reader wherever this departure occurs.

The following symbols and notation are used throughout the volume:

$\triangleq$ equal by definition
$\equiv$ identically equal

$\doteq$ approximately equal
$f(t) \approx g(t)$ means $f(t)/g(t) \rightarrow 1$ as t approaches some specified limit
$I \triangleq \{0, 1, 2, \ldots\}$
$I_+ \triangleq \{1, 2, \ldots\}$
$(a)^+ = \max\{a, 0\}$
$(a)^- = \max\{-a, 0\}$
$a \vee b = \max\{a, b\}$
$a \wedge b = \min\{a, b\}$
$f(t^+) = \lim_{s \downarrow t} f(s)$
$f(t^-) = \lim_{s \uparrow t} f(s)$
$\# \{\cdot\}$ = number of elements in the set $\{\cdot\}$
$\square$ = end of a proof or an example
$\mathbf{e}$ = column vector with 1 as every element
$\mathbb{R}$ = set of real numbers = $(-\infty, \infty)$
$\mathbb{R}_+$ = set of nonnegative real numbers = $[0, \infty)$
$\mathbb{R}^n$ = set of n-tuples of real numbers
$\mathbb{R}^n_+$ = set of n-tuples of nonegative real numbers

Also the identity matrix is denoted by I; it will be clear from the context which interpretation of I is intended.

The abbreviations r.v. and i.i.d. stand for random variable and independent and identically distributed.

PART C

CHAPTER

TWO

STOCHASTIC OPTIMIZATION: PROLOGUE

This volume presents methods to compare alternative stochastic processes. For example, consider a model of a production line and two alternative rules for deciding when to start and stop the line depending on the level of finished goods inventory at the end of the line. Suppose that the model is discrete in time and that demand for the finished goods is a stochastic process. Then the implementation of either production rule induces (in the model) a stochastic process of successive inventory levels, production rates, etc. The comparison of the two rules therefore requires the comparison of the stochastic processes induced by the rules.

A stochastic process has two features that complicate comparisons such as the one above. One complication is that the sample space of a (discrete-time) stochastic process contains sequences and it is not obvious how sequences should be compared. Section 2-3 discusses this complication. Even if the model were deterministic, the comparison of alternative rules would lead to the comparison of sequences. Section 2-5 presents an axiomatic foundation for the use of present-value formulas to compare time-indexed sequences of real numbers. A present-value formula reduces such a sequence to a single number.

The other difficulty in the comparison of stochastic processes stems from their probabilistic content. Even if each stochastic process is the degenerate case of a single r.v. (random variable), there remains the formidable task of comparing two distribution functions. The problem is somewhat similar to that of comparing alternative time-indexed sequences. Perhaps the simplest resolution of the problem is to replace all r.v.'s by measures of central tendency such as means or medians. Section 2-2 uses a simple specific model to explore the extent to which this simple resolution can be simplistic and misleading. Less naive resolutions of uncertainty are presented in Section 2-4. There we draw from the literature in decision theory to relate decision criteria and attitudes to risk. Indeed, Sections 2-2 through 2-5 are primarily concerned with the choice of criteria in stochastic optimization problems. The remaining chapters of the volume focus on the problem of determining a decision rule that optimizes a given criterion.

In Section 2-1, we discuss the elements of inventory and production models. In both volumes, we frequently use inventory and production models as expository vehicles.

The proofs of the major results in Sections 2-4 and 2-5 are presented in Sections 2-6 and 2-7, respectively.

2-1 ELEMENTS OF INVENTORY AND PRODUCTION MODELS

This section contains a rather lengthy discussion of the elements of inventory and production models. Why should you plow through this material before learning anything at all about stochastic optimization?

We chose inventory (and production) models as the prototype on which to base a discussion of model elements. Why? First, these models are important historically and in practice. They were the first class of stochastic operations research models whose optimization was investigated in detail. Moreover, inventory and production models are used widely in practice. Second, the elements of other kinds of models usually arise in inventory models too. The early practical usefulness and analytical elegance of inventory models made them prototypes when models for other purposes were formulated. Also, inventory and production situations are so varied that the considerations pertinent to other contexts usually are found in some inventory situations.

Any model is a compromise with realism. You strike the compromise in order to balance insight, tractability, and realism. For example, if you travel by car from New York to Atlanta, you do not select a road map with a scale of 1 centimeter of map per kilometer of terrain. The resulting "map" 1,500 centimeters long, would be too unwieldy to yield easily such insights as "where should we stop the first night?" On the other hand, if you are estimating the cost to widen a small portion of the same road, the 1,500-centimeter "map" is too small. This section is included so that you start to appreciate the balance between realism and tractability that is struck by inventory and production models. That appreciation, of course, will be enhanced by understanding the properties of models derived in the following chapters.

Most inventory models are discrete in time, and at first we shall honor tradition.

A Discrete-Time Model

Suppose there are stocks of several commodities that are periodically depleted by demands and replenished through purchases or production. The commodities may differ from one another, perhaps only because they are stored in different locations. Perhaps they are earmarked for different kinds of customers—or perhaps they are materially different. Of course, a single commodity stocked at one location is a special case. By "commodity" we encompass raw materials, work-in-process stored inside a factory, and goods whose manufacture is complete. Suppose that there are L commodities and that replenishment decisions must be made during a succession of N periods. In other words, the *planning horizon* is N periods in length.

Answers to two questions are needed for each commodity each period:

1. Should the commodity's stock level be altered?
2. If the stock level should be changed, what change is appropriate?

Let s_n^j denote the number of units of commodity j that are on hand at the beginning of period n ($1 \leq n \leq N$, $1 \leq j \leq L$). How is s_{n+1}^j related to s_n^j? They are connected by a conservation equation:

$$\begin{aligned} s_{n+1}^j = s_n^j &+ \text{amount of commodity } j \text{ added to stock in period } n \\ &- \text{amount of commodity } j \text{ issued from stock in period } n \end{aligned} \tag{2-1}$$

Let z_n^j denote the quantity of commodity j added to stock in period n. It includes

(i) Quantities returned by dissatisfied customers to whom stock was issued during earlier periods
(ii) Deliveries during period n of quantities ordered from suppliers during periods prior to n
(iii) Quantities whose manufacture is completed during period n but which was initiated prior to n
(iv) Deliveries during period n of quantities ordered from suppliers during period n
(v) Quantities whose manufacture is initiated and completed during period n

We defer until Section 3-2 and 7-2 the effects of the *delivery lag* in (ii) and (iii). Here we treat (i) as a negative component in the "amount ... issued from stock ..." in (2-1). The sum of (iv) and (v) is the "amount ... added to stock ..." in (2-1). Ignoring (ii) and (iii) as sources of additions to stock is called an *immediate delivery* assumption. Let z_n^j denote the sum of the increments due to (iv) and (v). The sources (iv) and (v) are indistinguishable in the effects of their deliveries on the stock level, and only the sum of goods that are manufactured and ordered is used to connect s_{n+1}^j to s_n^j. The sum

$$a_n^j = s_n^j + z_n^j \tag{2-2}$$

is the total amount of goods available to satisfy demand in period n.

In (2-1), the "amount ... issued from stock ..." includes

(vi) Quantities requested in earlier periods and issued (or delivered) in period n
(vii) Damaged or spoiled or outdated items that are discarded
(viii) Items that are stolen
(ix) Items that are misshelved or otherwise lost
(x) Algebraically negative amounts "issued," such as previously lost items that are found during period n and the returned items mentioned in (i)
(xi) Quantities requested and issued during period n

The stock level can be depleted because of (vi) only if some customers are willing to wait awhile for delivery to occur. The models discussed here are motivated by phenomena in which there is no reason to postpone delivery if sufficient stock is available to satisfy demand. Equivalently, here the "customers" are assumed to be indistinguishable. If the customers ought to be distinguished from one another and there is a possibility that the stock level will occasionally be zero—a *stockout*—a model ought to encompass the assignment of customer priority. By suppressing the distinguishability of customers, a previously requested quantity would be issued in period n only if a stockout occurred in the earlier period and the customer was willing to wait. In this case, the earliest demand is said to have been *back-ordered* or *backlogged*. The *lost sales* case occurs when no customer is willing to wait—so that demand is always "lost" if it cannot be satisfied immediately. Lost sales and back-ordering are extreme cases, and often neither is warranted; if demand exceeds the supplies on hand, some of it is lost and some may be back-ordered but for only a limited amount of time. Let D_n^j denote the algebraic sum of the quantities in (vii) to (xi). In the lost sales case,

$$s_{n+1}^j = \begin{cases} s_n^j + z_n^j - D_n^j & \text{if} \quad s_n^j + z_n^j \geq D_n^j \\ 0 & \text{if} \quad s_n^j + z_n^j < D_n^j \end{cases}$$

We use the notation $(u)^+$ for $\max\{0, u\}$ so

$$s_{n+1}^j = (a_n^j - D_n^j)^+ \qquad \text{(Lost sales)} \tag{2-3}$$

In the back-ordering case, $s_{n+1}^j = a_n^j - D_n^j$ if $a_n^j \geq D_n^j$. If $a_n^j < D_n^j$, the physical stock level drops to zero and $D_n^j - a_n^j$ is "owed" to the waiting customers. If the physical stock level is nonnegative, the quantity owed is zero—and conversely. In fact, if $a_n^j - D_n^j < 0$, $-(a_n^j - D_n^j)$ is the amount owed. Therefore, we extend the notation s_n^j to denote the difference between the stock level and the amount owed:

$$s_{n+1}^j = \begin{cases} a_n^j - D_n^j & \text{if} \quad a_n^j \geq D_n^j \\ -(D_n^j - a_n^j) & \text{if} \quad a_n^j < D_n^j \end{cases}$$

or $$s_{n+1}^j = a_n^j - D_n^j \qquad \text{(Back-ordering)} \tag{2-4}$$

If $s_n^j \geq 0$, there are s_n^j units in stock at the beginning of period n; if $s_n^j < 0$, $-s_n^j$ units of back-ordered demand are owed to consumers.

Sometimes the lost sales and back-ordering cases are treated simultaneously with intermediate possibilities (some excess demand lost and some back-ordered); so we use the notation

$$s_{n+1}^j = v_j(a_n^j, D_n^j)$$

for the rule that transforms a_n^j and D_n^j to s_{n+1}^j. We can treat all L items simultaneously with the vector notation

$$\mathbf{s}_{n+1} = \mathbf{v}(\mathbf{a}_n, \mathbf{D}_n) \tag{2-5}$$

Inventory and Shortage Costs

An enterprise incurs several costs when it keeps a commodity in storage. These costs often include rent or maintenance and amortization of costs of constructing the storage facility, heat, light, insurance against fire and other risks, obsolescence, and state or municipal taxes on the value of the stored commodity. However, the dominant cost of inventory is usually the opportunity cost of the capital that has been tied up in the procurement or production of the commodity. This unit cost is at least as high as the interest rate on liquid short-term secure investments such as 90-day U.S. Treasury bills.

The opportunity cost usually should be reckoned higher than the short-term interest rate because most enterprises have internal activities to which funds could be profitably diverted. The replacement of older production machinery with more efficient new machinery is a typical example in a manufacturing enterprise. In most well-managed enterprises, the marginal rate of return on internally invested capital is much higher than the short-term interest rate. In 1979, when the short-term interest rate was approximately 0.15, many firms used a figure between 0.25 and 0.30 for the overall unit cost of inventory.

We usually summarize the storage costs each period with a term such as a unit cost times the inventory level at the end† of the period. The inventory level at the end of one period is the level at the start of the next period; so we might use (2-5) to specify the contribution of period n to the aggregate inventory holding cost as

$$H(\mathbf{s}_{n+1}) \equiv H[\mathbf{v}(\mathbf{a}_n, \mathbf{D}_n)] \tag{2-6a}$$

where H is a real-valued function with domain $\mathbb{R}^L$ [L is the number of different commodities; so $\mathbf{v}(\cdot, \cdot)$ is an L-vector-valued function]. Usually, the role of

† Models could as easily be constructed so that the storage cost depends on the inventory level at the beginning of the period. Models where the cost depends on the inventory level in the middle of the period are harder to analyze. However, period lengths are relatively short in models used in practice; so basing the cost on the level at the start or end of the period is reasonable and adequate.

$\mathbf{v}(\cdot, \cdot)$ is implicit and we write

$$h(\mathbf{s}_{n+1}) \equiv h(\mathbf{a}_n, \mathbf{D}_n) \equiv H[\mathbf{v}(\mathbf{a}_n, \mathbf{D}_n)] \tag{2-6b}$$

as the inventory cost in period n. In practice, $h(\cdot, \cdot)$ is estimated separately for each commodity; so

$$\sum_{j=1}^{L} h_j(a_n^j, D_n^j) \equiv \sum_{j=1}^{L} h_j(s_{n+1}^j) \tag{2-6c}$$

is the storage cost incurred in period n.

Many inventory models assess a penalty in the event that demand exceeds supply in a period. This "shortage cost" is quite real in contexts where some shortages *must* be met. Then there may be costs of expediting delivery, of turning to nearby retail sources rather than to more distant wholesalers, and some of the excess demand may be met with a more costly substitute commodity at no extra price to the customers. For example, airlines often must pay a fee to overbooked passengers with "coach" tickets.

Penalty costs often are included in models for enterprises that do not respond to excess demand except, possibly, by back-ordering it when consumers are willing to wait. These penalty costs reflect the goodwill that is lost because some consumers cannot be satisfied immediately. It is difficult to estimate such penalty costs, but usually, it is even harder to model explicitly the dependence of the demand process on the degree to which demands do not exceed stock levels.

Whether shortage costs are real or surrogates, their detailed dependence on the length of time during which a shortage persists is difficult to model. Consider the backlogging case, for example, and a period n during which the backlog includes demands from several earlier periods. Then period n's shortage cost ought to depend on the aggregate amount of backlog *and* on the various "ages" of the backlogged demand. This dependence would complicate a model; so it is rarely made explicit. Instead, one usually finds a shortage cost for commodity j that is postulated to depend only on the aggregate excess demand $(D_n^j - a_n^j)^+$.

One usually thinks of holding and shortage costs as being physically dependent on the stock levels s_n^j. But our ultimate goal is to choose a_n^j shrewdly; so it is convenient to express these costs in terms of $\mathbf{a}_n$ and $\mathbf{D}_n$. Therefore, we shall represent shortage costs in period n as

$$w(\mathbf{a}_n, \mathbf{D}_n) \tag{2-7a}$$

to include the possibility of utilizing an excess supply of one commodity to satisfy the excess demand for a substitute commodity. When no such interaction exists, the shortage cost in period n decomposes into

$$\sum_{j=1}^{L} w_j[(D_n^j - a_n^j)^+] \tag{2-7b}$$

A period's storage costs in (2-6*a*), (2-6*b*), and (2-6*c*) and shortage costs in (2-7*a*) and (2-7*b*) are given by functions that depend only on $\mathbf{a}_n$ and $\mathbf{D}_n$. Usually, the sum of these two costs is more important than the individual components.

Let

$$g(\mathbf{a}, \mathbf{d}) = h(\mathbf{a}, \mathbf{d}) + w(\mathbf{a}, \mathbf{d}) \tag{2-8}$$

denote this sum ($\mathbf{a} \in \mathbb{R}^L$ and $\mathbf{d}$ is in the sample space of $\mathbf{D}_n$ for every n), which sometimes is called a *generalized inventory holding cost.*

Costs of Changing the Production Rate

The symbol z_n denotes the quantity added to stock during period n. If this quantity is manufactured by the same enterprise that stocks it, good management usually tries to avoid abrupt changes in the production rate. If z_n is much less than z_{n-1}, some workers may have to work for a smaller than usual number of hours each week or even be laid off—"furloughed"—and some production lines or factories be closed. If the workers have been employed under terms of a union contract, seniority provisions may permit the more senior furloughed workers to "bump" their unfurloughed coworkers having less seniority if they are doing less highly skilled work. Therefore, a forced contraction of the labor force immediately creates turbulence within the factory. The productive efficiency of the remaining workers may be affected by their heightened fears for their own job security. Also, some of the furloughed workers in whom much training may have been invested may accept permanent positions with other employers. Longer-range impacts include higher future unemployment insurance premiums and less community goodwill toward the employer so that in future years it may be more difficult—hence more expensive—to hire skilled workers.

If production increases abruptly, that is, z_n is much greater than z_{n-1}, there may be other kinds of significant expenses. The existing work force may be asked to work overtime while new workers are hired and trained. These steps cause direct expenses of hiring, training, and overtime. Moreover, there are indirect expenses due to the lower productive efficiency of the new workers while they become accustomed to their jobs and of the "old hands" whose overtime fatigues them.

The costs of changing the production rate often are called "smoothing costs" because they provide incentives for the pattern of production over time, namely, the graph of $\{(z_n, n): n = 1, 2, \dots\}$, to be less jagged than it might be in the absence of costs of change.

Setup Costs

Production and purchasing are the principal means to procure a stocked commodity. Here we discuss a cost which often pertains to both purchasing and production.

An enterprise may purchase or produce some commodities only occasionally. Suppose this is true of commodity j; in our notation, $z_n^j = 0$ for some n's and $z_n^j > 0$ for other n's. If n is a period in which $z_n^j > 0$, the enterprise may experience significant administrative expenses unrelated to the size of z_n^j (except that it is positive).

If the commodity is being purchased, a vendor must be selected, a purchase order has to be prepared and then transmitted to the selected vendor, the delivery may have to be expedited, a receiving report must be prepared upon delivery of the commodity from the vendor, and this may require checking that the delivered goods are, in fact, the same ones which were ordered. Then the delivered goods will be unpacked and shelved and their arrival noted on the commodity's stock record (possibly computerized) so that the stock level will be appropriately increased. Simultaneously, the vendor's invoice will be compared with the purchase order and the receiving report, and after any differences are reconciled, a check will be issued to the vendor. Issuance of the check will initiate a series of financial bookkeeping procedures. All these activities absorb the time of personnel, and some of them may utilize equipment such as computers, conveyor belts, telephones, and typewriters. Therefore, this administrative expense can be quite high.

Suppose that the commodity is procured by producing it. We assume that a production line is used for this purpose, sporadically, so that its purchase costs are not at issue. However, setting it up requires the transfer of workers and, perhaps, some general-purpose equipment (e.g., fork-lift trucks) from the production of other commodities. Beforehand, a check must be made that the necessary amount of raw material or components used in manufacturing the commodity are available. If any of their amounts are inadequate, they must be ordered and their deliveries expedited.

The name *setup cost* is given to the sum of the administrative costs of purchasing or the expense of setting up a production line.

When $L = 1$, so one commodity is unrelated to others, the setup cost is usually represented as

$$K\delta(z_n) \tag{2-9a}$$

where $K > 0$ and $\delta(z) = 0$ if $z \le 0$ and $\delta(z) = 1$ if $z > 0$. If $L > 1$, the appropriate setup cost is

$$\sum_{j=1}^{L} K_j \delta(z_n^j) \tag{2-9b}$$

in some contexts, and

$$K\delta\left(\sum_{j=1}^{L} z_n^j\right) \tag{2-9c}$$

in others. The setup cost appears as (2-9*c*) in contexts where payment must be made for the use of an entire freight car or truck to ship the purchased commodity. This situation arises, for example, when purchasing various radioactive isotopes, laboratory animals, gold, and diamonds. The setup cost appears as (2-9*b*) when administrative efforts for different commodities are independent of each other.

Other Purchasing and Production Costs

The enumeration of costs has not yet included payment of the vendor if the commodity is purchased, or the direct cost of wages and materials if the com-

Table 2-1 Example of a quantity discount

Order size Min	Order size Max	Unit price for all items ordered ($)
0	99	0.25
100	499	0.20
500	999	0.17
1,000	—	0.15

modity is produced. The purchase cost may be quoted by the vendor with *quantity discounts*; i.e., the price decreases as the order quantity increases. Table 2-1 presents an example of a quantity discount. For example, an order size of 200 units costs 200 × 0.20 = $40. Figure 2-1 graphs cost versus order quantity.

You may notice in Table 2-1 and Figure 2-1 that it is cheaper to buy 100 units than to by 81 through 99. Vendors are aware of such anomalies but sometimes use them as incentives to induce larger order quantities. Other vendors prefer to offer quantity discounts without these anomalies. Table 2-2 and Figure 2-2 present as example of this alternative form. For example, an order of 200 units costs 100 × 0.22 + 100 × 0.17 = $39. Figure 2-2 is based on Table 2-2 and displays total cost as a function of order size.

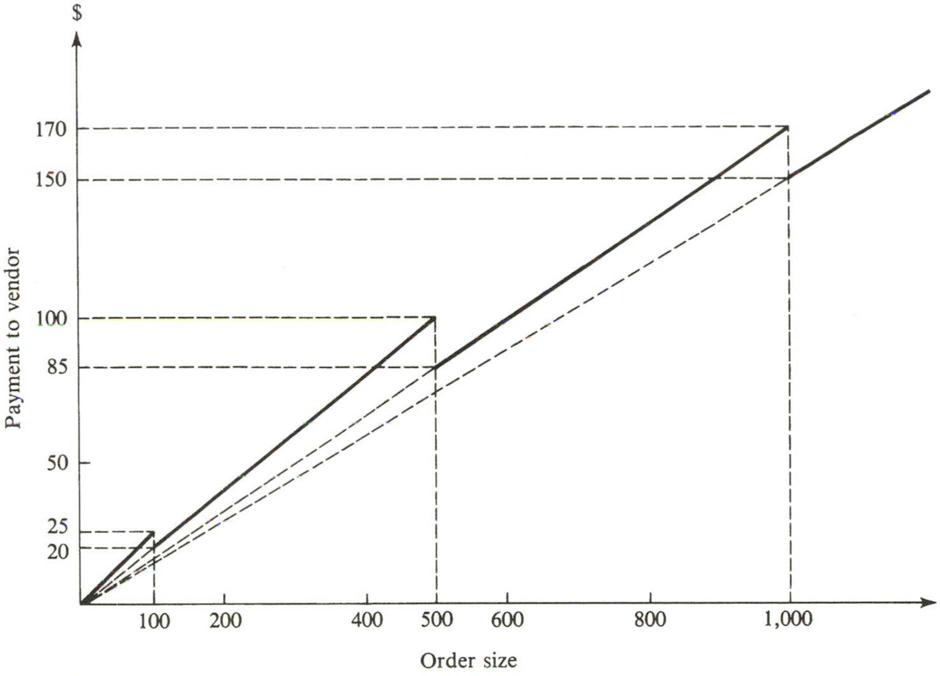

Figure 2-1 Example of a quantity discount.

Table 2-2 Another example of a quantity discount

Interval	Price ($)
First 100 units	0.22
Next 400 units	0.17
Next 500 units	0.14
All additional units	0.13

Vendors offer quantity discounts, also called *price breaks*, because they experience setup costs when filling orders and because they prefer to receive revenue sooner rather than later. The price break is an incentive to purchase a large quantity now rather than some now and the rest later. Of course, the saving in purchase cost (and, perhaps, subsequent setup costs) must be balanced against the higher storage cost entailed by larger purchase quantities.

Vendors discourage a purchase quantity from becoming too large, i.e., larger than they can readily handle, by spreading out the delivery in several shipments or by postponing the delivery date. A similar influence is exerted if the commodity is being produced by the same enterprise that "orders" it. If the production rate is very high, the enterprise must resort to overtime use of labor, it must expedite the delivery of materials, and it may even have to subcontract the

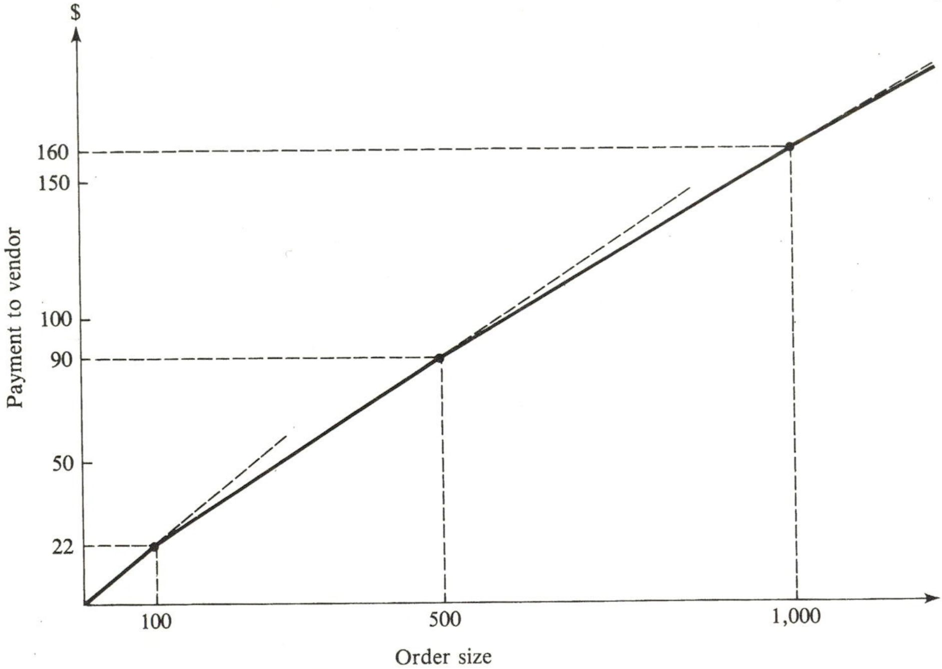

Figure 2-2 Another kind of quantity discount.

production of some portion of the "order." These costly consequences of high production rates or large purchase orders are called *bottleneck costs*.

Figure 2-3 illustrates the way in which purchasing or production costs, including setup cost, might typically (though not necessarily) depend on quantity. The marginal purchasing (or production) cost (slope of the tangent to the curve) starts to increase at quantity m, and at that point bottleneck costs intrude. In the interval from 0 to m, the marginal cost is decreasing because the setup cost and other factors offer economies of scale. The average cost per unit purchased is the ratio of the total cost to the quantity, for example, B/b. We see that the average cost (unit cost) is decreasing as the quantity increases to j (note $j > m$) and thereafter is increasing.

Our models usually contain simpler purchasing (or production) cost functions than the one in Figure 2-3. Perhaps the most frequently assumed cost function, exclusive of setup cost, is linear. Then there is an L-vector $\mathbf{c}$ (there are L commodities being stocked) so that the purchasing cost is

$$\mathbf{c}\mathbf{z}_n \qquad \mathbf{z}_n \geq \mathbf{0} \tag{2-10}$$

in period n plus setup cost when $\mathbf{z}_n$ is the vector of order quantities. Simple forms of functions are used for two main reasons: in practice, their parameters [for example, $\mathbf{c}$ in (2-10)] are relatively simple to estimate, and simple functions facilitate mathematical and numerical analysis. The models in any science trade off realistic detail in particular cases vs. generality of conclusion. The experience and shrewdness of the practitioner are reflected in the wisdom of the compromise struck among realism, simplicity, and generality.

A simple approximation of Figure 2-3, of the form

$$K\delta(z) + cz \qquad z \geq 0 \tag{2-11}$$

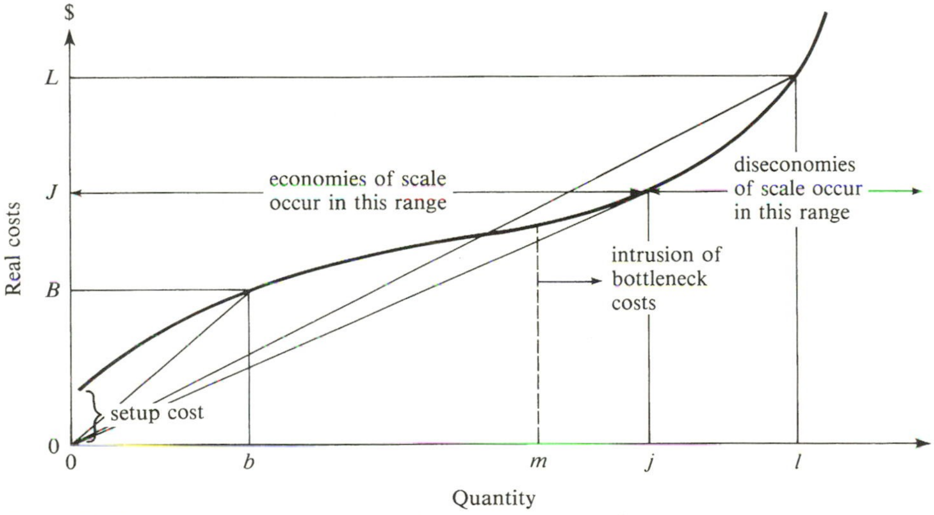

Figure 2-3 Dependence of purchasing or production costs on quantity.

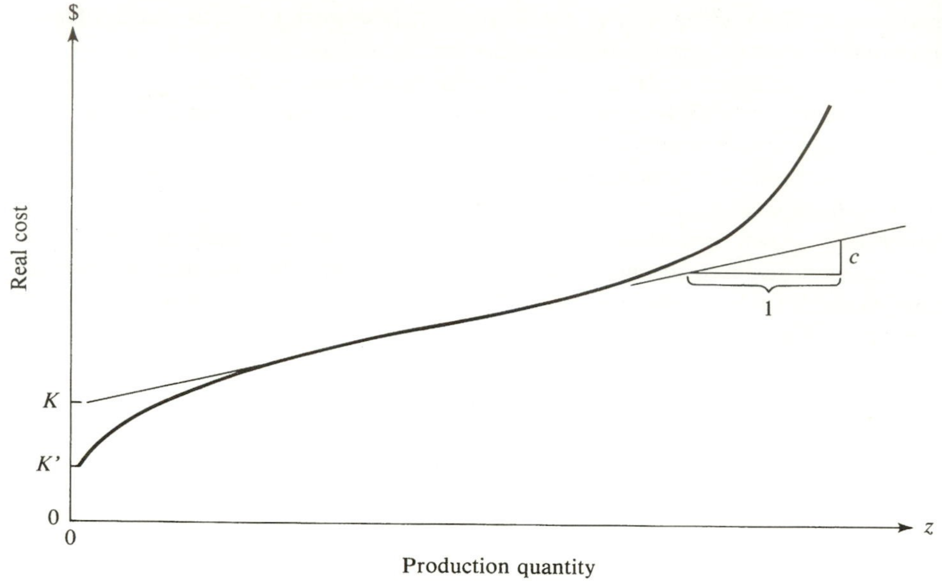

Figure 2-4 Simple approximation of purchasing or production costs.

is exhibited in Figure 2-4. The approximation is reasonably good except at very small and very large quantities. However, very small quantities are unlikely decisions using either the "true" or approximate cost functions. The setup cost, be it K or K', induces a choice of quantities sufficiently large to warrant paying the setup cost. Very large quantities are equally unlikely under normal conditions. Suppose "normal" conditions have changed so that there are incentives to choose quantities that heretofore were large enough to generate bottleneck costs. Then the enterprise's management will expand its manufacturing capacity or obtain additional suppliers or turn from retailers to wholesalers or from wholesalers to manufacturers. In other words, management will work to eliminate the constrictions of the bottlenecks! Therefore, cost functions such as (2-10) or (2-11) are frequently used in models of stockage under routine circumstances.

Salvage Values

Salvage value is the revenue you receive when an item is discontinued and either put on sale or sold to other enterprises, e.g., ski equipment at the end of winter. However, this notion must be modeled carefully to separate revenue due to usual demand from salvage-value revenue.

"Demand" in most models originates from consumers at a less aggregated stratum of the economy than the enterprise stocking the commodity. For example, the enterprise may be a retailer and its customers the general public, or it may be a wholesaler having retailers as customers, or it may be a manufacturer having wholesalers or its own production workshops as "customers." In these instances, the enterprise and its customers do their buying in different markets. We speak of salvage value, i.e., the revenue received from the disposal of a unit of the commodity, only if the markets are the same. In this way we distinguish between the revenue associated with demand (D_n^j) and the revenue received from disposal ($z_n^j < 0$).

The revenue received from disposal may depend nonlinearly on the quantity sold. Two nonlinear influences are setup costs and market saturation lowering the salvage value of a larger quantity. However, the disposed quantities are usually small, and in most models disposal is an option only when further sales to consumers (D_n^j) are no longer possible. Therefore, the salvage value is represented linearly in most cases.

The salvage value of an item is usually lower than the price to buy it. However, the gap is often small in markets for wholesale and raw commodities; so most models ignore the price gap and assume that the salvage value of an item is the same as the price to buy it. Then an extension of (2-10) is used as the purchase cost net of revenue from disposal (but not including setup cost, if any):

$$\mathbf{c}\mathbf{z}_n \qquad \mathbf{z}_n \in \mathbb{R}^L \tag{2-12}$$

Revenue

Revenue is treated cavalierly in most inventory models—partly because inventory theory and microeconomics have had little impact on each other. Also, most stockage models are used in contexts either where prices vary slowly or where no revenue at all is received. For example, revenues are absent when commodities are stocked by the same enterprise that utilizes them in production. An automobile manufacturer receives revenue for sales only of finished vehicles, but it must stock thousands of different raw materials and components which are consumed and assembled during the manufacturing process. Therefore, in many models it is sufficient and certainly convenient to assume that revenue is proportional to sales and that the constant of proportionality (price) is known in advance.

Another reason why revenues often do not appear explicitly in inventory models is illustrated in Exercise 2-3; minimizing expected inventory costs may be equivalent to maximizing expected net profits. If the management level at which inventory decisions are made does not control prices, it is appropriate to regard revenue from sales as a fixed selling price multiplied by the (random) amount sold. A typical example is the management of a single retail outlet of a chain of stores.

Let r denote the fixed selling price. With the earlier notation, net revenue is

$$B = r(D \wedge a) - (\text{purchase costs} + \text{holding costs} + \text{shortage cost})$$

Since $D \wedge a = D - (D - a)^+$,

$$B = rD - [r(D - a)^+ + \text{purchase costs} + \text{holding costs} + \text{shortage costs}]$$

When D is independent of a, $E(B)$ is maximized by minimizing the expected value of the bracketed terms, which are the inventory costs. Typically, the term $r(D - a)^+$ is included in the shortage costs.

There are two collections of models where we can abandon the foregoing assumption of proportionality. Some models having certain kinds of nonlinear purchase costs (exclusive of setup costs) can be analyzed mathematically; so the complexity of these models is not heightened by certain nonlinearities in the revenue terms. Another class of models retains the assumption that revenue is proportional to sales but abandons the assumption that the constant of proportionality, i.e., unit price, is known and set in advance. The latter class of models focuses on the interaction between marketing and stockage. It is directed at the occasionally encountered problem of jointly setting prices and production levels.

Given the cost and revenue structure in this section, the primary stochastic element in inventory and production models is the stochastic process $\mathbf{D}_1, \mathbf{D}_2, \ldots$ of demands.

2-2 DETERMINISTIC OPTIMIZATION VS. STOCHASTIC OPTIMIZATION

Stochastic optimization problems differ from deterministic optimizations problems in several ways that may not be immediately apparent. The major future consequences of present actions cannot be foretold with certainty. Since the resulting future options for choice are uncertain, a sensible decision rule should take into account whatever information is gained as time unfolds. Different information patterns may warrant taking different decisions in the future. This leads to the notion of a decision rule being a contingency plan. We use the term *policy* for such contingency plans.

Consider a problem with many alternative decisions and suppose that each one leads to a random payoff. Then maximization of the payoff is ill-defined. Should you maximize the mean or the median? Should you minimize the variance? Should you maximize the mode? There is further discussion of this matter in Section 2-3. For the moment, we use the criterion of expected value to elucidate other issues.

When is it prudent to ignore stochastic considerations? When should you replace random quantities in a model by their means? When by their medians? It is trite to assert that you should make such simplifications when their effects are minor. But how can you tell? You will see that the effects of misspecifying a stochastic model depend on economic parameters as well as on the degree of randomness.

The following illustration uses a simple specific inventory model. Starting in Chapter 3 you will learn how to analyze more complicated inventory models with less labor than is needed for the simple model here.

A Simple Inventory Model

Consider a simple stochastic model of a single inventory decision. You must choose the amount z of goods to be ordered. It will be delivered quickly, and thereafter a demand for D units will occur where D is an r.v. Let s denote the amount in stock before z is chosen and let $a = s + z$ denote the amount that is available, after ordering, to satisfy demand. Sales will be D if $a \geq D$ and a if $D > a$; hence sales will be $a \wedge D$ = minimum $\{a, D\}$ which satisfies

$$a \wedge D = a - (a - D)^+ = D - (D - a)^+$$

The residual stock, after demand occurs, will be $(a - D)^+$ and the sales lost to excess demand will be $(D - a)^+$.

Let r denote the selling price and let c denote the purchase cost. It is natural to assume $r \geq c$ and $r > 0$. Suppose that $c/2$ is the salvage value of the residual stock. What value of z maximizes the expected value of the net revenue? The net revenue B is the random variable given by

$$B = r[a - (a - D)^+] - cz + \frac{c}{2}(a - D)^+$$

$$= (r - c)a + cs - \left(r - \frac{c}{2}\right)(a - D)^+$$

You cannot maximize B because it is an r.v. Suppose, instead, you maximize its expected value, which is

$$E(B) = (r - c)a + cs - \left(r - \frac{c}{2}\right)E[(a - D)^+] \qquad \text{(2-13)}$$

In order to evaluate the objective, $E(B)$, the expected value of B, we must compute $E[(a - D)^+]$. To perform this computation, let $F(\cdot)$ be the d.f. of D. By the definition of $(a - D)^+$,

$$(a - D)^+ = \begin{cases} 0 & \text{if } D \geq a \\ a - D & \text{if } D < a \end{cases}$$

Taking expectations on both sides yields

$$E[(a - D)^+] = \int_0^a (a - x)\, dF(x) \qquad \text{(2-14a)}$$

Observe that $E[(a - D)^+]$ may depend on more of the function $F(\cdot)$ than just a few moments.

To make the model more explicit, assume that demand is uniformly distributed over $[\mu - v, \mu + v]$ where $0 < v < \mu$. Then the variance of D is an increasing function of v, $E(D) = \mu$, and (2-14a) becomes

$$E[(a-D)^+] = \begin{cases} 0 & a \le \mu - v \\ \dfrac{1}{2v}\displaystyle\int_{\mu-v}^{a} (a-x)\,dx = \dfrac{(a-\mu+v)^2}{4v} & \mu - v < a \le \mu + v \\ a - \mu & a > \mu + v \end{cases} \quad (2\text{-}14b)$$

To emphasize the dependence of B on s and a, write $\phi(s, a) = E(B)$. Substituting (2-14b) into (2-13) yields

$$\phi(s, a) = (r-c)a + cs - \left(r - \frac{c}{2}\right) \cdot \begin{cases} 0 & a \le \mu - v \\ \dfrac{(a-\mu+v)^2}{4v} & \mu - v < a \le \mu + v \\ a - \mu & \mu + v < a \end{cases} \quad (2\text{-}15)$$

What does $\phi(s, a)$ look like as a function of a? Of course, $a = s + z$ and $z \ge 0$ so $a \ge s$, but ignore this constraint at first. From (2-15),

$$\frac{\partial \phi(s, a)}{\partial a} = \begin{cases} r - c & a \le \mu - v \\ r - c - 2\left(r - \dfrac{c}{2}\right) \dfrac{a - \mu + v}{4v} & \mu - v < a \le \mu + v \\ -\dfrac{c}{2} & a > \mu + v \end{cases}$$

$$\frac{\partial^2 \phi(s, a)}{\partial a^2} = \begin{cases} 0 & a \le \mu - v \\ \dfrac{-(r - c/2)}{2v} & \mu - v < a \le \mu + v \\ 0 & a > \mu + v \end{cases} \quad (2\text{-}16)$$

The global maximum of $\phi(s, \cdot)$ occurs on $[\mu - v, \mu + v]$ if $r \ge c$, as is assumed. Set $\partial\phi(s, a)/\partial a = 0$ for $\mu - v < a \le \mu + v$, and call the solution a^*; thus† (assume $c > 0$)

$$a^* = \mu + \frac{v(2r - 3c)}{2r - c} \quad (2\text{-}17)$$

The graph of $\phi(s, \cdot)$ is presented in Figure 2-5.

Under the assumption $r \ge c$, it follows that $\mu - v < a^* \le \mu + v$. Therefore, the constraint $s \le a$ does not affect the choice of $a = a^*$ if $s \le a^*$. Suppose $s > a^*$. It is apparent from (2-16) and Figure 2-5 that a should be chosen as small as possible, that is, $a = s$. Therefore, the optimal choice of a is

$$a = a^* \vee s$$

† The notation a^* suppresses the dependence of the solution on various parameters, most notably s.

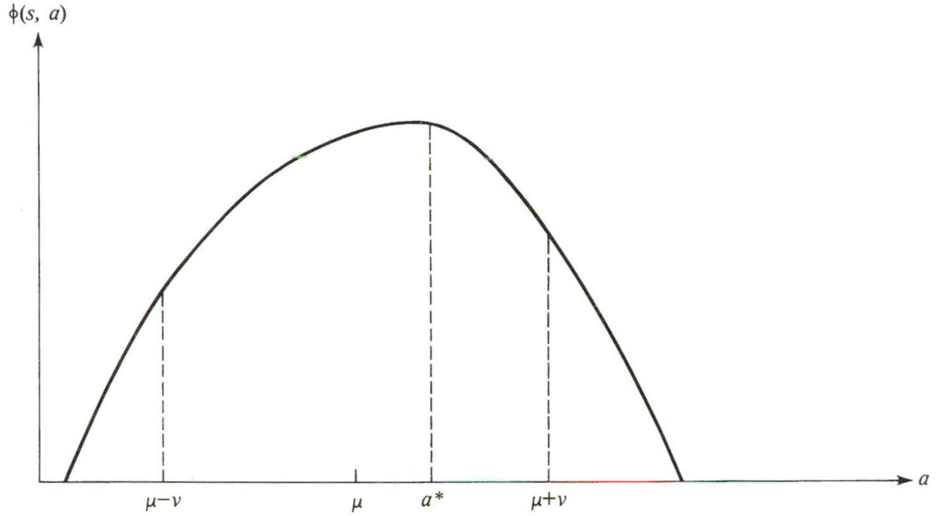

Figure 2-5 Expected net revenue as a function of available stock.

Let $Q(s)$ denote the maximum attainable value of the objective, that is, $Q(s) = \phi(s, a^* \vee s)$ and let $K = (2r - 3c)/(2r - c)$. Then $a^* = \mu + vK$ and, from (2-15) and (2-17),

$$Q(s) = \begin{cases} (r - c)\mu + cs - \dfrac{vc(r - c)}{2r - c} & s \le a^* \\ rs - \dfrac{(2r - c)(s - \mu + v)^2}{8v} & a^* < s \le \mu + v \\ r\mu + \dfrac{c(s - \mu)}{2} & \mu + v < s \end{cases} \tag{2-18}$$

Replacement of Random Demand by Its Mean

Suppose that D is replaced in (2-13) by $E(D) = \mu$. How much of an error is caused? Let $b(s, a)$ denote B in (2-13) if D is replaced by μ:

$$b(s, a) = (r - c)a + cs - \left(\frac{r - c}{2}\right)(a - \mu)^+$$

Then Exercise 2-1 asks you to verify that $b(s, \cdot)$ is maximized on $[s, \infty)$ [that is, on $(-\infty, \infty)$ subject to $a \ge s$] by

$$a = \mu \vee s$$

Suppose $2r \ge 3c$ so $0 \le K = (2r - 3c)/(2r - c)$ and $\mu \le a^* = \mu + Kv$. The degree of error in $\mu \vee s$ is

$$
\begin{aligned}
a^* \vee s - \mu \vee s &= Kv && \text{if } s \leq \mu \\
&= \mu + Kv - s && \text{if } \mu < s \leq a^* \\
&= 0 && \text{if } a^* < s
\end{aligned}
$$

Let $\Delta(s) = Q(s) - \phi(s, s \vee \mu)$; it is the change in the optimal value of the objective function when D is replaced by $E(D)$. From (2-15) and (2-18),

$$
\Delta(s) = \begin{cases} \dfrac{v(2r - 3c)^2}{8(2r - c)} & s \leq \mu \\[2ex] \dfrac{(2r - c)(s - \mu + v)^2}{8v} + \dfrac{vcr}{2r - c} & \mu < s \leq a^* \\[2ex] 0 & a^* \leq s \end{cases} \tag{2-19}
$$

If $s \leq \mu$, the relative error, namely, $\Delta(s)/Q(s)$, ranges from 0 when $v \to 0$ to 1 when $c = s = 0$ and $v \to \mu$. More generally, from (2-19) we conclude that the relative error of replacement of random quantities by their means depends on the extent of the randomness (v), the initial situation (s), the location of the probability distribution (μ), and economic parameters (r and c). It is insufficient to check only whether or not the variance of D, namely, $v^2/3$, is small. "Small" must be interpreted with respect to the considerations above. For example, choose a "small" variance, say $v^2 = 10^{-4}$; if $c = s = 0$ and $2\mu = v = 10^{-2}$, the relative error is 0.5.

EXERCISES

2-1 Show that for fixed s, max $\{b(s, a)\colon a \geq s\}$ is achieved by (2-7).

2-2 Suppose $2r < 3c$ but $r \geq c$ so $c \leq r < 3c/2$. What expression for $\Delta(s)$ replaces (2-19)?

2-3 Let the r.v. C be defined by

$$C \triangleq c(a - s) - \frac{c(a - D)^+}{2} + r(D - a)^+$$

The terms of C are the purchase costs, the salvage value, and the revenue foregone because more items are demanded than are stocked. The maximization of $E(B)$ in (2-13) with respect to s and a is equivalent to minimization of $E(C)$ with respect to s and a. Verify that (2-17) minimizes $E(C)$ for each fixed s.

2-3 EXPECTED PRESENT VALUE

Most of us prefer to receive revenues sooner rather than later and to defer the payment of expenses as long as possible. How does this attitude affect the amount that you would be willing to pay now for a guarantee of receiving \$1 in 1 year? A dollar in hand now can be invested in an insured savings account and earn a riskless return at interest rate i, say, annually. Your account balance after 1 year is $\$(1 + i)$; it increases to $\$(1 + i)^2$ after 2 years and to $\$(1 + i)^n$ after n years.

Let $\$A$ be the amount that you would be willing to pay now in order to receive \$1 in 1 year. If you did not make the payment, you could bank the $\$A$ and would be sure to have $\$A(1 + i)$ after 1 year. But you are receiving \$1 in 1 year in exchange for $\$A$ now; so you should insist that

$$\$A(1 + i) \le \$1$$

or $A \le \beta$ where

$$\beta = \frac{1}{1 + i} \tag{2-20}$$

Similarly, you should not be willing to pay more than $\$A\beta^n$ now in exchange for A dollars that you would receive n years from now. It is important to observe that $i > 0$ implies $\beta < 1$. The quantity β is called a *discount factor*.

More generally, how much might you be willing to pay in return for guaranteed payments of x_1 now, x_2 after 1 year passes, x_3 after 2 years pass, ..., x_N after $N - 1$ years pass? That is, x_n is the payment you are sure to receive at the beginning of the nth year; so for each n you are willing to spend perhaps as much as $\beta^{n-1}x_n$ now in return for the later payment. Summing over n,

$$\sum_{n=1}^{N} \beta^{n-1} x_n$$

is called the *present value* (sometimes called the *discounted value*) of the time stream $x_1, x_2, \ldots, x_N$. The present value is the most that you should be willing to pay now in return for guaranteed payments of $x_1, x_2, \ldots, x_N$ as time passes.

The foregoing discussion obeys tradition and ignores important practical considerations of inflation, income taxes, and transaction costs. For the remainder of this section, we suppress the effects of inflation with the assumption that all costs and revenues are adjusted for inflation. For example, suppose there is 10 percent inflation during the first year and 12 percent during the second. Then \$1 after 2 years has the same purchasing power as $\$1/(1.12 \times 1.10)$ today. So we replace $\$x_2$ in 2 years with $\$x_2/(1.12 \times 1.10)$. *Henceforth, we assume that costs and revenues have already been adjusted in this way.*

The discussion that leads to the present-value formula has a fundamental flaw. Most enterprises expect internal investments to be more lucrative than an insured savings account at a fixed interest rate. Examples of such investments for a private enterprise include advertising campaigns, expansion of the sales force, replacement of older machinery with newer and more efficient machinery, expansion of manufacturing capacity, increased support of trade association lobbyists, and addition of new product lines. Examples of such investments for a public housing authority, for example, include replacement of old maintenance vehicles with new ones, expansion of the number of maintenance workers to improve the quality of maintenance service to tenants, establishment of social programs for tenants, and modernization of older apartments.

In other words, most enterprises act as if their purposes are better served by internal allocations of resources than by long-term investments in riskless bank accounts. Their discount factors, determined as in (2-20) by $\beta = (1 + i)^{-1}$, should be based on a numerical value of i that is higher than the prevailing annual interest rate for riskless investments (but investments of what duration?). The determination of an appropriate value of β in a large enterprise can be quite difficult and involves tax considerations and concepts from economics, cost accounting, and finance.

Section 2-5 presents an axiomatic basis for comparing alternative investments via the present values of their time streams $(x_1, \ldots, x_N)$. The axiomatic approach yields sufficient conditions for a rational decision maker to choose an investment whose present value is maximal.

The literature in finance offers numerous alternatives to present value.† However, present value and its close relative, internal rate of return, are used in financial practice more frequently than other criteria. Moreover, present value is used in conjunction with operations research models far more often than any other criterion.

A shortcoming of the previous discussion is the deceptive manner in which effects of wealth were ignored. For example, it was asserted that "... you should not be willing to pay more than $\$A\beta^n$ now in exchange for A dollars that you would receive n years from now." But suppose your wealth totals only $\$(A\beta^n - 1)$. Should you be willing to spend all of it on this investment opportunity? If not, then how much? The answers to these questions depend not only on your present wealth but also on many other considerations such as your anticipated income from other sources during years $1, 2, \ldots, n-1, n, n+1, \ldots$, your notions of your own mortality, your bequest motives, and the amount of life insurance that you carry. A model in which investment decisions are influenced by levels of wealth is presented in Section 8-3.

The role of uncertainty in this discussion of time preferences is conspicuous by its absence. The time streams of actual costs and revenues which result from most investments are stochastic processes, inflation rates are uncertain, personal and organizational lifetimes have uncertain durations, tax laws may change as time passes, and unanticipated investment opportunities may be available in the future.

How should uncertainty influence the comparison of alternative investments? Let $X_1, X_2, \ldots, X_N$ be the sequence of r.v.'s with generic outcome $x_1, x_2, \ldots, x_N$. For each investment, we might construct the joint distribution (function) for $X_1, \ldots, X_N$. Then we might compare investments via comparisons of the joint distributions. In fact, there are at least three reasons for avoiding this procedure in practical problems.

† Solomon (1959) surveys other methods. See the Bibliographic Guide at the end of the chapter for the full citation and for other references.

The joint distribution of $X_1, \ldots, X_N$ is a function with N arguments; so it requires much computation and storage space to approximate the function reasonably well. For N larger than 3 or 4, the effort and space become prohibitively great. Keep in mind that the joint distribution would be needed for each of perhaps many alternative investments.

Another reason to avoid explicit computation of the joint distributions is that human beings have limited information-processing capabilities. If $N > 2$, we do not yet know how to present tabulations of functions with N argument so that humans would detect subtle differences between functions.

There is a third reason to avoid explicit computation of the joint distribution of $x_1, \ldots, x_N$ for each alternative investment. Each investment generates a probability space whose sample space consists of the sample paths of $x_1, \ldots, x_N$. Each outcome ω in the sample space is an N-tuple of real numbers $x_1(\omega)$, $x_2(\omega)$, $\ldots$, $x_N(\omega)$. We have already argued that a reasonable summary of this sequence is its present value

$$\sum_{n=1}^{N} \beta^{n-1} x_n(\omega)$$

Hence, the present value is an r.v. This observation suggests the comparison of alternative investments via the probability distributions of the present values of the investments. Each such distribution function is one-dimensional.

We have reduced the problem of comparing N-dimensional joint distributions to the problem of comparing one-dimensional distributions of the present value. Even the latter problem begs for simplification. The most frequent simplification is to represent each investment's distribution of present value by the expected value of the distribution.† We call this number the *expected present value* of the investment.

Expected present value is the most important criterion for the analysis of stochastic time-varying models of practical problems. Usually, we should wish to know also the variance (and other characteristics) of the probability distribution of present value. Nevertheless, probability distributions are more often compared via their means than with any other criterion. The next section presents an axiomatic justification for this criterion.

2-4 EXPECTED-UTILITY CRITERION

This section contains necessary and sufficient conditions for decisions based on expected values to be consistent with a decision maker's preferences. See DeGroot (1970) or Luce and Raiffa (1957) for other formal approaches to making decisions under uncertainty.

For simplicity, we consider a single-period investment whose profit can take only finitely many values. Specifically, let $\mathscr{P}$ denote the set of all probability

† For a probability distribution on $\mathbb{R}$ with distribution function $F(\cdot)$, we say that $\int_{-\infty}^{\infty} x\,dF(x)$ is the expected value of the *distribution*.

distributions on the sample space $\{1, 2, \ldots, K\}$ for some $K \in I_+$. Therefore, $\mathbf{p} = (p_1, \ldots, p_K) \in \mathscr{P}$ means that $\mathbf{p} \geq 0$ and $\sum_{i=1}^{K} p_i = 1$. Suppose that you face a decision problem whose alternatives will generate payoffs that are r.v.'s with distributions in $\mathscr{P}$. Then we present necessary and sufficient conditions for the existence of $\mathbf{u} = (u_1, \ldots, u_K) \in \mathbb{R}^K$ such that, for every $\mathbf{p} \in \mathscr{P}$ and $\mathbf{q} \in \mathscr{P}$, you prefer $\mathbf{p}$ to $\mathbf{q}$ if, and only if,

$$\sum_{i=1}^{K} p_i u_i > \sum_{i=1}^{K} q_i u_i \tag{2-21}$$

Such a K-vector $\mathbf{u}$ is called a *utility function.*

Here is an alternative interpretation of $\mathbf{u}$. Let X and Y be two r.v.'s representing payoffs with distributions $\mathbf{p}$ and $\mathbf{q}$ so $p_i = P\{X = i\}$ and $q_i = P\{Y = i\}$. Let $u(\cdot)$ denote the real-valued function on $\{1, \ldots, K\}$ with $u(i) = u_i$. Then (2-21) states that

$$E[u(X)] > E[u(Y)] \tag{2-22}$$

Hence, we present conditions for the existence of $u(\cdot)$ such that X is preferred to Y (equivalently, $\mathbf{p}$ is preferred to $\mathbf{q}$) if, and only if, the expected utility of X exceeds the expected utility of Y.

Such a result is interesting and useful for several reasons. First, it reduces the comparison of K-dimensional distributions to a one-dimensional comparison such as (2-21) and (2-22). This reduction is important in practice because it eliminates the actual need to compare large numbers of distributions. A second advantage of the reduction is that, as we shall see, it suggests procedures for estimating $\mathbf{u}$. Third, the reduction permits the analysis of decision problems as if the analyst had the same attitude to risk as the actual decision maker. For these reasons, and because many extant methods use expected-value criteria, most of this volume concerns the comparison of expected values of complicated r.v.'s. Also, the result illuminates normative theories in economics, political science, and psychology. The basic theorem is due to John von Neumann and Oskar Morgenstern,† who included it in their foundation of game theory.

During recent years, numerous applications of operations research have included assessments of decision makers' utility functions. Some of these applications are location decisions, capacity expansion, forest pest management, oil and natural gas drilling, new product development, oil tanker safety, and health care management. The term *decision analysis* refers to normative approaches to decision making which use utility functions, "judgmental" probabilities, or both. A recent‡ special issue of *Operations Research* reflects current research and applications of decision analysis.

In this volume, preference orderings are discussed in five places: here, in the next section, and in Chapters 4, 6, and 9. We shall briefly explain why we use

† J. von Neumann and O. Morgenstern, *Theory of Games and Economic Behavior*, Princeton University Press, Princeton, N.J. (1944).

‡ *Oper. Res.*, **28**(1) (1980).

binary relations† to describe preferences in each case. Suppose that we wish to describe a decision maker's preferences for elements of the set $\{a, b, c\}$. In this section, each element of the set is a probability distribution. A preference ordering based on a binary relation describes (at most) whether or not the decision maker thinks a is preferred to b, a is preferred to c, b is preferred to a, b is preferred to c, c is preferred to a, and c is preferred to b. More generally, if the set has L elements, the decision maker is asked to make at most $L(L - 1)$ comparisons.

Suppose, in addition, that we invite the decision maker to tell us whether or not a is preferred to the better of b and c. In other words, a is compared with $\{b, c\}$. By asking for all such responses, the six comparisons above must be augmented by a vs. $\{b, c\}$, b vs. $\{a, c\}$, and c vs. $\{a, c\}$. Hence, the number of comparisons has increased. Also, we have ruled out the possibility of obtaining an unambiguous rank ordering of the elements, such as a is preferred to b, b is preferred to c, and a is preferred to c. Then the rank ordering a, b, c would be natural.

Binary preference orderings are simpler and more natural than other kinds; so they underlie major normative theories in the social sciences. In this volume, preferences are described only with binary relations. Hence, we pause to introduce definitions.

Binary Relations

Let V be a nonempty set and $W \subset V \times V$. The set W of ordered pairs of elements of V is called a binary relation. If $(p, q) \in W$, it is convenient to use the notation $p \succ q$ and $q \prec p$. If $(p, q) \notin W$, we write $p \not\succ q$. Here are labels for binary relations with various properties.

Definition 2-1 Let V be a nonempty set with $W \subset V \times V$. Then W is a *binary relation* on V.

(*a*) W is *reflexive* if $p \succ p$ for all $p \in V$.
(*b*) W is *irreflexive* if $p \not\succ p$ for all $p \in V$.
(*c*) W is *transitive* if, for all p, q, and r in V, $p \succ q$ and $q \succ r$ implies $p \succ r$.
(*d*) W is *complete* if, for all p and q in V, either $p \succ q$ or $q \succ p$ or both $p \succ q$ and $q \succ p$.
(*e*) W is *symmetric* if, for all p and q in V, $p \succ q$ implies $q \succ p$.
(*f*) W is *antisymmetric* if, for all p and q in V, $p \succ q$ and $q \succ p$ implies $p = q$.
(*g*) W is *asymmetric* if, for all p and q in V, $p \succ q$ implies $q \not\succ p$.
(*h*) W is *negatively transitive* if, for all p, q, and r in V, $p \not\succ q$ and $q \not\succ r$ implies $p \not\succ r$.

Combinations of these properties lead to further labels.

† A binary relation is a set of ordered pairs.

Definition 2-2 Let W be a binary relation on V.

(*a*) W is a *preorder*† [and (W, V) is a *preordered set*] if W is transitive and reflexive.
(*b*) W is a *complete order* [and (W, V) is a *completely ordered set*] if W is transitive and complete.
(*c*) W is a *partial order* [and (W, V) is a *partially ordered set*] if W is an antisymmetric preorder.
(*d*) W is a *weak order* [and (W, V) is a *weakly ordered set*] if W is asymmetric and negatively transitive.
(*e*) W is an *equivalence relation* if W is reflexive, symmetric, and transitive.

Definition 2-3 Let (W, V) and (X, Y) be partially ordered sets and let f be a function with domain V and range Y. Then f is an *isotone function* if

$$p \succ q \Rightarrow [f(p), f(q)] \in X$$

And f is an *antitone function* if

$$p \succ q \Rightarrow [f(q), f(p)] \in X$$

In words, an isotone function takes ordered elements in V into ordered elements in Y and an antitone function reverses the order in Y.

We manipulate weak orders in this section; so the following result is useful.

Proposition 2-1 Suppose W is a weak order on V. Then:

(*a*) W is transitive.
(*b*) $Z \triangleq \{(p, q): p \not\succ q \text{ and } q \not\succ p\}$ is an equivalence relation.
(*c*) For all p, q, and r in V: (i) if $p \succ q$ and $(q, r) \in Z$, then $p \succ r$; and (ii) if $(p, q) \in Z$ and $q \succ r$, then $p \succ r$.
(*d*) Let $Y = W \cup Z$. Then Y is transitive and complete on V.

PROOF Left as Exercise 2-1. □

Preference Orderings for Distributions

Recall that $\mathscr{P}$ denotes the set of all probability distributions on a particular sample space with finitely many elements. Specifically, for a positive integer K, say, $\mathscr{P}$ contains all the nonnegative K-vectors whose components sum to 1. For $\mathbf{p} = (p_i) \in \mathscr{P}$, $\mathbf{q} = (q_i) \in \mathscr{P}$, and $0 \le \alpha \le 1$, let $\alpha\mathbf{p} + (1 - \alpha)\mathbf{q}$ denote the K-vector whose ith component is $\alpha p_i + (1 - \alpha)q_i$. Note that $\alpha\mathbf{p} + (1 - \alpha)\mathbf{q} \in \mathscr{P}$.

Let W be a binary relation on $\mathscr{P}$. We interpret $\mathbf{p} \succ \mathbf{q}$ as "distribution $\mathbf{p}$ is preferred to distribution $\mathbf{q}$."

† Some writers use the labels *partial order* and *quasi-order* for our preorder.

It is convenient to use the following alternative notation for pairs of distributions in W, Z [part (*b*) of Definition 2-1], and Y [part (*d*) of Definition 2-1]:

(*a*) If $\mathbf{p} \not\succ \mathbf{q}$ and $\mathbf{q} \not\succ \mathbf{p}$, then write $\mathbf{p} \sim \mathbf{q}$ or $\mathbf{q} \sim \mathbf{p}$.

(*b*) If $\mathbf{p} \succ \mathbf{q}$ or $\mathbf{p} \sim \mathbf{q}$, then write $\mathbf{p} \succsim \mathbf{q}$ or $\mathbf{q} \precsim \mathbf{p}$.

(*c*) If $\mathbf{p} \succ \mathbf{q}$ and $\mathbf{q} \succ \mathbf{r}$, then write $\mathbf{p} \succ \mathbf{q} \succ \mathbf{r}$ (similarly with $\succsim$ and $\sim$).

The pairs in Z and Y, respectively, correspond to $\mathbf{p} \sim \mathbf{q}$ and $\mathbf{p} \succsim \mathbf{q}$. Recall that we describe $(\mathbf{p}, \mathbf{q}) \in W$, hence $\mathbf{p} \succ \mathbf{q}$, as meaning that "distribution $\mathbf{p}$ is preferred to distribution $\mathbf{q}$." Therefore, we interpret $\mathbf{p} \sim \mathbf{q}$ as "indifference between $\mathbf{p}$ and $\mathbf{q}$" and $\mathbf{p} \succsim \mathbf{q}$ as "$\mathbf{p}$ is as good as $\mathbf{q}$."

We emphasize that $\mathbf{p} \succ \mathbf{q}$ implies that $\mathbf{p}$ and $\mathbf{q}$ are comparable. However, $\mathbf{p} \not\prec \mathbf{q}$ can be written either when $\mathbf{p} \succ \mathbf{q}$, hence $\mathbf{p}$ and $\mathbf{q}$ are comparable, or when $\mathbf{p}$ and $\mathbf{q}$ are not comparable. Hence, $\mathbf{p} \succ \mathbf{q} \Rightarrow \mathbf{p} \not\prec \mathbf{q}$ but not conversely.

Our definition of $\sim$ [Z in part (*b*) of Proposition 2-1] forces $\succsim$ to be complete even if $\succ$ is not complete. Some expositions of utility theory use $\succsim$ as the basic relation, assume completeness of $\succsim$, and then define $\succ$ and $\sim$ as derived relations. Their assumption of completeness for $\succsim$ is no stronger than our assumptions.

Example 2-1 If $K = 2$, then $\mathscr{P}$ is the set of distributions on $\{1, 2\}$. For $\mathbf{p} \in \mathscr{P}$ let $\mathbf{p}^x$ denote $(1 - x, x)$ so the mean and variance of $\mathbf{p}^x$ are $x + 1$ and $x(1 - x)$, respectively. Suppose that some of the distributions in $\mathscr{P}$ correspond to random payoffs in a decision problem. Decision makers usually behave as if they prefer high means and low variances; so $\mathbf{p}^x = (1 - x, x)$ would be preferred to $p^y = (1 - y, y)$ if $1/2 \le y < x \le 1$. In symbols,

$$1/2 \le y < x \le 1 \Rightarrow \mathbf{p}^x \succ \mathbf{p}^y \tag{2-23}$$

Let W consist of the pairs of distribution specified in (2-23):

$$W = \{(\mathbf{p}^x, \mathbf{p}^y)\colon 1/2 \le y < x \le 1\}$$

This is a weak order on $\mathscr{P}$. It is not reflexive; so it is neither a preorder nor a partial order. It is not complete on $\mathscr{P}$ (suppose $0 \le x < 1/2$); so it fails to be a complete order.

Let $\mathbf{u} = (u_1, u_2)$ be a vector with $u_1 < u_2$ and let $E[\mathbf{u}(\mathbf{p}^x)]$ denote $(1 - x)u_1 + xu_2$. Suppose also that $1/2 \le y < x \le 1$. Then

$$\mathbf{p}^x \succ \mathbf{p}^y \Leftrightarrow E[\mathbf{u}(\mathbf{p}^x)] > E[\mathbf{u}(\mathbf{p}^y)]$$

Of course, you do not need a utility function to mimic a decision maker's preferences in this case; $\mathbf{p}^x$ is preferred to $\mathbf{p}^y$ if $x > y \ge 1/2$.

Suppose we wish to describe preferences when x or y or both x and y are between 0 and 1/2. In this range, as x increases, the mean and variance of $\mathbf{p}^x$ both increase. Therefore, a description of preferences in this range must address the trade-off between mean and variance. Different decision makers may prefer markedly different trade-offs; so they may have different utility functions. □

It is convenient to extend the notation $E[\mathbf{u}(\mathbf{p})]$ introduced in Example 2-1 to the general case. Let $\mathbf{u} = (u_1, \ldots, u_K) \in \mathbb{R}^K$, i.e., $\mathbf{u}$ is any real-valued function on the sample space $\{1, \ldots, K\}$. For any $\mathbf{p} = (p_1, \ldots, p_K) \in \mathscr{P}$,

$$E[\mathbf{u}(\mathbf{p})] \triangleq \sum_{i=1}^{K} p_i u_i \tag{2-24}$$

Expected Utility

Here are the main results. Their proofs comprise Section 2-6.

Theorem 2-1 Suppose that $\succ$ is a binary relation on the set $\mathscr{P}$ of all distributions on a finite sample space. There exists $\mathbf{u}$ such that

$$\mathbf{p} \succ \mathbf{q} \Leftrightarrow E[\mathbf{u}(\mathbf{p})] > E[\mathbf{u}(\mathbf{q})] \tag{2-25}$$

for all $\mathbf{p}$ and $\mathbf{q}$ in $\mathscr{P}$ if, and only if, $\succ$ has the following properties:

(*a*) $\succ$ is asymmetric on $\mathscr{P}$ ($\mathbf{p} \succ \mathbf{q} \Rightarrow \mathbf{q} \not\succ \mathbf{p}$).
(*b*) $\succ$ is negatively transitive on $\mathscr{P}$ ($\mathbf{p} \not\succ \mathbf{q}$ and $\mathbf{q} \not\succ \mathbf{r} \Rightarrow \mathbf{p} \not\succ \mathbf{r}$).
(*c*) For all $\mathbf{p}$, $\mathbf{q}$, and $\mathbf{r}$ in $\mathscr{P}$ and $0 < \alpha < 1$,

$$\mathbf{p} \succ \mathbf{q} \Rightarrow \alpha\mathbf{p} + (1 - \alpha)\mathbf{r} \succ \alpha\mathbf{q} + (1 - \alpha)\mathbf{r} \tag{2-26}$$

(*d*) For all $\mathbf{p}$, $\mathbf{q}$, and $\mathbf{r}$ in $\mathscr{P}$, if $\mathbf{p} \succ \mathbf{q} \succ \mathbf{r}$ then there exists $\alpha \in (0, 1)$ such that

$$\alpha\mathbf{p} + (1 - \alpha)\mathbf{r} \succ \mathbf{q} \succ \alpha\mathbf{r} + (1 - \alpha)\mathbf{p} \tag{2-27}$$

The "if, and only if" in Theorem 2-1 yields two statements. First, if the preference ordering satisfies (*a*) through (*d*), then it can be represented by a utility function. Second, a utility function can be obtained only if the preference ordering satisfies (*a*) through (*d*). Incidentally, (2-25) does *not* imply "$\mathbf{p} \succ \mathbf{q}$ because $E[\mathbf{u}(\mathbf{p})] > E[\mathbf{u}(\mathbf{q})]$." We obtain from (2-25) that $E[\mathbf{u}(\cdot)]$ mimics the preference ordering; it does not *cause* the ordering to occur. Sometimes we say that "the utility function $u(\cdot)$ is consistent with the preference ordering $\succ$."

Theorem 2-1 establishes the existence of a utility function which permits distributions to be compared via their expected utilities. However, there is insufficient structure to determine the amount by which $\mathbf{p}$ is preferred to $\mathbf{q}$, when $\mathbf{p} \succ \mathbf{q}$. The underlying reason is that preferences are ordinal,† not cardinal.

Properties (*a*) through (*d*) in Theorem 2-1 are not innocuous. For example, $\alpha\mathbf{p} + (1 - \alpha)\mathbf{r}$ in (2-26) can be interpreted as the distribution induced by the following two-stage randomization. At the first stage, $\mathbf{p}$ or $\mathbf{r}$ is selected with respective probabilities α and $1 - \alpha$. At the second stage, whatever distribution was selected at the first stage describes the likelihood of outcomes in the sample space $\{1, \ldots, K\}$. The distribution $\alpha\mathbf{q} + (1 - \alpha)\mathbf{r}$ has a similar two-stage interpre-

† Intuitively, an ordinal relationship is a rank ordering while a cardinal relationship measures the "distances" between successively ranked items. For precise definitions, see pages 109 and 127 in P. Suppes, *Axiomatic Set Theory*, Van Nostrand, Princeton, N.J. (1960).

tation. Then (*c*) asserts that, if **p** is preferred to **q**, then every two-stage mixture of **p** and **r** is preferred to the comparable mixture of **q** and **r**. Moreover, this must be true for every **r**. In Example 2-2 we discover that W in Example 2-1 fails to satisfy (*c*). Following Example 2-2, we critique properties (*a*) through (*d*).

The next theorem shows that the utility function $u(\cdot)$ of Theorem 2-1 is not unique. Its scale and origin can be changed at will.

Theorem 2-2 If $\mathbf{u} \in \mathbb{R}^K$ satisfies (2-25), then $\mathbf{v} \in \mathbb{R}^K$ satisfies

$$\mathbf{p} \succ \mathbf{q} \Leftrightarrow E[\mathbf{v}(\mathbf{p})] > E[\mathbf{v}(\mathbf{q})]$$

for all **p** and **q** in $\mathscr{P}$ if, and only if, there are numbers $c > 0$ and b such that $\mathbf{v} = c\mathbf{u} + b$ (that is, $u_i = cv_i + b$ for all i).

The conclusion of Theorem 2-2 is sometimes described as *uniqueness up to a positive linear transformation.*

The case where $\mathbf{u} = (u_i)$ has $u_i = b + ci$ for some $c > 0$ and all i is particularly important. Then

$$E[\mathbf{u}(\mathbf{p})] = \sum_{i=1}^{K} u_i p_i = b + c \sum_{i=1}^{K} ip_i$$

so (2-25) $\{p \succ q \Leftrightarrow E[\mathbf{u}(\mathbf{p})] > E[\mathbf{u}(\mathbf{q})]\}$ becomes

$$\mathbf{p} \succ \mathbf{q} \Leftrightarrow \sum_{i=1}^{K} ip_i > \sum_{i=1}^{K} iq_i$$

In words, $\mathbf{p} \succ \mathbf{q}$ if, and only if, the mean of **p** exceeds the mean of **q**. This case is labeled *linear utility*. It reduces the comparison of random variables to the comparison of their mean values.

Example 2-2 In Example 2-1, any (u_1, u_2) such that $u_1 < u_2$ satisfies (2-25); so Theorem 2-2 does not apply. Why? Properties (*a*), (*b*), and (*d*) in Theorem 2-1 are satisfied. However, property (*c*) fails to hold. For example, take $\mathbf{p} = (0, 1)$, $\mathbf{q} = (1/2, 1/2)$, and $\mathbf{r} = (1, 0)$. Then (2-26) requires, *for all* $0 < \alpha < 1$,

$$(0, \alpha) + (1 - \alpha, 0) \succ \left(\frac{\alpha}{2}, \frac{\alpha}{2}\right) + (1 - \alpha, 0)$$

which is $(1 - \alpha, \alpha) \succ (1 - \alpha/2, \alpha/2)$. But $0 < \alpha < 1/2$ causes $[(1 - \alpha, \alpha), (1 - \alpha/2, \alpha/2)] \notin W$ so $(1 - \alpha, \alpha) \not\succ (1 - \alpha/2, \alpha/2)$.

Suppose that W in Example 2-1 is replaced by W^*, consisting of all pairs $(\mathbf{p}^x, \mathbf{p}^y)$ such that $\mathbf{p}^x \succ \mathbf{p}^y$ below:

$$\left.\begin{array}{l} 2x^2 - x > 2y^2 - y \\ 0 \le x \le 1,\ 0 \le y \le 1 \end{array}\right\} \Rightarrow \mathbf{p}^x \succ \mathbf{p}^y$$

This preference order corresponds to a criterion of (mean)-2 · (variance). Again, this preference order violates (2-26). Take $\mathbf{p} = (0, 1)$, $\mathbf{q} = (3/4, 1/4)$,

$\mathbf{r} = (1, 0)$, and $\alpha = 1/4$. Then $\alpha\mathbf{p} + (1 - \alpha)\mathbf{r} = \mathbf{q} \prec \alpha q + (1 - \alpha)\mathbf{r} \prec \mathbf{p}$, which is contrary to (2-26). Indeed, with W^*, no $\mathbf{u} = (u_1, u_2)$ can satisfy (2-25). From Theorem 2-2, $u_1 = 0$ without loss of generality; so we seek u_2 such that

$$2x^2 - x > 2y^2 - y \Leftrightarrow u_2 \cdot (x - y) > 0$$

for all x and y in $[0, 1]$. No such u_2 exists. In a few pages, we recall W^* and argue that (*c*) is reasonable whereas W^* is unreasonable.

Suppose that W is redefined, again, according to the criterion (mean) − (variance). Then $\mathbf{p}^x \succ \mathbf{p}^y \Leftrightarrow x^2 > y^2$. This preference order *does* satisfy (2-26) (and the other conditions in Theorem 2-1). Take $\mathbf{u} = (0, 1)$ so

$$E[\mathbf{u}(\mathbf{p}^x)] = x$$

and $\mathbf{p}^x \succ \mathbf{p}^y \Leftrightarrow E[\mathbf{u}(\mathbf{p}^x)] > E[\mathbf{u}(\mathbf{p}^y)]$, which is (2-25). But any (u_1, u_2) with $u_1 < u_2$ also satisfies (2-25) because

$$E[\mathbf{u}(p^x)] = (1 - x)u_1 + xu_2 = u_1 + (u_2 - u_1)x$$

Then every such (u_1, u_2) must be a positive linear transformation of (0, 1). In the notation of Theorem 2-2,

$$u_1 = c \cdot 0 + b \qquad \text{and} \qquad u_2 = c \cdot 1 + b$$

so $b = u_1$ and $c = u_2 - u_1 > 0$. □

Critique

How reasonable are the conditions of Theorem 2-1? The first condition, asymmetry of $\succ$, asserts: $\mathbf{p} \succ \mathbf{q} \Rightarrow \mathbf{q} \not\succ \mathbf{p}$. This is consistent with $\succ$ as "strong" preference in comparison with $\succsim$ as "weak" preference. The significance of the second condition, negative transitivity of $\succ$, is unclear. In conjunction with asymmetry, it asserts that $\succ$ is a weak order. From Proposition 2-1, the consequences include:

(*a*) $\succ$ is transitive.
(*b*) $\sim$ is reflexive, symmetric, and transitive.
(*c*) $\succsim$ is transitive and complete.

Transitivity is sometimes given as a definition of rational behavior. However, it leads to two kinds of problems. The first problem is that people sometimes exhibit *in*transitivities in $\succ$ on P. When these are pointed out, an individual typically revises $\succ$ to make it transitive. Nevertheless, there are theories of stochastic choice behavior which admit intransitivity. See DeGroot (1970) for an exposition and references; Fishburn (1978) and Whitt (1979) have recent contributions.

A second transitivity problem stems from (*b*) above, namely, transitivity of indifference. For example, I am indifferent between n and $n + 1$ grains of salt on my omelet. Therefore, *if* indifference is transitive, I am indifferent between having an omelet without any salt and having an omelet with a pound of salt! Another example, from Fishburn (1970), has $\mathbf{p}$ putting probability 1 on \$35, $\mathbf{q}$ putting probability 1 on \$36, and $\mathbf{r}$ putting equal probability on \$0 and \$100. Certainly $\mathbf{p} \prec \mathbf{q}$, but it is conceivable that $\mathbf{p} \sim \mathbf{r}$ and $\mathbf{q} \sim \mathbf{r}$, in which case $\sim$ is not transitive. Again, various theories have weakened this assumption. See Fishburn (1970) for an explanation and references.

Condition (*c*) in Theorem 2-1, namely, (2-26), is

$$\mathbf{p} \succ \mathbf{q} \Rightarrow \alpha\mathbf{p} + (1 - \alpha)\mathbf{r} \succ \alpha\mathbf{q} + (1 - \alpha)\mathbf{r}$$

for all $\mathbf{p}$, $\mathbf{q}$, and $\mathbf{r}$ and any $\alpha \in (0, 1)$. Suppose that we interpret $\alpha\mathbf{p} + (1 - \alpha)\mathbf{r}$ as the distribution which corresponds to a two-stage randomization [described below (2-27)]. Now (2-26) asserts that, if $\mathbf{p}$ is preferred to $\mathbf{q}$, the two-stage mixture of $\mathbf{p}$ and $\mathbf{r}$ should be preferred to the two-stage mixture of $\mathbf{q}$ and $\mathbf{r}$. Moreover, this should be true for every $\mathbf{r}$. In other words, $\mathbf{r}$ as an alternative to $\mathbf{p}$ or $\mathbf{q}$ should not affect the preference for $\mathbf{p}$ over $\mathbf{q}$. We call this an *independence condition* because (2-26) asserts that preferences should be independent of irrelevant alternatives.

But is $\mathbf{r}$ in (2-26) an "irrelevant alternative?" In Example 2-2, with W replaced by W^*, we saw that (2-26) is violated. Is W^* reasonable? The criterion implicit in W^*, (mean) $-2 \cdot$ (variance), is a plausible trade-off of mean vs. variance. However, it implies that $\mathbf{p}^0 \succ \mathbf{p}^x$ for all $x \in (0, 1/2)$. In words, a certain reward of 1 is preferred to an uncertain reward that is, nevertheless, at least 1. We believe that most people would revise their descriptions of preferences when they appreciate this consequence of W^*. That is, they would keep (*c*) and discard the criterion of mean $-2 \cdot$ (variance) for two-point distributions.

The fourth condition in Theorem 2-1 is that, if $\mathbf{p} \succ \mathbf{q} \succ \mathbf{r}$, there exist nontrivial mixtures of $\mathbf{p}$ and $\mathbf{r}$ which are less preferred and more preferred than $\mathbf{q}$. Specifically, there is $\alpha \in (0, 1)$ such that

$$\alpha\mathbf{p} + (1 - \alpha)\mathbf{r} \succ \mathbf{q} \succ \alpha\mathbf{r} + (1 - \alpha)\mathbf{p}$$

This *continuity* assumption, in effect, insists that outcomes in the sample space be comparable.

Consider the following example from Luce and Raiffa (1957). We prefer \$1 to \$0.01 and that to death. However, is there any probability $\alpha < 1$ such that the composite gamble with outcomes \$1, with probability α, and death, with probability $1 - \alpha > 0$, is preferred to \$0.01 (with probability 1)? Perhaps $\alpha = 1 - 10^{-100,000,000,000}$. Do we accept much higher probabilities of death each time we cross a traffic-laden street or drive a car? Fortunately, the sample spaces in most applications of operations research do not include such disparate outcomes.

Example 2-3 Your friend invites you to join him in a mining venture. The mine has \$$M$ worth of silver, with probability α, and \$0, with probability $1 - \alpha$. You can have a share of this mine for $B = M/10^3$. However, M is so large that, in order to raise B, you must refinance your house, sell all your other assets, and take out personal loans. If there is no silver, you will be ruined.

If your preference ordering satisfies the continuity assumption, there is a sufficiently large value of $\alpha < 1$ at which you prefer the investment to standing pat. Some people live in nice homes because other people believed the theorem! □

Construction of a Utility Function

Recall that $E[\mathbf{u}(\mathbf{p})]$ denotes $\sum_{i=1}^{K} u_i p_i$ where $\mathbf{u} = (u_i)$ and $\mathbf{p} = (p_i)$. Theorem 2-1 provides necessary and sufficient conditions for the existence of $\mathbf{u}$ such that

$$\mathbf{p} \succ \mathbf{q} \Leftrightarrow E[\mathbf{u}(p)] > E[\mathbf{u}(\mathbf{q})] \tag{2-28}$$

Theorem 2-2 asserts that $\mathbf{u}$ is unique up to a positive linear transformation. Here we describe a method to construct $\mathbf{u}$ which is based on the proof in Section 2-6 of Theorem 2-1.

Let $\mathbf{e}_i$ denote the ith unit vector in $\mathbb{R}^K$ (zero except for one in the ith coordinate). The unit vectors $\mathbf{e}_1, \ldots, \mathbf{e}_K$ are distributions in $\mathscr{P}$. Also, $E[u(\mathbf{e}_i)] = u_i$, $i = 1, \ldots, K$, which leads to a method for assessing $u_1, \ldots, u_K$. In this subsection, it simplifies the exposition if we assume that the outcomes in the sample space $\{1, \ldots, K\}$ are possible rewards \$1, ..., \$$K$ so $\mathbf{e}_1 \prec \mathbf{e}_2 \prec \cdots \prec \mathbf{e}_K$.

The proof of Theorem 2-1 shows that, for each $\mathbf{p} \in \mathscr{P}$, there is a unique number $\alpha(\mathbf{p}) \in (0, 1)$ such that

$$\begin{aligned} \mathbf{p} &\sim [1 - \alpha(\mathbf{p})]\mathbf{e}_1 + \alpha(\mathbf{p})\mathbf{e}_K \\ &= (1 - \alpha(\mathbf{p}), 0, 0, \ldots, 0, 0, \alpha(\mathbf{p})) \end{aligned} \tag{2-29}$$

Let α_i denote $\alpha(\mathbf{e}_i)$, $i = 1, \ldots, K$. Then (2-29) with $\mathbf{p}$ replaced by $\mathbf{e}_i$ is

$$\mathbf{e}_i \sim (1 - \alpha_i)\mathbf{e}_1 + \alpha_i \mathbf{e}_K = (1 - \alpha_i, 0, \ldots, 0, \alpha_i) \tag{2-30}$$

Moreover, $0 = \alpha_1 < \alpha_2 < \cdots < \alpha_K = 1$ for the following reasons. Since $\sim$ is reflexive (it is an equivalence relation due to Proposition 2-1), $\alpha_1 = 0$ and $\alpha_K = 1$. We obtain $\alpha_i < a_{i+1}$ for each i because (a) the proof in Section 2-6 shows that $\mathbf{p} \succ \mathbf{q}$ if, and only if, $\alpha(\mathbf{p}) > \alpha(\mathbf{q})$, and ($b$) we have assumed $\mathbf{e}_i \prec \mathbf{e}_{i+1}$ for each i.

It follows from (2-30) that

$$\begin{aligned} u_i &= E[u(\mathbf{e}_i)] = (1 - \alpha_i)E[u(\mathbf{e}_1)] + \alpha_i E[u(\mathbf{e}_K)] \\ &= (1 - \alpha_i) \cdot 0 + \alpha_i \cdot 1 = \alpha_i \end{aligned}$$

for each $i = 1, \ldots, K$.

Expression (2-30) describes the decision maker's indifference between receiving an amount i (with probability 1) or engaging in a gamble which pays 1 with probability $1 - \alpha_i$ and pays K with probability α_i. This description suggests the following method to construct an empirical utility function. Let M be a positive integer and $\Delta = M^{-1}$. Then present the decision maker with a description of the following gamble for each $j \in \{\Delta, 2\Delta, 3\Delta, \ldots, (M-1)\Delta\}$. The gamble pays K with probability j, and it pays 1 with probability $1 - j$. The decision maker is asked to specify an amount, say Q, which, if received with probability 1, is equivalent (for *that* decision maker) to the gamble based on j. Then j is used as $\alpha(Q)$.

The preceding method is feasible, in theory, because $\alpha_1 < \alpha_2 < \cdots < \alpha_K$; so, viewing the sequence $\alpha_1, \ldots, \alpha_K$ as a function on $\{1, \ldots, K\}$, strict monotonicity implies existence of an inverse function. The preceding method constructs the inverse function, and by inverting the inverse, the method constructs the utility function.

Example 2-4 Suppose the method is used with $K = 100$ and $M = 10$. Say that it generates the data in Table 2-3. These data are graphed in Figure 2-6, and a smooth line has been drawn through them.

The smooth-line approximation is used, at the points 1, 2, ..., 100, as an estimate of the utility function. For example, we estimate $\alpha_{41} = 0.73$. □

There are three practical obstacles to using this method. First, there are the empirical departures from the conditions assumed in Theorem 2-1. These departures have already been discussed briefly. In professional practice, they can be overcome. Second, the extraction of decision makers' preferences is an art. The art is based on experimental work in the behavioral sciences, but like any art, its effectiveness depends significantly on the creativity and experience of the "artist." Fishburn (1967), Keeney and Raiffa (1976), and Eliashberg (1980) survey the assessment of utility functions.

Figure 2-6 exemplifies the third obstacle to using the preceding method. The data in Table 2-3 are shown as points in the figure. How should we have interpolated between these points? We chose to interpolate so that the curve is everywhere strictly increasing but at a decreasing rate. That is, the first derivative is positive and the second derivative is negative. Is this reasonable? The next subsection briefly reviews the connection between the shape of the utility function and the decision maker's attitude to risk.

Table 2-3 Hypothetical data for the inverse of a utility function

j	Q	j	Q
0	1	0.6	23
0.1	2	0.7	33
0.2	4	0.8	48
0.3	7	0.9	68
0.4	11	1.0	100
0.5	16		

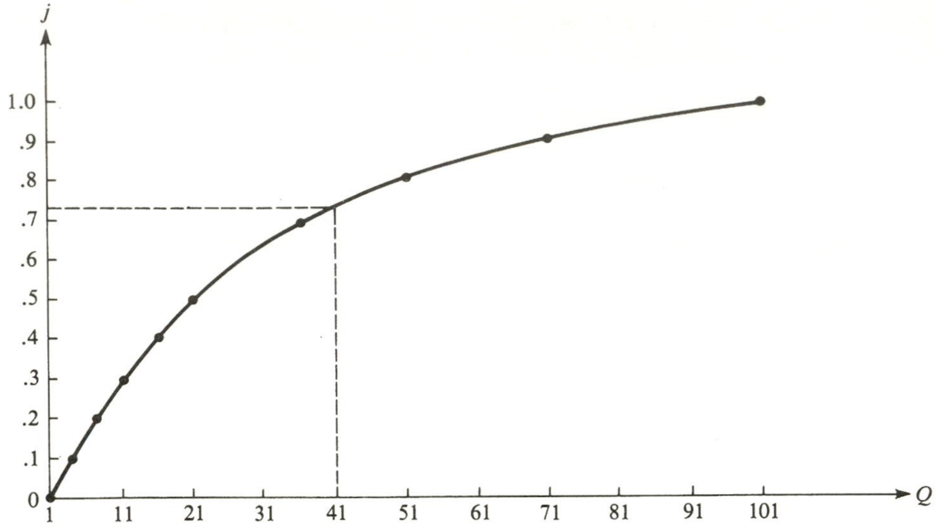

Figure 2-6 Hypothetical estimated utility function.

Risk Aversion

Theorems 2-1 and 2-2 are predicated on a finitely large sample space. This restriction is imposed merely to simplify the preceding exposition and to avoid technicalities in the proofs (in Section 2-6) of the theorems. The theorems are essentially true, as stated, without the restriction to a finite sample space. Then a utility function becomes a function $u(\cdot)$ from $\mathbb{R}$ to $\mathbb{R}$. See DeGroot (1970) or Fishburn (1970) for an exposition and references.

In this subsection we assume that $u(\cdot)$ is a twice differentiable function (from $\mathbb{R}$ to $\mathbb{R}$) with strictly positive first derivative. For an r.v. X with distribution function $F_X(\cdot)$, let γ_X denote the expected value of X. For all r.v.'s in this subsection, say X, we assume that $|\gamma_X| < \infty$ and that the integral†

$$E[u(X)] = \int_{-\infty}^{\infty} u(x)\, dF_X(x)$$

exists and is finite. Theorem 2-1, in its generalized form, essentially states necessary and sufficient conditions for

$$F_X(\cdot) \succ F_Y(\cdot) \Leftrightarrow E[u(X)] > E[u(Y)] \tag{2-31}$$

An individual's comparison of two risks usually is influenced by the total wealth prior to exposure to either risk. Therefore, we must specify whether $u(\cdot)$ is a utility function for total wealth or a utility function for changes in total wealth.

† See Definition A-6 in Appendix A for the definition of the Riemann-Stieltjes integral we use throughout the book.

The two interpretations are closely related, but we choose the latter. Suppose $U(\cdot)$ is a utility function for total wealth. Then $u(X) = U(z + X)$, where z denotes current total wealth. Since the notation $u(\cdot)$ suppresses the parametric dependence on z, in this book we shall make only a few comments about the connection between the shape of $u(\cdot)$ and the effects of changes in wealth on attitude to risk. See Arrow (1971), Pratt (1964), and Ross (1981) for lucid descriptions of the connection.

Suppose that $u(\cdot)$ describes your attitude to risk, in the sense of (2-31), and that you are presented with an opportunity having payoffs described by an r.v. X. What is the amount s, received for sure instead of X, such that you are indifferent between s and X? From (2-31),

$$u(s) = E[u(X)]$$

The quantity s is sometimes called the *certainty equivalent* of X. It is useful to describe s in terms of a departure from the mean of X, namely, γ_X; so we write $s = \gamma_X - \pi$ where π satisfies

$$u(\gamma_X - \pi) = E[u(X)] \tag{2-32}$$

A unique solution π to (2-32) necessarily exists because, by assumption, $u(\cdot)$ is continuous and strictly increasing. The solution π is called a *risk premium* because it is the amount you would pay in order to receive γ_X, for certain, in order to avoid the randomness in X.

The risk premium, π in (2-32), does not necessarily satisfy $\pi \geq 0$. Compulsive gamblers would wish to be compensated with an amount greater than γ_X in order to forgo X, i.e., $\pi < 0$. However, concavity† of $u(\cdot)$ is a simple necessary and sufficient condition for $\pi \geq 0$. It follows, essentially, from Exercise B-3 that $u(\cdot)$ is concave if, and only if,

$$u(\gamma_X) \geq E[u(X)] \tag{2-33}$$

for all r.v.'s X for which both sides are finite. But (2-33) and the fact that $u'(y) > 0$ for all y imply that $\gamma_X - \pi \leq \gamma_X$ in order for (2-32) to be satisfied. Therefore, the risk premium π is nonnegative for all r.v.'s X if, and only if, the utility function $u(\cdot)$ is concave.

A decision maker who will pay to avoid risks is said to be a *risk averter*. Therefore, risk averters have concave utility functions.

Example 2-5 Suppose X is uniformly distributed on [0, 1] and the portion of $u(\cdot)$ on [0, 1] is given by $u(x) = 3x - x^2$. Observe that $u'(x) > 0$ and

† A twice differentiable function $f(\cdot)$ is said to be *concave* on an interval (a, b) if $f''(y) \leq 0$ for all $y \in (a, b)$. Figure 2-5 shows a concave function on (1, 100). Appendix B presents many properties of concave functions.

$u''(x) < 0$. Here,

$$\gamma_X = E(X) = \int_0^1 x dx = 1/2$$

$$u(\gamma_X - \pi) = 3(1/2 - \pi) - (1/2 - \pi)^2 = 5/4 - 2\pi - \pi^2$$

$$E[u(X)] = \int_0^1 (3x - x^2)\, dx = 7/6$$

For this example, (2-32) yields

$$5/4 - 2\pi - \pi^2 = 7/6 \Leftrightarrow \pi = 0.04 \text{ or } \pi = -2.04,$$

so $\pi = 0.04$ because concavity of $u(\cdot)$ implies $\pi \geq 0$. □

Recall that the case of *linear utility* denotes preferences consistent with a linear utility function. Hence,

$$u(x) = b + cx \qquad c > 0 \qquad x \in \mathbb{R}$$

A linear utility function is indeed concave and, with $c > 0$, is strictly increasing. In this case, $E[u(X)] = b + c\gamma_X$ and (2-32) becomes

$$b + c \cdot (\gamma_X - \pi) = b + c\gamma_X$$

so $\pi = 0$. In words, a linear utility function causes the risk premium to be zero for all r.v.'s X. Therefore, an individual with a linear utility function is said to be *risk neutral.*

The next two subsections obtain further properties of linear utility functions and characterize exponential and quadratic utility functions.

Absolute and Relative Risk Aversion

Let X and Y be two r.v.'s and k a real number. How does the value of k affect a decision maker's preference for $X + k$ versus $Y + k$? This question leads to the issue of the effect of wealth on risk aversion. We seek a measure of the "local" risk aversion of $u(\cdot)$ at x, for each $x \in \mathbb{R}$

Suppose that two people have utility functions $u_1(\cdot)$ and $u_2(\cdot)$, respectively, and $u_1(x) = cu_2(x)$ for all x (with $c > 0$). Then, for all r.v.'s X and Y for which the following expectations are finite,

$$E[u_1(X)] - E[u_1(Y)] = c\{E[u_2(X)] - E[u_2(Y)]\}$$

so $u_1(\cdot)$ and $u_2(\cdot)$ induce exactly the same partial orderings of r.v.'s. Also, we know that the first person is a risk averter if, and only if, $u_1(\cdot)$ is concave. If $u_1(\cdot)$ is twice differentiable, concavity is equivalent to $u_1''(x) \leq 0$; so it is tempting to use $|u_1''(x)|$ as a measure of the "local" risk aversion at x. The same is true for the second person, but $|u_1''(x)| = c|u_2''(x)|$. For every r.v. X, both people would have the same risk premiums in (2-32); so a measure of local risk aversion should not vary with c. More generally, the measure should be invariant under positive linear transformations of a utility function.

Let $u(\cdot)$ be a strictly increasing twice differentiable utility function on $\mathbb{R}$. Define the *risk-aversion function*

$$\rho(x) = \frac{-u''(x)}{u'(x)} \tag{2-34}$$

The function $\rho(\cdot)$ is nonnegative if $u(\cdot)$ is concave (risk-averse decision maker), increases as $|u''(x)|$ increases, and $\rho_1(x) = \rho_2(x)$ where $\rho_j(\cdot)$ corresponds to utility function $u_j(\cdot)$ (in the preceding paragraph).

Sometimes $\rho(x)$ is called the *absolute* risk-aversion function and $x\rho(x)$ is called the *relative* risk-aversion function, $x \in \mathbb{R}$. Arrow (1971) shows that $x\rho(x)$ would decrease with x if an increase in an individual's wealth would induce an increase in the fraction of the individual's investments in risky (vs. riskless) assets.

If an individual has preferences consistent with a linear utility function, that is, $u(x) = b + cx$, $x \in \mathbb{R}$ $(c > 0)$, then $\rho(x) = 0$ for all x and we know that the risk premium $\pi = 0$ for all r.v.'s X. We shall see that an exponential utility function also has a constant $\rho(\cdot)$. Suppose

$$u(x) = be^{-cx} \qquad bc < 0 \tag{2-35}$$

where b and c have opposite signs and are nonzero. Then $u'(x) > 0$ and $\rho(x) = c$ for all x. Also, substitution of (2-35) in (2-32) yields

$$\pi = \gamma + \frac{\log\,[E(e^{-cX})]}{c}$$

so the risk premium is generally nonzero.

Only linear and exponential utility functions possess constant risk-aversion functions. From the definition of $\rho(\cdot)$ in (2-34), $\rho(x) = c$ for all x is equivalent to

$$u''(x) - cu'(x) = 0 \qquad x \in \mathbb{R} \tag{2-36}$$

Exercise 2-12 asks you to prove that the only nonzero solutions of (2-36) are $u(x) = g + be^{-cx}$ and $u(x) = b + gx$ for appropriate constants b and g.

Here is another property shared only by linear and exponential utility functions. Let X be an r.v., k a number, and $Y = X + k$. Let π_X and π_Y denote the risk premiums associated with X and Y, respectively. Since $\gamma_Y = k + \gamma_X$,

$$\begin{cases} u(\gamma_X + k - \pi_Y) = E[u(X + k)] \\ u(\gamma_X - \pi_X) = E[u(X)] \end{cases} \tag{2-37}$$

Invariance of the risk premium under translations of the r.v. means that $\pi_X = \pi_Y$ (for all $k \in \mathbb{R}$). Equivalently, if an r.v. increases by an amount k, its certainty equivalent increases by exactly the same amount. Hence, (2-37) and $\pi_X = \pi_{X+k}$ for all k is equivalent to the existence of $s = \gamma_X - \pi_X$ such that

$$u(s + k) = E[u(X + k)] \qquad k \in \mathbb{R} \tag{2-38}$$

Linear and exponential functions are the unique nonzero functions which satisfy (2-38) [cf. Rothblum (1975)].

Mean-Variance Trade-offs

A quadratic utility function is

$$u(x) = x - cx^2 \qquad x \in \mathbb{R} \qquad c > 0 \tag{2-39}$$

Since $u'(x) < 0$ if $x > 1/2c$, (2-39) is used only as an approximation to the utility function for r.v.'s with a limited range. Let X be an r.v. with mean γ and variance σ^2. From (2-39),

$$E[u(X)] = \gamma - cE(X^2) = \gamma - c(\sigma^2 + \gamma^2) \tag{2-40}$$

Therefore, a quadratic utility function implies that a "rational" decision maker compares r.v.'s only according to their means and variances. It is a surprising and important fact that the converse is true:

A "rational" decision maker who compares r.v.'s only according to their means and variances must have preferences consistent with a quadratic utility function.

Now we prove† the converse.

Let X and Y be two r.v.'s with distribution functions $F_X(\cdot)$ and $F_Y(\cdot)$ between which a decision maker is indifferent. Let (γ_X, σ_X) and (γ_Y, σ_Y) be the respective means and standard deviations. It follows from the assumptions of Theorem 2-1 that the decision maker is indifferent among X, Y, and any randomization‡ between X and Y. Suppose X is chosen with probability p and Y with probability $1 - p$. Call Z the resulting r.v. and let γ and σ^2 denote its mean and variance. The resulting distribution function is

$$F(z) = pF_X(z) + (1 - p)F_Y(x)$$

and it has mean and mean square

$$\gamma = p\gamma_X + (1 - p)\gamma_Y = \gamma_Y + p(\gamma_X - \gamma_Y) \tag{2-41a}$$

$$E(Z)^2 = \gamma^2 + \sigma^2 = p(\gamma_X^2 + \sigma_X^2) + (1 - p)(\gamma_Y^2 + \sigma_Y^2) \tag{2-41b}$$

We assume $E(X) \neq E(Y)$ and $E(X^2) \neq E(Y^2)$. From (2-41a, 2-41b),

$$p = \frac{\gamma - \gamma_Y}{\gamma_X - \gamma_Y}$$

$$\sigma^2 = p\sigma_X^2 + (1 - p)\sigma_Y^2 + p(1 - p)(\gamma_X - \gamma_Y)^2$$

Substitution of the first expression in the second one and manipulating terms yields

$$\gamma - c(\gamma^2 + \sigma^2) = M \tag{2-42}$$

† Our proof follows Mossin (1973).

‡ You are asked to prove this intuitive result in part C of Exercise 2-17.

where $c = \dfrac{\gamma_X - \gamma_Y}{\gamma_X^2 + \sigma_X^2 - \gamma_Y^2 - \sigma_Y^2}$ and $M = \dfrac{\gamma_X \gamma_Y(\gamma_X - \gamma_Y) + \gamma_Y \sigma_X^2 - \gamma_X \sigma_Y^2}{\gamma_X^2 + \sigma_X^2 - \gamma_Y^2 - \sigma_Y^2}$

Equation (2-42) specifies all the r.v.'s Z such that the decision maker is indifferent among X, Y, and Z. By Theorem 2-1, all such r.v.'s Z must have the same expected utility; so (2-42) implies

$$E[u(Z)] = \gamma_z - c(\gamma_z^2 + \sigma_z^2).$$

Therefore,

$$u(z) = z - cz^2 \tag{2-43}$$

as was to be proved.

EXERCISES

2-4 Prove Proposition 2-1. [Hint: Prove and use the fact that W is negatively transitive on V if, and only if, for all p, q, and r in V, $(p, q) \in W$ implies either $(p, r) \in W$ or $(r, q) \in W$ or both].

2-5 Prove or find a counterexample:

(a) If W is complete, it is reflexive.

(b) If W is reflexive, it is complete.

(c) If W is asymmetric, it is irreflexive [i.e., $(p, p) \notin W$ for all $p \in V$].

2-6 Verify that the risk premium π associated with an r.v. and a utility function $u(\cdot)$ is invariant with respect to all positive linear transformations of $u(\cdot)$.

2-7 According to (2-23), $(10^{-10}, 1 - 10^{-10})$ is not necessarily preferred to (1, 0). Is this reasonable? If not, how would you augment (2-23)?

2-8 Let $K = 3$ and $u(1) = 0$, $u(2) = v$, and $u(3) = v_3$ with $0 < v < v_3 < 2v$ so $u(\cdot)$ is strictly risk-averse. Give an example of two distributions, $\mathbf{p}$ and $\mathbf{q}$, such that $\mathbf{p} \succ \mathbf{q}$ is consistent with $u(\cdot)$, $\Sigma_i ip_i = \Sigma_i iq_i$, and $q_3 > p_3$.

2-9 In many factories, *bills of materials* are used to describe product structures in terms of subproducts. For example, an automobile's bill of materials might show that the subproducts are transmission, chassis, engine, controls, and trim. The engine's subproducts may include engine block, distributor, carburetor, spark plugs, etc. The spark plugs may be purchased (not having subproducts) while the carburetor has subproducts. Let a *component* be a final product or any subproduct and let V be the set of all components in a factory. Let $A = \{(x, y): x \in V, y \in V$, and x is a subproduct of $y\}$. Discuss whether or not A generally has each of the properties (a) through (h) in Definition 2-1.

2-10 Suppose that a utility function on $\mathbb{R}$ is $u(x) = be^{-cx}$ with $c > 0$ and $b < 0$. Verify the following properties:

(a) $u(\cdot)$ is a concave function.

(b) $u(\cdot)$ has a constant risk-aversion function.

(c) $u(\cdot)$ exhibits increasing relative risk aversion.

(d) If X is a Poisson r.v. with $E(X) = \lambda$ and c is small, the risk premium is approximately $\lambda c/2$.

How do these properties change if $c < 0$ and $b > 0$?

2-11 Suppose that a utility function is $u(x) = x - x^2/2$ with $c > 0$ and domain $(-\infty, 1)$. Verify the following properties:

(a) $u(\cdot)$ is a concave function.

(b) $u(\cdot)$ has an increasing risk-aversion function.

(c) $u(\cdot)$ has increasing relative risk aversion.

(d) If X is uniformly distributed on (0, 1), the risk premium is 0.077 (to the nearest 0.001).

2-12 Prove that the only nonzero solutions of (2-36) have the form $u(x) = g + be^{-cx}$ and $u(x) = b + cx$ for appropriate constants b and g. Hence, only linear and exponential utility functions possess constant risk-aversion functions.

2.5 JUSTIFICATION FOR PRESENT-VALUE FORMULAS

This section contains an axiomatic basis for the comparison of alternative time streams of cash via their respective present values. We use binary relations to describe preference orderings of alternative cash flows. This approach is both natural and convenient for the same reasons given in Section 2-4.

Consider a planning horizon of N periods and a cash-flow vector $\mathbf{p} = (p_1, p_2, \ldots, p_N)$ where $p_n > 0$ and $p_n < 0$ indicate profit and loss, respectively. Each cash flow $\mathbf{p}$ is a vector of real numbers. In Section 2-4, $\mathbf{p}$ is a probability distribution; so its components are nonnegative and sum to 1. Neither constraint is applicable in this section. However, we use binary relations in both sections; so the notation is similar.

We ignore the important issue of whether there is a feasible decision rule that generates the cash flow $\mathbf{p} = (p_1, \ldots, p_N)$. Instead, we analyze the logical consequences of a few axioms concerning the comparison of any two cash flows. The feasibility issue is addressed in all subsequent chapters. The most important result below is Corollary 2-3, which asserts that a present-value formula is the only method of comparing alternative cash flows that is consistent with the axioms. More precisely, any increasing function of a present-value formula also satisfies the axioms.

Let V denote the nonempty set of all conceivable cash flows (vectors) in a given context; $V \subset \mathbb{R}^N$. It is convenient to use the terminology of binary relations introduced in Section 2-4 to describe a preference ordering on V. Let $W \subset V \times V$ describe cash-flow preferences in the following manner. If $\mathbf{p} \in V$ and $\mathbf{q} \in V$, we interpret $(\mathbf{p}, \mathbf{q}) \in W$ as "cash flow $\mathbf{p}$ is at least as desirable as cash flow $\mathbf{q}$." Therefore, $(\mathbf{p}, \mathbf{q}) \in W$ denotes "weak" preference, whereas it denoted "strong" preference in Section 2-4. Hence, we use the $\succsim$ notation from Section 2-4: $\mathbf{p} \succsim \mathbf{q}$ whenever $(\mathbf{p}, \mathbf{q}) \in W$. In words, $\mathbf{p} \succsim \mathbf{q}$ if cash-flow vector $\mathbf{p}$ is *as good as* cash-flow vector $\mathbf{q}$. We write $\mathbf{p} \succsim \mathbf{q}$ and $\mathbf{q} \precsim \mathbf{p}$ interchangeably. If W is complete and $(\mathbf{q}, \mathbf{p}) \notin W$, then $\mathbf{p} \succsim \mathbf{q}$ must be true (valid). In this negation of $\mathbf{q} \succsim \mathbf{p}$, as in Section 2-4 we write $\mathbf{p} \succ \mathbf{q}$ (interchangeably $\mathbf{q} \prec \mathbf{p}$) and say $\mathbf{p}$ *is preferred to* $\mathbf{q}$. If $\mathbf{p} \succsim \mathbf{q}$ and $\mathbf{q} \succsim \mathbf{p}$, as in Section 2-4 we write $\mathbf{p} \sim \mathbf{q}$ (interchangeably $\mathbf{q} \sim \mathbf{p}$) and say that $\mathbf{p}$ and $\mathbf{q}$ *are indifferent*. Alternatively, we say that $\mathbf{p}$ and $\mathbf{q}$ *are equivalent* (with respect to the binary relation $\succsim$) if $\mathbf{p} \sim \mathbf{q}$.

It is convenient to assume $\mathbf{0} = (0. \ldots, 0) \in V$ and also that V is large enough to include all sums, differences, and scalar products of its elements. In other words, we assume that V is a vector space. If $\mathbf{p} = (p_1, \ldots, p_N)$, $\mathbf{q} = (q_1, \ldots, q_N)$, and $b \in \mathbb{R}$, then $\mathbf{p} + \mathbf{q}$, $\mathbf{p} - \mathbf{q}$, and $b\mathbf{p}$ denote the vectors with nth components $p_n + q_n$, $p_n - q_n$, and bp_n, respectively.

The following notation is used for elements of V: $\mathbf{p} = (p_1, \ldots, p_N) = \mathbf{0} \Leftrightarrow p_n = 0$ for all n; $\mathbf{p} > \mathbf{0} \Leftrightarrow p_n > 0$ for all n; $\mathbf{p} \geq \mathbf{0} \Leftrightarrow \mathbf{p} \neq \mathbf{0}$ and $p_n \geq 0$ for all n; $\mathbf{p} \geq \mathbf{q} \Leftrightarrow \mathbf{p} - \mathbf{q} \geq \mathbf{0}$; $\mathbf{p} > \mathbf{q} \Leftrightarrow \mathbf{p} - \mathbf{q} > \mathbf{0}$. If $\{\mathbf{p}^i\}$ is a countable sequence of cash-flow vectors, then

$$\mathbf{p} = \lim_{i \to \infty} \mathbf{p}^i \Leftrightarrow p_n = \lim_{i \to \infty} p_n^i, \; n = 1, \ldots, N$$

The Axioms

The first axiom is innocuous.

Axiom 2-1: Completeness

$$\succsim \text{ is a complete order, i.e., transitive and complete} \tag{2-44}$$

Completeness is the same as comparability of all pairs of cash-flow vectors. Transitivity is a widely applied version of rationality.

The next axiom asserts that more cash is better than less.

Axiom 2-2: Greed

$$\text{If} \quad \mathbf{p} \geq \mathbf{q} \quad \text{then} \quad \mathbf{p} \succ \mathbf{q} \tag{2-45}$$

The desirability of hastening the receipt of revenue and delaying expenditure is at the heart of discounting and is reflected in the next axiom.

Axiom 2-3: Impatience For all numbers $\xi > 0$, vectors $\mathbf{p} = (p_1, \ldots, p_N) \in V$, and $n = 2, \ldots, N$,

$$(p_1, \ldots, p_{n-2}, p_{n-1} + \xi, p_n - \xi, p_{n+1}, \ldots, p_N) \succ \mathbf{p} \tag{2-46}$$

Axioms 2-1, 2-2, and 2-3 are relatively innocuous, but the next axiom merits close scrutiny.

It is convenient to reduce the comparison of any two cash-flow vectors $\mathbf{p}$ and $\mathbf{q}$ to the comparison of their difference $\mathbf{p} - \mathbf{q}$ with $\mathbf{0} \in V$:

Axiom 2-4: Marginal consistency

$$\mathbf{p} \succsim \mathbf{q} \quad \text{if, and only if,} \ \mathbf{p} - \mathbf{q} \succsim 0 \tag{2-47}$$

For example, $(4, 3) \succsim (3, 4)$ if, and only if, $(1, -1) \succsim (0, 0)$. This axiom is more troublesome than any other and will be discussed shortly.

We assume that V is dense enough in the region of $\mathbf{0}$ and the ordering $\succsim$ is "smooth" enough near $\mathbf{0}$ for the following continuity axiom to be valid:

Axiom 2-5: Continuity

$$\text{If} \quad \mathbf{p} \succ \mathbf{0} \quad \text{there is } \mathbf{q} > \mathbf{0} \text{ with } \mathbf{p} - \mathbf{q} \succ \mathbf{0} \tag{2-48}$$

One consequence of (2-47) is that, with (2-48), it implies continuity everywhere on V (not only at $\mathbf{0}$). Exercise 2-13 asks you to prove that (2-47) and (2-48) are equivalent to the combination of (2-47) with

$$\text{If} \quad \mathbf{p} \succ \mathbf{q} \quad \text{there is } \mathbf{r} > \mathbf{0} \text{ with } \mathbf{p} - \mathbf{r} \succ \mathbf{q} \tag{2-48$'$}$$

Axiom 2-5 merely asserts that there are "many" cash flows near $\mathbf{0}$ and that the preference ordering $\succsim$ is not too coarse in its discrimination among them.

Axioms 2-1 through 2-5 imply that the preference ordering is consistent with a present-value criterion but the discount factors are not necessarily the same fraction raised to successively higher powers. The latter result requires another axiom.

Axiom 6: Persistence of preference

$$\text{If} \quad (p_1, p_2, \ldots, p_{N-1}, 0) \sim \mathbf{0} \quad \text{then} \quad (0, p_1, p_2, \ldots, p_{N-1}) \sim \mathbf{0} \tag{2-49}$$

Roughly speaking, (2-49) postulates that if you are indifferent between receiving a cash flow and not receiving it, then you are indifferent to whether or not you receive the one-period postponement of that cash flow. The effect of (2-49) is to eliminate the beginning of period 1 as an absolute reference for time. A situation where (2-49) is unlikely to hold is when the remaining life of the enterprise is known to be $N - 1$ periods.

To appreciate the force of (2-47), consider two cash flows $\mathbf{q}$ and $\mathbf{q} + \mathbf{r}$. Then (2-47) asserts that $\mathbf{q} + \mathbf{r}$ is as good as $\mathbf{q}$ if, and only if, $\mathbf{r}$ is as good as $\mathbf{0}$. Moreover, the comparison of the incremental cash flow $\mathbf{r}$ with $\mathbf{0}$ does not depend on the base-level cash flow $\mathbf{q}$. In other words, the absolute cash flow does not affect the assessment of whether or not an incremental cash flow is worthwhile. This assumption seems unreasonable for incremental flows that would materially alter an individual's economic consumption possibilities. Suppose there is an opportunity to receive \$10,000 five years hence in return for an investment of \$5,000 now. Would you make the investment vs. the option of making none, i.e., do you prefer $(-5{,}000, 0, 0, 0, 10{,}000)$ to $(0, 0, 0, 0, 0)$? For most of us, the answer would be influenced by whether the first year's total money income is high enough to permit a reasonable standard of living even after it is reduced by \$5,000. On the other hand, suppose this option is faced by a large corporation in which you hold a small fraction of the shares of stock. Then your personal consumption possibilities during the next 5 years will be affected only minutely according to whether or not the corporation makes the investment. In the latter case, Axiom 2-4 is a reasonable characteristic of your preferences.

Principal results

The principal results are stated and discussed here. The proofs are deferred to Section 2-7, which includes a dozen intermediate propositions.

Theorem 2-3 Axioms 2-1 through 2-5 imply existence of numbers $\gamma_1, \ldots, \gamma_N$ such that $1 = \gamma_1 < \gamma_2 < \cdots < \gamma_N$ and

$$\mathbf{p} \sim \mathbf{0} \Leftrightarrow \sum_{n=1}^{N} \frac{p_n}{\gamma_n} = 0$$

Corollary 2-3 Axioms 2-1 through 2-5 imply

$$\mathbf{p} \begin{Bmatrix} \succ \\ \sim \\ \prec \end{Bmatrix} \mathbf{0} \Leftrightarrow \sum_{n=1}^{N} \frac{p_n}{\gamma_n} \begin{Bmatrix} > \\ = \\ < \end{Bmatrix} 0$$

Theorem 2-4 Axioms 2-1 through 2-6 imply the existence of a number β, $0 < \beta < 1$, such that $\gamma_n^{-1} = \beta^{n-1}$, $n = 1, \ldots, N$.

The most important consequence of the axioms is Corollary 2-3, which asserts that $\mathbf{p} = (p_1, \ldots, p_N)$ is preferred to $\mathbf{0}$ if, and only if, $\sum_{n=1}^{N} \gamma_n^{-1} p_n > 0$, where the γ_n's are numbers that satisfy $1 = \gamma_1 < \gamma_2 < \cdots < \gamma_N$. Hence, $1 = \gamma_1^{-1} > \gamma_2^{-1} > \cdots > \gamma_N^{-1} > 0$. Theorem 2-4 provides sufficient conditions for there to be a number β, $0 < \beta < 1$, such that $\gamma_n^{-1} = \beta^{n-1}$. In words, the corollary asserts that behavior consistent with the axioms must lead to cash flow $\mathbf{p}$ being preferred to cash flow $\mathbf{q}$ if, and only if, the present value of $\mathbf{p}$ is higher than that of $\mathbf{q}$. However, the discount factors may differ from one period to the next. In terms of the γ_n's above, the factor for year n is $\beta_n \triangleq \gamma_{n-1}/\gamma_n$ for $n > 1$ and $\beta_1 = 1$. Then $\gamma_n^{-1} = \prod_{n=1}^{N} \beta_i$. Theorem 2-4 provides conditions for $\beta_2 = \beta_3 = \cdots = \beta_N \triangleq \beta$.

It is important to recognize that the corollary does not assert that a cash flow $\mathbf{p}$ is "worth" its present value in any absolute sense. The present-value formula emerges as a measure of relative worth. The preference ordering $\succsim$ is ordinal, not cardinal; so the axioms lead to a formula with only ordinal significance. Indeed, let $Y(\cdot)$ be any strictly increasing function from $\mathbb{R}$ to $\mathbb{R}$, say, $Y(x) = 2x^3 + x - 10$. Let $\mathbf{p}$ and $\mathbf{q}$ be cash flows in V and let $U(\cdot)$ be the function from V to $\mathbb{R}$ given by $U(\mathbf{p}) = \sum_{n=1}^{N} \gamma_n^{-1} p_n$. Then $U(\mathbf{p}) > U(\mathbf{q}) \Leftrightarrow Y[U(\mathbf{p})] > Y[U(\mathbf{q})]$. Hence the composite function $Y[U(\cdot)]$ has the same ordinal properties as $U(\cdot)$.

EXERCISES

2-13 Prove that the pair (2-47) and (2-48) is equivalent to the pair (2-47) and (2-48′).

2-14 Prove that (2-47) is equivalent to the following statement: if $\mathbf{p} \succsim \mathbf{q}$ and $\mathbf{r} \succsim \mathbf{s}$, then $\mathbf{p} + \mathbf{r} \succsim \mathbf{q} + \mathbf{s}$.

2-15 Fill in the details of the following proof that axioms 2-2 through 2-6 are independent (i.e., that none of them is implied by the rest).

(i) The lexicographic ordering $\mathbf{p} \succ \mathbf{q} \Leftrightarrow p_n > q_n$ for some n and $p_j = q_j$ for $j < n$ satisfies all the axioms except 2-5.

(ii) Let

$$U(\mathbf{p}) = -\sum_{n=1}^{N} \beta^{n-1} p_n$$

with $\beta > 1$. The ordering induced by

$$\mathbf{p} \succsim \mathbf{q} \Leftrightarrow U(\mathbf{p}) \geq U(\mathbf{q})$$

satisfies all the axioms except 2-2.

(iii) If in (ii), $U(\mathbf{p}) = \sum_{n=1}^{N} \beta^{n-1} p_n$ and $\beta \geq 1$, then all the axioms are satisfied except 2-3.

(iv) If in (ii), $U(\mathbf{p}) = \sum_{n=1}^{N} [\exp(\sum_{j=1}^{n} p_j) - 1]^+ + \sum_{n=1}^{N} \beta^{n-1} p_n$, then only 2-4 is violated.

(v) If in (ii), $U(\mathbf{p}) = \sum_{n=1}^{N} (\prod_{j=1}^{n} \beta_j) p_n$ where $\beta_1 = 1$ and $0 < \beta_n < 1$ for $n > 1$ but not all the β_n's are equal ($n \geq 2$), then only 2-6 is violated.

2-16 Use an axiom to prove: For all $\mathbf{p} \in V$ and $\gamma > \mathbf{0}$ there is a number b_0 such that $b \geq b_0$ implies $b\gamma \succ \mathbf{p}$.

2-6* PROOFS FOR EXPECTED-UTILITY CRITERION

This section and Sections 2-7, 4-4, and 9-7 contain proofs based on axioms that concern binary preference relations. The four axiom systems and proofs are somewhat similar to one another. Recall the notation

$$E[\mathbf{u}(\mathbf{p})] \triangleq \sum_{i=1}^{K} u_i p_i \tag{2-50}$$

for $\mathbf{u} = (u_1, \ldots, u_K)$.

Theorem 2-1 Suppose that $\succ$ is a binary relation on the set $\mathscr{P}$ of probability distributions on a finite sample space. There exists $\mathbf{u}$ such that

$$\mathbf{p} \succ \mathbf{q} \Leftrightarrow E[\mathbf{u}(\mathbf{p})] > E[\mathbf{u}(\mathbf{q})] \tag{2-51}$$

for all $\mathbf{p}$ and $\mathbf{q}$ in $\mathscr{P}$ if, and only if, $\succ$ has the following properties:

(*a*) $\succ$ is asymmetric on $\mathscr{P}$ ($\mathbf{p} \succ \mathbf{q} \Rightarrow \mathbf{q} \not\succ \mathbf{p}$),
(*b*) $\succ$ is negatively transitive on $\mathscr{P}$ ($\mathbf{p} \not\succ \mathbf{q}$ and $\mathbf{q} \not\succ \mathbf{r} \Rightarrow \mathbf{p} \not\succ \mathbf{r}$),
(*c*) For all $\mathbf{p}$, $\mathbf{q}$, and $\mathbf{r}$ in $\mathscr{P}$ and $0 < \alpha < 1$,

$$\mathbf{p} \succ \mathbf{q} \Rightarrow \alpha\mathbf{p} + (1 - \alpha)\mathbf{r} \succ \alpha\mathbf{q} + (1 - \alpha)\mathbf{r} \tag{2-52}$$

(*d*) For all $\mathbf{p}$, $\mathbf{q}$, and $\mathbf{r}$ in $\mathscr{P}$, if $\mathbf{p} \succ \mathbf{q} \succ \mathbf{r}$, then there exists $\alpha \in (0, 1)$ such that

$$\alpha\mathbf{p} + (1 - \alpha)\mathbf{r} \succ \mathbf{q} \succ \alpha\mathbf{r} + (1 - \alpha)\mathbf{p} \tag{2-53}$$

Theorem 2-2 If $\mathbf{u} \in \mathbb{R}^K$ satisfies (2-51), then $\mathbf{v} \in \mathbb{R}^K$ satisfies

$$\mathbf{p} \succ \mathbf{q} \Leftrightarrow E[\mathbf{v}(\mathbf{p})] > E[\mathbf{v}(\mathbf{q})] \tag{2-54}$$

for all $\mathbf{p}$ and $\mathbf{q}$ in $\mathscr{P}$ if, and only if, there are numbers $c > 0$ and b such that $\mathbf{v} = c\mathbf{u} + b$ (i.e., $v_i = cu_i + b$ for all i).

Proofs To prove the "if" part of Theorem 2-2, it follows from (2-50) that, if $c > 0$ and $b \in \mathbb{R}$,

$$E[\mathbf{u}(\mathbf{p})] > E[\mathbf{u}(\mathbf{q})] \Leftrightarrow b + cE[\mathbf{u}(\mathbf{p})] > b + cE[\mathbf{u}(\mathbf{q})]$$

Therefore, if $\mathbf{v} = b + c\mathbf{u}$,

$$E[\mathbf{u}(\mathbf{p})] > E[\mathbf{u}(\mathbf{q})] \Leftrightarrow E[\mathbf{v}(\mathbf{p})] > E[\mathbf{v}(\mathbf{q})]$$

and if $\mathbf{u}$ satisfies (2-51), then $\mathbf{v}$ satisfies (2-54). We prove the "only if" part of Theorem 2-2 last of all.

The "if" part of Theorem 2-2 permits the assumption $\mathbf{u} \geq 0$ without loss of generality. This assumption is made in the following proof that (2-51) implies conditions (*a*) through (*d*). Since $>$ is irreflexive on the set of real numbers ($x \not> x$ for all $x \in \mathbb{R}$), (*a*) is implied by

$$\mathbf{p} \succ \mathbf{q} \Rightarrow E[\mathbf{u}(\mathbf{p})] > E[\mathbf{u}(\mathbf{q})] \Rightarrow E[\mathbf{u}(q)] \leq E[\mathbf{u}(\mathbf{p})] \Rightarrow \mathbf{q} \not\succ \mathbf{p}$$

For (*b*), use (2-51) and transitivity of $\geq$ on $\mathbb{R}$ to obtain

$$\mathbf{p} \not\succ \mathbf{q} \not\succ \mathbf{r} \Leftrightarrow E[\mathbf{u}(\mathbf{p})] \leq E[\mathbf{u}(\mathbf{q})] \leq E[\mathbf{u}(\mathbf{r})]$$

$$\Rightarrow E[\mathbf{u}(\mathbf{p})] \leq E[\mathbf{u}(\mathbf{r})] \Rightarrow \mathbf{p} \not\succ \mathbf{r}$$

For (*c*), $\mathbf{u} \geq 0$ implies

$$\mathbf{p} \succ \mathbf{q} \Leftrightarrow E[\mathbf{u}(\mathbf{p})] > E[\mathbf{u}(q)] \Leftrightarrow \sum_{i=1}^{K} u_i p_i > \sum_{i=1}^{K} u_i q_i$$

$$\Leftrightarrow M + \alpha \sum_{i=1}^{K} u_i p_i > M + \alpha \sum_{i=1}^{K} u_i q_i$$

for all $\alpha > 0$ and $M \in \mathbb{R}$. Take $\alpha < 1$ and $M = (1 - \alpha) \sum_{i=1}^{K} u_i r_i$ for any $\mathbf{r} = (r_1, \ldots, r_K) \in \mathscr{P}$. Therefore,

$$\mathbf{p} \succ \mathbf{q} \Leftrightarrow \sum_{i=1}^{K} u_i [(1 - \alpha) r_i + \alpha p_i] > \sum_{i=1}^{K} u_i [(1 - \alpha) r_i + \alpha q_i]$$

$$\Leftrightarrow \alpha\mathbf{p} + (1 - \alpha)\mathbf{r} \succ \alpha\mathbf{q} + (1 - \alpha)\mathbf{r}$$

For (*d*), $\mathbf{u} \geq \mathbf{0}$ implies

$$\mathbf{p} \succ \mathbf{q} \succ \mathbf{r} \Leftrightarrow \sum_{i=1}^{K} u_i p_i > \sum_{i=1}^{K} u_i q_i > \sum_{i=1}^{K} u_i r_i$$

For $\alpha \in (0, 1)$ sufficiently close to 1 and real numbers γ_1 and γ_2, the right side is equivalent to

$$\alpha \sum_{i=1}^{K} u_i p_i + (1 - \alpha)\gamma_1 > \sum_{i=1}^{K} u_i q_i > \alpha \sum_{i=1}^{K} u_i r_i + (1 - \alpha)\gamma_2$$

Take $\gamma_1 = \sum_{i=1}^{K} u_i r_i$ and $\gamma_2 = \sum_{i=1}^{K} u_i p_i$. Then

$$E\{\mathbf{u}[\alpha\mathbf{p} + (1 - \alpha)\mathbf{r}]\} > E[\mathbf{u}(\mathbf{q})] > E\{\mathbf{u}[\alpha\mathbf{r} + (1 - \alpha)\mathbf{p}]\}$$

$$\Leftrightarrow \alpha\mathbf{p} + (1 - \alpha)\mathbf{r} \succ \mathbf{q} \succ \alpha\mathbf{r} + (1 - \alpha)\mathbf{p}$$

Conditions (*a*) *through* (*d*) *imply* (2-51)

The half of the proof that shows that conditions (*a*) through (*d*) imply (2-51) begins with preliminary consequences of (*a*) through (*d*). The proof of Lemma 2-1 appears after the proofs of the theorems.

Lemma 2-1 Suppose $\succ$ is a binary relation on the set $\mathscr{P}$ on a finite sample space such that conditions (*a*) through (*d*) of Theorem 2-1 are satisfied. Then for all $\mathbf{p}$, $\mathbf{q}$, and $\mathbf{r}$ in $\mathscr{P}$:

(i) $\mathbf{p} \succ \mathbf{q}$ and $0 \le \alpha < \beta \le 1 \Rightarrow \beta\mathbf{p} + (1 - \beta)\mathbf{q} \succ \alpha\mathbf{p} + (1 - \alpha)\mathbf{q}$.
(ii) $\mathbf{p} \succsim \mathbf{q} \succsim \mathbf{r}$ and $\mathbf{p} \succ \mathbf{r} \Rightarrow$ there is a unique $\alpha \in [0, 1]$ such that $\mathbf{q} \sim \alpha\mathbf{p} + (1 - \alpha)\mathbf{r}$.
(iii) $\mathbf{p} \sim \mathbf{q}$ and $0 \le \alpha \le 1 \Rightarrow \alpha\mathbf{p} + (1 - \alpha)\mathbf{r} \sim \alpha\mathbf{q} + (1 - \alpha)\mathbf{r}$.

Now Lemma 2-1 is used to show that conditions (*a*) through (*d*) imply (2-51). First, suppose that $\mathscr{P}$ does not include any $\mathbf{p}$ and $\mathbf{q}$ such that $\mathbf{p} \succ \mathbf{q}$. Then, for all $\mathbf{p}$ and $\mathbf{q}$, $\mathbf{p} \not\succ \mathbf{q}$ and $\mathbf{q} \not\succ \mathbf{p}$ so $\mathbf{p} \sim \mathbf{q}$ so $\mathbf{u} = \mathbf{0}$ suffices in (2-51).

Suppose there exists $\mathbf{x}$ and $\mathbf{z}$ in $\mathscr{P}$ such that $\mathbf{z} \succ \mathbf{x}$. Let $\mathscr{D} = \{\mathbf{p}\colon \mathbf{p} \in \mathscr{P}$ and $\mathbf{z} \succsim \mathbf{p} \succsim \mathbf{x}\}$. For each $\mathbf{p} \in \mathscr{D}$, part (ii) of the lemma asserts that there is a unique $\alpha(\mathbf{p}) \in [0, 1]$ such that

$$\mathbf{p} \sim [1 - \alpha(\mathbf{p})]\mathbf{x} + \alpha(\mathbf{p})\mathbf{z} \qquad \mathbf{p} \in \mathscr{D} \tag{2-55}$$

If $\mathbf{p}$ and $\mathbf{q}$ are in $\mathscr{D}$ and $\alpha(\mathbf{p}) > \alpha(\mathbf{q})$, then part (i) of the lemma implies

$$[1 - \alpha(\mathbf{p})]\mathbf{x} + \alpha(\mathbf{p})\mathbf{z} \succ [1 - \alpha(\mathbf{q})]\mathbf{x} + \alpha(\mathbf{q})\mathbf{z}$$

Therefore, (2-55) and parts (*c*) and (*d*) of Proposition 2-1 imply

$$\mathbf{p} \sim [1 - \alpha(\mathbf{p})]\mathbf{x} + \alpha(\mathbf{p})\mathbf{z} \succ [1 - \alpha(\mathbf{q})]\mathbf{x} + \alpha(\mathbf{q})\mathbf{z} \sim \mathbf{q}$$

so $\mathbf{p} \succ \mathbf{q}$. If $\mathbf{p}$ and $\mathbf{q}$ are in $\mathscr{D}$ and $\alpha(\mathbf{p}) = \alpha(\mathbf{q})$, then (2-55) and the fact that $\sim$ is an equivalence relation [part (*b*) of Proposition 2-1] imply $\mathbf{p} \sim \mathbf{q}$. Therefore,

$$\mathbf{p} \succ \mathbf{q} \Leftrightarrow \alpha(\mathbf{p}) > \alpha(\mathbf{q}) \qquad \mathbf{p} \in \mathscr{D}, \mathbf{q} \in \mathscr{D} \tag{2-56}$$

The remainder of the proof has three parts. The first part shows that $\alpha(\cdot)$, the function in (2-56) from $\mathscr{D}$ to $[0, 1]$, satisfies

$$\alpha[\beta\mathbf{p} + (1 - \beta)\mathbf{q}] = \beta\alpha(\mathbf{p}) + (1 - \beta)\alpha(\mathbf{q}) \qquad \mathbf{p} \in \mathscr{D}, \mathbf{q} \in \mathscr{D}, 0 \le \beta \le 1 \tag{2-57}$$

The second part extends (2-56) and (2-57) from $\mathscr{D} \subset \mathscr{P}$ to all of $\mathscr{P}$. The third part shows that (2-57) (and its extension from $\mathscr{D}$ to $\mathscr{P}$) implies that $\alpha(\cdot)$ must have the form

$$\alpha(\mathbf{p}) = \sum_{i=1}^{K} \alpha_i p_i \qquad \mathbf{p} = (p_i) \in \mathscr{P} \tag{2-58}$$

From the extension of (2-56) to $\mathscr{P}$ and (2-58),

$$\alpha(\mathbf{p}) > \alpha(\mathbf{q}) \Leftrightarrow \sum_{i=1}^{K} \alpha_i p_i > \sum_{i=1}^{K} \alpha_i q_i$$
$$\Leftrightarrow E[\boldsymbol{\alpha}(\mathbf{p})] > E[\boldsymbol{\alpha}(\mathbf{q})]$$

where $\boldsymbol{\alpha} = (\alpha_1, \ldots, \alpha_K)$. This verifies (2-51) because $\boldsymbol{\alpha}$ is formally identical to $\mathbf{u}$.

In order to verify (2-57), we first show that $\mathbf{p} \in \mathscr{D}$, $\mathbf{q} \in \mathscr{D}$, and $0 \le \beta \le 1$ implies $\beta\mathbf{p} + (1 - \beta)\mathbf{q} \in \mathscr{D}$. This is immediately true if $\beta = 0$ or $\beta = 1$. If

$0 < \beta < 1$, then

$$\mathbf{x} = \beta\mathbf{x} + (1 - \beta)\mathbf{x} \precsim \beta\mathbf{p} + (1 - \beta)\mathbf{x}$$

by condition (*c*) ("independence") if $\mathbf{x} \prec \mathbf{p}$, and by part (iii) of Lemma 2-1 if $\mathbf{x} \sim \mathbf{p}$. Repetition of the same arguments and transitivity of $\precsim$ [part (*d*) of Proposition 2-1] gives

$$\begin{aligned}\mathbf{x} &\precsim \beta\mathbf{p} + (1 - \beta)\mathbf{x} \precsim \beta\mathbf{p} + (1 - \beta)\mathbf{q}\\ &\precsim \beta\mathbf{z} + (1 - \beta)\mathbf{q} \precsim \beta\mathbf{z} + (1 - \beta)\mathbf{z} = \mathbf{z} \Rightarrow \beta\mathbf{p} + (1 - \beta)\mathbf{q} \in \mathscr{D}\end{aligned}$$

Application of (2-55) to $\beta\mathbf{p} + (1 - \beta)\mathbf{q}$ yields

$$\beta\mathbf{p} + (1 - \beta)\mathbf{q} \sim \{1 - \alpha[\beta\mathbf{p} + (1 - \beta)\mathbf{q}]\}\mathbf{x} + \alpha[\beta\mathbf{p} + (1 - \beta)\mathbf{q}]\mathbf{z} \tag{2-59}$$

Repeated application of (2-55) to $\mathbf{p}$ and $\mathbf{q}$ and transitivity of $\sim$ [part (*b*) of Proposition 2-1] gives

$$\begin{aligned}\beta\mathbf{p} + (1 - \beta)\mathbf{q} &\sim \beta\{[1 - \alpha(\mathbf{p})\mathbf{x} + \alpha(\mathbf{p})\mathbf{z}\} + (1 - \beta)\{[1 - \alpha(\mathbf{q})]\mathbf{x} + \alpha(\mathbf{q})\mathbf{z}\}\\ &\sim [1 - \beta\alpha(\mathbf{p}) - (1 - \beta)\alpha\mathbf{q})]\mathbf{x} + [\beta\alpha(\mathbf{p}) + (1 - \beta)\alpha(\mathbf{q})]\mathbf{z}\end{aligned} \tag{2-60}$$

Transitivity of $\sim$, (2-59), and (2-60) imply

$$\begin{aligned}&\{1 - \alpha[\beta\mathbf{p} + (1 - \beta)\mathbf{q}]\}\mathbf{x} + \alpha[\beta\mathbf{p} + (1 - \beta)\mathbf{q}]\mathbf{z}\\ &\qquad\sim [1 - \beta\alpha(\mathbf{p}) - (1 - \beta)\alpha(\mathbf{q})]\mathbf{x} + [\beta\alpha(\mathbf{p}) + (1 - \beta)\alpha(\mathbf{q})]\mathbf{z}\end{aligned}$$

It follows from part (i) of Lemma 2-1 that if the coefficients of $\mathbf{x}$ (or of $\mathbf{z}$) in the two expressions were different, then $\sim$ would have to be replaced by $\succ$ or by $\prec$. Therefore, the coefficients are equal, so

$$\alpha[\beta\mathbf{p} + (1 - \beta)\mathbf{q}] = \beta\alpha(\mathbf{p}) + (1 - \beta)\alpha(\mathbf{q})$$

which verifies (2-57).

Now we extend (2-56) and (2-57) from $\mathscr{D}$ to $\mathscr{P}$. Every $\mathbf{p} \in \mathscr{P}$ must lie in a set such as $\mathscr{D}$; i.e., there exists $\mathbf{x} \prec \mathbf{z}$ such that $\mathbf{x} \precsim \mathbf{p} \precsim \mathbf{z}$ (why?). The problem is that $\mathbf{p}$ may (usually does) lie in more than one such set and our construction of $\alpha(\cdot)$ seems to depend on the *particular* $\mathscr{D}$ in use. To extend (2-56) and (2-57), it is sufficient to prove that, if $\mathbf{p}$ lies in two such sets, then $\alpha(\mathbf{p})$ is the same numerical value (up to a positive linear transformation) regardless of which set was used to construct $\alpha(\cdot)$.

Fix $\mathbf{x} \prec \mathbf{z}$ and $\mathscr{D}$ as defined previously and suppose there are two other subsets of $\mathscr{P}$, $\mathscr{D}_1$, and $\mathscr{D}_2$, such that $\mathscr{D} \subset \mathscr{D}_1 \cap \mathscr{D}_2$. It follows from the application of (2-57) to $\mathscr{D}_j$ that there is a function $\alpha_j^*(\cdot)$ that satisfies (2-56) and (2-57) with $\mathscr{D}$ replaced by $\mathscr{D}_j$, $j = 1, 2$. Let $\alpha_j(\cdot)$ be the positive linear transformation of $\alpha_j^*(\cdot)$ such that $\alpha_j(\mathbf{x}) = 0$ and $\alpha_j(\mathbf{z}) = 1$, that is,

$$\alpha_j(\mathbf{p}) = \frac{\alpha_j^*(\mathbf{p}) - \alpha_j(\mathbf{x})}{\alpha_j^*(\mathbf{z}) - \alpha_j^*(\mathbf{x})} \qquad \mathbf{p} \in \mathscr{D}_j,\ j = 1, 2 \tag{2-61}$$

Suppose $\mathbf{p} \in \mathscr{D}$. If $\mathbf{p} \sim \mathbf{x}$ or $\mathbf{p} \sim \mathbf{z}$, then $\alpha_1(\mathbf{p}) = \alpha_2(\mathbf{p})$ by construction (2-61).

Otherwise, there are three possibilities: $\mathbf{p} \prec \mathbf{x} \prec \mathbf{z}$, or $\mathbf{x} \prec \mathbf{p} \prec \mathbf{z}$, or $\mathbf{x} \prec \mathbf{z} \prec \mathbf{p}$. In these cases, strict preference and part (ii) of Lemma 2-1 implies existence of unique numbers ξ, γ, and λ in (0, 1) such that, respectively,

$$\mathbf{x} \sim \xi\mathbf{p} + (1 - \xi)\mathbf{z}$$

$$\mathbf{p} \sim \gamma\mathbf{x} + (1 - \gamma)\mathbf{z}$$

$$\mathbf{z} \sim \lambda\mathbf{x} + (1 - \lambda)\mathbf{p}$$

The application of (2-56) and (2-57) to these statements for $j = 1$ and 2 yields

$$\alpha_j(\mathbf{x}) = \xi\alpha_j(\mathbf{p}) + (1 - \xi)\alpha_j(\mathbf{z}) \Leftrightarrow 0 = \xi\alpha_j(\mathbf{p}) + 1 - \xi$$

$$\alpha_j(\mathbf{p}) = \gamma\alpha_j(\mathbf{x}) + (1 - \gamma)\alpha_j(\mathbf{z}) \Leftrightarrow \alpha_j(\mathbf{p}) = 1 - \gamma$$

$$\alpha_j(z) = \lambda\alpha_j(\mathbf{x}) + (1 - \lambda)\alpha_j(\mathbf{p}) \Leftrightarrow 1 = (1 - \lambda)\alpha_j(\mathbf{p})$$

so that $\alpha_1(\mathbf{p}) = \alpha_2(\mathbf{p})$ in each case.

To complete the proof, we show that the extension of (2-57) to $\mathscr{P}$, namely,

$$\alpha[\beta\mathbf{p} + (1 - \beta)\mathbf{q}] = \beta\alpha(\mathbf{p}) + (1 - \beta)\alpha(\mathbf{q}) \qquad \mathbf{p} \in \mathscr{P}, \mathbf{q} \in \mathscr{P}, 0 \le \beta \le 1 \quad \text{(2-62)}$$

implies (2-58), that is,

$$\alpha(\mathbf{p}) = \sum_{i=1}^{K} \alpha_i p_i \qquad \mathbf{p} = (p_i) \in \mathscr{P} \quad \text{(2-63)}$$

Let $\mathbf{q}^1, \ldots, \mathbf{q}^K$ be a sequence of distributions in $\mathscr{P}$ and let $\mathbf{p} = (p_1, \ldots, p_K) \in \mathscr{P}$. Then induction on (2-57) yields

$$\alpha\left(\sum_{i=1}^{K} p_i \mathbf{q}^i\right) = \sum_{i=1}^{K} p_i \alpha(\mathbf{q}^i) \quad \text{(2-64)}$$

For each i, let $\mathbf{q}^i = \mathbf{e}_i$ ($\mathbf{e}_i$ denotes the ith unit vector in $\mathbb{R}^K$, i.e., the K-vector which is all 0 except for one in its ith coordinate). Let α_i denote $\alpha(\mathbf{q}^i)$. Then (2-64) becomes

$$\alpha(\mathbf{p}) = \sum_{i=1}^{K} \alpha_i p_i$$

which is (2-63) and (2-58).

The remaining task is the "only if" part of the proof of Theorem 2-2. Suppose both (2-51) and (2-54) are valid so

$$E[\mathbf{u}(\mathbf{p})] > E[\mathbf{u}(\mathbf{q})] \Leftrightarrow \mathbf{p} \succ \mathbf{q} \Leftrightarrow E[\mathbf{v}(\mathbf{p})] > E[\mathbf{v}(\mathbf{q})]$$

If $\mathbf{p} \sim \mathbf{q}$ for all $\mathbf{p}$ and $\mathbf{q}$ in $\mathscr{P}$, then there are numbers μ and ν such that $u_i = \mu$ and $v_i = \nu$ for all i [because of (2-51) and (2-54)]. Hence, $\mathbf{v} = b + c\mathbf{u}$ with $c = 1$ and $b = \mu - \nu$.

Suppose that there are $\mathbf{x}$ and $\mathbf{z}$ in $\mathscr{P}$ such that $\mathbf{x} \prec \mathbf{z}$. The following argument is similar to the material between (2-61) and (2-62). Let

$$w_u(\mathbf{p}) = \frac{E[\mathbf{u}(\mathbf{p})] - E[\mathbf{u}(\mathbf{x})]}{E[\mathbf{u}(\mathbf{z})] - E[\mathbf{u}(\mathbf{x})]}$$
$$w_v(\mathbf{p}) = \frac{E[\mathbf{v}(\mathbf{p})] - E[\mathbf{v}(\mathbf{x})]}{E[\mathbf{v}(\mathbf{z})] - E[\mathbf{v}(\mathbf{x})]} \tag{2-65}$$

with the denominators positive due to (2-51), (2-54), and $\mathbf{x} \prec \mathbf{z}$. Observe that $w_u(\cdot)$ is a positive linear transformation of $\mathbf{u}$ so the "if" part of Theorem 2-2 implies that $w_u(\cdot)$ satisfies (2-51) with u_i replaced by $b + cu_i$ where

$$b = -\frac{E[\mathbf{u}(\mathbf{x})]}{E[\mathbf{u}(\mathbf{z})] - E[\mathbf{u}(\mathbf{x})]} \qquad \text{and} \qquad c = \frac{1}{E[\mathbf{u}(\mathbf{z})] - E[\mathbf{u}(\mathbf{x})]}$$

A similar statement applies to $w_v(\cdot)$. Also,

$$w_u(\mathbf{x}) = w_v(\mathbf{x}) = 0 \qquad \text{and} \qquad w_v(\mathbf{z}) = w_u(\mathbf{z}) = 1$$

so, if $\mathbf{p} \sim \mathbf{x}$ or $\mathbf{p} \sim \mathbf{z}$, then $w_u(\mathbf{p}) = w_v(\mathbf{p})$. If $\mathbf{p} \not\sim \mathbf{x}$ and $\mathbf{p} \not\sim \mathbf{z}$, then there are the same three possibilities that followed (2-61): $\mathbf{p} \prec \mathbf{x} \prec \mathbf{z}$, or $\mathbf{x} \prec \mathbf{p} \prec \mathbf{z}$, or $\mathbf{x} \prec \mathbf{z} \prec \mathbf{p}$. Essentially the same argument used between (2-61) and (2-62) yields $w_u(\mathbf{p}) = w_v(\mathbf{p})$ here. Therefore, $w_u(\cdot) \equiv w_v(\cdot)$ so we equate the two expressions in (2-65). Rearrangement of terms results in

$$E[\mathbf{v}(\mathbf{p})] = cE[u(\mathbf{p})] + b$$

where $\quad c = \dfrac{E[\mathbf{v}(\mathbf{z})] - E[\mathbf{v}(\mathbf{x})]}{E[\mathbf{u}(\mathbf{z})] - E[\mathbf{u}(\mathbf{x})]} \quad$ and $\quad b = E[\mathbf{v}(\mathbf{x})] - cE[\mathbf{u}(\mathbf{x})]$

Therefore, $\mathbf{v} = b + c\mathbf{u}$. □

Lemma 2-1 and Its Proof

Lemma 2-1 Suppose $\succ$ is a binary relation on the set $\mathscr{P}$ of distributions on a finite sample space and conditions (*a*) through (*d*) of Theorem 2-1 are satisfied. Then for all $\mathbf{p}$, $\mathbf{q}$, and $\mathbf{r}$ in $\mathscr{P}$:

(i) $\mathbf{p} \succ \mathbf{q}$ and $0 \le \alpha < \beta \le 1 \Rightarrow \beta\mathbf{p} + (1-\beta)\mathbf{q} \succ \alpha\mathbf{p} + (1-\alpha)\mathbf{q}$,

(ii) $\mathbf{p} \succsim \mathbf{q} \succsim \mathbf{r}$ and $\mathbf{p} \succsim \mathbf{r} \Rightarrow$ there is a unique $\alpha \in [0, 1]$ such that $\mathbf{q} \sim \alpha\mathbf{r} + (1-\alpha)\mathbf{p}$.

(iii) $\mathbf{p} \sim \mathbf{q}$ and $0 \le \alpha \le 1 \Rightarrow \alpha\mathbf{p} + (1-\alpha)\mathbf{r} \sim \alpha\mathbf{q} + (1-\alpha)\mathbf{r}$.

Proof (i) If $\beta = 1$,

$$\mathbf{p} \succ \mathbf{q} = \beta\mathbf{q} + (1-\beta)\mathbf{p}$$

If $\beta < 1$, condition (*c*) implies

$$\mathbf{p} = \beta\mathbf{p} + (1-\beta)\mathbf{p} \succ \beta\mathbf{q} + (1-\beta)\mathbf{p}$$

Therefore, if $\alpha > 0$, condition (*c*) implies

$$\frac{\alpha}{\beta}[\beta\mathbf{q} + (1-\beta)\mathbf{p}] + \left(1 - \frac{\alpha}{\beta}\right)[\beta\mathbf{q} + (1-\beta)\mathbf{p}]$$

$$\prec \frac{\alpha}{\beta}[\beta\mathbf{q} + (1-\beta)\mathbf{p}] + \left(1 - \frac{\alpha}{\beta}\right)\mathbf{p} \quad (2\text{-}66)$$

Note that $0 \le \alpha < \beta < 1 \Rightarrow \beta > 0$. Straightforward algebra on both sides of (2-66) yields

$$\alpha\mathbf{q} + (1-\alpha)\mathbf{p} \prec \beta\mathbf{q} + (1-\beta)\mathbf{p}$$

If $\alpha = 0$, condition (*c*) implies

$$\alpha\mathbf{q} + (1-\alpha)\mathbf{p} = \mathbf{p} \succ \beta\mathbf{q} + (1-\beta)\mathbf{p}$$

(iii) If $\alpha = 0$, the result is implied by reflexivity of $\sim$ [part (*b*) of Proposition 2-1]. If $\alpha = 1$, the result is true by assumption. Henceforth, suppose $0 < \alpha < 1$ and $\mathbf{p} \sim \mathbf{q}$. If $\mathbf{r} \sim \mathbf{p}$, transitivity of $\sim$ and part *C* of Exercise 2-17 implies the desired result via

$$\alpha\mathbf{p} + (1-\alpha)\mathbf{r} \sim \mathbf{p} \sim \mathbf{q} \sim \alpha\mathbf{q} + (1-\alpha)\mathbf{r}$$

Now suppose $\mathbf{r} \prec \mathbf{p}$. We shall prove

$$\alpha\mathbf{p} + (1-\alpha)\mathbf{r} \succ \alpha\mathbf{q} + (1-\alpha)\mathbf{r} \Rightarrow \alpha\mathbf{p} + (1-\alpha)\mathbf{r} \precsim \alpha\mathbf{q} + (1-\alpha)\mathbf{r}$$

$$\alpha\mathbf{p} + (1-\alpha)\mathbf{r} \prec \alpha\mathbf{q} + (1-\alpha)\mathbf{r} \Rightarrow \alpha\mathbf{p} + (1-\alpha)\mathbf{r} \succsim \alpha\mathbf{q} + (1-\alpha)\mathbf{r}$$

Since both of these statements are absurd, (iii) is valid. There is a similar argument when $\mathbf{p} \succ \mathbf{r}$.

From (i) and $\alpha > 0$,

$$\mathbf{r} = 0 \cdot \mathbf{p} + 1 \cdot \mathbf{r} \prec \alpha\mathbf{p} + (1-\alpha)\mathbf{r}$$

If $\alpha\mathbf{p} + (1-\alpha)\mathbf{r} \prec \alpha\mathbf{q} + (1-\alpha)\mathbf{r}$, then (ii) implies existence of a unique $\beta \in (0, 1)$ such that

$$\alpha\mathbf{p} + (1-\alpha)\mathbf{r} \sim \beta[\alpha\mathbf{q} + (1-\alpha)\mathbf{r}] + (1-\beta)\mathbf{r}$$

$$= \alpha\beta\mathbf{q} + (1-\alpha\beta)\mathbf{r} \quad (2\text{-}67)$$

From part (*c*) of Proposition 2-1, $\mathbf{r} \prec \mathbf{p}$ and $\mathbf{p} \sim \mathbf{q} = \mathbf{r} \prec \mathbf{q}$. Therefore, (i) implies

$$(1-\beta)\mathbf{r} + \beta\mathbf{q} \prec 0 \cdot \mathbf{r} + 1 \cdot \mathbf{q} = \mathbf{q}$$

Then $\mathbf{q} \sim \mathbf{p}$ and part (*c*) of Proposition 2-1 yield

$$\beta\mathbf{q} + (1-\beta)\mathbf{r} \prec \mathbf{p}$$

Hence, condition (b) implies

$$\alpha\mathbf{p} + (1-\alpha)\mathbf{r} \succ \alpha[\beta\mathbf{q} + (1-\beta)\mathbf{r}] + (1-\alpha)\mathbf{r} = \alpha\beta\mathbf{q} + (1-\alpha\beta)\mathbf{r}$$

which is contrary to (2-67). Therefore, $\alpha\mathbf{p} + (1-\alpha)\mathbf{r} \succsim \alpha\mathbf{q} + (1-\alpha)\mathbf{r}$. A similar argument with the hypothesis $\alpha\mathbf{p} + (1-\alpha)\mathbf{r} \succ \alpha\mathbf{q} + (1-\alpha)\mathbf{r}$ implies $\alpha\mathbf{p} + (1-\alpha)\mathbf{r} \precsim \alpha\mathbf{q} + (1-\alpha)\mathbf{r}$. Therefore,

$$\alpha\mathbf{p} + (1-\alpha)\mathbf{r} \sim \alpha\mathbf{q} + (1-\alpha)\mathbf{r}$$

(ii) We have $\mathbf{p} \succsim \mathbf{q} \succsim \mathbf{r}$ and $\mathbf{p} \succ \mathbf{r}$. Define

$$U = \{\alpha: \alpha\mathbf{r} + (1-\alpha)\mathbf{p} \succ \mathbf{q}, \qquad 0 \le \alpha \le 1\}$$

$$L = \{\alpha: \alpha\mathbf{r} + (1-\alpha)\mathbf{p} \prec \mathbf{q}, \qquad 0 \le \alpha \le 1\}$$

$$H = \{\alpha: \alpha\mathbf{r} + (1-\alpha)\mathbf{p} \sim \mathbf{q}, \qquad 0 \le \alpha \le 1\}$$

If $\mathbf{q} \sim \mathbf{p}$, then $0 \in H$ so part (i) of the lemma implies $\alpha \in U$ if $\alpha > 0$. Therefore, $U = (0, 1]$ and $H = \{0\}$. Similarly, if $\mathbf{q} \sim \mathbf{r}$, then $L = [0, 1)$ and $H = \{1\}$.

Now suppose $\mathbf{p} \succ \mathbf{q} \succ \mathbf{r}$. Part (i) of the lemma implies U and L are intervals; they are nonempty because $0 \in U$ and $1 \in L$. Therefore, part (i) of the lemma implies that there must be $\beta \in (0, 1)$ such that

$$\alpha > \beta \Rightarrow \alpha \in L \qquad \textit{and} \qquad \alpha < \beta \Rightarrow \alpha \in U \tag{2-68}$$

Hence, either H is empty or $H = \{\beta\}$. The remainder of the proof shows that H cannot be empty.

If H were empty, then either $\beta \in L$ or $\beta \in U$. Suppose $\beta \in L$ so

$$\beta\mathbf{r} + (1-\beta)\mathbf{p} \prec \mathbf{q} \prec \mathbf{p}$$

Condition (*d*) of Theorem 2-1 ("continuity") implies existence of $\alpha \in (0, 1)$ such that

$$\alpha[\beta\mathbf{r} + (1-\beta)\mathbf{p}] + (1-\alpha)\mathbf{p} \prec \mathbf{q} \Leftrightarrow \alpha\beta\mathbf{r} + (1-\alpha\beta)\mathbf{p} \prec \mathbf{q} \tag{2-69}$$

However, $\alpha\beta < \beta$ and (2-68) imply $\alpha\beta \in U$ so

$$\alpha\beta\mathbf{r} + (1-\alpha\beta)\mathbf{p} \succ \mathbf{q}$$

which contradicts (2-69). Therefore, $\beta \in L$ is false. The initial assumption $\beta \in U$ and an analogous argument leads to a similar contradiction; so $\beta \in H$ and H is nonempty. □

EXERCISES

2-17 For $\mathbf{p}$, $\mathbf{q}$, $\mathbf{r}$, and $\mathbf{s}$ in $\mathscr{P}$ and $0 \le \alpha \le 1$, prove the following assertions under conditions (*a*) through (*d*) of Theorem 2-1.

A. $\alpha\mathbf{p} + (1-\alpha)\mathbf{r} \sim \alpha\mathbf{q} + (1-\alpha)\mathbf{r} \Rightarrow \mathbf{p} \sim \mathbf{q}$

B. $\mathbf{p} \succ \mathbf{q}$ and $\mathbf{r} \succ \mathbf{s} \Rightarrow \alpha\mathbf{p} + (1-\alpha)\mathbf{r} \succ \alpha\mathbf{q} + (1-\alpha)\mathbf{s}$

C. $\mathbf{p} \sim \mathbf{q} \Rightarrow \alpha\mathbf{p} + (1-\alpha)\mathbf{q} \sim \mathbf{p}$

[Hint for *C*: Suppose $\mathbf{p} \succ \alpha\mathbf{p} + (1-\alpha)\mathbf{q}$. Use part (*c*) of Proposition 2-1 and *B* to prove $\alpha\mathbf{p} + (1-\alpha)\mathbf{q} \succ \alpha\mathbf{p} + (1-\alpha)\mathbf{q}$, which is false because $\succ$ is irreflexive.]

2-18 Give an intuitive explanation of Lemma 2-1.

2-7* PROOFS FOR JUSTIFICATION OF PRESENT-VALUE FORMULAS

Before indulging in axiomantics, a word on tactics is in order. You might properly interpret Corollary 2-3 as the following hierarchy of results. A preference ordering satisfies the axioms if, and only if:

(i) There is a real-valued function $U(\cdot)$ defined on V such that $U(\mathbf{p}) > 0 \Leftrightarrow \mathbf{p} \succ \mathbf{0}$.

(ii) There are N real-valued functions $u_n(\cdot)$ defined on $\mathbb{R}$ such that $U(\mathbf{p}) = \sum_{n=1}^{N} u_n(p_n)$.

(iii) For each n there is a number $\gamma_n \geq 1$ such that $u_n(p_n) = \gamma_n^{-1} p_n$ and $1 = \gamma_1 < \gamma_2 < \cdots < \gamma_N$.

The hierarchy of (i), (ii), and (iii) is nòt the route followed here. It would require substantially deeper and more intricate mathematical analysis than we shall need with a more circuitous path.

The following reasoning frequently uses a *contrapositive* method of proof. A statement "A implies B" is logically equivalent to its contrapositive, which is "the negation of B implies the negation of A." Therefore, the original statement can be proved by proving its contrapositive. For example, the first proposition below is $\mathbf{p} \succsim 0$ implies $\mathbf{p} + \mathbf{r} \succsim \mathbf{r}$ for all $\mathbf{r}$. This is proved by establishing the contrapositive, which is $\mathbf{p} + \mathbf{r} \prec \mathbf{r}$ implies $\mathbf{p} \prec 0$.

The argument proceeds in several stages. The first stage shows† that $\mathbf{p} \succsim 0$ implies $b\mathbf{p} \succsim 0$ for all $b \geq 0$. The second stage establishes that there are unique numbers $\gamma_1 \ldots, \gamma_N$ such that $\mathbf{p} \sim \mathbf{0}$ if, and only if, $\sum_{n=1}^{N} \gamma_n^{-1} p_n = 0$. The third stage provides a sufficient condition for γ_n/γ_{n+1} to be the same for all $n \geq 2$.

Until the proof of Theorem 2-4 is reached, we invoke Axioms 2-1 through 2-5. In the proof of Theorem 2-4 we also invoke Axiom 2-6. The axioms are listed below.

2-1 $\succsim$ is a complete order, i.e., transitive and complete.

2-2 $\mathbf{p} \geq \mathbf{q} \Rightarrow \mathbf{p} \succ \mathbf{q}$.

2-3 For all $\xi > 0$, $\mathbf{p} = (p_1, \ldots, p_N) \in V$, and $n = 2, \ldots, N$, $(p_1, \ldots, p_{n-2}, p_{n-1} + \xi, p_n - \xi, p_{n+1}, \ldots, p_N) \succ \mathbf{p}$.

2-4 $\mathbf{p} \succsim \mathbf{q} \Leftrightarrow \mathbf{p} - \mathbf{q} \succsim \mathbf{0}$.

2-5 $\mathbf{p} \succ \mathbf{0} \Rightarrow$ there exists $\mathbf{q} > 0$ with $\mathbf{p} - \mathbf{q} \succ \mathbf{0}$.

2-6 $(p_1, p_2, \ldots, p_{N-1}, 0) \sim \mathbf{0} \Rightarrow (0, p_1, p_2, \ldots, p_{N-1}) \sim \mathbf{0}$.

Preliminary Results

Proposition 2-2 $\mathbf{p} \succsim \mathbf{0} \Leftrightarrow \mathbf{p} + \mathbf{r} \succsim \mathbf{r}$ for all $\mathbf{r} \in V$.

PROOF The contrapositive of Axiom 2-4 is $\mathbf{p} \prec \mathbf{q} \Leftrightarrow \mathbf{p} - \mathbf{q} \prec \mathbf{0}$. Therefore, $\mathbf{p} + \mathbf{r} \prec \mathbf{r} \Leftrightarrow -\mathbf{p} \succ \mathbf{0}$. Therefore, $\mathbf{p} + \mathbf{r} \prec \mathbf{r} \Leftrightarrow -\mathbf{p} \succ \mathbf{0} \Leftrightarrow \mathbf{p} \prec \mathbf{0}$. □

† Review our notation for vector inequalities on page 44.

Proposition 2-3 $\mathbf{p} \succsim (\prec)\ \mathbf{0}$ and $\mathbf{r} \succsim (\prec)\ \mathbf{0}$ implies $\mathbf{p} + \mathbf{r} \succ (\prec)\ \mathbf{0}$.

PROOF Proposition 2-2 and $\mathbf{p} \succsim \mathbf{0}$ yield $\mathbf{p} + \mathbf{r} \succsim \mathbf{r}$. Then $\mathbf{r} \succsim \mathbf{0}$ and transitivity (Axiom 2-1) imply $\mathbf{p} + \mathbf{r} \succsim \mathbf{0}$. The contrapositive of Proposition 2-2 and transitivity establish the result for $\prec \mathbf{0}$. □

Proposition 2-4 $\mathbf{p} \sim \mathbf{0} \Leftrightarrow \mathbf{p} + \mathbf{r} \sim \mathbf{r}$ for all $\mathbf{r} \in V$.

PROOF $\mathbf{p} \sim \mathbf{0} \Leftrightarrow \mathbf{p} \succsim \mathbf{0}$ and $\mathbf{p} \precsim \mathbf{0} \Leftrightarrow$ (Proposition 2-2 and Axiom 2-4) $\mathbf{p} + \mathbf{r} \succsim \mathbf{r}$ and $\mathbf{p} + \mathbf{r} \precsim \mathbf{r}$ for all $\mathbf{r}$. Hence, $\mathbf{p} + \mathbf{r} \sim \mathbf{r}$ for all $\mathbf{r}$. □

Proposition 2-5 If $\langle \mathbf{p}^i \rangle$ is a countable sequence of elements of V that converge to $\mathbf{p} \in V$ and $\mathbf{p}^i \succsim \mathbf{0}$ for all i, then $\mathbf{p} \succsim \mathbf{0}$.

PROOF A contrapositive argument is $\mathbf{p} \prec \mathbf{0} \Leftrightarrow \mathbf{0} \prec -\mathbf{p}$, which implies (due to Axiom 2-5) that there exists $\gamma > \mathbf{0}$ such that $-\mathbf{p} - \gamma \succ \mathbf{0}$. But $-\mathbf{p} - \gamma \succ \mathbf{0} \Leftrightarrow \mathbf{0} \succ \mathbf{p} + \gamma$. Since $\gamma_n > 0$ for $n = 1, \ldots, N$ and $\mathbf{p}^i \to \mathbf{p}$, there exists i^* such that $i \geq i^*$ implies $\mathbf{p}^i \prec \mathbf{0}$. □

Recall the notation $I = \{0, 1, 2, \ldots\}$ and $I_+ = \{1, 2, \ldots\}$.

Proposition 2-6 $\mathbf{p} \succsim (\prec)\ \mathbf{0} \Leftrightarrow b\mathbf{p} \succsim (\prec)\ \mathbf{0}$ for all $b \in I$.

PROOF The claim is true for $b = 0$ by reflexivity and for $b = 1$ by assumption. The inductive assumption is $(b - 1)\mathbf{p} \succsim \mathbf{0}$, and $\mathbf{p} \succsim \mathbf{0}$ is given. Therefore, Proposition 2-3 yields $b\mathbf{p} \succsim \mathbf{0}$. The same argument with $\mathbf{p} \prec \mathbf{0}$ implies $b\mathbf{p} \prec \mathbf{0}$.

The converse is $b\mathbf{p} \succsim (\prec)\ \mathbf{0}$ for all $b \in I$ implies $\mathbf{p} \succsim (\prec)\ 0$. The contrapositive to the case with $\succsim$ is $\mathbf{p} \prec \mathbf{0}$ implies existence of some $b \in I$ such that $b\mathbf{p} \prec \mathbf{0}$. Then $b = 1$ is sufficient. The case with $\prec$ is proved similarly. □

Proposition 2-7 If $\mathbf{p} \sim \mathbf{0}$ and $\mathbf{r} \sim \mathbf{0}$, then $\mathbf{p} + \mathbf{r} \sim \mathbf{0}$.

PROOF Proposition 2-4 and transitivity of $\succsim$ (Axiom 2-1) implies transitivity of $\sim$.

□

Proposition 2-8 $\mathbf{p} \succsim \mathbf{0} \Leftrightarrow b\mathbf{p} \succsim \mathbf{0}$ for all rational numbers $b \geq 0$.

PROOF Let $b = M/T$ with $M \in I$ and $T \in I_+$. The claim is $\mathbf{p} \succsim \mathbf{0}$ implies $(M/T)\mathbf{p} \succsim \mathbf{0}$, whose contrapositive is $(M/T)\mathbf{p} \prec 0$ only if $\mathbf{p} \prec 0$. However, Proposition 2-6 and $(M/T)\mathbf{p} \prec \mathbf{0}$ imply $T(M/T)\mathbf{p} \prec \mathbf{0}$ or $M\mathbf{p} \prec \mathbf{0}$. Then the contrapositive of Proposition 2-6 implies $\mathbf{p} \prec \mathbf{0}$. The converse is proved similarly.

□

Proposition 2-9 $\mathbf{p} \succsim \mathbf{0} \Leftrightarrow b\mathbf{p} \succsim \mathbf{0}$ for all $b \geq 0$.

PROOF For any number $b \geq 0$ there is a sequence $b_1, b_2, \ldots$ of nonnegative rational numbers with $b_i \to b$. Proposition 2-8 and $\mathbf{p} \succsim \mathbf{0}$ imply $b_i \mathbf{p} \succsim \mathbf{0}$ for all i. Also, $b_i \mathbf{p} \to b\mathbf{p}$ as $i \to \infty$. Therefore, Proposition 2-5 implies $b\mathbf{p} \succsim \mathbf{0}$. The converse is proved similarly. □

Proposition 2-10 $\mathbf{p} \sim \mathbf{0}$ implies $b\mathbf{p} \sim \mathbf{0}$ for all $b \in \mathbb{R}$. Conversely, if $b \neq 0$ and $b\mathbf{p} \sim \mathbf{0}$, then $\mathbf{p} \sim \mathbf{0}$.

PROOF Left as Exercise 2-19. □

Proposition 2-11 Let $\mathbf{p}^k \in V$ satisfy $\mathbf{p}^k \sim \mathbf{0}$ and $b_k \in \mathbb{R}$, $k = 1, \ldots, K$. Then $\sum_{k=1}^{K} b_k \mathbf{p}^k \sim \mathbf{0}$.

PROOF Left as Exercise 2-20. □

Proposition 2-12 If $\mathbf{q} \prec \mathbf{p}$ and $\mathbf{p} \prec \mathbf{r}$, then there is a number b^*, $0 < b^* < 1$, with $\mathbf{p} \sim b^*\mathbf{q} + (1 - b^*)\mathbf{r}$.

PROOF Dedekind's theorem† asserts: If $A \subset \mathbb{R}$ and $B \subset \mathbb{R}$ are both nonempty, $A \cup B = \mathbb{R}$, and $a \in A$ and $b \in B$ implies $a < b$, then there is a number c such that $b > c$ implies $b \in B$ and $a < c$ implies $a \in A$. To apply this theorem to Proposition 2-12, let $A = \{b: \mathbf{p} \precsim b\mathbf{q} + (1 - b)\mathbf{r},\ b \in \mathbb{R}\}$ and $B = \{b: \mathbf{p} \succ b\mathbf{q} + (1 - b)\mathbf{r},\ b \in \mathbb{R}\}$. Both sets are nonempty because $1 \in B$ and $0 \in A$. Completeness (Axiom 2-1) implies $A \cup B = \mathbb{R}$. Suppose $a \in A$ and $b \in B$. Then

$$b\mathbf{q} + (1 - b)\mathbf{r} \prec \mathbf{p} \qquad \mathbf{p} \precsim a\mathbf{q} + (1 - a)\mathbf{r}$$

so Axiom 2-4 yields $(b - a)(\mathbf{r} - \mathbf{q}) \succ \mathbf{0}$. Transitivity of $\succsim$ implies transitivity of $\succ$ so $\mathbf{r} \succ \mathbf{q}$, which implies (Axiom 2-4) $\mathbf{r} - \mathbf{q} \succ \mathbf{0}$. Therefore, Proposition 2-9 implies $b > a$; so the conditions of Dedekind's theorem are satisfied and $0 \leq c \leq 1$. Exercise 2-22 asks you to prove that continuity (Axiom 2-5) implies $0 < c < 1$. □

Proposition 2-13 For each $n = 2, \ldots, N$, there is a unique number γ_n for which

$$\mathbf{0} \sim (-1, 0, \ldots, 0, \gamma_n, 0, \ldots, 0) \tag{2-70}$$

Also, $1 < \gamma_2 < \ldots < \gamma_N$.

PROOF For each $n = 2, \ldots, N$, let $\mathbf{e}_n$ denote the N-vector whose entries are all 0 except for unity in the nth coordinate. Then Axiom 2-2 implies $\mathbf{e}_n \succ \mathbf{0}$. Also, Axiom 2-2 implies $\mathbf{0} \succ (-1, 0, \ldots, 0) = -\mathbf{e}_1$. Then Proposition 2-12 with $-\mathbf{e}_1 = \mathbf{q}$, $\mathbf{0} = \mathbf{p}$, and $\mathbf{e}_n = \mathbf{r}$ implies there exists a number b_n, $0 < b_n < 1$,

† G. H. Hardy, *A Course of Pure Mathematics*, Cambridge University Press, Cambridge (1963).

such that

$$\mathbf{0} \sim b_n(-\mathbf{q}) + (1 - b_n)\mathbf{e}_n = b_n\left(-1, 0, \ldots, 0, \frac{1-b_n}{b_n}, 0, \ldots, 0\right)$$

Therefore, Proposition 2-10 implies (2-70) with $\gamma_n = (1 - b_n)/b_n$. For uniqueness of γ_n, suppose γ_n is some number that satisfies (2-70). Then Axiom 2-2 implies

$$(-1, 0, \ldots, 0, \gamma, 0, \ldots, 0) \succ (\prec)\, \mathbf{0} \text{ if } \gamma > (<) \gamma_n$$

Also, $\gamma_n > 1$ because $0 < b_n < 1$. Lastly, Axiom 2-3 and

$$(-1, 0, \ldots, \gamma_n, 0, \ldots, 0) \sim \mathbf{0} \sim (-1, 0, \ldots, 0, \gamma_{n+1}, 0, \ldots, 0)$$

imply $\gamma_n < \gamma_{n+1}$. □

It is convenient to define $\gamma_1 = 1$.

The Theorems

Theorem 2-3 Axioms 2-1 through 2-5 imply existence of numbers $\gamma_1, \ldots, \gamma_N$ such that $1 = \gamma_1 < \gamma_2 < \ldots < \gamma_N$ and

$$\mathbf{p} \sim \mathbf{0} \Leftrightarrow \sum_{n=1}^{N} \gamma_n^{-1} p_n = 0$$

Proof For $n = 2, \ldots, N$ let $\mathbf{r}_n = (-1, 0, \ldots, \gamma_n, 0, \ldots, 0)$. Then $\mathbf{r}_n \sim \mathbf{0}$ from (2-70) and $\sum_{n=2} b_n \mathbf{r}_n \sim \mathbf{0}$ (from Proposition 2-11) for any $(b_2, \ldots, b_N) \in \mathbb{R}^{N-1}$. Take $b_n = \gamma_n^{-1} p_n$; then $\sum_{n=2}^{N} \gamma_n^{-1} p_n \mathbf{r}_n \sim \mathbf{0}$. Observe that

$$\sum_{n=2}^{N} \gamma_n^{-1} p_n \mathbf{r}_n = \left(-\sum_{n=2}^{N} p_n \gamma_n^{-1}, p_2, p_3, \ldots, p_N\right)$$

Then Axiom 2-2 implies $\mathbf{p} = (p_1, p_2, \ldots, p_N) \sim \mathbf{p}' \Leftrightarrow p_1 = -\sum_{n=2}^{N} p_n \gamma_n^{-1}$, which is the same as

$$0 = \sum_{n=1}^{N} \gamma_n^{-1} p_n \tag{2-71}$$

However, $\mathbf{p}' \sim \mathbf{0}$; so transitivity of $\sim$ implies $\mathbf{p} \sim \mathbf{0} \Leftrightarrow$(2-71) □

The most important result is Corollary 2-3 which justifies using present-value formulas to compare alternative cash flows.

Corollary 2-3 $\mathbf{p} \succ (\prec) \mathbf{0} \Leftrightarrow \sum_{n=1}^{N} \gamma_n^{-1} p_n > (<)\, 0$.

Proof As in the proof of Theorem 2-3, $\mathbf{p} \succ (\prec) \mathbf{0} \Leftrightarrow p_1 > (<) -\sum_{n=2}^{N} \gamma_n^{-1} p_n$. □

The next result presents conditions under which there is a number β, $0 < \beta < 1$, such that $\gamma_n^{-1} = \beta^{n-1}$ for each $n = 1, \ldots, N$.

Theorem 2-4 Axioms 2-1 through 2-6 imply there exists a number β, $0 < \beta < 1$, such that $\gamma_n^{-1} = \beta^{n-1}$, $n = 1, 2, \ldots, N$.

PROOF From (2-70) and Axiom 2-6 and transitivity of $\sim$, for each $n = 2, \ldots, N-1$,

$$\mathbf{0} \sim (0, -1, 0, \ldots, 0, \gamma_n, 0, \ldots, 0) \tag{2-72}$$

where γ_n is the $(n+1)$th coordinate in the right side of (2-72). Theorem 2-3 implies (2-72) is valid $\Leftrightarrow -\gamma_2^{-1} + \gamma_n \gamma_{n+1}^{-1} = 0$. Therefore, $\gamma_2^{-1} = \gamma_n/\gamma_{n+1}$ for all $n = 2, \ldots, N-1$. Hence, $\gamma_n^{-1} = \gamma_2^{-n+1}$, $n = 2, \ldots, N$. Let $\beta \triangleq \gamma_2^{-1}$; then $\gamma_n^{-1} = \beta^{n-1}$, $n = 2, \ldots, N$. Lastly, $\gamma_1 = 1$ so $\beta^{n-1} = \gamma_n^{-1}$ if $n = 1$. $\square$

EXERCISES

2-19 Suppose $\mathbf{p}^1, \mathbf{p}^2, \mathbf{p}^3, \ldots$ is a countable sequence of elements of V that converges to $\mathbf{p} \in V$ and $\mathbf{p}^i \sim \mathbf{0}$ for all i. Prove that $\mathbf{p} \sim \mathbf{0}$.

2-20 Prove Proposition 2-10

2-21 Prove Proposition 2-11.

2-22 Complete the proof of Proposition 2-12 by showing that $b^* \neq 0$ and $b^* \neq 1$.

2-23 Prove that Proposition 2-12 is equivalent to the following claim: if $\mathbf{0} \prec \mathbf{p} \prec \mathbf{r}$, then there is a number b, $0 < b < 1$, such that $\mathbf{p} \sim (1 - b)\mathbf{r}$.

2-24 How is $N < \infty$ used in this section? Let $N \to \infty$ so $V \subset \mathbb{R}^\infty$ and $\mathbf{p} = (p_1, p_2, \ldots)$. Among Exercises 2-13 through 2-16 and Exercises 2-19 find one that is false and offer a counterexample.

Bibliographic Guide

Expected-Value Criteria and Expected Utility

A large literature concerns criteria for comparing alternative decisions having uncertain consequences. It dates at least from the eighteenth century when Daniel Bernoulli considered the question of a "fair" entry fee for the opportunity to participate in a gamble whose payoff has an infinite expected value. The books by DeGroot (1970), Fishburn (1970), and Luce and Raiffa (1957) have good expositions and many references. Our exposition is based on Fishburn (1970).

Several important topics are not discussed in Sections 2-3, 2-4, or 2-6. First, the actual use of stochastic models requires the specification of numerical probability distributions. Classical inferential statistical theory provides many methods with which distributions can be estimated from objective data. However, objective data are sometimes sparse; so probabilities are subjectively estimated. The book by Savage (1954) is the classic axiomatic foundation for this approach; also, see part III of Fishburn (1970).

Behavioral scientists now understand some of the systematic errors in subjective estimates of probabilities and have devised methods to elicit "good" estimates [Estes (1972) and Tversky and Kahneman (1974)]. Similarly, behavioral scientists have investigated the experimental process of eliciting preferences (in order to estimate a utility function as described in Section 2-4). See Fishburn (1967), Keeny and Raiffa (1976), and Eliashberg (1980) for discussions of how to assess a utility function.

The modern study of risk aversion was initiated by Arrow (1971, an expansion of a 1964 book) and Pratt (1964). Since, then, several writers have explored the connection between the shape of a utility function and partial orderings of distribution functions. See Bawa (1982), Bessler and Veinott (1969), Mossin (1973), Nachman (1979), Ross (1981), Rothschild and Stiglitz (1970, 1971), Veinott (1965), Whitmore and Findlay (1978), and Ziemba and Vickson (1975), and their references.

Another direction of study of risk aversion concerns multivariate utility functions; i.e., each outcome in the sample space is a vector. See Duncan (1977), Rothblum (1975), and their references.

The remainder of this volume exemplifies the fact that many decision problems require the comparison of alternative time streams (or of probability distributions of time streams). Section 2-5 provides an axiomatic foundation for discounting when uncertainty does not intrude. Several authors have recently asked risk-aversion questions for general time streams. See the following articles and their references: Bell (1974), Kreps and Porteus (1978), Nachman (1975), Rossman and Selden (1978), and Selden (1978).

References for Sections 2-3, 2-4, and 2-6

Arrow, K. J.: *Essays in the Theory of Risk-Bearing*, Markham Publishing Company, Chicago, Ill. (1971).

Bawa, V. S.: "Stochastic Dominance: A Research Bibliography," *Manage. Sci.* **28**: 698–712 (1982).

Bell, D. E.: "Evaluating Time Streams of Income," *OMEGA, Int. J. Manage. Sci.* **2**(5): 691–699 (1974).

Bessler, S. A., and A. F. Veinott, Jr.: "Optimal Policy for a Dynamic Multi-Echelon Inventory Model," *Naval Res. Logistics Quart.* **13**: 355–389 (1966).

DeGroot, M. H.: *Optimal Statistical Decisions*, McGraw-Hill, New York (1970).

Duncan, G. T.: "A Matrix Measure of Multivariate Local Risk Aversion," *Econometrica* **45**(4): 895–903 (May 1977).

Eliashberg, J.: "Consumer Preference Judgements: An Exposition with Empirical Applications," *Manage. Sci.* **26**(1): 60–77 (1980)

Estes, W. K.: "Research and Theory on the Learning of Probabilities," *J. Am. Stat. Assoc.* **67**(337): 91–102(1972).

Fishburn, P. C.: "Methods of Estimating Additive Utilities," *Manage. Sci.* **13**(7): 435–453 (1967).

———: *Utility Theory for Decision Making*, Wiley, New York, Publications in Operations Research No. 18 (1970).

———: "Stochastic Dominance without Transitive Preference," *Manage. Sci.* **24**: 1268–1277 (1978).

Keeney, R. L., and H. Raiffa: *Decisions with Multiple Objectives: Preferences and Value Tradeoffs*, Wiley, New York (1976).

Kreps, D. M., and Evans L. Porteus: "Temporal Resolution of Uncertainty and Dynamic Choice Theory," *Econometrica*, **46**(1): 185–200 (January 1978).

———: "Temporal von Neumann-Morgenstern and Induced Preferences," *J. Econ. Theory* **20**: 81–109 (1979).

Luce, R. D., and H. Raiffa: *Games and Decisions*, Wiley, New York (1957).

Mossin, J.: *Theory of Financial Markets*, Prentice-Hall, Englewood Cliffs, N.J. (1973).

Nachman, D. C.: "Risk Aversion, Impatience, and Optimal Timing Decisions," *J. Econ. Theory* **11** (2) 196–246 (1975).

———: "On the Theory of Risk Aversion and the Theory of Risk," *J. Econ. Theory*, **(1979)**: 1–19.

Pratt, J. W.: "Risk Aversion in the Small and in the Large," *Econometrica* **32**(1-2): 122–135 (January-April 1964).

Raiffa, H.: *Decision Analysis*, Addison-Wesley, Reading, Mass. (1968).

Ross, S. A.: "Some Stronger Measures of Risk Aversion in the Small and the Large with Applications." *Econometrica* **49**: 621–638 (1981).

Rossman, M., and L. Selden: "Time Preferences, Conditional Risk Preferences, and Two Period Cardinal Utility," *J. Econ. Theory*, **19**(1): 64–83 (1978).

Rothblum, U. G.: "Multivariate Constant Risk Posture," *J. Econ. Theory*, **10**: 309–332 (1975).

Rothschild, M., and J. E. Stiglitz: "Increasing Risk: I. A Definition," *J. Econ. Theory* **2**: 225–243, (1970).

——— and ———: "Increasing Risk: II. Its Economic Consequences," *J. Econ. Theory*, **3**: 66–84 (1971).

Savage, L. J.: *The Foundations of Statistics*, New York (1954).

Selden, L.: "A New Representation of Preferences Over "Certain × Uncertain" Consumption Pairs: the "Ordinal Certainty Equivalent" Hypothesis," *Econometrica* **46**(5): 1045–1060 (September 1978).

Tversky, A., and D. Kahneman: "Judgement under Uncertainty: Heuristics and Biases," *Science* **185**(4157): 1124–1131 (1974).

Veinott, A. F., Jr.: "Optimal Policy in a Dynamic, Single Product, Non-Stationary Inventory Model with Several Demand Classes," *Oper. Res.* **13**(5): 761–778 (1965).

Whitmore, G. A., and M. C. Findlay: *Stochastic Dominance*, Lexington Books, Lexington, Mass. (1978).

Whitt, W.: "Comparing Probability Measures on a Set with an Intransitive Preference Relation," *Manage. Sci.*, **25**: 505–511 (1979).

Ziemba, W. T., and R. G. Vickson: *Stochastic Optimization Models in Finance*, Academic Press, New York, 1975.

Present-Value Formulas

Several scholars during the 1960s explored properties for preference orderings for time streams that induce present-value formulas. Particularly good references are Diamond (1965), Koopmans (1972), and Williams and Nasser (1966). The article by Koopmans has an extensive list of references.

The article by Williams and Nasser is the basis for our presentation of Sections 2-5 and 2-7. This article seems to be disjoint from the economics literature which is referenced by Koopmans (1972).

The fields of finance and accounting contain a large literature that compares alternative methods for assessing time streams of revenues and costs. There are numerous methods other than present-value formulas, but present-value formulas are used in practice far more often than any other method. Moreover, present-value methods are particularly compatible with stochastic models in operations research. For other methods, see Solomon (1959).

Diamond, P. A.: "The Evaluation of Infinite Utility Streams," *Econometrica* **33** (1): 170–177 (1965).

Koopmans, T. C.: "Representation of Preference Orderings over Time," chap. 4 in *Decisions and Organizations*, edited by C. B. McGuire and R. Radner, North-Holland Publishing Co., Amsterdam (1972).

Solomon, E.: *The Management of Corporate Capital*, Free Press, Glencoe, Ill. (1959).

Williams, A. C., and J. I. Nassar: "Financial Measurement of Capital Investments," *Manage. Sci.* **12** (11): 851–864 (1966).

CHAPTER

THREE

MYOPIC OPTIMAL POLICIES

Temporal dependence usually complicates a model's optimization. You will see in Chapter 4 that the optimization of an n-period model typically consumes more time and requires more computer storage than n separate one-period problems. This painful inflation of a problem's difficulty would be curbed if the future consequences of present decisions could safely be ignored. Then *myopic* (or "greedy") behavior would be optimal. For each dynamic problem in this chapter, greediness *is* optimal; the solution is obtained by optimizing a static (i.e., one-period) problem. The static problem is derived easily from the data of the n-period problem. A static problem suppresses temporal dependence; so its solution is necessarily myopic. Therefore, this chapter's dynamic problems can be said to possess myopic optimal solutions.

Myopia can be optimal for any of several separate reasons. Sections 3-1 through 3-5 investigate models where the initial specification lacks a myopic optimum but a transformation of the model possesses a myopic optimal policy. Applications in those sections include inventories, productive capacity, maintenence, and advertising policy. This chapter's Bibliographic Guide cites several references with different justifications for optimality of myopic optima.

The inventory model in Section 3-1 may seem absurdly simple, but it is quite useful in practice. Also, some of its limitations are removed with the embellishments presented in Section 3-2.

3-1 MYOPIC INVENTORY POLICIES

Consider the problem of a retailer who sells many products. Among these products is the sole one ordered from a particular wholesaler. The wholesaler makes deliveries in the region once a month; so the retailer must decide each month how much to order. Suppose that the retail sales of the product are random, that the profits on sales are appreciable, that the wholesale unit cost is high, and that the retailer uses debt financing. The debt financing ensures that (1) there are real costs of storing unsold items and (2) the retailer prefers to incur costs in the future rather than in the present.

Specifically, suppose that the demands $D_1, D_2, \ldots$ in successive months are i.i.d. (independent and identically distributed) nonnegative r.v.'s (random variables). Let s_n denote the number of items on hand at the beginning of month n when the quantity z_n is purchased from the wholesaler. We assume that excess demand cannot be backlogged; so successive months' inventory levels are related by $s_{n+1} = (s_n + z_n - D_n)^+$. Let $a_n = s_n + z_n$ denote the total number of items available to satisfy demand in month n so

$$s_{n+1} = (a_n - D_n)^+$$

Let r denote the fixed (known) retail price so the revenue received during month n is†

$$r(a_n \wedge D_n) = rD_n - r(D_n - a_n)^+ = ra_n - r(a_n - D_n)^+$$

Suppose that the relevant retail costs are of two kinds—purchasing and storage. Let c denote the wholesale unit cost so $c \cdot z_n$ is the total purchasing cost during month n when z_n items are purchased by the retailer. We assume that the storage cost, or *holding cost*, is proportional to the number of items left over at the end of the period; i.e., there is a parameter h such that the holding cost in month n is $h \cdot (a_n - D_n)^+$. In the notation of (2-5) and (2-8), $v(a, d) = (a - d)^+$ and $g(a, d) = -r(a \wedge d) + h \cdot (a - d)^+$.

Suppose that the retailer's time preference for money is embodied in a discount factor β, $0 < \beta < 1$. The retailer expects to operate the store for many more years; so the criterion, we assume, is to maximize the expectation‡ of the sum of the discounted revenues net of costs. In our notation, therefore, the problem is to choose $z_n \geq 0$, for each $n \in I_+$, to maximize $E(B)$ where

$$B = \sum_{n=1}^{\infty} \beta^{n-1}[r(a_n \wedge D_n) - cz_n - h \cdot (a_n - D_n)^+] \tag{3-1}$$

† Recall that $a \wedge b$ denotes min $\{a, b\}$ and $a \vee b$ denotes max $\{a, b\}$.

‡ This criterion is equivalent to assuming that the retailer is risk neutral, i.e., that the retailer's attitude to risk is consistent with a linear utility function. This is the usual assumption in the analysis of operations research models, but see Section 2-4 for a discussion of attitude toward risk.

Substitute $a_n - s_n$ for z_n and $(a_{n-1} - D_{n-1})^+$ for s_n if $n > 1$. Then

$$B = \sum_{n=2}^{\infty} \beta^{n-1}\{r[a_n - (a_n - D_n)^+] - h(a_n - D_n)^+ - c[a_n - (a_{n-1} - D_{n-1})^+]\}$$

$$+ ra_1 - r(a_1 - D_1)^+ - h(a_1 - D_1)^+ - c(a_1 - s_1)$$

$$= cs_1 - \sum_{n=1}^{\infty} \beta^{n-1}[(c - r)a_n + (a_n - D_n)^+(r + h - \beta c)] \qquad (3\text{-}2)$$

Let $w(a_n, D_n)$ denote the following quantity in (3-2):

$$w(a_n, D_n) = (c - r)a_n + (a_n - D_n)^+(r + h - \beta c)$$

Then $$E(B) = cs_1 - \sum_{n=1}^{\infty} \beta^{n-1}E[w(a_n, D_n)]$$

Observe that a_n and D_n are stochastically independent† because the order quantity z_n, hence a_n, can depend only on the information available at the start of period n:

$$s_1, a_1, D_1, s_2, a_2, D_2, \ldots, s_{n-1}, D_{n-1}, \text{ and } s_n$$

But $D_1, D_2, \ldots, D_{n-1}, D_n$ are stochastically independent. As a result of the independence of a_n and D_n,

$$E[w(a_n, D_n)] = E\{E[w(a_n, D_n) \mid a_n]\}$$

For $a \geq 0$, define

$$G(a) = (c - r)a + (r + h - \beta c)E[(a - D_1)^+] \qquad a \geq 0 \qquad (3\text{-}3)$$

Therefore, $E[w(a_n, D_n)] = E[G(a_n)]$ so

$$E(B) = cs_1 - \sum_{n=1}^{\infty} \beta^{n-1}E[G(a_n)] \qquad (3\text{-}4)$$

Since $a_n = s_n + z_n$, $E(B)$ is maximized over choices of $z_n \geq 0$, $n \in I_+$, if, and only if,

$$\sum_{n=1}^{\infty} \beta^{n-1}E[G(a_n)]$$

is minimized over choices of $a_n \geq s_n$, $n \in I_+$.

Suppose $r > c$. You should verify that there necessarily is a number a^* at which $G(\cdot)$ attains its minimum on $\mathbb{R}_+$.

Assumption There exists $a^* \geq 0$ such that $G(a^*) \leq G(a)$ for all $a \geq 0$.

† z_1 can depend only on s_1 so it has the form $u_1(s_1)$ and $a_1 = s_1 + z_1 = s_1 + u_1(s_1)$ which has the form $h_1(s_1)$. Then z_2 can depend only on s_1, a_1, D_1, and s_2 so it has the form $z_2 = u_2[s_1, h_1(s_1), D_1, s_2]$ but $s_2 = (a_1 - D_1)^+ = [h_1(s_1) - D_1]^+$. Thus, $z_2 = u_2\{s_1, h_1(s_1), D_1, [h_1(s_1) - D_1]^+\}$ and $a_2 = s_2 + z_2 = [h_1(s_1) - D_1]^+ + z_2$ which has the form $a_2 = h_2(s_1, D_1)$. Continuing, we see that a_n has the form $h_n(s_1, D_1, \ldots, D_{n-1})$. Thus, a_n and D_n are independent.

Proposition 3-1 If $s_1 \leq a^*$, then $a_n = a^*$ for all $n \in I_+$ is feasible and optimal.

PROOF First, for every n, $E[G(a_n)] \geq G(a^*)$ because $G(a^*) \leq G(a)$ for all $a \in \mathbb{R}_+$. Therefore,

$$\sum_{n=1}^{\infty} \beta^{n-1} E[G(a_n)] \geq \sum_{n=1}^{\infty} \beta^{n-1} G(a^*)$$

Second, if $s_1 \leq a^*$, the following argument shows that $a_n = a^*$ for all n is feasible, hence optimal. If $s_1 \leq a^*$, the constraint $a_n \geq s_n$ allows $a_1 = a^*$. If $a_n = a^*$ for some n, then $s_{n+1} = (a_n - D_n)^+ = (a^* - D_n)^+ \leq a^*$ because $a^* \geq 0$ and $D_n \geq 0$. Hence, $s_{n+1} \leq a^*$ and $a_{n+1} = a^*$ is feasible. □

Suppose $s_1 \leq a^*$ so $a_n = a^*$ for all n is indeed optimal. What are the resulting order quantities, namely, the z_n's? Remember $z_n = a_n - s_n$ so $z_n = a^* - s_n$ for all n. Therefore, for $n > 1$, $z_n = a^* - (a_{n-1} - D_{n-1})^+ = a^* - (a^* - D_{n-1})^+ = a^* \wedge D_{n-1}$.

Suppose that the monthly demand r.v. D_1 has a density function $q(\cdot)$ and a distribution function $Q(\cdot)$ with $Q(0) = 0$. Then (3-3) is

$$G(a) = (c - r)a + (r + h - \beta c) \int_0^a (a - u)q(u)\, du \tag{3-5}$$

The derivative of $G(a)$ is obtained from Leibnitz' rule, which states that

$$\frac{d}{dt} \int_{x(t)}^{y(t)} f(u, t)\, du = f[y(t), t]y'(t) - f[x(t), t]x'(t) + \int_{x(t)}^{y(t)} \frac{\partial f(u, t)}{\partial t}\, du \tag{3-6}$$

when $f(u, t)$, $\partial f(u, t)/\partial t$, $x(t)$, $x'(t)$, $y(t)$, and $y'(t)$ exist and are continuous on suitable domains (cf. Proposition A-9). The application of this rule to the integral in (3-5) yields

$$\frac{d}{da} \int_0^a (a - u)q(u)\, du = 1 \cdot (a - a)q(a) - 0 \cdot (a - 0)q(0) + \int_0^a 1 \cdot q(u)\, du = Q(a)$$

As a result,

$$G'(a) = c - r + (r + h - \beta c)Q(a) \tag{3-7}$$

$$G''(a) = (r + h - \beta c)q(a) \tag{3-8}$$

Therefore, $G''(a) \geq 0$ for all $a \geq 0$ if $r + h > \beta c$. It is natural to assume that the parameters satisfy this inequality because $r + h - \beta c \geq r - c$ and the item should not be offered for sale if the selling price is less than the purchase cost. Now $G''(a) \geq 0$ for all $a \geq 0$ implies that a^* minimizes $G(\cdot)$ on $[0, \infty)$ if $a^* \geq 0$ and $G'(a^*) = 0$. From (3-7), $G'(a) = 0$ if, and only if,

$$Q(a) = \frac{r - c}{r + h - \beta c} \tag{3-9}$$

There is a unique value of a that satisfies equation (3-9) unless either (1) no value of a satisfies (3-9), or (2) $(r - c)/(r + h - \beta c)$ is a flat point of $Q(\cdot)$. However, (1) is impossible because $Q(\cdot)$ has a density function and is therefore continuous. We resolve the nonuniqueness problem (2) in the following manner: Let $Q^{-1}(\cdot)$ be the inverse† of $Q(\cdot)$: if $0 \le b \le 1$, then

$$Q^{-1}(b) \triangleq \min \{u: Q(u) \ge b\}$$

Thus $Q^{-1}(b)$ is the smallest value of u at which $P\{D_1 \le u\} \ge b$. One solution of (3-9) is

$$Q^{-1}\left(\frac{r - c}{r + h - \beta c}\right)$$

By assumption, the numerator and denomenator are positive. Therefore, $P\{D_1 \ge 0\} = 1$ implies that this expression is a nonnegative number. Since a^* minimizes $G(\cdot)$, it must be a solution of (3-9); as a result,

$$a^* = Q^{-1}\left(\frac{r - c}{r + h - \beta c}\right) \tag{3-10}$$

The right side of (3-10) is sometimes called a *critical fractile formula.*

Example 3-1 Suppose demand is an exponential r.v., that is, for $a \ge 0$, $Q(a) = 1 - e^{-a/\mu}$ $(\mu > 0)$. Then $Q^{-1}(x) = -\mu \log(1 - x)$ so (3-10) yields

$$a^* = -\mu \log \frac{h + c(1 - \beta)}{r + h - \beta c} \tag{3.11}$$

In this case, the amount a^* of goods available to satisfy demand is proportional to μ, the mean demand (and to $\mu \log 2$, the median demand). □

Example 3-2 If D is a uniform r.v. on the interval $[y, b + y]$, then $\mu = y + b/2$ so $b = 2(\mu - y)$ and $Q(a) = (a - y)/b$ if $y \le a \le y + b$. Therefore,

$$a^* = y + \frac{b(r - c)}{r + h - \beta c} \tag{3-12}$$

Here, a^* is not proportional to μ. □

You may have noticed that an entire contingency plan, a *policy*, has not yet been specified. If $s_1 > a^*$, you do not yet know how to assign values to $a_1, a_2, \ldots$. Recall that $s_1 \le a^*$ leads to the order quantity $z_n = a^* - s_n$ for all n so $z_n \downarrow 0$ as $s_n \uparrow a^*$. You might conjecture that $z_n = 0$ is optimal if $s_n > a^*$. Indeed it is, and $z_n = (a^* - s_n)^+$ or

$$a_n = a^* \vee s_n \qquad n = 1, 2, \ldots \tag{3-13}$$

completes the specification of an optimal policy.

† $Q(u) = P\{D_1 \le u\}$ is continuous from the right so $P\{0 \le D_1 < \infty\} = 1$ and $0 \le b < 1$ imply that the minimum is attained.

Proposition 3-2 An optimal policy is specified by (3-13).

PROOF To verify optimality of (3-13) when $s_1 > a^*$, let N be the first time that (3-13) causes $s_n < a^*$, that is,

$$N = \inf \left\{ n \colon \sum_{i=1}^{n} D_i > s_1 - a^* \right\}$$

Let $s'_1, s'_2, \ldots$ and $a'_1, a'_2, \ldots$ denote the inventory levels and decisions caused by (3-13) so $s'_1 = s_1$,

$$s'_n = s_1 - \sum_{i<n} D_i \qquad n = 1, 2, \ldots, N$$

and $a'_n = s'_n$ if $n < N$. Consider any other decision rule and let $s_1, s_2, \ldots$ and $a_1, a_2, \ldots$ denote its inventory levels and decisions. It will be shown that $G(a'_n) \le G(a_n)$ for all n [although $E[G(a'_n)] \le E[G(a_n)]$ would be sufficient].

For $n \ge N$, $a'_n = a^*$ and $G(a^*) \le G(a)$ for all a so $G(a'_n) \le G(a_n)$. For $n < K$,

$$a'_n = s'_n = s_1 - \sum_{i<n} D_i \le s_1 - \sum_{i<n} [D_i - (a_i - s_i)] = s_n \le a_n$$

Observe from (3-3) that $G(\cdot)$ has a right-hand derivative which is nondecreasing on $\mathbb{R}_+$. Hence, $G(\cdot)$ is nondecreasing on $[a^*, \infty)$. Therefore, $a'_n \le a_n$ implies $G(a'_n) \le G(a_n)$ if $n < N$, which completes the proof that (3-13) is optimal in (3-4). □

Notice that we have proved a stronger result than was needed. *On every sample path,* $G(a'_n) \le G(a_n)$ for all n; so (3-13) is optimal for other criteria than merely (3-4). For example, (3-13) minimizes $\sum_{n=1}^{\infty} \beta^{n-1} G(a_n)$ with probability one!† Also with probability 1, it minimizes the long-run "average cost per period"

$$\lim_{n\to\infty} \frac{1}{n} \sum_{i=1}^{n} G(a_i)$$

This is not the usual average cost per period because the discount factor β is a parameter in the specification of $G(\cdot)$ in (3-3).

Value of the Objective‡

What value does $E(B)$ take with an optimal policy? From (3-4), if $s_1 \le a^*$,

$$E(B) = cs_1 - \frac{G(a^*)}{1-\beta} \tag{3-14}$$

† A statement is said to be *true with probability* 1, abbreviated w.p. 1, if the subset of the sample space where it is true has probability 1. Of course, if Z is an event with $P\{Z\} = 0$, then Z is not necessarily impossible. For example, if X is an exponential r.v., $P\{X = x\} = 0$ for all $x \ge 0$ although $P\{X \ge 0\} = 1$.

‡ The remainder of this section should be skipped by readers who are unfamiliar with elementary properties of renewal functions (cf. Section 5-4 in Volume I).

If $s_1 > a^*$, the answer is more complicated because no orders are placed until the inventory level drops below a^*. With the policy (3-13), let $s = s_1$ and let N denote the number of periods until the inventory level drops below a^*:

$$N = \inf\left\{n\colon \sum_{i=1}^{n} D_i > s - a^*\right\} \tag{3-15}$$

so $s_1 \geq s_2 \geq \ldots \geq s_N \geq a^* > s_{N+1}$ and $z_1 = z_2 = \ldots = z_N = 0 < z_{N+1}$. Since D_1, D_2, ... are i.i.d. nonnegative r.v.'s, in (3-15) we may interpret N as "the number of renewals at 'time' $s - a^*$." See Figure 3-1. With the notation N,

$$E(B) = cs_1 - E\left[\sum_{n=1}^{N} \beta^{n-1} G\left(s_1 - \sum_{i=1}^{n-1} D_i\right) + \beta^N \frac{G(a^*)}{1-\beta}\right] \tag{3-16}$$

The quantity $s_1 - \sum_{i=1}^{n-1} D_i$ is the inventory level at the beginning of period

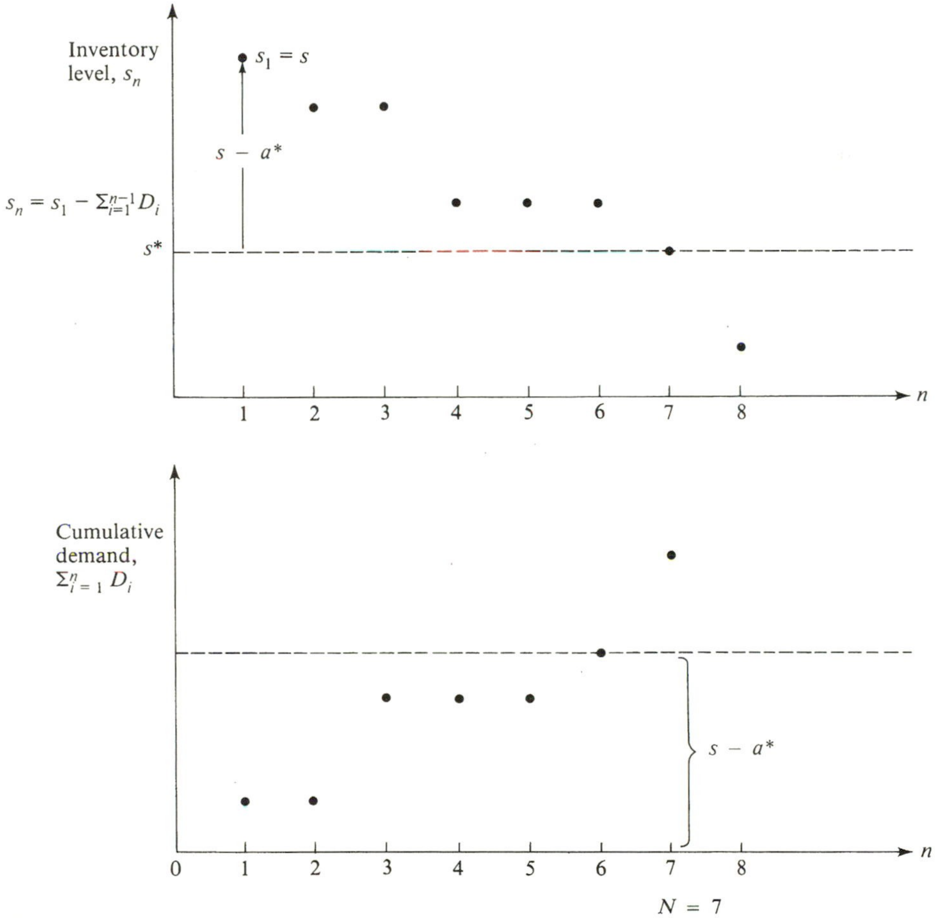

Figure 3-1 N is the "number of renewals" at "time" $s - a^*$.

$n \le N$. The program now is to evaluate $E(\beta^N)$ and

$$E\left[\sum_{n=1}^{N} \beta^{n-1} G\left(s_1 - \sum_{i=1}^{n-1} D_i\right)\right]$$

For simplicity, suppose D_1 is a discrete r.v. with $q_j = P\{D = j\}$, $j \in I$. Let $\Delta = s - a^*$ and $Q^n(j) = P\{\sum_{i=1}^n D_i \le j\}$ with $Q^1(\cdot) = Q(\cdot)$ and $Q^0(\cdot) \equiv 1$. Also, let $q^n(j) = P\{\sum_{i=1}^n D_i = j\}$ and

$$M_\beta(j) = \sum_{n=1}^{\infty} \beta^n Q^n(j) \qquad m_\beta(j) = \sum_{n=1}^{\infty} \beta^n q^n(j) \tag{3-17}$$

From page 120 in Volume I, $M_\beta(\cdot)$ can be computed recursively via

$$M_\beta(j) = (1 - \beta q_0)^{-1} \beta \left[Q(j) + \sum_{i=0}^{j-1} M_\beta(i) q_{j-i}\right]$$

The computation of $m_\beta(j)$ is simply $m_\beta(0) = M_\beta(0)$ and $m_\beta(j) = M_\beta(j) - M_\beta(j-1)$ if $j \in I_+$. From (3-15),

$$\begin{aligned} P\{N = n\} &= P\left\{\sum_{i=1}^{n-1} D_i \le \Delta < \sum_{i=1}^{n} D_i\right\} = P\left\{\sum_{i=1}^{n-1} D_i \le \Delta\right\} - P\left\{\sum_{i=1}^{n} D_i \le \Delta\right\} \\ &= Q^{n-1}(\Delta) - Q^n(\Delta) \end{aligned}$$

From (3-17),

$$\begin{aligned} E(\beta^N) &= \sum_{n=1}^{\infty} \beta^n [Q^{n-1}(\Delta) - Q^n(\Delta)] \\ &= \beta[1 + M_\beta(\Delta)] - M_\beta(\Delta) = \beta - (1-\beta) M_\beta(\Delta) \end{aligned} \tag{3-18}$$

Observe that

$$\{N > n\} = \left\{\sum_{i=1}^{n} D_i \le \Delta\right\}$$

so (3-17) yields

$$\begin{aligned} &E\left[\sum_{n=1}^{N} \beta^{n-1} G\left(s - \sum_{i=1}^{n-1} D_i\right)\right] \\ &\quad = G(s) + E\left[\sum_{n=2}^{N} \beta^{n-1} G\left(s - \sum_{i=1}^{n-1} D_i\right)\right] \\ &\quad = G(s) + E\left[\sum_{n=1}^{N-1} \beta^{n} G\left(s - \sum_{i=1}^{n} D_i\right)\right] \\ &\quad = G(s) + \sum_{n=1}^{\infty} \beta^n E\left[G\left(s - \sum_{i=1}^{n} D_i\right) \mid N > n\right] P\{N > n\} \\ &\quad = G(s) + \sum_{n=1}^{\infty} \beta^n \sum_{k=0}^{\Delta} G(s-k) q^n(k) \\ &\quad = G(s) + \sum_{k=0}^{\Delta} G(s-k) m_\beta(k) \end{aligned} \tag{3-19}$$

When $s_1 = s > a^*$, the combination of (3-16), (3-18), and (3-19) is

$$E(B) = cs - G(s) - \sum_{k=0}^{\Delta} G(s-k)m_\beta(k) - [\beta - (1-\beta)M_\beta(\Delta)] \frac{G(a^*)}{1-\beta} \quad (3\text{-}20)$$

Two uses can be made of (3-15) through (3-20). First, (3-20) itself can be used to decide how much of an excess initial inventory to sell. Let $b(s)$ denote the dependence of $E(B)$ in (3-20) on $s_1 = s$ (recall $\Delta = s - a^*$ also depends on s). Suppose there is a salvage value v for each unit sold. Then it is optimal to sell $s - s^*$ units where s^* maximizes $b(a) + v(s - a)$ on $\{0, 1, \ldots, s - a^*\}$. The second use of (3-15) through (3-20) is in an algorithm to find an optimal policy in inventory problems having setup costs. See page 314 and the Bibliographic Guide for Chapter 7 for a brief discussion and references.

Summary

The initial formulation (3-1) does not possess a myopic optimum. The nth term of the objective in (3-1) is

$$\beta^{n-1}[r(a_n \wedge D_n) - c(a_n - s_n) - h(a_n - D_n)^+] \quad (3\text{-}21)$$

which includes s_n in a nontrivial way. Also, for a given value of s_n, what value of a_n maximizes the expectation of (3-21) conditional on s_n? In general, it is different from a^* that minimizes $G(\cdot)$ in (3-3).

The procedure that transformed (3-1) to (3-4) substituted $a_n - s_n$ for z_n and $(a_{n-1} - D_{n-1})^+$ for s_n. Then, for each n, the coefficients of all the terms involving a_n were collected and their expectations were evaluated [in (3-3)] as a function of a_n. This resulted in (3-4), which had a sum that depended only on the a_n's and had another term that did not depend on the s_n's (except for s_1, which is known), a_n's, or z_n's. This other term was extraneous because its value was not influenced by the decision variables (the a_n's).

Minimization of the sum in (3-4) became the objective. Except for the discount factor β^{n-1}, the functional forms of the terms in the sum were the same, namely, $G(\cdot)$. Let a^* minimize $G(\cdot)$ on $[0, \infty)$. The constraints $a_n \geq s_n$ might have prevented the assignment $a_n = a^*$ for all n. The assignment $a_n = a$ for all n might be regarded as an optimal solution to an unconstrained optimization problem. The same solution is optimal for the constrained problem, that is, $a_n \geq s_n$ for all n, if the constraints are not binding. We saw that if the initial inventory level is sufficiently low, that is, $s_1 \leq a^*$, then indeed the constraints are not binding. If $s_n > a^*$ so the constraint $a_n \geq s_n$ is binding, then the myopic constrained optimization

$$\min \{G(a): a \geq s_n\}$$

has the solution $a = s_n$ because $G(s_n) \leq G(a)$ for all $a \geq s_n$ is implied by $G(\cdot)$ being nondecreasing on $[a^*, \infty)$.

This line of argument is made general in Section 3-3 and the general results are applied in Sections 3-4 and 3-5. Before turning to the general case, we extend the conversion of (3-1) to (3-4) to other inventory problems.

EXERCISES

3-1 Let $b(\cdot)$ be given by (3-14) on $[0, a^*) \cap I$ and by (3-20) on $[a^*, \infty) \cap I$. (*a*) Show that $b(s+1) - b(s) \leq c$ if $s \in [a^*, \infty) \cap I$. (*b*) How do you interpret the inequality in (*a*)? (Hint: From (3-14), $b(s+1) - b(s) = c$ if $s \in [0, a^* - 1) \cap I$.)

3-2 Let $b(\cdot)$ denote the function specified in Exercise 3-1. Consider a salvage value v for each unit of the initial inventory that is sold as "excess." Then the expected profit net of salvage revenue is $b(a) + v(s - a)$ when $s - a$ units are sold for salvage. Assume $v \leq c$ and let s^* denote the largest element a of $\mathbb{R}_+$ that maximizes $b(a) + v(s - a)$ on $\mathbb{R}_+$ (assume s^* exists). Prove that $s^* \geq a^*$.

3-3 Suppose D_1 is a discrete r.v. with $q_j = P\{D_1 = j\}$ $j \in I$. What form does (3-10) take?

3-4 The discrete analog of Example 3-1 occurs when D_1 has a geometric distribution, that is, $q_j = P\{D_1 = j\} = (1 - \lambda)\lambda^j$ for some $0 < \lambda < 1$. In this case, prove that a^* is the smallest nonnegative integer x which satisfies

$$x \geq \frac{\log \{[h + c(1 - \beta)]/(r + h - \beta c)\}}{\log \lambda} - 1$$

3-2 MULTIPLE PRODUCTS AND OTHER GENERALIZATIONS

Single-Item Models

More elaborate inventory models than the one in Section 3-1 have myopic optima. First, the structure $s_{n+1} = (a_n - D_n)^+$ is just one of many ways in which "tomorrow's" inventory level might depend on "today's" demand and the total amount of goods available for sale. More generally, suppose

$$s_{n+1} = v(a_n, D_n) \tag{3-22}$$

For example, if $s_{n+1} = a_n - D_n$ so excess demand is back-ordered, then $v(a, u) = a - u$.

Second, $h \cdot (a_n - D_n)^+ - r(a_n \wedge D_n)$ in Section 3-1 is just one of many plausible models of inventory-related costs. Generally, denote this cost by

$$g(a_n, D_n) \tag{3-23a}$$

An example that sometimes occurs in practice is

$$g(a, u) = h \cdot (a - u)^+ - r(a \wedge u) + \pi \cdot (u - a)^+$$

where π is the unit cost of delayed receipts due to excess demand.

Use (3-22) and (3-23a) to alter (3-1) to

$$\begin{aligned} -B &= \sum_{n=1}^{\infty} \beta^{n-1}[g(a_n, D_n) + cz_n] \\ &= g(a_1, D_1) + c(a_1 - s_1) + \sum_{n=2}^{\infty} \beta^{n-1}\{g(a_n, D_n) \\ &\qquad + c[a_n - v(a_{n-1}, D_{n-1})]\} \\ &= -cs_1 + \sum_{n=1}^{\infty} \beta^{n-1}[ca_n + g(a_n, D_n) - \beta c v(a_n, D_n)] \end{aligned} \tag{3-23b}$$

The definition

$$G(a) = ca + E[g(a, D) - \beta cv(a, D)] \tag{3-24}$$

yields

$$cs_1 - E(B) = \sum_{n=1}^{\infty} \beta^{n-1} E[G(a_n)]$$

Constraints other than $a_n \geq s_n$ are sometimes appropriate. For example, suppose disposal of excess stock, say up to U units, is possible. Then $a_n \geq s_n - U$ is an appropriate constraint. Generally, we impose the constraint $a_n \geq w(s_n)$; in the instances above, $w(s) \equiv s$ if the constraint is $a_n \geq s_n$ and $w(s) \equiv s - U$ if the constraint is $a_n \geq s_n - U$.

Sometimes, a_n is constrained in ways unrelated to s_n. For example, suppose there is space to store at most m items. Then $a_n \leq m$ is necessary. Generally, let A denote the set of values to which each a_n is constrained without regard to s_n. Then the full constraint on each a_n is $a \in A \cap [w(s_n), \infty)$.

The following theorem can be proved by using essentially the same argument as in Section 3-1.

Theorem 3-1 Suppose

(*a*) G attains its minimum on A at a^*.
(*b*) $w[v(a^*, u)] \leq a^*$ for all $u \geq 0$.
(*c*) For all $u \geq 0$, $G(\cdot)$, $w(\cdot)$, and $v(\cdot, u)$ are nondecreasing on $[a^*, \infty) \cap A$.
(*d*) If $a^* < w(s)$, then $w(s) \in A$.

Then $a_n = a^* \vee s_n$ for all n is optimal in

$$\min \left\{ \sum_{n=1}^{\infty} \beta^{n-1} E[G(a_n)] \colon a_n \geq w(s_n),\ a_n \in A, \text{ for all } n \right\}$$

PROOF Left as Exercise 3-11. □

Corollary 3-1 Suppose D_1 has a mean μ, density function $q(\cdot)$, and distribution function $Q(\cdot)$ with $Q(0) = 0$, excess demand is back-ordered so $v(a, u) = a - u$, and there are linear holding and penalty costs so

$$g(a, u) = h \cdot (a - u)^+ + \pi \cdot (u - a)^+ - ru$$

Then

$$a^* = Q^{-1} \left[\frac{\pi - c(1 - \beta)}{h + \pi} \right] \tag{3-25}$$

PROOF In this case, (3-24) is

$$G(a) = ca + h \int_0^a (a - x)q(x)\, dx + \pi \int_a^{\infty} (x - a)q(x)\, dx - r\mu - \beta c(a - \mu)$$

so Leibnitz' rule [cf. (3-6) and Proposition A-9] yields

$$G'(a) = c + hQ(a) - \pi[1 - Q(a)] - \beta c$$

$$G''(a) = (h + \pi)q(a) \geq 0$$

Therefore, $G'(a^*) = 0$ implies (3-25). □

Example 3-3 Suppose, as in Example 3-1, that demand is exponentially distributed so $Q(x) = 1 - e^{-x/\mu}$. Then (3-25) is

$$1 - e^{-a^*/\mu} = \frac{\pi - c(1 - \beta)}{h + \pi}$$

so

$$a^* = \mu \log \frac{h + \pi}{h + c(1 - \beta)} \tag{3-26}$$

and, as in (3-9), a^* is proportional to mean demand μ. Comparison of (3-9) with (3-26) shows that a^* is the same in the two cases if $\pi = r - \beta c$. This result agrees with intuition. In the lost-sales case of Section 3-1, you lose revenue r for every unit of excess demand but, unlike the back-order case here, you need not purchase the unit next period (suppose $a^* > 0$). Hence, you save βc. □

Example 3-4 If D is uniform on $[y, b + y]$, as in Example 3-2, then (3-26) is

$$a^* = y + \frac{2(\mu - y)[\pi - c(1 - \beta)]}{h + \pi} \tag{3-27}$$

This value of a^* is the same as (3-10) if, as in Example 3-3, $\pi = r - \beta c$. □

Multiple Products

Suppose that several products are stocked simultaneously. Their costs may be interdependent if the products include complementary or substitute goods or if it is cheaper to order several products at one time than to order each of them separately. Exercise 3-12 asks you to prove the following version of Theorem 3-1: $a_n = a^*$ for all n is optimal under conditions (a) and (b) of Theorem 3-1 if s_n, a_n, D_n, and a^* are interpreted as vectors with one component for each different product and $\mathbf{s}_1 \leq \mathbf{a}^*$.

Example 3-5 Suppose L products are stocked in the same location, excess demand is lost, and the storage cost depends only on the aggregate number of items in stock. This is a reasonable approximation when, for example, the products are different colors of the same model automobile. For each product j let c_j and r_j be the unit cost and unit revenue, respectively. Let $h(x)$ denote the storage cost if x items remain in storage at the end of a period. Then the expected discounted cost is

$$-E(B) = \sum_{n=1}^{\infty} \beta^{n-1} E\left(\sum_{j=1}^{L} \{c_j(a_n^j - s_n^j) - r_j[a_n^j - (a_n^j - D_n^j)^+]\} + h\left[\sum_{j=1}^{L} (a_n^j - D_n^j)^+ \right] \right)$$

because the aggregate amount stored at the end of period n is

$$\sum_{j=1}^{L} (a_n^j - D_n^j)^+$$

Therefore,
$$-E(B) = -\sum_{j=1}^{L} c_j s_1^j + E\left[\sum_{n=1}^{\infty} \beta^{n-1} G(\mathbf{a}_n)\right]$$

where

$$G(\mathbf{a}) = \sum_{j=1}^{L} \{(c_j - r_j)a^j + (r_j - \beta c_j)E[(a^j - D^j)^+]\} + E\left\{h\left[\sum_{j=1}^{L} (a^j - D^j)^+\right]\right\} \tag{3-28}$$

Suppose that $h(x) = x$ and for all j, $c_j = c$, $r_j = r$, and the marginal distribution of D^j is uniform on $[0, 1]$, $j = 1, \ldots, L$. (When might a continuous distribution like this be reasonable?) Then the components of $\mathbf{a}^*$ are all the same and Exercise 3-13 asks you to verify that they are given by the expression

$$\frac{r - c}{r - \beta c + 1} \qquad \square \tag{3-29}$$

Delivery Delay

Consider the following alteration of the general single-item problem which led to (3-24). Suppose excess demand is back-ordered and there is a delay of ν periods between ordering goods and receiving them. We assume ν to be constant but comment later on the random case. Thus far, $\nu = 0$. In general, z_n is not available to satisfy demand until period $\nu + n$.

It is convenient to redefine s_n as the amount of goods on hand *plus on order* (i.e., plus $z_{n-\nu+1} + z_{n-\nu+2} + \cdots + z_{n-1}$) after $z_{n-\nu}$ is delivered at the beginning of period n but before z_n has been ordered and before demand D_n has been experienced. As before, let

$$a_n = s_n + z_n \tag{3-30a}$$

but a_n now denotes the amount on hand plus on order after z_n is ordered but before D_n is experienced. It is still true that

$$s_{n+1} = v(a_n, D_n) = a_n - D_n \tag{3-30b}$$

Also, the ordering cost $c \cdot z_n$ is still given by $c \cdot (a_n - s_n)$.

So far, so good. The costs in (3-23b) are those of ordering and of inventory-related expenses. How do we represent the latter? Consider a specific case to appreciate the general situation. Suppose there are unit costs h, for holding inventory, and π, for excess demand. When $\nu = 0$, this resulted in the expression

$$h \cdot (a_n - D_n)^+ + \pi \cdot (D_n - a_n)^+$$

If $v > 0$, the inventory level at the end of period n is given by†

$$\left(a_n - \sum_{i=0}^{v-1} z_{n-i} - D_n\right)^+ \tag{3-31}$$

The middle term is the amount of goods on order that is included in a_n. But (3-31) is related to a_{n-v} because the amount of goods on hand at the beginning of period n is precisely the amount on hand plus on order in period $n - v$ less the demand during the intervening periods:

$$s_n - \sum_{i=1}^{v-1} z_{n-i} = s_{n-v} + z_{n-v} - \sum_{i=n-v}^{n-1} D_i$$

Therefore, substituting (3-30*a*) and (3-30*b*) in (3-31) implies

$$\begin{aligned} a_n - \sum_{i=0}^{v-1} z_{n-i} - D_n &= s_n + z_n - \sum_{i=0}^{n-1} z_{n-i} - D_n = s_n - \sum_{i=1}^{v-1} z_{n-i} - D_n \\ &= s_{n-v} + z_{n-v} - \sum_{i=n-v}^{n-1} D_i - D_n = a_{n-v} - \sum_{i=n-v}^{n} D_i \end{aligned}$$

so the inventory cost in period n is

$$h \cdot \left(a_{n-v} - \sum_{i=n-v}^{n} D_i\right)^+$$

Similarly, the penalty cost for excess demand is

$$\pi \cdot \left(\sum_{i=n-v}^{n} D_i - a_{n-v}\right)^+$$

This example suggests replacement of (3-23*a*) with the following general expression for inventory-related costs incurred during period n:

$$g\left(a_{n-v}, \sum_{i=n-v}^{n} D_i\right) \tag{3-32}$$

By assumption, the demands $\{D_i\}$ are not influenced by the decisions $z_1, z_2, \ldots$ or the inventory levels $s_1, s_2, \ldots$; so (3-32) implies that the inventory-related costs incurred during periods $1, 2, \ldots, v$ are beyond control. The costs in period $v + 1$ will be influenced by the choice of a_1. Therefore, in place of (3-23*b*), a reasonable measure of performance is

$$\sum_{n=1}^{\infty} \beta^{n-1} \left[cz_n + \beta^v g\left(a_n, \sum_{i=n}^{n+v} D_i\right)\right]$$

† We use the convention that $\sum_{i=b}^{c} = 0$ if $c < b$.

Substitution of (3-30a) and (3-30b) yields

$$\sum_{n=2}^{\infty} \beta^{n-1} \left\{c \cdot [a_n - (a_{n-1} - D_{n-1})] + \beta^v g\left(a_n, \sum_{i=n}^{n+v} D_i\right)\right\}$$

$$+ c \cdot (a_1 - s_1) + \beta^v g\left(a_1, \sum_{i=1}^{v+1} D_i\right)$$

$$= \sum_{n=1}^{\infty} \beta^{n-1} \left[c(1-\beta)a_n + \beta^v g\left(a_n, \sum_{i=n}^{n+v} D_i\right)\right] + \sum_{n=1}^{\infty} \beta^n D_n - cs_1$$

The following line of reasoning should now be familiar. Observe that

$$E\left(\sum_{n=1}^{\infty} \beta^n D_n - cs_1\right)$$

is not influenced by the decisions $a_1, a_2, \ldots$; so a decision rule is optimal if, and only if, it minimizes

$$\sum_{n=1}^{\infty} \beta^{n-1} E\left[c(1-\beta)a_n + \beta^v g\left(a_n, \sum_{i=n}^{n+v} D_i\right)\right] \tag{3-33}$$

The expression $\sum_{i=n}^{n+v} D_i$ in (3-33) is the sum of $v + 1$ i.i.d. r.v.'s. Therefore, $\sum_{i=n}^{n+v} D_i$ and $\sum_{i=1}^{v+1} D_i$ have the same distribution for every n, and given a_n in (3-33),

$$g\left(a_n, \sum_{i=n}^{n+v} D_n\right)$$

has a conditional distribution that depends on n only through a_n. Hence, (3-33) can be written as

$$\sum_{n=1}^{\infty} \beta^{n-1} E[G(a_n)]$$

where
$$G(a) = c(1-\beta)a + \beta^v E\left[g\left(a, \sum_{i=1}^{v+1} D_i\right)\right] \tag{3-34}$$

It is straightforward to verify that Theorem 3-1 applies here too.

Suppose D has a density function q and distribution function Q and $g(\cdot, \cdot)$ is given by linear holding and penalty costs:

$$g(a, x) = h \cdot (a - x)^+ + \pi \cdot (x - a)^+ - rx$$

Let $\mu = E(D_1)$, $q_{v+1}(\cdot)$ denote the density function for $\sum_{i=1}^{v+1} D_i$ (given by the $v + 1$-fold convolution of q), and let Q_{v+1} denote the corresponding distribution function. Then (3-34) is

$$G(a) = c(1-\beta)a + \beta^v \left[h \int_0^a (a - x)\, q_{v+1}(x)\, dx\right.$$

$$\left. + \pi \int_a^{\infty} (x - a) q_{v+1}(x)\, dx\right] - r(v + 1)$$

Here,
$$G'(a) = c(1-\beta) + \beta^{v}\{hQ_{v+1}(a) - \pi[1 - Q_{v+1}(a)]\}$$
$$G''(a) = \beta^{v}(h+\pi)q_{v+1}(a) \geq 0$$
so
$$a^* = Q_{v+1}^{-1}\left[\frac{\pi - c(1-\beta)/\beta^{v}}{h+\pi}\right] \tag{3-35}$$
which is the same as (3-25) except that $Q_{v+1}(\cdot)$ replaces $Q(\cdot)$ and $c(1-\beta)/\beta^{v}$ replaces $c(1-\beta)$.

What is the effect of v in (3-35)? Exercise 3-9 asks you to verify that $P\{D \geq 0\} = 1$ implies $Q_i(x) \geq Q_{i+1}(x)$ for all $i = 1, 2, \ldots$ and $x \in \mathbb{R}$. Therefore, $Q_{i+1}^{-1}(b) \geq Q_i^{-1}(b)$ for all $i = 1, 2, \ldots$ and $b \in [0, 1]$. Let y_{v+1} denote the argument of $Q_{v+1}^{-1}(\cdot)$ in (3-35). Then y_{v+1} is a decreasing function of v. Since $Q_{v+1}^{-1}(\cdot)$ is a nondecreasing function, a^* may either increase or decrease with v.

Let $a^*(v)$ indicate the dependence of (3-35) on v. But $a^*(v)$ is not the inventory level! The justification of (3-32) showed that the inventory level at the end of period n (possibly negative if there is excess demand) is
$$a_{n-v} - \sum_{i=n-v}^{n} D_i \tag{3-36}$$
If $n > v$ and $s_1 \leq a^*(v)$, the optimality of a myopic policy and the use of that policy implies $a_{n-v} = a^*(v)$. Then $a^*(v) - \sum_{i=n-v}^{n} D_i$ is the inventory level. The demands are i.i.d.; so (3-36) has the same distribution as
$$U_v \triangleq a^*(v) - \sum_{i=1}^{v+1} D_i \tag{3-37}$$
To appreciate the effect of v in (3-37),
$$E(U_{v+1} - U_v) = a^*(v+1) - a^*(v) - \mu$$
$$= Q_{v+1}^{-1}(y_{v+1}) - Q_v^{-1}(y_v) - \mu$$
where $y_v = [\pi - c(1-\beta)\beta^{v}]/(h+\pi)$. Exercise 3-9 asks you to prove that $Q_v^{-1}(y_{v+1}) \leq E(U_{v+1} - U_v) + Q^{-1}(y_v) + \mu \leq Q_{v+1}^{-1}(y_v)$.

You should review this section on the inclusion of a delivery delay to appreciate its dependence on the back-ordering assumption (3-30*b*). It was essential in the justification of (3-32) as the form of the inventory-related costs. This form includes certain cases where inflation is included in the model.

If the delivery delay is random, the entire problem of choosing a suitable model becomes complicated.† For example, if the delays of various deliveries are modeled as i.i.d. r.v.'s, then it is possible for orders to "cross"; i.e., an order placed in period $n+1$ may be delivered sooner than one placed in period n. This

† An exception to this complexity occurs if the delays are i.i.d. geometric r.v.'s. This case was encountered in Example 7-15 of Volume I. The following references use other probabilistic models for delays: Kaplan, R. S.: "A Dynamic Inventory Model with Stochastic Lead Times," *Manage. Sci.* **16**: 491–507 (1970); and Nahmias, S.: "Simple Approximations for a Variety of Dynamic Leadtime Lost-Sales Inventory Models," *Oper. Res.* **27**: 904–924 (1979).

phenomenon rarely occurs in commercial practice; so a more complicated stochastic model is needed. The extant theory is incomplete here.

Finite Planning Horizon

Suppose the retailer in Section 3-1 has a store in a building that is scheduled to be demolished in N periods. Suppose that N is small enough so that β^N is not nearly zero. An infinite planning horizon no longer seems appropriate. Let us replace the infinite planning horizon in (3-23*b*) with N and then use $z_n = a_n - s_n$ and $s_{n+1} = v(a_n, D_n)$:

$$B_N \triangleq \sum_{n=1}^{N} \beta^{n-1}[g(a_n, D_n) + cz_n]$$

$$= g(a, D_1) + c(a_1 - s_1) + \sum_{n=2}^{N} \beta^{n-1}\{g(a_n, D_n) + c[a_n - v(a_{n-1}, D_{n-1})]\} \tag{3-38a}$$

$$= -cs_1 + \sum_{n=1}^{N} \beta^{n-1}[g(a_n, D_n) + ca_n - \beta cv(a_n, D_n)] + \beta^N cv(a_N, D_N) \tag{3-38b}$$

This formulation *lacks* a myopic optimum because of the appendage $\beta^N cv(a_n, D_N)$ in (3-38*b*). We realize that it would have been convenient to add the term $-\beta\, cs_{N+1}$ to (3-38*a*). It would have canceled the unwanted $+\beta^N cv(a_N, D_N)$ in (3-38*b*). What justification could there be for the term $-\beta^N cs_{N+1}$ being added to (3-38*a*)?

When the retailer's store must be closed, what is the value of the unsold merchandise; i.e., what is its *salvage value*? Suppose, for the present, that the retailer would be able to resell it wholesale at its original cost; i.e., its unit salvage value is c. Usually, the salvage value would be less than c, and this flexibility will be encompassed by the next generalization.

Suppose that the salvage value is received just after the last period, say, at the beginning of period $N + 1$ so its present value (at the start of the first period) is $\beta^N cs_{N+1}$. If excess demand is backlogged so $s_{N+1} < 0$ is possible, then it is assumed that the backlogged demand must be met with a concomitant unit purchase cost c. Therefore, to (3-38*a*) and (3-38*b*) we add $-\beta^N cs_{N+1} = -\beta^N cv(a_N, D_N)$. This leads to the criterion

$$B_N^* \triangleq \sum_{n=1}^{N} \beta^{n-1}[g(a_n, D_n) + cz_n] - \beta^N cv(a_N, D_N) \tag{3-39}$$

Therefore,
$$E(B_N^*) = -cs_1 + \sum_{n=1}^{N} \beta^{n-1} E[G(a_n)] \tag{3-40}$$

where $G(\cdot)$ is given by (3-24). As in Theorem 3-1, if $s_1 \le a^*$, then $a_1 = a_2 = \cdots = a_N = a^*$ is optimal for the minimization of (3-40) subject to $a_n \ge s_n$, $n = 1, \ldots, N$.

Nonstationary Model

In many contexts, perhaps most, you expect conditions to change as time passes. Perhaps you expect unit costs and prices to rise with inflation but the pertinent discount factor to rise more slowly. Consider a nonstationary version of (3-38) with (3-39) in which every entity is subscripted with the period to which it pertains. Suppose that the demands $D_1, D_2, \ldots, D_N$ are independent r.v.'s with, perhaps, different distributions. In place of β^{n-1}, let β_n denote the factor to discount period n's cash flow back to the beginning of period 1. Let $\alpha_0 = 1$ and $\alpha_1, \ldots, \alpha_N$, all nonnegative, satisfy $\beta_{n+1} = \alpha_n \beta_n$ so

$$\beta_n = \prod_{i=0}^{n-1} \alpha_i \qquad n = 1, \ldots, N+1$$

With this notation,

$$\beta_n c_n z_n = \beta_n c_n(a_n - s_n) = \beta_n c_n a_n - \beta_{n-1}\alpha_{n-1} c_n v_{n-1}(a_{n-1}, D_{n-1})$$

so B_N^*, as in (3-39), becomes

$$B_N^* = \sum_{n=1}^{N} \beta_n[g_n(a_n, D_n) + c_n z_n] - \beta_{N+1} c_{N+1} s_{N+1}$$

$$= -c_1 s_1 + \sum_{n=1}^{N} \beta_n \left[g_n(a_n, D_n) + c_n a_n - \alpha_n c_{n+1} v_n(a_n, D_n)\right] \qquad (3\text{-}41)$$

and
$$E(B_N^*) = -c_1 s_1 + \sum_{n=1}^{N} \beta_n E[G_n(a_n)]$$

where
$$G_n(a) = c_n a + E[g_n(a, D_n) - \alpha_n c_{n+1} v_n(a, D_n)] \qquad (3\text{-}42)$$

Finally, suppose a is constrained to be in $A_n \cap [w_n(s_n), \infty)$ and $G_n(\cdot)$ is minimized on A_n at a_n^*. Then Exercise 3-10 asks you to verify that $a_n = a_n^* \vee w_n(s_n)$, $n = 1, 2, \ldots, N$ minimizes the expectation of (3-41), subject to $a_n \geq w_n(s_n)$ and $a_n \in A_n$ for $n = 1, \ldots, N$, if

$$a_1^* \leq a_2^* \leq \cdots \leq a_{N-1}^* \leq a_N^* \qquad (3\text{-}43)$$

and conditions (*a*), (*b*), and (*c*) of Theorem 3-1 are satisfied for each n.

Condition (3-43) is trivial in the stationary case but nontrivial otherwise.

Example 3-6 Here is an example where $a_N^* < a_n^*$ for $n < N$. Suppose that a retailer has to close down and that the unit salvage value in period $N + 1$ is much less than c, say $c/10$ ("ten cents on the dollar!"). You might conjecture that it would be prudent to let inventories diminish as period N approaches, at least $a_N^* < a_1^*$. Consider the back-order case with $c_n = c$ and $\alpha_n = \beta$ for $n = 1, 2, \ldots, N$. For $n = 1, 2, \ldots, N - 1$, let

$$g_n(a, x) = h \cdot (a - x)^+ + \pi \cdot (x - a)^+$$

Suppose the "penalty" for excess demand in period N is the necessity of meeting it in period $N + 1$ so

$$g_N(a, x) = h \cdot (a - x)^+ + 9\beta c \cdot \frac{(x - a)^+}{10}$$

so, if $s_{N+1} < 0$, the cost to meet it is $-9\beta c s_{N+1}/10$ included in $g_N(\cdot, \cdot)$ plus $-\beta^N c s_{N+1}/10$ as a salvage "value." Suppose $D_1, D_2, \ldots, D_N$ are i.i.d. r.v.'s with density and distribution functions q and Q and suppose $c_{N+1} = c/10$. From (3-25),

$$a_n^* = Q^{-1}\left[\frac{\pi - c(1 - \beta)}{h + \pi}\right] \qquad n = 1, \ldots, N - 1$$

For $n = N$, (3-42) is

$$G_N(a) = ca + h \int_0^a (a - x)q(x)\, dx + \frac{9\beta c}{10} \int_a^x (x - a)q(x)\, dx - \beta \frac{c}{10}(a - \mu)$$

where μ is the mean demand. Therefore,

$$G_N'(a) = c\left(1 - \frac{\beta}{10}\right) + \left(h + \frac{9\beta c}{10}\right)Q(a) - \frac{9\beta c}{10}$$

$$G_N''(a) = \left(h + \frac{9\beta c}{10}\right)q(a) \geq 0$$

$$Q(a_N^*) = \frac{9\beta c/10 - c(1 - \beta/10)}{h + 9\beta c/10} = \frac{9\beta c - c(10 - \beta)}{10h + 9\beta c}$$

$$= \frac{10c(\beta - 1)}{10h + 9\beta c} < 0$$

because $\beta < 1$. Hence, $a_N^* = -\infty$ so (3-43) is violated. □

On the other hand, (3-43) *is* satisfied in many nonstationary models. Suppose D_n has density and distribution functions q_n and Q_n and mean μ_n, $v_n(a, x) = a - x$, and

$$h_n(a, x) = h_n \cdot (a - x)^+ + \pi_n \cdot (x - a)^+$$

for all $n = 1, \ldots, N$. Then (3-42) yields

$$G_n(a) = c_n a + h_n \int_0^a (a - x)q_n(x)\, dx + \pi_n \int_a^\infty (x - a)q_n(x)\, dx$$

$$-\alpha_n c_{n+1}(a - \mu_n)$$

$$G_n'(a) = (c_n - \alpha_n c_{n+1}) - \pi_n + (h_n + \pi_n)Q_n(a)$$

$$G_n''(a) = (h_n + \pi_n)q_n(a) \geq 0$$

$$a_n^* = Q_n^{-1}\left(\frac{\pi_n - c_n + \alpha_n c_{n+1}}{h_n + \pi_n}\right) \tag{3-44}$$

Let
$$u_n = \frac{\pi_n - c_n + \alpha_n c_{n+1}}{h_n + \pi_n}$$
denote the argument of $Q_n^{-1}(\cdot)$ in (3-44). If demand is stochastically increasing with n, that is, $Q_n(x) \geq Q_{n+1}(x)$ for all $x \in \mathbb{R}$ and each n, then $Q_n^{-1}(y) \leq Q_{n+1}^{-1}(y)$ for all $y \in [0, 1]$. Therefore, in this case, (3-43) is satisfied if $u_n \leq u_{n+1}$ for each n.

For example, suppose for all n that $\alpha_n = \beta$, $\pi_n = b^n\pi$, $c_n = b^n c$, and $h_n = b^n h$, where $b > 1$ is the rate at which costs are rising each period. Then
$$u_1 = u_2 = \cdots = u_N = \frac{\pi - c(1 - b\beta)}{h + \pi} \tag{3-45}$$
so (3-43) is satisfied if demand is stochastically increasing. Observe in (3-45) that $u > 1$ is possible if $b\beta$ exceeds 1 by a large enough margin. However, this is unlikely if the entire economy is expected to experience a high inflation rate; then β would ordinarily be a commensurately smaller fraction.

In summary, the most important instance of (3-43) is stochastic growth in demand with cost parameters changing relatively slowly.

EXERCISES

3-5 Consider an infinite-horizon single-item inventory model with nonnegative i.i.d. r.v. demands, back-ordering of excess demands, and linear purchase costs. Suppose the delivery delay is zero, i.e., ordered goods are delivered immediately, a penalty π is incurred to expedite delivery per unit of excess demand, and the holding cost incurred in period n is proportional to the average of the inventory levels at the beginnings of periods $n - 1$ and n. Let c and h denote the respective unit costs of ordering goods and of holding inventory. (*a*) Verify that this model has a myopic optimum by specifying its $G(\cdot)$. (*b*) Derive an expression for a^* under the assumption that demand has distribution function $Q(\cdot)$ with density function $q(\cdot)$. (*c*) Compute a^* if D is uniformly distributed on $[0, 10]$, $h = 1$, $\pi = 9$, $c = 10$, and $\beta = 0.9$. (*d*) Compute a^* if D is exponentially distributed with mean 5 and the other parameters are as in (*c*). Compare your answers to (*c*) and (*d*).

3-6 Repeat Exercise 3-5 with excess demand being lost instead of back-ordered.

3-7 Repeat Exercise 3-5 with a two-period delivery delay.

3-8 Prove that the expected discounted costs in the nonstationary model are minimized by $a_n = a_n^*$, $n = 1, \ldots, N$, if $s_1 \leq a_1^*$ and (3-43) is valid.

3-9 (*a*) Suppose $Q(\cdot)$ is the distribution function of a nonnegative random variable D, i.e., $P\{D \geq 0\} = 1$. Let $Q_i(\cdot)$ denote the i-fold convolution of $Q(\cdot)$. Prove $Q_i(x) \geq Q_{i+1}(x)$ and $Q_{i+1}^{-1}(b) \geq Q_i^{-1}(b)$ for all $i = 1, 2, \ldots$, $x \in \mathbb{R}$, and $b \in [0, 1]$. (*b*) Use the notation following (3-37) and prove $Q_v^{-1}(y_{v+1}) \leq E(U_{v+1} - U_v) + Q_v^{-1}(y_v) + \mu \leq Q_{v \times 1}^{-1}(y_v)$

3-10 Repeat parts (*c*) and (*d*) of Exercise 3-5 if each of the following complications is separately applicable: (*a*) there is space to store at most 6 items; (*b*) for each n, a_n must be an even integer, i.e., $A = \{\ldots, -2, 0, 2, \ldots\}$.

3-11 Prove Theorem 3-1 (in the single-item case).

3-12 Prove the version of Theorem 3-1 stated for the multiple-product model described below (3-27).

3-13 Verify that each component of a^* is given by (3-29) in the special case of Example 3-5 described after (3-28).

3-14 Following (3-23), there is an example of $g(\cdot, \cdot)$ where the retail price r is assumed to be fixed. Instead, let r_n denote the price in period n and suppose that (D_1, r_1), (D_2, r_2), $\ldots$ are independent and identically distributed random vectors. Assume that (0, 1), (1, 1), (2, 1), (0, 2), and (1, 2) are all the outcomes of (D_1, r_1) and that they are equally likely. Let $c = 2$, $h = 1$, $\pi = 3$, $\beta = 0.9$, $s_1 = 0$, and assume an infinitely long planning horizon. What is an optimal sequence of order quantities when excess demand is lost?

3-3 SUFFICIENT CONDITIONS FOR A MYOPIC OPTIMUM

Pause here to conjecture the reasons why the model in Section 3-1 and its generalizations in Section 3-2 possessed myopic optima. The linear purchase cost facilitated the transformation from $\{(s_n, z_n)\}$ variables to $\{a_n\}$ variables, but that property, alone, did not ensure myopic optima.

This section presents a class of general sequential decision problems and deduces sufficient conditions for such a problem to have a myopic optimum. The general problems are called *Markov decision processes*, abbreviated MDPs, or *dynamic programming problems*, abbreviated DPs. Chapter 4 presents elements of the theory and computational characteristics of MDPs. An MDP is defined here merely to acquire notation concerning myopic optima.

Markov Decision Process

The elements of an MDP are a set $\mathcal{S}$ of *states*, a set A_s of *actions* at each state $s \in \mathcal{S}$, a probability measure on a (suitable) collection of subsets of $\mathcal{S}$ defined for each pair $(s, a) \in \mathscr{C}$,

$$\mathscr{C} \triangleq \{(s, a): a \in A_s \text{ and } s \in \mathcal{S}\}$$

a *single-stage reward* whose expected value $r(\cdot, \cdot)$ is defined on $\mathscr{C}$, and if the planning horizon N is finite, a *salvage-value* function $L(\cdot)$ on $\mathcal{S}$. These elements have the following interpretations. At each point in time, the state of affairs is described by an element $s \in \mathcal{S}$ assumed known by the "decision maker." A_s is an array of alternative actions, and exactly one of them must be taken immediately. The notation A_s emphasizes the dependence of the set of alternative actions on the current state s. Selection of $a \in A_s$ has two consequences. First, there is an immediate scalar reward $r(s, a)$ whose value (expected value if the reward is an r.v.), in general, depends on the current state s and on the action a chosen from A_s, but on nothing else. Second, the state of affairs changes, at least partly influenced by s and a, to another state in $\mathcal{S}$. The probability measure mentioned above governs this transition. The Markovian assumption is that neither the reward nor the probability measure depends on events (states of affairs and actions taken) prior to the epoch when state s was occupied. Finally, if the planning horizon N is finite, and the state of affairs after the ultimate period N is s, then $L(s)$ is the salvage value, namely, the reward received in period $N + 1$.

All the inventory models in Sections 3-1 and 3-2 are MDPs. In them, the set of states $\mathcal{S}$ is the set of all possible inventory levels s_n, and the set of actions A_s is the set of all possible values of z_n, namely, $\mathbb{R}_+$. Equivalently, A_s is the set of possible values of a_n, $\{a: a \geq s_n\}$, which is contingent on the state s_n. The probability measures that determine successive states are induced by the distribution of D and of $v(\cdot, \cdot)$ [in $s_{n+1} = v(a_n, D_n)$]:

$$P\{s_{n+1} \leq m \mid s_n = s, a_n = a\} = P\{v(a, D) \leq m\}$$

For example, if $v(a, d) = a - d$ as in the back-order case, then

$$P\{s_{n+1} \leq m \mid s_n = s, a_n = a\} = P\{D \geq a - m\} \tag{3-46}$$

This section concerns only discounted MDPs. Let $\beta \geq 0$ be a discount factor; if the planning horizon is infinite, $\beta < 1$ is assumed too. First, consider a finite planning horizon. Let B_N denote the discounted sum of the rewards net of salvage value:

$$B_N \triangleq \sum_{n=1}^{N} \beta^{n-1} r(s_n, a_n) + \beta^N L(s_{N+1}) \tag{3-47}$$

where s_n and a_n denote the state occupied and action taken in period n. If N is infinite, B denotes the sum of the discounted rewards:

$$B \triangleq \sum_{n=1}^{\infty} \beta^{n-1} r(s_n, a_n) \tag{3-48}$$

Both B_N and B are r.v.'s, and the notion of optimum used here depends on the expectations $E(B_N)$ and $E(B)$. Precise definitions of "optimum" for MDPs will be given in Chapter 4. The criterion used most widely in practice, and the basic idea here, is to maximize $E(B)$ and $E(B_N)$.

It is convenient to define

$$A = \bigcup_{s \in \mathcal{S}} A_s$$

and

$$S(a) = \{s\colon a \in A_s \text{ and } s \in \mathcal{S}\} \qquad a \in A$$

Hence $S(a)$ is the set of states from which action a is feasible.

Assumptions

The following assumptions are sufficient for a myopic optimum.

Assumption I

$$r(s, a) = K(a) + L(s) \qquad (s, a) \in \mathscr{C} \tag{3-49}$$

Assumption II For each $a \in A$ there is an r.v. $\xi(a)$ such that†

$$s_{n+1} \sim \xi(a_n) \qquad n = 1, 2, \ldots \tag{3-50}$$

Let

$$\gamma(a) \triangleq K(a) + \beta E\{L[\xi(a)]\} \qquad a \in A \tag{3-51}$$

† For two r.v.'s Y and Z on the same probability space, the notation $Y \sim Z$ means that they have the same distribution.

Assumption III

$$\text{There exists } a^* \text{ such that } \gamma(a^*) \geq \gamma(a) \text{ for all } a \in A \tag{3-52}$$

Assumption IV

$$P\{\xi(a^*) \in S(a^*)\} = 1\dagger \tag{3-53}$$

Assumption I asserts that the expected single-stage reward depends additively on the state s and action a. Moreover, the dependence on s_n is given by the salvage-value function $L(\cdot)$. Assumption II concerns the transition probabilities. In an MDP, we have already assumed that these probabilities depend only on the current state and current action; i.e., s_{n+1} has a distribution that depends only on s_n and a_n. Assumption II further reduces the dependence to a_n, alone, so that the distribution of s_{n+1} does not depend on the state s_n from which the transition starts. In (3-50) we use the notation $\xi(a_n)$ to denote an r.v. with the same distribution as s_{n+1}.

Assumption III asserts existence of an action a^* that maximizes a function defined in (3-51), $\gamma(\cdot)$, whose role is analogous to that of $G(\cdot)$ in Section 3-1. Last, suppose that action a^* is taken at some state s where it is feasible, that is, $s \in \mathcal{S}(a^*)$. Then Assumption IV ensures (with probability one) that a^* will again be feasible next period because it is feasible at every state to which the process can transit.

MDP models which are built in professional practice usually contain only finitely many states and actions. That is, $\mathcal{S}$ is a finite set as is A_s for each $s \in \mathcal{S}$. Hence, $\mathscr{C}$ and A are finite sets. Then there necessarily exists an element a^* which maximizes $\gamma(\cdot)$ on A so Assumption III is satisfied.

When $\mathcal{S}$ is a countable set (or a finite set), the probability measure which describes the movements from one period to the next can be described with transition probabilities. There are nonnegative numbers $\{p_{sj}^a\colon (s, a) \in \mathscr{C}, j \in \mathcal{S}\}$ such that

$$p_{sj}^a = P\{s_{n+1} = j \mid s_n = s, a_n = a\}$$

so

$$\sum_{j \in \mathcal{S}} p_{sj}^a = 1 \qquad (s, a) \in \mathscr{C}$$

Assumption II states that p_{sj}^a can depend only on j and a but not on s. That is, for each $j \in \mathcal{S}$ and $a \in A$ there is a nonnegative number ρ_j^a such that

$$p_{sj}^a = \rho_j^a \qquad (s, a) \in \mathscr{C}, j \in \mathcal{S}$$

so

$$\sum_{j \in \mathcal{S}} \rho_j^a = 1 \qquad a \in A$$

† Let $(\Omega, \mathscr{B}, P)$ be the probability space involved here. Then by (3-53) we mean $P\{\xi(a^*) \in H\} = 1$ for all $H \in \mathscr{B}$ that have $S(a^*) \subset H$.

In this case, (3-51) can be written

$$\gamma(a) = K(a) + \beta \sum_{j \in \mathcal{S}} \rho_j^a L(j) \tag{3-51$'$}$$

Compatibility with Earlier Inventory Models

In (3-23), the single-stage reward is $-g(a_n, D_n) - c(a_n - s_n)$ so

$$r(s, a) = cs_n - ca_n - E[g(a_n, D)] \tag{3-54}$$

which satisfies (3-49) with $L(s)$ given by cs_n and $K(a)$ given by $-ca_n - E[g(a, D)]$. Hence, $L(s) = cs$ is necessary and, in (3-38) to (3-40), is precisely the salvage-value term that is used. From (3-22), it is apparent that the transition probabilities do not depend on the state s_n.

One of the ways that a model can satisfy (3-53) is $\xi(a) \in S(a)$ for all $a \in A$. For example, suppose $\mathcal{S} \subset \mathbb{R}^b$, $A \subset \mathbb{R}^n$, $n < \infty$, $b < \infty$, and

$$A_s = \{a\colon a \geq M(s) \text{ and } a \in A\} \qquad s \in \mathcal{S} \tag{3-55}$$

with M a nondecreasing function from $\mathcal{S}$ to A. Suppose also that $a \geq M(s)$ if s is in the sample space of $\xi(a)$. In many inventory models with s as the inventory level (a vector in the multi-item case) and a as the decision, we have $A_s = \{a\colon\ a \geq s\}$, which satisfies (3-55) with $M(s) \equiv s$. Also, recall (3-22):

$$a - \xi(a) = a - v(a, D) \geq 0$$

so $a_n = a$ implies $s_{n+1} \leq a$ and $a \in A_{s_{n+1}}$.

The Assumptions Yield a Myopic Optimal Policy

Substitution of (3-49) in (3-47) yields

$$\begin{aligned} B_N &= \sum_{n=1}^{N} \beta^{n-1} r(s_n, a_n) + \beta^N L(s_{N+1}) \\ &= \sum_{n=1}^{N} \beta^{n-1} [K(a_n) + L(s_n)] + \beta^N L(s_{N+1}) \end{aligned}$$

Substitution of (3-50) gives

$$\begin{aligned} B_N &\sim K(a_1) + L(s_1) + \sum_{n=2}^{N} \beta^{n-1} \{K(a_n) + L[\xi(a_{n-1})]\} + \beta^N L[\xi(a_N)] \\ &= L(s_1) + \sum_{n=1}^{N} \beta^{n-1} \{K(a_n) + \beta L[\xi(a_n)]\} \end{aligned}$$

so

$$E(B_N) = L(s_1) + E\left[\sum_{n=1}^{N} \beta^{n-1} \gamma(a_n)\right] \tag{3-56}$$

where $\gamma(\cdot)$ is defined by (3-51).

Suppose $s_1 \in S(a^*)$. From (3-56) and (3-52), for any method of choosing a_1, $a_2, \ldots, a_N$,

$$E(B_N) - L(s_1) = E\left[\sum_{n=1}^{N} \beta^{n-1}\gamma(a_n)\right] \leq \gamma(a^*)\sum_{n=1}^{N} \beta^{n-1} \tag{3-57}$$

The right side of (3-57) would result from $a_1 = a_2 = \cdots = a_N = a^*$ if these actions were feasible. When are they feasible? First, it is necessary that $a_1 = a^*$ be feasible, that is, $s_1 \in S(a^*)$. For some n suppose $s_n \in S(a^*)$ and $a_n = a^*$. Then (3-53) implies (with probability 1) $s_{n+1} \in S(a^*)$ so $a_{n+1} = a^*$ is again feasible. This argument is a proof of the following principal result in Theorem 3-2.

Theorem 3-2 Assumptions I through IV and $s_1 \in S(a^*)$ imply $a_n = a^*$, $n = 1, 2, \ldots, N$, maximizes $E(B_N)$ subject to the constraints $P\{a_n \in A_{s_n}, n = 1, 2, \ldots, N\} = 1$.

When the goal is maximization of the infinite-horizon objective $E(B)$ [where B is the present value defined by (3-48)], essentially the same argument as used above shows

$$E(B) = \sum_{n=1}^{\infty} \beta^{n-1} E[\gamma(a_n)] + L(s_1)$$

Corollary 3-2 Assumptions I through IV and $s_1 \in S(a^*)$ imply $a_n = a^*$ for all $n = 1, 2, \ldots,$ maximizes $E(B)$ subject to the constraints $P\{a_n \in A_{s_n}, n = 1, 2, \ldots\} = 1$

You saw in Section 3-2 that a nonstationary inventory model had a myopic optimal policy. A straightforward nonstationary version of Theorem 3-2 is stated in Exercise 3-18.

Example 3-7 Here is a prototype of a model of a reservoir† which is used for multiple purposes. Suppose that the reservoir's capacity is 4 units of water and that discretization yields a state space $\mathcal{S} = \{0, 1, 2, 3, 4\}$. We interpret $s_n \in \mathcal{S}$ as the level, i.e., the amount of water in the reservoir, at the beginning of period n. Suppose that the inflows, i.e., the amounts of water which flow into the reservoir in successive periods, are‡ i.i.d. r.v.'s with the following

† See Section 2-5 in Volume I for a brief discussion of reservoir modeling. Section 8-3 in this volume has more elaborate reservoir models than the one in this example.

‡ The i.i.d. assumption is made for simplicity of exposition. If inflows are independent but not necessarily identically distributed r.v.'s, the qualitative results may remain valid. See pages 321 to 324 in Section 7-2 for a discussion of this point. Also, see Exercise 3-9. Dropping the independence assumption leads to an enlargement of the MDP state space, and the enlarged model may not have a myopic optimum. See Exercise 7-13 for an enlarged inventory model.

distribution of, say, an r.v. D:

d	0	1	2	3	4	5
$P\{D = d\}$	0.3	0	0.3	0.2	0.1	0.1

We assume that the discount factor β is 0.95, there is \$50 profit from each unit of water which is discharged, and profits accrue from recreational uses of the reservoir. Suppose that $l(s)$ is the profit in a period from recreational uses if s is the level at the beginning of the period. The following table specifies $l(\cdot)$:

s	0	1	2	3	4
$l(s)$	-50	200	300	250	100

Let D_n be the amount of the inflow and let z_n be the amount of the discharge in period n. We assume that the discharge must be chosen before the amount of the inflow is known. Hence, $0 \leq z_n \leq s_n$ because the inflow might be zero, in which case the maximum discharge is the level at the beginning of the period. An overflow occurs if the inflow exceeds the freeboard, i.e., the space available, so

$$s_{n+1} = \begin{cases} s_n - z_n + D_n & \text{if } s_n - z_n + D_n \leq 4 \\ 4 & \text{if } s_n - z_n + D_n > 4 \end{cases}$$

Letting $a_n = s_n - z_n$ permits rewriting this expression as

$$s_{n+1} = (a_n + D_n) \wedge 4$$

The constraint $0 \leq z_n \leq s_n$ is equivalent to $0 \leq a_n \leq s_n$. Letting $\rho_j^a = P\{s_{n+1} = j \mid a_n = a\}$ and using the distribution of D yields the following table for $\{\rho_j^a\}$.

		Next state				
	$a \backslash j$	0	1	2	3	4
Level after discharge	0	0.3	0	0.3	0.2	0.2
	1	0	0.3	0	0.3	0.4
	2	0	0	0.3	0	0.7
	3	0	0	0	0.3	0.7
	4	0	0	0	0	1

For example, $\rho_4^2 = P\{D_n \geq 2\} = 0.7$ and $\rho_3^0 = P\{D_n = 3\} = 0.2$.

The profit in period n is $l(s_n) + 50z_n$ or $l(s_n) + 50(s_n - a_n)$ so the single-period reward function is $r(s, a) = l(s) + 50(s - a)$. Thus $r(s, a) = K(a) + L(s)$, where $K(a) = -50a$ and $L(s) = l(s) + 50s$. The combination of these functions, $\{\rho_k^a\}$, and definition (3-51) or (3-51′) for $\gamma(a)$ yields the following values for $\{\gamma(a)\}$:

a	0	1	2	3	4
$\gamma(a)$	232.75	249.25	213.5	163.5	85

For example,

$$\gamma(1) = K(1) + \beta \sum_{j=0}^{4} \rho_j^1 \, L(j)$$

$$= -50(1) + 0.95 \sum_{j=0}^{4} \rho_j^1 \, [l(j) + 50j]$$

$$= -50(1) + 0.95\,[(0.3)(200 + 50) + (0.3)(250 + 150) + (0.4)(100 + 200)]$$

In this example, $\gamma(1) \geq \gamma(a)$ for all values of a so $a^* = 1$. If $s_n \geq 1$, then $a_n = 1$ is feasible because the contraint is $0 \leq a_n \leq s_n$ so $S(a^*) = \{1, 2, 3\}$. Hence, $s_1 \geq 1$ implies $a_n = 1$ for all n is optimal. However, $s_n = 0$ implies that $a_n = 0$ is the only possible choice of a_n; so an optimal decision rule is as follows. If $s_1 = 0$, then necessarily discharge nothing until the first time that the reservoir level is positive. Thereafter, discharge just enough to reduce the level to 1. □

A Generalization

Theorem 3-2 and its corollary provide sufficient conditions for $s_1 \in S(a^*)$ to imply existence of an optimal rule that is not only myopic but whose decisions are invariant with respect to successive states occupied. The decisions $a^* = a_1 = a_2 = a_3 = \cdots$ are optimal a priori. However, the myopic property is shared by any decision rule that chooses a_n based on a numerical procedure that involves only s_n and $\gamma(\cdot)$. Additional assumptions yield a complete prescription of an optimal myopic decision rule.

Assumption V For every $s \in \mathcal{S}$ there is $b_s \in A_s$ such that

$$\gamma(b_s) \geq \gamma(a) \qquad a \in A_s \tag{3-58}$$

In words, at each state s there is a feasible action b_s that maximizes $\gamma(\cdot)$ on A_s. If $s \in S(a^*)$, note that $\gamma(b_s) = \gamma(a^*)$ so $a^* = b_s$ without loss of optimality. Dirickx and Jennergren (1975) present sufficient conditions for the myopic rule $a_n = b_{s_n}$, $n = 1, 2, \ldots$, to be optimal.

Suppose now that states and actions all comprise real numbers, that is,†

$$A = \mathcal{S} = \mathbb{R} \tag{3-59}$$

Assumption VI

$$\gamma(\cdot) \text{ is nonincreasing on } [a^*, \infty) \tag{3-60}$$

Assumption VII

$$a^* \le a \le a' \qquad \text{implies} \qquad P\{\xi(a) \le \xi(a')\} = 1 \tag{3-61}$$

Assumption VIII There is a nondecreasing function $M(\cdot)$ such that‡

$$M(s) = \min\{a: a \in A_s\} \qquad \text{for all} \qquad s \notin S(a^*) \tag{3-62}$$

Assumption IX

$$s \notin S(a^*) \qquad \text{and} \qquad a \in A_s \qquad \text{implies} \qquad a > a^* \tag{3-63}$$

Assumption VI is satisfied if $\gamma(\cdot)$ is unimodal. Assumption II asserts that $s_{n+1} \sim \xi(a_n)$. Then Assumption VII causes s_{n+1} to be a monotone function of a_n if $a^* \le a_n$. For example, in an inventory model where excess demand is backlogged, $s_{n+1} = a_n - D_n$. If a and a' are two candidates for a_n with $a < a'$, then $\xi(a') - \xi(a) = (a' - D_n) - (a - D_n) = a' - a \ge 0$ which satisfies (3-61). Assumption VIII asserts that, if a^* is not feasible from state s, then the smallest feasible action is a monotone function of s. For example, $M(s) = s$ in an inventory model with $A_s = [s, \infty)$ for all $s \in \mathcal{S}$. From Assumption IX, if a^* is not feasible in state s, then all feasible actions are bigger than a^*.

From (3-60) and (3-62), if $s \notin S(a^*)$, then $b_s = M(s)$.

Theorem 3-3 Assumptions I through IX imply

$$a_n = \begin{cases} a^* & \text{if } s_n \in S(a^*) \\ M(s_n) & \text{if } s_n \notin S(a^*) \end{cases} \tag{3-64}$$

for all $n = 1, 2, \ldots$ maximizes $E(B)$ subject to the restriction $P\{a_n \in A_{s_n}$ for all $n = 1, 2, \ldots\} = 1$.

PROOF If $s_1 \in S(a^*)$, Corollary 3-2 yields optimality of (3-64). Suppose $s_1 \notin S(a^*)$ and policy (3-64) is used for all n; let $s_1^*, a_1^*, s_2^*, a_2^*, \ldots,$ denote the associated sample path of states and actions. Let T be the earliest period in which a^* is feasible, that is, $T = \inf\{n: s_n^* \in S(a^*)\}$. Consider any other

† If A or $\mathcal{S}$ are proper subsets of $\mathbb{R}$, it is merely more cumbersome to state Assumption VI.

‡ Our use of min rather than inf implies that the infimum is attained so $M(s) \in A_s$.

policy and its associated sample path $s_1, a_1, s_2, a_2 \ldots$. It will be shown that

$$a^* \le a_n^* \le a_n \qquad n = 1, 2, \ldots, T-1 \tag{3-65}$$

(with probability 1) so (3-60) and (3-63) imply $\gamma(a_n^*) \ge \gamma(a_n)$ for $n = 1, 2, \ldots, T-1$. Then optimality of (3-64) follows from $n \ge T$ implying $a_n^* = a^*$ and $\gamma(a^*) \ge \gamma(a_n)$ for all $a_n \in A$. To establish (3-65) and the theorem, observe that $s_1 \equiv s_1^* \notin S(a^*)$, (3-62), and (3-64) imply $a_1^* = M(s_1) \le a$ for all $a \in A_{s1}$. Therefore, $a_1^* \le a_1$. Suppose $a_{n-1}^* \le a_{n-1}$ for some $n \le T-1$. Then (3-61) implies $P\{s_n^* \le s_n\} = 1$ so (3-62) and (3-64) yield $a_n^* = M(s_n^*) \le M(s_n) \le a_n$ with probability 1. □

Example 3-8 Single-product inventory model in Section 3-1 The single-product inventory model in Section 3-1 has a lost-sales assumption so $\mathcal{S} = \mathbb{R}^+$. Action a denotes an order-up-to level so $A_s = [s, \infty)$. From Section 3-1, if $s \le a^*$, then $b_s = a^*$ minimizes $G(\cdot)$ on A_s (assuming $r + h > \beta c$). If $s > a^*$, then (3-7), (3-9), and (3-10) imply that $G'(a) > 0$ for all $a \in A_s$ so $a = s$ minimizes $G(\cdot)$ on A_s. Hence, $b_s = s$ if $s > a^*$ so Assumptions V and VI are satisified.

In order to verify that Assumption VII is satisfied, let $\xi(a) = (a - D_1)^+$ and let $a \le a'$. If $D_1 \le a$,

$$\xi(a') - \xi(a) = (a' - D_1 - (a - D_1) \ge 0$$

If $a' \le D_1$, then $\xi(a) = \xi(a') = 0$ so $\xi(a') - \xi(a) = 0$. If $a < D_1 < a'$, then $\xi(a) = 0$ and $\xi(a') = a' - D_1$ so

$$\xi(a') - \xi(a) = a' - D_1 > 0$$

and Assumption VII is satisfied.

For all $s \in \mathcal{S}$, $A_s = [s, \infty)$ so s is the minimal element in A_s. Then $M(s) = s$ is nondecreasing and Assumption VIII is satisfied.

In this model $S(a^*) = [0, a^*]$ so $s \notin S(a^*)$ implies $s > a^*$. With $A_s = [s, \infty)$ this implies $a > a^*$ for all $a \in A_s$; so Assumption IX is satisfied.

Theorem 3-3 asserts that $a_n = M(s_n)$ is best if $s_n \notin S(a^*)$, that is, $a_n = s_n$ if $s_n > a^*$. This is exactly the conclusion we reached in Proposition 3-2. □

Discussion

Assumptions I, II, and IV genuinely restrict the applicability of Theorem 3-2 and its corollary. Nevertheless Sections 3-1 and 3-2 (and 3-4 and 3-5 to come) demonstrate that there are important applications of these general results.

In Sections 3-1 and 3-2 you saw that a change of decision variable sometimes induced a myopic optimum that failed to exist in the original variables. Such a change cannot always be made. For example, consider the inclusion of a setup cost in the inventory model in (3-23*b*). The expected single-stage reward is

$$r(s, a) = -c \cdot (a - s) - E[g(a, D)] - k\delta(a - s) \tag{3-66}$$

where k, the setup cost, is incurred if $a - s > 0$ so $\delta(a - s) = 1$. The term $-k\delta(a - s)$ is not additively separable in terms of a and s as specified in (3-49). However, (3-50) *is* satisfied. The change of variable $z = a - s$ remedies the difficulty with $-k\delta(a - s)$ but creates a new problem with $-E[g(a, D)]$ which becomes $-E[g(s + z, D)]$, a term that is not (ordinarily) additively separable.

The next two sections concern myopic optima in models of capacity expansion and advertising expenditures.

EXERCISES

3-15 Suppose the performance of a system depends on two stochastically independent and identical modules. A working module has probability 1/2 of becoming defective at the end of the day. It costs \$1,000 to replace a module. A replacement module is sure to be working properly at the time of replacement (but with probability $\frac{1}{2}$ it will become defective at day's end). The system cannot function without at least one module working properly. However, if *only* one module is working, the system's performance is degraded by \$500 per day. It is obligatory that at least one module be working after replacements, if any, have been made.

Let s be the number of modules *not* working at the beginning of a day; $s \in S = \{0, 1, 2\}$. Let action a denote the number of modules not working after replacements, if any, are made. Then $A_1 = A_2 = \{0, 1\}$ and $A_0 = \{0\}$. Let p_{sj}^a denote the *transition probability* $P\{s_{n+1} = j \mid s_n = s, a_n = a\}$. The single-stage returns and transition probabilities do not depend on s and are:

$r(s, a)$

s \ a	0	1
0	0	—
1	−1,000	−500
2	−2,000	−1,500

p_{sj}^a

a \ j	0	1	2
0	0.25	0.5	0.25
1	0	0.5	0.5

The single day's discount factor is 0.9. Suppose that the initial state is 2; i.e., both modules are defective. Find a decision rule which minimizes the expected present value of the costs during an infinite planning horizon.

3-16 Consider a small commercial catfish farming pond. Harvests occur annually and the harvest takes only a few days. The farmer can accurately estimate the amount (tons) of fish in the pond and then decide on the harvest quantity. The farmer's experience and U.S. Department of Agriculture data lead to the following probabilities:

		Tons of fish at harvest time next year					
		0	1	2	3	4	5
Tons of fish remaining after harvest this year	0	1	0	0	0	0	0
	1	0	0.9	0.1	0	0	0
	2	0	0	0.8	0.1	0.1	0
	3	0	0	0	0.7	0.2	0.1
	4	0	0	0	0	0.6	0.4
	5	0	0	0	0	0	1

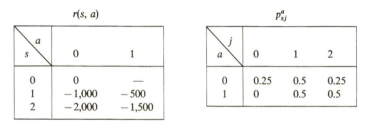

The farmer's annual discount factor is 0.9 and the net profit (i.e., revenue − cost) is \$500 per ton. The farmer is young and optimistic and uses an infinitely long planning horizon. There are 5 tons in the pond at the present harvest time. How many tons should be harvested now to maximize the expected present value of the net profit? What will the expected present value of the profit be when an optimal policy is used? Justify your answer.

3-17 In Exercise 3-16, suppose, instead, that the farmer has only 1 ton in the pond now and has decided always to use a policy of harvesting $(s-2)^+$ tons (where s denotes the tons of fish in the pond before harvesting occurs). What is the expected present value of the profit of this policy?

3-18 A nonstationary MDP with a finite horizon N consists of (i) sets of states $S^1, \ldots, S^{N+1}$; (ii) a nonempty set of actions A_s^n for each $s \in S^n$ and $n = 1, \ldots, N$; (iii) a reward function $r_n(s, a)$ for each $a \in A_s^n$, $s \in S^n$, and $n = 1, \ldots, N$; (iv) a law of motion in which s_{n+1} depends only on s_n and a_n and $P\{s_{n+1} \in S^{n+1}\} = 1$; (v) a salvage-value function $L_{N+1}(\cdot)$ on S^{N+1}; and (vi) a sequence of nonnegative single-period discount factors $\alpha_1, \ldots, \alpha_N$. Let $\alpha_0 = 1$. In place of Assumptions I through IV, suppose for each $n = 1, \ldots, N$;

(I′) $$r_n(s, a) = K_n(a) + L_n(s) \qquad a \in A_s^n, \qquad s \in S^n$$

(II′) $$s_{n+1} \sim \xi_n(a_n) \qquad a_n \in A^n \triangleq \bigcup_{s \in S^n} A_s^n$$

Let

$$\gamma_n(a) \triangleq K_n(a) + \alpha_n E\{L_{n+1}[\xi_n(a)]\} \qquad a \in A^n$$

(III′) There exists a_n^* which maximizes $\gamma_n(\cdot)$ on A^n.
Let

$$S_n^* \triangleq \{s: a_n^* \in A_s\}$$

(IV′) $$P\{\xi_n(a_n^*) \in S_{n+1}^*\} = 1 \qquad (n < N)$$

Prove that assumptions (I′) through (IV′) and $s_1 \in S_1^*$ imply $a_n = a_n^*$ for all n is optimal.

3-4 ADVERTISING POLICY

The demand for a firm's products is often a complicated phenomenon. It may depend significantly on phenomena beyond the firm's immediate control. For example, competitors' prices may influence demand greatly, and this suggests a game-like perspective which is adopted in Section 9-5. Empirical studies† support the intuition that a firm can affect demand through advertising expenditures. Most of this section focuses on a model to optimize advertising outlays without regard to other decisions a firm may make. The end of the section expands the advertising model to include decisions on price and inventory level.

Empirical demand functions show that the impact of advertising diminishes as time passes.‡ Here we assume that the sales effect of advertising depreciates at a constant rate θ per time period, $0 \le \theta \le 1$. Let z_n denote the firm's advertising expenditure in period n, $n = 1, 2, \ldots$, so $z_n \theta^j$ is the impact of z_n on demand in period $n + j$. We use the label *goodwill* for the aggregate impact of past advertis-

† J. D. C. Little, "Aggregate Advertising Models: The State of the Art," *Oper. Res.* **27** (4):629–667 (1979).

‡ Little, op. cit.

ing expenditures. Then the firm's goodwill in period n, denoted a_n, is

$$a_n = z_1\theta^{n-1} + z_2\theta^{n-2} + \cdots + z_{n-1}\theta + z_n$$

so
$$a_n = \sum_{k=0}^{n-1} z_{n-k}\theta^k = z_n + \theta a_{n-1} = z_n + s_n \tag{3-67}$$

where s_n denotes θa_{n-1}.

Let $D_1, D_2, \ldots$ be the demands for the firm's product in successive periods. We assume for each n that the distribution of D_n depends only on a_n. Let $\mu(a) = E(D_n | a_n = a)$, $a \geq 0$, which is assumed to be the same for all n.

Suppose r is the firm's gross profit per unit of demand not including advertising expenditures. Then $rD_n - z_n$ is the gross profit in period n. Let β denote the firm's single-period discount factor so

$$B = \sum_{n=1}^{\infty} \beta^{n-1}(rD_n - z_n) \tag{3-68}$$

is the sum of discounted gross profits. Let $s_1 = a_0\theta$ denote the initial goodwill, "early" in period 1, so $a_1 = z_1 + s_1$ (which includes the case $a_0 = 0$ and $a_1 = z_1$) and (3-67) is valid for all $n = 1, 2, \ldots$. Substitution of $z_n = a_n - \theta a_{n-1}$ in (3-68) yields

$$B = \sum_{n=1}^{\infty} \beta^{n-1}(rD_n - a_n + \theta a_{n-1}) = \theta a_0 + \sum_{n=1}^{\infty} \beta^{n-1}[rD_n - (1 - \beta\theta)a_n]$$

Let

$$G(a) = r\mu(a) - (1 - \beta\theta)a \qquad a \geq 0 \tag{3-69}$$

so
$$E(B) = \theta a_0 + E\left[\sum_{n=1}^{\infty} \beta^{n-1}G(a_n)\right] \tag{3-70}$$

The constraint on a_n is $0 \leq z_n = a_n - \theta a_{n-1} \leq a_n$, which is not stochastic. Therefore, an a priori choice of $a_1, a_2, \ldots$ is possible. In other words, the realized values of $D_1, D_2, \ldots$ do not affect feasibility of $a_1, a_2, \ldots$.

We assume

Assumption X

$$\text{There exists } a^* \text{ that maximizes } G(\cdot) \text{ on } \mathbb{R}_+ \tag{3-71}$$

Assumption XI

$$G(\cdot) \text{ is nonincreasing on } [a^*, \infty) \tag{3-72}$$

Exercise 3-19 asks you to verify that the MDP thus far meets the conditions of Theorem 3-3. Hence, an optimal policy is specified by

$$a_n = a^* \vee \theta a_{n-1} = a^* \vee s_n \qquad n = 1, 2, \ldots \tag{3-73}$$

The advertising expenditures that correspond to (3-73) are $z_n = (a^* - \theta a_{n-1})^+$. Therefore, $\theta a_0 \le a^*$ implies $z_n = (1-\theta)a^*$ for all $n > 1$. If $\theta a_0 > a^*$, then $z_n = 0$ for all $n < N$, where N is the smallest integer with $\theta^N \le a^*/a_0$. Then $z_N = a^* - \theta^N a_0$ and $z_n = (1-\theta)a^*$ for all $n > N$.

Assumption XI, that $G(\cdot)$ is nonincreasing on $[a^*, \infty)$, is a restriction on the rate at which $\mu(\cdot)$ can increase. From (3-69), if $\mu(\cdot)$ is differentiable on (a^*, ∞), then $\mu'(a) \le (1 - \beta\theta)$ ensures (3-72). The following example has this property.

Example 3-9 Suppose $\mu(a) = k\Phi[(a-\lambda)/\sigma]$ where $\Phi(\cdot)$ is the distribution function of a standard normal r.v. (zero mean and unit variance) and $\sigma > 0$ and λ are parameters. This form of $\mu(\cdot)$ exhibits the market-saturation effects often encountered in practice. Let $\phi(\cdot)$ denote the density function for a standard normal r.v., $\phi(x) = (2\pi)^{-1/2} \exp(-x^2/2)$, $x \in \mathbb{R}$. Then

$$G(a) = rk\Phi\left(\frac{a-\lambda}{\sigma}\right) - (1-\beta\theta)a \tag{3-74a}$$

$$G'(a) = \frac{rk}{\sigma}\,\phi\left(\frac{a-\lambda}{\sigma}\right) - (1-\beta\theta) \tag{3-74b}$$

and

$$G''(a) = \frac{rk}{\sigma^2}\,\phi'\left(\frac{a-\lambda}{\sigma}\right) \tag{3-74c}$$

From (3-74*b*), $G'(a) = 0$ if, and only if,

$$(2\pi)^{-1/2} \exp\left[\frac{[-(a-\lambda)/\sigma]^2}{2}\right] = \frac{\sigma(1-\beta\theta)}{rk}$$

or

$$\left(\frac{a-\lambda}{\sigma}\right)^2 = -2\log\left[\frac{\sigma(1-\beta\theta)\sqrt{2\pi}}{rk}\right] \tag{3-75}$$

Substitute

$$H \triangleq -2\sigma^2 \log\left[\frac{\sigma(1-\beta\theta)\sqrt{2\pi}}{rk}\right]$$

in (3-75) to obtain

$$(a-\lambda)^2 = H \tag{3-76a}$$

We already have $0 \le \beta\theta \le 1$ so (3-76*a*) has solutions only if $H \ge 0$, that is,

$$\sigma(1-\beta\theta)\sqrt{2\pi} \le rk$$

which we now assume. Then (3-76*a*) implies

$$a = \lambda \pm H^{1/2} \tag{3-76b}$$

Exercise 3-20 asks you to verify that

$$a^* = \lambda + H^{1/2} \qquad \square \tag{3-76c}$$

Several Decision Variables

Suppose the firm not only chooses its advertising expenditures but also sets its product's price and orders additional goods each period. Let the model in Section 3-1 describe inventory costs and revenues (excess demand is lost and ordered goods are delivered immediately). Let p_n denote the price charged in period n and let x_n, y_n, and w_n carry the same definitions, respectively, as s_n, a_n, and w_n in Section 3-1. In words, x_n is the inventory on hand at the beginning of period n, y_n is the inventory on hand after ordered goods are delivered, and w_n is the amount ordered. Recall that the inventory holding cost in period n is $h \cdot (y_n - D_n)^+$. Suppose that the demand D_n is an r.v. whose distribution depends only on goodwill and price, which are denoted by a_n and p_n, respectively.

The sum of discounted revenues net of costs is

$$B = \sum_{n=1}^{\infty} \beta^{n-1}[p_n D_n - p_n(D_n - y_n)^+ - h(y_n - D_n)^+ - cw_n - z_n]$$

$$= \theta a_0 + cx_1 + \sum_{n=1}^{\infty} \beta^{n-1}[p_n D_n - p_n(D_n - y_n)^+$$

$$- h(y_n - D_n)^+ - cy_n + \beta c(y_n - D_n)^+ - a_n(1 - \beta\theta)]$$

where the substitutions $w_n = y_n - x_n$, $x_n = (y_{n-1} - D_{n-1})^+$ if $n > 1$, and $z_n = a_n - \theta a_{n-1}$ have been made. Therefore,

$$E(B) = \theta a_0 + cx_1 + E\left[\sum_{n=1}^{\infty} \beta^{n-1} G(p_n, y_n, a_n)\right]$$

where

$$G(p, y, a) = E[pD_1 - p(D_1 - y)^+ - h(y - D_1)^+ + \beta c(y - D_1)^+ | p_1 = p, a_1 = a]$$
$$- cy - (1 - \beta\theta)a$$

Suppose (p^*, y^*, a^*) maximizes $G(\cdot, \cdot, \cdot)$ on $\mathbb{R}^3_+$. Then Corollary 3-1 implies $(p_n, y_n, a_n) = (p^*, y^*, a^*)$ for all n is optimal if $x_1 \le x^*$ and $\theta a_0 \le a^*$.

EXERCISES

3-19 Verify that the MDP described in this section [up to and including (3-72)] satisfies the conditions of Theorem 3-3.

3-20 Verify (3-76*c*).

3-21 Suppose that demand exhibits a lagged response to advertising via ($n > 1$)

$$a_n = z_1\theta^{n-2} + z_2\theta^{n-3} + \cdots + z_{n-2}\theta + z_{n-1} \tag{3-77}$$

instead of (3-67). Still assume $\mu(a) = E(D_n | a_n = a)$ with a_n specified by (3-77) instead of (3-67); let s_1 be given.

(*a*) Show

$$E(B) = \beta^{-1}a_1 + \sum_{n=1}^{\infty} \beta^{n-1}G(a_n)$$

where $$G(a) = r\mu(a) - (\beta^{-1} - \theta)a$$

(*b*) In this case, does an optimal policy specify higher or lower advertising levels than with (3-69) and (3-70)? Make explicit your assumptions concerning $\mu(\cdot)$.

3-5* Capacity Expansion

Industrial capacity is a major element of many firms' long-range planning processes. Nevertheless, it is difficult to measure the capacity of a plant, firm, or industry. We think of capacity as a maximum production rate per unit time, but the "maximum" is an elusive number. It is a measure that is relative rather than absolute. For example, during the 1978 U.S. coal strike many firms in the midwest found that their output was constrained by their fuel reserves. A longer-range perspective would lead to other constraints, such as physical plant size, being dominant.

Here, the perspective is exceedingly distant from measurement problems. Instead, we use a model that appears to be complex but is useful for "quick and dirty" analyses. The model's principal deficiency is its lack of the economies of scale often associated with capacity expansions. Its major positive feature is the inclusion of the delay between the time a firm opts for an expansion and the date that it can utilize the expansion. These delays are central to the problem of planning capital improvements.

Let K_n be the capacity at the beginning of period n and let D_n be the demand (amount of product that could be sold if it could be made) during period n. Suppose there is a delay of L periods between the time that it is decided to augment capacity and the time that the added capacity is usable. Let z_n be the amount of capacity that is "ordered" in period n and usable in period $n + L$. Then

$$K_n = K_{n-1} + z_{n-L} \tag{3-78}$$

The decision problem is to choose $z_1, z_2, z_3, \ldots$. Note that (3-78) implies that the decisions $z_1, z_2, \ldots$ will first affect capacity in period $n = L + 1$. Until then, the capacities are beyond control because $K_1, K_2, \ldots, K_L$ are determined by earlier decisions (i.e., prior to period 1). It is convenient to define $z_n = K_{n+L-1}$ if $-L + 1 \le n \le 0$.

Define $X_n = D_n - D_{n-1}$ so that

$$D_n = D_0 + \sum_{i=1}^{n} X_i \qquad n = 1, 2, \ldots \tag{3-79}$$

We assume that $X_1, X_2, \ldots$ are nonnegative i.i.d. r.v.'s.

Let p be the unit profit per item sold. An essential feature of capacity is its constraining effect on production; so we assume that operating profit, exclusive of costs of augmenting capacity, is

$$pD_n \quad \text{if} \quad D_n \leq K_n$$

and

$$pK_n \quad \text{if} \quad K_n < D_n$$

which is written concisely as

$$pD_n - p(D_n - K_n)^+ \tag{3-80}$$

Discounted Profit

Suppose that the cost of adding capacity z_n is linear, namely, cz_n, and that this cost is incurred in period n (rather than period $n + L$). Then the total discounted profit, net of costs of capacity expansion, is

$$B = \sum_{n=1}^{\infty} \beta^{n-1}[pD_n - p(D_n - K_n)^+ - cz_n] \tag{3-81}$$

An infinite number of periods is used in (3-81) as a surrogate for a large and uncertain finite number of periods.

In order to simplify (3-81), let

$$a_n = K_n + \sum_{i=n-L+1}^{n} z_i - D_{n-1} \tag{3-82}$$

Then a_n is the amount by which period n's capacity, plus the aggregrate increments to capacity that have been ordered but not yet delivered, exceeds the most recent demand. But L periods later, the increments $z_{n-L+1}, z_{n-L+2}, \ldots, z_n$ will be usable. Therefore, in period $n + L$ the actual excess capacity will be a_n less whatever increments to demand occur during the periods between n and $n + L$. In symbols,

$$K_{n+L} - D_{n+L} = a_n - \sum_{i=n}^{n+L} X_i$$

or

$$D_n - K_n = \sum_{i=n-L}^{n} X_i - a_{n-L} \tag{3-83}$$

To derive (3-83) formally, use (3-78), (3-79), and (3-82):

$$a_n - a_{n-1} = K_n - K_{n-1} + \sum_{i=n-L+1}^{n} z_i - D_{n-1} + D_{n-2}$$

$$= z_{n-L} + z_n - z_{n-L} - X_{n-1} = z_n - X_{n-1} \tag{3-84}$$

Recursive use of (3-84) yields

$$a_n = a_{n-1} + z_n - X_{n-1} = a_{n-2} + z_n - z_{n-1} - X_{n-1} - X_{n-2}$$
$$= \ldots = a_{n-L} + \sum_{i=n-L+1}^{n} z_i - \sum_{i=n-L}^{n-1} X_i$$

Using (3-82):

$$K_n - D_n = a_{n-L} + \sum_{i=n-L+1}^{n} z_i - \sum_{i=n-L}^{n-1} X_i - X_n$$
$$= a_{n-L} - \sum_{i=n-L}^{n} X_i$$

Now we put B into a convenient form. Substitution of (3-80), (3-83), and (3-84) in (3-81) gives

$$B = \sum_{n=1}^{\infty} \beta^{n-1}\left[pD_n - p\left(\sum_{i=n-L}^{n} X_i - a_{n-L}\right)^+ - c(a_n - a_{n-1} + X_n)\right]$$

in which (3-84) has been used to substitute $a_n - a_{n-1} + X_n$ for z_n. Collecting terms involving a_n yields

$$B = \sum_{n=1}^{\infty} \beta^{n-1}\left[pD_n - p\beta^L\left(\sum_{i=n}^{n+L} X_i - a_n\right)^+ - c(1-\beta)a_n - cX_n\right]$$
$$+ ca_0 - \sum_{n=1}^{L} \beta^{n-1} p\left(\sum_{i=-L+n}^{n} X_i - a_{-L+n}\right)^+$$
$$= \sum_{n=1}^{\infty} \beta^{n-1}(pD_n - cX_n) + ca_0 - \sum_{n=1}^{L} \beta^{n-1} p\left(\sum_{i=-L+n}^{n} X_i - a_{-L+n}\right)^+$$
$$- \sum_{n=1}^{\infty} \beta^{n-1}\left[c(1-\beta)a_n + \beta^L p\left(\sum_{i=n}^{n+L} X_i - a_n\right)^+\right] \qquad (3\text{-}85)$$

Our choice of capacity increments $z_1, z_2, \ldots$ or, equivalently, of $a_1, a_2, \ldots$ influences only the third sum. Also, the X_i's are i.i.d.; so $\sum_{i=n}^{n+L} X_i$ has the same distribution function as $\sum_{i=1}^{L+1} X_i$. As a consequence, $E(\sum_{i=1}^{n+L} X_i - a_n)^+$ depends on n only via a_n. Let

$$G(a) = c(1-\beta)a + \beta^L pE\left[\left(\sum_{i=1}^{L+1} X_i - a\right)^+\right] \qquad (3\text{-}86)$$

Then

$$E(B) = \begin{pmatrix}\text{terms that do not} \\ \text{depend on the} \\ \text{decisions } z_i, z_2, \ldots \\ \text{or } a_1, a_2, \ldots\end{pmatrix} - E\left[\sum_{n=1}^{\infty} \beta^{n-1} G(a_n)\right]$$

Optimization

The expected present value of the operating profit (minus expansion costs), namely, $E(B)$, is maximized if, and only if,

$$E\left[\sum_{n=1}^{\infty} \beta^{n-1} G(a_n)\right] \tag{3-87}$$

is minimized. Suppose a^* minimizes $G(\cdot)$ on $\mathbb{R}$. Then, as in Corollary 3-2, if

$$P\{X_1 \geq 0\} = 1 \tag{3-88a}$$

and

$$K_1 + \sum_{i=2-L}^{0} z_i \leq a^* + D_0 \tag{3-88b}$$

then

$$z_1 = a^* + D_0 - K_1 - \sum_{i=2-L}^{0} z_i \tag{3-89a}$$

and

$$z_n = X_{n-1} \qquad \text{for all } n \geq 2 \tag{3-89b}$$

minimizes $E(B)$ subject to the constraint $P\{z_n \geq 0 \text{ for all } n\} = 1$. Exercise 3-22 asks you to verify this claim.

The model just analyzed lacks economies of scale. Therefore, it may be more useful for planning the aggregate capacity of numerous facilities than for the detailed planning of increments at a single facility.

Suppose X_1 has a density function $q(\cdot)$ and distribution function $Q(\cdot)$. You should verify

$$Q_{L+1}(a^*) = 1 - \frac{c}{p}\frac{1-\beta}{\beta^L} \tag{3-90}$$

where Q_i denotes the i-fold convolution [i.e., convolving $Q(\cdot)$ with itself i times].

How does the average capacity $E(K_n)$ depend on L?

From (3-78) and (3-89),

$$K_n = K_1 + \sum_{i=-L+2}^{n-L} z_i = K_1 + \sum_{i=-L+2}^{0} z_i + \sum_{i=2}^{n-L} z_i$$

$$= a^* + D_0 - z_1 + \sum_{i=1}^{n-L-1} X_i + z_1 = a^* + D_0 + \sum_{i=1}^{n-L+2} X_i$$

Therefore,

$$E(K_n) = a^* + D_0 + (n - L)E(X_1) \qquad n \geq L \tag{3-91}$$

However, a^* depends on L; so the effect of L on $E(K_n)$ is not clear.

Table 3-1 Effect of lag L on operating characteristics

	Lag L			
Item	1	2	3	4
a^*	3.24	4.26	5.13	5.91
z_1	6.24	7.26	8.13	8.91
$E(K_{10})$	90.24	90.26	90.13	89.91

Example 3-10 Suppose $c/p = 3$, $\beta = 0.9$, $D_0 = 78$, $K_1 = 75$, and the X_n's are exponentially distributed with $E(X_1) = 1$. Then the distribution function of $\sum_{i=1}^{L+1} X_i$ is†

$$Q_{L+1}(x) = 1 - \sum_{i=1}^{L+1} e^{-x} \frac{x^i}{i!} \qquad (x \geq 0) \tag{3-92}$$

so (3-90) becomes

$$e^{a^*} = \sum_{i=1}^{L+1} \frac{(a^*)^i}{i!} \frac{0.9^L}{0.3} \tag{3-93}$$

Table 3-1 shows the effect of L on a^*, z_1, and $E(K_{10})$ [via (3-93), (3-89), and (3-91)]. In this table, a^* and z_1 are concave increasing functions of the lag L. However, $E(K_{10})$ is not a monotone function of L. □

EXERCISES

3-22 Identify the elements in the model in this section which correspond to the elements of the MDP in Corollary 3-2. Verify that the elements here satisfy the conditions of that corollary.

3-23 Suppose that the cost of additional capacity is incurred at the time that the capacity first becomes usable; i.e., cz_n is paid in period $L + n$ instead of in period n as in (3-81). (*a*) Derive $G(\cdot)$ in this case [instead of (3-86), which is based on (3-81)]. Then specify a relation analogous to (3-90) in this case and compare your relation with (3-90). (*b*) Compute the entries in Table 3-1 if cz_n is paid in period $n + L$.

3-24 Suppose $G(\cdot)$ is nondecreasing on $[a^*, \infty)$ and (3-88*b*) is violated; that is,

$$K_1 + \sum_{i=2-L}^{0} z_i > a^* + D_0$$

† If $X_1, \ldots, X_{L+1}$ are i.i.d. exponential r.v.'s with $E(X_1) = \lambda^{-1}$, the r.v. $\sum_{i=1}^{L+1} X_i$ has a gamma distribution with parameters $L + 1$ and λ. Therefore,

$$Q_{L+1}(x) = \frac{1}{L!} \int_0^x y^L \lambda^{L+1} e^{-\lambda y}\, dy$$

Repeated integration by parts and $\lambda = 1$ yields (3-92). See Section A-2 for further properties of exponential r.v.'s. Also, see Exercise 5-5 on page 111 of Volume I.

(*a*) Specify an optimal policy and substantiate your claim of optimality. (*b*) Use (3-20) as a guide to specify $E(B)$ if your policy in (*a*) is used. Assume that X_1 is a discrete r.v.

3-25 Verify (3-90) under the assumption that X_1 has density $q(\cdot)$ and distribution function $Q(\cdot)$.

BIBLIOGRAPHIC GUIDE

Arrow, Beckmann, and Karlin (1958) deduced a myopic optimum in a deterministic capacity expansion model† different from the one in Section 3-5. Then Arrow (1962) obtained similar results in a deterministic capital-accumulation model which, at the time, appeared fundamentally different from the capacity-expansion model. A model structurally similar to capital accumulation arises in the context of advertising and was analyzed by Nerlove and Arrow (1962). Dhrymes (1962) augmented the advertising model to include research expenditures. The models in the aforementioned articles satisfy the conditions in Section 3-3.

The advertising model in Section 3-4 is a discrete-time stochastic version of the model presented by Nerlove and Arrow (1962). Sethi (1977) has surveyed the literature on normative advertising models.

Veinott (1965), then Ignall and Veinott (1969), explored the extent to which dynamic versions of "newsboy" inventory models possess myopic optima. Veinott (1965) references root ideas in earlier work by S. Karlin. The reasoning in Veinott's paper is used in Section 3-3, and some of his models are presented in Section 3-2. Frey and Nemhauser (1972) apply the myopic result in Veinott's paper to a model of capacity expansion in a transportation system. For a different approach to myopic optima in multiechelon production and inventory systems see Schwarz and Schrage (1975) and Szendrovits (1981).

Dirickx and Jennergren (1975) and Hakansson (1971) present MDPs with myopic optima. Their notions of "myopic" differ from those in this chapter.

Several models are "nearly" myopic and closely related to the general structure in Section 3-3. Denardo (1968) presents conditions for "separable" MDPs (which we discuss in Section 7-3). Adjoining (3-53) to Denardo's conditions yields our conditions in Section 3-3. Also, Denardo and Rothblum (1979)'s "affine" MDPs and Spence and Starrett (1975)'s "most rapid approach paths" are related to the model in Section 3-3.

Sections 3-1 and 3-2 are among six sections in this volume which draw on the huge literature on inventory and production models. The following critiques, reviews, books, and surveys are useful for planning forays into the literature: Clark (1972), Graves (1981), Hadley and Whitin (1963), Iglehart (1967), Johnson and Montgomery (1974), Peterson and Silver (1979), Scarf (1963), Silver (1981), Veinott (1966), and Wagner (1980).

† Their model is continuous in time, but its discrete-time analog satisfies the conditions in Section 3-3.

REFERENCES

Arrow, K. J.: "Optimal Capital Adjustment," in *Studies in Applied Probability and Management Science*, edited by K. J. Arrow, S. Karlin, and H. Scarf, Stanford University Press, Stanford, Calif. (1962), pp. 1–17.

———, M. J. Beckmann, and S. Karlin: "The Optimal Expansion of the Capacity of a Firm," in *Studies in the Mathematical Theory of Inventory and Production*, by K. J. Arrow, S. Karlin, and H. Scarf, Stanford University Press, Stanford, Calif. (1958), pp. 92–105.

Clark, A. J.: "An Informal Survey of Multi-Echelon Inventory Theory," *Nav. Res. Logistics Quart.* **19**: 621–650 (1972).

Denardo, E. V.: "Separable Markovian Decision Problems," *Manage. Sci.* **14**: 451–462 (1968).

——— and U. G. Rothblum: "Affine Dynamic Programming," in *Dynamic Programming and Its Applications*, edited by M. L. Puterman, Academic Press, New York (1979).

Dhrymes, P. J.: "On Optimal Advertising Capital and Research Expenditures under Dynamic Conditions," *Economica* **39**: 275–279 (1962).

Dirickx, Y. M. I., and L. P. Jennergren: "On the Optimality of Myopic Policies in Sequential Decision Problems," *Manage. Sci.* **21**: 550–556 (1975).

Frey, S. C., Jr., and G. L. Nemhauser: "Temporal Expansion of a Transportation Network," *Transp. Sci.* **6**: 306–323, 395–406 (1972).

Graves, S. C.: "A Review of Production Scheduling," *Oper. Res.* **29**: 646–675 (1981).

Hadley, G., and T. M. Whitin: *Analysis of Inventory Systems*, Prentice-Hall, Englewood Cliffs, N.J. (1963).

Hakansson, N. H.: "On Optimal Myopic Portfolio Policies, with and without Serial Correlation on Yields," *J. Business* **44**: 324–334 (1971).

Iglehart, D.: "Recent Results in Inventory Theory," *J. Ind. Eng.* **18**: 48–51 (1967).

Ignall, E., and A. F. Veinott, Jr.: "Optimality of Myopic Inventory Policies for Several Substitute Products," *Manage. Sci.* **15**: 284–304 (1969).

Johnson, L. A., and D. C. Montgomery: *Operations Research in Production Planning, Scheduling, and Inventory Control*, Wiley, New York (1974).

Nerlove, M., and K. J. Arrow: "Optimal Advertising Policy under Dynamic Conditions," *Economica* **22**: 129–142 (1962).

Peterson, R., and E. A. Silver: *Decision Systems for Inventory Management and Production Planning*, Wiley, New York (1979).

Scarf, H.: "A Survey of Analytic Techniques in Inventory Theory," chap. 7 in *Multistage Inventory Models and Techniques*, edited by H. Scarf, D. Gilford, and M. Shelly, Stanford University Press, Stanford (1963).

Schwarz, L. B., and L. Schrage: "Optimal and Systems Myopic Policies for Multi-Echelon Production/Inventory Assembly Systems," *Manage. Sci.* **21**: 1285–1294 (1975).

Sethi, S. P.: "Dynamic Optimal Control Models in Advertising: A Survey," *SIAM Rev.* **19**: 685–725 (1975).

Silver, E. A.: "Operations Research in Inventory Management: A Review and Critique," *Oper. Res.* **29**: 628–645 (1981).

Sobel, M. J.: "Myopic Solutions of Markov Decisions Processes and Stochastic Games," *Oper. Res.* **29**: 995–1009 (1981).

Spence, M., and D. Starrett: "Most Rapid Approach Paths in Accumulation Problems," *Int. Econ. Rev.* **16**: 388–403 (1975).

Szendrovits, A. Z.: "Comments on the Optimality in Optimal and System Myopic Policies for Multi-Echelon Production Inventory Systems," *Manage. Sci.* **27**: 1081–1087 (1981).

Veinott, A. F., Jr.: "Optimal Policy for a Multi-Product, Dynamic Nonstationary Inventory Problem," *Manage. Sci.* **12**: 206–222 (1965).

———: "The Status of Mathematical Inventory Theory," *Manage. Sci.* **12**: 745–777 (1966).

Wagner, H. M.: "Research Portfolio for Inventory Management and Production Planning Systems," *Oper. Res.* **28**: 445–475 (1980).

CHAPTER

FOUR

MARKOV DECISION PROCESSES

Sequential decision processes differ from static problems in three essential ways. An action taken now has the following influences: the amounts of rewards that can be earned in the future, whether or not an action will be feasible in the future, and the information that will be available in the future. The notion of a *state* is fundamental in the distinction between static and dynamic problems. In Chapter 3 little attention is paid to this notion because dynamic problems there had the happy property of being transformable to static problems. Most of the problems in the remainder of the book are *not* equivalent to static problems; so this chapter develops general results for dynamic decision problems.

Two broad classes of questions should be asked of a sequential decision process: (1) what are its qualitative properties, and (2) how can these properties be exploited computationally? This chapter answers many such questions for the most important kind of sequential decision process, the discrete-time Markov decision process with the criterion of expected present value. The final section concerns the Markov decision process with the criterion of average reward per period.

4-1 THE MODEL

Most of the sequential decision models in this book (and in professional practice) are Markov decision processes (MDPs). Here we expand the description in Section 3-3 of the canonical form of an MDP. An MDP is a *model*; in spite of the presence of the word "process," an MDP is not a stochastic process.

The canonical elements of an MDP include a set $\mathcal{S}$ of *states* and, for each $s \in \mathcal{S}$, a set A_s of alternative *actions*. The other canonical elements are a *transition function* and a *single-stage reward function*. The general idea is that a sequence of decisions must be made and that the state of affairs just prior to making a decision is adequately summarized by some state $s \in \mathcal{S}$. "Adequately" has three qualifications. First, the constraining effects of the past history on the array of feasible alternative actions are completely specified by s in A_s. Second, the subsequent state of affairs is conditionally independent† of the past history given the current state and current action. Third, the immediate reward, possibly an r.v., also is conditionally independent of the past history given the current state and action. These are Markovian assumptions. We show in Chapters 7 and 8 in Volume I that essentially every stochastic process can be viewed as a Markov process (with a suitably large state space). The art, in practice, is to formulate a process so that the state space is large enough to induce a Markov process but not so large that it masks the essential properties of the underlying phenomena. The same remarks apply to MDPs.

The sample path (or outcome) of an MDP specifies the successive states and actions. Let $\mathbf{s}_n$ and $\mathbf{a}_n$ denote the nth state and action, respectively. In this‡ section and Section 4-2 we use **boldface** type and write $\mathbf{s}_n$ and $\mathbf{a}_n$ for the r.v.'s while s_n and a_n denote the values taken by these r.v.'s. The identity of the $(n + 1)$th state $\mathbf{s}_{n+1}$ is assumed to depend stochastically only on s_n and a_n.

Definition 4-1 The *history* up to the time at which the nth action is taken is

$$H_n \triangleq (s_1, a_1, s_2, a_2, \ldots, s_{n-1}, a_{n-1}, s_n) \tag{4-1}$$

At time $j < n$, H_j is known but $\mathbf{s}_n$ and $\mathbf{a}_n$ are not known. They are r.v.'s because they depend on how the process evolves during periods $j + 1, j + 2, \ldots, n$. The sample path (or outcome) of an MDP is generated in the following way. When action a_n is taken in state s_n, two things happen in this order. First, state $\mathbf{s}_{n+1}$ is determined in a manner which depends only on s_n and a_n. That is, assume that, for any set of states J,

$$P\{\mathbf{s}_{n+1} \in J \mid \mathbf{H}_n, \mathbf{a}_n\} = P\{\mathbf{s}_{n+1} \in J \mid \mathbf{s}_n, \mathbf{a}_n\} \tag{4-2}$$

This is a Markovian assumption. It says that $\mathbf{s}_{n+1}$ is conditionally independent of $\mathbf{H}_n$ given $\mathbf{s}_n$ and $\mathbf{a}_n$. If the conditional probabilities do not depend on

† See page 501 for a definition of conditional independence.

‡ In most of this book, **boldface** type denotes a vector.

n, the assumption implies existence of a transition function $p(\cdot\,|\,\cdot,\cdot)$ defined as follows.

Definition 4-2 The *transition function* is $p(J\,|\,s, a) \triangleq P\{\mathbf{s}_{n+1} \in J\,|\,\mathbf{s}_n = s, \mathbf{a}_n = a\}, n \in I_+$. The equality in (4-2) is an *assumption* of an MDP.

Second, a reward $\mathbf{X}_n$ (possibly random) that depends only on $\mathbf{s}_n$ and $\mathbf{a}_n$ is determined. *Assume*

$$E(\mathbf{X}_n\,|\,\mathbf{H}_n, \mathbf{a}_n) = E(\mathbf{X}_n\,|\,\mathbf{s}_n, \mathbf{a}_n) \tag{4-3}$$

Definition 4-3 The *single-stage reward function* is

$$r(s, a) \triangleq E(\mathbf{X}_n\,|\,\mathbf{s}_n = s, \mathbf{a}_n = a)$$

Definition 4-4 A *Markov decision process* (an MDP) is a model which consists of the sets $\mathcal{S}$ and A_s, $s \in \mathcal{S}$, the transition function, and the single-stage reward function, and which satisfies (4-2) and (4-3) for all n.

The domain of $r(\cdot, \cdot)$ and $p(J\,|\,\cdot,\cdot)$ is the set of feasible state-action pairs labeled $\mathscr{C}$:

$$\mathscr{C} = \{s, a): a \in A_s, s \in \mathcal{S}\}$$

When $\mathcal{S}$ is a discrete set, we write p_{sj}^a for $p(\{j\}\,|\,s, a)$. When $\mathscr{C}$ is a finite set, the model is called a *finite* MDP.

The duration of an MDP is called a *planning horizon, N*. It is either finite or infinite according to the number of periods in the model. If $N < \infty$, we also specify a *salvage-value function* $L(\cdot)$ on S which determines a terminal reward $L(s)$ if the state is s after period N ends. In practice it may be quite difficult to choose a particular value N and a corresponding salvage-function.

Example 4-1: Gambler's ruin Consider two players D and B who start a coin-matching game with two coins each. At each play of the game, each player flips a fair coin on a table (so two coins are flipped). Player D wins, i.e., gains a coin from B, if both coins match; otherwise player B wins a coin from D. All the plays are independent of one another. Suppose that B is willing to play until either he or D has all four coins. Suppose that D can stop whenever he wishes to do so (or when he has either all four coins or none). In other words, the game *must* stop if either player wins all the coins; in addition, player D can stop the game any time he pleases.† When the game stops, both players retain whatever coins they hold.

Player D faces the following MDP. The set of states is $\mathcal{S} = \{0, 1, 2, 3, 4\}$, where $s = 1, 2,$ or 3 denotes the number of coins held by D prior to his decision whether or not to stop before the next play. Our labels for actions

† This example differs from the Markov-chain model in Example 7-1 on page 208 in Volume I; there the game stops only if states 0 or 4 are reached.

are arbitrary. Let $a = 0$ denote the "continue" action and let $a = 1$ denote "stop." Let $s = 4$ indicate that D has just won all the coins; the game must then stop. We use $s = 0$ to indicate either that B has just won all the coins and the game must then stop or that the game has already stopped. Then the sets of actions are

$$A_1 = A_2 = A_3 = \{0, 1\} \qquad \text{and} \qquad A_0 = A_4 = \{1\}$$

The set of feasible pairs of states and actions is $\mathscr{C} = \{(0, 1), (1, 0), (1, 1), (2, 0), (2, 1), (3, 0), (3, 1), (4, 1)\}$.

A possible sequence of states is $s_1 = 2$, $s_2 = 3$, $s_3 = 2$, $s_4 = 3$, $s_5 = 4$, $s_6 = s_7 = \cdots = 0$. This sequence occurs if D always elects to continue and the following sequence of coin comparisons occurs: match, no match, match, and match. Another possible sequence is $s_1 = 2$, $s_2 = 3$, $s_3 = 0$, $s_4 = s_5 = \cdots = 0$. This sequence occurs if D elects to continue in the first period, a match occurs, and then D elects to stop.

If D has s coins (D is in state s), $0 < s < 4$, and continues (D chooses action 0), then the next state is $s - 1$ or $s + 1$, each with probability 1/2. Therefore,

$$p^0_{s,\,s+1} = p^0_{s,\,s-1} = 1/2 \qquad 0 < s < 4$$

The game remains stopped once it stops; so

$$p^1_{s0} = 1 \qquad 0 \le s \le 4$$

Let

$$\mathbf{X}_n = \begin{cases} 0 & \text{if } \mathbf{a}_n = 0 \\ \mathbf{s}_n & \text{if } \mathbf{a}_n = 1 \end{cases}$$

and $L(s) = s$, $0 \le s \le 4$. In order to check that this specification is sensible, we adopt the convention that $a_{N+1} = 1$ and define

$$\mathbf{T} = \min\,\{n\colon \mathbf{a}_n = 1,\, n \le N + 1\}$$

Then $\mathbf{T}$ is the period in which player D elects to stop if there is such a period. Otherwise, $\mathbf{T} = N + 1$. In either case, $L(\mathbf{s_T})$ is the number of coins D holds when the game stops. The specification of $\mathbf{X}_n$ and $L(\cdot)$ yields

$$\mathbf{B}_N = \sum_{n=1}^{N} \beta^{n-1}\mathbf{X}_n + \beta^N L(\mathbf{s}_{N+1}) = \beta^{\mathbf{T}-1}\mathbf{s_T}$$

This quantity is the present value of the number of coins held by player D at the time that the game stops. Therefore,

$$r(s, 0) = 0 \qquad 1 \le s \le 3$$

$$r(s, 1) = 1 \qquad 0 \le s \le 4$$

□

Example 4-2: Machine maintenance Consider a machine that is in one of three states: working perfectly ($s = 0$), working with a minor defect ($s = 1$), and totally failed but being repaired ($s = 2$). Therefore, $\mathcal{S} = \{0, 1, 2\}$. When it

Table 4-1 p_{sj}^0 **in Example 4-2**

s \ j	0	1	2
0	u	$(1-u)(1-g)$	$(1-u)g$
1	0	d	$1-d$

works perfectly, it should be left alone; so $A_0 = \{0\}$, where $a = 0$ denotes "do nothing." When it is totally failed, it must be repaired, so $A_2 = \{1\}$, where $a = 1$ denotes "under repair." When it works with a minor defect, it can be left alone or repaired, so $A_1 = \{0, 1\}$. The set of feasible state-action pairs is $\mathscr{C} = \{(0, 0), (1, 0), (1, 1), (2, 1)\}$.

Example 7-2 on page 210 of Volume I presents a Markov-chain model of the machine's condition if it is always left alone when it has a minor defect. The transition matrix in that example is

$$\begin{pmatrix} u & (1-u)(1-g) & (1-u)g \\ 0 & d & 1-d \\ (1-\rho)\phi & (1-\rho)(1-\phi) & \rho \end{pmatrix}$$

and the parameters have the following interpretations: ρ is the probability that the machine will still be under repair next period, u is the probability that a machine that works perfectly will continue to do so next period, d is the probability that a machine with a minor defect will not fail totally, g is the conditional probability that a machine that works perfectly and develops some defect will fail totally, and ϕ is the probability that a newly repaired machine will work perfectly. If $a = 0$ when $s = 1$, these parameters lead to the transition probabilities in Table 4-1.

The matrix in Table 4-1 is not square (as in Markov chains) because action 0 is not available in state $s = 2$. Suppose that the effect of taking action $a = 1$ in state $s = 1$ is to bring the machine to the repair process. Then the transition probabilities from states $s = 1$ and $s = 2$ are presented in Table 4-2.

As in Example 7-20 on page 255 of Volume I, suppose that the expected profit per period from the machine is \$10 when it is operating perfectly and

Table 4-2 p_{sj}^1 **in Example 4-2**

s \ j	0	1	2
1	0	0	1
2	$(1-\rho)\phi$	$(1-\rho)(1-\phi)$	ρ

Table 4-3 Single-stage rewards $r(s, a)$ in Example 4-2

s \ a	0	1
0	10	—
1	5	0
2	—	0

\$5 when it has a minor defect. If the cost of repair is primarily the lost profit from the forgone production, the single-stage rewards are given in Table 4-3. □

Example 4-3: Inventory model Consider the single-product inventory model in Section 3-1 where demands in successive months, $\mathbf{D}_1, \mathbf{D}_2, \ldots$ are i.i.d. nonnegative r.v.'s and excess demand is lost. The state in Section 3-1 is $\mathbf{s}_n$, the amount of goods on hand at the beginning of period n. Ordered goods are delivered immediately, and the action $\mathbf{a}_n$ is the amount of goods on hand after the goods are ordered and delivered. Therefore, $A_s = \{a: a \geq s\}$ because $a - s \geq 0$ is the amount of goods ordered. Since excess demand is lost, $\mathcal{S} = \{s: s \geq 0\} = \mathbb{R}_+$. Another consequence of lost excess demand is $\mathbf{s}_{n+1} = (\mathbf{a}_n - \mathbf{D}_n)^+$ so

$$p((-\infty, j] \mid s, a) = P\{(a - \mathbf{D}_1)^+ \leq j\} = \begin{cases} P\{\mathbf{D}_1 \geq a - j\} & \text{if } j \geq 0 \\ 0 & \text{if } j < 0 \end{cases}$$

From (3-1), the net revenue in period n is†

$$\rho(\mathbf{a}_n \wedge \mathbf{D}_n) - c(\mathbf{a}_n - \mathbf{s}_n) - h(\mathbf{a}_n - \mathbf{D}_n)^+$$

so

$$r(s, a) = E[\rho(a \wedge \mathbf{D}_1) - ca + cs - h(a - \mathbf{D}_1)^+] \qquad \square \quad (4\text{-}4)$$

The single-stage reward in (4-4) is itself an expectation of an r.v., and occasionally in an MDP the randomness is explicit.

Now we present a general causal model for the transition from $\mathbf{s}_n$ to $\mathbf{s}_{n+1}$. Suppose that the MDP model induces a sequence of r.v.'s $\boldsymbol{\xi}_1, \boldsymbol{\xi}_2, \ldots$ as well as the sequence of states and actions. Amend (4-1) to

$$H_n = (s_1, a_1, \xi_1, s_2, a_2, \xi_2, \ldots, s_{n-1}, a_{n-1}, \xi_{n-1}, s_n) \tag{4-5}$$

† In this example we use ρ, whereas r is employed in Section 3-1.

From (4-3) and Definition 4-3, recall that $\mathbf{X}_n$ is the reward received in period n. With the more detailed specification in (4-5) of the history of an MDP, we assume that there are mappings M and g through which $\mathbf{s}_{n+1}$ and $\mathbf{X}_n$ are determined by $\mathbf{s}_n$, $\mathbf{a}_n$, and ξ_n:

$$\mathbf{s}_{n+1} = M(\mathbf{s}_n, \mathbf{a}_n, \xi_n) \tag{4-6}$$

$$\mathbf{X}_n = g(\mathbf{s}_n, \mathbf{a}_n, \xi_n) \tag{4-7}$$

Then assumptions (4-2) and (4-3) require that ξ_n be conditionally independent of $\mathbf{H}_n$ given $\mathbf{s}_n$ and $\mathbf{a}_n$ because

$$r(s, a) = E[g(\mathbf{s}_n, \mathbf{a}_n, \xi_n) \mid \mathbf{s}_n = s, \mathbf{a}_n = a] \tag{4-8}$$

$$p(J \mid s, a) = P\{M(\mathbf{s}_n, \mathbf{a}_n, \xi_n) \in J \mid \mathbf{s}_n = s, \mathbf{a}_n = a\} \tag{4-9}$$

That is, ξ_n is conditionally independent of the past history given the current state and action.

Proposition 4-1 In an MDP, for each $n = 1, 2 \ldots, N$, ξ_n is conditionally independent of $\mathbf{H}_n$ given $\mathbf{s}_n$ and $\mathbf{a}_n$.

In Example 4-1 (Gambler's Ruin), if $0 < \mathbf{s}_n < 4$, then $\mathbf{s}_{n+1} = \mathbf{s}_n \pm 1$ so $\mathbf{s}_{n+1} = \mathbf{s}_n + \xi_n$ with

$$P\{\xi_n = 1\} = P\{\xi_n = -1\} = 0.5$$

Then the mapping M in Example 4-1 is

$$\begin{cases} M(s, 0, \xi_n) = s + \xi_n & 0 < s < 4 \\ M(s, 1, \cdot) \equiv 0 \equiv M(0, 0, \cdot) & 0 \le s \le 4 \end{cases} \tag{4-10}$$

In the inventory model of Example 4-3, $\xi_n \equiv \mathbf{D}_n$. Then $M(\cdot, a, \xi_n) \equiv (a - \xi_n)^+$, and

$$g(s, a, \xi_n) = \rho(a \wedge \xi_n) - c(a - s) - h(a - \xi_n)^+$$

Criteria

Several criteria are used to evaluate decision rules in an MDP. First, suppose that the planning horizon N is finite. The total reward, denoted B_N, is the r.v.

$$\mathbf{B}_N = \sum_{n=1}^{N} \mathbf{X}_n + L(\mathbf{s}_{N+1}) \tag{4-11}$$

Different decision rules induce different distributions for $\mathbf{B}_N$, and in Section 2-3 we observe that this situation introduces embarrassingly many criteria of optimization. The formal criterion most commonly applied to (4-11) is $E(\mathbf{B}_N)$.† However-

† This criterion implies risk neutrality unless monetary units are replaced by utility values. See Section 2-4 for a discussion of attitude to risk.

ever, (4-11) is a special case of the discounted sum

$$\mathbf{B}_N = \sum_{n=1}^{N} \beta^{n-1}\mathbf{X}_n + \beta^N L(\mathbf{s}_{N+1}) \tag{4-12}$$

where $\beta = 1$. Usually $0 < \beta < 1$ in (4-12), but instances of $\beta \geq 1$ occur in plans for economic development of regions and nations. Again, $E(\mathbf{B}_N)$ (expected present value) is the formal criterion most often applied to (4-12).

What happens in (4-11) as $N \to \infty$? Suppose in (4-7) that $g(\cdot, \cdot, \cdot) \geq 0$. Moreover, suppose the event $\{\mathbf{X}_n \geq m > 0$ for infinitely many $n\}$ has positive probability for every decision rule. This condition is usually met in practical models. Then, in (4-11), $E(\mathbf{B}_N) \to \infty$ as $N \to \infty$. This suggests the criterion of the *rate* at which $E(\mathbf{B}_N)$ grows with N, that is, some limit point of $E(\mathbf{B}_N)/N$ as $N \to \infty$.

In Section 4-6 we shall discuss which limit point seems appropriate, distinguish between the limits of $E(\mathbf{B}_N)/N$ and $\mathbf{B}_N/N$ when these limits (the second with probability 1) both exist, and construct some algorithms to find a "good" limit point. Maximization of a limit point of $E(\mathbf{B}_N)/N$ as $N \to \infty$ is called the *average-return, average-reward*, or *gain-rate* criterion; some writers label it *gain optimality*.

The criterion in Sections 4-2 through 4-5 is maximization of expected present value. The optimization of MDPs with the average-return criterion is contrasted in Section 4-6 with discounted criteria. Other criteria are optimized in Section 6-3.

EXERCISES

4-1 Formulate the capacity expansion model in Section 3-5 as an MDP; specify the canonical elements $\mathcal{S}$, $\{A_s: s \in \mathcal{S}\}$, $\mathscr{C}$, $p(\cdot|\cdot,\cdot)$, and $r(\cdot,\cdot)$.

4-2 Specify the MDP canonical elements in the simple inventory model in Section 2-2.

4-3 Suppose that the costs in Example 4-2 are modified as follows. If the machine has been working with a minor defect for b consecutive periods, the reward in such a period is $r_1 - b$. Specify the MDP canonical elements of the modified example.

4-4 Suppose that the transition probabilities in Example 4-2 are modified as follows. When a machine has a minor defect but is not shifted to the maintenance shop, the number of periods $\mathbf{T}$ until it breaks down completely has the following distribution:

$$P\{\mathbf{T} = 1\} = 0.3, \; P\{\mathbf{T} = 2\} = P\{\mathbf{T} = 3\} = \cdots = P\{\mathbf{T} = 8\} = 0.1$$

Specify the MDP canonical elements of the modified example.

4-5 Prove Proposition 4-1.

4-6 Let $\delta_1, \delta_2, \ldots$ be a sequence of single-stage decision rules used via $a_n = \delta_{N-n+1}(s_n)$ for each n. Suppose $\mathbf{X}_n = h(\mathbf{s}_n, \mathbf{a}_n, \mathbf{s}_{n+1})$ for each n and let $r_{sj}(n) = h[s, \delta_{N-n+1}(s), j]$ and $p_{sj}(n) = p_{sj}^{\delta_{N-n+1}(s)}$, respectively. When $s_1 = s$, let $F_s^N(\cdot)$, v_s^N, and σ_{Ns}^2 denote the distribution function, mean, and variance of $\mathbf{B}_N$ defined in (4-12). Let $v_s^0 = L(s)$ and $\sigma_{0s}^2 = 0$, $s \in \mathcal{S}$. Derive the following formulas:

(a) $F_s^N(x) = \sum_{j \in \mathcal{S}} p_{sj}(N) F_j^{N-1}\left[\dfrac{x - r_{sj}(N)}{\beta}\right] \qquad (0 < \beta)$

(b) $v_{sN} = \sum_{j \in \mathcal{S}} p_{sj}(N)[r_{sj}(N) + \beta v_{j,N-1}]$

(c) $\sigma_{Ns}^2 = \beta^2 \sum_{j \in \mathcal{S}} p_{sj}(N)\sigma_{j,N-1}^2 + \sum_{j \in \mathcal{S}} p_{sj}(N)[r_{sj}(N) + \beta v_{j,N-1}]^2 - (v_{sN})^2$

4-2 FINITE-HORIZON MDPs

Let $\mathbf{B}_N(n)$ denote the present value at the beginning of period n of the rewards from period n through period N, $N \geq n$:

$$\mathbf{B}_N(n) = \sum_{\tau=n}^{N} \beta^{\tau-n} g(\mathbf{s}_\tau, \mathbf{a}_\tau, \xi_\tau) + \beta^{N-n+1} L(\mathbf{s}_{N+1})$$

We write $\mathbf{B}_N$ for $\mathbf{B}_N(1)$ so

$$\mathbf{B}_N = \sum_{\tau=1}^{N} \beta^{\tau-1} g(\mathbf{s}_\tau, \mathbf{a}_\tau, \xi_\tau) + \beta^N L(\mathbf{s}_{N+1})$$

is the present value of the rewards from an MDP whose planning horizon N is finite. In pursuit of the objective of maximizing $E(\mathbf{B}_N)$ we shall also maximize $E[\mathbf{B}_N(n)]$ for each $n = 1, 2, \ldots, N$. The maximization is with respect to alternative *policies*, namely, decision rules for choosing $a_1, \ldots, a_N$.

A policy is a contingency plan for choosing actions. The contingencies are the r.v.'s $\xi_1, \xi_2, \ldots, \xi_N$. Therefore, a policy is a function that depends on the histories $\mathbf{H}_1, \mathbf{H}_2, \ldots, \mathbf{H}_N$. Let $I_N = \{1, 2, \ldots, N\}$. For each $n \in I_N$, we consider only deterministic rules† π_n for choosing a_n. In other words, π_n assigns an element of A_s to every history $\mathbf{H}_n$ in (4-1) having $\mathbf{s}_n = s$ as the terminal entry. Equivalently, π_n must satisfy $[s, \pi_n(\mathbf{H}_n)] \in \mathscr{C}$ if $\mathbf{H}_n$ specifies the state $\mathbf{s}_n = s$ at the beginning of period n.

Consider a sequence of rules $\pi_1, \pi_2, \ldots, \pi_N$.

> **Definition 4-5** $\Pi_1 = (\pi_1, \pi_2, \ldots, \pi_N)$ is a *policy* if $[\mathbf{s}_n, \pi_n(\mathbf{H}_n)] \in \mathscr{C}$ for all possible $\mathbf{H}_n$ and $n \in I_N$.

For any $n \in I_N$ let $\Pi_n = (\pi_n, \pi_{n+1}, \ldots, \pi_N)$. Then Π_n is the segment of Π_1 from period n onward.

The distribution of the r.v. $\mathbf{B}_N$ depends on Π_1 and $H_1 = (s_1)$, which we assume known. Similarly, the distribution of $\mathbf{B}_N(n)$ depends on Π_n and H_n, which we assume known in period n. Let the notation $\mathbf{B}_N(n, \Pi_n)$ make the influence of Π_n explicit. For each $n \in I_N$, Π_n, and H_n we define

$$F_n(\Pi_n \mid H_n) \triangleq E[\mathbf{B}_N(n, \Pi_n) \mid \mathbf{H}_n = H_n] \tag{4-13}$$

which is the expected value‡ conditioned on $\mathbf{H}_n = H_n$, of the present value of the

† The exclusion of randomized policies (which arise in game theory) is without loss of optimality in this chapter.

‡ See Section A-3 for a brief review of integration. In (4-13), we assume that the expectation exists. Three requirements are implicit in this assumption: (i) $g(\cdot, \cdot, \cdot)$ and $L(\cdot)$ cannot be too irregular; (ii) the conditional distribution of ξ_n, given $\mathbf{s}_n$ and $\mathbf{a}_n$, cannot be too diffuse; and (iii) for each $\tau, \Pi_\tau(H_\tau)$ cannot be too irregular as a function of H_τ. If (iii) is violated, $\mathbf{B}_N(n)$ may not be a bona fide r.v.; i.e., its distribution may be ill-defined. If (i) or (ii) are violated, the distribution of $\mathbf{B}_N(n)$ may not be integrable. The general question of the existence of r.v.'s in MDPs is investigated in detail in Bertsekas and Shreve (1978) and Hinderer (1970).

rewards starting in period n, if Π_n is used to choose actions $\mathbf{a}_n, \mathbf{a}_{n+1}, \ldots, \mathbf{a}_N$. Often we use the notation

$$E_{\Pi_n|H_n}[\mathbf{B}_N(n)] \triangleq E[\mathbf{B}_N(n, \Pi_n) \mid \mathbf{H}_n = H_n]$$

Proposition 4-1 asserts that ξ_n, given $\mathbf{s}_n$ and $\mathbf{a}_n$, is conditionally independent of $(\mathbf{H}_{n-1}, \mathbf{a}_{n-1}, \xi_{n-1})$ so Definition 4-3 of $r(s, a)$, the single-stage reward function, yields

$$E_{\Pi_n|H_n}[g(\mathbf{s}_\tau, \mathbf{a}_\tau, \xi_\tau)] = E_{\Pi_n|H_n}[r(\mathbf{s}_\tau, \mathbf{a}_\tau)] \qquad \text{for } \tau \geq n$$

Therefore,

$$F_n(\Pi_n \mid H_n) = E_{\Pi_n|H_n}\left[\sum_{\tau=n}^{N} \beta^{\tau-n} r(\mathbf{s}_\tau, \mathbf{a}_\tau) + \beta^{N-n+1} L(\mathbf{s}_{N+1})\right] \tag{4-14}$$

Suppose s_1 is the initial state. It is tempting to define a policy Π_1^* as optimal if $F_1(\Pi_1^* \mid s_1) \geq F_1(\Pi_1 \mid s_1)$ for all policies Π_1. However, a policy is a contingency plan that covers all possible histories and states; so we use the following more stringent notion† of optimality.

Definition 4-6 $\Pi_1^* = (\pi_1^*, \pi_2^*, \ldots, \pi_N^*)$ is an *optimal policy* if, and only if,

$$F_n(\Pi_n^* \mid H_n) \geq F_n(\Pi_n \mid H_n) \qquad \text{for all } H_n, \Pi_n, \text{ and } n \in I_N \tag{4-15}$$,

where $\Pi_n^* = (\pi_n^*, \pi_{n+1}^*, \ldots, \pi_N^*)$.

This notion of optimality leads naturally (later in this section) to a recursive algorithm to compute an optimal policy.

For each n, the distribution of ξ_n is determined by $(s_n, a_n) \in \mathscr{C}$. Let Ω be the union over all $(s, a) \in \mathscr{C}$ of the possible sample spaces of the ξ_n. Then H_n in (4-13) is an element of

$$\times_{j=1}^{n-1}(\mathscr{C} \times \Omega)) \times \mathbb{S} \tag{4-16}$$

The definition of optimality requires that (4-15) be valid over all elements H_n of the set in (4-16) and that this be true for every $n \in I_N$. Hence, Π_1^* must dominate every policy from every conceivable intermediate point. Consider a history in which an optimal policy was not used for $n - 1$ periods and foolish decisions were made during the first $n - 1$ periods. Then the definition requires that Π_1^* (via Π_n^*) optimally extricate the history from a situation that would never be encountered under a sensible policy.

† This definition of optimality is similar to the basis for many algorithms which compute shortest routes in networks. The algorithms compute lengths of shortest routes not only between the source and the sink but between all pairs of nodes.

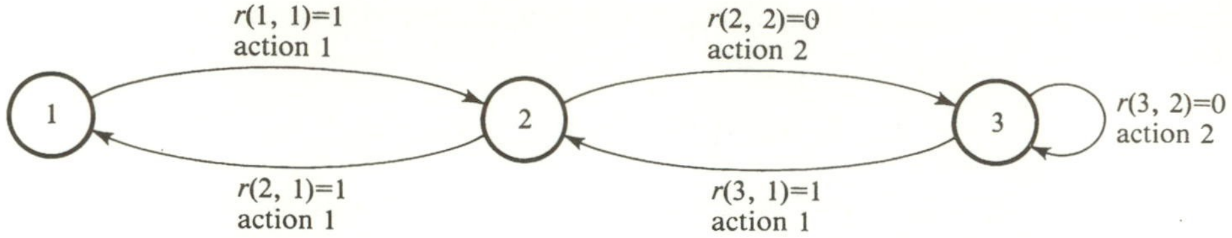

Figure 4-1 A deterministic MDP.

Example 4-4 Consider the deterministic MDP in Figure 4-1 with $\mathcal{S} = \{1, 2, 3\}$ and $L(\cdot) \equiv 0$. Nodes in the figure represent states, and arcs correspond to actions and their consequent transitions. The data for this MDP are

$$A_1 = \{1\}, A_2 = A_3 = \{1, 2\}$$

$$p_{12}^1 = p_{21}^1 = p_{23}^2 = p_{32}^1 = p_{33}^2 = 1$$

$$r(1, 1) = r(2, 1) = r(3, 1) = 1, \qquad \text{and} \qquad r(2, 2) = r(3, 2) = 0$$

Starting in either state 1 or state 2, it is best to shuttle back and forth between those states. Starting in state 3, it is best to move immediately to state 2 and then shuttle between states 1 and 2. Consider the policy Π', which always takes action 1 in states 1 and 2 but takes action 1 in state 3 only in the first period. In any period $n > 1$, if $\mathbf{s}_n = 3$, then Π' takes action 2. According to Definition 4-6, Π' is not optimal although it attains $F_1(\Pi' \mid s_1) \geq F_1(\Pi \mid s_1)$ for all Π and s_1. Suppose $N > 2$ and consider $\mathbf{H}_2 = H_2 \triangleq (3, 2, 3)$ so $\mathbf{s}_1 = \mathbf{s}_2 = 3$ and $\mathbf{a}_1 = 2$. Of course, Π' precludes $\mathbf{s}_n = 3$ for any $n > 1$. Nevertheless, $F_2(\Pi' \mid H_2) < F_2(\Pi^* \mid H_2)$, where Π^* is the same as Π' except that it always takes action 1 from state 3. □

The importance of Definition 4-5 is that optimality is equivalent to recursive optimization in a sense made precise in Theorem 4-1, below. From Proposition 4-1, the distribution of ξ_n depends only on $\mathbf{s}_n$ and $\mathbf{a}_n$. Let $\xi_{s,a}$ be an r.v. with the same distribution as ξ_n when $\mathbf{s}_n = s$ and $\mathbf{a}_n = a$.

We shall invoke Theorem 4-1 frequently in the remainder of the book, and at such times the label *principal of optimality* may be used. However, this "principle" is a mathematical theorem rather than an axiom. A host of books and articles contain dynamic programs, and too many authors justify their recursive equations by appealing to the "principle" as if it were both self-evident and impossible to prove from fundamental assumptions.

Theorem 4-1 The principle of optimality Consider an MDP with reward $\mathbf{X}_n = \beta^{n-1} g(\mathbf{s}_n, \mathbf{a}_n, \xi_n)$ in period n, $\mathbf{s}_{n+1} = M(\mathbf{s}_n, \mathbf{a}_n, \xi_n)$, and ξ_n conditionally independent of $\mathbf{H}_n$ given $\mathbf{s}_n$ and $\mathbf{a}_n$.

(*a*) If there is an optimal policy $\Pi_1^* = (\pi_1^*, \ldots, \pi_N^*)$, then there are N real-valued function $f_1, f_2, \ldots, f_N$ on $\mathcal{S}$ which satisfy

$$f_n(s) = F_n(\Pi_n^* \mid H_n) \tag{4-17}$$

where H_n assigns the value s to s_n, $s \in \mathcal{S}$, $n \in I_N$. Furthermore,

$$f_n(s) = \sup \{r(s, a) + \beta E(f_{n+1}[M(s, a, \xi_{s,a})]): a \in A_s\} \quad s \in \mathcal{S},\ n \in I_N \tag{4-18}$$

with the supremum attained and $f_{N+1}(s) = L(s)$ for all $s \in \mathcal{S}$.

(*b*) Conversely, suppose $f_1, \ldots, f_N$ is a sequence of functions that satisfy (4-18) with the supremum attained for all n and s. Then there is an optimal policy $\Pi_1^* = (\pi_1^*, \ldots, \pi_N^*)$ with $\pi_n^*(H_n) \equiv a_n^*(s_n)$ being any element $a \in A_{s_n}$ that attains the maximum in (4-18).

PROOF See page 123. □

Comments

The theorem asserts that the optimization of dynamic processes is equivalent to the solution of recursive equation such as (4-18). Some of the labels given to the recursion (4-18) are *dynamic program, dynamic programming problem,* and *optimality equation.* The part of operations research concerned with MDPs is sometimes called *dynamic programming.*

The numerical value of $f_n(s)$ in (4-18) is the value of an optimal compromise between the two quantities in braces on the right side of (4-18). One quantity is the immediate reward $r(s, a)$; the other quantity is the discounted expected present value of an optimal policy starting next period. An optimal decision must optimally balance these short-run and long-run benefits.

The numerical value of $f_n(s)$ in (4-18) has the following interpretation. It is the maximal attainable expected value of $\mathbf{B}_N(n)$ (the present value in period n of the rewards from period n onward) if the MDP is in state s at the start of period n, that is, $\mathbf{s}_n = s$. Similarly, for two states s and s', $f_n(s') - f_n(s)$ is the relative advantage of being in state s' rather than s in period n. Therefore, we often refer to $f_n(s)$ as the *value of an optimal policy* starting from state s in period n. We call $f_1(\cdot)$, $f_2(\cdot), \ldots, f_N(\cdot)$ *value functions.*

The last assertion of the theorem is that the nth action a_n can be chosen at the beginning of period n without considering any of the data in the history H_n except the current state s_n. In practice, this result enormously reduces the amount of data that must be stored and retrieved whenever a decision is to be made. It permits a degree of retrospective myopia when future decisions are being weighed. It is one form of the economist's aphorism, "Let bygones be bygones," which reflects the admonition to disregard "sunk" costs when deciding what action to take at present.

The proofs of both parts of the theorem are straightforward, established by induction, and tedious. However, we include the entire proof at the end of this section because the theorem is important and its proof is not available in other references.

When $\mathcal{S}$ is a discrete set, it is sensible to use the notation

$$p_{sj}^a = P\{M(s, a, \xi_{s,a}) = j\}$$

Then (4-18) becomes

$$f_n(s) = \max \{r(s, a) + \beta \sum_{j \in \mathcal{S}} p_{sj}^a f_{n+1}(j)\colon a \in A_s\} \qquad s \in \mathcal{S}, n \in I_N \qquad (4\text{-}18')$$

Example 4-5: Gambler's ruin—Example 4-1 continued This example is a straightforward application of (4-18′) and (4-18). In Example 4-1, the salvage-value function is $L(s) = s$, $0 \le s \le 4$, the actions are $a = 0$ and $a = 1$, which denote "continue" and "stop," the transition probabilities are

$$p_{s0}^1 = 1 \qquad 0 \le s \le 4$$

$$p_{s,\,s-1}^0 = p_{s,\,s+1}^0 = 1/2 \qquad 1 \le s \le 3$$

The discount factor is $0 \le \beta \le 1$, and the single-stage rewards are

$$r(s, 0) = 0 \qquad \text{and} \qquad r(s, 1) = s$$

Therefore, (4-18) for $n = N$ is

$$f_N(s) = \max \left\{r(s, 0) + \frac{\beta[(s+1) + (s-1)]}{2}, r(s, 1) + \beta \cdot 0\right\}$$

$$= \max \{\beta s, s\} = s \qquad 0 < s < 4 \qquad (4\text{-}19)$$

so it is optimal to stop in the last period. Also, $f_N(4) = r(4, 1) + \beta \cdot 0 = 4$ and $f_N(0) = 0$ so

$$f_N(s) = s \qquad 0 \le s \le 4 \qquad (4\text{-}20)$$

Continuing with (4-18), if $n < N$

$$f_n(s) = \max \left\{\frac{\beta[f_{n+1}(s-1) + f_{n+1}(s+1)]}{2}, s\right\} \qquad 0 < s < 4 \qquad (4\text{-}21)$$

and $f_n(0) = 0$, and $f_n(4) = 4$. We use (4-20) to initiate an inductive proof that $f_n(s) = s$, $0 \le s \le 4$, for all $n < N$. Suppose in (4-21) that $f_{n+1}(s-1) = s - 1$ and $f_{n+1}(s+1) = s + 1$. Then (4-21) becomes

$$f_n(s) = \max \left\{\frac{\beta[(s-1) + (s+1)]}{2}, s\right\}$$

$$= \max \{\beta s, s\} = s \qquad (4\text{-}22)$$

and it is optimal to stop however many periods $N - n + 1$ remain and however many coins s you hold.

When $p = 1/2$, the roles of players B and D are symmetric;† so it is not surprising that there is no advantage in continuing to play the "fair" coin-matching game. On the other hand, suppose that the game is more favorable† for player D, the one with the option to stop whenever he wishes, than

† The reader familiar with martingales will recognize that when $p = 1/2$, successive states comprise a symmetric random walk which is a martingale. If $p \ge 1/2$, the process is a submartingale. These terms are defined on page 44 of Volume I; however, we do not use any martingale theory in this book.

it is for player B. Then

$$p^0_{s,s+1} = p \qquad p^0_{s,s-1} = 1 - p \qquad 0 < s < 4$$

with $p \geq 1/2$. Consequently, in place of (4-19) and (4-21), player D has the dynamic program $f_N(0) = 0, f_N(4) = 4$,

$$f_N(s) = \max \{0 + \beta[p(s+1) + (1-p)(s-1)], s\}$$
$$= \max \{\beta(2p + s - 1), s\} \qquad 1 \leq s \leq 3$$

and for $n < N$ and $1 \leq s \leq 3$,

$$f_n(s) = \max \{\beta[pf_{n+1}(s+1) + (1-p)f_{n+1}(s-1)], s\} \tag{4-23}$$

As in the symmetric case, $f_n(0) = 0$ and $f_n(4) = 4$. For a particular β and p, whether it is optimal in (4-23) to continue ($a = 0$) or to stop ($a = 1$) depends on s. Exercise 4-7 asks you to obtain some properties of (4-23). We shall see in Section 8-2 that if it is optimal to continue in state s, then it is necessarily optimal to continue in all states s' with $s' \leq s$. □

Example 4-6 This example shows that the recursive optimization equation (4-18) can fail to hold if Definition 4-6 is replaced with maximization of expected present value starting in period one, namely, $F_1(\cdot \mid s_1)$. Consider the deterministic MDP in Figure 4-2 with $N = 2$, $\mathcal{S} = \{1, 2, 3\}$, $s_1 = 1$, $\beta = 1$, and $L(\cdot) \equiv 0$. Nodes in the figure represent states and arcs correspond to actions and their consequent transitions.

However many periods remain, it is obvious that it is best to choose $a = 1$ in all states.

Consider the policy Π which chooses $a = 1$ in states $s = 1$ and $s = 3$ and chooses $a = 2$ in state $s = 2$. If $s_n = 1$ or $s_n = 3$, then $F_n(\Pi \mid H_n) = 3\text{-}n$. If $s_n = 2$, then $F_n(\Pi \mid H_n) = 0$. Therefore, $F_n(\Pi \mid H_n)$ depends only on n and s_n; so let $f_n(s)$ denote $F_n(\Pi \mid H_n)$ when $s_n = s$. The right side of (4-18) when $n = 1$

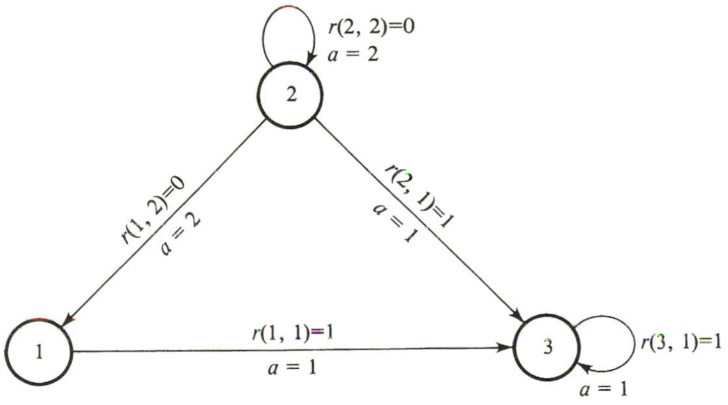

Figure 4-2 A deterministic MDP.

and $s = 2$ is

$$\begin{aligned} &\max \{r(2, 1) + \beta f_2(3), r(2, 2) + \beta f_2(2)\} \\ &= \max \{1 + 1, \quad 0 + 0\} = 2 \end{aligned}$$

However, the left side of (4-18) is $f_1(2)$, which is $F_1(\Pi \mid 2)$, $= 0$ so (4-18) is not valid. On the other hand, Π is optimal with respect to the notion of maximizing $F_1(\cdot \mid s_1)$. Recall that this notion is weaker than Definition 4-6. □

Example 4-7: machine maintenance—Example 4-2 continued The probabilistic structure of this example is more complicated than that of Example 4-6. Recall from Example 4-2 that $s = 0$, 1, and 2 indicate that a machine is working perfectly, has a minor defect, and is being repaired, respectively. The actions $a = 0$ and $a = 1$ denote "do nothing" and "under repair," respectively; so $A_0 = \{0\}$, $A = \{0, 1\}$, and $A_2 = \{1\}$. The transition probabilities and rewards are given in Tables 4-1, 4-2, and 4-3 (repeated here as Tables 4-4, 4-5, and 4-6).

Suppose in Example 4-2 that the salvage-value function is $L(s) = v_s$, $s = 0, 1, 2$, with $v_0 > v_1 > v_2$. From Example 4-2 (4-18) for $n = N$ is

$$\begin{aligned} f_N(0) &= 10 + \beta[uv_0 + (1 - u)(1 - g)v_1 + (1 - u)gv_2] \\ f_N(1) &= \max \{5 + \beta[dv_1 + (1 - d)v_2], 0 + \beta v_2\} \\ f_N(2) &= 0 + \beta[(1 - \rho)\phi v_0 + (1 - \rho)(1 - \phi)v_1 + \rho v_2] \end{aligned} \tag{4-24}$$

For $n < N$, (4-18) is

$$\begin{aligned} f_n(0) &= 10 + \beta[uf_{n+1}(0) + (1 - u)(1 - g)f_{n+1}(1) + (1 - u)gf_{n+1}(2)] \\ f_n(1) &= \max \{5 + \beta[df_{n+1}(1) + (1 - d)f_{n+1}(2)], 0 + \beta f_{n+1}(2)\} \\ f_n(2) &= 0 + \beta[(1 - \rho)\phi f_{n+1}(0) + (1 - \rho)(1 - \phi)f_{n+1}(1) + \rho f_{n+1}(2)] \end{aligned} \tag{4-25}$$

Suppose $N = 2$, $\beta = 1$, and

$$\begin{aligned} u &= 0.7 & v_0 &= 5 \\ g &= 0.333 & v_1 &= 2 \\ d &= 0.4 & v_2 &= 0 \\ \rho &= 0.5 & & \\ \phi &= 0.8 & & \end{aligned}$$

Table 4-4 p_{sj}^0 **in Example 4-7**

$s \backslash j$	0	1	2
0	u	$(1 - u)(1 - g)$	$(1 - u)g$
1	0	d	$1 - d$

Table 4-5 p^1_{sj} **in Example 4-7**

$s \backslash j$	0	1	2
1	0	0	1
2	$(1-\rho)\phi$	$(1-\rho)(1-\phi)$	ρ

Then (4-24) and (4-25) are (to the nearest one-tenth)

$$f_2(0) = 10 + 1.[0.7(5) + 0.3(0.667)(2) + 0.3(0.333)(0)] = 13.9$$

$$f_2(1) = \max\{5 + 1 \cdot [0.7(5) + 0.3(0.667)(2) + 0.3(0.333)(0)], 0\} = 9.2$$

$$f_2(2) = 0 + 1 \cdot [0.5(0.8)5 + 0.5(0.2)(2) + 0.5(0) = 2.2$$

$$f_1(0) = 10 + 0.7(13.9) + 0.3(0.667)(9.2) + 0.3(0.333)(2.2) = 21.7$$

$$f_1(1) = \max\{\mathbf{5 + 0.4(9.2) + 0.6(2.2)}, 2.2\} = 10.0$$

$$f_1(2) = 0.5(0.8)(13.9) + 0.5(0.2)(9.2) + 0.5(2.2) = 2.6$$

Terms that attain maxima are written in boldface type. From the second and fifth equations, the ones for $f_2(1)$ and $f_1(1)$, we reach the following conclusion: It is better not to move a machine with a minor defect to the maintenance shop when the machine faces disposal within two periods. □

Usually, the significance of "period n" depends only on $N - n$. Therefore, dynamic programming recursions often are written after making the following transformation: $f_N \to f_1^*$, $f_{N-1} \to f_2^*$, ..., $f_2 \to f_{N-1}^*$, $f_1 \to f_N^*$. Then the recursive optimization equation (4-18) is

$$f_n^*(s) = \max\{r(s, a) + \beta E(f_{n-1}^*[M(s, a, \xi_{s,a})]) : a \in A_s\} \qquad s \in \mathcal{S}, n \in I_N \qquad (4\text{-}26)$$

with $f_0^*(s) = L(s)$ for all $s \in \mathcal{S}$. The subscript n in (4-26) measures the number of periods *remaining* until the end of the planning horizon.

Of course, (4-18) and (4-26) are equivalent, but we are usually interested in the effects of the number of remaining periods. Therefore, we use (4-26) rather

Table 4-6 Single-stage rewards $r(s, a)$ **in Example 4-7**

$s \backslash a$	0	1
0	10	—
1	5	0
2	—	0

than (4-18). Moreover, the restriction $n \in I_N$ is unnecessary in (4-26). If $n > N$, then $f_n^*(s)$ still is the value of an optimal policy when n periods remain and the present state is s. Finally, we shall abandon the superscript * on f_n^* and replace (4-26) with

$$f_n(s) = \max \{r(s, a) + \beta E[f_{n-1}(M[s, a, \xi_{s,a}]): a \in A_s\} \qquad s \in \mathcal{S}, n \in I_+ \qquad (4\text{-}27)$$

and $f_0(s) = L(s)$, $s \in \mathcal{S}$. When you see an equation such as (4-18) or (4-27), you can determine the proper interpretation of n as follows. If the subscript on $f(\cdot)$ on the right is $n - 1$, then n denotes the number of periods remaining. If the subscript is $n + 1$, then n denotes chronological time.

Just as (4-18′) can replace (4-18) if $\mathcal{S}$ is a discrete set, (4-27) can be replaced with

$$f_n(s) = \max \left\{r(s, a) + \beta \sum_{j \in \mathcal{S}} p_{sj}^a f_{n-1}(j): a \in A_s\right\} \qquad s \in \mathcal{S}, n \in I_+ \qquad (4\text{-}27')$$

Example 4-8: Inventory model Consider a single-product inventory model. Suppose that ordered goods are delivered immediately, excess demand is lost, inventory holding cost is proportional to inventory level at the end of the period, and demands in successive periods are i.i.d. nonnegative r.v.'s. Thus far, the model is the same as in Example 1-3. However, suppose here that the purchase cost consists of a setup cost as well as a unit cost and that there is limited storage capacity. Say $\mathcal{S} = \{0, 1, \ldots, 4\}$ where $\mathbf{s}_n \in \mathcal{S}$ denotes the inventory level at the beginning of period n. If $\mathbf{a}_n$ denotes the inventory level after ordered goods are delivered but before demand $\mathbf{D}_n$ occurs, then $\mathbf{a}_n \in A_{\mathbf{s}_n} = \{\mathbf{s}_n, \mathbf{s}_n + 1, \ldots, 4\}$ because $\mathbf{a}_n - \mathbf{s}_n$ is the amount ordered and $\mathbf{s}_{n+1} = (\mathbf{a}_n - \mathbf{D}_n)^+ \le \mathbf{a}_n$ so $\mathbf{a}_n \le 4$.

To be specific, suppose that the net cost in period n is

$$\delta(\mathbf{a}_n - \mathbf{s}_n) + 10(\mathbf{a}_n - \mathbf{s}_n) + (\mathbf{a}_n - \mathbf{D}_n)^+ + 12(\mathbf{a}_n - \mathbf{D}_n)^+ - 12\mathbf{a}_n$$

[where $\delta(u) = 1$ if $u > 0$ and $\delta(u) = 0$ if $u \le 0$]. This expression corresponds to a setup cost of 1, a unit purchase cost of 10, a unit holding cost of 1, and a unit revenue of 12. Suppose also that the single-period discount factor is 0.9,

$$P\{\mathbf{D}_1 = j\} = 0.25 \qquad j = 0, 1, 2, 3 \qquad (4\text{-}28)$$

and the salvage value of unsold items is the same as their purchase cost, namely, 10.

Using the same reasoning as in Section 3-1, the total discounted cost for N periods, net of salvage value, is

$$\sum_{n=1}^{N} 0.9^{n-1}[\delta(\mathbf{a}_n - \mathbf{s}_n) + 10(\mathbf{a}_n - \mathbf{s}_n) + (\mathbf{a}_n - \mathbf{D}_n)^+ + 12(\mathbf{a}_n - \mathbf{D}_n)^+ - 12\mathbf{a}_n] - 0.9^N(10)\mathbf{s}_{N+1}$$

In $10(\mathbf{a}_n - \mathbf{s}_n)$ and $\mathbf{s}_{N+1}$ we substitute $\mathbf{s}_n = (\mathbf{a}_{n-1} - \mathbf{D}_{n-1})^+$ for $n > 1$ and rearrange terms to obtain

$$\sum_{n=1}^{N} 0.9^{n-1}[\delta(\mathbf{a}_n - \mathbf{s}_n) - 2\mathbf{a}_n + (12 + 1 - 10 \times 0.9)(\mathbf{a}_n - \mathbf{D}_n)^+] - 10s_1 \qquad (4\text{-}29)$$

Table 4-7 Single-period Costs $r(s, a)$ in Example 4-8

Inventory after ordering

	$s \backslash a$	0	1	2	3	4
	0	0	0	0	1	3
Inventory	1	—	−1	0	1	3
before	2	—	—	−1	1	3
ordering	3	—	—	—	0	3
	4	—	—	—	—	2

The constant $-10s_1$ is not influenced by actions $\mathbf{a}_1, \ldots, \mathbf{a}_n$; so we delete this term from the objective. The modified objective is

$$E\left[\sum_{n=1}^{N} 0.9^{n-1} r(\mathbf{s}_n, \mathbf{a}_n)\right]$$

where $r(\cdot, \cdot)$ is the following expectation of the bracketed term in (4-29):

$$r(s, a) = \delta(a - s) - 2a + 4E[(a - \mathbf{D}_1)^+] \tag{4-30}$$

The substitution of (4-28) in (4-30) yields the numerical values of $r(\cdot, \cdot)$ in Table 4-7.

Observe that $r(s, a)$ here is the expected single-period *cost* rather than reward. Hence, the recursive equations will have "min" rather than "max." You see in (4-18′) and (4-27′) that the transition probability p_{sj}^a and discount factor β always appear as a product in the recursive equations. Here,

$$p_{sj}^a = P\{(a - \mathbf{D}_1)^+ = j\}$$

Substitution of (4-28) leads to the numerical values of βp_{sj}^a displayed in Table 4-8.

Each row of entries in Table 4-9 sums to $\beta = 0.9$. Hence Table 4-9 is a *substochastic matrix*: its entries are nonnegative and its row sums do not exceed unity.

To find an optimal policy with n periods remaining, we solve (4-27′) recursively. Let $a_n(s)$ denote a feasible action that attains the optimum in

Table 4-8 $\beta p_{sj}^a = \beta P\{(a - \mathbf{D}_1)^+ = j\}$ in Example 4-8

Subsequent inventory level

	$a \backslash j$	0	1	2	3	4
	0	0.9	0	0	0	0
Inventory	1	0.675	0.225	0	0	0
after	2	0.45	0.225	0.225	0	0
ordering	3	0.225	0.225	0.225	0.225	0
	4	0	0.225	0.225	0.225	0.225

Table 4-9 $f_1(s)$ **and** $a_1(s)$ **in Example 4-8**

s	$f_1(s)$	$a_1(s)$
0	0	0, 1, or 2
1	-1	1
2	-1	2
3	0	3
4	2	4

(4-27) [or (4-27′)]. Recall that the actual salvage values are built into Table 4-7; so let $f_0(\cdot) \equiv 0$. Then (4-27′) with $n = 1$ (and min instead of max) is

$$f_1(s) = \min \{r(s, a): s \le a \le 4\} \tag{4-31}$$

From Table 4-7 and (4-31), we obtain the values of $f_1(s)$ and $a_1(s)$ in Table 4-9. Therefore, in the final period, order zero or $(1 - s)^+$ units when the initial inventory level is s.

We discover what to do when two periods remain by solving (4-27′) for $n = 2$. For example, if $s = 3$,

$$\begin{aligned} f_2(3) &= \min \left\{ r(s, a) + \beta \sum_{j=0}^{4} p_{3j}^a f_1(j): a \in \{3, 4\} \right\} \\ &= \min \left\{ 0 + 0.225 \sum_{j=0}^{3} f_1(j),\ 3 + 0.225 \sum_{j=1}^{4} f_1(j) \right\} \\ &= \min \{0 - 0.45, 3 - 0\} = \min \{-0.45, 3\} \\ &= -0.45 \qquad \text{and} \qquad a_2(3) = 3 \end{aligned}$$

Thus it is best not to order anything if there are three items in inventory and two periods to go. The results when $n = 2$ periods remain are summarized in Table 4-10.

Table 4-10 $f_2(s)$ **and** $a_2(s)$ **in Example 4-8**

s	$f_2(s)$	$a_2(s)$
0	-0.45	2
1	-1.225	1
2	-1.45	2
3	-0.45	3
4	2	4

□

Proof of Theorem 4-1

Theorem 4-1 (*a*) If there is an optimal policy $\Pi_1^* = (\pi_1^*, \ldots, \pi_N^*)$, then there are N real-valued functions $f_1, f_2, \ldots, f_N$ on $\mathcal{S}$ which satisfy

$$f_n(s) = F_n(\Pi_n^* \mid H_n) \tag{4-17}$$

when H_n assigns the value s to s_n, $s \in S$, $n \in I_N$. Furthermore,

$$f_n(s) = \max\,\{r(s, a) + \beta E(f_{n+1}[M(s, a, \xi_{s,a})]) : a \in A_s\} \qquad s \in \mathcal{S}, n \in I_n \tag{4-18}$$

with $f_{N+1}(s) = L(s)$ for all $s \in \mathcal{S}$.

(*b*) Conversely, if $f_1, \ldots, f_N$ is a sequence of functions that satisfy (4-18), then there is an optimal policy $\Pi_1^* = (\pi_1^*, \ldots, \pi_N^*)$ with $\pi_n^*(H_n) \equiv a_n^*(s_n)$ being any element $a \in A_{s_n}$ that attains the maximum in (4-18).

PROOF The proof of (*a*) is by induction starting with $n = N$, and it exploits $f_{N+1}(\cdot) \equiv L(\cdot)$. If there is an optimal policy $\Pi_1^* = (\pi_1^*, \ldots, \pi_N^*)$, then (4-14) and Definition 4-6 yield

$$E_{\pi_{N*}|H_N}\{r(\mathbf{s}_N, \mathbf{a}_N) + \beta L[M(\mathbf{s}_N, \mathbf{a}_N, \xi_{\mathbf{s}_N, \mathbf{a}_N})]\}$$
$$\geq E_{\pi_N|H_N}\{r(\mathbf{s}_N, \mathbf{a}_N) + \beta L[M(\mathbf{s}_N, \mathbf{a}_N, \xi_{s_N, a_N})]\}$$

for all H_N and π_N; i.e., for all H_N (and s_N specified by H_N),

$$r[s_N, \pi_N^*(H_N)] + \beta E(L\{M[s_N, \pi_N^*(H_N), \xi_{sN, \pi N*(H_N)}]\})$$
$$= \max\,\{r[s_N, \pi_N(H_N)] + \beta E(L\{M[s_N, \pi_N(H_N), \xi_{sN, \pi N*(H_N)}]\} : \pi_N(H_N) \in A_{sN}\}$$
$$= \max\,\{r(s_N, a) + \beta E(L[M(s_N, a, \xi_{s,a})]) : a \in A_{sN}\} = f_N(s_N)$$

which verifies (4-17) and (4-18) for $n = N$.

An expansion of (4-14) yields

$$F_n(\Pi_n \mid H_n)$$
$$= r[s_n, \pi_n(H_n)] + E_{\Pi_n|H_n}\left[\sum_{j=n+1}^{N} \beta^{j-n} r(\mathbf{s}_j, \mathbf{a}_j) + \beta^{N-n+1} L(\mathbf{s}_{N+1})\right]$$
$$= r[s_n, \pi_n(H_n)] + E\{F_{n+1}[\Pi_{n+1} \mid H_n, \pi_n(H_n), \xi_n, M[s_n, \pi_n(H_n), \xi_n]\} \tag{4-32}$$

for $n \in I_N$ with $F_{N+1}(\cdot \mid H_{N+1}) \triangleq L(s_{N+1})$. Suppose now that Π_1^* is optimal and (4-17) and (4-18) are valid for $n + 1$. Then

$$F_n(\Pi_n^* \mid H_n)$$
$$= \sup\,\{F_n(\Pi_n \mid H_n) : \Pi_n \text{ is a policy}\}$$
$$= \sup\,\{F_n(\pi_n, \Pi_{n+1} \mid H_n) : \pi_n, \Pi_{n+1}\}$$
$$= \sup\,\{r[s_n, \pi_n(H_n)] + \beta E\{F_{n+1}[\Pi_{n+1} \mid H_n, \pi_n(H_n), \xi_n, M[s_n, \pi_n(H_n), \xi_n]\} :$$
$$\pi_n, \Pi_{n+1}]\}$$

$$= \sup \{r[s_n, \pi_n(H_n)] + \beta E(\sup \{F_{n+1}\{\Pi_{n+1} \mid H_n, \pi_n(H_n), \xi_n, M[s_n, \pi_n(H_n), \xi_n]\}: \Pi_{n+1}\}: \pi_n(H_n) \in A_{s_n}\})$$

$$= \sup \{r[s_n, \pi_n(H_n)] + \beta E(f_{n+1}\{M[s_n, \pi_n(H_n), \xi_n]\}): \pi_n(H_n) \in A_{s_n}\}$$

$$= \max \{r(s_n, a) + \beta E\{f_{n+1}[M(s_N, a, \xi_n)]\}: a \in A_{s_n}\} \triangleq f_n(s_n)$$

which verifies (4-17) and (4-18) for n and completes the inductive proof of (*a*).

To prove (*b*), suppose $f_1, \ldots, f_N$ satisfy (4-18) with $f_{N+1}(s) = L(s)$ for all $s \in \mathcal{S}$ and $a_n^*(s) \in A_s$ attains the maximum in (4-18). Also suppose that Π_1^* satisfies (4-15) for $n \geq j + 1$, that is,

$$F_n(\Pi_n^* \mid H_n) \geq F_n(\Pi_n \mid H_n) \qquad \text{for all } H_n, \Pi_n, \text{ and } n \in I_N$$

Then the definition of $F_n(\Pi_n \mid H_n)$ [(4-14)], (4-15), and (4-32), $s \in \mathcal{S}$, $a \in A_s$, and arbitrary $\Pi_j = (\pi_j, \Pi_{j+1})$ imply

$$\begin{aligned} f_j(s) &= r[s, a_j^*(s)] + E(f_{j+1}\{M[s, a_j^*(s), \xi_j]\}) \\ &\geq r(s, a) + E\{f_{j+1}[M(s, a, \xi_j)]\} \\ &= r(s, a) + E(\max \{F_{j+1}[\Pi_{j+1} \mid H_j, a, \xi_j, M(s, a, \xi_j)]: \Pi_{j+1}\}) \\ &\geq r(s, a) + E\{F_{j+1}[\Pi_{j+1} \mid H_j, a, \xi_j, M(s, a, \xi_j)]\} \end{aligned} \tag{4-33}$$

The last inequality is true, in particular, if $a = \pi_j(H_j)$:

$$\begin{aligned} f_j(s) &\geq r[s, \pi_j(H_j)] + E(F_{j+1}\{\Pi_{j+1} \mid H_j, \pi_j(H_j), \xi_j, M[s, \pi_j(H_j), \xi_j]\}) \\ &= F_j(\pi_j, \Pi_{j+1} \mid H_j) = F_j(\Pi_j \mid H_j) \end{aligned}$$

With the inductive assumption and (4-33), this inequality implies

$$\begin{aligned} F_j(\Pi_j \mid H_j) &\leq f_j(s) \\ &= r[s, a_j^*(s)] + E(f_{j+1}\{M[s, a_j^*(s), \xi_j]\}) \\ &= r[s, a_j^*(s)] + E(F_{j+1}\{\Pi_{j+1}^* \mid H_j, a_j^*(s), \xi_j, M[s, a_j^*(s), \xi_j]\}) \\ &= F_j[a_j^*(\cdot), \Pi_{j+1}^* \mid H_j] \end{aligned}$$

so $\Pi_j^* = [a_j^*(\cdot), \Pi_{j+1}^*]$ satisfies (4-15) for $n = j$.

The inductive proof of (*b*) will be completed by establishing (*b*) for $n = N$. Suppose $f_N(\cdot)$ satisfies (4-18) with $f_{N+1}(\cdot) \equiv L(\cdot)$ and there exists $a_N^*(s) \in A_s$, $s \in \mathcal{S}$, such that

$$r[s, a_N^*(s)] \geq r(s, a) \qquad a \in A_s, s \in \mathcal{S}$$

In particular, let H_N specify $s_N = s$, $a = \pi_N(H_N)$ and use (4-28):

$$\begin{aligned} F_N[a_N^*(\cdot) \mid H_N] &= r[s, a_N^*(s)] + \beta E(L\{M[s, a_N^*(s), \zeta_{s, a_N*(s)}]\}) \\ &\geq r[s, \pi_N(H_N)] + \beta E(L\{M[s, \pi_N(H_N), \zeta_{s, \pi_N(H_N)}]\}) \\ &= F_N(\pi_N \mid H_N) \end{aligned}$$

so (4-15) is satisfied for $n = N$ with $a_N^*(\cdot) = \pi_N^*(\cdot) = \Pi_N^*$. □

EXERCISES

4-7 Consider the dynamic program (4-23) associated with the favorable game ($p \geq 1/2$) version of the gambler's ruin model in Example 4-5. Let $N = 2$.

(*a*) Suppose $\beta = 0.9$ and $p = 7/12$. Find $f_1(s)$ and $f_2(s)$ for $0 \leq s \leq 4$ and specify an optimal policy.

(*b*) Suppose $\beta < 1$. In the last period prove that action $a = 0$ ("continue") is best if, and only if,

$$s \leq \frac{(2p-1)\beta}{1-\beta}$$

4-8 Alter Example 4-5 so that the salvage value of s coins is 0, that is, $L(s) = 0$, $0 \leq s \leq 4$.

(*a*) What form does $f_N(s)$ take in place of (4-20)? What is the associated optimal action?

(*b*) Is (4-22) still valid? Justify your answer.

(*c*) Repeat part (*a*) of Exercise 4-7 in this case.

4-9 In the maintenance model of Example 4-7, what is an optimal policy when $N = 3$? Use (4-18′) and the numerical values following (4-25).

4-10 (*a*) Write (4-27) for the MDP in Example 4-3.

(*b*) Suppose $P\{\mathbf{D}_1 = 0\} = P\{\mathbf{D}_1 = 1\} = 0.3$, $P\{\mathbf{D}_1 = 2\} = 0.4$, $\rho = 4$, $c = 2$, $h = 1$, and $\beta = 0.9$. Assume $\mathcal{S} = I$ and $L(s) = 0$ for all $s \in \mathcal{S}$. What is an optimal inventory rule when two periods remain and 0 units are in inventory? When 2 units are in inventory? When 4 units are in inventory?

4-11 Repeat part (*b*) of Exercise 4-10 with $L(s) = 2s$, $s \in \mathcal{S}$. Contrast your answer with a solution based on a myopic optimum from Section 3-2.

4-12 A *nonstationary* MDP is a model whose elements may depend on the period n in an essential way. For each $n \in I_+$ let $\mathcal{S}_n$ and A_{sn} be nonempty sets, $s \in \mathcal{S}_n$, and let

$$\mathscr{C}_n = \{(s, a):\quad a \in A_{sn} \text{ and } s \in \mathcal{S}_n\}$$

Let $r_n(\cdot\,, \cdot)$ be a real-valued function on $\mathscr{C}_n$ such that

$$r_n(s, a) = E(X_n \mid s_n = s,\ a_n = a)$$

For each $(s, a) \in \mathscr{C}_n$ let $\xi_{s,a,n}$ be an r.v. and $M_n(s, a, \cdot)$ a mapping such that

$$P\{M_n(s, a, \xi_{s,a,n}) \in \mathcal{S}_{n+1}\} = 1$$

Let $\beta_0, \beta_1, \beta_2, \ldots$ be a sequence of positive numbers such that $1 = \beta_0 \geq \beta_1 \geq \beta_2 \geq \ldots$. Review the proof of Theorem 4-1 to verify that it remains valid for a nonstationary MDP if (4-18) is replaced by

$$f_n(s) = \max\left\{r_n(s, a) + \frac{\beta_n}{\beta_{n-1}} E(f_{n+1}[M_n(s, a, \xi_{s,a,n})]):\quad a \in A_{sn}\right\},\qquad s \in \mathcal{S}_n,\ n \in I_N$$

with $f_{N+1}(s) = L(s)$ for all $s \in \mathcal{S}_1$.

4-3 INFINITE-HORIZON MDPs AND STATIONARY POLICIES

Consider an inventory problem in a retail store which will close permanently in two years. Suppose replenishment can occur daily and the store is open every day of the year. Do you think that today's replenishment, as part of an optimal policy for this problem, would change much if the store were to remain open for 735 days instead of 730? Suppose costs and revenues are discounted at a daily rate $\beta < 1$. It is our intuition that the first period's action is insensitive to the horizon length N if N is large. In fact, as N tends to infinity, the optimal rule for selecting the first period's action loses its dependence on N (Theorem 4-2 below). Then the

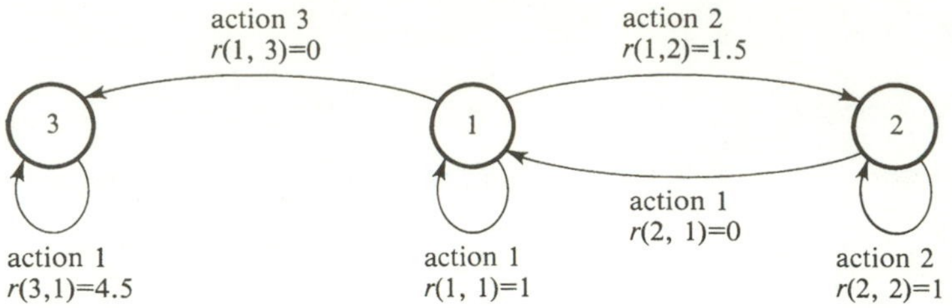

Figure 4-3 A deterministic MDP.

same decision rule (not the same action) is optimal in all succeeding periods too. A time-invariant decision rule is called a *stationary policy*.

Example 4-9 Consider the deterministic MDP in Figure 4-3 with state space $\mathcal{S} = \{1, 2, 3\}$ and salvage-value function $L(\cdot) \equiv 0$. Nodes in the figure represent states and arcs correspond to actions and their consequent transitions. The data are $p_{11}^1 = p_{12}^2 = p_{13}^3 = p_{33}^1 = p_{21}^1 = p_{22}^2 = 1$, $r(1, 1) = 1$, $r(1, 2) = 3/2$, $r(1, 3) = 0$, $r(2, 1) = 0$, $r(2, 2) = 1$, $r(3, 1) = 9/2$, and $\beta = 1/2$.

The dynamic programming recursion (4-27′) takes the following form here: $f_0(s) = 0$ for $s = 1, 2$, and 3 and, for $n \in I_+$,

$$f_n(1) = \max\,\{1 + 0.5\,f_{n-1}(1),\ 1.5 + 0.5f_{n-1}(2),\ 0.5f_{n-1}(3)\}$$

$$f_n(2) = \max\,\{0.5f_{n-1}(1),\ 1 + 0.5f_{n-1}(2)\}$$

$$f_n(3) = 4.5 + 0.5f_{n-1}(3)$$

Let $a_n(s)$ denote any value of a that causes the "max" in $f_n(s)$ to be attained. For example, $a_n(3) = 1$ for all $n \in I_+$. Then the recursion yields the results in Table 4-11.

As n increases, $[a_n(1), a_n(2), a_n(3)]$ moves through (2, 2, 1), (3, 2, 1), and finally (3, 1, 1). As you might guess, the vector remains at (3, 1, 1) no matter how much larger n gets. Is this behavior true in general? Also, for each $s \in \mathcal{S}$, the value of $f_n(s)$ seems to be converging as n gets larger. Is that behavior true in general? These questions contribute to the motivation for Sections 4-3 through 4-6. As you may have come to expect, we provide answers in a roundabout manner. □

Table 4-11 An optimal policy and its values in Example 4-9

n	$a_n(1)$	$a_n(2)$	$a_n(3)$	$f_n(1)$	$f_n(2)$	$f_n(3)$
1	2	2	1	1.5	1	4.5
2	3	2	1	2.25	1.5	6.75
3	3	2	1	3.375	1.75	7.875
4	3	2	1	3.9375	1.875	8.4375
5	3	1	1	4.2188	1.9688	8.7188
6	3	1	1	4.3594	2.1094	8.8594
7	3	1	1	4.4297	2.1797	8.9297

Preliminaries

For the remainder of this chapter, we abandon the notation in Sections 4-1 and 4-2 of boldface type for r.v.'s and ordinary typeface for values taken by r.v.'s. Whether a symbol represents an r.v. or a value taken by an r.v. should be clear in each context. Henceforth, **boldface** type denotes a vector.

Suppose that the planning horizon is infinite, that $\beta < 1$, and the objective is to maximize $E(B)$ where

$$B = \sum_{\tau=1}^{\infty} \beta^{\tau-1} g(s_\tau, a_\tau, \xi_\tau) \tag{4-34}$$

Let $B(n)$ denote the present value of the rewards from period n onward:

$$B(n) = \sum_{\tau=n}^{\infty} \beta^{\tau-n} g(s_\tau, a_\tau, \xi_\tau)$$

Recall the notation $I_+ = \{1, 2, \ldots\}$ and $\mathscr{C} = \{(s, a): a \in A_s, s \in \mathcal{S}\}$. As in Section 4-2, π_n is a feasible decision rule in period n if π_n assigns a feasible action to each possible history H_n. Consider a sequence of rules $\pi_1, \pi_2, \ldots$.

Definition 4-7 $\Pi_1 = (\pi_1, \pi_2, \ldots)$ is a *policy* if $[s_n, \pi_n(H_n)] \in \mathscr{C}$ for all possible H_n and $n \in I_+$.

For $n \in I_+$, let $\Pi_n = (\pi_n, \pi_{n+1}, \ldots)$ denote the segment of Π_1 from period n onward. The distribution of the r.v. B depends on Π_1, which is influenced by $H_1 = (s_1)$. Similarly, the distribution of $B(n)$ depends on Π_n, which is influenced by H_n. The notation $B(n, \Pi_n)$ and

$$F_n(\Pi_n \mid H_n) \triangleq E_{\Pi_n \mid H_n}[B_N(n)]$$

make the dependence explicit.

Proposition 4-1 asserts that ξ_n, given s_n and a_n, is conditionally independent of $(H_{n-1}, a_{n-1}, \xi_{n-1})$; so Definition 4-3 of $r(s, a)$ yields

$$F_n(\pi_n \mid H_n) = E_{\Pi_n \mid H_n}\left[\sum_{\tau=n}^{\infty} \beta^{\tau-n} r(s_\tau, a_\tau)\right]$$

The following definition of optimality is essentially the same stringent notion as in Section 4-2.

Definition 4-8 $\Pi_1^* = (\pi_1^*, \pi_2^*, \ldots)$ is an *optimal policy* if, and only if [where $\Pi_n^* = (\pi_n^*, \pi_{n+1}^* \ldots)$],

$$F_n(\Pi_n^* \mid H_n) \geq F_n(\Pi_n \mid H_n) \text{ for all } H_n, \Pi_n, \text{ and } n \in I_+ \tag{4-35}$$

Is the infinite-horizon analog of Theorem 4-1 valid? Further explanation is needed to pose the inquiry properly, and at first we shall not be rigorous.

Recall the recursion in (4-27). Let $f_0(\cdot) \equiv L(\cdot)$ and

$$f_n(s) = \max \{r(s, a) + \beta E\{f_{n-1}[M(s, a, \xi_{s,a})]\} : a \in A_s\} \qquad s \in \mathcal{S}, n \in I_+ \tag{4-36}$$

What happens as n gets larger? If B in (4-34) is replaced with

$$B_N \triangleq \sum_{n=1}^{N} \beta^{n-1} g(s_n, a_n, \xi_n) + \beta^N L(s_{N+1})$$

and $\beta < 1$, then $\beta^N L(s_{N+1})$ should tend to zero as $N \to \infty$ [if $|L(s_{N+1})|$ does not get too large]; so B_N should tend to B. Therefore, we might as well use $L(\cdot) \equiv 0$ to investigate infinite-horizon problems. This argument is made rigorous in Section 5-4, but until then we use (4-36) with $L(\cdot) \equiv 0$.

Suppose there is a number $u < \infty$ such that

$$0 \le r(s, a) \le u \qquad (s, a) \in \mathscr{C} \tag{4-37}$$

Exercise 4-13 asks you to prove that assumption (4-37) is no worse than assuming that $|r(\cdot, \cdot)|$ is uniformly bounded† on $\mathscr{C}$. Exercise 4-14 asks you to show that (4-37) implies

$$f_1(s) \le f_2(s) \le f_3(s) \le \cdots \le \frac{u}{1-\beta} \qquad s \in \mathcal{S}$$

Therefore, there exists

$$f(s) \triangleq \lim_{n \to \infty} f_n(s) \qquad s \in \mathcal{S} \tag{4-38}$$

As a result, $f_n(s) \to f(s)$ on the left side of (4-36) and $f_{n-1}[M(s, a, \xi_{s,a})] \to f[M(s, a, \xi_{s,a})]$ on the right side of (4-36). Hence, you might conjecture that the limit function $f(\cdot)$ [whose point values are defined by (4-38)] must satisfy the following variant of (4-36) in which the subscripts n and $n - 1$ are deleted:

$$f(s) = \max \{r(s, a) + \beta E\{f[M(s, a, \xi_{s,a})]\} : a \in A_s\} \qquad s \in \mathcal{S} \tag{4-39}$$

In Sections 4-5 and 5-4 we investigate the connection between (4-38) and (4-39) and present sufficient conditions for (4-39) to have a unique solution that is the same as (4-38). Here, we merely note that (4-39) is the important *functional equation of dynamic programming*.‡ The following conjecture is an infinite-horizon version of Theorem 4-1.

Conjecture (*a*) If there is an optimal policy $\Pi_1^* = (\pi_1^*, \pi_2^*, \ldots)$, there is a real-valued function f on $\mathcal{S}$ which satisfies (4-39) and

$$f(s) = F(\Pi_n^* \mid H_n)$$

when H_n assigns the value s to s_n, $s \in \mathcal{S}$, $n \in I_+$. (*b*) Conversely, if f satisfies (4-39), there is an optimal policy $\Pi_1^* = (\pi^*, \pi^*, \pi^*, \ldots)$ with $\pi^*(H_n) \equiv a^*(s_n)$ being any element $a \in A_{s_n}$ that attains the maximum in (4-39).

† The function $r(\cdot, \cdot)$ is uniformly bounded on $\mathscr{C}$ if there is a number $u < \infty$ such that $|r(s, a)| \le u$ for all $(s, a) \in \mathscr{C}$.

‡ Some of the important functional equations in Volume I are $\boldsymbol{\pi} = \boldsymbol{\pi}P$, $\Sigma_i \pi_i = 1$, and $\boldsymbol{\pi} \ge \mathbf{0}$ for Markov chains [(7-45*a*), (7-45*b*), and (7-45*c*) on page 242] and the renewal equation [(5-26) on page 119].

The inductive proof of each part of Theorem 4-1 was initiated by $f_0(s) = L(s)$ for all $s \in \mathcal{S}$. In the present infinite-horizon case, there is no comparably convenient starting point for an inductive proof. The conjecture is true if additional assumptions are made, but the proofs are indirect. The second part of Conjecture (*b*) asserts that there is a stationary optimal policy if there is any optimal policy. This claim is proved in Theorem 4-3 below. Then Section 4-4 uses the machinery introduced for the proof to construct an algorithm that finds an optimal policy. Part of Section 4-5 and all of Section 5-4 concern the remainder of the conjecture.

The interpretation of $f(s)$ is similar to that of $f_n(s)$. Hence, $f(s)$ is the value of an optimal policy in the infinite-horizon model if the initial state is s. It is the maximal expected present value if the initial state is s. For two states s and s', $f(s) - f(s')$ is the relative value of starting from state s instead of s'. It is the most you should pay to have the initial state shifted from s' to s. We call $f(\cdot)$ the *value function.*

A solution $f(\cdot)$ to (4-39) is a *vector maximum* in the sense that the same function appears on both sides of the equation. If there is a sample path emanating from state s that eventually passes through s', then $f(s)$ is affected by $f(s')$. The recursion (4-36) is not a vector-maximization problem because $f_n(\cdot)$ depends on $f_{n-1}(\cdot)$ rather than on $f_n(\cdot)$ itself.

Stationary Policies

This subsection defines the sets of Markov and stationary policies. Then it shows that a restriction to the set of Markov policies is without loss of optimality. The fundamental assumptions are that $\mathcal{S}$ is either a finite or countably infinite set and $0 \le \beta < 1$.

Definition 4-9 A *single-stage decision rule* is a function on $\mathcal{S}$ that assigns to each $s \in \mathcal{S}$ an element of A_s.

Let Δ denote the set of all single-stage decision rules, so

$$\Delta = \underset{s \in \mathcal{S}}{\times} A_s$$

If $\delta \in \Delta$ is used to select the nth period's action a_n, then $a_n = \delta(s_n)$. In this case, the action is selected by a procedure that depends on the past history only through the current state.

Definition 4-10 A *Markov policy* is a sequence $(\delta_1, \delta_2, \ldots)$ of single-stage decision rules.

In a Markov policy $a_n = \delta_n(s_n)$ for all $n \in I_+$; so in every period, the action is selected by a procedure δ_n that depends on the past history only through the current state. However, different procedures may be used at different times because $\delta_i \neq \delta_j$ is possible if $i \neq j$. The set Y of all Markov policies is

$$Y = \mathop{\times}_{n=1}^{\infty} \Delta = \mathop{\times}_{n=1}^{\infty} \mathop{\times}_{s \in \mathcal{S}} A_s$$

Definition 4-11 A Markov policy $\pi = (\delta_1, \delta_2, \ldots)$ is a *stationary policy* if δ_n is the same for all n.

Therefore, Δ indexes the set of stationary policies. It is convenient to write δ^∞ for the stationary policy $\pi = (\delta, \delta, \delta, \ldots)$. Let Z denote the set of all stationary policies so

$$Z = \{\delta^\infty : \delta \in \Delta\}$$

Example 4-10: Continuation of Example 4-9 Example 4-9 is a deterministic MDP with three states whose data are exhibited in Figure 4-3 and $\beta = 0.5$. In this example, for any specific stationary policy σ^∞, let $[\sigma(1), \sigma(2), \sigma(3)]$ specify the corresponding single-stage decision rule. Then Δ, the set of single-stage decision rules, is $\{(1, 1, 1), (1, 2, 1), (2, 1, 1), (2, 2, 1), (3, 1, 1), (3, 2, 1)\}$. Let $\delta = (3, 1, 1)$ and $\gamma = (2, 2, 1)$. An example of a policy that is *nonstationary* (i.e., not stationary) but Markov is $(\gamma, \delta, \gamma, \delta, \ldots) \in Y - Z$.

An example of a policy that is *not* Markov, i.e., not in Y, is to use δ^∞ if $s_1 = 1$ and use γ^∞ if $s_1 \neq 1$. If $s_1 = 1$, the sequence $(s, a_1, s_2, a_2, \ldots)$ of states and actions induced by this policy is $(1, 3, 3, 1, 3, 1, 3, 1, \ldots)$; if $s_1 = 2$ it is $(2, 2, 2, 2 \ldots)$; and if $s_1 = 3$ it is $(3, 1, 3, 1, 3, 1, \ldots)$. These are exactly the same sequences that would be induced by the Markov policy $[(3, 2, 1), \gamma, \gamma, \gamma, \ldots]$.

□

The remainder of this subsection justifies our restriction of policies to the set Y of Markov policies. Suppose that $\mathcal{S}$ is a finite or countably infinite set. Let $\pi = (\delta_1, \delta_2, \ldots)$ be a Markov policy, $v_s(\pi)$ be the expected present value if π is used and the initial state is s. Suppress the dependence on s and let

$$w_n(j) = P\{s_n = j \mid s_1 = s\}$$

and $\mathbf{w}_n = [w_n(i), i \in \mathcal{S}]$. The vector $\mathbf{w}_n$ is a probability distribution on $\mathcal{S}$. Let W be the set of all probability distributions on $\mathcal{S}$. Then $\mathbf{w} = [w(i), i \in \mathcal{S}] \in W$ if, and only if, $w(i) \geq 0$ for all $i \in \mathcal{S}$ and $\sum_{i \in \mathcal{S}} w(i) = 1$.

With the preceding notation,

$$v_s(\pi) = E\left\{\sum_{n=1}^{\infty} \beta^{n-1} r[s_n, \delta_n(s_n)] \mid s_1 = s\right\}$$

More generally, if the initial state has the probability distribution $\mathbf{w} = [w(i), i \in \mathcal{S}]$, let $v_{\mathbf{w}}(\pi)$ denote the expected present value if policy π is used. Then

$$v_{\mathbf{w}}(\pi) = \sum_{n=1}^{\infty} \beta^{n-1} \sum_{j \in \mathcal{S}} w_n(j) r[j, \delta_n(j)]$$

where $w_1(j) = w(j)$ for each $j \in \mathcal{S}$. Let

$$\rho(\mathbf{w}, \delta) = \sum_{i \in \mathcal{S}} w(i) r[i, \delta(i)] \qquad w \in W \qquad \delta \in \Delta \tag{4-40}$$

This is the expected value of the single-stage reward in a period when the state has the distribution $\mathbf{w}$ and the decision rule is δ. Then

$$v_{\mathbf{w}}(\pi) = \sum_{n=1}^{\infty} \beta^{n-1} \rho(\mathbf{w}_n, \delta_n) \tag{4-41}$$

Now we show that $\mathbf{w}_1, \mathbf{w}_2, \ldots$ can be computed recursively. The first element $\mathbf{w}_1$ has $w_1(s) = 1$ and $w_1(i) = 0$ for all $i \neq s$. For each $n \geq 1$ and j,

$$P\{s_{n+1} = j \mid s_1 = s\} = \sum_{i \in \mathcal{S}} P\{s_n = i \mid s_1 = s\} P\{s_{n+1} = j \mid s_n = i\}$$

so

$$w_{n+1}(j) = \sum_{i \in \mathcal{S}} w_n(i) p_{ij}^{\delta_n(i)} \tag{4-42}$$

Let

$$M(\mathbf{w}, \delta)(j) = \sum_{i \in \mathcal{S}} w(i) p_{ij}^{\delta(i)} \qquad \mathbf{w} \in W \qquad \delta \in \Delta \qquad j \in \mathcal{S} \tag{4-43}$$

Thus $M(\mathbf{w}, \delta)$ is the mapping from $W \times \Delta$ to W which takes the value $M(\mathbf{w}, \delta)(j)$ at j. This value is the conditional probability of being in state j next period if $\mathbf{w}$ is the distribution of the state this period and we use decision rule δ. Hence, $w_{n+1}(j) = M(w_n, \delta_n)(j)$ and

$$\mathbf{w}_{n+1} = M(\mathbf{w}_n, \delta_n) \qquad n = 1, 2, \ldots \tag{4-44}$$

It follows from (4-40) through (4-44) that an MDP with the expected-present-value criterion and a Markov policy is completely characterized by the sequence $(\mathbf{w}_1, \delta_1, \mathbf{w}_2, \delta_2, \mathbf{w}_3, \delta_3, \ldots)$.

Definition 4-12 A *posterity* is a sequence $(\mathbf{w}_1, \delta_1, \mathbf{w}_2, \delta_2, \ldots)$ such that $\mathbf{w}_{n+1} = M(\mathbf{w}_n, \delta_n)$ for all $n \in I_+$. The set $\mathbf{\Phi}_{\mathbf{w}}$ of all posterities with the initial distribution $\mathbf{w}$ is

$$\mathbf{\Phi}_{\mathbf{w}} = \{(\mathbf{w}_1, \delta_1, \mathbf{w}_2, \delta_2, \ldots) : \mathbf{w}_{n+1} = M(\mathbf{w}_n, \delta_n) \text{ for all } n, \text{ and } \mathbf{w}_1 = \mathbf{w}\}$$

The expected-present-value criterion is

$$E\left[\sum_{n=1}^{\infty} \beta^{n-1} r(s_n, a_n) \mid s_1 = s\right] = \sum_{n=1}^{\infty} \beta^{n-1} \sum_{j \in \mathcal{S}} w_n(j) E[r(j, a_n) \mid s_1 = s, s_n = j]$$

We consider only deterministic policies; so it is sufficient in $E[r(j, a_n) \mid s_1 = s, s_n = j]$ to consider a_n as a function of at most s and j. A Markov policy has this property because $a_n = \delta_n(s_n) = \delta_n(j)$ is a function only of j. Therefore, for fixed initial state $s \in \mathcal{S}$ it is sufficient to compare alternative posterities in $\mathbf{\Phi}_{\mathbf{w}}$, where

$w(s) = 1$ and $w(i) = 0$ for all $i \neq s$. However, each posterity $(\mathbf{w}, \delta_1, \mathbf{w}_2, \delta_2, \ldots) \in \Phi_\mathbf{w}$ corresponds to a unique policy $(\delta_1, \delta_2, \ldots) \triangleq \pi \in Y$. Conversely, each policy $(\delta_1, \delta_2, \ldots) \triangleq \pi \in Y$ and $\mathbf{w} \in W$ corresponds to a unique posterity $(\mathbf{w}, \delta_1, \mathbf{w}_2, \delta_2, \ldots) \triangleq \tau_\mathbf{w}(\pi)$ such that $\mathbf{w}_{n+1} = M(\mathbf{w}_n, \delta_n)$ for all n; i.e., policy π generates posterity $\tau_\mathbf{w}(\pi)$. Therefore, different posterities in $\Phi_\mathbf{w}$ are equivalent to different policies in Y.

Deterministic MDPs

From (4-42), (4-43), and (4-44), an initial distribution $\mathbf{w}_1 = \mathbf{w}$ and Markov policy π induce a unique posterity $\tau_\mathbf{w}(\pi)$. For the remainder of this section we view the optimization of an MDP as the problem of comparing alternative posterities. This point of view yields a general foundation for the optimization of discounted MDPs. We build on that foundation in this section to prove the existence of an optimal stationary policy. The foundation is used in the next section to develop an algorithm which identifies an optimal policy, in Section 6-3, which discusses MDPs with multiple criteria, and in Section 9-7, which concerns sequential games. A unified approach to these topics requires notions of preference orderings.

Expressions (4-41) and (4-44) are

$$v_\mathbf{w}(\pi) = \sum_{n=1}^{\infty} \beta^{n-1} \rho(\mathbf{w}_n, \delta_n) \qquad \text{where } \mathbf{w}_{n+1} = M(\mathbf{w}_n, \delta_n)\ n \in I_+ \tag{4-45}$$

and $\mathbf{w}_1 = \mathbf{w}$. We can interpret (4-45) as a deterministic MDP with w_n and δ_n as the respective state and action in period n, state space W, feasible action set Δ for every state $\mathbf{w} \in W$, single-stage reward function $\rho(\cdot, \cdot)$, and dynamics given by $M(\cdot, \cdot)$. If $\tau = \tau_\mathbf{w}(\pi) = (\mathbf{w}_1, \delta_1, \mathbf{w}_2, \delta_2, \ldots)$ is the posterity generated by the policy $\pi = (\delta_1, \delta_2, \ldots)$ from the initial state $\mathbf{w}_1 = \mathbf{w}$, we often write $v(\tau)$ instead of $v_\mathbf{w}(\pi)$ for the expected present value

$$v(\tau) = \sum_{n=1}^{\infty} \beta^{n-1} \rho(\mathbf{w}_n, \delta_n) \tag{4-46}$$

We assume that the sum is well defined and that the objective (a precise specification of the objective is given below) is to maximize $v(\tau)$.

Definition 4-13 Let

$$\Theta_\mathbf{w} = \{(\tau, \tau') : (\tau, \tau') \in \Phi_\mathbf{w} \times \Phi_\mathbf{w} \text{ and } v(\tau) \geq v(\tau')\} \tag{4-47}$$

We say τ is *as good as* τ' if, and only if, $(\tau, \tau') \in \Theta_s$.

Therefore, τ is as good as τ' if, and only if, both posterities have the same initial state and $v(\tau) \geq v(\tau')$.

Recall that $\tau_\mathbf{w}(\pi)$ denotes the posterity generated by policy π from the initial "state" $\mathbf{w}$ and $v_\mathbf{w}(\pi)$ denotes the value of this posterity:

$$v_\mathbf{w}(\pi) \triangleq v[\tau_\mathbf{w}(\pi)]$$

The value function $v(\cdot)$ can be used to compare policies in the obvious way. We *might* say that π' is at least as good as π in "state" $\mathbf{w}$ if $v_{\mathbf{w}}(\pi') \geq v_{\mathbf{w}}(\pi)$. If we were to compare policies in this way (we won't!), the best policy in one "state" might not be best in another "state." However, we use a more ambitious criterion of optimality and seek a policy that is best from every state. Indeed, this more ambitious criterion is necessarily a property of a policy that is optimal in the sense of Definition 4-8. A more fundamental justification of the ambitious criterion is that it is a "free good" in every computationally significant problem.

For the remainder of this section, we use the following definition of optimality.

Definition 4-14′ If π and π' are Markov policies, then π is *as good as* π', written $\pi \gtrsim \pi'$ or $\pi' \lesssim \pi$, if, and only if, $[\tau_{\mathbf{w}}(\pi), \tau_{\mathbf{w}}(\pi')] \in \mathbf{\Theta}_{\mathbf{w}}$, that is,

$$v_{\mathbf{w}}(\pi) \geq v_{\mathbf{w}}(\pi') \qquad \text{for all } \mathbf{w} \in W \tag{4-48}$$

If $\pi^* \in Y$, then π^* is *optimal* if, and only if, $[\tau_{\mathbf{w}}(\pi^*), \tau_{\mathbf{w}}(\pi)] \in \Theta_{\mathbf{w}}$ for all $\mathbf{w}$ and π, that is,

$$v_{\mathbf{w}}(\pi^*) \geq v_{\mathbf{w}}(\pi) \qquad \text{for all } \mathbf{w} \in W \text{ and } \pi \in Y \tag{4-49}$$

In words, an optimal policy π^*, if one exists, must be as good as every other Markov policy in every state. If $\pi \gtrsim \pi'$ and there is at least one state $\mathbf{w} \in W$ with a strict inequality in (4-48), then we write $\pi > \pi'$ or $\pi' < \pi$ and say that π is *better than* π'. Here is a restatement of Definition 4-14′ in the $\gtrsim$ notation.

Definition 4-14 π is *as good as* $\pi' \Leftrightarrow \pi \gtrsim \pi'$.
π is *better than* $\pi' \Leftrightarrow \pi > \pi'$.
π^* is *optimal* $\Leftrightarrow \pi^* \gtrsim \pi$ for all $\pi \in Y$.

Definition 4-14 of optimality may seem weaker than Definition 4-8; however, Exercise 4-15 asks you to prove that they are equivalent for stationary policies. Usually, Definition 4-8 is more convenient in proofs.

Binary Relations

The binary relation $\gtrsim$ on Y has important properties which are exploited below. First, recall some terminology from Section 2-4. Let D be a nonempty set and $L \subset D \times D$. Then (L, D) is *transitive* if $(a, b) \in L$ and $(b, c) \in L$ implies $(a, c) \in L$; (L, D) is *reflexive* if $(b, b) \in L$ for all $b \in D$; and L is a *preorder* and (L, D) is a *preordered set* if L is transitive and reflexive. A mapping f from D to D is *isotone* if $(a, b) \in L$ implies $[f(a), f(b)] \in L$.

The remainder of this section exploits the following properties of $\{(\Theta_{\mathbf{w}}, \Phi_{\mathbf{w}})$: $\mathbf{w} \in W\}$.

Lemma 4-1 Suppose $0 \le \beta < 1$ and $\mathcal{S}$ contains at most countably many states. Then:

$(\boldsymbol{\Theta}_{\mathbf{w}}, \boldsymbol{\Phi}_{\mathbf{w}})$ is a preordered set for each $\mathbf{w} \in W$ (4-50′)

For all $\mathbf{w} \in W$, if $(\tau, \tau') \in \Theta_{\mathbf{w}}$, then $[(\mathbf{y}, \delta, \tau), (\mathbf{y}, \delta, \tau')] \in \Theta_{\mathbf{y}}$ for all $\mathbf{y} \in W$ and $\delta \in \Delta$ with $\mathbf{w} = M(\mathbf{y}, \delta)$ (4-51′)

For any $\mathbf{w} \in W$, let $\tau^j = (\mathbf{w}, \delta_1^j, \mathbf{w}_2^j, \delta_2^j, \ldots) \in \Phi_{\mathbf{w}}$, $j \in I$, with τ^j matching τ^0 up to $\mathbf{w}_j^j$ and δ_j^j (that is, $\mathbf{w}_n^0 = \mathbf{w}_n^j$ and $\delta_n^0 = \delta_n^j$ for all $n \le j$) (4-52′)

(*a*) If $(\tau^{j+1}, \tau^j) \in \Theta_{\mathbf{w}}$ for all $j \in I_+$, then $(\tau^0, \tau^j) \in \Theta_{\mathbf{w}}$ for all $j \in I_+$.
(*b*) If $(\tau^j, \tau^{j+1}) \in \Theta_{\mathbf{w}}$, for all $j \in I_+$, then $(\tau^j, \tau^0) \in \Theta_{\mathbf{w}}$ for all $j \in I_+$.

PROOF Left as Exercise 4-16. □

To interpret (4-51′), suppose two posterities τ and τ' both emanate from initial "state" $\mathbf{w}$ and suppose τ is as good as τ'. Now consider one-period delays in τ and τ' caused by starting at "state" $\mathbf{y}$ and taking "action" δ so that $\mathbf{w} = M(\mathbf{y}, \delta)$ and the second period's "state" is $\mathbf{w}$. Thereafter, pursue either τ or τ'. Then (4-51′) asserts that the one-period delay should not reverse the preference; that is, $(\mathbf{y}, \delta, \tau)$ is at least as desirable as $(\mathbf{y}, \delta, \tau')$. For this reason, (4-51′) is called *consistent choice* or *temporal persistence of preference.*

The consequence of (4-51′) is that preference persists as time passes. If an action should not be taken when $\mathbf{w}_1 = \mathbf{w}$, it should not be taken when $\mathbf{w}_n = \mathbf{w}$ for any $n > 1$. An infinite planning horizon is essential to this argument. If a policy should be postponed at all, it should be postponed forever. This is the essence of Theorem 4-2 below.

Property (4-52′) might be called *continuity*. It can be viewed as a continuity† property in the sense that $\tau^0 \lesssim \tau^j \lesssim \tau^{j+1}$ for all j implies $\tau^0 \lesssim \lim_{j\to\infty} \tau^j$ (and similarly when $\gtrsim$ replaces $\lesssim$). Alternatively and equally loosely, for posterity $\tau = (\mathbf{w}_1, d_1, \mathbf{w}_2, d_2, \ldots)$ let $K_n(\tau) = (\mathbf{w}_n, d_n, \mathbf{w}_{n+1}, d_{n+1}, \ldots)$ for $n \in I_+$ and label an invariant characteristic in the sequence $K_1(\tau)$, $K_2(\tau)$, $K_3(\tau)$, ... a *tail property* of τ. Then (4-52′) asserts that the preference ordering does not depend on tail properties.

Property (4-50′) is a widely invoked assumption of "rationality." It is a reasonable feature of criteria in most operations research models.

Properties (4-51′) and (4-52′) can be restated more compactly, but additional notation is needed.

Let γ be a single-period decision rule and π be a Markov policy. Then $T_\gamma^n \pi \in Y$ denotes the Markov policy which delays the use of π for n periods during which γ is the decision rule. In symbols, we have this definition.

† A formal justification of this claim would require topological assumptions.

Definition 4-15 Let $\pi = (\delta_1, \delta_2, \ldots) \in Y$ and $\gamma \in \Delta$. Then

$$T_\gamma \pi \triangleq (\gamma, \delta_1, \delta_2, \delta_3, \ldots)$$
$$T_\gamma^1 \pi \triangleq T_\gamma \pi, \qquad \text{and}$$
$$T_\gamma^{n+1} \pi \triangleq T_\gamma T_\gamma^n \pi$$

In Definitions 4-14′ and 4-14, the binary relations $\{(\Theta_{\mathbf{w}}, \Phi_{\mathbf{w}}): \mathbf{w} \in W\}$ induce the binary relation $(\gtrsim, Y)$. Therefore, the properties of $\{(\Theta_{\mathbf{w}}, \Phi_{\mathbf{w}}): \mathbf{w} \in W\}$ asserted in Lemma 4-1 induce properties of $(\gtrsim, Y)$.

Lemma 4-2 Suppose $0 \le \beta < 1$ and $\mathcal{S}$ contains at most countably many states. Then:

$(\gtrsim, Y)$ is a preordered set (4-50)

For every $\delta \in \Delta$, T_δ is an isotone function on $(\gtrsim, Y)$ (4-51)

For any $\pi = (\delta_1, \delta_2, \ldots) \in Y$ (4-52)

(*a*) If $T_{\delta_1} T_{\delta_2} \ldots T_{\delta_{K+1}} \pi' \lesssim T_{\delta_1} T_{\delta_2} \ldots T_{\delta_K} \pi'$ for all $K \in I$, then $\pi \lesssim \pi'$.
(*b*) If $T_{\delta_1} T_{\delta_2} \ldots T_{\delta_{K+1}} \pi' \gtrsim T_{\delta_1} T_{\delta_2} \ldots T_{\delta_K} \pi'$ for all $K \in I$, then $\pi \gtrsim \pi'$.

Exercise 4-16 asks you to prove Lemma 4-2. One method of proof exploits order properties of the real numbers, the fact that $v(\tau)$ is a real number for each posterity τ, and the fact that the notation $\pi \gtrsim \pi'$ means that $v_{\mathbf{w}}(\pi) \ge v_{\mathbf{w}}(\pi')$ for all $\mathbf{w} \in W$. There is a less direct proof via Lemma 4-1. It depends neither on the definition of $\Theta_{\mathbf{w}}$ nor on properties of real numbers.

Proposition 4-2 Suppose $0 \le \beta < 1$ and $\mathcal{S}$ contains at most countably many states. For each $\mathbf{w} \in W$, let $\Theta_{\mathbf{w}}$ be an arbitrary subset of $\Phi_{\mathbf{w}} \times \Phi_{\mathbf{w}}$ and define $\pi \gtrsim \pi' \Leftrightarrow [\tau_{\mathbf{w}}(\pi), \tau_{\mathbf{w}}(\pi')] \in \Theta_{\mathbf{w}}$ for all $\mathbf{w} \in W$. Then (4-50′), (4-51′), and (4-52′) imply (4-50), (4-51), and (4-52).

Exercise 4-17 asks you to prove this proposition, which is exploited in Sections 6-3 and 9-7. Also, in the next section, it permits statements of results in terms of either policies or posterities. We sometimes use the label "continuity" for (4-52) because (4-52′) implies (4-52).

Example 4-11: Deterministic average-return MDPs Suppose that the original MDP with canonical elements $\mathcal{S}$, $\{A_s\}$, $r(\cdot, \cdot)$, etc., has finitely many states and is deterministic; i.e., for each i and j in $\mathcal{S}$ and $a \in A_i$ either $p_{ij}^a = 1$ or 0. Then if $w_1(s_1) = 1$ and $w_1(i) = 0$ for all $i \ne s_1$, it follows that a Markov policy causes $w_n(i) = 1$ or 0 for all i and n. Hence, s_n is deterministic and $s_n = i$ where $w_n(i) = 1$. A posterity in this case yields a deterministic sequence $s_1, a_1, s_2, a_2, \ldots$ of states and actions. Let

$$g(\tau) = \limsup_{N \to \infty} \frac{1}{N} \sum_{n=1}^{N} r(s_n, a_n)$$

denote the long-run average reward per period. Then define $\pi \gtrsim \pi' \Leftrightarrow g[\tau_s(\pi)] \geq g[\tau_s(\pi')]$ for all $s \in \mathcal{S}$. Exercise 4-18 asks you to prove that this preference ordering has properties (4-50) and (4-51) but not (4-52). □

Example 4-12: Denardo's counterexample As in Example 4-11, suppose that the original MDP is deterministic. This example demonstrates that a preference ordering can lack countable transitivity (4-52) even though $\mathscr{C}$ is a simple set. Let $\mathcal{S} = \{0\}$ and $A_0 = \{0, 1\}$ so $\mathscr{C} = \{(0, 0), (0, 1)\}$ and $M(0, \cdot) \equiv 0$. For a posterity $\tau = (0, a_1, 0, a_2, \ldots)$ let $J(\tau) = \sup \{n: a_n = 1\}$ denote the last period in which action 1 is taken. Define

$$g(\tau) = \begin{cases} 0 & \text{if } J(\tau) = \infty \text{ or } a_n = 0 \text{ for all } n \\ 1 - 2^{-J(\tau)} & \text{if } J(\tau) < \infty \end{cases}$$

Then define $\pi \gtrsim \pi' \Leftrightarrow g[\tau(\pi)] \geq g[\tau(\pi')]$. This example lacks properties (4-51) and (4-52), but it satisfies (4-50). Verification is left to you. □

Example 4-13: Lexicographical orderings As in Example 4-11 and 4-12, suppose that the original MDP is deterministic. Let $r_1(\cdot, \cdot)$ and $r_2(\cdot, \cdot)$ be real-valued functions on $\mathscr{C}$ and suppose that

$$v^j(\tau) \triangleq \sum_{n=1}^{\infty} \beta^{n-1} r_j(s_n, a_n) \qquad j = 1, 2 \ (0 \leq \beta < 1)$$

is well defined for each τ. We interpret $r_1(\cdot, \cdot)$ and $r_2(\cdot, \cdot)$ as different single-stage reward functions and $v^1(\cdot)$ and $v^2(\cdot)$ as different present-value functions (of the posterity). In general, there are trade-offs between $v^1(\cdot)$ and $v^2(\cdot)$. Then define $\pi \gtrsim \pi' \Leftrightarrow$ for all $s \in \mathcal{S}$ either $v_s^1(\pi) > v_s^1(\pi')$ or $v_s^1(\pi) = v_s^1(\pi')$ and $v_s^2(\pi) \geq v_s^2(\pi')$. Hence, $v^1(\cdot)$ is of primary importance and $v^2(\cdot)$ is used to resolve ties. This preference ordering is *not* induced by a real-valued criterion, but Exercise 4-19 asks you to prove that it does possess properties (4-50), (4-51), and (4-52). This criterion is called a *lexicographical* ordering.

□

Example 4-14: Nonstationary rewards Again, suppose that the original MDP is deterministic. Let $\mathcal{S} = \{0\}$, $A_0 = \{0, 1\}$, and for $\tau = (0, a_1, 0, a_2, \ldots)$ let

$$g(\tau) = -a_1 + \sum_{n=2}^{\infty} \beta^{n-1} a_n \qquad (0 < \beta < 1)$$

Define $\pi \gtrsim \pi' \Leftrightarrow g[\tau(\pi)] \geq g[\tau(\pi')]$. This example satisfies (4-50) and (4-52) but not (4-51): Let $\tau = (0, a_1, 0, a_2, \ldots)$, $\tau' = (0, a_1', 0, a_2', \ldots)$, with $a_1 = 0$, $a_n = 1$ for all $n > 1$, and $a_n' = 1$ for all $n \geq 1$. Then $g(\tau) > g(\tau')$ but $g(0, 1, \tau) < g(0, 1, \tau')$, which violates (4-51). Generally, an MDP with nonstationary single-stage reward functions lacks property (4-51). □

Solutions in Stationary Policies

Definition 4-14 of an optimal policy means that it performs at least as well as any other policy from *every* initial state. An optimum, if it exists, attains a vector maximum (a vector with $\#\mathcal{S}$ coordinates).

Theorem 4-3 below asserts that there is an optimal stationary policy if there is an optimal Markov policy. Theorem 4-2 below is used in the proof and leads in Section 4-4 to an algorithm for finding an optimal policy. Theorem 4-2 states that a policy should be postponed forever if it should be postponed at all.

Theorem 4-2 Let $\pi \in Y$ and $\delta \in \Delta$ and suppose $0 \le \beta < 1$ and $\mathcal{S}$ contains at most countably many states. If $T_\delta \pi \gtrsim \pi$, then $\delta^\infty \gtrsim \pi$.

PROOF Let $\pi = (\gamma_1, \gamma_2, \ldots)$ and for $n \ge 1$, $\pi_n = T_\delta^n \pi$; let $\pi_0 \equiv \pi$. The hypothesis is $\pi_1 \gtrsim \pi_0$, which initiates an inductive proof. For some n, suppose $\pi_n \gtrsim \pi_{n-1}$. Then isotonicity (4-51) implies $T_\delta \pi_n \gtrsim T_\delta \pi_{n-1}$ or $\pi_{n+1} \gtrsim \pi_n$. Hence $\pi_n \gtrsim \pi_{n-1}$ for all n so $\pi_0 \lesssim \pi_1 \lesssim \pi_2 \lesssim \ldots$. Then part (*b*) of (4-52) ("continuity") yields $\delta^\infty \gtrsim \pi$. □

Theorem 4-3 Suppose $0 \le \beta < 1$ and $\mathcal{S}$ contains at most countably many states. There is an optimal stationary policy if (and only if) there is an optimal Markov policy.

PROOF The Markov policies contain the stationary policies, that is, $Z \subset Y$; so there is an optimal policy in Y if there is one in Z. Suppose now there is an optimal policy $\pi^* \in Y$. We shall identify $\delta \in \Delta$ such that $\delta^\infty \gtrsim \pi^*$. But $\pi^* \gtrsim \pi$ for all π so transitivity will imply $\delta^\infty \gtrsim \pi$ for all π; that is, δ^∞ will be optimal.

Let $\pi^* = (\delta, \delta_1, \delta_2, \ldots)$ and $\pi' = (\delta_1, \delta_2, \ldots)$ so $\pi^* = T_\delta \pi'$. Since $\pi^* \gtrsim \pi$ for all π, that is, $T_\delta \pi' \ge \pi$ for all π, then $T_\delta \pi' \gtrsim \pi'$ so $\delta^\infty \gtrsim \pi'$ by Theorem 4-2. Hence, isotonicity (4-51) yields

$$\delta^\infty = T_\delta\, \delta^\infty \gtrsim T_\delta \pi' = \pi^* \qquad \square$$

The next corollary is an immediate consequence of the preceding proof.

Corollary 4-3 If $\pi^* = (\delta, \delta_1, \delta_2, \ldots)$ is an optimal policy, then δ^∞ is an optimal policy.

Theorem 4-3 justifies consideration only of stationary policies when seeking an optimal policy. Several algorithms can be used to conduct the search, and one of them is developed in the next section. However, the catch in applying Theorem 4-3 is to prove that *some* Markov policy is optimal.

Example 4-15: Continuation of Example 4-12 In this deterministic MDP, let $\mathcal{S} = \{0\}$ and $A_0 = \{0, 1\}$. For a posterity $\tau = (0, a_1, 0, a_2, \ldots)$ let $J(\tau) =$

sup $\{n: a_n = 1\}$, and

$$g(\tau) = \begin{cases} 0 & \text{if } J(\tau) = \infty \text{ or } a_n = 0 \text{ for all } n \\ 1 - 2^{-J(\tau)} & \text{if } J(\tau) < \infty \end{cases}$$

If an optimal policy π_0 were to exist, let τ_0 denote the posterity it would induce so $g(\tau_0) \geq g(\tau)$ for all τ. Then $J(\tau_0) < \infty$; but there is a posterity τ with $J(\tau) = J(\tau_0) + 1$ so $g(\tau) = g(\tau_0) + 2^{-J(\tau_0)-1}$. Let π be the policy that induces τ so $\tau = \tau(\pi)$. Then $\pi \gtrsim \pi_0$ but $\pi_0 \not\gtrsim \pi$ so π_0 is not optimal. In words, there is no optimal policy. □

Example 4-16 Here is a trivial deterministic MDP that lacks an optimal stationary policy for a different reason than in Example 4-15. Let $\beta = 0.5$, $\mathcal{S} = \{1\}$, $A_1 = I$, $M(1, \cdot) \equiv 1$, and $r(1, a) = a$ for all $a \in I$. Notice that $|r(1, \cdot)|$ cannot be bounded. Let a^∞ denote the stationary policy of always taking action a, $a \in I$. Then

$$v_1(a^\infty) = 2a < 2(a + 1) = v_1[(a + 1)^\infty] \qquad a \in I$$

so every stationary policy is strictly dominated by another one. Then Theorem 4-3 implies that there is no optimal policy (of any kind). □

EXERCISES

4-13 We say that two MDPs are *equivalent* if their sets of optimal policies are the same. Suppose an MDP has $|r(s, a)| \leq u' < \infty$ for all $(s, a) \in \mathscr{C}$. Show that this MDP is equivalent to another MDP whose single-period reward function satisfies $0 \leq r(s, a) \leq u$ for all $(s, a) \in \mathscr{C}$, for some $u < \infty$. What is the connection between the functions $r(\cdot, \cdot)$ in the two MDPs? What is the connection between the functions $f_n(\cdot)$ in the two MDPs?

4-14 Prove that $0 \leq r(s, a) \leq u < \infty$, $(s, a) \in \mathscr{C}$, and $f_0(\cdot) \equiv 0$ imply $f_1(s) \leq f_2(s) \leq \ldots \leq u/(1 - \beta)$ for all $s \in \mathcal{S}$.

4-15 For a deterministic MDP, prove that Definitions 4-8 and 4-14 are equivalent among stationary policies.

4-16 Prove Lemmas 4-1 and 4-2 (do not invoke Lemma 4-1 in your proof of Lemma 4-2).

4-17 Prove Proposition 4-2.

4-18 Show, with examples, that Example 4-11 lacks property (4-52) and Example 4-12 lacks properties (4-51) and (4-52). To appreciate why *lim sup* is used in Example 4-11, construct a deterministic MDP where property (4-50) is not valid when *lim sup* is replaced by *lim.*

4-19 Prove that the preference ordering in Example 4-13 has properties (4-50), (4-51), and (4-52).

4-20 Properties (4-50), (4-51), and (4-52) are independent. Example 4-11 has properties (4-50) and (4-51) but not (4-52). Example 4-14 has (4-50) and (4-52) but not (4-51). Construct an example that has (4-51) and (4-52) and reflexivity but not transitivity [hence not (4-50)]. (Hint: make the example simple. It is possible to construct an example with $\#\mathscr{C} = 6$.)

4-21 Where are the assumptions that $0 \leq \beta < 1$ and $\mathcal{S}$ is at most countably infinite used in the proofs of Theorems 4-2 and 4-3?

4-22 A plausible alternative to (4-52) is:

For any $\pi = (\delta_1, \delta_2, \ldots) \in Y$ (4-53)

(*a*) If $\pi' \gtrsim T_{\delta_1} T_{\delta_2} \ldots T_{\delta_K} \pi'$ for all $K < \infty$, then $\pi' \gtrsim \pi$.
(*b*) If $\pi' \lesssim T_{\delta_1} T_{\delta_2} \ldots T_{\delta_K} \pi'$ for all $K < \infty$, then $\pi' \lesssim \pi$.

Notice that (4-51) with (4-53) appears to be weaker than (4-51) with (4-52) because transitivity plus the hypotheses of (4-52*a*) and (4-52*b*) imply the hypotheses of (4-53*a*) and (4-53*b*). It *is* weaker. C. Blair (1981) observes† that (4-53) excludes lexicographical orderings as in Example 4-13. [Exercise 4-19 asked you to verify that such orderings do satisfy (4-52)]. Construct a counterexample for (4-53) based on Example 4-13. Use $\mathcal{S} = \{0, 1\}$, $A_0 = A_1 = \{0, 1\}$, $\beta = 1/2$, $M(s, a) = a$ for all $(s, a) \in \mathscr{C}$, $r_1(0, 0) = r_2(0, 1) = 1$, $r_1(1, 1) = 2$, and $r_1(0, 1) = r_1(1, 0) = r_2(0, 0) = r_2(1, 1) = r_2(1, 0) = 0$.

4-4 POLICY IMPROVEMENT

Many algorithms consist of (*a*) an initial trial solution followed by (*b*) a series of *iterations* which terminates at an acceptable solution. If the purpose of the algorithm is to optimize some criterion, "acceptable" means "optimal" or at least approximately optimal (or a signal that an optimal solution does not exist). Step *a* for some problems and algorithms is as much a chore as *b*.

An iteration in many algorithms consists of the following steps. First, test the current trial solution for acceptability. If it is acceptable, stop. Otherwise, convert the current trial solution to a new and improved trial solution.

The *policy-improvement* algorithm for MDPs consists of an optimality test (Theorem 4-4 below), and if the test result is negative, it provides an improvement procedure (Theorem 4-5 below). Stationary policies are the objects being tested and improved.

We continue to use the notation and approach in Section 4-3. For any Markov policy π and single-stage decision rule γ, recall that $T_\gamma \pi$ denotes the Markov policy where γ is used for the first period and the one-period deferral of π is used thereafter. Theorem 4-2 asserts that a policy worth postponing for one period is worth postponing forever. The next theorem says that a policy is optimal if it is at least as good as any one-period postponement.

Theorem 4-4 Suppose $0 \le \beta < 1$ and $\mathcal{S}$ contains at most countably many states. If π^* is a Markov policy which satisfies

$$\pi^* \gtrsim T_\delta \pi^* \qquad \text{for all } \delta \in \Delta \tag{4-54}$$

then π^* is optimal.

PROOF Let $\pi = (\delta_1, \delta_2, \ldots)$ be any Markov policy and let $k \in I_+$. Then (4-54) implies

$$\pi^* \gtrsim T_{\delta_k} \pi^*$$

so isotonicity (4-51) and transitivity (4-50) yield

$$\pi^* \gtrsim T_{\delta_k} \pi^* \gtrsim T_{\delta_k} T_{\delta_k} \pi^* \tag{4-55}$$

Now use $T_{\delta_{k-1}}$ on (4-55) and $\pi^* \gtrsim T_{\delta_{k-1}} \pi^*$ from (4-54):

$$\pi^* \gtrsim T_{\delta_{k-1}} \pi^* \gtrsim T_{\delta_{k-1}} T_{\delta_k} \pi^*$$

† Blair, C. E.: "Axioms and Examples Related to Ordinal Dynamic Programming," *Math. of Oper. Res.*, to appear.

Then use $T_{\delta_{k-2}}, T_{\delta_{k-3}}, \ldots, T_{\delta_2}$, and T_{δ_1} to obtain

$$\pi^* \gtrsim T_{\delta_1}\pi^* \gtrsim T_{\delta_1}T_{\delta_2}\pi^* \gtrsim T_{\delta_1}T_{\delta_2}T_{\delta_3}\pi^* \gtrsim \cdots \gtrsim T_{\delta_1}T_{\delta_2}\ldots T_{\delta_{k-1}}T_{\delta_k}\pi^*$$

so $\pi^* \gtrsim T_{\delta_1}T_{\delta_2}\ldots T_{\delta_k}\pi^*$ for every $k \in I_+$. Hence, continuity (4-52) implies $\pi^* \gtrsim \pi$; so π^* is optimal. □

Theorem 4-4 implies that a sufficient condition for a stationary policy γ^∞ to be optimal is $\gamma^\infty \gtrsim T_\delta \gamma^\infty$ for all $\delta \in \Delta$. Theorem 4-3 asserts that only stationary policies need to be tested, but these results do *not* imply that γ^∞ cannot be optimal if it fails the test. This conclusion will follow from Theorem 4-5 below.

The Optimality Test in Detail

Recall that $v_s(\pi)$ denotes the expected present value if π is used from the initial state s.

Corollary 4-4a A stationary policy γ^∞ is optimal for an MDP with at most countably many states and $0 \le \beta < 1$ if

$$v_s(\gamma^\infty) \ge r(s, a) + \beta \sum_{j \in \mathcal{S}} p_{sj}^a v_j(\gamma^\infty) \qquad \text{for all } (s, a) \in \mathscr{C} \tag{4-56}$$

PROOF First we show that $\pi \gtrsim \pi'$ if, and only if, $v_s(\pi) \ge v_s(\pi')$ for all $s \in \mathcal{S}$. By the definition of $\gtrsim$,

$$\pi \gtrsim \pi' \Leftrightarrow v_{\mathbf{w}}(\pi) \ge v_{\mathbf{w}}(\pi') \qquad \text{for all } \mathbf{w} \in W$$

where W consists of the probability distributions on $\mathcal{S}$. Therefore, $\pi \gtrsim \pi'$ if, and only if,

$$0 \le \sum_{s \in \mathcal{S}} w(s)[v_s(\pi) - v_s(\pi')] \tag{4-57}$$

for all $\mathbf{w}$ such that $\sum_i w(i) = 1$ and $w(i) \ge 0$ for all i. Taking $w(s) = 1$ and $w(i) = 0$ if $i \ne s$ implies $v_s(\pi) \ge v_s(\pi')$ so $\pi \gtrsim \pi'$ implies $v_s(\pi) \ge v_s(\pi')$ for all s. Suppose $v_s(\pi) \ge v_s(\pi')$ for all s. Then $w(s)[v_s(\pi) - v_s(\pi')] \ge 0$ if $w(s) \ge 0$. Hence (4-57) is valid for all $\mathbf{w} \in W$.

From Theorem 4-4, γ^∞ is optimal if $\gamma^\infty \gtrsim T_\delta \gamma^\infty$ for all $\delta \in \Delta$. The latter is valid if, and only if,

$$v_s(\gamma^\infty) \ge v_s(T_\delta \gamma^\infty) \qquad \text{for all } \delta \in \Delta \text{ and } s \in \mathcal{S}$$

Now we obtain an expression for $v_s(T_\delta \gamma^\infty)$.

Let $\pi = (\delta_1, \delta_2, \ldots)$. By the definition of $v_s(\pi)$,

$$v_s(T_\delta \pi) = E\left\{r[s, \delta(s)] + \sum_{n=2}^{\infty} \beta^{n-1} r[s_n, \delta_{n-1}(s_n)] \mid s_1 = s\right\}$$

$$= r[s, \delta(s)] + \sum_{j \in \mathcal{S}} p_{sj}^{\delta(s)} E$$

$$\cdot \left\{ \sum_{n=2}^{\infty} \beta^{n-1} r[s_n, \delta_{n-1}(s_n)] \mid s_2 = j \right\}$$

$$= r[s, \delta(s)] + \beta \sum_{j \in \mathcal{S}} p_{sj}^{\delta(s)} E \left\{ \sum_{n=1}^{\infty} \beta^{n-1} r[s_{n+1}, \delta_n(s_{n+1})] \mid s_2 = j \right\}$$

$$= r[s, \delta(s)] + \beta \sum_{j \in \mathcal{S}} p_{sj}^{\delta(s)} v_j(\pi)$$

Letting $\pi = \gamma^\infty$ yields

$$v_s(T_\delta \gamma^\infty) = r[s, \delta(s)] + \beta \sum_{j \in \mathcal{S}} p_{sj}^{\delta(s)} v_j(\gamma^\infty) \tag{4-58}$$

Therefore, γ^∞ is optimal if, and only if,

$$v_s(\gamma^\infty) \geq r[s, \delta(s)] + \beta \sum_{j \in \mathcal{S}} p_{sj}^{\delta(s)} v_j(\gamma^\infty) \qquad \text{for all } \delta \in \Delta \text{ and } s \in \mathcal{S} \tag{4-59}$$

In (4-59) fix $s \in \mathcal{S}$ and let δ range over Δ. Then $\delta(s)$ ranges over A_s so (4-59) is equivalent to

$$v_s(\gamma^\infty) \geq r(s, a) + \beta \sum_{j \in \mathcal{S}} p_{sj}^{a} v_j(\gamma^\infty) \qquad \text{for all } a \in A_s \text{ and } s \in \mathcal{S}$$

This collection of inequalities is (4-56). □

For any $\gamma \in \Delta$ and $s \in \mathcal{S}$, let

$$G(s, \gamma) = \Big\{ a\colon a \in A_s \text{ and} \tag{4-60a}$$

$$r(s, a) + \beta \sum_{j \in \mathcal{S}} p_{sj}^{a} v_j(\gamma^\infty) > v_s(\gamma^\infty) \Big\} \tag{4-60b}$$

$G(s, \gamma)$ is the set of actions whose use in state s to delay γ^∞ for one period is strictly preferred to immediate use of γ^∞. Note that $a = \gamma(s)$ fails to satisfy (4-60b) [use (4-57) and $\gamma^\infty = T_\gamma \gamma^\infty$]; so† $G(s, \gamma) = \phi$ is possible. Therefore, Theorem 4-4 has the following corollary.

Corollary 4-4*b* Suppose an MDP has at most countably many states and $0 \leq \beta < 1$.

$$\bigcup_{s \in \mathcal{S}} G(s, \gamma) = \phi \quad \Rightarrow \gamma^\infty \text{ is optimal} \tag{4-61}$$

This equivalence is useful in the proof of Theorem 4-5.

Examples

Example 4-17: Continuation of Examples 4-9 and 4-10 In Example 4-10, $\mathscr{C} = \{(1, 1), (1, 2), (1, 3), (2, 1), (2, 2), (3, 1)\}$ so Δ, the set of single-stage rules, contains six elements. Hence, there are six stationary policies. Let $\gamma \in \Delta$

† We use ϕ to denote the empty set.

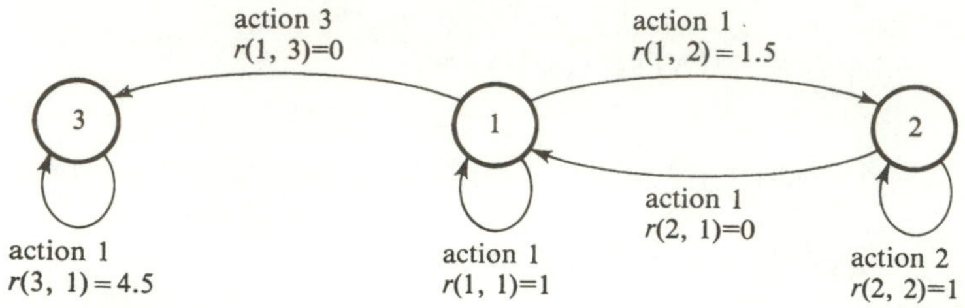

Figure 4-4 A deterministic MDP.

denote the rule $\gamma(1) = 3$, $\gamma(2) = 2$, $\gamma(3) = 1$. In Table 4-11 on page 126, we see that γ is the best rule to use if there are two, three, or four periods remaining but not if there are five or six. We might guess that γ is not best if there are five or more periods left, and that γ is not best for an infinite horizon. We shall use (4-56) to test the optimality of γ^∞.

The basic data are $\beta = 0.5$ and Figure 4-4. To appreciate the effects of γ, delete arcs in Figure 4-4 that correspond to actions that γ does *not* take as shown in Figure 4-5.

In this deterministic MDP, let $m(s, a)$ denote the j which satisfies $p_{sj}^a = 1$. Thus, $m(1, 1) = m(2, 1) = 1$, $m(2, 2) = m(1, 2) = 2$, and $m(1, 3) = m(3, 1) = 3$. From (4-58) and $\gamma^\infty = T_\gamma \gamma^\infty$,

$$v_s(\gamma^\infty) = r[s, \gamma(w)] + \beta \sum_{j \in S} p_{sj}^{\gamma(s)} v_j(\gamma^\infty) \tag{4-62}$$

so

$$v_s(\gamma^\infty) = r[s, \gamma(s)] + \beta v_{m[s, \gamma(s)]}(\gamma^\infty) \tag{4-63}$$

Therefore,

$$v_1(\gamma^\infty) = 0 + 0.5\, v_3(\gamma^\infty)$$

$$v_2(\gamma^\infty) = 1 + 0.5\, v_2(\gamma^\infty)$$

$$v_3(\gamma^\infty) = 4.5 + 0.5\, v_3(\gamma^\infty)$$

so $v_1(\gamma^\infty) = 4.5$, $v_2(\gamma^\infty) = 2$, and $v_3(\gamma^\infty) = 9$. Alternatively, using Figure 4-5, geometric series could have been summed directly to obtain the same results:

$$v_1(\gamma^\infty) = 0 + 4.5(0.5 + 0.5^2 + \cdots) = 4.5$$

$$v_2(\gamma^\infty) = 1 + 1(0.5 + 0.5^2 + \cdots) = 2.0$$

$$v_3(\gamma^\infty) = 4.5 + 4.5(0.5 + 0.5^2 + \cdots) = 9.0$$

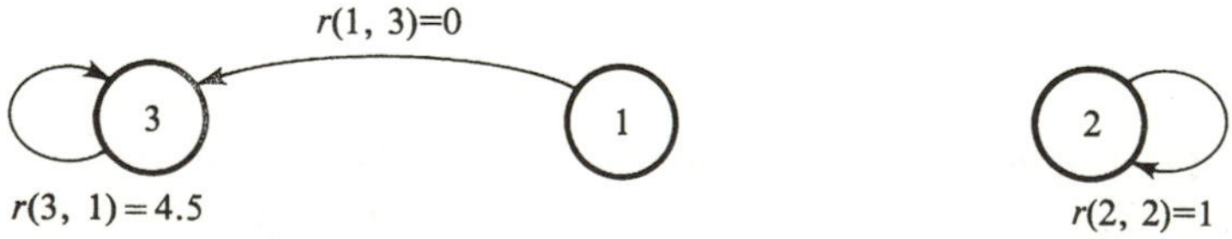

Figure 4-5 Network induced by γ.

These values of $v_s(\gamma^\infty)$ are used in (4-56). First, we compare $v_1(\gamma^\infty) = 4.5$ with $r(1, a) + \beta v_{m(1,a)}(\gamma^\infty)$, for $a = 1$ and $a = 2$. Note that $a = 3$ is the action which γ^∞ stipulates in state $s = 1$ so

$$v_1(\gamma^\infty) = r(1, 3) + \beta v_{m(1,3)}(\gamma^\infty)$$

In words, γ^∞ never fails the optimality test (4-56) due to $a \in A_s$ such that $\gamma(s) = a$. Here are the test results:

$$a = 1\colon r(1, 1) + \beta v_{m(1,1)}(\gamma^\infty) = 1 + 0.5v_1(\gamma^\infty) = 3.25$$

$$a = 2\colon r(1, 2) + \beta v_{m(1,2)}(\gamma^\infty) = 1.5 + 0.5v_2(\gamma^\infty) = 2.5$$

Therefore, γ^∞ satisfies (4-56) when $s = 1$. For $s = 2$, compare $v_2(\gamma^\infty) = 2$ with $r(2, a) + \beta v_{m(2,a)}(\gamma^\infty)$ for $a = 1$ (why is it unnecessary to test with $a = 2$?):

$$a = 1\colon r(2, 1) + \beta v_{m(2,1)}(\gamma^\infty) = 0 + 0.5v_1(\gamma^\infty) = 2.25$$

Therefore, γ^∞ fails to pass the optimality test (4-56) at $s = 2$ and $a = 1$. □

Example 4-18: Continuation of machine-maintenance Examples 4-2 and 4-7 There are three states: 0, 1, and 2. Action 0 is available in state 0, action 1 is available in state 2, and actions 0 and 1 are available in state 1. The data are the same as are listed below (4-25) in Example 4-7:

$$r(0, 0) = 10 \qquad r(1, 0) = 5 \qquad r(1, 1) = 0,\ r(2, 1) = 0$$

$$p_{00}^0 = 0.7 \qquad p_{01}^0 = 0.2 \qquad p_{02}^0 = 0.1$$

$$p_{10}^0 = 0 \qquad p_{11}^0 = 0.4 \qquad p_{12}^0 = 0.6$$

$$p_{10}^1 = 0 \qquad p_{11}^1 = 0 \qquad p_{12}^1 = 1$$

$$p_{20}^1 = 0.4 \qquad p_{21}^1 = 0.1 \qquad p_{22}^1 = 0.5$$

Also, let $\beta = 0.5$ and $\mathcal{S} = \{0, 1, 2\}$ where $s = 0$, $s = 1$, and $s = 2$ denote, respectively, perfect operation, a minor defect, and obligatory repair. Let $a = 0$ indicate continuing production and $a = 1$ indicate that the machine is in the repair process. Then $A_0 = \{0\}$, $A_2 = \{1\}$, and $A_1 = \{0, 1\}$. The only real choice arises with a minor defect. There are two stationary policies. Let b denote the rule $b(0) = 0$, $b(1) = 0$, and $b(2) = 1$. The other single-stage rule, of course, differs only in state 1 where action 1 is taken instead of action 0.

The first step in the optimality test (4-56) is to use (4-62) with $\gamma^\infty = b^\infty$ to compute $v_s = v_s(b^\infty)$ for $s = 0$, 1, and 2. From (4-62) and the data above,

$$v_0 = 10 + 0.5(0.7v_0 + 0.2v_1 + 0.1v_2)$$

$$v_1 = 5 + 0.5(0v_0 + 0.4v_1 + 0.6v_2)$$

$$v_2 = 0 + 0.5(0.4v_0 + 0.1v_1 + 0.5v_2)$$

The solution is

$$v_0 = 17.031 \qquad v_1 = 8.157 \qquad \text{and} \qquad v_2 = 5.085$$

Is b^∞ optimal? From (4-56) the test is $v_1(b^\infty) = v_1 = 8.157$ versus

$$r(1, 1) + \beta \sum_j p_{1j}^1 v_j \tag{4-64}$$

[why not also test v_1 versus $r(1, 0) + \beta \sum_j p_{1j}^0 v_j$?]. The numerical value of (4-64) is

$$0 + 0.5(0v_0 + 0v_1 + 1v_2) = 2.542 < v_1$$

so b^∞ passes the test and is indeed optimal (see Exercise 4-23). □

Policy Improvement

Lemma 4-2 and Theorem 4-4 are combined in Theorem 4-5 below to show that any stationary policy that fails the optimality test must be dominated by another stationary policy. Thus it is necessary as well as sufficient that an optimal policy pass the test. The proof is constructive; it identifies a better stationary policy.

Theorem 4-5: Policy improvement Suppose an MDP has $0 \leq \beta < 1$ and at most countably many states. For every $\gamma \in \Delta$, either γ^∞ is optimal or there is $\delta \in \Delta$ with

$$\delta^\infty > \gamma^\infty \tag{4-65}$$

PROOF Recall the definition (4-60) of the sets $G(s, \gamma)$ for $s \in \mathcal{S}$. If

$$\bigcup_{s \in \mathcal{S}} G(s, \gamma) = \phi \tag{4-66}$$

then Corollary 4-4*b* asserts that γ^∞ is optimal.

If (4-66) is false so

$$\bigcup_{s \in \mathcal{S}} G(s, \gamma) \neq \phi \tag{4-67}$$

then specify $\delta \in \Delta$ in the following way:

$$\delta(s) = \begin{cases} \gamma(s) & \text{if} \quad G(s, \gamma) = \phi \\ \text{any } a \in G(s, \gamma) & \text{if} \quad G(s, \gamma) \neq \phi \end{cases} \tag{4-68}$$

From (4-67), (4-68), and $\gamma(s) \notin G(s, \gamma)$,

$$\delta \neq \gamma \tag{4-69}$$

From (4-60*b*), $a \in G(s, \gamma)$ implies

$$r(s, a) + \beta \sum_{j \in \mathcal{S}} p_{sj}^a v_j(\gamma^\infty) > v_s(\gamma^\infty)$$

so

$$v_s(T_\delta \gamma^\infty) \begin{Bmatrix} = \\ > \end{Bmatrix} v_s(\gamma^\infty) \qquad \text{if } G(s, \gamma^\infty) \begin{Bmatrix} = \\ \neq \end{Bmatrix} \phi \tag{4-70}$$

Therefore, $T_\delta \gamma^\infty > \gamma^\infty$ so isotonicity of T_δ [(4-51) in Lemma 4-2] implies

$T_\delta^2 \gamma^\infty \gtrsim T_\delta \gamma^\infty$; hence

$$T_\delta^2 \gamma^\infty \gtrsim T_\delta \gamma^\infty > \gamma^\infty$$

Theorem 4-2 asserts $T_\delta \pi \gtrsim \pi$ implies $\delta^\infty \gtrsim \pi$. Let $\pi = T_\delta \gamma^\infty$ so the hypothesis of the theorem is $T_\delta^2 \gamma^\infty \gtrsim T_\delta \pi$, which is satisfied. Therefore, $\delta^\infty \gtrsim T_\delta \gamma^\infty > \gamma^\infty$ so $\delta^\infty > \gamma^\infty$. □

An algorithm is implicit in the preceding proof. Start with any stationary policy, say δ_1^∞, test it for optimality, and compute† $\mathbf{v}(\delta_1^\infty) = (v_s(\delta_1^\infty),\ s \in \mathcal{S})$. Use $\mathbf{v}(\delta_1^\infty)$ in (4-66) to check whether (4-66) is satisfied; i.e., $G(s, \delta_1) = \phi$ for all s. If so, stop because δ_1^∞ is optimal. If not, use $\{G(s, \delta_1): s \in \mathcal{S}\}$ to construct a better policy, say δ_2^∞. As in (4-68), if $G(s, \delta_1) = \phi$, let $\delta_2(s) = \delta_1(s)$. If $G(s, \delta_1) \neq \phi$, let $\delta_2(s)$ be any element of $G(s, \delta_1)$, i.e., any action a whose one-period delay of δ_1^∞ from state s is better than using δ_1^∞ in state s. As (4-65) asserts, the new policy δ_2^∞ is better than δ_1^∞. Next, relabel δ_2^∞ as δ_1^∞ and begin another iteration by testing δ_1^∞ (previously labeled δ_2^∞) for optimality.

Notice that the algorithm does not require that you know all the elements in $G(s, \delta_1)$ if $G(s, \delta_1)$ is nonempty. In that case, it is sufficient to know just a single element $a \in G(s, \delta_1)$. Here is the algorithm in schematic form.

Discounted-Return Policy-Improvement Algorithm

(*a*) Let $i = 1$ and select $\delta_1 \in \Delta$.

(*b*) Compute† $\mathbf{v}^i \triangleq \mathbf{v}(\delta_i^\infty)$. Let v_s^i denote the sth component of $\mathbf{v}^i$.

(*c*) For each $s \in \mathcal{S}$ compute

$$\psi_s^i \triangleq \max\left\{r(s, a) + \beta \sum_{j \in \mathcal{S}} p_{sj}^a v_j^i:\quad a \in A_s - \{\delta_i(s)\}\right\} - v_s^i$$

If $\psi_s^i > 0$, then $G(s, \delta_i) \neq \phi$; so let $\delta_{i+1}(s)$ be any $a \in A_s$ that causes

$$r(s, a) + \beta \sum_{j \in \mathcal{S}} p_{sj}^a v_j^i - v_s^i > 0$$

If $\psi_s^i \leq 0$, then $G(s, \delta_i) = \phi$; so let $\delta_{i+1}(s) = \delta_i(s)$.

(*d*) If $\delta_{i+1} = \delta_i$, stop, and δ_{i+1}^∞ is optimal. Otherwise replace i with $i + 1$ and return to (*b*).

The special case of finite MDPs, i.e., MDPs with finitely many states and finitely many actions in each state, is very important. If $\#\mathcal{S} < \infty$ and $\#A_s < \infty$ for each $s \in \mathcal{S}$, then $\#\mathscr{C} < \infty$. This is the case toward which algorithms are ordinarily directed, and it enjoys the following major property.

† Sections 4-5 and 4-6 discuss how the computation might actually be accomplished. We shall see in Theorem 4-7 that $\mathbf{v}(\delta^\infty) = (I - \beta P_\delta)^{-1}\mathbf{r}$, where $p_{sj}^{\delta(s)}$ is the (s, j)th element of P_δ and $r[s, \delta(s)]$ is the sth element of $\mathbf{r}_\delta$.

Theorem 4-6 In an MDP with $0 \leq \beta < 1$, if $\#\mathscr{C} < \infty$, there is an optimal stationary policy.

PROOF Label any element of Δ as δ_1. From Theorem 4-5, either δ_1^∞ is optimal or there is $\delta_2 \in \Delta$ such that $\delta_2^\infty > \delta_1^\infty$. The construction (4-68) generates a sequence $\delta_1, \delta_2, \ldots$ of better and better policies, that is, $\delta_n^\infty < \delta_{n+1}^\infty$ with $\delta_n \neq \delta_{n+1}$ [from (4-69)] for each n (until the sequence terminates if it does). The hypothesis

$$\infty > \#\mathscr{C} = \sum_{s \in \mathcal{S}} \#A_s$$

implies

$$\#\Delta = \prod_{s \in \mathcal{S}} \#A_s < \infty$$

so the sequence $\delta_1, \delta_2, \ldots$ must terminate in at most $\#\Delta$ iterations of (4-68). Let $\delta_* \in \Delta$ denote the terminal element of the sequence. Theorem 4-5 asserts δ_*^∞ is optimal. □

The hypothesis of Theorem 4-6, $\#\mathscr{C} < \infty$, is merely sufficient. In Section 5-4 we shall encounter processes with infinitely many stationary policies but which nevertheless possess an optimal stationary policy.

Example 4-19: Continuation of Example 4-17 The basic data are $\beta = 0.5$ and Figure 4-6. Let $\gamma \in \Delta$ denote the rule $\gamma(1) = 3$, $\gamma(2) = 2$, and $\gamma(3) = 1$. In Example 4-17 we found that γ^∞ failed the optimality test and $v_1(\gamma^\infty) = 4.5$, $v_2(\gamma^\infty) = 2$, and $v_3(\gamma^\infty) = 9$. We use these results in (4-60) to test whether $2 \in G(1, \gamma)$, $1 \in G(1, \gamma)$, and $1 \in G(2, \gamma)$. Recall the notation $m(s, a)$ for the j which satisfies $p_{sj}^a = 1$.

Test if $2 \in G(1, \gamma)$:

$$r(1, 2) + \beta v_{m(1,2)}(\gamma^\infty) - v_1(\gamma^\infty)$$
$$= 1.5 + 0.5 v_2(\gamma^\infty) - 4.5 = -2 < 0 \Rightarrow 2 \notin G(1, \gamma)$$

Test if $1 \in G(1, \gamma)$:

$$r(1, 1) + \beta v_{m(1,1)}(\gamma^\infty) - v_1(\gamma^\infty)$$
$$= 1 + 0.5 v_1(\gamma^\infty) - 4.5 = -1.25 < 0 \Rightarrow 1 \notin G(1, \gamma)$$

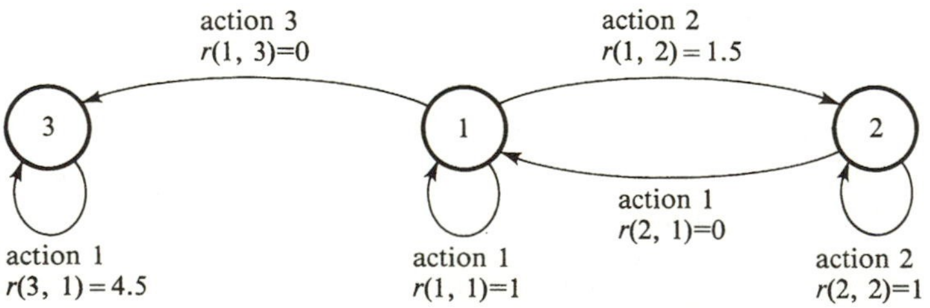

Figure 4-6 A deterministic MDP.

Test if $1 \in G(2, \gamma)$:

$$r(2, 1) + \beta v_{m(2,\,1)}(\gamma^{\infty}) - v_2(\gamma^{\infty}) = 0 + 0.5v_1(\gamma^{\infty}) - 2 = 0.5 > 0 \Rightarrow 1 \in G(2, \gamma)$$

Therefore, $G(1, \gamma) = \phi$ and $G(2, \gamma) = \{1\}$; $G(3, \gamma) = \phi$ is trivially true. The construction (4-68) in the proof of Theorem 4-5 is

$$\delta(1) = \gamma(1) = 3, \quad \delta(2) = 1, \quad \text{and} \quad \delta(3) = \gamma(3) = 1$$

Theorem 4-5 asserts $\delta^{\infty} > \gamma^{\infty}$; that is, for each $s \in \mathcal{S}$, $v_s(\gamma^{\infty}) \geq v_s(\gamma^{\infty})$ and for at least one $s \in \mathcal{S}$ the inequality is strict. We compute $\{v_s(\delta^{\infty})\colon s \in \mathcal{S}\}$ as in Example 4-18.

$$v_1(\delta^{\infty}) = 0 + 0.5v_3(\delta^{\infty})$$

$$v_2(\delta^{\infty}) = 0 + 0.5v_1(\delta^{\infty})$$

$$v_3(\delta^{\infty}) = 4.5 + 0.5v_3(\delta^{\infty})$$

so $v_1(\delta^{\infty}) = 4.5$, $v_2(\delta^{\infty}) = 2.25$, and $v_3(\delta^{\infty}) = 9$. Observe that $v_s(\delta^{\infty}) - v_s(\gamma^{\infty}) \geq 0$ for each s with $v_2(\delta^{\infty}) - v_2(\gamma^{\infty}) > 0$.

Is δ^{∞} optimal or should (4-60) and the construction in (4-68) be used again to improve δ^{∞}? Remember that the optimality test (4-54), with δ in place of γ, is equivalent to $G(s, \delta) = \phi$ for all $s \in \mathcal{S}$; so we test whether or not $1 \in G(1, \delta)$, $2 \in G(1, \delta)$, and $2 \in G(2, \delta)$.

Test if $1 \in G(1, \delta)$:

$$r(1, 1) + \beta v_{m(1,\,1)}(\delta^{\infty}) - v_1(\delta^{\infty})$$
$$= 1 + 0.5v_1(\delta^{\infty}) - 4.5 = -1.25 < 0 \Rightarrow 1 \notin G(1, \delta)$$

Test if $2 \in G(1, \delta)$:

$$r(1, 2) + \beta v_{m(1,\,2)}(\delta^{\infty}) - v_1(\delta^{\infty})$$
$$= 1.5 + 0.5v_2(\delta^{\infty}) - 4.5 = -1.875 < 0 \Rightarrow 2 \notin G(1, \delta)$$

Test if $2 \in G(2, \delta)$:

$$r(2, 2) + \beta v_{m(2,\,2)}(\delta^{\infty}) - v_2(\delta^{\infty})$$
$$= 1 + 0.5v_2(\delta^{\infty}) - 2.25 = -0.125 < 0 \Rightarrow 2 \notin G(2, \delta)$$

Therefore, $G(s, \delta) = \phi$ for all $s \in \mathcal{S}$; so δ^{∞} passes the optimality test. δ^{∞} is an optimal policy. (Cheers heard in the background.) □

Example 4-20: Continuation of Example 4-8 The basic data for this inventory model are $\beta = 0.9$ and Tables 4-12 and 4-13 on the next page. Recall that $\mathcal{S} = \{0, 1, \ldots, 4\}$ and $A_s = \{s, s + 1, \ldots, 4\}$.

We shall begin with the initial policy of ordering $(3 - s)^{+}$ items, which corresponds to $a = \max\{3, s\}$. Therefore, we start with b^{∞}, where $b(0) = b(1) = b(2) = b(3) = 3$ and $b(4) = 4$. For simplicity, let v_s denote $v_s(b^{\infty})$.

Table 4-12. Single-period costs $r(s, a)$ in Example 4-20

	$s \backslash a$	Inventory after ordering: 0	1	2	3	4
Inventory before ordering	0	0	0	0	1	3
	1	—	−1	0	1	3
	2	—	—	−1	1	3
	3	—	—	—	0	3
	4	—	—	—	—	2

Then (4-62) becomes

$$v_0 = 1 + 0.225v_0 + 0.225v_1 + 0.225v_2 + 0.225v_3$$

$$v_1 = 1 + 0.225v_0 + 0.225v_1 + 0.225v_2 + 0.225v_3$$

$$v_2 = 1 + 0.225v_0 + 0.225v_1 + 0.225v_2 + 0.225v_3$$

$$v_3 = 0 + 0.225v_0 + 0.225v_1 + 0.225v_2 + 0.225v_3$$

$$v_4 = 2 + 0.225v_1 + 0.225v_2 + 0.225v_3 + 0.225v_4$$

This causes $v_0 = v_1 = v_2 = v_3 + 1$; so the unique solution is

$$v_0 = v_1 = v_2 = 7.75, \quad v_3 = 6.75, \quad \text{and} \quad v_4 = 9.04$$

The next step is to form $\{G(s, b)\colon s \in \mathcal{S}\}$. For example, for $G(2, b)$ we test $2 \in G(2, b)$ and $4 \in G(2, b)$.

$2 \in G(2, b)$?

$$r(2, 2) + \beta \sum_{j \in \mathcal{S}} p_{2j}^2 v_j - v_2 = -1 + 6.975 - 7.75 = -1.775 < 0 \Rightarrow 2 \in G(2, b)$$

$4 \in G(2, b)$?

$$r(2, 4) + \beta \sum_{j \in \mathcal{S}} p_{2j}^4 v_j - v_2 = 3 + 7.04 - 7.75 = 2.29 \geq 0 \Rightarrow 4 \notin G(2, b)$$

Table 4-13. $\beta p_{sj}^a = \beta P\{(a - D_1)^+ = j\}$ in Example 4-20

	$a \backslash j$	Subsequent inventory level: 0	1	2	3	4
Inventory after ordering	0	0.9	0	0	0	0
	1	0.675	0.225	0	0	0
	2	0.45	0.225	0.225	0	0
	3	0.225	0.225	0.225	0.225	0
	4	0	0.225	0.225	0.225	0.225

The results are

$$G(s, b) = \{2\} \qquad \text{for } s = 0, 1, 2$$

and $G(3, b) = G(4, b) = \phi$. Therefore, b^∞ is not optimal and $a^\infty > b^\infty$, where $a(s) = 2$ for $s = 0, 1, 2$, $a(3) = 3$, and $a(4) = 4$.

The next step is to compute $\{v_s(a^\infty)\}$ in order to construct $\{G(s, a): s \in \mathcal{S}\}$. Now, let v_s denote $v_s(a^\infty)$. Then (4-62) is

$$v_0 = v_1 = 0 + 0.45v_0 + 0.225(v_1 + v_2)$$

$$v_2 = v_0 - 1$$

$$v_3 = 0 + 0.225 \sum_{j=0}^{3} v_j$$

$$v_4 = 2 + 0.225 \sum_{j=1}^{4} v_j$$

Therefore, $v_0 = v_1 = -2.25$, $v_2 = -3.25$, $v_3 = -2.25$, and $v_4 = 0.33$. We find $G(s, a) = \phi$ for $s = 0, 2, 3$, and 4 and $G(1, a) = \{1\}$. For example,

$4 \in G(0, a)$?

$$r(0, 4) + \beta \sum_{j \in \mathcal{S}} p_{0j}^4 v_j - v_0 = 3 - 1.41 + 2.25 = 3.84 \geq 0 \Rightarrow 4 \notin G(0, a)$$

$1 \in G(1, a)$?

$$r(1, 1) + \beta \sum_{j \in \mathcal{S}} p_j^1 v_j - v_1 = -1 - 2.25(0.9) + 2.25 = -7.75 < 0 \Rightarrow 1 \in G(1, a)$$

Therefore, a^∞ is not optimal and $c^\infty > a^\infty$, where $c(1) = 1$, $c(0) = c(2) = 2$, $c(3) = 3$, and $c(4) = 4$.

Now, let v_s denote $v_s(c^\infty)$. Then (4-62) is

$$v_0 = 0.45v_0 + 0.225(v_1 + v_2)$$

$$v_2 = v_0 - 1$$

$$v_1 = -1 + 0.675v_0 + 0.225v_1$$

$$v_3 = 0.225 \sum_{j=0}^{3} v_j$$

$$v_4 = 2 + 0.225 \sum_{j=1}^{4} v_j$$

whose unique solution is

$$v_0 = -4, \quad v_1 = -4.774, \quad v_2 = -5, \quad v_3 = -4, \quad \text{and} \quad v_4 = -1.419$$

Finally, testing for elements in $G(s, c)$ yields $G(s, c) = \phi$ for all s; so c^∞ is optimal.

Exercise 4-27 asks you to analyze the effects on an optimal policy of changes in the cost data. □

Generalizations†

Our principal interest is discounted MDPs; so we use the definition $\pi \gtrsim \pi' \Leftrightarrow v_{\mathbf{w}}(\pi) \geq v_{\mathbf{w}}(\pi')$ for all $\mathbf{w} \in W \Leftrightarrow v_s(\pi) \geq v_s(\pi')$ for all $s \in \mathcal{S}$. Recall that Y denotes the set of all Markov policies. Lemma 4-2 asserts that $(\gtrsim, Y)$ has properties (4-50), (4-51), and (4-52): reflexivity, transitivity, isotonicity, and continuity. The subsequent formal results (Theorems 4-2, 4-3, 4-4, 4-5, and 4-6 and Corollaries 4-3, 4-4*a*, and 4-4*b*) are justified solely‡ through properties (4-50), (4-51), and (4-52) defined in (4-41) and $\pi \gtrsim \pi' \Leftrightarrow v_{\mathbf{w}}(\pi) \geq v_{\mathbf{w}}(\pi')$ for all $\mathbf{w} \in W$ rather than through properties of $v(\cdot)$. Therefore, the formal results are valid for any other preference ordering on Y that has properties (4-50), (4-51), and (4-52). More generally, (4-50), (4-51), and (4-52) can be treated as axioms and the formal results can be augmented with the initial statement "Invoke axioms (4-50), (4-51), and (4-52)."

The power of this generality is suggested in Example 4-13, where we observe that lexicographical preference orderings have properties (4-50), (4-51), and (4-52). Therefore, if $\#\mathscr{C} < \infty$, there is a stationary policy that is lexicographically optimal (due to Theorem 4-6).§ Also, an optimum can be computed via the policy-improvement algorithm (due to Theorem 4-5).

Theorem 4-2, namely, $T_\delta \pi \gtrsim \pi$ implies $\delta^\infty \gtrsim \pi$, is fundamentally important to the proofs of Theorem 4-3 and Theorems 4-4, 4-5, and 4-6. The proof of Theorem 4-2 exploits (4-51), which we have called *temporal persistence of preference*. The force of (4-51) is that preference is time-invariant. However, none of the proofs (of the lemmas or theorems) depend on time invariance of the dynamics. At the outset, we could have permitted a different mapping M_n for each n in (4-44), that is, $w_{n+1} = M_n(w_n, \delta_n)$. For each $n \in I_+$, M_n would have been a mapping from $\{(w, z): z \in Z_n \text{ and } w \in W\}$ to W. Then $\mathbf{\Phi}_w$ would have been defined as

$$\Phi_w = \{(w_1, z_1, w_2, z_2, \ldots): z_n \in Z_{w_n}, w_{n+1} = M_n(w_n, z_n) \text{ for all } n \in I_+, w_1 = w\}$$

Indeed, each set of actions Z_w also could have depended on n, and the proofs would have been unaffected. However, Example 4-14 shows that the preference ordering must remain time-invariant in order to retain (4-45).

There are several generalizations of the policy-improvement algorithm, and we shall describe one of them. For $\delta \in \Delta$ and $\pi \in Y$, recall the notation $T_\delta^1 \pi =$

† Skipping the remainder of this section will not interfere with understanding later unstarred sections.

‡ The proof of Theorem 4-5 seems to depend on (4-60), hence on (4-58), which depends on (4-41). But (4-58) is not implied by (4-50), (4-51), and (4-52). However, the definition (4-60) of $G(s, \gamma)$ can be freed of $v(\cdot)$ as follows. Let

$$\Theta_s = \{(\tau, \tau'): \text{there exist } \pi \text{ and } \pi' \text{ such that } \tau = \tau_s(\pi), \tau' = \tau_s(\pi'), \text{ and } \pi \gtrsim \pi'\}$$

When $\gtrsim$ is defined with $v(\cdot)$, Θ_s assumes the form given after Lemma 4-1. Generally, we have a nonempty set W, nonempty sets Z_w for all $w \in W$, and a mapping M from $\{(w, z): z \in Z_w, w \in W\}$ to W. Define

$$G(w, \gamma) = \{z: z \in Z_w, \{[w, z, \tau_{M(w,z)}(\gamma^\infty)], \tau_w(\gamma^\infty)\} \in \Theta_w, \{\tau_w(\gamma^\infty), [w, z, \tau_{M(w,z)}(\gamma^\infty)]\} \notin \Theta_w\}$$

Then (4-60) is the special case when $\gtrsim$ is induced by $v(\cdot)$ in (4-41).

§ For conditions where C is countable and there is a stationary policy that is lexicographically optimal. M. Henig, (unpublished).

$T_\delta \pi$ and $T_\delta^{k+1} = T_\delta T_\delta^k \pi$ for any $k \in I_+$. Thus $T_\delta^k \pi$ defers the use of π for k periods during which δ is used. The policy-improvement algorithm in the proof of Theorem 4-5 finds a one-period delay of γ^∞, namely, $T_\delta \gamma^\infty$, which is strictly better than γ^∞ (unless γ^∞ is optimal). Then δ^∞ replaces γ^∞ in the next iteration. However, δ^∞ is equivalent to an infinite-period deferral of γ^∞, namely $T_\delta^\infty \gamma^\infty$. In the following modification of the algorithm, a policy π leads to a search for a one-period delay, namely, $T_\delta \pi$, which is strictly better than π. If $\pi = \gamma^\infty$, this step in the unmodified and modified algorithms is the same. However, for some fixed $k \in I_+$, π is replaced by $T_\delta^k \pi$ in the modified algorithm. This is a k-period deferral of π rather than an infinite-period deferral as would be true in the unmodified algorithm.

The modified algorithm begins with a policy $\pi_1 \in Y$. At the nth iteration, let π_n denote the policy† being tested for optimality. The test consists of checking whether or not $G(\pi_n)$ is empty where

$$G(\pi) \triangleq \{\delta\colon \delta \in \Delta \text{ and } T_\delta \pi > \pi\}$$
$$= \times_{s \in \mathcal{S}} \{a\colon a \in A_s \text{ and } r(s, a) + \beta \sum_{j \in \mathcal{S}} p_{sj}^a v_j(\pi) > v_s(\pi)\} \qquad (4\text{-}71)$$

is analogous to $\times_{s \in \mathcal{S}} G(s, \gamma)$ in (4-60). This is the same optimality test as in Theorem 4-4. If $G(\pi_n)$ is nonempty,

$$\pi_{n+1} = T_{\delta_{n+1}}^k \pi_n \qquad k \in I_+ \qquad (4\text{-}72)$$

where δ_{n+1} is an element of $G(\pi_n)$. Then the $n + 1$st iteration occurs, etc. The unmodified algorithm is the special case where $k = \infty$.

The motivation for modifying the algorithm is numerical, and it is discussed in Section 6-2. However, the legitimacy of the modification depends on Proposition 4-3 below. Proofs of the following corollaries to Theorem 4-3 and Theorems 4-4 and 4-5 comprise Exercise 4-29.

Proposition 4-3 Let $\pi \in Y$, $\delta \in \Delta$, and $k \in I_+$ and suppose $0 \le \beta < 1$ and there are at most countably many states.

(*a*) If $T_\delta \pi \gtrsim \pi$, then $T_\delta^k \pi \gtrsim \pi$.
(*b*) If $T_\delta \pi > \pi$, then $T_\delta^k \pi > \pi$.

Proposition 4-4 Suppose an MDP has $0 < \beta < 1$, there are at most countably many states, and there exists an optimal policy. For any $k \in I_+$, $\pi^* \in Y$ is optimal if

$$\pi^* \gtrsim T_\delta^k \pi^* \qquad \delta \in \Delta \qquad (4\text{-}73)$$

Proposition 4-5 Suppose an MDP has $0 \le \beta < 1$ and at most countably many states. For any $k \in I_+$ and $\pi \in Y$, either π is optimal or there is $\delta \in \Delta$ with

$$T_\delta^k \pi > \pi \qquad (4\text{-}74)$$

† The notation π_n in this section and π_n in Section 4-2 are not related.

In general, the sequence $\pi_{n+1} = T^k_{\delta_{n+1}}\pi_n$ does not terminate in a finite number of iterations. Nevertheless, under computationally reasonable assumptions, in Section 6-2 we show that it does converge (in a sense made precise there), and to an optimum.

EXERCISES

4-23 In Example 4-18, verify the optimality of b^∞ by enumeration. In other words, let c denote the rule

$$c(0) = 0,\ c(1) = 1,\ c(2) = 1$$

Compute $v_s(c^\infty)$ for each s and check $v_s(b^\infty) \geq v_s(c^\infty)$.

4-24 Find an optimal infinite-horizon policy for the coin-matching game in Example 4-5. Use $\beta = 0.9$ and $p = 7/12$.

4-25 Use policy improvement to find a policy which minimizes the expected discounted cost in the maintenance model in Exercise 3-15.

4-26 Use policy improvment to find a policy which maximizes the expected present value in the fishery model in Exercise 3-16.

4-27 Example 4-20 concerns the inventory model introduced in Example 4-8. We find in Example 4-20 that c^∞ is optimal, where $c(0) = 2$ and $c(s) = s$ if $s \neq 2$. This is an example of a (σ, Σ) inventory policy, namely, $a_n = \sigma$ if $s_n < \sigma$ and $a_n = s_n$ if $s_n \geq \sigma$. Here, $\sigma = 1$ and $\Sigma = 2$. See Sections 7-1, 7-2, and 7-4 for conditions under which an inventory model has an optimal (σ, Σ) policy and for algorithms which exploit the (σ, Σ) structure.

(*a*) What are the smallest and largest numerical values of K, respectively, K_* and K^*, for which c^∞ remains optimal?

(*b*) Suppose $K = K^* + \epsilon$ where $\epsilon > 0$ is suitably small. Specify a policy d^∞ which dominates c^∞.

4-28 Here is a notion of policy improvement that is different from the construction in the proof of Theorem 4-5. Let $\gamma_j \in \Delta, j = 1, 2$, and let

$$\mathcal{S}_0 = \{s\colon s \in \mathcal{S},\ v_s(\gamma_1^\infty) > v_s(\gamma_2^\infty)\}$$

so that $\mathcal{S}_0$ is the subset of states from which γ_1^∞ is better than γ_2^∞. Then define $\delta_- \in \Delta$ and $\delta_+ \in \Delta$ with

$$\delta_-(s) = \gamma_2(s) \text{ and } \delta_+(s) = \gamma_1(s) \text{ if } s \in \mathcal{S}_0$$

$$\delta_-(s) = \gamma_1(s) \text{ and } \delta_+(s) = \gamma_2(s) \text{ if } s \notin \mathcal{S}_0$$

Prove that $\delta_+^\infty \succsim \gamma_j^\infty$ and $\gamma_j^\infty \succsim \delta_-^\infty$ for $j = 1, 2$.

4-29 Prove Propositions 4-3, 4-4, and 4-5. [Hint for Proposition 4-4: show (4-66) implies $\pi^* \succsim \delta^\infty$ for all $\delta \in \Delta$ and use Theorem 4-2.] [Hint for Proposition 4-5: show $G_1(\pi) \subset G_k(\pi)$ and use Theorem 4-5.]

4-30 Throughout most of Sections 4-3 and 4-4 we assume that $\mathcal{S}$ is a countable (or finite) set in order to avoid technical difficulties when we specify W as the set of all probability distributions on W and

$$w_{n+1}(j) = M(\mathbf{w}_n, \delta_n)(j) = \sum_{i \in \mathcal{S}} w_n(i) p_{ij}^{\delta(i)}$$

The potential difficulties are absent in a deterministic MDP. Suppose for each n that s_{n+1} is a deterministic function $m(\cdot, \cdot)$ of s_n and a_n. That is, $s_{n+1} = m(s_n, a_n)$ for all n where $m\colon \mathscr{C} \to \mathcal{S}$. We assume merely that $\mathcal{S}$ is nonempty and A_s is nonempty for each $s \in \mathcal{S}$. Verify that the following results are valid in this case if W, M, and $W \times \Delta$ are replaced with $\mathcal{S}$, m, and $\mathscr{C}$, respectively: Lemma 4-2, Propositions 4-2 and 4-3, Theorems 4-2, 4-3, 4-4, and 4-5, and Corollaries 4-4*a* and 4-4*b*.

4-5 A CORNUCOPIA OF ALGORITHMS

This section presents algorithms for stochastic MDPs with a criterion of infinite-horizon expected discounted reward and with finitely many states and actions.† Let $\pi \triangleq (\delta_1, \delta_2, \ldots)$ be a Markov policy. The objective of the algorithms in this section is to find a policy π^* that attains

$$\max_{\pi \in Y} E\left\{ \sum_{n=1}^{\infty} \beta^{n-1} r[s_n, \delta_n(s_n)] \mid s_1 = s \right\} \qquad s \in \mathcal{S} \tag{4-75}$$

with $0 \leq \beta < 1$. This is a vector-maximization problem because the same π^* must attain the maximum for all initial states $s \in \mathcal{S}$.

Throughout the section, we assume

$$\#\mathscr{C} < \infty \quad \text{where} \quad \mathscr{C} = \{(s, a): a \in A_s, s \in \mathcal{S}\} \tag{4-76}$$

In words, there are only finitely many states and actions. We call this a *finite MDP*. Let S denote the number of states, so $S \triangleq \#\mathcal{S}$.

From Theorem 4-3, there is no loss of optimality in (4-75) if the maximization is restricted to the set Δ of stationary policies. From Theorem 4-6, there is a stationary optimal policy when $\#\mathscr{C} < \infty$. Therefore, an optimal policy can be found in finitely many steps by computing

$$E\left(\sum_{n=1}^{\infty} \beta^{n-1} r[s_n, \delta(s_n)] \mid s_1 = s \right) \qquad s \in \mathcal{S} \tag{4-77}$$

for each of the finitely many single-period rules $\delta \in \Delta$. You might guess that this enumeration algorithm has some drawbacks.

A possible drawback, you might imagine, is that we may be unable to evaluate (4-77) numerically. We shall see that (4-77) is well defined and enjoys desirable properties.

A first severe drawback is that the computation in (4-77), for a *single* $\delta \in \Delta$ and for *all* $s \in \mathcal{S}$, is comparable with inversion of an $S \times S$ matrix. Since S is large in practical problems ($S \leq 100$ is unusual), the labor involved in (4-77) is significant. A second drawback is that Δ usually contains prodigiously many elements:

$$\#\Delta = \prod_{s \in \mathcal{S}} \#A_s$$

so $S = 100$ and $\#A_s = 2$ for all s yields $\#\Delta = 2^{100}$. Hence, enumeration is not feasible in practice.

You already know one algorithm, namely, policy improvement. It requires markedly fewer computations than does enumeration. All the algorithms in this section are much better than enumeration, but like policy improvement, their computational labor is roughly proportional to S^3. Therefore, large models (say, with $S > 10^4$) invite approximate rather than exact solution. Some approximate methods for large MDPs are presented in Section 6-4.

† Section 6-2 presents methods which accelerate some of the algorithms in this section.

The notation in this section is the same as in Sections 4-1 through 4-4 except that we write $v_s(\delta)$, rather than $v_s(\delta^\infty)$, for the expected present value in (4-95).

Recall from Section 4-3 that nonnegativity of $r(\cdot, \cdot)$ is equivalent to $r(\cdot, \cdot)$ being bounded below on $\mathscr{C}$. Since $\#\mathscr{C} < \infty$, of course, $r(\cdot, \cdot)$ is bounded below. Also, it is bounded above; so we may assume, without loss of generality, that there is $u < \infty$ such that

$$0 \le r(s, a) \le u \qquad (s, a) \in \mathscr{C} \tag{4-78}$$

The Value Function

Let

$$v_s(\delta) \triangleq E\left(\sum_{n=1}^{\infty} \beta^{n-1} r[s_n, \delta(s_n)] \mid s_1 = s \right) \qquad s \in \mathcal{S},\ \delta \in \Delta \tag{4-79}$$

Is (4-79) well defined? The computation of $v_s(\delta)$ is fundamentally important in the policy-improvement algorithms in Section 4-4, and definition (4-79) is basic for all our algorithms. We had no difficulty with the computations in the examples in Section 4-4, but how easy is it to compute $v_s(\delta)$ in general? Theorem 4-7 below asserts that $v_s(\delta)$ *is* well defined and is part of the solution to a system of linear algebraic equations with a nonsingular matrix of coefficients. Much of the remainder of this section and all of Section 6-2 are concerned with algorithms to compute $v_s(\delta)$ for an optimal policy δ^∞.

Policy δ^∞ induces a Markov chain with stationary transition probabilities

$$P\{s_{n+1} = j \mid s_n = s\} = p_{sj}^{\delta(s)} \qquad s \in \mathcal{S},\ j \in \mathcal{S}$$

Let P_δ and $Q_\delta \triangleq \beta P_\delta$ denote the $S \times S$ matrices whose (s, j)th entries are $p_{sj}^{\delta(s)}$ and $q_{sj}^{\delta(s)} = \beta p_{sj}^{\delta(s)}$, respectively. Let $\mathbf{r}_\delta$ and $\mathbf{v}(\delta)$ denote the column vectors whose sth entries are $r[s, \delta(s)]$ and $v_s(\delta)$, respectively. We use the notation I for the identity matrix.

Recall the notation X_n for the reward earned in period n. Then

$$r(s, a) = E(X_n \mid s_n = s, a_n = a) \qquad (s, a) \in \mathscr{C}$$

Let

$$B_s(\delta) = \sum_{n=1}^{\infty} \beta^{n-1} X_n \qquad \text{where } a_n = \delta(s_n) \text{ for all } n$$

The r.v. $B_s(\delta)$ is the present value of the sequence of single-stage rewards. Let the distribution function, mth moment, and variance of $B_s(\delta)$ be labeled

$$F_s^\delta(x) = P\{B_s(\delta) \le x\}$$

$$v_m(s, \delta) = E\{[B_s(\delta)]^m\} = \int_{-\infty}^{\infty} x^m dF_s^\delta(x) \qquad v_s(\delta) = v_s^{(1)}(\delta)$$

and

$$\sigma_s^2(\delta) = v_s^{(2)}(\delta) - [v_s(\delta)]^2$$

Let $\boldsymbol{\sigma}^2(\delta)$ denote the vector whose sth component is $\sigma_s^2(\delta)$.

It is convenient to assume for each n that X_n has a discrete sample space $\mathscr{K}$ and to use the notation

$$p^a_{sjk} = P\{s_{n+1} = j,\ X_n = k \mid s_n = s,\ a_n = a\}$$

for the joint conditional probability distribution of the next state and the current reward. In our usual notation,

$$p^a_{sj} = \sum_{k \in \mathscr{K}} p^a_{sjk}$$

Here is a system of equations for the distribution functions. Let p_{sjk} and $F_s(\cdot)$ denote $p^{\delta(s)}_{sjk}$ and $F^\delta_s(\cdot)$, respectively.

Lemma 4-3 $F_s(x) = \sum_{j \in \mathcal{S}} \sum_{k \in \mathscr{K}} p_{sjk} F_j\left(\frac{x-k}{\beta}\right) \qquad s \in \mathcal{S} \quad x \in \mathbb{R}$

PROOF By definition of $B_s(\delta)$,

$$B_s(\delta) = X_1 + \beta\left(X_2 + \sum_{n=2}^{\infty} \beta^{n-1} X_{n+1}\right) \sim X_1 + \beta B_{s_2}(\delta)$$

where $X \sim Y$ denotes two r.v.'s with the same distribution. Therefore,

$$\begin{aligned} F_s(x) &= P\{X_1 + \beta B_{s_2}(\delta) \le x\} \\ &= \sum_{j \in \mathcal{S}} \sum_{k \in \mathscr{K}} P\{X_1 + \beta B_{s_2}(\delta) \le x \mid s_2 = j,\ X_1 = k\} p_{sjk} \\ &= \sum_{j \in \mathcal{S}} \sum_{k \in \mathscr{K}} F_j\left(\frac{x-k}{\beta}\right) p_{sjk} \end{aligned}$$

□

It follows from (4-80) and (4-81) in Theorem 4-7 below that the various moments of $B_s(\delta)$ are well defined and computable as solutions of linear algebraic equations. Let $\boldsymbol{\theta}_\delta$ and $\mathbf{r}_\delta$ denote the vectors whose sth components are

$$\theta_s(\delta) = \sum_{j \in \mathcal{S}} \sum_{k \in \mathscr{K}} p^{\delta(s)}_{sjk} [k + \beta v_j(\delta)]^2 - [v_s(\delta)]^2$$

and $r[s, \delta(s)]$, respectively.

Theorem 4-7 If $0 \le \beta < 1$ and $S < \infty$, then

$$\mathbf{v}(\delta) = \mathbf{r}(\delta) + Q_\delta \mathbf{v}(\delta) = (I - Q_\delta)^{-1} \mathbf{r}_\delta \tag{4-80}$$

$$\boldsymbol{\sigma}^2(\delta) = \boldsymbol{\theta}_\delta + \beta^2 P_\delta \boldsymbol{\sigma}^2(\delta) = (I - \beta^2 P_\delta)^{-1} \boldsymbol{\theta}_\delta \tag{4-81}$$

and

$$v_m(s, \delta) = \sum_{i=0}^{m} \binom{m}{i} \beta^i \sum_{j \in \mathcal{S}} v_i(j, \delta) \sum_{k \in \mathscr{K}} k^{m-i} P_{sjk} \tag{4-82}$$

PROOF Exercise 4-44 asks you to use the equations of Lemma 4-3 for the distribution functions to derive (4-80) and (4-81). We use alternative arguments here to prove (4-80) and (4-81).

As in the proof of Lemma 4-3, δ is suppressed in the notation. Let $r_s = r[s, \delta(s)]$. For (4-80), as in the proof of Lemma 4-3, $B_s \sim X_1 + \beta B_{s_2}$; so

$$\begin{aligned} v_s = E(B_s) &= \sum_{j \in \mathcal{S}} \sum_{k \in \mathcal{K}} E(B_s \mid s_2 = j, X_1 = k) p_{sjk} \\ &= \sum_{j \in \mathcal{S}} \sum_{k \in \mathcal{K}} (k + \beta v_j) p_{sjk} \\ &= r_s + \beta \sum_{j \in \mathcal{S}} p_{sj} v_j \end{aligned}$$

which is $\mathbf{v} = \mathbf{r} + \beta P \mathbf{v} = \mathbf{r} + Q\mathbf{v}$ in (4-80)

Also, $\mathbf{v} = \mathbf{r} + Q\mathbf{v}$ yields $(I - Q)\mathbf{v} = \mathbf{r}$; so $\mathbf{v} = (I - Q)^{-1}\mathbf{r}$ if $I - Q$ is nonsingular. The assumption $0 \le \beta < 1$ implies $\beta^n P^n = Q^n \to 0$ as $n \to \infty$ (that is, each element of Q^n converges to the scalar 0). Theorem 7-3 on page 223 of Volume I states that if a square matrix Q has the property $Q^n \to 0$ as $n \to \infty$, then $I - Q$ is nonsingular and

$$(I - Q)^{-1} = \sum_{n=0}^{\infty} Q^n$$

where $Q^0 \triangleq I$. Therefore, $I - Q_\delta$ is nonsingular for each $\delta \in \Delta$, and

$$\sum_{n=0}^{\infty} Q_\delta^n \mathbf{r}_\delta = (I - Q)^{-1} \mathbf{r}_\delta$$

which completes the proof of (4-80).

For (4-81),

$$B_s \sim X_1 + \beta B_{s_2} \qquad \text{and} \qquad \sigma_s^2 = E[(B_s)^2] - v_s^2$$

implies

$$\begin{aligned} \sigma_s^2 &= E[(X_1 + \beta B_{s_2})^2] - v_s^2 \\ &= \sum_{j \in \mathcal{S}} \sum_{k \in \mathcal{K}} p_{sjk}[k^2 + \beta^2 E(B_j^2) + 2\beta k v_j] - v_s^2 \\ &= \sum_{j \in \mathcal{S}} \sum_{k \in \mathcal{K}} p_{sjk}[k^2 + \beta^2(\sigma_j^2 + v_j^2) + 2\beta k v_j] - v_s^2 \\ &= \sum_{j \in \mathcal{S}} \sum_{k \in \mathcal{K}} p_{sjk}(k + \beta v_j)^2 - v_s^2 + \beta^2 \sum_{j \in \mathcal{S}} p_{sj} \sigma_j^2 \\ &= \theta_s + \beta^2 \sum_{j \in \mathcal{S}} p_{sj} \sigma_j^2 \end{aligned}$$

Therefore, $\boldsymbol{\sigma}^2 = \beta^2 P \boldsymbol{\sigma}^2 + \boldsymbol{\theta}$ so $(I - \beta^2 P)\boldsymbol{\sigma}^2 = \boldsymbol{\theta}$. The same argument that leads to nonsingularity of $I - \beta P$ yields nonsingularity of $I - \beta^2 P$; so $\boldsymbol{\sigma}^2 = (I - \beta^2 P)^{-1} \boldsymbol{\theta}$.

The proof of (4-82) is left as Exercise 4-34. □

Example 4-21 Suppose that $\beta = 0.5$, $\mathcal{S} = \{1, 2, 3\}$ with states 1 and 2 being absorbing, and $A_1 = A_2 = \{1\}$. From state 3, $A_3 = \{1, 2\}$ with $a = 2$ causing a certain transition to state 2. Using $a = 1$ in state 3 results in either a transition to state 1, with probability .2, or staying in state 3, with probability .8. The single-stage rewards are 10 and 1.5 in states 1 and 2, respectively. For state 3, $r(3, 1) = 0$ and $r(3, 2) = 1.5$.

The high-gain and high-risk policy is to choose action 1 in state 3 and to hope that it results in a transition to state 1 where the high rewards are earned. Until such a transition occurs, the single-stage rewards are zero. The modest-gain but low-risk policy is to choose action 2 in state 3. It yields a certain reward of 1.5 in every period thereafter.

Let δ denote the policy which takes action 1 in state 3. Let γ denote the policy which takes action 2 in state 3. Then

$$v_1(\delta) = v_1(\gamma) = 10(1 + \beta + \beta^2 + \cdots) = \frac{10}{1 - 0.5} = 20$$

$$v_3(\gamma) = v_2(\gamma) = v_2(\delta) = 1.5(1 + \beta + \beta^2 + \cdots) = 3$$

$$v_3(\delta) = 0 + 0.2\beta(20) + (0.2\beta)^2(20) + \cdots = 3.333$$

We shall verify these numbers with formula (4-80).

$$P_\delta = \begin{pmatrix} 1 & 0 & 0 \\ 0 & 1 & 0 \\ 0.2 & 0 & 0.8 \end{pmatrix} \qquad P_\gamma = \begin{pmatrix} 1 & 0 & 0 \\ 0 & 1 & 0 \\ 0 & 1 & 0 \end{pmatrix}$$

$$(I - \beta P_\delta)^{-1} = \begin{pmatrix} 2 & 0 & 0 \\ 0 & 2 & 0 \\ 0.333 & 0 & 1.666 \end{pmatrix} \qquad \mathbf{r}_\delta = \begin{pmatrix} 10 \\ 1.5 \\ 0 \end{pmatrix}$$

$$(I - \beta P_\gamma)^{-1} = \begin{pmatrix} 2 & 0 & 0 \\ 0 & 2 & 0 \\ 0 & 1 & 1 \end{pmatrix} \qquad \mathbf{r}_\gamma = \begin{pmatrix} 10 \\ 1.5 \\ 1.5 \end{pmatrix}$$

$$(I - \beta P_\delta)^{-1}\mathbf{r}_\delta = \begin{pmatrix} 20 \\ 3 \\ 3.333 \end{pmatrix} \qquad (I - \beta P_\gamma)^{-1}\mathbf{r}_\gamma = \begin{pmatrix} 20 \\ 3 \\ 3 \end{pmatrix}$$

These numbers confirm the direct calculations.

In order to compute the variance with (4-81), we must calculate $(I - \beta^2 P_\delta)^{-1}$, $(I - \beta^2 P_\gamma)^{-1}$, $\boldsymbol{\theta}_\delta$, and $\boldsymbol{\theta}_\gamma$. Using P_δ and P_γ above and the identity for θ_s in Exercise 4-41,

$$(I - \beta^2 P_\delta)^{-1} = \begin{pmatrix} \frac{4}{3} & 0 & 0 \\ 0 & \frac{4}{3} & 0 \\ \frac{1}{12} & 0 & \frac{5}{4} \end{pmatrix} \qquad (I - \beta^2 P_\gamma)^{-1} = \begin{pmatrix} \frac{4}{3} & 0 & 0 \\ 0 & \frac{4}{3} & 0 \\ 0 & \frac{1}{3} & 1 \end{pmatrix}$$

$$\theta_1(\delta) = \theta_1(\gamma) = (0.5)^2(20)^2 - (20 - 10)^2 = 0$$

$$\theta_3(\gamma) = \theta_2(\delta) = \theta_2(\gamma) = (0.5)^2(3)^2 - (3 - 1.5)^2 = 0$$

$$\theta_3(\delta) = (0.5)^2[(0.2)(20)^2 + (0.8)(3.333)^2] - (4.333 - 0)^2 = 11.111$$

From (4-81),

$$\boldsymbol{\sigma}^2(\gamma) = (I - \beta^2 P_\gamma)^{-1}\boldsymbol{\theta}_\gamma = \begin{pmatrix} 0 \\ 0 \\ 0 \end{pmatrix}$$

$$\boldsymbol{\sigma}^2(\delta) = (I - \beta^2 P_\delta)^{-1}\boldsymbol{\theta}_\delta = \begin{pmatrix} 0 \\ 0 \\ 13.888 \end{pmatrix}$$

In order to confirm that these variances are correct, observe that

$$P\{B_3(\gamma) = B_2(\delta) = B_2(\gamma) = 3\} = P\{B_1(\delta) = B_1(\gamma) = 20\} = 1$$

so

$$\sigma_3^2(\gamma) = \sigma_1^2(\delta) = \sigma_1^2(\gamma) = \sigma_2^2(\delta) = \sigma_2^2(\gamma) = 0$$

which agrees with the formula (4-81) calculations. If $s_1 = 3$ and δ is used, let T be the number of periods in which the state is 3; that is, $T + 1$ is the first period in which the state is 1. From $B_3^\delta = 20(0.5)^T$ and the geometric distribution of T,

$$P\{B_3^\delta = 20(0.5)^k\} = (0.2)(0.8)^{k-1} \qquad k = 1, 2, \ldots$$

so

$$v_2(3, \delta) = E\{[B_3(\delta)]^2 = \sum_{k=1}^{\infty} (0.2)(0.8)^{k-1}[20(0.5)^k]^2$$

$$= (0.2)(20)^2(0.5)^2 \sum_{k=1}^{\infty} [(0.8)(0.5)^2]^{k-1} = 25$$

$$\sigma_3^2(\delta) = 25 - (3.333)^2 = 13.888$$

which agrees with the calculation based on (4-81). □

A square matrix is said to be *substochastic* if its entries are nonnegative and its rows sum to at most 1. For each δ, $Q_\delta \geq 0$ and $\sum_j q_{sj}^{\delta(s)} = \beta < 1$; so Q_δ is substochastic. Most of this section does not depend on the further property that the row sums are all the same, that is, β. Instead, we exploit (4-80), (4-81), and (4-82), for which $Q_\delta^n \to 0$ is sufficient. From Section 7-2 in Volume I, $Q_\delta^n \to 0$ if the largest eigenvalue of Q_δ^n is strictly less than 1. Several sections in Chapter 5 contain models with the formal appearance of (4-80), (4-81), and (4-82) but where the row sums of Q_δ are not all the same. Since $Q_\delta^n \to 0$ for those models, we shall be able to use the algorithms in this section.

Successive Approximations

How rapidly does the sum in (4-79) [which defines $v_s(\delta)$] converge? Repeating (4-79),

$$v_s(\delta) \triangleq E\left\{\sum_{n=1}^{\infty} \beta^{n-1} r[s_n, \delta(s_n)] \mid s_1 = s\right\} \qquad s \in \mathcal{S} \qquad \delta \in \Delta$$

Let

$$v_s^N(\delta) \triangleq E\left(\sum_{n=1}^{N} \beta^{n-1} r[s_n, \delta(s_n)] \mid s_1 = s\right) \qquad s \in \mathcal{S},\ N \in I_+$$

Nonnegativity of $r(s, a)$ for all $(s, a) \in \mathscr{C}$ implies $v_s^N(\delta) \leq v_s^{N+1}(\delta)$ for all s, N, and δ. Hence the definition of $v_s(\delta)$ and $r(s, a) \leq u$ for all $(s, a) \in \mathscr{C}$ yield

$$v_s^N(\delta) \leq v_s(\delta) = v_s^N(\delta) + E\left(\sum_{n=N+1}^{\infty} \beta^{n-1} r[s_n, \delta(s_n)] \mid s_1 = s\right)$$

$$\leq v_s^N(\delta) + \frac{u\beta^N}{1-\beta}$$

so

$$0 \leq v_s(\delta) - v_s^N(\delta) \leq \frac{\beta^N u}{1-\beta} \qquad N \in I_+,\ s \in \mathcal{S},\ \delta \in \Delta \tag{4-83}$$

and convergence is at least geometrically fast.†

The following recursion is called *successive approximations* (or *value iteration*):

$$f_n(s) = \max\ \{r(s, a) + \sum_{j \in \mathcal{S}} q_{sj}^a f_{n-1}(j)\colon a \in A_s\} \tag{4-84}$$

This expression is used in Section 4-2 with $f_0(\cdot)$ given by $L(\cdot)$, a real-valued salvage function on $\mathcal{S}$. In Section 4-3 we suggest that $f_n(s)$ may approximate an infinite-horizon optimum if n is sufficiently large. Now we obtain bounds on the approximation.

Definition 4-16 A *norm* on $\mathbb{R}^S$ is a real-valued function $\|\cdot\|$ on $\mathbb{R}^S$ with three properties:

(i) $\|\mathbf{w}\| \geq 0$ for all $\mathbf{w} \in \mathbb{R}^S$
(ii) $\|\mathbf{w}\| = 0 \Leftrightarrow \mathbf{w} = \mathbf{0}$
(iii) $\|\mathbf{w} + \mathbf{v}\| \leq \|\mathbf{w}\| + \|\mathbf{v}\|$ for all $\mathbf{w}$ and $\mathbf{v}$

A *metric* on $\mathbb{R}^S$ is a real-valued function $d(\cdot, \cdot)$ on $\mathbb{R}^{2S}$ with three properties for all $\mathbf{u}$, $\mathbf{v}$, and $\mathbf{w}$ in $\mathbb{R}^S$:

(i) $d(\mathbf{u}, \mathbf{v}) \geq 0$
(ii) $d(\mathbf{u}, \mathbf{v}) = 0 \Leftrightarrow \mathbf{u} = \mathbf{v}$
(iii) $d(\mathbf{u}, \mathbf{v}) + d(\mathbf{v}, \mathbf{w}) \geq d(\mathbf{u}, \mathbf{w})$

Because of these properties, we think of $d(\mathbf{u}, \mathbf{v})$ as a measure of the distance between $\mathbf{u}$ and $\mathbf{v}$. It follows from the definitions that a norm $\|\cdot\|$ induces the metric $d(\mathbf{u}, \mathbf{v}) = \|\mathbf{u} - \mathbf{v}\|$.

† For methods to expedite the computational solution of (4-80) and (4-81), see E. L. Porteus and J. C. Totten, "Accelerated Computation of the Expected Discounted Return in a Markov Chain," *Oper. Res.* **26**(2): 350–358 (1978).

In this section, we use the norm

$$\|\mathbf{w}\| \triangleq \max \{|w_s|: s \in \mathcal{S}\} \qquad \mathbf{w} \in \mathbb{R}^S$$

which induces the metric $d(\mathbf{v}, \mathbf{w}) \triangleq \max \{|v_s - w_s|: s \in \mathcal{S}\}$.

Definition 4-17 Let

$$f(s) \triangleq \max \{v_s(\delta): \delta \in \Delta\} \qquad s \in \mathcal{S}$$

The *value of an optimal policy* is the S-vector

$$\mathbf{f} \triangleq [f(s), s \in \mathcal{S}]$$

The *n-period value function* is the S-vector

$$\mathbf{f}_n \triangleq [f_n(s), s \in \mathcal{S}] \qquad n \in I_+$$

where $f_n(s)$ is given by (4-84).

We say that a sequence of S-vectors $\mathbf{x}_1, \mathbf{x}_2, \ldots$ converges to $\mathbf{y}$ if the jth components $x_1(j), x_2(j), \ldots$ converge to $y(j)$ for each $j = 1, \ldots, S$. We show now that $\mathbf{f}_n \to \mathbf{f}$ geometrically fast, for any initial $\mathbf{f}_0$, in the sense that $d(\mathbf{f}, \mathbf{f}_n) \le K\beta^n$ for some $K < \infty$.

Let $\mathbf{L}$ denote the mapping† from $\mathbb{R}^S$ to $\mathbb{R}^S$ such that $\mathbf{v} \in \mathbb{R}^S$ causes $\mathbf{L}v$ to have the sth coordinate

$$\mathbf{Lv}(s) = \max \{r(s, a) + \sum_{j \in \mathcal{S}} q_{sj}^a v_j: a \in A_s\} \tag{4-85'}$$

where v_j is the jth coordinate of $\mathbf{v}$. We interpret $\mathbf{Lv}(s)$ as the optimal value of a one-period optimization problem which starts in state s and has salvage value function $\mathbf{v}$. It is convenient to write (4-85′) as

$$\mathbf{Lv} = \text{vmax} \{\mathbf{r}_\delta + Q_\delta \mathbf{v}: \delta \in \Delta\} \tag{4-85}$$

where "vmax" stands for "vector maximum" in the sense of (4-85′) at each coordinate s.

Theorem 4-8 Suppose $0 \le \beta < 1$ and $S < \infty$ and let $\mathbf{f}$ be the value of an optimal policy.

(*a*) For all $\mathbf{f}_0 \in \mathbb{R}^S$, $\mathbf{f}_n \to \mathbf{f}$, and

$$\|\mathbf{f} - \mathbf{f}_n\| \le \frac{\|\mathbf{f}_{n+1} - \mathbf{f}_n\|}{1 - \beta} \le \beta^n \frac{\|\mathbf{f}_1 - \mathbf{f}_0\|}{1 - \beta}$$

(*b*) $\mathbf{f}$ is the unique solution of

$$\mathbf{Lf} = \mathbf{f}$$

† Properties of **L** are derived in Section 5-4 with less effort but deeper mathematics than is used in the forthcoming proof of Theorem 4-8. The mapping **L** has no connection with the notation $L(\cdot)$ for the salvage value function.

PROOF The property (see Exercise 4-32)

$$\| \mathbf{f}_{i+1} - \mathbf{f}_i \| \le \beta \| \mathbf{f}_i - \mathbf{f}_{i-1} \| \qquad i \in I_+ \tag{4-86}$$

implies $\| \mathbf{f}_{i+1} - \mathbf{f}_i \| \le \beta^i \| \mathbf{f}_1 - \mathbf{f}_0 \|$. Therefore, for each $s \in \mathbb{S}$, $f_0(s)$, $f_1(s)$, $f_2(s)$, ... is a Cauchy sequence of real numbers, hence convergent. Therefore, $\mathbf{f}_0$, $\mathbf{f}_1$, $\mathbf{f}_2$, ... is convergent in $\mathbb{R}^S$. Let $\mathbf{F}$ denote the limit function so $\mathbf{f}_n \rightarrow \mathbf{F}$. First we show $d(\mathbf{f}_n, \mathbf{F}) \le K\beta^n$ for some $K < \infty$ and then $\mathbf{f} = \mathbf{F}$. From (4-86), property (iii) of a metric, and

$$\mathbf{F} = \mathbf{f}_n - \sum_{i=n}^{\infty} (\mathbf{f}_{i+1} - \mathbf{f}_i)$$

we obtain

$$\begin{aligned}
\| \mathbf{F} - \mathbf{f}_n \| &= \left\| \sum_{i=n}^{\infty} (\mathbf{f}_{i+1} - \mathbf{f}_i) \right\| \\
&\le \sum_{i=n}^{\infty} \| \mathbf{f}_{i+1} - \mathbf{f}_i \| \le \sum_{i=n}^{\infty} \beta^{i-n} \| \mathbf{f}_{n+1} - \mathbf{f}_n \| \\
&= \frac{\| \mathbf{f}_{n+1} - \mathbf{f}_n \|}{1-\beta} \le \frac{\beta \| \mathbf{f}_n - \mathbf{f}_{n-1} \|}{1-\beta} \le \cdots \le \beta^n \frac{\| \mathbf{f}_1 - \mathbf{f}_0 \|}{1-\beta}
\end{aligned}$$

Therefore, $d(\mathbf{f}_n, \mathbf{F}) \le K\beta^n$ with $K = \| \mathbf{f}_1 - \mathbf{f}_0 \| /(1-\beta)$.

To establish $\mathbf{F} = \mathbf{f}$ and complete the proof of part (*a*) of the theorem, we prove that $\mathbf{F}$ is the unique function for which $d(\mathbf{f}_n, \mathbf{F}) \rightarrow 0$ regardless of $\mathbf{f}_0$. Define $\mathbf{L}^n$ recursively with $\mathbf{L}^1 = \mathbf{L}$ and $\mathbf{L}^{n+1} = \mathbf{L}(\mathbf{L}^n)$. Then $\mathbf{f}_n = \mathbf{L}^n\mathbf{f}_0$, $n \in I_+$; so $d(\mathbf{L}^n\mathbf{f}_0, \mathbf{F}) \rightarrow 0$ as $n \rightarrow \infty$. Also, (4-86) is an instance of (Exercise 4-32)

$$d(\mathbf{Lu}, \mathbf{Lv}) \le \beta d(\mathbf{u}, \mathbf{v}) \qquad u \in \mathbb{R}^S,\ v \in \mathbb{R}^S \tag{4-87}$$

It follows from (4-87) and $0 \le \beta < 1$ that $\mathbf{L}$ is continuous on $\mathbb{R}^S$ (Exercise 4-32). Therefore,

$$\mathbf{LF} = \mathbf{L}\left(\lim_{n\rightarrow\infty} \mathbf{f}_n\right) = \lim_{n\rightarrow\infty} \mathbf{Lf} = \lim_{n\rightarrow\infty} \mathbf{f}_{n+1} = \mathbf{F}$$

so $\mathbf{LF} = \mathbf{F}$, that is, $\mathbf{F}$ is a *fixed point* of $\mathbf{L}$. Also, $\mathbf{F}$ is the unique fixed point: If $\mathbf{F}'$ also is a fixed point so $\mathbf{LF}' = \mathbf{F}'$, then $d(\mathbf{F}, \mathbf{F}') = d(\mathbf{LF}, \mathbf{LF}') \le \beta \mathbf{d}(\mathbf{F}, \mathbf{F}')$; so $0 \le \beta < 1$ implies $d(\mathbf{F}, \mathbf{F}') = 0 \Leftrightarrow \mathbf{F}' = \mathbf{F}$.

To complete the proof that $\mathbf{f} = \mathbf{F}$, observe that $\mathbf{LF} = \mathbf{F}$ and (4-85) imply existence of $\gamma \in \Delta$ such that

$$\mathbf{r}_\gamma + Q_\gamma \mathbf{F} = \text{vmax}\ \{\mathbf{r}_\delta + Q_\delta \mathbf{F} : \delta \in \Delta\}$$

Let

$$\mathbf{f} \triangleq \text{vmax}\ \{\mathbf{v}(\delta) : \mathbf{v}(\delta) = \sum_{i=0}^{\infty} Q_\delta^i \mathbf{r}_\delta,\ \delta \in \Delta\}$$

if the second (vector) maximum is achieved. Then $\mathbf{F} = \sum_{i=0}^{\infty} Q_\gamma^i \mathbf{r}_\gamma \le \mathbf{f}$. To prove $\mathbf{F} \ge \mathbf{f}$, let

$$\mathbf{G}_\delta^k \mathbf{w} = \sum_{i=0}^{k-1} Q_\delta^i \mathbf{r}_\delta + Q_\delta^k \mathbf{w}$$

for any $\mathbf{w} \in \mathbb{R}^S$, $\delta \in \Delta$, and $k \in I_+$. We interpret $\mathbf{G}_\delta^k \mathbf{w}(s)$ as the expected present value of using δ for k periods, starting in state s, and ending with the salvage-value function $\mathbf{w}$. Suppose $\mathbf{w} \le \mathbf{v}$. Then $Q_\delta \ge \mathbf{0}$ implies $\mathbf{G}_\delta^k \mathbf{w} \le \mathbf{G}_\delta^k \mathbf{v}$. Observe that $\lim_{k \to \infty} \mathbf{G}_\delta^k \mathbf{w} = \mathbf{v}(\delta)$ in (4-80), regardless of $\mathbf{w}$. Therefore, $\mathbf{F} \le \mathbf{F}$, $\mathbf{LF} = \mathbf{F}$, and induction on k implies $\mathbf{G}_\delta^k \mathbf{F} \le \mathbf{F}$, for all $\delta \in \Delta$ and k; so $\mathbf{v}(\delta) \le \mathbf{F}$. As a result, $\mathbf{v}(\delta) \le \sup \{\mathbf{v}(\delta)\colon \delta \in \Delta\} = \mathbf{f} \le \mathbf{F}$. Hence, $\mathbf{F} \le \mathbf{f} \le \mathbf{F}$; so $\mathbf{f} = \mathbf{F}$, which completes the proof of part (*a*) of the theorem.

Also, we have proved $\mathbf{f}$ is the unique solution of $\mathbf{Lf} = \mathbf{f}$, which is part (*b*).

□

In the special case $f_0(\cdot) \equiv 0$, $0 \le r(s, a) \le u$ for all $(s, a) \in \mathscr{C}$ causes $\mathbf{f}_n \le \mathbf{f}$ for all n and $\| \mathbf{f}_1 - \mathbf{f}_0 \| \le u$. Then part (*a*) of the theorem implies $0 \le f(s) - f_n(s) \le \beta^n u/(1 - \beta)$.

Example 4-22 Table 4-11 on page 126 presents $f_n(\cdot)$ for a simple deterministic MDP with $\beta = 0.5$ and $f_0(\cdot) \equiv 0$. From that table, $\| \mathbf{f}_{n+1} - \mathbf{f}_n \| = 0.5 \| \mathbf{f}_n - \mathbf{f}_{n-1} \|$ for $1 \le n \le 6$:

n	$\| f_n - f_{n-1} \|$
1	4.5
2	2.25
3	1.125
4	0.5625
5	0.2813
6	0.1406

Convergence of $f_n(s)$ to $f(s)$ is at least geometrically fast. □

From part (*a*) of Theorem 4-8, successive approximations is a *bona fide* algorithm to find an approximately optimal policy, i.e., a policy whose value is close to $f(s)$ for every $s \in S$. Here is the basic idea. For $n \in I_+$, construct a Markov policy π_n as follows. Solve for $\mathbf{f}_1, \mathbf{f}_2, \ldots, \mathbf{f}_n$ and let δ_i be a single-period decision rule which attains the i-period maximum, $i = 1, \ldots, n$. Let $\pi_n = (\delta_n, \delta_{n-1}, \ldots, \delta_2, \delta_1, \gamma_1, \gamma_2, \ldots)$, where $\gamma_1, \gamma_2, \ldots$ is arbitrary. Then, for all $\epsilon > 0$, there is a sufficiently large $n < \infty$ such that $\| \mathbf{f} - \mathbf{v}(\pi_n) \| < \epsilon$.

Section 6-2 suggests ways to accelerate the convergence of $\mathbf{f}_n$ to $\mathbf{f}$. Successive approximations in a setting more general than $\#\mathscr{C} < \infty$ is examined in Section 5-4.

Necessary and Sufficient Conditions for Optimality

From (4-70), recall the notation

$$G(s, \delta) = \{a: a \in A_s, r(s, a) + \sum_{j \in \mathcal{S}} q_{sj}^a v_j(\delta^\infty) > v_s(\delta^\infty)\}$$

Corollary 4-4*b* and Theorem 4-5 imply that δ^∞ is optimal if, and only if, $G(s, \delta) = \phi$ for all $s \in \mathcal{S}$. Therefore, $\delta(s) \in A_s$ implies

$$G(s, \delta) = \phi \Leftrightarrow v_s(\delta) = \max \{r(s, a) + \sum_{j \in \mathcal{S}} q_{sj}^a v_j(\delta): a \in A_s\} \tag{4-88}$$

Here we obtain groups of conditions that are equivalent to $\bigcup_{s \in \mathcal{S}} G(s, \delta) = \phi$, and each group is the basis of an algorithm.

Several algorithms are based on the following result.

Theorem 4-9 $\mathbf{y}$ is a solution of

$$y_s = \max \{r(s, a) + \sum_{j \in \mathcal{S}} q_{sj}^a y_j: a \in A_s\} \qquad s \in \mathcal{S} \tag{4-89a}$$

if, and only if, there is $\delta \in \Delta$ such that $\mathbf{y} = \mathbf{v}(\delta)$ and δ^∞ is optimal.

It is convenient to rewrite (4-89*a*) as

$$\mathbf{y} = \text{vmax} \{\mathbf{r}_\delta + Q_\delta \mathbf{w}: \delta \in \Delta\} \tag{4-89b}$$

where "vmax" connotes component-by-component maximization as in (4-89*a*). We refer to both (4-89*a*) and (4-89*b*) as (4-89).

Proof of Theorem 4-9 (Informal) Observe that $G(s, \delta)$ depends on δ only via $\mathbf{v}(\delta)$. Let $\mathbf{y} = (w_s) \in \mathbb{R}^S$ and define

$$\mathcal{G}(s, \mathbf{y}) = \{a: a \in A_s, r(s, a) + \sum_{j \in \mathcal{S}} q_{sj}^a y_j > y_s\}$$

Then $G(s, \delta) = \mathcal{G}[s, \mathbf{v}(\delta)]$, $\delta \in \Delta$. In general,

$$\bigcup_{s \in \mathcal{S}} \mathcal{G}(s, \mathbf{y}) = \phi \Leftrightarrow r(s, a) + \sum_{j \in \mathcal{S}} q_{sj}^a y_j \le y_s \qquad (s, a) \in \mathscr{C} \tag{4-90}$$

Suppose $\mathbf{y}$ satisfies (4-90). Is $\mathbf{y} = \mathbf{v}(\delta)$ for some optimal policy δ? Actually, it sufficies to determine if $\mathbf{y} = \mathbf{v}(\delta)$ for *any* (rather than *optimal*) policy $\delta \in \Delta$. If so, then (4-88) implies $G(s, \delta) = \phi$ for all $s \in \mathcal{S}$ so δ is optimal.

From (4-82), for every $\delta \in \Delta$, $v_s(\delta) = r[s, \delta(s)] + \sum_{j \in \mathcal{S}} q_{sj}^{\delta(s)} v_j(\delta)$, $s \in \mathcal{S}$. Therefore, if $\mathbf{y} = \mathbf{v}(\delta)$ for some $\delta \in \Delta$ and $\mathbf{y}$ satisfies (4-90), there must be equality in (4-90) when $a = \delta(s)$. In other words, for each $s \in \mathcal{S}$, in (4-90) the inequality cannot be strict for all $a \in A_s$. Therefore, $\mathbf{y}$ satisfies (4-90) and $\mathbf{y} = \mathbf{v}(\delta)$ for some $\delta \in \Delta$ if, and only if, (4-89) is satisfied. □

Theorem 4-6 asserts existence of an optimal policy (if $\#\mathscr{C} < \infty$ and $0 \le \beta < 1$). Therefore, we should be able to claim that (4-89) has a solution.

Theorem 4-10 Equation (4-89) has a solution $\mathbf{y}^*$, and it is unique.

Proof Existence of a solution is implied by existence of an optimal policy (Theorem 4-6). From (4-89*b*), if $\mathbf{y}^*$ is a solution, then $\mathbf{y}^* \geq \mathbf{r}_\delta + Q_\delta \mathbf{y}^*$ for all $\delta \in \Delta$ with equality for some δ. Therefore,

$$\mathbf{y}^* \geq (\mathbf{I} - Q_\delta)^{-1}\mathbf{r}_\delta \qquad \delta \in \Delta$$

and
$$\mathbf{y}^* = \text{vmax}\,\{(\mathbf{I} - Q_\delta)^{-1}\mathbf{r}_\delta \colon \delta \in \Delta\}$$

with the "vmax" understood, as in (4-89*b*), to apply to each component. Let $[(\mathbf{I} - Q_\delta)^{-1}\mathbf{r}_\delta]_s$ denote the sth component of the S-vector $(\mathbf{I} - \mathbf{Q}_\delta)^{-1}\mathbf{r}_\delta$. For each $s \in \mathcal{S}$, $\mathbf{y}_s^*$ is the maximum of the set of numbers $\{[(\mathbf{I} - Q_\delta)^{-1}\mathbf{r}_\delta]_s \colon \delta \in \Delta\}$; so it is unique. □

Identification of an Optimal Policy

Policy improvement is an algorithm that solves (4-89). The iterations continue until $\bigcup_{s \in \mathcal{S}} G(s, \delta) = \phi$; so at that point, $\mathbf{y} = \mathbf{v}(\delta)$ satisfies (4-89). Also, (4-84) and part (*a*) of Theorem 4-7 imply that successive approximations solves (in the limit) (4-89). Suppose you solve (4-89) with one of these algorithms or some other one. Then you know $y_s = f(s)$, which is the value of an optimal policy from the initial state s, for each $s \in \mathcal{S}$. The values y_s, $s \in \mathcal{S}$, can be used in the following way to identify an optimal policy, i.e., specify $\delta \in \Delta$ such that $\mathbf{v}(\delta) = \mathbf{y} = \mathbf{f}$.

Corollary 4-10 For each $s \in \mathcal{S}$, let $\delta(s)$ be some $a \in A_s$ that causes the maximum in (4-89) to be attained. Then δ^∞ is optimal.

Proof (4-89) and $\mathbf{y} = \mathbf{v}(\delta)$ implies $G(s, \delta) = \phi$ for each $s \in \mathcal{S}$; so δ^∞ is optimal.

□

Linear Programming

Another class of algorithms for finite MDPs uses linear programs. Let $\alpha_s > 0$ be arbitrary but positive for each $s \in \mathcal{S}$. Consider the linear-programming problem of choosing variables y_s, $s \in \mathcal{S}$, in order to

$$\text{Minimize } \sum_{s \in \mathcal{S}} \alpha_s y_s$$

$$\text{subject to} \tag{4-91}$$

$$y_s - \sum_{j \in \mathcal{S}} q_{sj}^a y_j \geq r(s, a) \qquad (s, a) \in \mathscr{C}$$

whose constraints are the same as (4-90). The coefficient of y_j in the (s, a) constraint in (4-91) is $1 - q_{ss}^a$ if $j = s$, and is $-q_{sj}^a$ if $j \neq s$. There is an important relationship between (4-89) and (4-91).

Theorem 4-11 $\mathbf{y}$ is optimal in (4-91) $\Leftrightarrow$ $\mathbf{y}$ is the solution of (4-89a), i.e.,

$$y_s = \max \{r(s, a) + \sum_{j \in \mathcal{S}} q_{sj}^a y_j : a \in A_s\} \qquad s \in \mathcal{S} \qquad \text{(4-89}a\text{)}$$

PROOF Suppose $\mathbf{y}$ is optimal, hence feasible in (4-91), i.e.,

$$0 \le \xi_{sa} \triangleq y_s - \sum_{j \in \mathcal{S}} q_{sj}^a y_j - r(s, a) \qquad (s, a) \in \mathscr{C} \qquad \text{(4-92)}$$

but assume $\mathbf{y}$ is not a solution of (4-89a). We shall obtain a contradiction. Then there is at least one state $k \in \mathcal{S}$ and $m > 0$ such that $0 < m \le \xi_{ka}$ for all $a \in A_k$. Specify $\mathbf{y}'$ as $y'_j = y_j$ if $j \ne k$ and $y'_k = y_k - m$. Let ξ'_{sa} denote ξ_{sa} when $\mathbf{y}'$ replaces $\mathbf{y}$. Then

$$\xi'_{sa} = \xi_{sa} + mq_{sk}^a \ge \xi_{sa} \ge 0 \qquad s \ne k,\ a \in A_S$$

$$\xi'_{ka} = \xi_{ka} - (1 - q_{kk}^a)m \ge \xi_{ka} - m \ge 0 \qquad a \in A_k$$

so $\mathbf{y}'$ is feasible in (4-91). However,

$$\sum_{j \in \mathcal{S}} \alpha_j (y_j - y'_j) = \alpha_k m > 0$$

so $\mathbf{y}$ is not optimal in (4-91).

Now we use a contrapositive proof to show that if $\mathbf{y}$ is the solution of (4-89a), then it must be optimal in (4-91). If $\mathbf{y}$ is feasible but not optimal in (4-91), either (i) there is at least one state k and $m > 0$ such that $0 < m \le \xi_{ka}$ for all $a \in A_k$ or (ii) $\min \{\xi_{sa} : a \in A_s\} = 0$ for all $s \in \mathcal{S}$. Case (i) denies that $\mathbf{y}$ solves (4-89a). Case (ii) is impossible unless there are at least two solutions of (4-89a) [we have already proved that an optimal solution of (4-91) is a solution of (4-89)]. Hence, (ii) is precluded by Theorem 4-10. □

It follows from Corollary 4-10 and Theorem 4-11 that it is simple to identify an optimal policy in an optimal solution to the linear program (4-91). For each state s, identify an action a whose "slack" variable ξ_{sa} [defined in (4-92)] is zero in the optimal solution. Then assign $\delta(s) = a$ because a causes the maximum in (4-89) to be attained.

Example 4-23: Continuation of Example 4-20 The MDP in Example 4-8 and 4-20 is an inventory model with $\mathcal{S} = \{0, \ldots, 4\}$ and $A_s = \{s, s+1, \ldots, 4\}$ so $\#\mathscr{C} = 15$. The basic data are $\beta = 0.9$ and Tables 4-14 and 4-15.

In Example 4-20, an optimal policy is found to be $a = 2 \vee s$ if $s \ne 1$ and $a = 1$ if $s = 1$ with concomitant values $y_0 = y_3 = y_2 + 1 = -4$, $y_1 = -4.77$, and $y_4 = -1.42$. The optimality test performed in Example 4-20 verifies (4-89a).

The constraints for the linear program (4-91) for this example are presented in Table 4-16. The first two columns indicate the values of s and a to which the row corresponds. The next five columns present the coefficients of $y_0, y_1, \ldots, y_4$ in the (s, a) constraint, and the last column specifies the

Table 4-14. Single-period expected costs, $r(s, a)$

Inventory after ordering

	$s \backslash a$	0	1	2	3	4
	0	0	0	0	1	3
Inventory	1	...	−1	0	1	3
before	2	...	...	−1	1	3
ordering	3	...	...	...	0	3
	4	...	...	...	...	2

"right-hand side" $r(s, a)$. The linear program with the constraints in Table 4-16 and the objective

$$\text{Maximize } \sum_{s=0}^{4} \alpha_s y_s$$

for any positive numerical values of $\alpha_0, \ldots, \alpha_4$ has optimal values of y_s mentioned above. The objective has "maximize" instead of "minimize" because Table 4-14 presents costs instead of rewards.

Maximizing $\sum_{s=0}^{4} \alpha_s y_s$ (with $\alpha_s > 0$ for each s) subject to the constraints in Table 4-16 yields a solution in which the slack variables ξ_{sa} are zero for $(s, a) \in \{(0, 2), (1, 1)\ (2, 2), (3, 3), (4, 4)\}$. Therefore, an optimal policy is δ^∞, where $\delta(0) = 2$, $\delta(1) = 1$, $\delta(2) = 2$, $\delta(3) = 3$, and $\delta(4) = 4$. This is precisely the policy which is optimal via the policy-improvement algorithm in Example 4-20. Also, an optimal linear-program solution yields

$$y_0 = -4,\ y_1 = -4.774,\ y_2 = -5,\ y_3 = -4, \text{ and } y_4 = -1.419$$

which were the optimal values found in Example 4-20. □

The linear program which is dual to (4-91) is

$$\text{Maximize } \sum_{(s, a) \in \mathscr{C}} r(s, a) x_{sa}$$

Table 4-15. β × transition probabilities, q_{sj}^a

Subsequent inventory level

	$a \backslash j$	0	1	2	3	4
	0	0.9	0	0	0	0
Inventory	1	0.675	0.225	0	0	0
after	2	0.45	0.225	0.225	0	0
ordering	3	0.225	0.225	0.225	0.225	0
	4	0	0.225	0.225	0.225	0.225

Table 4-16 Constraints for the linear program in Example 4-23

s	a	y_0	y_1	y_2	y_3	y_4		$r(s, a)$
0	0	0.1	0	0	0	0		0
0	1	0.325	−0.225	0	0	0		0
0	2	0.55	−0.225	−0.225	0	0		0
0	3	0.775	−0.225	−0.225	−0.225	0		1
0	4	1.0	−0.225	−0.225	−0.225	−0.225	$\begin{bmatrix} y_0 \\ y_1 \\ y_2 \\ y_3 \\ y_4 \end{bmatrix} \le$	3
1	1	−0.675	0.775	0	0	0		−1
1	2	−0.45	0.775	−0.225	0	0		0
1	3	−0.255	0.775	−0.225	−0.225	0		1
1	4	0	0.775	−0.225	−0.225	−0.225		3
2	2	−0.45	−0.225	0.775	0	0		−1
2	3	−0.225	−0.225	0.775	−0.225	0		1
2	4	0	−0.225	0.775	−0.225	−0.225		3
3	3	−0.225	−0.225	−0.225	0.775	0		0
3	4	0	−0.225	−0.225	0.775	−0.225		3
4	4	0	−0.225	−0.225	−0.225	0.775		2

subject to $x_{sa} \geq 0$, $(s, a) \in \mathscr{C}$, and

$$\sum_{a \in A_j} x_{ja} - \beta \sum_{(s,\, a) \in \mathscr{C}} p_{sj}^{a} x_{sa} = \alpha_j \qquad j \in \mathcal{S}$$

In Section 6-1 we analyze this dual problem under the assumption that $0 < \alpha_j = P\{s_1 = j\}$ for all j (so $\sum_{j \in \mathcal{S}} \alpha_j = 1$). Then the conclusion is that an optimal solution $\{x_{sa}\}$ has the interpretation

$$x_{sa} = \sum_{n=1}^{\infty} \beta^{n-1} P\{s_n = s,\, a_n = a\}$$

Example 4-24 Continuation of Example 4-23 If $\alpha_j = 0.2$ for $j = 0, 1, \ldots, 4$ in Example 4-23, then the dual linear program has the optimal solution

$$x_{02} = 5.060,\ x_{11} = 2.450,\ x_{22} = 1.899,\ x_{33} = 0.333,\ x_{44} = 0.258$$

and $x_{sa} = 0$ for all other (s, a). Therefore, if $P\{s_1 = j\} = 0.2$ for all j, then

$$\sum_{n=1}^{\infty} (0.9)^{n-1} P\{s_n = 2\} = \sum_{a \in A_2} x_{2a} = x_{22} = 1.899$$

Observe that

$$\sum_{(s,\, a) \in \mathscr{C}} x_{sa} = 10 = (1 - 0.9)^{-1} = (1 - \beta)^{-1}$$

which is true in general (see Section 6-1). □

Complementarity and Fixed-Point Approaches

Yet another route to a solution of (4-89) is provided by (4-92) and an alternative condition for the maximum in (4-90) to be achieved; i.e., for each s there is at least one $a \in A_s$ with $\xi_{sa} = 0$. Let $\mathbf{D} = (D_{sa}, a \in A_s, s \in \mathcal{S})$. You may interpret D_{sa} as the stationary conditional probability of taking action a given that the state is s.

Theorem 4-12 $\mathbf{y}$ is a solution of

$$y_s = \max \{r(s, a) + \beta \sum_{j \in \mathcal{S}} q_{sj}^a y_j : a \in A_s\} \quad s \in \mathcal{S} \tag{4-89}$$

if, and only if, there exist $\mathbf{D}$ and $\boldsymbol{\xi}$ such that

$$y_s = r(s, a) + \sum_{j \in \mathcal{S}} q_{sj}^a y_j + \xi_{sa} \qquad (s, a) \in \mathscr{C} \tag{4-93a}$$

$$\boldsymbol{\xi} \geq \mathbf{0} \qquad \mathbf{D} \geq \mathbf{0} \tag{4-93b}$$

$$\sum_{a \in A_s} D_{sa} = 1 \qquad s \in \mathcal{S} \tag{4-93c}$$

$$\sum_{s \in \mathcal{S}} \sum_{a \in A_s} D_{sa} \xi_{sa} = 0 \tag{4-93d}$$

PROOF Equation (4-93a) and $\boldsymbol{\xi} \geq \mathbf{0}$ comprise (4-90). Then $\mathbf{D} \geq \mathbf{0}$, and (4-93$c$) and (4-93$d$) ensure for each $s \in \mathcal{S}$ that there is at least one $a \in A_s$ such that $\xi_{sa} = 0$. Therefore, $\mathbf{y}$ is a solution of (4-90) if $(\mathbf{D}, \boldsymbol{\xi}, \mathbf{y})$ is a solution of (4-93). Conversely, if $\mathbf{y}$ is a solution of (4-89), then $\boldsymbol{\xi}$ defined by (4-92) satisfies (4-93a) and $\boldsymbol{\xi} \geq \mathbf{0}$. For each $s \in \mathcal{S}$ let $D_{sz} = 1$ for some $z \in A_s$ with $\xi_{sz} = 0$ and $D_{sa} = 0$ for all $a \in A_s - \{z\}$. Then $\mathbf{D}$ satisfies (4-93c) and $(\mathbf{D}, \boldsymbol{\xi})$ satisfies (4-93d). □

You may interpret (4-93a-d) as requiring that the conditional probability of taking action a in state s be zero unless action $a \in A_s$ attains the maximum in (4-90a).

Condition (4-93d) is termed *complementary slackness* and (4-93) is a *linear complementarity problem*. There are several algorithms for such problems† but they have yet to be shown effective on large MDPs.

Complementarity approaches other than (4-93) have been proposed for MDPs in Eaves (1977) and Koehler (1979). Eaves presents a special complementarity algorithm for MDPs. Koehler shows that the linear program (4-91) is equivalent to a convex quadratic program to which he applies complementarity algorithms.

A complementarity problem also can be interpreted as a fixed-point problem or as the search for a zero of a mapping. For $\mathbf{y} \in \mathbb{R}^S$, in notation similar to (4-86), let

$$\mathbf{L}^*\mathbf{y} \triangleq \text{vmax} \{\mathbf{r}_\delta + (Q_\delta - \mathbf{I})\mathbf{y} : \delta \in \Delta\}$$

† B. C. Eaves and H. Scarf, "The Solution of Systems of Piecewise Linear Equations," *Math. Oper. Res.* **1**: 1–27 (1976).

Then $\mathbf{y}$ solves (4-89*b*) if, and only if,

$$\mathbf{L}\mathbf{y} = \mathbf{y}$$

so

$$\mathbf{L}^*\mathbf{y} = \mathbf{0}$$

We say that $\mathbf{y}$ is a fixed point of $\mathbf{L}$ and a zero of $\mathbf{L}^*$. The fixed-point property has already been asserted in part (*b*) of Theorem 4-8. See the books† by Scarf and Zangwill and Garcia for algorithms which are avowedly directed at the computation of approximate fixed points.

Example 4-25: Continuation of Example 4-23 The linear complementarity problem (4-93) consists of a modification of Table 4-16 plus

$$\sum_{a=0}^{4} D_{0a} = \sum_{a=1}^{4} D_{1a} = \sum_{a=2}^{4} D_{2a} = \sum_{a=3}^{4} D_{3a} = D_{44} = 1$$

$$\boldsymbol{\xi} \geq \mathbf{0},\ \mathbf{D} \geq \mathbf{0}$$

$$\sum_{s=0}^{4} \sum_{a=s}^{4} D_{sa}\, \xi_{sa} = 0$$

The modification of Table 4-16 is to subtract ξ_{sa} from the (s, a) constraint. Then $D_{02} = D_{12} = D_{22} = D_{33} = D_{44} = 1$, all other $D_{sa} = 0$, and $\mathbf{y} = (-2.25, -2.25, -3.25, -2.25, 0.33)$ satisfies this system because the optimality test in Example 4-20 determines $\boldsymbol{\xi} \geq 0$ with $\xi_{sa} = 0$ whenever $D_{sa} > 0\ (=1)$ above.

□

EXERCISES

4-31 Under assumption (4-76), that is, $\#\mathscr{C} < \infty$, let $\pi \triangleq (\delta_1, \delta_2, \ldots, \delta_N)$, and consider the problem

$$\sup_{\pi} E\left(\sum_{n=1}^{N} \beta^{n-1} r[s_n, \delta_n(s_n)] \mid s_1 = s \right) \qquad s \in \mathbb{S} \tag{4-94}$$

where $N < \infty$. Prove that there is an optimal policy for this finite-horizon problem. [Hint: There are at least two proofs. (i) For each fixed N, show that this problem is a special case of an infinite-horizon problem; then invoke Theorem 4-6. (ii) Use an inductive proof that starts with $N = 1$. Exploit $\#\mathscr{C} < \infty$ and construct $f_N(\cdot)$.]

4-32 (*a*) For any $\mathbf{f}_0 \in \mathbb{R}^S$, prove that the recursion (4-84) satisfies (4-87), hence (4-86).
(*b*) Use (4-87) and $0 \leq \beta < 1$ to prove that $\mathbf{L}$ is a continuous mapping on $\mathbb{R}^S$.
(*c*) Explain in detail how (4-86) is a special case of (4-87).

4-33 Consider a finite-horizon version (4-94) of Example 4-23. Let $N = 2$. Formulate this problem as a special case of an infinite-horizon problem and specify a linear program (4-91) (in detail) which would solve it.

4-34 Prove that the moment formula (4-82) in Theorem 4-7 is valid.

† Scarf H., *The Computation of Economic Equilibria* (in collaboration with T. Hansen), Yale University Press, New Haven, Conn. (1973). W. I. Zangwill, and C. B. Garcia, *Pathways to Solutions, Fixed Points, and Equilibria*, Prentice-Hall, Englewood Cliffs, N.J. (1981).

4-35 Use linear programming to find a policy which minimizes the expected present value (infinite planning horizon), with $\beta = 0.9$, in the maintenance model in Exercise 3-15.

4-36 Let δ denote the policy you offer as a solution to Exercise 4-35. Compute $\boldsymbol{\sigma}^2(\delta)$.

4-37 Repeat Exercise 4-35 using the policy-improvement algorithm.

4-38 Show that your solution to Exercise 4-35 is associated with a solution to the linear-complementarity problem (4-93).

4-39 (It is advisable to work this exercise on a computer.) Use linear programming to find a policy which maximizes the present-value (infinite-planning-horizon) criterion, with $\beta = 0.85$, for the fishery model in Exercise 3-16. Check the correctness of your solution by identifying the myopic optimum.

4-40 (It is advisable to work this exercise on a computer.) Repeat Exercise 4-39 using the policy-improvement algorithm.

4-41 Suppose for each s, k, and a that $\sum_{j \in \mathcal{S}} p_{sjk}^a$ is either zero or one.

(*a*) Interpret this assumption.

(*b*) Let $r_s = r[s, \delta(s)]$ and suppress other notational dependence on δ. Prove the identity

$$\theta_s = \sum_{j \in \mathcal{S}} p_{sj}(r_s + \beta v_j)^2 - v_s^2 = \beta^2 \sum_{j \in \mathcal{S}} p_{sj} v_j^2 - (v_s - r_s)^2$$

4-42 In the inventory model in Section 3-1, suppose that there is room to hold at most 4 items in inventory and let $r = 10$, $c = 5$, $h = 0.5$, $\beta = 0.9$, and $\mathbf{P}\{D_1 = i\} = 0.5^{i+1}$, $i \in I$. Formulate a linear program for the problem of maximizing the expected present value of the net profits.

4-43 In the notation of Lemma 4-3, let

$$\tilde{F}_s(z) = \int_0^\infty e^{-zt}\, dF_s(t)$$

(*a*) Prove that

$$\tilde{F}_s(z) = \sum_{j \in \mathcal{S}} \sum_{k \in \mathcal{K}} p_{sjk} e^{-zk} \tilde{F}_j(z\beta)$$

(*b*) Use (*a*) to obtain another proof of (4-80) in Theorem 4-7.

(*c*) Use (*a*) to obtain another proof of (4-81). (Hint: This exercise uses properties of Laplace-Stieltjes transforms. See pages 516 to 529 in Volume I for a summary of properties of transforms.)

4-44 Use Lemma 4-3 to obtain another proof of (4-80) and (4-81) in Theorem 4-7.

4-45 The conditional variance formula is (cf. Exercise 6-1 on page 173 of Volume I)

$$\text{Var}(X) = \text{Var}[E(X \mid Y)] + E[\text{Var}(X \mid Y)]$$

Use this formula to obtain another proof of (4-81). Hint: $\theta_s = \text{Var}[E(B_s \mid s_2)]$.

4-6 AVERAGE-REWARD CRITERION

The objective in Sections 4-3, 4-4, and 4-5 is maximization of the expected discounted return. This section concerns the maximization of the long-run average of the expected reward per period. Our objective is (1) to prove that some stationary policy is optimal, and (2) to present algorithms (policy improvement and linear programming) for computing an optimal policy.

The mathematics of MDPs with the average-reward criterion is richer and more delicate than that of MDPs with the discounted criterion. Hence, throughout this section we assume that the MDP has only finitely many states, and only finitely many actions in each state; i.e., we assume $\#\mathscr{C} < \infty$.

Suppose a discounted criterion is based on an annual opportunity cost of 25 percent (due to interest and other considerations) so that the annual discount factor is $1/(1 + 0.25) = 0.8$. If decisions are made daily, the appropriate daily discount factor β satisfies $\beta^{365} = 0.8$ so $\beta \doteq 0.9994$. There are two reasons why a discounted criterion might not be used in this situation. First, if β is close to 1, the algorithms in Section 4-5 are either slow to converge, numerically unstable, or both. Second, if β is close enough to 1, then $\sum_{n=1}^{N} \beta^{n-1} x_n$ is close to $\sum_{n=1}^{N} x_n = N(\sum_{n=1}^{N} x_n/N)$; so maximization of the long-run average reward is a reasonable alternative criterion.

Recall the terminology and notation in Section 4-3. A single-stage decision rule δ is a function on $\mathcal{S}$ that assigns $\delta(s) \in A_s$ for all $s \in \mathcal{S}$. A Markov policy $\pi = (\delta_1, \delta_2, \ldots)$ is a sequence of single-stage decision rules. A stationary policy is a Markov policy in which δ_n is the same for all n. Let Δ, Y, and Z denote the sets of single-stage decision rules, Markov policies, and stationary policies, respectively.

For $\pi = (\delta_1, \delta_2, \ldots) \in Y$, let $B(\pi, N)$ denote the total reward during the first N periods:†

$$B(\pi, N) = \sum_{n=1}^{N} r(s_n, a_n) = \sum_{n=1}^{N} r[s_n, \delta_n(s_n)] \tag{4-95}$$

Let $b_s(\pi, N)$ denote the expected total reward during the first N periods if the initial state is s:

$$b_s(\pi, N) = E[B(\pi, N) \mid s_1 = s] = E\left\{\sum_{n=1}^{N} r[s_n, \delta_n(s_n)] \mid s_1 = s\right\} \tag{4-96}$$

The average reward (or average return) is the asymptotic rate at which $b_s(\pi, N)$ grows with N. Unfortunately,

$$\lim_{N \to \infty} \frac{b_s(\pi, N)}{N} \tag{4-97}$$

usually does not exist for some policies $\pi \in Y$; so we must settle on a limit point of $b_s(\pi, N)/N$. For a technical reason,‡ we choose to maximize

$$\liminf_{N \to \infty} \frac{b_s(\pi, N)}{N}$$

In any case, we shall see that the limit exists (so lim inf = lim sup) if π is a stationary policy.

† It is implicit in (4-95) that the salvage-value function is identically zero. None of our results depend on that assumption, which is made for expository convenience.

‡ The reason for our use of lim inf stems from part (*a*) of Lemma 4-4 forthcoming.

For $\pi \in Y$, let $g_s(\pi)$ denote the average reward per period if π is the policy and the initial state is s:

$$g_s(\pi) \triangleq \liminf_{N\to\infty} \frac{b_s(\pi, N)}{N}$$
$$= \liminf_{N\to\infty} \frac{E\{\sum_{n=1}^{N} r[s_n, \delta_n(s_n)] \mid s_1 = s\}}{N} \tag{4-98}$$

Some writers call $g_s(\pi)$ the *gain rate* of policy π from initial state s. We use the labels *average return per period* and *average reward per period*. Let u_s denote the maximal average return per period from initial state s:

$$u_s \triangleq \sup \{g_s(\pi): \pi \in Y\}$$

Let $\mathbf{b}(\pi, N)$, $\mathbf{g}(\pi)$, and $\mathbf{u}$ be the vectors with sth components $b_s(\pi, N)$, $g_s(\pi)$, and u_s, respectively. If π is a stationary policy, that is, $\pi = \delta^\infty$ for some $\delta \in \Delta$, then we simplify the notation by writing $\mathbf{b}(\delta, N)$, $b_s(\delta, N)$, $\mathbf{g}(\delta)$, and $g_s(\delta)$ instead of $\mathbf{b}(\delta^\infty, N)$, $b_s(\delta^\infty, N)$, $\mathbf{g}(\delta^\infty)$, and $g_s(\delta^\infty)$.

A policy is said to be *optimal* if its average return per period is at least as great as that of every other policy from every initial state.

Definition 4-18 Policy $\pi_* \in Y$ is *optimal* if

$$\mathbf{u} = \mathbf{g}(\pi_*)$$

The proof that some stationary policy is optimal and the justification of the algorithms to find an optimal policy depend on the connection between a policy's average return per period and its discounted return as the discount factor tends to 1. Let P be a stochastic matrix and let $P^0 \triangleq I$, the identity matrix. The proofs depend too on the behavior of $\sum_{n=0}^{N-1} P^n/N$ as $N \to \infty$. Both limiting results use the following mathematical prerequisite.

A Tauberian Theorem

Abelian and Tauberian theorems address conditions under which sums and integrals defined on open sets retain their values at the boundaries of those sets. Here is the Tauberian theorem, which we use to relate the discounted and average-return criteria.

Proposition 4-6 (*a*) If $K(\cdot)$ is nondecreasing on $[0, \infty)$ and

$$f(x) = \int_0^\infty e^{-xt}\, dK(t) \qquad x > 0$$

is convergent, then

$$\liminf_{t\to\infty} \frac{K(t)}{t} \le \liminf_{x\downarrow 0} xf(x) \le \limsup_{x\downarrow 0} xf(x) \le \limsup_{t\to\infty} \frac{K(t)}{t}$$

(*b*) If $\lim_{t\to\infty} K(t)/t$ exists, then $\lim_{x\downarrow 0} xf(x)$ exists and

$$\lim_{x\downarrow 0} xf(x) = \lim_{t\to\infty} \frac{K(t)}{t}$$

PROOF See pages 181 and 182 in D. V. Widder, *The Laplace Transform*, Princeton University Press, Princeton, N.J. (1941). □

The following result is a corollary of Proposition 4-6.

Proposition 4-7 (*a*) Let $0 \le c_n \le m < \infty$ for all $n \in I_+$. Then

$$\liminf_{N\to\infty} \frac{1}{N} \sum_{n=1}^{N} c_n \le \liminf_{\beta\uparrow 1} (1-\beta) \sum_{n=1}^{\infty} \beta^{n-1} c_n$$

$$\le \limsup_{\beta\uparrow 1} (1-\beta) \sum_{n=1}^{\infty} \beta^{n-1} c_n \le \limsup_{N\to\infty} \frac{1}{N} \sum_{n=1}^{N} c_N.$$

(*b*) If $\lim_{N\to\infty} \sum_{n=1}^{N} c_n/N$ exists, then $\lim_{\beta\uparrow 1} (1-\beta) \sum_{n=1}^{\infty} \beta^{n-1} c_n$ exists and

$$\lim_{\beta\uparrow 1} (1-\beta) \sum_{n=1}^{\infty} \beta^{n-1} c_n = \lim_{N\to\infty} \frac{1}{N} \sum_{n=1}^{N} c_N$$

PROOF Left as Exercise 4-54. □

Application of Proposition 4-7

For a Markov policy $\pi = (\delta_1, \delta_2, \ldots)$, let the notation $v_s(\pi, \beta)$ make explicit the dependence of the expected present value on the discount factor β:

$$v_s(\pi, \beta) \triangleq E\left\{ \sum_{n=1}^{\infty} \beta^{n-1} r[s_n, \delta_n(s_n)] \mid s_1 = s \right\} \qquad s \in \mathcal{S},\ 0 \le \beta < 1,\ \pi \in Y$$

As in Section 4-5, the assumption $\#\mathscr{C} < \infty$ permits us to assume that $r(\cdot, \cdot)$ is nonnegative and bounded on $\mathscr{C}$. Therefore, Proposition 4-7 has the following consequence.

Lemma 4-4 Suppose $\#\mathscr{C} < \infty$. Then

(*a*) $$\liminf_{\beta\uparrow 1} (1-\beta) v_s(\pi, \beta) \ge g_s(\pi) \qquad s \in \mathcal{S},\ \pi \in Y \tag{4-99}$$

(*b*) If $\lim_{N\to\infty} b_s(\pi, N)/N$ exists, then $\lim_{\beta\uparrow 1}(1-\beta) v_s(\pi, \beta)$ exists and

$$g_s(\pi) = \lim_{\beta\uparrow 1} (1-\beta) v_s(\pi, \beta)$$

PROOF Left as Exercise 4-55. □

Lemma 4-5 Let P be the matrix of transition probabilities for a Markov chain with S states. That is, P is $S \times S$, has nonnegative entries, and each row sums to 1. (*a*) There exists the limit

$$P^* = \lim_{N \to \infty} \frac{1}{N} \sum_{n=0}^{N-1} P^n$$

where $P^0 \triangleq I$ is the identity matrix.

(*b*) $P^* = PP^* = P^*P = P^*P^*$
(*c*) There exists $(I - P + P^*)^{-1}$
(*d*) $P^* = \lim_{\beta \uparrow 1} (1 - \beta) \sum_{n=0}^{\infty} \beta^n P^n$
(*e*) Rank $(I - P) = S -$ rank $P^* = S -$ number of communicating classes in P

Proof Parts (*a*), (*b*), (*c*), and (*e*) are the following results in Volume I: Propositions 7-6 and 7-7 on page 259 and Lemmas 7-1 and 7-2 on page 256. Part (*d*) is implied by (*a*) above and part (*b*) of Proposition 4-7. □

Suppose $\pi = \delta^\infty$ for some $\delta \in \Delta$. Let P_δ denote the matrix whose (s, j)th entry is $p_{sj}^{\delta(s)}$, and let

$$P_\delta^* = \lim_{N \to \infty} \frac{1}{N} \sum_{n=0}^{N-1} P_\delta^n \qquad \delta \in \Delta \tag{4-100}$$

whose existence is asserted in (*a*) of Lemma 4-5. Let $\mathbf{r}_\delta$ denote the vector whose sth entry is $r[s, \delta(s)]$.

The following theorem is fundamentally important. It is the basis of policy-improvement algorithms for the average-reward criterion.

Theorem 4-13 For each $\delta \in \Delta$:

(*a*)
$$\mathbf{g}(\delta) = P_\delta^* \mathbf{r}_\delta \tag{4-101}$$

(*b*) The number of linearly independent equations in

$$(I - P_\delta)\mathbf{w} = \mathbf{r}_\delta - \mathbf{g}(\delta)$$

is S minus the number of communicating classes in P_δ.

(*c*) The equations

$$(I - P_\delta)\mathbf{w} = \mathbf{r}_\delta - \mathbf{g} \tag{4-102}$$

$$P_\delta^* \mathbf{w} = \mathbf{0} \tag{4-103}$$

have the solution $\mathbf{g} = \mathbf{g}(\delta)$ and $\mathbf{w} = \mathbf{w}(\delta)$ where

$$\mathbf{w}(\delta) \triangleq (I - P_\delta + P_\delta^*)^{-1}(I - P_\delta^*)\mathbf{r}_\delta \tag{4-104}$$

(*d*) $\mathbf{g} = \mathbf{g}(\delta)$ and $\mathbf{w} = \mathbf{w}(\delta)$ is the unique solution to (4-102) and (4-103) for which $g_s = g_j$ if s and j are in the same communicating class of P_δ, and $g_s = g_s(\delta)$ if state s is transient in P_δ.

Lemma 4-5 is the principal tool in the proof below of Theorem 4-13. By using properties of regenerative reward processes (Section 6-4 in Volume I) and further Markov-chain results (from Section 7-7 in Volume I), a more intuitive proof than the one below can be constructed. We outline this proof in Exercises 4-56, 4-57, and 4-58. In particular, Exercise 4-58 asks you to show that (4-103) follows from properties of the matrix $F_\delta \triangleq (I - P_\delta + P_\delta^*)^{-1}$ given in Section 7-7 of Volume I and $\mathbf{w}(\delta) = (F_\delta - P_\delta^*)\mathbf{r}_\delta$ via Corollary 7-15 on page 257 of Volume I. It follows from that corollary and Exercise 4-57 that

$$b_s(N) \triangleq E\left\{\sum_{n=1}^{N} r[s_n, \delta(s_n) \mid s_1 = s\right\} \approx Ng_s(\delta) + w_s(\delta)$$

if all states are in the same communicating class. Then

$$w_s(\delta) - w_j(\delta) = \lim_{N\to\infty} [b_s(N) - b_j(N)]$$

measures the asymptotic advantage of starting in state s rather than state j. Also,

$$g_s(\delta) = \lim_{N\to\infty} \frac{b_s(N)}{N}$$

which justifies the label *gain rate* for $g_s(\delta)$. It is the long-run average reward per unit time.

Proof of Theorem 4-13 Using the notation of (4-96), let $\mathbf{b}(\pi, N)$ denote the vector whose sth entry is $b_s(\pi, N)$. Then

$$\mathbf{b}(\delta, N) = \mathbf{r}_\delta + P_\delta \mathbf{r}_\delta + \cdots + P_\delta^{N-1}\mathbf{r}_\delta = \sum_{n=0}^{N-1} P_\delta^n \mathbf{r}_\delta$$

Therefore, (4-100) implies that there exists

$$\lim_{N\to\infty} \frac{b(\delta, N)}{N} = \lim_{N\to\infty} \frac{1}{N} \sum_{n=0}^{N-1} P_\delta^n \mathbf{r}_\delta = P_\delta^* \mathbf{r}_\delta$$

which proves (*a*). Part (*b*) is the result of part (*e*) of Lemma 4-5.

In order to prove part (*c*), we shall verify that (4-104) satisfies (4-102) and (4-103). From (4-104),

$$(I - P_\delta + P_\delta^*)\mathbf{w}(\delta) = (I - P_\delta^*)\mathbf{r}_\delta \tag{4-105}$$

Premultiplication of the left side of (4-105) by P_δ^* and part (*b*) of Lemma 4-5 yield

$$P_\delta^*(I - P_\delta + P_\delta^*)\mathbf{w}(\delta) = (P_\delta^* - P_\delta^* + P_\delta^*)\mathbf{w}(\delta) = P_\delta^* \mathbf{w}(\delta)$$

Premultiplication of the right side of (4-105) by P_δ^* yields

$$P_\delta^*(I - P_\delta^*)\mathbf{r}_\delta = (P_\delta^* - P_\delta^*)\mathbf{r}_\delta = \mathbf{0}$$

Therefore, $P_\delta^* \mathbf{w}(\delta) = \mathbf{0}$, which is (4-103) with $\mathbf{w} = \mathbf{w}(\delta)$.

To show that (4-102) is satisfied with $\mathbf{g} = \mathbf{g}(\delta)$ and $\mathbf{w} = \mathbf{w}(\delta)$, expand (4-105) and write

$$\mathbf{w}(\delta) - P_\delta \mathbf{w}(\delta) + P_\delta^* w(\delta) = \mathbf{r}_\delta - P_\delta^* \mathbf{r}_\delta$$

Use (4-103) to write this as

$$\mathbf{w}(\delta) - P_\delta \mathbf{w}(\delta) + \mathbf{0} = \mathbf{r}_\delta - \mathbf{g}(\delta)$$

which is (4-102).

In order to prove part (*d*), suppose s and j are in the same class of P_δ and let $\boldsymbol{\lambda}$ be the stationary probability vector common to all the rows of P_δ^* whose states are in the same class as s and j. Then (4-101) ($\mathbf{g}_\delta = P_\delta^* \mathbf{r}_\delta$) implies

$$g_s(\delta) - g_j(\delta) = \boldsymbol{\lambda}\mathbf{r}_\delta - \boldsymbol{\lambda} r_\delta = 0$$

Suppose ($\mathbf{g}$, $\mathbf{w}$) is a solution to (4-102) and (4-103) and $g_s = g_j$ if s and j are in the same communicating class of P_δ. Arguing as in the proof that (4-104) satisfies (4-103), we premultiply the left side of (4-102) by P_δ^* and use part (*b*) of Lemma 4-5 ($P^* = PP^* = P^*P = P^*P^*$):

$$P_\delta^*(I - P_\delta)\mathbf{w} = (P_\delta^* - P_\delta^*)\mathbf{w} = \mathbf{0}$$

Premultiplication of the right side of (4-102) by P_δ^* and the use of (4-103) yield

$$P_\delta^*(\mathbf{r}_\delta - \mathbf{g}) = \mathbf{g}(\delta) - P_\delta^* \mathbf{g}$$

so

$$P_\delta^* \mathbf{g} = \mathbf{g}(\delta) \tag{4-106}$$

Suppose state s is in the (recurrent) class $\mathcal{S}'$ with $\boldsymbol{\lambda}^\delta$ common to the rows of P_δ^* corresponding to states in $\mathcal{S}'$. Then $\lambda_j^\delta = 0$ if $j \notin \mathcal{S}'$; so (4-106) yields

$$g_s(\delta) = \sum_{j \in \mathcal{S}'} \lambda_j^\delta g_j = g_s$$

because, by assumption, g_j is the same for all $j \in \mathcal{S}'$.

To prove $w_s = w_s(\delta)$ for all $s \in \mathcal{S}$, let $d_s = w_s(\delta) - w_s$ and $\mathbf{d} = (d_s, s \in \mathcal{S})$. Write (4-102) for each of the two solutions [using $\mathbf{g} = \mathbf{g}(\delta)$ proved above] and subtract:

$$(I - P_\delta)\mathbf{d} = \mathbf{0}$$

Therefore, $\mathbf{d} = P_\delta \mathbf{d}$. Iterating, $\mathbf{d} = P_\delta^2 \mathbf{d} = P_\delta^n \mathbf{d}$ for all $n \in I_+$. As a result,

$$\mathbf{d} = \frac{1}{N} \sum_{n=0}^{N-1} P_\delta^n \mathbf{d} \rightarrow P_\delta^* \mathbf{d}$$

so

$$\mathbf{d} = P_\delta^* \mathbf{d} = P_\delta^*[\mathbf{w}(\delta) - \mathbf{w}] = \mathbf{0} - \mathbf{0} = \mathbf{0}$$

due to (4-103). □

Example 4-26: Continuation of machine-maintenance Examples 4-2, 4-7, and 4-18 There are three states: 0, 1, and 2. Action 0 is available in states 0 and 1 and action 1 is available in states 1 and 2. The following data are the same as those listed below (4-25) in Example 4-7:

$$r(0, 0) = 10 \qquad r(1, 0) = 5 \qquad r(1, 1) = 0 \qquad r(2, 1) = 0$$

$$p_{00}^0 = 0.7 \qquad p_{01}^0 = 0.2 \qquad p_{02}^0 = 0.1$$

$$p_{10}^0 = 0 \qquad p_{11}^0 = 0.4 \qquad p_{12}^0 = 0.6$$

$$p_{10}^1 = 0 \qquad p_{11}^1 = 0 \qquad p_{12}^1 = 1$$

$$p_{20}^1 = 0.4 \qquad p_{21}^1 = 0.1 \qquad p_{22}^1 = 0.5$$

State 0 connotes perfect operation, 1 connotes a minor defect, and 2 connotes obligatory repair. Action $a = 0$ indicates a continuation of production, and $a = 1$ indicates that the machine is in the repair process. The decision problem is whether or not to repair a minor defect. Let δ denote the policy where $\delta(1) = 1$.

Because

$$P_\delta = \begin{pmatrix} 0.7 & 0.2 & 0.1 \\ 0 & 0 & 1 \\ 0.4 & 0.1 & 0.5 \end{pmatrix}$$

is the transition matrix of an irreducible Markov chain, each row of P_δ^* consists of the stationary distribution $\boldsymbol{\lambda}$ satisfying

$$\boldsymbol{\lambda} \geq \mathbf{0} \qquad \boldsymbol{\lambda}\mathbf{e} = 1 \qquad \text{and} \qquad \boldsymbol{\lambda} = \boldsymbol{\lambda} P_\delta$$

where $\mathbf{e}$ is the column vector with $+1$ in each component. Here, $\boldsymbol{\lambda} =$ (40/81, 11/81, 30/81) and the transpose of $\mathbf{r}_\delta$ is (10, 0, 0). In this example and the following one, we write $\mathbf{g}$ and $\mathbf{w}$ instead of $\mathbf{g}(\delta)$ and $\mathbf{w}(\delta)$. Therefore, each component of $\mathbf{g}$ is $\boldsymbol{\lambda}\mathbf{r}_\delta = 400/81 = 4.9383$. Hence, (4-102) is

$$\begin{aligned} 4.9383 + w_0 &= 10 + 0.7w_0 + 0.2w_1 + 0.1w_2 \\ 4.9383 + w_1 &= w_2 \\ 4.9383 + w_2 &= 0.4w_0 + 0.1w_1 + 0.5w_2 \end{aligned}$$

which, as asserted in part (*b*) of Theorem 4-13, contains *only two independent equations but three variables.* We include (4-103), which is equivalent to

$$40w_0 + 11w_1 + 30w_2 = 0$$

to obtain the solution

$$w_0 = 6.1886 \qquad w_1 = -9{,}6510 \qquad \text{and} \qquad w_2 = -4.7127 \quad \square$$

Example 4-27: Continuation of Examples 4-9 and 4-17 In this deterministic MDP with three states, one of the stationary policies induces the network shown in Figure 4-7.

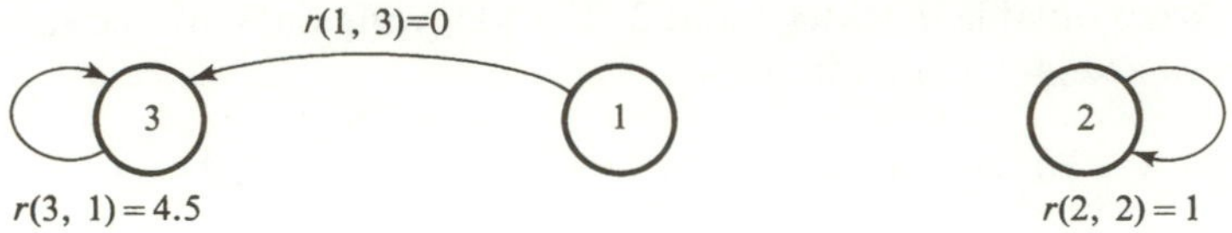

Figure 4-7 Network induced by a stationary policy.

Here, the stationary policy is γ^∞, where $\gamma(1) = 3$, $\gamma(2) = 2$, and $\gamma(3) = 1$. The transition matrix

$$P_\gamma = \begin{pmatrix} 0 & 0 & 1 \\ 0 & 1 & 0 \\ 0 & 0 & 1 \end{pmatrix}$$

has two absorbing states, states 2 and 3; state 1 is transient. Therefore,

$$P_\gamma^* = \begin{pmatrix} 0 & 0 & 1 \\ 0 & 1 & 0 \\ 0 & 0 & 1 \end{pmatrix}$$

so (4-101) yields

$$\mathbf{g} = P_\gamma^* \mathbf{r}_\gamma = \begin{pmatrix} 4.5 \\ 1 \\ 4.5 \end{pmatrix}$$

The equations $P_\gamma^* \mathbf{w} = \mathbf{0}$ yield $w_2 = w_3 = 0$. The first equation in $\mathbf{g} + \mathbf{w} = \mathbf{r}_\gamma + P_\gamma \mathbf{w}$ is $4.5 + w_1 = 0 + w_3$; so $w_1 = -4.5$. □

An Optimal Stationary Policy

The next result is that there is a stationary policy δ_*^∞ which maximizes the average reward per period.

Theorem 4-14 If $\mathscr{C}$ is a finite set, there is a stationary policy δ_*^∞ with

$$\mathbf{u} = \mathbf{g}(\delta_*) \geq \mathbf{g}(\pi) \qquad \pi \in Y \tag{4-107}$$

PROOF Let $\beta_1, \beta_2, \ldots$ be a sequence such that $0 < \beta_k < 1$ for each k and $\beta_k \to 1$ as $k \to \infty$. From Theorem 4-6, the restriction $\#\mathscr{C} < \infty$ implies for each k that some stationary policy δ_k^∞ is optimal for the expected present-value criterion when β_k is the discount factor. That is, there is a single-period decision rule $\delta_k \in \Delta$ such that

$$\mathbf{v}(\delta_k, \beta_k) \geq \mathbf{v}(\pi, \beta_k) \qquad \pi \in Y \tag{4-108}$$

The restriction $\#\mathscr{C} < \infty$ implies that Δ is a finite set; so the sequence $\delta_1, \delta_2, \ldots$ must contain at least one element δ_* which occurs infinitely often. Let $k(1), k(2), \ldots$ index a subsequence with $\delta_* = \delta_{k(1)} = \delta_{k(2)} = \ldots$.

From (4-108), for all $\pi \in Y$,

$$(1 - \beta_{k(j)})\mathbf{v}(\delta_*, \beta_{k(j)}) \geq (1 - \beta_{k(j)})\mathbf{v}(\pi, \beta_{k(j)}) \qquad j = 1, 2, \ldots$$

Therefore, $\beta_{k(j)} \uparrow 1$ as $j \to \infty$, Theorem 4-13, and part (*a*) of Lemma 4-4 imply

$$\mathbf{g}(\delta_*) = \lim_{\beta \uparrow 1}(1 - \beta)\mathbf{v}(\delta_*, \beta) = \lim_{j \to \infty}(1 - \beta_{k(j)})\mathbf{v}(\delta_*, \beta_{k(j)})$$

$$\geq \liminf_{j \to \infty} (1 - \beta_{k(j)})\mathbf{v}(\pi, \beta_{k(j)}) \geq \mathbf{g}(\pi)$$

Hence, $\mathbf{g}(\delta_*) = \mathbf{u}$. □

The Unichain Case

The following simplifying assumption is sometimes appropriate in practice.

Unichain assumption Every single-stage decision rule $\delta \in \Delta$ has a transition matrix P_δ that induces a Markov chain with one communicating class of states and a (possibly empty) set of transient states.

The unichain assumption and finitely many states (because $\#\mathscr{C} < \infty$) implies† existence of a unique stationary distribution which is the same regardless of the initial state. That is, for each $\delta \in \Delta$ there is a unique probability vector $\boldsymbol{\lambda}^\delta$ with components λ_s^δ, $s \in \mathcal{S}$, such that

$$\boldsymbol{\lambda}^\delta \geq \mathbf{0} \qquad \boldsymbol{\lambda}^\delta \mathbf{e} = 1 \qquad \text{and} \quad \lambda_j^\delta = \sum_{s \in \mathcal{S}} \lambda_s^\delta p_{sj}^{\delta(s)} \qquad j \in \mathcal{S} \tag{4-109}$$

where $\mathbf{e}$ denotes the column vector whose components are all 1. Without the unichain assumption, we should have to partition $\mathcal{S}$ into its communicating classes. Since each row‡ of P_δ^* consists of $\boldsymbol{\lambda}^\delta$, each component of $P_\delta^* \mathbf{r}_\delta$ is $\boldsymbol{\lambda}^\delta \cdot \mathbf{r}_\delta$. Let $g_\delta = \boldsymbol{\lambda}^\delta \cdot \mathbf{r}_\delta$. Then equations (4-101), (4-102), and (4-103) reduce to

$$g_\delta = \boldsymbol{\lambda}^\delta \cdot \mathbf{r}_\delta \tag{4-110}$$

$$\boldsymbol{\lambda}^\delta \cdot \mathbf{w} = 0 \tag{4-111}$$

$$\mathbf{e} \cdot g_\delta + \mathbf{w} = \mathbf{r}_\delta + P_\delta \mathbf{w} \tag{4-112}$$

$$\boldsymbol{\lambda}^\delta \geq \mathbf{0} \tag{4-113}$$

$$\boldsymbol{\lambda}^\delta \cdot \mathbf{e} = 1 \tag{4-114}$$

$$\boldsymbol{\lambda}^\delta = \boldsymbol{\lambda}^\delta P_\delta \tag{4-115}$$

† Let k be the period of the communicating class and $p_{ij}(n) = P\{s_n = j \mid s_1 = i\}$. Then $\#\mathcal{S} < \infty$, Corollary 7-9 and Theorem 7-10 on pages 241 and 242, respectively, of Volume I imply $\lambda_j^\delta = \lim_{n \to \infty} p_{jj}(nk)$ is the unique solution of (4-109) (even if j is transient).

‡ If $k = 1$ is the period of the communicating class, P^n has a limit, say P^∞, and $P^* = P^\infty$. Then the result is asserted by Proposition 7-5 on page 240 of Volume I. If $k > 1$, the definition of P_δ^* and Proposition 7-4 on page 230 of Volume I imply that each row of P_δ^* is $\boldsymbol{\lambda}^\delta$

It follows from part (*b*) of Theorem 4-13 and the unichain assumption that (4-112) contains $S - 1$ independent equations. Therefore, (4-110) and (4-112) are inadequate to determine the $S + 1$ variables g_δ and $w_s(\delta)$, $s \in \mathbb{S}$. From parts (*c*) and (*d*) of Theorem 4-13, including (4-111) with (4-110) and (4-112) yields a solution. However, (4-111) requires first solving (4-113), (4-114), and (4-115) for $\boldsymbol{\lambda}^\delta$. In practice, when the unichain assumption is valid, (4-111) is *not* added to (4-110) and (4-112). Instead, a state s^* is chosen and $w_{s*} \equiv 0$ is added to (4-110) and (4-112). The resulting solution $(g, \mathbf{w})$ satisfies $g = g_\delta$, although $\mathbf{w} \neq \mathbf{w}(\delta)$.

Proposition 4-8 If the unichain assumption is valid and $w_{s*} \equiv 0$ is added to (4-110) and (4-112), there is a unique solution $(g, \mathbf{w})$ and $g = g_\delta$.

PROOF The argument above justifies the claim of uniqueness and $g = g_\delta$ due to (4-110). □

Linear Programming in the Unichain Case

Here we seek $\delta_* \in \Delta$, which maximizes g_δ over $\delta \in \Delta$. We first replace Δ with a larger set, then obtain a linear program, and finally argue that the linear program has an optimal solution corresponding to some $\delta_* \in \Delta$.

The pertinent set which contains Δ is the set of *stationary randomized policies.*

Definition 4-19 Policy π is a *stationary randomized policy* if, for each $(s, a) \in \mathscr{C}$, $P\{a_n = a \mid s_n = s\}$ is the same for all $n \in I_+$.

To each stationary randomized policy there correspond numbers D_{sa}, $(s, a) \in \mathscr{C}$, such that

$$D_{sa} = P\{a_n = a \mid s_n = s\}$$

Therefore,

$$D_{sa} \geq 0 \qquad (s, a) \in \mathscr{C} \tag{4-116}$$

and

$$\sum_{a \in A_s} D_{sa} = 1 \qquad s \in \mathbb{S} \tag{4-117}$$

A stationary policy δ^∞ is the randomized stationary policy obtained by letting $D_{s,\,\delta(s)} = 1$ and $D_{sa} = 0$ if $a \neq \delta(s)$ for each $s \in \mathbb{S}$.

Let $\mathbf{D}$ denote $\{D_{sa}: (s, a) \in \mathscr{C}\}$, which is assumed to satisfy (4-116) and (4-117). The stationary Markov chain induced by $\mathbf{D}$ has a transition matrix $P_{\mathbf{D}}$ with (s, j)th component

$$\begin{aligned} \rho_{sj}^{\mathbf{D}} &= P\{s_{n+1} = j \mid s_n = s\} \\ &= \sum_{a \in A_s} P\{s_{n+1} = j \mid s_n = s,\, a_n = a\} P\{a_n = a \mid s_n = a\} \\ &= \sum_{a \in A_s} p_{sj}^a D_{sa} \end{aligned} \tag{4-118}$$

Therefore, $\rho_{sj}^{\mathbf{D}}$ is the weighted average of transition probabilities, where $\{D_{sa},\ a \in A_s\}$ provides the weights.

Exercise 4-47 asks you to prove that the unichain assumption implies that $P_{\mathbf{D}}$ satisfies the unichain assumption. Therefore, there exists a stationary distribution $\boldsymbol{\lambda}^{\mathbf{D}}$ such that

$$\boldsymbol{\lambda}^{\mathbf{D}} \geq \mathbf{0} \tag{4-119a}$$

$$\boldsymbol{\lambda}^{\mathbf{D}}\mathbf{e} = 1 \tag{4-119b}$$

$$\boldsymbol{\lambda}^{\mathbf{D}} = \boldsymbol{\lambda}^{\mathbf{D}} P_{\mathbf{D}} \tag{4-119c}$$

Now we obtain a system of equations which is equivalent to (4-119*a*), (4-119*b*), and (4-119*c*).

Henceforth, we suppress the superscript **D**. Let

$$x_{sa} \triangleq D_{sa} \lambda_s \qquad (s, a) \in \mathscr{C} \tag{4-120}$$

From (4-117) and (4-119*b*),

$$\sum_{a \in A_s} x_{sa} = \lambda_s \qquad s \in \mathcal{S} \tag{4-121}$$

$$\sum_{(s, a) \in \mathscr{C}} x_{sa} = 1 \tag{4-122}$$

Substitution of (4-118) and (4-120) in (4-119*c*) yields

$$\begin{aligned} \lambda_j &= \sum_{s \in \mathcal{S}} \lambda_s p_{sj} \\ &= \sum_{s \in \mathcal{S}} \lambda_s \sum_{a \in A_s} p_{sj}^a D_{sa} \\ &= \sum_{(s, a) \in \mathscr{C}} p_{sj}^a x_{sa} \end{aligned}$$

so (4-121) implies

$$\sum_{a \in A_j} x_{ja} = \sum_{(s, a) \in \mathscr{C}} p_{sj}^a x_{sa} \qquad j \in \mathcal{S} \tag{4-123}$$

This completes the proof of the next result.

Proposition 4-9 The system of equations (4-119*a*), (4-119*b*), and (4-119*c*) is equivalent to (4-122), (4-123), and

$$x_{sa} \geq 0 \qquad (s, a) \in \mathscr{C} \tag{4-124}$$

The variables of the linear program will be $\{x_{sa}\colon (s, a) \in \mathscr{C}\}$ and the constraints will be (4-122), (4-123), and (4-124). What is the "objective function" of the linear program?

If **D** is used, the expected reward of a transition starting in state s is given by

$$\begin{aligned} z_s &\triangleq E[r(s_n, a_n) \mid s_n = s] \\ &= \sum_{a \in A_s} P\{a_n = a \mid s_n = s\} E[r(s_n, a_n) \mid s_n = s, a_n = a] \\ &= \sum_{a \in A_s} D_{sa} r(s, a) \qquad s \in \mathcal{S} \end{aligned}$$

Therefore, the expected value of the sum of the rewards in the first N transitions is

$$b_s(\mathbf{D}, N) = E\left[\sum_{n=1}^{N} r(s_n, a_n) \mid s_1 = s, \mathbf{D}\right]$$

$$= \sum_{n=1}^{N} \sum_{j \in S} E[r(s_n, a_n) \mid s_n = j] P\{s_n = j \mid s_1 = s\}$$

$$= \sum_{n=1}^{N} \sum_{j \in S} z_j P\{s_n = j \mid s_1 = s\}$$

$$= \sum_{(j, a) \in \mathscr{C}} D_{ja} r(j, a) \sum_{n=1}^{N} P\{s_n = j \mid s_1 = s\} \tag{4-125}$$

Let $\mathbf{c}$ denote the column vector whose jth component is

$$c_j = \sum_{a \in A_s} D_{ja} r(j, a)$$

Then (4-125) is the system

$$\mathbf{b}(\mathbf{D}, N) = \mathbf{c} + P_{\mathbf{D}}\mathbf{c} + \cdots + P_{\mathbf{D}}^{N-1}\mathbf{c}$$

because $P\{s_n = j \mid s_1 = s\}$ in (4-125) is the (s, j)th element of $P_{\mathbf{D}}^{n-1}$. Using part (a) of Lemma 4-5, let

$$P_{\mathbf{D}}^* = \lim_{N \to \infty} \frac{1}{N} \sum_{n=0}^{N-1} P_{\mathbf{D}}^n$$

Then

$$\mathbf{g}(\mathbf{D}) = \lim_{N \to \infty} \frac{\mathbf{b}(\mathbf{D}, N)}{N} = P_{\mathbf{D}}^* \mathbf{c}$$

Under the unichain assumption, each row of $P_{\mathbf{D}}^*$ is $\boldsymbol{\lambda}$; so each component of the average-return vector $\mathbf{g}(\mathbf{D})$ is

$$\boldsymbol{\lambda}\mathbf{c} = \sum_{j \in S} \lambda_j c_j = \sum_{(j, a) \in \mathscr{C}} \lambda_j D_{ja} r(j, a) = \sum_{(j, a) \in \mathscr{C}} x_{ja} r(j, a)$$

The objective of the linear program corresponds to maximizing the average return per period, i.e.,

$$\text{Maximize} \sum_{(j, a) \in \mathscr{C}} r(j, a) x_{ja} \tag{4-126}$$

The linear-programming problem is to find values of $\{x_{ja}: (j, a) \in \mathscr{C}\}$ *which attain* (4-126) *subject to* (4-122), (4-123), *and* (4-124). Suppose the linear program is solved and $\{x_{ja}: (j, a) \in \mathscr{C}\}$ is an optimal solution. What is the corresponding decision rule $\mathbf{D}$? From (4-120) and (4-121), if $\pi_s = \sum_{a \in A_s} x_{sa} > 0$,

$$D_{sa} = \frac{x_{sa}}{\sum_{j \in A_s} x_{ja}} \qquad a \in A_s \tag{4-127}$$

Now suppose $\lambda_s = \sum_{a \in A_s} x_{sa} = 0$. Then state s is transient under an optimal policy and the rule for choosing an action in state s does not affect the average return per period (because, with probability 1, state s is visited at most finitely many times). Therefore, assign D_{sa}, $a \in A_s$, to satisfy

$$D_{sa} \geq 0 \qquad a \in A_s \qquad \text{and} \qquad \sum_{a \in A_s} D_{sa} = 1 \tag{4-128}$$

in any manner (Exercise 4-49) which preserves s as a transient state.

The procedure above leads to a "pure" (i.e., unrandomized) stationary policy only if (4-127) yields $D_{sa} \in \{0, 1\}$ [(4-128) permits $D_{sa} \in \{0, 1\}$, $a \in A_s$ if $\lambda_s = \sum_{a \in A_s} x_{sa} = 0$]. In (4-127), $D_{sa} \in \{0, 1\}$ for all $a \in A_s$ if, and only if, there is exactly one $a_s \in A_s$ such that $x_{sa_s} > 0$ and $x_{sa} = 0$ for all $a \neq a_s$. It is unnecessary to invoke the unichain assumption to prove that the procedure yields a pure stationary policy.

Proposition 4-10 The following linear program has an optimal solution in which, for each $s \in \mathcal{S}$, there is at most one $a \in A_s$ such that $x_{sa} > 0$:

$$\text{Maximize} \sum_{(s, a) \in \mathscr{C}} r(s, a) x_{sa} \tag{4-129a}$$

$$\text{subject to } x_{sa} \geq 0 \qquad (s, a) \in \mathscr{C} \tag{4-129b}$$

$$\sum_{(s, a) \in \mathscr{C}} x_{sa} = 1 \tag{4-129c}$$

$$\sum_{a \in A_j} x_{ja} = \sum_{(s, a) \in \mathscr{C}} p_{sj}^a x_{sa} \qquad j \in \mathcal{S} \tag{4-129d}$$

PROOF A linear program with a nonempty and bounded set of feasible solutions is the optimization of a continuous function on a compact set; so there exists an optimal solution. The set of $\mathbf{x}$ which satisfy (4-129*b*), (4-129*c*), and (4-129*d*) is nonempty and bounded; so there is an optimal solution, say $\{x_{sa}^0 : (s, a) \in \mathscr{C}\}$. There are two cases.

Suppose $\sum_{a \in A_s} x_{sa}^0 > 0$ for every $s \in \mathcal{S}$ so $\mathbf{x}^0$ contains at least $S = \#\mathcal{S}$ positive variables. Summing the S equations in (4-129*b*) yields (4-129*c*); so the $S + 1$ equations in (4-129*c*) and (4-129*d*) contain at most S independent equations. Now invoke the property of a linear program that, if there is an optimum, then there is an optimum with at most as many positive variables as there are independent constraints.† Without loss of generality, assume that $\mathbf{x}^0$ has this property. Therefore, $\mathbf{x}^0$ contains at most S positive variables; so it contains exactly S positive variables. Finally, $\sum_{a \in A_s} x_{sa}^0 > 0$ for all $s \in \mathcal{S}$ implies for each $s \in \mathcal{S}$ that there is at least one $a \in A_s$ such that $x_{sa}^0 > 0$. Therefore, for each $s \in \mathcal{S}$ there is exactly one $a \in A_s$ with $x_{sa}^0 > 0$.

† Chapter 6 in G. B. Dantzig, *Linear Programming and Extensions*, Princeton University Press, Princeton, N.J. (1963).

Suppose

$$\mathscr{T} \triangleq \left\{ s\colon \sum_{a \in A_s} x_{sa}^0 = 0, \qquad s \in \mathcal{S} \right\}$$

is nonempty. Then replace (4-129a), (4-129b), (4-129c), and (4-129d) with the following smaller linear program. First eliminate all columns corresponding to $(s, a) \in \{(j, k)\colon k \in A_j, j \in \mathscr{T}\}$. Then eliminate all rows $j \in \mathscr{T}$ from (4-129d). The reduced linear program has the same structure† as (4-129a), (4-129b), (4-129c), and (4-129d) but with $\mathcal{S}$ replaced by $\mathcal{S} - \mathscr{T}$. Moreover, $\sum_{a \in A_j} x_{sa}^0 > 0$ if $s \in \mathcal{S} - \mathscr{T}$; so the preceding argument applies to the reduced linear program. That is, there is an optimal solution $\mathbf{x}^0$ such that, for all $s \in \mathcal{S} - \mathscr{T}$, there is exactly one $s \in A_s$ with $x_{sa}^0 > 0$. If $s \notin \mathcal{S} - \mathscr{T}$, then $x_{sa}^0 = 0$ for all $a \in A_s$. □

Example 4-28: Continuation of Example 4-26 This stochastic MDP has three states, and its data are repeated below:

$r(0, 0) = 10$	$r(1, 0) = 5$	$r(1, 1) = 0$	$r(2, 1) = 0$
$p_{00}^0 = 0.7$	$p_{01}^0 = 0.2$	$p_{02}^0 = 0.1$	
$p_{10}^0 = 0$	$p_{11}^0 = 0.4$	$p_{12}^0 = 0.6$	
$p_{10}^1 = 0$	$p_{11}^1 = 0$	$p_{12}^1 = 1$	
$p_{20}^1 = 0.4$	$p_{21}^1 = 0.1$	$p_{22}^1 = 0.5$	

The unichain assumption is satisfied. The linear program (4-129a) to (4-129d) is given below:

$$\begin{aligned}
&\text{Maximize} \quad 10x_{00} + 5x_{10} \\
&\text{subject to} \quad x_{00}, x_{10}, x_{11}, \text{ and } x_{21} \geq 0 \\
&x_{00} + x_{10} + x_{11} + x_{21} = 1 \\
&0.3x_{00} - 0.4x_{21} = 0 \\
&-0.2x_{00} + 0.6x_{10} + x_{11} - 0.1x_{21} = 0 \\
&-0.1x_{00} - 0.6x_{10} - x_{11} + 0.5x_{21} = 0
\end{aligned}$$

The optimal solution is

$$x_{00} = 0.4528 \qquad x_{10} = 0.2075 \qquad x_{11} = 0 \qquad \text{and} \qquad x_{21} = 0.3396$$

and the value of an optimal solution is 5.5655. Therefore, (4-120) yields

† States s in $\mathscr{T}$ are transient under $\mathbf{x}^0$; so

$$\sum_{j \in \mathscr{T}} \sum_{a \in A_s} x_{sa}^0 p_{sj}^a = 0 \qquad s \in \mathcal{S} - \mathscr{T}$$

which justifies the claim that elimination of the rows preserves the structure.

$\lambda_0 = 0.4528$, $\lambda_1 = 0.2075$, and $\lambda_2 = 0.3396$. The optimal policy specifies action 0 in state 1. □

Policy Improvement in the Unichain Case

Let

$$\gamma_{sj} = \begin{cases} 1 & \text{if } s = j \\ 0 & \text{if } s \neq j \end{cases}$$

The dual to the linear program (4-129*a*) to (4-129*d*) is to find variables g and $w_j, j \in \mathcal{S}$, in order to

$$\text{Minimize } g \qquad \text{(4-130}a\text{)}$$

$$\text{subject to } g + \sum_{j \in \mathcal{S}} (\gamma_{sj} - p_{sj}^a) w_j \geq r(s, a) \qquad (s, a) \in \mathscr{C} \qquad \text{(4-130}b\text{)}$$

The functional equation

$$g + w_s = \max \left\{ r(s, a) + \sum_{j \in \mathcal{S}} p_{sj}^a w_j : a \in A_s \right\} \qquad s \in \mathcal{S} \qquad \text{(4-131)}$$

with variables g and w_s, $s \in \mathcal{S}$, is nearly equivalent to (4-130*a*) and (4-130*b*).

Theorem 4-15 Suppose the unichain assumption is valid. If $(g^0, \mathbf{w}^0)$ is optimal in (4-130*a*) and (4-130*b*) and $(g^1, \mathbf{w}^1)$ solves (4-131), then $g^0 = g^1$. Also, $u = g^0 = g^1$.

Proof Let $(g^0, \mathbf{w}^0)$ be optimal in (4-130*a*) and (4-130*b*) and let $(g^1, \mathbf{w}^1)$ satisfy (4-131). From (4-131), for each $s \in \mathcal{S}$,

$$g^1 + w_s^1 \geq r(s, a) + \sum_{j \in \mathcal{S}} p_{sj}^a w_j \qquad a \in A_s$$

with equality for at least one $a \in A_s$. This inequality is the same as (4-130*b*); so $(g^1, \mathbf{w}^1)$ is feasible in (4-130*a*) and (4-130*b*), which implies $g^0 \leq g^1$.

From (4-130*b*),

$$g^0 \geq r(s, a) + \sum_{j \in \mathcal{S}} (p_{sj}^a - \gamma_{sj}) w_j^0 \qquad (s, a) \in \mathscr{C}$$

so the objective (4-130*a*) yields

$$g^0 = \max \left\{ r(s, a) + \sum_{j \in \mathcal{S}} (p_{sj}^a - \gamma_{sj}) w_j^0 : \ (s, a) \in \mathscr{C} \right\} \qquad \text{(4-132)}$$

From (4-131),

$$g^1 = \max \left\{ r(s, a) + \sum_{j \in \mathcal{S}} (p_{sj}^a - \gamma_{sj})\, w_j^1 : \ a \in A_s \right\} \qquad s \in \mathcal{S}$$

so $(g^1, \mathbf{w}^1)$ is optimal in (4-130*a*) and (4-130*b*) with the added constraints that

$$\max\left\{r(s, a) + \sum_{j \in \mathcal{S}} (p_{sj}^a - \gamma_{sj})w_j : \quad a \in A_s\right\}$$

be the same value for all $s \in \mathcal{S}$. The added constraints imply $g^1 \leq g^0$. Therefore, $g^1 = g^0$. □

The solution of

$$g + w_s = \max\left\{r(s, a) + \sum_{j \in \mathcal{S}} p_{sj}^a w_j : \quad a \in A_s\right\} \qquad s \in \mathcal{S} \tag{4-133}$$

yields the value of the maximal average reward per period. *With the following procedure,* (4-133) *also yields an optimal policy.* Let $(g, \mathbf{w})$ be a solution of (4-133) and for each $s \in \mathcal{S}$ assign to $\delta(s)$ an $a \in A_s$ that attains the maximum in (4-133). That is,

$$g + w_s = r[s, \delta(s)] + \sum_{j \in \mathcal{S}} p_{sj}^{\delta(s)} w_j \qquad s \in \mathcal{S}$$

which is (4-102):

$$\mathbf{e}g + \mathbf{w} = \mathbf{r}_\delta + P_\delta \mathbf{w}$$

Theorem 4-15 implies $g = g_s(\delta)$ for each $s \in \mathcal{S}$; so

$$\mathbf{e}g = P_\delta^* \mathbf{r}_\delta$$

from (4-101). Now we explain how to solve (4-133) with a policy-improvement algorithm.

In order to compare the discounted- and average-reward algorithm, first we summarize the discounted-reward policy-improvement algorithm in Section 4-4. In the notation below, v_s^i is the sth component of the S-vector $\mathbf{v}^i$. Also, we write $\mathbf{r}_i$ and P_i instead of $\mathbf{r}_{\delta_i}$ and P_{δ_i}, respectively.

Discounted-Reward Policy-Improvement Algorithm

(*a*) Let $i = 1$ and select $\delta_1 \in \Delta$.

(*b*) Solve the following equation for $\mathbf{v}^i$:

$$\mathbf{v}^i = (I - \beta P_i)^{-1} \mathbf{r}_i$$

(*c*) For each $s \in \mathcal{S}$, compute

$$\psi_s^i = \max\left\{r(s, a) + \beta \sum_{j \in \mathcal{S}} p_{sj}^a v_j^i : a \in A_s - \{\delta_i(s)\}\right\} - v_s^i \tag{4-134}$$

If $\psi_s^i > 0$, let $\delta_{i+1}(s)$ be any $a \in A_s$ that causes

$$r(s, a) + \beta \sum_{j \in \mathcal{S}} p_{sj}^a v_j^i - v_s^i > 0$$

If $\psi_s^i \leq 0$, let $\delta_{i+1}(s) = \delta_i(s)$.

(*d*) If $\delta_{i+1} = \delta_i$, stop, and δ_{i+1}^∞ is optimal. Otherwise replace i with $i + 1$ and return to (*b*).

Average-Reward Policy-Improvement Algorithm—Unichain Case

The following average-reward algorithm has the same structure as (*a*) through (*d*) above.

(*A*) Let $i = 1$ and select $\delta_1 \in \Delta$. Choose $s^* \in \mathcal{S}$ and let $w_{s*}^i \equiv 0$ for all i.
(*B*) Solve the following equations for g^i and $\mathbf{w}^i$:

$$\mathbf{e}g^i + \mathbf{w}^i = \mathbf{r}_i + P_i \mathbf{w}^i \tag{4-135}$$

(*C*) For each $s \in \mathcal{S}$ compute

$$\psi_s^i \triangleq \max \left\{ r(s, a) + \sum_{j \in \mathcal{S}} p_{sj}^a w_j^i : a \in A_s - \{\delta_i(s)\} \right\} - g^i - w_s^i \tag{4-136}$$

If $\psi_s^i > 0$, let $\delta_{i+1}(s)$ be any $a \in A_s$ that causes

$$r(s, a) + \sum_{j \in \mathcal{S}} p_{sj}^a w_j^i - g^i - w_s^i > 0$$

If $\psi_s^i \leq 0$, let $\delta_{i+1}(s) = \delta_i(s)$
(*D*) If $\delta_{i+1} = \delta_i$, stop. Otherwise replace i with $i + 1$ and return to (*B*).

Below we question whether or not the algorithm converges to an optimal policy in a finite number of iterations. Recall from part (*b*) of Theorem 4-13 that the solution of (4-135) in (*B*) is not unique. The nonuniqueness concerns $\mathbf{w}^i$, which we know from part (*c*) of Theorem 4-13 can be constrained to satisfy $P_{\delta_i}^* \mathbf{w}^i = \mathbf{0}$. Instead, the usual practice is to select a particular state s^* and let $w_{s*}^i \equiv 0$, as we do in (*A*). Exercise 4-59 asks you to prove that (4-135) then has a unique solution under the unichain assumption. Uniqueness is not used in the proof of Theorem 4-16 below.

In (*C*), it is unnecessary to find the maximum in (4-136). It is sufficient to find some $a \in A_s - \{\delta_i(s)\}$ that causes the test quantity

$$r(s, a) + \sum_{j \in \mathcal{S}} p_{sj}^a w_j^i - g^i - w_s^i$$

to be strictly positive. Then $\delta_{i+1}(s)$ can be any such a.

Theorem 4-16 Suppose the unichain assumption is valid. (*a*) Then $g^i \leq g^{i+1}$ for each i until the policy-improvement algorithm terminates. (*b*) If also, for every $\delta \in \Delta$, P_δ corresponds to an irreducible Markov chain (i.e., there are no transient states), then $g^i < g^{i+1}$ until termination, which occurs after finitely many iterations.

PROOF To simplify the notation, we write $\mathbf{r}_i$ and P_i instead of $\mathbf{r}_{\delta_i}$ and P_{δ_i}. For vectors $\mathbf{x} = (x_j)$ and $\mathbf{y} = (y_j)$ of the same size, we write $\mathbf{x} < \mathbf{y}$ if $x_j \leq y_j$ for $\mathbf{x} \neq \mathbf{y}$. From (*B*) and (*C*) of the algorithm, if i is not the final iteration, then

$$\mathbf{e}g^i + \mathbf{w}^i = \mathbf{r}^i + P_i \mathbf{w}^i < \mathbf{r}_{i+1} + P_{i+1} \mathbf{w}^i \tag{4-137}$$

Hence,

$$\mathbf{e}g^i < \mathbf{r}_{i+1} + (P_{i+1} - I)\mathbf{w}^i$$

Premultiplication by P_{i+1}^n preserves the inequality (P_{i+1}^n is nonnegative and nonzero); so

$$P_{i+1}^n \mathbf{e} g^i = \mathbf{e} g^i < P_{i+1}^n \mathbf{r}_{i+1} + (P_{i+1}^{n+1} - P_{i+1}^n)\mathbf{w}^i$$

Summing over n from 0 to $N-1$ and dividing by N yields

$$\mathbf{e} g^i < \frac{1}{N} \sum_{n=0}^{N-1} P_{i+1}^n \mathbf{r}_{i+1} + (P_{i+1}^N - I)\frac{\mathbf{w}^i}{N}$$

As $N \to \infty$, the second term tends to $\mathbf{0}$ and the first to $P_{i+1}^* \mathbf{r}_{i+1}$ (part (*a*) of Lemma 4-5). Therefore, Theorem 4-13 yields

$$\mathbf{e} g^i \le P_{i+1}^* \mathbf{r}_{i+1} = \mathbf{e} g^{i+1}$$

so $g^i \le g^{i+1}$, which proves (*a*).

The hypothesis of (*b*) implies that the rows of P_{i+1}^* are all the same, say $\boldsymbol{\lambda}^{i+1}$, and that $\lambda_s^{i+1} > 0$ for all $s \in \mathcal{S}$. Premultiplying (4-137) by $\boldsymbol{\lambda}^{i+1}$ and using Lemma 4-14 and Theorem 4-13 yields

$$\boldsymbol{\lambda}^{i+1}\mathbf{e} g^i + \boldsymbol{\lambda}^{i+1}\mathbf{w}^i < \boldsymbol{\lambda}^{i+1}\mathbf{r}^{i+1} + \boldsymbol{\lambda}^{i+1} P_{i+1} \mathbf{w}^i$$

Consequently, $\boldsymbol{\lambda}^{i+1} = \boldsymbol{\lambda}^{i+1} P_{i+1}$ implies

$$g^i + \boldsymbol{\lambda}^{i+1}\mathbf{w}^i < g^{i+1} + \boldsymbol{\lambda}^{i+1}\mathbf{w}^i$$

so

$$g^i < g^{i+1}$$

The assumption $\#\mathscr{C} < \infty$ causes Δ to have only finitely many single-stage rules δ; so the sequence $g^1 < g^2 < \ldots$ corresponding to $\delta_1, \delta_2, \ldots$ must terminate finitely soon. □

Example 4-29: Continuation of Examples 4-26 and 4-28 The data are repeated below:

$r(0, 0) = 10$	$r(1, 0) = 5$	$r(1, 1) = 0$	$r(2, 1) = 0$
$p_{00}^0 = 0.7$	$p_{01}^0 = 0.2$	$p_{02}^0 = 0.1$	
$p_{10}^0 = 0$	$p_{11}^0 = 0.4$	$p_{12}^0 = 0.6$	
$p_{10}^1 = 0$	$p_{11}^1 = 0$	$p_{12}^1 = 1$	
$p_{20}^1 = 0.4$	$p_{21}^1 = 0.1$	$p_{22}^1 = 0.5$	

From Example 4-28, we know action 0 in state 1 is optimal. Ignoring this information, suppose for (*A*) of the algorithm we select $s^* = 0$, $\delta_1(0) = 0$, $\delta_1(1) = 1$, and $\delta_1(2) = 1$. Then (4-135), together with $w_0^1 \equiv 0$, is

$$\begin{aligned} g^1 \quad &= 10 + 0.2w_1^1 + 0.1w_2^1 \\ g^1 + w_1^1 &= \qquad\qquad w_2^1 \\ g^1 + w_2^1 &= \quad 0.1w_1^1 + 0.5w_2^1 \end{aligned}$$

The solution is

$$g^1 = 4.9383 \qquad w_0^1 = 0 \qquad w_1^1 = 18.3333 \qquad \text{and} \qquad w_2^1 = -13.3950$$

which completes (*B*) when $i = 1$. For (*C*), we compute ψ_1^1; $0 = \psi_0^i = \psi_2^i$ for each i because $\# A_0 = \# A_2 = 1$.

$$\psi_1^1 = \max\left\{r(1, a) + \sum_{j=0}^{2} p_{1j}^a w_j^1 : a \in \{0, 1\} - \{1\}\right\} - g^1 - w_1^1$$
$$= 5 + 0.4(18.3333) + 0.6(-13.395) - 4.9383 + 18.3333$$
$$= -10.3703 + 13.395 = 3.0247 > 0$$

Since $\psi_1^1 > 0$, let $\delta_2(1) = 0$ because $a = 0$ attains the maximum in ψ_1^1. Now $\delta_2 \neq \delta_1$; so return to (*B*) with $i = 2$. Then (4-135), together with $w_0^2 \equiv 0$, is

$$g^2 \qquad = 10 + 0.2w_1^2 + 0.1w_2^2$$
$$g^2 + w_1^2 = 5 + 0.4w_1^2 + 0.6w_2^2$$
$$g^2 + w_2^2 = 0.1w_1^2 + 0.5w_2^2$$

The solution is

$$g^2 = 5.5655 \qquad w_0^2 = 0 \qquad w_1^2 = -15.0940 \qquad \text{and} \qquad w_2^2 = -14.1509$$

At (*C*), we compute ψ_1^2 to test the optimality of δ_2^∞:

$$\psi_1^2 = \max\left\{r(1, a) + \sum_{j=0}^{2} p_{1j}^a w_j^2 : a \in \{0, 1\} - \{0\}\right\} - g^2 - w_1^2$$
$$= 0 + 1(-14.1508) - 5.5655 + 15.0940$$
$$= -4.6223 < 0$$

Therefore, $\delta_3 = \delta_2$; so stop because δ_2^∞ is optimal. □

Secondary Criteria

When the stationary policy δ^∞ is used, the vector (due to various initial states) of the expected total reward during N periods is

$$\mathbf{b}(\delta, N) = \sum_{n=0}^{N-1} P_\delta^n \mathbf{r}_\delta = \mathbf{r}_\delta + P_\delta \mathbf{b}(\delta, N-1)$$

It follows from Exercise 4-57 that

$$\mathbf{b}(\delta, N) = N\mathbf{g}(\delta) + \mathbf{w}(\delta) + \boldsymbol{\epsilon}(N, \delta) \tag{4-138}$$

where $\boldsymbol{\epsilon}(N, \delta) \to \mathbf{0}$ as $N \to \infty$. The vector $\mathbf{w}(\delta)$ in (4-138) is the (unique) solution of $P^*\mathbf{w} = \mathbf{0}$ and $\mathbf{g} + \mathbf{w} = \mathbf{r} + P\mathbf{w}$ rather than the vector obtained by joining $w_{s*} = 0$ to $\mathbf{g} + \mathbf{w} = \mathbf{r} + P\mathbf{w}$. It follows from (4-138) that

$$b_s(\delta, N) - b_j(\delta, N) \to w_s(\delta) - w_j(\delta) \text{ as } N \to \infty$$

Therefore, if δ^∞ is an optimal policy, $w_s(\delta) - w_j(\delta)$ measures the relative advantage of having s rather than j as the initial state.

Two policies, say δ and δ', can both be optimal; so $\mathbf{g}(\delta) = \mathbf{g}(\delta')$. However, $\mathbf{w}(\delta)$ and $\mathbf{w}(\delta')$ may differ; so (4-138) suggests that this difference may be useful in resolving ties. This point is pursued further in the references cited in the Bibliographic Guide in Chapter 5 under Average-Reward and Sensitive-Discount Criteria.

EXERCISES

4-46 Prove $\mathbf{g}(\delta) = P_\delta \mathbf{g}(\delta)$ for each $\delta \in \Delta$.

4-47 Suppose the unichain assumption is satisfied. For each stationary randomized policy **D**, prove that $P_{\mathbf{D}}$ has one communicating class and a (possibly empty) set of transient states.

4-48 Use the policy-improvement algorithm to find a policy which minimizes the average-return criterion for the inventory model in Example 4-23.

4-49 Suppose $\mathcal{T} \subset \mathcal{S}$, $\mathcal{T} \neq \phi$, and for each $s \in \mathcal{S} - \mathcal{T}$ that $\delta(s) \in A_s$ with

$$\sum_{j \in \mathcal{T}} p_{sj}^{\delta(s)} = 0 \qquad s \in \mathcal{S} - \mathcal{T}$$

A *completion* of δ is any specification of $\delta(s) \in A_s$ for each $s \in \mathcal{T}$. Suppose that there is a completion of δ such that all states in $\mathcal{T}$ are transient under P_δ. Construct an algorithm to find a completion with this property.

4-50 Use linear programming to find a policy which minimizes the long-run average cost per unit time in the maintenance model in Exercise 3-15.

4-51 Repeat Exercise 4-50 using the policy-improvement algorithm.

4-52 Use policy improvement to find a policy which maximizes the average-return criterion for the fishery model in Exercise 3-16.

4-53 Repeat Exercise 4-52 using a linear-programming algorithm.

4-54 Prove Proposition 4-7.

4-55 Prove Lemma 4-4.

4-56 Theorem 4-13 and regenerative reward processes†. Fill in the details of the following outline of an alternative proof of (4-101), that is,

$$\mathbf{g}(\delta) = P_\delta^* \mathbf{r}_\delta$$

Some state, 0 say, is ergodic (why?); so $\{B(\delta, n), n \in I_+\}$ is a regenerative stochastic process (see Section 6-4 in Volume I) where the r.v. $B(\delta, N)$ is given by $\sum_{n=1}^{N} r[s_n, \delta(s_n)]$. Let p_{ij}^* denote the (i, j)th element of P_δ^*. Then p_{ij}^* is the proportion of transitions starting in state i that enter state j. From a version of the regenerative theorem (Corollary 6-8*a* on page 185 of Volume I),

$$\text{Average return per period} = \sum_{j \in \mathcal{S}} (\text{proportion of time in state } j) \times E(\text{return in } j) \qquad (4\text{-}139)$$

When there is only one communicating class, (4-139) is (4-101). If there are several communicating classes, use (4-139) and average over which class is entered to obtain (4-101).

4-57 Alternative derivation‡ of (4-102). Make rigorous the following heuristic derivation of (4-102), that is, $(I - P_\delta)\mathbf{w}(\delta) = \mathbf{r}_\delta - \mathbf{g}(\delta)$. For simplicity, suppose that P_δ has one communicating class and,

† This exercise depends on Section 6-4 in Volume I.

‡ This exercise depends on Section 7-7 in Volume I.

perhaps, transient states (average over the classes if there are several classes). Also, we drop δ from the notation because it is fixed. Equation (7-76) in Corollary 7-15 on page 257 of Volume I implies

$$\mathbf{b}(n) \approx nP^*\mathbf{r} + (F - P^*)\mathbf{r} \tag{4-140}$$

where $b_s(n) \triangleq E[B(n) \mid s_1 = s]$ and $F \triangleq (I - P + P^*)^{-1}$[$\mathbf{b}(n)$ corresponds to $E[R(n+1)]$ in (7-76) in Volume I because of different numbering conventions]. $P^* = P^\infty \triangleq Q$ and each row of Q is the limiting (= stationary) probability vector $\boldsymbol{\lambda}$. Thus,

$$g = \lim_{n\to\infty} \frac{b_s(n)}{n} = \boldsymbol{\lambda}\mathbf{r} \qquad \text{for each } s \tag{4-141}$$

Using (4-141), write (4-140) as

$$b_s(n) \approx ng + w_s \tag{4-142}$$

where w_s is the sth element of $(F - P^*)\mathbf{r}$. Therefore,

$$b_s(n+1) \approx r_s + \sum_{j\in \mathcal{S}} p_{sj} b_j(n) \tag{4-143}$$

where r_s denotes $r[s, \delta(s)]$. Substituting (4-142) into (4-143) yields

$$b_s(n+1) \approx r_s + \sum_{j\in\mathcal{S}} p_{sj}(ng + w_j) = r_s + ng + \sum_{j\in\mathcal{S}} p_{sj} w_j$$

so

$$\mathbf{b}(n+1) \approx \mathbf{r} + n\mathbf{g} + P\mathbf{w} \tag{4-144}$$

Putting (4-142) in vector form and replacing n with $n + 1$ yields

$$\mathbf{b}(n+1) \approx (n+1)\mathbf{g} + \mathbf{w} \tag{4-145}$$

Equating the right sides of (4-144) and (4-145) implies

$$\mathbf{r} + P\mathbf{w} = \mathbf{g} + \mathbf{w}$$

which is (4-102).

4-58 Alternative derivation† of (4-103). Make rigorous the following alternative derivation of (4-103), that is, $P_\delta^* w(\delta) = 0$. For simplicity, suppose that P_δ has one communicating class. If there are several classes, average over them as in Exercises 4-56 and 4-57. Since δ is fixed here, we suppress δ in the notation. From Corollary 7-15 on page 257 of Volume I,

$$\mathbf{w} = (F - P^*)\mathbf{r} \tag{4-146}$$

where $F \triangleq (I - P + P^*)^{-1}$. Therefore,

$$P^*F^{-1} = P^* \tag{4-147}$$

From

$$F = I + \sum_{n=1}^{\infty} (P^n - Q)$$

where $Q \triangleq P^\infty = P^*$, we obtain $FP^* = P^*$; so (4-146) can be written as

$$\mathbf{w} = F(I - P^*) \tag{4-148}$$

Premultiplying (4-148) by P^*F^{-1} and using (4-147) yields $P^*\mathbf{w} = \mathbf{0}$.

4-59 Justification of (*B*) in the policy-improvement algorithm. Let P be an $S \times S$ stochastic matrix corresponding to a unichain Markov chain (including, perhaps, some transient states). Let $P^\#$ denote the $S \times S - 1$ matrix, which is P with its Sth column deleted. Prove that, for each $\mathbf{b} \in \mathbb{R}^S$,

$$\mathbf{e}g + (I - P^\#)\mathbf{w} = \mathbf{b}$$

† This exercise depends on Section 7-7 in Volume I.

has a unique solution consisting of a number g and a column vector $\mathbf{w} = (w_1, \ldots, w_{S-1})$. [Hint: Use part (b) of Theorem 4-13 and the following† fundamental theorem concerning linear inequalities. Let A be an $m \times n$ matrix and $\mathbf{c}$ an $m \times 1$ vector; then exactly one of the following alternatives holds. Either

$$A\mathbf{y} \geq \mathbf{c}$$

has a solution or

$$\mathbf{x}A = 0 \qquad \mathbf{xc} = 1 \qquad \mathbf{x} \geq 0$$

has a solution.]

BIBLIOGRAPHIC GUIDE

The earliest antecedents of MDPs are problems and methods in the calculus of variations developed in the seventeenth and eighteenth centuries; cf. Dreyfus (1965). The modern heritage begins with Wald's development of sequential statistical analysis during the 1930s and 1940s; cf. Wald (1947). Not much later, Massé devised what we now recognize as dynamic programming methods to analyze water-resource problems; cf. Massé (1946).

The modern foundations were laid between 1949 and 1953 by people who spent at least a part of that period as staff members at the RAND Corporation in Santa Monica, Calif. Dates of actual publication are not reliable guides to the order in which ideas were discovered during this period. One group of people studied sequential game models. The special case of a one-player game is an MDP (or a continuous-time deterministic analog of an MDP). The contributions included Bellman and Blackwell (1949), Bellman and LaSalle (1949), Shapley (1953), and early work by Isaacs summarized in a later book Isaacs (1965). Isaacs' work initiated the literature in differential games. Shapley's paper contains several ideas basic to Chapter 4, 5, and 9. During this period, an epochal paper by Arrow, Harris, and Marschak (1951) initiated optimization approaches to dynamic inventory models.

In the midst of this activity, Richard Bellman appreciated the broad applicability and mathematical content of sequential decision models. Bellman (1957) summarized some of his extraordinarily numerous and successful efforts to popularize dynamic programming. His book included the notion of a policy-improvement algorithm, but Ronald Howard was first to present it explicitly; cf. Howard (1960).

The algorithms in Howard's book had a major impact. De Ghellinck (1960) recognized that policy improvement was equivalent to block pivoting in a linear program, Manne (1960) presented a linear-programming formulation for the unichain case in the average-return criterion, Wagner (1960) showed that an extreme-point optimum to the linear program corresponded to an unrandomized stationary policy, and d'Epenoux (1960) presented a linear-programming formulation for the discounted case. Then Blackwell (1962) and Derman (1962) connec-

† Page 46 of D. Gale, *The Theory of Linear Economic Models*, McGraw-Hill, New York (1960).

ted the discounted- and average-return criteria. Blackwell presented a proof, different from Howard's, that policy improvement converged in the discounted case. Blackwell's arguments were the precursor of the results in Sections 4-3 and 4-4. Derman was the first to use the Tauberian theorem on which we rely in Section 4-6.

Howard (1964) and Jewell (1963) show that most of the properties of MDPs are shared by more general sequential decision processes based on semi-Markov processes and Markov renewal processes, an idea we pursue in Sections 5-1, 5-2, and 5-3. Blackwell (1965) and Denardo (1967) use the theory of contraction mappings (already in Shapley's 1953 paper!) in the discounted case. Denardo's paper influences Sections 4-5 and 5-4 and books and articles by many people since 1967. Veinott (1966) answers a question concerning the connection between discounted and undiscounted problems first raised by Blackwell (1962).

Our approach in Section 4-3 follows Sobel (1975), which was motivated by Mitten (1974). The earlier work by Brown and Strauch (1964) was appreciated only more recently.

Since the mid 1960s, research on MDPs has been very active. We particularly like the books Bertsekas (1976), Bertsekas and Shreve (1978), Denardo (1982), Derman (1970), and Dubins and Savage (1965), and the conference proceedings Hartley, Thomas, and White (1980), Puterman (1978), and Tijms and Wessels (1977).

References

Arrow, K. J., T. E. Harris, and J. Marschak: "Optimal Inventory Policy," *Econometrica*, **19**: 250–272 (1951)

Bellman, R.: *Dynamic Programming*, Princeton University Press, Princeton, N.J. (1957).

——— and D. Blackwell: "On a Particular Non-Zero Sum Game," RM-250, RAND Corp., Santa Monica, Calif. (1949).

——— and J. P. LaSalle: "On Non-Zero Sum Games and Stochastic Processes," RM-212, RAND Corp., Santa Monica, Calif. (1949).

Bertsekas, D. P.: *Dynamic Programming and Stochastic Control*, Academic Press, New York (1976).

——— and S. E. Shreve: *Stochastic Optimal Control. The Discrete Time Case*, Academic Press, New York (1978).

Blackwell, D.: "Discrete Dynamic Programming," *Ann. Math. Stat.* **33**: 719–726 (1962).

———: "Discounted Dynamic Programming," *Ann. Math. Stat.* **36**: 226–235 (1965).

Brown, T. A., and R. E. Strauch: "Dynamic Programming in Multiplicative Lattices," *J. Math. Anal. Appl.* **12**: 364–370 (1964).

de Ghellinck, G.: "Les Problemes de Decisions Sequentielles," *Cashiers Centre d'Etudes de Recherche Operationelle* **2**: 161–179 (1960).

Denardo, E. V.: "Contraction Mappings in the Theory Underlying Dynamic Programming," *SIAM Rev.* **9**: 165–177 (1967).

———: *Dynamic Programming*, Prentice-Hall, Englewood Cliffs, N.J. (1982).

d'Epenoux, F.: "Sur un Problème de Production et de Stockage dans l'Aléatoire," *Revue Francaise de Informatique et Recherche Opérationnelle* **14**: 3–16 (1960). [English translation: *Manag. Sci.* **10**: 98–108 (1963)].

Derman, C.: "On Sequential Decisions and Markov Chains," *Manag. Sci.* **9**: 16–24 (1962).

———: *Finite State Markovian Decision Processes*, Academic Press, New York (1970).

Dreyfus, S. E.: *Dynamic Programming and the Calculus of Variations*, Academic Press, New York (1965).

Dubins, L. E., and L. J. Savage: *How to Gamble if You Must*, McGraw-Hill, New York (1965).
Dynkin, E.: "Controlled Random Sequences," *Theory of Prob. and Appls.* **10**: 1–14 (1965).
———, and A. A. Yushkevich: *Controlled Markov Processes*, Springer-Verlag, Berlin (1979).
Eaton, J. H., and L. A. Zadeh: "Optimal Pursuit Strategies in Discrete State Probabilistic Systems," *J. Basic Eng., Ser. D* **84**: 23–29 (1961).
Eaves, B. C.: "Complementary Pivot Theory and Markovian Decision Chains," in *Fixed Points: Algorithms and Applications*, edited by S. Karamardian, Academic Press, New York (1977), pp. 59–85.
Hartley, R., L. C. Thomas, and D. J. White, eds.: *Recent Developments in Markov Decision Processes*, Academic Press, New York (1980).
Hinderer, K.: *Foundations of Non-Stationary Dynamic Programming with Discrete Time Parameter*, 33, Lecture Notes in Operations Research and Mathematical Systems, Springer-Verlag, Berlin (1970).
Howard, R. A.: *Dynamic Programming and Markov Processes*, MIT Press, Cambridge, Mass. (1960).
———: "Research in Semi-Markovian Decision Structures," *J. Oper. Res. Soc. Japan* **6**: 163–199 (1964).
Isaacs, R.: *Differential Games*, Wiley, New York (1965).
Jewell, W. S.: "Markov-Renewal Programming: I and II," *Oper. Res.* **11**: 938–971 (1963).
Karlin, S.: "The Structure of Dynamic Programming Models," *Naval Logistics Res. Quart.* **2**: 285–294 (1955).
Koehler, G. J.: "A Complementarity Approach for Solving Leontief Substitution Systems and (Generalized) Markov Decision Processes," *R.A.I.R.O. Recherche Opérationnelle/Operations Research* **13**: 75–80 (1979).
Kreps, D. M.: "Decision Problems with Expected Utility Criteria, I: Upper and Lower Convergent Utility," *Math. Oper. Res.* **2**: 45–53 (1977).
———: "Decision Problems with Expected Utility Criteria II: Stationarity," *Math. Oper. Res.* **2**: 266–274 (1977).
———: "Sequential Decision Problems with Expected Utility Criteria, III: Upper and Lower Transience," *SIAM J. Control Optimization*, **16**: 420–428 (1978).
——— and E. L. Porteus: "Dynamic Choice Theory and Dynamic Programming," *Econometrica* **47**: 91–100 (1979).
Manne, A. S.: "Linear Programming and Sequential Decisions," *Manage. Sci.* **6**: 259–267 (1960).
Massé, P.: *Les Réserves et la Régulation de l'Avenir dans la Vie Économique*, 2 vols., Hermann, Paris (1946).
Mitten, L. G.: "Preference Order Dynamic Programming," *Manag. Sci.* **21**: 43–46 (1974).
Puterman, M. L., ed.: *Dynamic Programming and Its Applications*, Academic Press, New York (1978).
Shapley, L. S.: "Stochastic Games," *Proc. Natl. Acad. Sci. USA* **39**: 1095–1100 (1953).
Sirjaev, A.: "Some New Results in the Theory of Controlled Random Processes," *Selected Translations in Math. Stat. and Prob.*, **8**: 49–130 (1970).
Sobel, M. J.: "Ordinal Dynamic Programming," *Manag. Sci.* **21**: 967–975 (1975).
Tijms, H. C., and J. Wessels, eds.: *Markov Decision Theory*, Tract 93, Mathematics Centrum, Amsterdam (1977).
Veinott, A. F., Jr.: "On Finding Optimal Policies in Discrete Dynamic Programming with No Discounting," *Ann. Math. Stat.* **37**: 1284–1294 (1966).
Wagner, H. M.: "On the Optimality of Pure Strategies," *Manage. Sci.* **6**: 268–269 (1960).
Wald, A.: *Sequential Analysis*, Wiley, New York (1947).

CHAPTER

FIVE

GENERALIZATIONS OF MDPs

Chapter 4 presents fundamental properties of discrete-time MDPs. This chapter presents continuous-time models and other generalizations of MDPs. Some phenomena are most naturally described with continuous-time models. The first three sections concern semi-Markov decision processes and continuous-time Markov decision chains. These are important continuous-time models. We conclude that the optimization of such models is equivalent to the optimization of discrete-time MDPs. Many of the results in Chapter 4 can be applied to these continuous-time models. The exposition in Sections 5-1 and 5-2 assumes a modest knowledge of semi-Markov and Markov renewal processes (cf. Sections 9-1 and 9-2 in Volume I). Section 5-3 assumes a modest knowledge of continuous-time Markov chains (cf. Chapter 8 in Volume I).

Section 5-4 addresses the connection between the expected-present-value criterion and contraction mappings via S. Banach's fixed-point theorem for contraction mappings. The presentation in Section 5-4 does not depend on prior knowledge of topology. The contraction-mapping approach is important because it elegantly encompasses many generalizations of MDPs including semi-Markov decision processes.

5-1* SEMI-MARKOV DECISION PROCESSES

The semi-Markov decision process (SMDP) is a generalization of the MDP model where the times between transitions are allowed to be r.v.'s. The r.v.'s may depend on the current state, the action taken, and possibly the next state. This generalization includes important operations research models of queues, maintenance, reliability, inventory, and health care.

The SMDP is a continuous-time model in which control can be exerted only at a denumerable sequence of epochs.† The MDP is a discrete-time model. Nevertheless, the SMDP and MDP share certain characteristics. The infinite-horizon discounted case is essentially the same for both models. We shall see that only trivial changes are needed in Sections 4-3, 4-4, and 4-5 for them to encompass SMDPs. The infinite-horizon average-reward case is similar in the two models, but the differences are less trivial. The two models are least similar in the finite-horizon case.

Here is a simple illustration of the manner in which continuous-time models arise in practice. Sections 3-1 and 3-2 concern discrete-time inventory models in which the stock levels of items are reviewed periodically at fixed intervals of time. However, the increasingly widespread implementation of nearly real-time computer systems has made periodic review less common. Instead, review procedures are increasingly "driven" by demands. Whenever a demand occurs, the diminished stock level is reviewed. The times separating successive demands are not usually fixed intervals; so a continuous-time model is apt.

The section begins with the definition of an SMDP. Then it considers the optimization of SMDPs with the expected-present-value criterion. The average-reward criterion is addressed in Section 5-2. Section 5-3 briefly considers the special case of controlled continuous-time Markov chains.

The Model

An MDP is a model obtained by appending actions and rewards to a Markov chain. An SMDP is obtained by appending actions and rewards to a semi-Markov process. Section 9-1 of Volume I presents important properties of the semi-Markov process, a generalization of a discrete-time Markov chain. Section 9-2 of Volume I concerns the Markov renewal process, a close relative of the semi-Markov process. As in those sections,‡ we consider a sequence of pairs (s_n, τ_n) of r.v.'s in which s_n is interpreted as the state of the SMDP at the epoch of its nth transition and τ_n as the elapsed time between the nth and $(n+1)$th transitions. For a semi-Markov process, Definition 9-2 on page 317 of Volume I implies that τ_n is conditionally independent of $\tau_0, \tau_1, \ldots, \tau_{n-1}$ given s_n. There, $\{s_n; n \in I\}$ is a Markov chain and the distribution of τ_n depends in an essential

† When discussing continuous-time models, we usually use *epoch* for a point in time, and *time* for an interval of elapsed time.

‡ The notation (s_n, τ_n) in this section is written (Y_n, X_n) in Sections 9-1 and 9-2 of Volume I.

way only on s_n and perhaps also on s_{n+1}. Here, we introduce action a_n taken at the epoch of the nth transition, permit the distribution of τ_n to depend on a_n as well as on s_n and s_{n+1}, and permit the conditional distribution of s_{n+1} to depend on both a_n and s_n. As with an MDP, a_n may be constrained by s_n.

The following definition of an SMDP mimics the definition of an MDP in Section 4-1. As in that section, we briefly use **boldface** type for r.v.'s and ordinary typeface for their realizations. Let $\mathcal{S}$ be a set of states and, for each $s \in \mathcal{S}$, let A_s be a set of actions. We assume $\mathcal{S} \subset I$. Let

$$\mathscr{C} \triangleq \{(s, a)\colon \quad a \in A_s \quad \text{and} \quad s \in \mathcal{S}\}$$

Definition 5-1 The *history* up to the epoch at which the nth action is taken is

$$H_n \triangleq (s_1, \tau_1, a_1, s_2, \tau_2, a_2, \ldots, s_{n-1}, \tau_{n-1}, a_{n-1}, s_n) \tag{5-1}$$

At the epoch of the jth transition, for $j < n$, H_j is known but $\boldsymbol{\tau}_{n-1}$, $\mathbf{s}_n$, and $\mathbf{a}_n$ are not known. There are semi-Markovian assumptions concerning the dynamics and rewards.

Definition 5-2 The *transition distribution* is a collection $\{Q_{sj}^a(\cdot)\colon (s, a) \in \mathscr{C}, j \in \mathcal{S}\}$ of functions on $[0, \infty)$ which satisfy

(i) $Q_{sj}^a(t) = 0$ for $t < 0$
(ii) $Q_{sj}^a(\cdot)$ is monotone
(iii) $\lim_{t\to\infty} \sum_{j\in\mathcal{S}} Q_{sj}^a(t) = 1 \qquad (s, a) \in \mathscr{C}$

For each n, x, j, H_n, and $(s, a) \in \mathscr{C}$,

$$P\{\mathbf{s}_{n+1} = j, \boldsymbol{\tau}_n \le x \mid \mathbf{H}_n = H_n, \mathbf{s}_n = s, \mathbf{a}_n = a\}$$
$$= P\{\mathbf{s}_{n+1} = j, \boldsymbol{\tau}_n \le x \mid \mathbf{s}_n = s, \mathbf{a}_n = a\} = Q_{sj}^a(x) \tag{5-2}$$

It follows from (iii) above (cf. Proposition 9-1 on page 317 of Volume I) that, if a stationary policy is used to choose actions, $\{\mathbf{s}_n;\ n \in I\}$ is a Markov chain. The transition probabilities are obtained from

$$p_{sj}^a \triangleq P\{\mathbf{s}_{n+1} = j \mid \mathbf{s}_n = s, \mathbf{a}_n = a\} = Q_{sj}^a(\infty) \tag{5-3}$$

A reward $\mathbf{X}_n$ is received during each transition of an SMDP. The probability distribution of $\mathbf{X}_n$ is assumed to depend on s_n, $\mathbf{s}_{n+1}$, $\boldsymbol{\tau}_n$, and $\mathbf{a}_n$. It is convenient to represent the expected values of rewards with a collection $\{\xi_{sj}^a(\cdot, \cdot)\}$ of functions on $\{(x, y)\colon 0 \le y \le x < \infty\}$. Let $\mathbf{t}_n$ denote the epoch of the nth transition,

$$\mathbf{t}_n = \sum_{l=1}^{n-1} \boldsymbol{\tau}_l$$

Then $\xi_{sj}^a(x, y)$ is the expected value of the cumulative reward during $[t_n, t_n + y)$ if $\mathbf{s}_n = s$, $\mathbf{s}_{n+1} = j$, $\mathbf{a}_n = a$, and $\boldsymbol{\tau}_n = x$.

Definition 5-3 The *reward function* is a collection $\{\xi_{sj}^a(\cdot,\cdot): (s, a) \in \mathscr{C}, j \in \mathcal{S}\}$ of functions on $\{(x, y): 0 \le y \le x < \infty\}$ such that

$$\xi_{sj}^a(x, x) = E(\mathbf{X}_n \mid \mathbf{s}_n = s, \mathbf{s}_{n+1} = j, \boldsymbol{\tau}_n = x, \mathbf{a}_n = a) \tag{5-4}$$

From Exercises 5-1 and 5-3,

$$P\{\boldsymbol{\tau}_n \le x \mid \mathbf{s}_n = s, \mathbf{s}_{n+1} = j, \mathbf{a}_n = a\} = \frac{Q_{sj}^a(x)}{Q_{sj}^a(\infty)} = \frac{Q_{sj}^a(x)}{p_{sj}^a}$$

whose right side we define to be zero if $Q_{sj}^a(\infty) = 0$. Therefore, using (5-2), (5-3), and (5-4),†

$$r(s, a) \triangleq \sum_{j \in \mathcal{S}} \frac{1}{p_{sj}^a} \int_0^\infty \xi_{sj}^a(x, x)\, dQ_{sj}^a(x) = E(\mathbf{X}_1 \mid \mathbf{s}_1 = s, \mathbf{a}_1 = a) \tag{5-5}$$

is the expected reward during any transition which begins in state s with action a being taken.

Definition 5-4 A *semi-Markov decision process* (*SMDP*) is a model which consists of nonempty sets $\mathcal{S}$ and A_s, $s \in \mathcal{S}$, a transition distribution which satisfies (5-2), and a reward function which satisfies (5-4).

An Example

Example 5-1 Regional incinerators, traffic lights at intersections, food canneries, and equipment maintenance are among the phenomena which have been described with models of a queue having a controllable service rate. Here, we use the terminology of equipment maintenance. Suppose the epochs at which units of equipment fail constitute a Poisson process with intensity λ. The repairer, the person who can repair the failed units, can also be used on other tasks. At any epoch, the repairer is either busy on other tasks or is repairing a failed unit. A switch from the former to the latter can occur at any epoch when a piece of equipment fails. A switch from the latter to the former can occur at any epoch when the repair of a failed unit has been completed. Suppose that the times taken to repair failed units are i.i.d. r.v.'s with the distribution function $G(\cdot)$. Let ρ be the net profit per unit time when the server is busy on other tasks, c be the material cost per unit time of engaging in repairs, h be the holding cost per unit time per failed unit, and ζ be the revenue received when a repair is completed.

In order to specify the SMDP, let $\mathbf{t}_1, \mathbf{t}_2, \ldots$ be the sequence of epochs at which either (*a*) the repairer is busy with other tasks and a unit fails, or (*b*) the repairer completes the repair of a failed unit. Let $\mathbf{s}_n$ be the number of failed units at epoch $\mathbf{t}_n + 0$ [that is, including the newly failed unit in case (*a*) and not including the newly repaired unit in case (*b*)]. Here, $\mathcal{S} = I$. Let $a = 0$

† For the integral to exist, it is sufficient that $\xi_{sj}^a(x, x)$ have bounded variation for $x \in [0, \infty)$.

denote "busy with other tasks" and $a = 1$ denote "repairing a failed unit." Then $A_0 = \{0\}$ and $A_s = \{0, 1\}$ for all $s > 0$.

If $\mathbf{a}_n = 0$, then $\mathbf{s}_{n+1} = \mathbf{s}_n + 1$ and $\boldsymbol{\tau}_n$ has an exponential distribution with mean λ^{-1} (the interarrival times of a Poisson process have an exponential distribution and exponential r.v.'s are "memoryless"; cf. Corollary 4-6 in Volume I and Proposition A-2). If $\mathbf{a}_n = 1$, then $\boldsymbol{\tau}_n$ has the distribution function $G(\cdot)$ and $\mathbf{s}_{n+1}$ has the same distribution as $\mathbf{s}_n - 1 + \mathbf{X}$, where $\mathbf{X}$ denotes the number of units which fail while the defective unit is being repaired. The distribution of $\mathbf{X}$ is given by

$$P\{\mathbf{X} = i\} = \frac{1}{i!}\int_0^\infty e^{-\lambda y}(\lambda y)^i \, dG(y) \qquad i \in I \tag{5-6}$$

Therefore,

$$Q_{sj}^0(x) = \begin{cases} 1 - e^{-\lambda x} & \text{if } j = s + 1 \\ 0 & \text{if } j \neq s + 1 \end{cases} \tag{5-7a}$$

$$Q_{s,\, s+i}^1(x) = \begin{cases} \displaystyle\int_0^x e^{-\lambda y}(\lambda y)^{i+1} \, dG(y)/(i+1)! & \text{if } i \geq -1 \text{ and } s \geq 1 \\ 0 & \text{otherwise} \end{cases} \tag{5-7b}$$

Recall that $\xi_{sj}^a(x, y)$ is the cumulative expected reward during the first y time units of a transition if $\boldsymbol{\tau}_n = x$, $\mathbf{s}_n = s$, $\mathbf{s}_{n+1} = j$, and $\mathbf{a}_n = a$. Let $\mathbf{W}(y)$ be the number of failed units at time $\mathbf{t}_n + y$, $y \geq 0$. Notice that $\{\mathbf{W}(y) - \mathbf{W}(\mathbf{t}_n);\ y \geq 0\}$ is a Poisson process with rate λ. Then

$$\xi_{s,\, s+1}^0(x, y) = \rho y - hE\left[\int_0^y \mathbf{W}(z)\, dz\right] = (\rho - hs)y$$

because $\mathbf{a}_n = 0$ implies $\mathbf{W}(y) = \mathbf{W}(\mathbf{t}_n)$ for $0 \leq y < \boldsymbol{\tau}_n$. The case $\mathbf{a}_n = 1$ is more complicated.

If $y < x$

$$\xi_{sj}^1(x, y) = -cy - hE\left[\int_0^y \mathbf{W}(z)\, dz\right]$$

Let $\{\mathbf{N}(t);\ t \geq 0\}$ be a Poisson process with intensity λ. Then $\mathbf{W}(y) - \mathbf{W}(\mathbf{t}_n)$ has the same distribution as $\mathbf{N}(y)$. If $\mathbf{a}_n = 1$, $\mathbf{s}_n = 2$, and $\mathbf{s}_{n+1} = j$, then $\mathbf{W}(\tau_n - 0) = j + 1$, so

$$\begin{aligned} E\left[\int_0^y \mathbf{W}(z)\, dz\right] &= E\left(\int_0^y \{\mathbf{W}(\mathbf{t}_n) + [\mathbf{W}(\mathbf{z}) - \mathbf{W}(\mathbf{t}_n)]\}\, dz\right) \\ &= sy + E\left[\int_0^y \mathbf{N}(z)\, dz \,|\, \mathbf{N}(x) = j + 1 - s\right] \\ &= sy + \int_0^y \sum_{i=0}^{j+1-s} iP\{\mathbf{N}(z) = i \,|\, \mathbf{N}(x) = j + 1 - s\}\, dz \end{aligned}$$

Let $J = j + 1 - s$ and $p = z/x$, Then Exercise 5-56 on page 155 of Volume I asserts that

$$P\{\mathbf{N}(z) = i \mid \mathbf{N}(x) = J\} = \binom{J}{i} p^i (1-p)^{J-i}$$

This is a probability from a binomial distribution with parameters J and p, so its expected value is $Jp = Jt/x$. Therefore,

$$\sum_{i=0}^{J} iP\{\mathbf{N}(z) = i \mid \mathbf{N}(x) = J\} = \frac{Jz}{x}$$

and

$$E\left[\int_0^y \mathbf{W}(z)\,dz\right] = \int_0^y \frac{Jz}{x}\,dz = \frac{Jy^2}{2x}$$

Similarly,

$$E\left[\int_0^x \mathbf{W}(z)\,dz\right] = \frac{Jx}{2}$$

so

$$\xi_{sj}^1(x, y) = \begin{cases} -cy - h\left[sy + \dfrac{(j+1-s)y^2}{2x}\right] & 0 \le y < x \\ \zeta - cx - hx\left(\dfrac{s+j+1}{2}\right) & y = x \end{cases} \tag{5-8}$$

Let

$$\mu \triangleq \int_0^\infty y\,dG(y) \qquad \text{and} \qquad \mu_{(2)} \triangleq \int_0^\infty y^2\,dG(y)$$

denote the first and second moments of the repair time. Then the expected rewards during transitions, using (5-5), (5-7*a*), (5-7*b*), and (5-8) are

$$r(s, 0) = \frac{\rho - hs}{\lambda}$$

$$r(s, 1) = \sum_{i=0}^{\infty} \int_0^\infty \left[\theta - cx - hx\left(s + \frac{i}{2}\right)\right] \frac{e^{-\lambda x}(\lambda x)^i}{i!}\,dG(x)$$

In order to evaluate $r(s, 1)$, observe that

$$\sum_{i=0}^{\infty} \int_0^\infty e^{-\lambda x} \frac{(\lambda x)^i}{i!}\,dG(x) = 1$$

$$\sum_{i=0}^{\infty} \int_0^\infty xe^{-\lambda x} \frac{(\lambda x)^i}{i!}\,dG(x) = \int_0^\infty x\,dG(x) = \mu$$

$$\sum_{i=0}^{\infty} \int_0^\infty xie^{-\lambda x} \frac{(\lambda x)^i}{i!}\,dG(x) = \int_0^\infty \lambda x^2\,dG(x) = \lambda\mu_{(2)}$$

Therefore,

$$r(s, 1) = \zeta - (c + hs)\mu - \frac{h}{2}\lambda\mu_{(2)}$$

□

The remainder of this section refrains from the use of boldface type to denote r.v.'s. Instead, as in most of the book, boldface type denotes vectors.

Expected Present Value

Let $w(t)$ be the rate at which rewards will be earned t time units hence, $t \geq 0$. The present value of $w(\cdot)$ is

$$\int_0^\infty w(t)e^{-\alpha t}\, dt \tag{5-9}$$

where α is a parameter called the *continuous-time interest rate.* Observe in (5-9) that $w(t)$ is discounted by the factor β^t where $\beta = e^{-\alpha}$ is an instantaneous discount factor. Since $\alpha = -\log \beta$, $\alpha > 0 \Leftrightarrow \beta < 1$. Similarly, if $W(t)$ is the cumulative reward at epoch t, then the present value is

$$\int_0^\infty e^{-\alpha t}\, dW(t) \tag{5-10}$$

which is the Laplace-Stieltjes transform of $W(\cdot)$. In Volume I see pages 503, 504, 523, and 524 for a summary of properties of Laplace-Stieltjes transforms. The parameter $\alpha > 0$ is called the *continuous-time interest rate.*

Definition 4-7 asserts that a *policy* for an MDP is a nonanticipative contingency plan. For each n, a policy specifies the action a_n as some function of the history H_n which is known at the time action a_n is taken. From (4-5) and Proposition 4-1, H_n in an MDP has the form $(s_1, a_1, \theta_1, s_2, a_2, \theta_2, \ldots, s_{n-1}, a_{n-1}, \theta_{n-1}, s_n)$† where, for each n, θ_n is conditionally independent of H_n given s_n and a_n. For an SMDP, the history in Definition 5-1 has exactly this form, with $\tau_n = \theta_n$, because (2) implies that τ_n is conditionally independent of H_n given s_n and a_n. It follows (below) that a trivial generalization of the principle of optimality (Theorem 4-3) is valid for SMDPs with appropriate finite horizon criteria.

There are two major candidates for specifying the length of a finite planning horizon in an SMDP. They are (i) the number N of transitions which remain to be taken, and (ii) the clock time T which remains. If T is small and specific, alternative (ii) is aesthetically superior in most phenomena. If T is large (and vague), an analysis based on suitably large values of N gives approximately the same results for (i) and (ii). However, (ii) leads to more intricate analysis, while (i) leads more easily to infinite-horizon criteria; so we shall use (i) here.

As with the MDP, suppose that there is a salvage-value function $L(\cdot)$ so $L(s_{N+1})$ is the salvage value of ending in state s_{N+1}. In order to simplify the notation, let $\xi_n(x, y)$ denote $\xi_{ij}^a(x, y)$, where $i = s_n$, $j = s_{n+1}$, and $a = a_n$. Then the present value of the rewards during $[0, t_{N+1}]$ is the r.v.

$$\begin{aligned} B_N &\triangleq \sum_{n=1}^{N} \int_{t_n}^{t_{n+1}} e^{-\alpha y}\, d_y\, \xi_n(\tau_n, y) + \exp(-\alpha t_{N+1}) L(s_{N+1}) \\ &= \sum_{n=1}^{N} e^{-\alpha t_n} \int_0^{\tau_n} e^{-\alpha y}\, d_y\, \xi_n(\tau_n, y) + \exp(-\alpha t_{N+1}) L(s_{N+1}) \end{aligned}$$

† The symbol ξ in Section 4-1 is replaced by θ in this section.

Let

$$\beta(s, a) \triangleq E[\exp(-\alpha\tau_1) \mid s_1 = s, a_1 = a]$$
$$= \sum_{j \in \mathcal{S}} \int_0^\infty e^{-\alpha x} \, dQ_{sj}^a(x) \tag{5-11a}$$

and

$$r(s, a) \triangleq E\left[\int_0^{\tau_1} e^{-\alpha y} \, d_y \xi_1(\tau_1, y) \mid s_1 = s, a_1 = a\right]$$
$$= \sum_{j \in \mathcal{S}} \int_0^\infty \int_0^x e^{-\alpha y} \, d_y \xi_{sj}^a(x, y) \, dQ_{sj}^a(x) \tag{5-11b}$$

This notation correctly suggests that an SMDP can be construed as an MDP whose single-period reward is the expected present value of the reward function and whose discount factor depends on state and action. Note that (5-5) is the special case of (5-11b) with $\alpha = 0$.

Exercise 5-6 asks you to prove that

$$E(B_N) = E\left[\sum_{n=1}^{N} \prod_{l=1}^{n-1} \beta(s_l, a_l) r(s_n, a_n) + \prod_{l=1}^{N} \beta(s_l, a_l) L(s_{N+1})\right] \tag{5-12}$$

Expression (5-12) can be viewed as a generalization of an MDP with transition probabilities $\{p_{sj}^a = Q_{sj}^a(\infty)\}$ and single-period discount factors depending on the state and action. Expression (5-12) is formally the same as the expected present value of a finite-horizon MDP except that β^{n-1} in the MDP expression is replaced in (5-12) by

$$\prod_{l=1}^{n-1} \beta(s_l, a_l)$$

From Exercise 4-12, the principle of optimality (Theorem 4-2) is valid for the SMDP if β is replaced by $\beta(s, a)$. That is, the recursive optimality equations for an SMDP are

$$f_0(s) = L(s) \qquad s \in \mathcal{S}$$
$$f_n(s) = \sup\left\{r(s, a) + \beta(s, a) \sum_{j \in \mathcal{S}} p_{sj}^a f_{n-1}(j) : a \in A_s\right\} \qquad s \in \mathcal{S}, n \in I_+ \tag{5-13}$$

It is valid to interpret $f_n(s)$ as the maximal value of $E(B_n \mid s_1 = s)$ when $N = n$. Thus $f_n(s)$ is the maximal expected present value of the rewards which accrue during n transitions, if s is the initial state.

Another Example

Example 5-2: Continuation of Examples 4-2, 4-7, 4-18, 4-26, 4-28, and 4-29 Let $\mathcal{S} = \{0, 1, 2\}$ in a machine-maintenance model with states 0, 1, and 2 signifying "like new," "minor defect," and "being repaired," respectively.

The actions are 0, which indicates "let the machine continue to operate" and 1, which indicates "under repair." The sets of available actions are $A_0 = \{0\}$, $A_1 = \{0, 1\}$, and $A_2 = \{1\}$. The decision problem is whether or not to repair a machine with a minor defect.

Suppose that the machine is inspected at epochs separated by a unit of time while it is operational, i.e., in states 0 or 1. However, if the machine develops a minor defect or "fails" in use, i.e., moves to state 1 from state 0 or to state 2 from either state 0 or state 1, then the change in status is instantly recognized. We assumed that the failure epoch of a "like new" machine, given that it fails before it is next inspected, is uniformly distributed on [0, 1]. Similarly, if a machine has a minor defect and action 0 is taken ("let the machine continue to operate"), and if the machine fails before it is next inspected, then the failure epoch is uniformly distributed on [0, 1]. Finally, the time needed to repair a machine (regardless of the state from which it entered state 2) is an exponential r.v. with mean 0.5.

We choose $\{p_{sj}^a\}$ to be consistent with the MDP version (Example 4-7) of this model. For example, in the MDP version, $P\{s_{n+1} = 0 \mid s_n = 0, a_n = 0\} = 0.7$, $P\{s_{n+1} = 1 \mid s_n = 0, a_n = 0\} = 0.2$, and $P\{s_{n+1} = 2 \mid s_n = 0, a_n = 0\} = 0.1$. Therefore, we use $p_{00}^0 = 0.7$, $p_{01}^0 = 0.2$, and $p_{02}^0 = 0.1$. With the preceding description, this yields

$$Q_{00}^0(x) = \begin{cases} 0 \\ 0.7 \end{cases} \qquad Q_{01}^0(x) = \begin{cases} 0 \\ 0.2 \end{cases} \qquad \text{and} \qquad Q_{02}^0(x) = \begin{cases} 0.1x & 0 \le x < 1 \\ 0.1 & x \ge 1 \end{cases}$$

Similarly,

$$Q_{11}^0(x) = \begin{cases} 0 \\ 0.4 \end{cases} \qquad \text{and} \qquad Q_{12}^0(x) = \begin{cases} 0.6x & 0 \le x < 1 \\ 0.6 & x \ge 1 \end{cases}$$

If action 1 ("under repair") is taken in state 1 ("minor defect"), then the machine immediately enters state 2 so $Q_{12}^1(x) = 1$ for $x \ge 0$. This exemplifies a time between transitions which is 0 (with probability 1). Finally, the exponential repair times and the transition probabilities in Example 4-7 yield $Q_{20}^1(x) = 0.8(1 - e^{-2x})$ and $Q_{21}^1(x) = 0.2(1 - e^{-2x})$, $x \ge 0$, and $Q_{22}^1(\cdot) \equiv 0$.

Letting $x \to \infty$ in $Q_{sj}^a(x)$ yields Table 5-1.

Table 5-1 Transition probabilities in Example 5-2

s \ j	0	1	2
0	0.7	0.2	0.1
1	0	0.4	0.6

p_{sj}^0

s \ j	0	1	2
1	0	0	1
2	0.8	0.2	0

p_{sj}^1

Table 5-2 $\beta(s, a)$ **in Example 5-2**

$s \backslash a$	0	1
0	0.5188	—
1	0.6301	1
2	—	0.7407

In Example 4-7, $\beta = 0.5$; so here $\alpha = -\log \beta = 0.7$. Substitution of $\{Q_{sj}^{a}(\cdot)\}$ in (5-11a) yields $\{\beta(s, a)\}$ as follows:

$$\beta(0, 0) = 0.7e^{-0.7} + 0.2e^{-0.7} + 0.1 \int_0^1 e^{-0.7x}\, dx = 0.5188$$

$$\beta(1, 0) = 0.4e^{-0.7} + 0.6 \int_0^1 e^{-0.7x}\, dx = 0.6301$$

$$\beta(1, 1) = 1$$

and
$$\beta(2, 1) = \int_0^\infty e^{-0.7x} 2e^{-2x}\, dx = 0.7407$$

The salvage values are the same as in Example 4-7, namely, $L(0) = 5$, $L(1) = 2$, and $L(2) = 0$.

The SMDP reward structure which corresponds to Example 4-7 is

$$\xi_{0j}^{0}(x, y) = 10y \qquad \text{for all } j \text{ and } 0 \le y \le x$$

$$\xi_{1j}^{0}(x, y) = 5y \qquad \text{for all } j \text{ and } 0 \le y \le x$$

and
$$\xi_{sj}^{1}(x, y) = 0 \qquad \text{for all } s, j, x, \text{ and } y$$

Using (5-11b), the consequent values of $\{r(s, a)\}$ are

$$r(0, 0) = 10\left[0.7 \int_0^1 e^{-0.7x}\, dx + 0.2 \int_0^1 e^{-0.7x}\, dx + 0.1 \int_0^1 \int_0^x e^{-0.7y}\, dy dx\right] = 6.8737$$

Table 5-3 $r(s, a)$ **in Example 5-2**

$s \backslash a$	0	1
0	6.8737	—
1	2.6417	0
2	—	0

Table 5-4 Dynamic programming recursion in Example 5-2

n	$f_n(0)$	$f_n(1)$	$f_n(2)$	$a_n(1)$
1	8.8970	3.1458	3.2591	0
2	10.6002	4.6667	5.7380	0
3	11.5052	4.9872	6.9726	0
4	12.0349	6.9726	7.7045	1
5	12.3675	7.7045	8.2727	1

$$r(1, 0) = 5\left[0.4 \int_0^1 e^{-0.7x}\, dx + 0.6 \int_0^1 \int_0^x e^{-0.7y}\, dydx\right] = 2.6417$$

$$r(1, 1) = 0 \qquad \text{and} \qquad r(2, 1) = 0$$

Tables 5-2 and 5-3 summarize $\{\beta(s, a)\}$ and $\{r(s, a)\}$.

Table 5-4 summarizes the results when Tables 5-1, 5-2, and 5-3 and $L(\cdot)$ are substituted in the recursive optimality equations (5-13). The last column in Table 5-4, labeled $a_n(1)$, shows which action is optimal in state 1 when n transitions remain to be made. □

Expected Present Value—Infinite Horizon

Sections 4-3, 4-4, and 4-5 concern the infinite-horizon discounted MDP. The results in those sections depend on $\beta^N \to 0$ as $N \to \infty$. Here is an analog for an SMDP.

Assumption 5-1 For every initial state and every policy, if $\alpha > 0$ in (5-11*a*),

$$\lim_{N \to \infty} \prod_{n=1}^{N} \beta(s_n, a_n) = 0 \tag{5-14}$$

is valid with probability 1.

Referring to definition (5-11*a*) of $\beta(s, a)$, (5-14) fails to be valid if there is an initial state and a policy for which $P\{\tau_n = 0\} = 1$ for all n. Although $P\{\tau_n > 0\} = 1$ for most n in most models, in Example 5-2, $P\{\tau_n = 0 \mid s_n = 1, a_n = 1\}$ exemplifies a model with zero transition times. Nevertheless, Assumption 5-1 is valid in Example 5-2. For (5-14) to be valid, in general, it is sufficient that there be $\gamma > 0$ such that, with probability 1, $\{n\colon \tau_n > \gamma\}$ is countably infinite for every initial state and every policy. We do not know of any well-posed SMDP model of a real phenomenon that fails to satisfy Assumption 5-1.

Assumption 5-1 guarantees that an infinite-horizon SMDP with the expected present-value criterion and $\alpha > 0$ has the continuity property (4-52). Therefore, the proof of Theorem 4-3 can be modified to yield the following result.

Theorem 5-1 Suppose $\mathcal{S} \subset I$, $\alpha > 0$, and Assumption 5-1 is valid in an infinite-horizon SMDP with expected-present-value criterion. If there is an optimal Markov policy, there is an optimal stationary policy.

PROOF Left as Exercise 5-7. □

It should be apparent that $\mathcal{S} \subset I$, Assumption 5-1, and $\alpha > 0$ are sufficient for the results in Sections 4-4 and 4-5 to be valid for SMDPs. The results include Corollaries 4-4*a* and 4-4*b*, Theorems 4-4, 4-5, 4-6, 4-8, 4-9, 4-10, 4-11, and 4-12. The remainder of this subsection presents SMDP counterparts to some of these MDP results.

For the remainder of this section, we assume $\#\mathscr{C} < \infty$. Let δ be a single-period decision rule so $\delta(s) \in A_s$ for all $s \in \mathcal{S}$. Let $\mathbf{r}_\delta$ be the vector whose sth component is $r[s, \delta(s)]$, and $B_s(\delta)$ be the present value of the infinite-horizon rewards when s is the initial state and stationary policy δ^∞ is used. That is, $B_s(\delta)$ is the following r.v.:

$$B_s(\delta) \triangleq \sum_{n=1}^{\infty} e^{-\alpha t_n} \int_0^{\tau_n} e^{-\alpha y}\, d_y\, \xi_n(\tau_n, y) \tag{5-15}$$

where $\xi_n(\cdot,\cdot)$ denotes $\xi_{ij}^a(\cdot,\cdot)$ with $i = s_n$, $j = s_{n+1}$, and $a = \delta(s_n)$. Let the distribution function, mth moment, and variance of $B_s(\delta)$ be labeled

$$F_s^\delta(x) = P\{B_s(\delta) \le x\}$$

$$v_m(s, \delta) = E\{[B_s(\delta)]^m\} \qquad v_s(\delta) = v_1(s, \delta)$$

and

$$\sigma_s^2(\delta) = v_2(s, \delta) - [v_s(\delta)]^2$$

The vectors with components $v_s(\delta)$ and $\sigma_s^2(\delta)$ are denoted $\mathbf{v}(\delta)$ and $\boldsymbol{\sigma}^2(\delta)$, respectively. Finally, let $Z_{sj}^a(x)$ (an r.v.) be the discounted reward during a transition from state s to state j whose elapsed time is x, while action a is taken:

$$Z_{sj}^a(x) = \int_0^x e^{-\alpha y}\, d_y\, \xi_{sj}^a(x, y)$$

Reasoning as in the proof of Lemma 4-3, if $s_1 = s$, $s_2 = j$, and $\tau_1 = y$,

$$B_s(\delta) \sim Z_{sj}^{\delta(s)}(y) + e^{-\alpha y} B_j(\delta) \tag{5-16}$$

where the notation $X \sim Y$ indicates that r.v.'s X and Y have the same probability distribution. Therefore,

$$F_s^\delta(x \mid s_2 = j, \tau_1 = y) = F_j^\delta\{e^{\alpha y}[x - Z_{sj}^{\delta(s)}(y)]\}$$

Unconditioning on $s_2 = j$ and $\tau_1 = y$ above and in (5-16) yields the SMDP counterpart of Lemma 4-3, namely,

$$F_s^\delta(x) = \sum_{j \in \mathcal{S}} \int_0^\infty F_j\{e^{\alpha y}[x - Z_{sj}^{\delta(s)}(y)]\}\, dQ_{sj}^{\delta(s)}(y) \tag{5-17}$$

and

$$v_s(\delta) = r[s, \delta(s)] + \beta[s, \delta(s)] \sum_{j \in \mathcal{S}} p_{sj}^{\delta(s)} v_j(\delta) \tag{5-18a}$$

Let M_δ be the matrix whose (s, j)th element is $\beta[s, \delta(s)]p_{sj}^{\delta(s)}$. Then (5-18*a*) in matrix-vector notation is

$$\mathbf{v}(\delta) = \mathbf{r}_\delta + M_\delta \mathbf{v}_\delta \tag{5-18b}$$

Exercise 5-10 asks you to prove that $I - M_\delta$ is nonsingular so the analog of (4-80) is

$$\mathbf{v}(\delta) = (I - M_\delta)^{-1}\mathbf{r}_\delta \tag{5-19}$$

Let $R_{sj}(x)$ denote the total reward during the nth transition if $s_n = s$, $s_{n+1} = j$, $a_n = \delta(s_n)$, and $\tau_n = x$. Let $\mathscr{B}_\delta$ denote the matrix whose (s, j)th element is

$$\beta_{sj} \triangleq \int_0^\infty e^{-2\alpha y}\, dQ_{sj}^{\delta(s)}(y) \qquad \mathscr{B}_\delta = (\beta_{sj}) \tag{5-20a}$$

and define $\boldsymbol{\psi}_\delta$ as the vector whose sth element is

$$\psi_s \triangleq \sum_{j \in \mathcal{S}} \int_0^\infty [R_{sj}(x)]^2\, dQ_{sj}^{\delta(s)}(x) + \sum_{j \in \mathcal{S}} \beta_{sj}[v_j(\delta)]^2$$
$$+ 2\sum_{j \in \mathcal{S}} v_j(\delta)\int_0^\infty e^{-\alpha x}R_{sj}(x)\, dQ_{sj}^{\delta(s)}(x) - [v_s(\delta)]^2 \tag{5-20b}$$

Exercise 5-9 asks you to derive recursive equations for $v_m(s, \delta)$ and to prove the following generalization of (4-81):

$$\boldsymbol{\sigma}^2(\delta) = (I - \mathscr{B}_\delta)^{-1}\boldsymbol{\psi}_\delta \tag{5-20c}$$

Algorithms for Discounted Rewards

Here is the policy-improvement algorithm for SMDPs based on Corollaries 4-4*a* and 4-4*b*. Let v_s^n be the sth component of the vector $\mathbf{v}^n$ and let M_n and $\mathbf{r}_n$ denote M_{δ_n} and $\mathbf{r}_{\delta_n}$, respectively.

Discounted-Reward Policy-Improvement Algorithm

(*a*) Let $n = 1$ and select a single-period decison rule δ_1.
(*b*) Solve the equation $\mathbf{v}^n = \mathbf{r}_n + M_n \mathbf{v}^n$ for $\mathbf{v}_n$ by computing

$$\mathbf{v}^n = (I - M_n)^{-1}\mathbf{r}_n \tag{5-21a}$$

(*c*) For each $s \in \mathcal{S}$, determine whether

$$\psi_s^n \triangleq \max\left\{r(s, a) + \beta(s, a)\sum_{j \in \mathcal{S}} p_{sj}^a v_j^n:\quad a \in A_s - \{\delta_n(s)\}\right\} - v_s^n \tag{5-21b}$$

is positive or nonpositive. If $\psi_s^n > 0$, let $\delta_{n+1}(s)$ be any $a \in A_s$ that causes the maximum in (5-21*b*) to be attained. If $\psi_s^n \le 0$, let $\delta_{n+1}(s) = \delta_n(s)$.

(*d*) If $\delta_{n+1} = \delta_n$, stop, and δ_n^∞ is optimal. Otherwise, replace n with $n + 1$ and return to (*b*).

Theorem 5-2 If $\#\mathscr{C} < \infty$, $\alpha > 0$, and Assumption 5-1 is valid, then procedures (*a*) through (*d*) above converge to an optimum after finitely many iterations.

Proof Left as Exercise 5-7. □

The SMDP analog of the MDP infinite-horizon expected-present-value functional equation (4-89) is

$$y_s = \max \left\{ r(s, a) + \beta(s, a) \sum_{j \in \mathcal{S}} p_{sj}^a y_j : \quad a \in A_s \right\} \qquad s \in \mathcal{S} \tag{5-22}$$

Exercise 5-10 uses (5-22) to state the SMDP version of Theorem 4-9.

Here is the linear program for SMDPs, which is analogous to (4-91).

Discounted-Reward Linear Program

$$\begin{aligned} &\text{Minimize } \sum_{s \in \mathcal{S}} \gamma_s y_s \\ &\text{subject to} \\ &y_s - \beta(s, a) \sum_{j \in \mathcal{S}} p_{sj}^a y_j \geq r(s, a) \qquad (s, a) \in \mathscr{C} \end{aligned} \tag{5-23}$$

where $\gamma_s > 0$ is arbitrary for each $s \in \mathcal{S}$. Exercise 5-11 states the equivalence between (5-22) and (5-23) that is presented in Theorem 4-11 for MDPs.

Example 5-3: Continuation of Example 5-2 The data of this SMDP are summarized in Tables 5-5, 5-6, and 5-7. There are only two stationary policies; we start with $\delta_1(1) = 1$ [necessarily $\delta_1(0) = 0$ and $\delta_1(1) = 1$]. From Tables 5-5 and 5-6, M_1, the matrix with entries $p_{sj}^a \beta(s, a)$, where $a = \delta_1(s)$, is

$$M_1 = \begin{pmatrix} 0.3632 & 0.1038 & 0.0519 \\ 0 & 0 & 1 \\ 0.5926 & 0.1481 & 0 \end{pmatrix}$$

From Table 5-7, the transpose of $\mathbf{r}_1$ is (6.8737, 0, 0). Therefore, using (5-21*a*),

$$\mathbf{v} = (I - M_1)^{-1}\mathbf{r}_1 = \begin{pmatrix} 13.0038 \\ 9.0457 \\ 9.0457 \end{pmatrix}$$

Table 5-5 Transition probabilities in Example 5-2

$s \backslash j$	0	1	2
0	0.7	0.2	0.1
1	0	0.4	0.6

p_{sj}^0

$s \backslash j$	0	1	2
1	0	0	1
2	0.8	0.2	0

p_{sj}^1

Table 5-6 $\beta(s, a)$ **in Example 5-2**

$s \backslash a$	0	1
0	0.5188	—
1	0.6301	1
2	—	0.7407

The optimality test (5-21*b*) reduces to computing

$$r(1, 0) + \beta(1, 0) \sum_{j \in \mathcal{S}} p_{1j}^{0} v_j^1 - v_1^1$$

$$= 2.6417 + 0.2520(9.0457) + 0.3781(9.0457) - 9.0456 = -0.7043$$

This quantity is nonpositive; so δ_1^∞ is optimal.

For the linear program (5-23), let $\gamma_0 = \gamma_1 = \gamma_2 = 1$; then (5-23) is

$$\text{Minimize } y_0 + y_1 + y_2$$

subject to

$$y_0 - 0.3632y_0 - 0.1038y_1 - 0.0519y_2 \geq 6.8737$$

$$y_1 - 0.2520y_1 - 0.3781y_2 \geq 2.6417$$

$$y_1 - y_2 \geq 0$$

$$y_2 - 0.5926y_0 - 0.1481y_1 \geq 0$$

The optimal solution is $y_0 = 13.0038$ and $y_1 = y_2 = 9.0457$. The third constraint is satisfied as an equality; so taking action 1 in state 1 is optimal. The left side of the second constraint takes the value 3.3460 (at the optimal values of $\{y_s\}$), which exceeds the constraining value 2.6417 by an amount 0.7043. The optimality test in the policy-improvement algorithm yielded the quantity -0.7043.

Now we use (5-20*a*), (5-20*b*), and (5-20*c*) to compute the vector of variances of present values induced by policy δ_1^∞. From the specification of

Table 5-7 $r(s, a)$ **in Example 5-2**

$s \backslash a$	0	1
0	6.8737	—
1	2.6417	0
2	—	0

$\{Q^a_{sj}(\cdot)\}$ in Example 5-2, the definition in (5-20*a*), $\delta(0) = 0$, $\delta(1) = 1$, $\delta(2) = 1$, and $\alpha = 0.7$,

$$\beta_{00} = 0.73^{-1.4} = 0.1726 \qquad \beta_{01} = 0.2e^{-1.4} = 0.0493$$

$$\beta_{02} = 0.1 \int_0^1 e^{-1.4y}\, dy = 0.0538$$

$$\beta_{10} = 0 \qquad \beta_{11} = 0 \qquad \beta_{12} = 1$$

$$\beta_{20} = 0.8 \int_0^\infty e^{-1.4y} 2e^{-2y}\, dy = 0.4706 \qquad \beta_{22} = 0$$

and

$$\beta_{21} = 0.2 \int_0^\infty e^{-1.4y} 2e^{-2y}\, dy = 0.1176$$

Therefore,

$$\mathscr{B}_{\delta_1} = \begin{pmatrix} 0.8274 & -0.0493 & -0.0538 \\ 0 & 1 & -1 \\ -0.4706 & -0.1176 & 1 \end{pmatrix}$$

From the specification of $\{\xi^a_{sj}(\cdot,\cdot)\}$ in Example 5-2 and the definition

$$R_{sj}(x) = \xi^{\delta(s)}_{sj}(x, x) \qquad \delta = \delta_1$$

it follows that

$$R_{0j}(x) = 10x \qquad R_{1j}(x) = 0 \qquad \text{and} \qquad R_{2j}(x) = 0$$

Combining $\{Q^a_{sj}(\cdot)\}$ [with $a = \delta(s)$] from Example 5-2, the calculations above, and (5-20*b*) yields

$$\psi_0 = 0.9(100) + 0.1 \int_0^1 100x^2\, dx + (0.1726)(13.0038)$$

$$+ (0.0493)(9.0456) + 0.0538(9.0456) + 2\Big\{(13.0038)(10)e^{-0.7}$$

$$+ (9.0457)(10)e^{-0.7} + 0.1 \int_0^1 10xe^{-0.7x}\, dx\Big\} - (13.0038)^2 = 147.0367$$

$$\psi_1 = 0 + 0 + v_2^2 - v_1^2 = 0$$

and

$$\psi_2 = 0 + 0 + (0.4706)(13.0038) + (0.1176)(9.0457) - (9.0456)^2$$

$$= -74.6413$$

Using (5-20*c*),

$$\begin{pmatrix} \sigma_0^2(\delta_1) \\ \sigma_1^2(\delta_1) \\ \sigma_2^2(\delta_1) \end{pmatrix} = \begin{pmatrix} 0.8274 & -0.0493 & -0.0538 \\ 0 & 1 & -1 \\ -0.4706 & -0.1176 & 1 \end{pmatrix}^{-1} \begin{pmatrix} 147.0367 \\ 0 \\ -74.6413 \end{pmatrix}$$

$$= \begin{pmatrix} 179.0690 \\ 10.9118 \\ 10.9118 \end{pmatrix}$$

Since $v_1(\delta_1) = v_2(\delta_1)$ and δ_1 applied to state 1 causes an immediate and costless shift to state 2, we should have expected $\sigma_1^2(\delta_1) = \sigma_2^2(\delta_1)$. □

EXERCISES

5-1 Prove that an SMDP has the property

$$P\{\boldsymbol{\tau}_n \le x \mid \mathbf{s}_n = s, \mathbf{s}_{n+1} = j, \mathbf{a}_n = a, \mathbf{H}_n = H\} = \frac{Q_{sj}^a(x)}{Q_{sj}^a(\infty)}$$

whose right side is defined to be zero if $Q_{sj}^a(\infty) = 0$.

5-2 Suppose Example 5-1 is altered so that a setup cost K is incurred when an item's repair begins. Specify $\{r(s, a): (s, a) \in \mathscr{C}\}$.

5-3 Suppose Example 5-1 is altered so that a start-up cost K is incurred at epoch t_n if $a_{n-1} = 0$ and $a_n = 1$. Specify an SMDP model which encompasses the alteration. Include $\mathscr{C}$, $\{Q_{sj}^a(\cdot)\}$, $\{\xi_{sj}^a(\cdot, \cdot)\}$, and $\{r(s, a)\}$. [Hint: Let $s_n = (x_n, z_n)$, where x_n denotes the number of failed units and $z_n = a_{n-1}$.]

5-4 Suppose in Example 5-1 that $G(x) = 1 - e^{-\mu x}$, $x \ge 0$. Find $Q_{sj}^1(x)$, $x \ge 0$, and $r(s, 1)$.

5-5 Suppose in Example 5-1 that the epochs at which equipment fails constitute a renewal process with interarrival times which are not exponentially distributed. Explain why the model is no longer an SMDP.

5-6 Prove that (5-12) is valid.

5-7 Prove Theorems 5-1 and 5-2.

5-8 If $\alpha > 0$, $\#\mathscr{C} < \infty$, and Assumption 5-1 is valid, prove that $I - M_\delta$ is nonsingular for every $\delta \in \Delta$ (Δ is the set of single-stage decision rules).

5-9 (*a*) If $\alpha > 0$, $\#\mathscr{C} < \infty$, and Assumption 5-1 is valid, prove that $I - \mathscr{B}_\delta$ is nonsingular and

$$\boldsymbol{\sigma}^2(\delta) = \boldsymbol{\psi}_\delta + \mathscr{B}_\delta \boldsymbol{\sigma}_\delta^2 = (I - \mathscr{B}_\delta)^{-1} \boldsymbol{\psi}_\delta$$

(*b*) Derive the following generalization of (4-82) in which δ is suppressed in the notation:

$$v_m(s) = \sum_{i=0}^{m-1} \binom{m}{i} \sum_{j \in S} v_j(i) \int_0^\infty e^{-\alpha t} [Z_{sj}(t)]^{m-i} \, dQ_{sj}(t) + \sum_{j \in S} v_m(j) \int_0^\infty e^{-m\alpha t} \, dQ_{sj}(t)$$

5-10 Here is the SMDP version of Theorem 4-9:

w is a solution of (5-22) if, and only if, there is an optimal policy δ^∞ such that $\mathbf{w} = \mathbf{v}(\delta)$

Review the proof of Theorem 4-9 to verify that it applies to the above assertion.

5-11 The SMDP version of Theorem 4-11 is

$$\mathbf{w} \text{ is optimal in (5-23)} \Leftrightarrow \mathbf{w} \text{ is the solution of (5-22)}$$

Review the proof of Theorem 4-6 to verify that it applies to the above assertion.

5-12 Let δ be the optimal policy identified in Example 5-3. Compute $\boldsymbol{\sigma}^2(\delta)$.

5-13 Consider a firm that rents equipment to its customers. When a customer terminates service, the equipment is returned. A returned item may be repaired so that it is "as good as new," placed in inventory, and then used to satisfy future demand. Computer terminals are an example of this type of equipment.

For this type of inventory system, a natural question to ask is "when is the inventory level too high?" When is it uneconomical to retain another returned item because the cost of repairing the item and holding it in inventory until it can be used is greater than the savings that can be obtained by satisfying a future demand with a repaired item rather than with a new item? (*a*) Construct an SMDP† to answer this question in the following context.

The demand process is Poisson. The times between returns are i.i.d. rv.'s and independent of the demand process. Repair times and purchase lead times are negligible. If the on-shelf inventory level is zero at a demand epoch, the demand is immediately satisfied by a purchase. Let r be the repair cost per item, p the purchase cost per item, h the cost rate for holding an item in inventory, and α the continuous-time interest rate. (*b*) What is the significance of assuming that the demand process is Poisson?

5-14 (Continuation) (*a*) Show that an optimal stationary policy is a single critical number policy. That is, there is an inventory level N^* such that returns are accepted when the inventory is less than N^* and returns are junked otherwise. (*b*) Prove that there is a stationary optimal policy.

5-15 Section 7-3 describes a model of a production process in which items are made one at a time, demand is a renewal process, and excess demand is backlogged. Sections 4-6 and 11-6 in Volume I are related to this model. Suppose that the demand process is Poisson with rate λ and production times are i.i.d. r.v.'s which are independent of the demand process. Assume that a dormant production process can be started up at any time, but an active process can be shut down only at epochs when an item's manufacture has just been completed. That is, interruptions of partially completed items are prohibited. In the terminology of Section 7-3, let K_1 be the start-up cost, K_2 the shut-down cost, c the production cost per unit time when production is active, h the cost per unit out of stock, and μ^{-1} the expected value of the time to produce an item. Assume $0 < \lambda < \mu < \infty$. The criterion is expected present value with continuous-time interest rate α.

(*a*) Construct an MDP to answer the question: Under what conditions should production be started or halted?

(*b*) Repeat (*a*) under the assumption that excess demand is lost.

5-16 (Continuation) Suppose that demand is a renewal process, but not necessarily Poisson, and that production times are exponentially distributed. Assume that an interruption of a partially completed item is permitted but that, upon resumption, no credit is received for the work to date on the item.‡ More precisely, assume that start-ups and shut-downs are permissible only at epochs when a demand occurs.

(*a*) Repeat part (*a*) of Exercise 5-15.

(*b*) Repeat part (*b*) of Exercise 5-15.

5-17 In Example 5-1, let

$$\mu_{(-1)} = \int_0^\infty \frac{1}{x}\, dG(x) \qquad \text{and} \qquad \tilde{G}(\gamma) = \int_0^\infty e^{-\gamma x}\, dG(x) \qquad (\gamma > 0)$$

† See Examples 7-25, 7-26, and 8-4 on pages 270, 271, and 297 of Volume I for probabilistic analyses of closely related models.

‡ This is similar to a preempt-repeat discipline in a queueing model. See page 431 in Volume I for a description of this discipline.

and assume $\mu_{(-1)} < \infty$. Use definition (5-11*b*) of $r(s, a)$ to obtain the following formulas:

$$r(s, 0) = \frac{\rho - hs}{\lambda + \alpha}$$

$$r(s, 1) = \frac{h}{\alpha^2}(s\mu_{(-1)} - \lambda\mu) - \frac{c + hs}{\alpha} - \frac{sh}{\alpha^2}\int_{\alpha}^{\infty} \tilde{G}(x)\,dx$$

$$+ \left(\zeta + \frac{c + 2hs}{\alpha}\right)\tilde{G}(\alpha) - \frac{h\lambda}{\alpha^2}\frac{d\tilde{G}(\alpha)}{d\alpha} - \frac{h\lambda}{\alpha}\frac{d^2\tilde{G}(\alpha)}{d\alpha^2}$$

(Hint: See the properties of Laplace transforms and Laplace-Stieltjes transforms in Appendix A of Volume I.)

5-2* SMDP—Average-Reward Criterion

This section continues the contrast, first drawn in Section 5-1, between an SMDP (semi-Markov decision process) and a discrete-time MDP. We define the average-reward criterion, show that some stationary policy is optimal, and specify algorithms to find an optimal policy. As in Section 5-1, t_n denotes the epoch at which the nth transition begins (with $t_1 \equiv 0$). For an interval $[0, T)$, Let $N(T)$ denote the number of transitions which have been completed by epoch T:

$$N(T) = \sup\,\{n\colon t_{n+1} \le T\}$$

Recall that $\xi_n(x, y)$ denotes $\xi_{ij}^a(x, y)$, where $i = s_n$, $j = s_{n+1}$, and $a = \delta_n(s_n)$. Then the total reward induced by policy $\pi = (\delta_1, \delta_2, \ldots)$ during $[0, T]$ can be written as

$$B(\pi, T) \triangleq \sum_{n=1}^{N(T)} X_n + \xi_{N(T)}(\tau_{N(T)+1}, T - t_{N(T)}) \tag{5-24}$$

The second term on the right is the cumulative reward of the incomplete transition still in progress at epoch T. Let $b_s(\pi, T)$ denote the expected value of $B(\pi, T)$ if the initial state is s:

$$b_s(\pi, T) \triangleq E[B(\pi, T)\,|\,s_1 = s]$$

In (5-11*b*), $r(s, a)$ is defined as the expected discounted reward during a transition which starts in state s where action a is taken. If $\alpha = 0$, then $r(s, a)$ is the expected total cost during a transition as in (5-5). In this section, $r(s, a)$ denotes this latter special case of (5-11*b*):

$$r(s, a) = E[X_1\,|\,s_1 = s,\, a_1 = a] = \sum_{j \in S} \frac{1}{p_{sj}^a}\int_0^{\infty} R_{sj}^a(x)\,dQ_{sj}^a(x)$$

where we use the convention that the jth term is zero when $p_{sj}^a = 0$. Then

$$b_s(\pi, T) = E\left\{\sum_{n=1}^{N(T)} r[s_n, \delta_n(s_n)] + \varepsilon\,|\,s_1 = s\right\} \tag{5-25}$$

where ε is the second term on the right side of (5-24).

The average reward per unit time criterion is $\lim_{T \to \infty} b_s(\pi, T)/T$ when the limit exists. As in Section 4-6, we choose the limit point

$$h_s(\pi) \triangleq \liminf_{T \to \infty} \frac{b_s(\pi, T)}{T} \tag{5-26}$$

in the general case. If π is a stationary policy δ^∞, we write $h_s(\delta)$ instead of $h_s(\delta^\infty)$. As with MDPs, it turns out that the limit exists if π is a stationary policy and $\#\mathscr{C} < \infty$ (and if Assumption 5-3 below is valid). We interpret (5-26) as the long-run average reward per unit time. Sometimes, $h_s(\pi)$ is called a *gain rate*.

The discussion between (5-10) and (5-11*a*) corresponds to the choice between $\lim_{T \to \infty} b_s(\pi, T)/T$ versus $\lim_{N \to \infty} b_s(\pi, t_{N+1})/N$ as the appropriate specification of the long-run average reward per unit time. We see shortly that these two measures of effectiveness are very different.

Assumptions

Recall the definition

$$\beta(s, a) = E[\exp(-\alpha\tau_1) \mid s_1 = s, a_1 = a]$$

and the following assumption made in Section 5-1.

Assumption 5-2 For every initial state and every policy, if $\alpha > 0$ then

$$\lim_{N \to \infty} \prod_{n=1}^{N} \beta(s_n, a_n) = 0 \tag{5-27}$$

is valid with probability 1.

From (5-24) and (5-25), the average reward per unit time during $[0, T)$ is

$$\frac{b_s(\pi, T)}{T} = \frac{E\{\sum_{n=1}^{N(T)} r[s_n, \delta_n(s_n)] \mid s_1 = s\}}{T} + \frac{E(\varepsilon \mid s_1 = s)}{T}$$

In order to conclude that $E(\varepsilon \mid s_1 = s)/T \to 0$ as $T \to \infty$, we make the following assumption, which uniformly bounds the reward function.

Assumption 5-3 $\#\mathscr{C} < \infty$ and there exists $m < \infty$ such that, for all s, a, j, and y,

$$0 \le E[\xi_{s_1 s_2}^{a_1}(\tau_1, y \wedge \tau_1) \mid s_1 = s, a_1 = a, s_2 = j] \le m$$

The consequences of this assumption† include $0 \le r(s, a) \le m$ for all $(s, a) \in \mathscr{C}$ and $E(\varepsilon \mid s_1 = s)/T \to 0$ as $T \to \infty$. Therefore,

$$h_s(\pi) = \liminf_{T \to \infty} E\left\{ \frac{\sum_{n=1}^{N(T)} r[s_n, \delta_n(s_n)]/N(T)}{T/N(T)} \,\middle|\, s_1 = s \right\} \tag{5-28}$$

† Equation (5-28) is valid with less restrictive assumptions. A method of proof based on limit theorems for renewal-reward processes (cf. Section 6-3 in Volume I) leads to (5-28).

The numerator is the average reward per transition, and the denominator is the average duration of a transition. In the formulas below, the long-run average reward per unit time is the ratio of the expected reward per transition to the expected duration of a transition.†

Assumption 5-2 implies $P\{N(T) \to \infty \text{ as } T \to \infty\} = 1$ for every policy and initial state; so the numerator in (5-28) behaves as

$$\frac{1}{N} \sum_{n=1}^{N} r[s_n, \delta_s(s_n)]$$

as $N \to \infty$. Therefore, only the denominator of (5-28) obstructs the direct transfer of most of the results in Section 4-6 (which concerns the average-reward criterion for MDPs). In particular, Theorem 4-14 asserts that there is a stationary policy which maximizes the average reward per transition. Theorem 5-3 below makes the same assertion for the average reward per unit time. However, a stationary policy which optimizes one of the criteria need not optimize the other.

Suppose that π is a stationary policy δ^∞. Then $\{s_{N(t)};\ t \geq 0\}$ is a finite-state semi-Markov process; so‡ for each initial state s and single-stage decision rule δ, there is a number $\mu_s(\delta)$ such that (with probability 1)

$$\frac{t}{N(t)} \to \mu_s(\delta) \tag{5-29}$$

Assumption 5-2 implies $\mu_s(\delta) > 0$ for each s and δ.

Let P_δ be the matrix whose (s, j)th entry is $p_{sj}^{\delta(s)}$ and let $\mathbf{g}(\delta)$ be the vector whose sth component is

$$g_s(\delta) \triangleq \lim_{N \to \infty} \frac{1}{N} E\{\textstyle\sum_{n=1}^{N} r[s_n, \delta(s_n)] \mid s_1 = s\} \tag{5-30}$$

Lemma 4-5 guarantees that the limit exists. As in Section 4-6, we interpret $g_s(\delta)$ as the average reward per transition. However, transitions may have variable durations in an SMDP; so $g_s(\delta)$ and $h_s(\delta)$ may differ. Let $\mathbf{r}_\delta$ denote the vector whose sth component is $r[s, \delta(s)]$. Theorem 4-13 asserts

$$\mathbf{g}(\delta) = P_\delta^* \mathbf{r}_\delta \tag{5-31}$$

in which

$$P_\delta^* \triangleq \lim_{N \to \infty} \frac{1}{N} \sum_{n=0}^{N-1} P_\delta^n$$

where the limit exists due to Lemma 4-5. Therefore,

$$h_s(\delta) = \frac{g_s(\delta)}{\mu_s(\delta)} \tag{5-32}$$

† Compare (5-28) and (5-33) with Exercise 9-24 on page 339 of Volume I.

‡ Proposition 5-1 will justify this claim.

Let $v(s, a)$ be the expected duration of a transition starting from state s in which action a is taken:

$$v(s, a) \triangleq E[\tau_1 \mid s_1 = s, a_1 = a] = \sum_{j \in \mathcal{S}} \int_0^\infty x \, dQ_{sj}^a(x)$$

Let $\mathbf{v}_\delta$ and $\boldsymbol{\mu}(\delta)$ denote the vectors whose sth components are $v_s^{\delta(s)}$ and $\mu_s(\delta)$, respectively.

Proposition 5-1 Assumptions 5-2 and 5-3 imply

$$\boldsymbol{\mu}(\delta) = P_\delta^* \mathbf{v}_\delta \tag{5-33}$$

Proof If $\xi_{sj}^a(x, y) = y$ for all s, j, x, a, and y, then $r(s, a) = v(s, a)$; so $\mu_s(\delta) = g_s(\delta)$ and (5-31) yields (5-33). □

Exercise 5-18 asks you to explain how the proof of (5-33) depends on Assumptions 5-2 and 5-3.

The following result is an analog of Theorem 4-14, which states that a finite MDP with the average-return criterion has a stationary policy which is optimal. Recall from Chapter 4 that Δ and Y denote the sets of single-stage decision rules and Markov policies, respectively.

Theorem 5-3 Assumptions 5-2 and 5-3 imply that there is a stationary policy δ_*^∞ with

$$h_s(\delta_*) = \sup \{h_s(\pi): \pi \in Y\} \qquad s \in \mathcal{S}$$

Proof We only sketch a proof; Exercise 5-19 asks you to fill in the details. The general argument is the same as the proof of Theorem 4-14. That proof invokes part (*b*) of Lemma 4-4, parts (*a*) and (*d*) of Lemma 4-5, and part (*a*) of Theorem 4-4. Lemmas 4-4 and 4-5 depend on Proposition 4-7, which, in turn, depends on Proposition 4-6. We shall mimic the proof of Theorem 4-14 and identify the gaps.

Let $\alpha_1, \alpha_2, \ldots$ be a sequence such that $\alpha_k > 0$ for all k and $\alpha_k \to 0$ as $k \to \infty$. Let δ_k^∞ be optimal for the expected-present-value criterion when α_k is the continuous-time interest rate. Let $\mathbf{v}(\delta, \alpha)$ make explicit the dependence on α of the expected present value $v(\delta)$. Then

$$\mathbf{v}(\delta_k, \alpha_k) \geq \mathbf{v}(\pi, \alpha_k) \qquad \pi \in Y \qquad k \in I$$

Since $\#\mathscr{C} < \infty$, Δ is a finite set; so $\delta_1, \delta_2, \ldots$ contains a subsequence $\delta_{k(1)}, \delta_{k(2)}, \ldots$ such that $\delta_* = \delta_{k(j)}$ for all $j \in I$. Therefore,

$$\mathbf{v}(\delta_*, \alpha_{k(j)}) \geq \mathbf{v}(\pi, \alpha_{k(j)}) \qquad \pi \in Y \qquad j \in I \tag{5-34}$$

If

$$\mathbf{h}(\delta) = \alpha \lim_{\alpha \downarrow 0} \mathbf{v}(\delta, \alpha) \qquad \delta \in \Delta \tag{5-35}$$

then $$\mathbf{h}(\delta_*) = \lim_{\alpha \downarrow 0} \alpha \mathbf{v}(\delta_*, \alpha) = \lim_{j \to \infty} \alpha_{k(j)} v(\delta_*, \alpha_{k(j)}) \tag{5-36}$$

The step in the proof of Theorem 4-5 which is analogous to (5-35) depends on part (*b*) of Lemma 4-4, parts (*a*) and (*d*) of Lemma 4-5, and part (*a*) of Theorem 4-13.

If

$$\mathbf{h}(\pi) \le \liminf_{\alpha \downarrow 0} \alpha \mathbf{v}(\pi, \alpha) \qquad \pi \in Y \tag{5-37}$$

then (5-34) and (5-36) yield

$$\begin{aligned}\mathbf{h}(\delta_*) &= \lim_{j \to \infty} \alpha_{k(j)} \mathbf{v}(\delta_*, \alpha_{k(j)}) \\ &\ge \liminf_{j \to \infty} \alpha_{k(j)} \mathbf{v}(\pi, \alpha_{k(j)}) \ge \mathbf{h}(\pi)\end{aligned}$$

In the proof of Theorem 4-5, the step comparable with (5-37) depends on part (*a*) of Lemma 4-4.

Thus (5-35) and (5-37) must be proved in order to complete the argument above. □

Average-Return Criterion—Algorithms

Let $(\mathbf{u})_s$ denote the sth component of a vector $\mathbf{u}$. From (5-31), (5-32), (5-33), and Theorem 5-3, there is a single-stage decision rule δ_* which maximizes

$$h_s(\delta) = \frac{(P_\delta^* \mathbf{r}_\delta)_s}{(P_\delta^* \mathbf{v}_\delta)_s} \tag{5-38}$$

for all $s \in \mathcal{S}$ and $\delta \in \Delta$. We present linear programming and policy-improvement algorithms which are similar to those in Section 4-6. Therefore, there is an analogous unichain assumption.

Assumption 5-4 Every single-stage decision rule δ has a matrix P_δ that induces a Markov chain with one communicating class of states and a (possibly empty) set of transient states.

As in Section 4-6, this assumption implies for each $\delta \in \Delta$ (cf. Section 7-6 in Volume I) existence of a stationary distribution $\boldsymbol{\lambda}^\delta$ which is the same regardless of the initial state. Each row of P_δ^* is $\boldsymbol{\lambda}^\delta$, which satisfies

$$\boldsymbol{\lambda}^\delta \ge \mathbf{0} \qquad \boldsymbol{\lambda}^\delta \mathbf{e} = 1 \qquad \text{and} \qquad \boldsymbol{\lambda}^\delta = \boldsymbol{\lambda}^\delta P_\delta \tag{5-39}$$

Therefore,

$$(P_\delta^* \mathbf{r}_\delta)_s = \boldsymbol{\lambda}^\delta \mathbf{r}_\delta \qquad \text{and} \qquad (P_\delta^* \mathbf{v}_\delta)_s = \boldsymbol{\lambda}^\delta \mathbf{v}_\delta \qquad s \in \mathcal{S}$$

so, from (5-38),

$$h_s(\delta) = \frac{\boldsymbol{\lambda}^\delta r_\delta}{\boldsymbol{\lambda}^\delta v_\delta} \qquad s \in \mathcal{S} \qquad \delta \in \Delta \tag{5-40}$$

Exercise 5-21 asks you to use Lemma 5-1 below and the material between (4-110) and (4-130) to prove the following result.

Proposition 5-2 Under Assumptions 5-2, 5-3, and 5-4, the following mathematical program has an optimal solution in which, for each $s \in \mathcal{S}$, there is at most one $a \in A_s$ such that $z_{sa} > 0$:

$$\text{Maximize } \frac{\sum_{(s,\,a) \in \mathscr{C}} r(s, a) z_{sa}}{\sum_{(s,\,a) \in \mathscr{C}} v(s, a) z_{sa}} \tag{5-41a}$$

$$\text{subject to } z_{sa} \geq 0 \qquad (s, a) \in \mathscr{C} \tag{5-41b}$$

$$\sum_{(s,\,a) \in \mathscr{C}} z_{sa} = 1 \tag{5-41c}$$

$$\sum_{a \in A_j} z_{ja} = \sum_{(s,\,a) \in \mathscr{C}} p_{sj}^a z_{sa} \qquad j \in \mathcal{S} \tag{5-41d}$$

The system (5-41a) to (5-41d) is analogous to the linear program (4-129a) to (4-129d). Here, z_{sa} can be interpreted in terms of the embedded Markov chain $\{s_n; n \in I_+\}$ induced by a stationary policy. For this Markov chain, z_{sa} is the stationary joint probability of being in state s and taking action a.

The apparent obstacle to solving (5-41a) to (5-41d) is that the objective (5-41a) is nonlinear. However, there is an equivalent linear program due to the following result,† which you are asked to prove in Exercise 5-22.

Lemma 5-1 Let M be a matrix and $\mathbf{b}$ and $\mathbf{d}$ vectors such that

(i) $M\mathbf{z} = \mathbf{0}$ and $\mathbf{z} \geq \mathbf{0} \Rightarrow \mathbf{z} = \mathbf{0}$

(ii) if

$$\mathbf{z} \geq \mathbf{0} \qquad \text{and} \qquad M\mathbf{z} = \mathbf{b} \tag{5-42}$$

then $\mathbf{d}z > 0$

Then the transformation

$$\mathbf{x} = \frac{\mathbf{z}}{\mathbf{dz}} \qquad \text{and} \qquad y = (\mathbf{dz})^{-1} \tag{5-43}$$

is one-to-one between (5-42) and

$$\mathbf{x} \geq \mathbf{0} \qquad \mathbf{dx} = 1 \qquad \text{and} \qquad M\mathbf{x} - \mathbf{b}y = \mathbf{0} \tag{5-44}$$

As a result, a linear program with its constraints given by (5-42) and whose objective is to maximize $\mathbf{cx}/\mathbf{dx}$ is equivalent to the following linear program:

$$\text{Maximize } c\mathbf{x}$$

$$\text{subject to } \mathbf{x} \geq \mathbf{0} \qquad M\mathbf{x} = \mathbf{b} \qquad \text{and} \qquad \mathbf{dx} = 1$$

† C. Derman, "On Sequential Decisions and Markov Chains," *Manage. Sci.* **9**: 16–24 (1962). Also, see A. Charnes and W. W. Cooper, "Programming with Linear Fractional Functionals," *Nav. Res. Logistics Quart.* **9**: 181–186 (1962).

Assumption 5-4 implies that $\mathbf{z} = \mathbf{0}$ is the unique solution of (5-41*b*) and (5-41*d*) and (5-41*c*) replaced by $\mathbf{ez} = 0$; therefore, condition (i) in the lemma is satisfied by (5-41*b*) to (5-41*d*). Assumption 5-2 implies that the denominator of (5-41*a*) is positive for all solutions to (5-41*b*) to (5-41*d*) so condition (ii) of the lemma is satisfied. The transformation (5-43) applied to (5-41*a*) to (5-41*d*) is

$$y = \frac{1}{\sum_{(j,k) \in \mathscr{C}} v(j, k) z_{jk}} \qquad \text{and} \qquad x_{sa} = z_{sa} y \qquad (s, a) \in \mathscr{C} \tag{5-45}$$

Therefore, (5-41*a*) to (5-41*d*) is equivalent to the following linear program.

Average-Reward Linear Program

$$\text{Maximize} \sum_{(s,a) \in \mathscr{C}} r(s, a) x_{sa} \tag{5-46a}$$

$$\text{subject to } x_{sa} \geq 0 \qquad (s, a) \in \mathscr{C} \tag{5-46b}$$

$$\sum_{a \in A_j} x_{ja} = \sum_{(s,a) \in \mathscr{C}} p_{sj}^a x_{sa} \qquad j \in \mathcal{S} \tag{5-46c}$$

$$\sum_{(s,a) \in \mathscr{C}} v(s, a) x_{sa} = 1 \tag{5-46d}$$

If $v(\cdot, \cdot) \equiv 1$, then (5-46*a*) to (5-46*d*) is identical to (4-129*a*) to (4-129*d*), the linear program for the average-reward criterion in an MDP with the unichain assumption. The obvious analogs of Section 4-6 are valid here. In particular, the material between (4-126) and (4-128) describes the extraction of an optimal policy from a basic optimal solution of the linear program. Also, (5-46*c*) contains a redundant constraint so one of the $S = \#\mathcal{S}$ constraints in (5-46*c*) may be deleted.

The linear program which is dual to (5-46*a*) to (5-46*d*) is to find variables h and $w_j, j \in \mathcal{S}$, in order to

$$\text{Minimize } h$$

$$\text{subject to} \qquad hv(s, a) + w_s \geq r(s, a) + \sum_{j \in \mathcal{S}} p_{sj}^a w_j \qquad (s, a) \in \mathscr{C} \tag{5-47}$$

Just as linear programs (4-130*a*) and (4-130*b*) suggest the functional equation (4-131), so also (5-47) suggests

$$w_s = \max \left\{ r(s, a) - hv(s, a) + \sum_{j \in \mathcal{S}} p_{sj}^a w_j : a \in A_s \right\} \qquad s \in \mathcal{S} \tag{5-48}$$

Exercise 5-23 asks you to prove the following connection between (5-47) and (5-48).

Theorem 5-4 Suppose Assumptions 5-2, 5-3, and 5-4 are valid and $(h^0, \mathbf{w}^0, k^0)$ and $(h^1, \mathbf{w}^1)$ are optimal in (5-47) and (5-48), respectively. Then $h^0 = h^1 \geq h_s(\pi)$ for all s and π.

The following policy-improvement algorithm for the SMDP with the average-reward criterion is only slightly more laborious than for the MDP with the average-reward criterion. We write u_s^i for the sth component of an S-vector $\mathbf{u}^i$ and $\mathbf{r}^i$, $\boldsymbol{\mu}_i$, $\mathbf{v}_i$, and P_i instead of $\mathbf{r}_{\delta_i}$, $\boldsymbol{\mu}_{\delta_i}$, $\mathbf{v}_{\delta_i}$, and P_{δ_i}, respectively.

Average-Reward Policy-Improvement Algorithm

(*A*) Let $i = 1$ and select $\delta_1 \in \Delta$. Choose $s^* \in \mathcal{S}$ and let $w_{s*}^i \equiv 0$ for all i.
(*B*) Solve the following equation for h^i and $\mathbf{w}^i$:

$$\mathbf{v}_i h^i + \mathbf{w}^i = \mathbf{r}_i + P_i \mathbf{w}^i \tag{5-49}$$

(*C*) For each $s \in \mathcal{S}$, determine whether

$$\psi_s^i \triangleq \max \left\{ r(s, a) + \sum_{j \in \mathcal{S}} p_{sj}^a w_j^i - h^i v(s, a) \colon a \in A_s - \{\delta_i(s)\} \right\} - w_s^i \tag{5-50}$$

is positive or nonpositive. If $\psi_s^i > 0$, let $\delta_{i+1}(s)$ be any $a \in A_s$ that causes the right side of (5-50) to be positive. If $\psi_s^i \geq 0$, let $\delta_{i+1}(s) = \delta_i(s)$.
(*D*) If $\delta_{i+1} = \delta_i$, stop. Otherwise, replace i with $i + 1$ and return to (*B*).

Exercise 5-24 asks you to prove the following analog of Theorem 4-16.

Theorem 5-5 Suppose Assumptions 5-2, 5-3, and 5-4 are valid. (*a*) Then $h^i \leq h^{i+1}$ for each i until the policy-improvement algorithm terminates. (*b*) If also, for every $\delta \in \Delta$, P_δ corresponds to an irreducible Markov chain (i.e., there are no transient states), then $h^i < h^{i+1}$ until termination, which occurs after finitely many iterations.

Example 5-4: Continuation of Example 5-3 The data for this SMDP are presented in Example 5-2, and some data are repeated below: $\xi_{0j}^0(x, y) = 10y$, $\xi_{1j}^0(x, y) = 5y$, and $\xi_{sj}^1(x, y) = 0$. From those data and Table 5-8,

$$r(0, 0) = 10(0.7 + 0.2 + 0.1 \times \tfrac{1}{2}) = 9.5$$

$$r(1, 0) = 5(0.4 + 0.6 \times \tfrac{1}{2}) = 3.5$$

$$r(1, 1) = r(2, 1) = 0$$

Table 5-8 Transition probabilities in Example 5-4

s \ j	0	1	2
0	0.7	0.2	0.1
1	0	0.4	0.6

p_{sj}^0

s \ j	0	1	2
1	0	0	1
2	0.8	0.2	0

p_{sj}^1

Let $\delta_1(1) = 1$, [$\delta_1(0) = 0$ and $\delta_1(2) = 1$ necessarily]; then

$$P_1 = \begin{pmatrix} 0.7 & 0.2 & 0.1 \\ 0 & 0 & 1 \\ 0.8 & 0.2 & 0 \end{pmatrix}$$

so (5-39) yields $\boldsymbol{\lambda}^1 = (0.606, 0.167, 0.227)$. Also, $\mathbf{r}_1 = (9.5, 0, 0)$ and $\mathbf{v}_1 = (0.95, 0, 0.5)$ so (5-31), (5-32), and (5-33) produce $g^1 = 5.757$, $\mu^1 = 0.689$, and $h^1 = 8.353$. Retaining $h^1 = 8.353$ as a check, (5-49) is

$$\begin{aligned} h^1 + w_0^1 &= 9.5 + 0.7w_0^1 + 0.2w_1^1 + 0.1w_2^1 \\ h^1 + w_1^1 &= w_2^1 \\ h^1 + w_2^1 &= 0.8w_0^1 + 0.2w_1^1 \end{aligned}$$

Let $s^* = 1$ so $w_1^1 = 0$; then the solution is $h^1 = 8.353$ and $\mathbf{w}^1 = (14.394, 0.5, 5.758)$. The test statistic (5-50) is $\psi_0^1 = \psi_2^1 = 0$ and

$$\begin{aligned} \psi_1^1 &= r(1, 0) + \sum_{j \in S} p_{1j}^0 w_j^1 - h^1 - w_1^1 \\ &= 3.5 + 0.4w_1^1 + 0.6w_2^1 - h^1 - w_1^1 = 1.197 > 0 \end{aligned}$$

(where $\{p_{1j}^0\}$ is taken from Table 5-8). Completing step (*C*), $\delta_2(0) = 0$, $\delta_2(2) = 1$, and $\delta_2(1) = 0$. Since δ_1^∞ is not optimal and $\Delta = \{\delta_1, \delta_2\}$, of course, δ_2^∞ is optimal. Exercise 5-25 asks you to verify the optimality of δ_2^∞ via the test (5-50). □

EXERCISES

5-18 How does the proof of (5-33) depend on Assumptions 5-2 and 5-3?

5-19 Prove Theorem 5-3 in detail.

5-20 Prove (5-35) and (5-37), thus completing the proof of Theorem 5-3.

5-21 Prove Proposition 5-2 (Hint: Proposition 4-9).

5-22 Prove Lemma 5-1.

5-23 Prove Theorem 5-4.

5-24 Prove Theorem 5-5.

5-25 In Example 5-1, let $\delta(0) = 0$, $\delta(1) = 0$, and $\delta(2) = 1$. Use (5-48) to verify that δ^∞ is optimal for the criterion of average reward per unit time.

5-26 In Example 5-1, find the policy which maximizes the long-run average reward per transition. Notice that the resulting value of the objective is different from the optimal value of the objective in Exercise 5-25.

5-27 Suppose a stationary policy δ^∞ is used in an SMDP. Let $\rho_s(t)$ denote the expected reward during $[0, t \wedge \tau_1]$ if $s_1 = s$:

$$\rho_s(t) = \sum_{j \in S} \left[\int_0^t \xi_{sj}^{\delta(s)}(y, y)\, dQ_{sj}^{\delta(s)}(y) + \int_t^\infty \xi_{sj}^{\delta(s)}(y, t)\, dQ_{sj}^{\delta(s)}(y) \right]$$

Let $b_s(t)$ denote the expected total reward during $[0, t]$ if $s_1 = s$ and policy δ^∞ is used. Prove that the following equation is valid:

$$b_s(t) = \rho_s(t) + \sum_{j \in \mathcal{S}} \int_0^t b_s(t-x)\, dQ_{sj}^{\delta(s)}(x)$$

5-28 Exercises 5-13 and 5-14 model the amount of equipment stocked by a firm that rents equipment to customers. Under the assumptions of Exercise 5-14 and with its notation, suppose that the demand and the return processes are independent Poisson processes with the same rate λ. We assume that $p - r - d > 0$ to avoid trivialities. Show that the asymptotic cost rate is minimized by the largest value of N which satisfies

$$\frac{h}{\lambda} \frac{N(N+1)}{2} \leq p - r - d$$

(Do not use the probabilistic methods in Example 7-26 on pages 271 and 272 of Volume I.)

5-3* CONTINUOUS-TIME MARKOV DECISION CHAINS

A continuous-time Markov chain can be viewed as a generalization of a discrete-time Markov chain in which the times to move from state to state are exponential r.v.'s. These processes are the subject of Chapter 8 in Volume I. A semi-Markov process is a further generalization in which the times to move from state to state may have arbitrary distributions (not necessarily exponential distributions). Alternatively, a continuous-time Markov chain may be viewed as a special kind of semi-Markov process. In comparison, a continuous-time Markov decision chain, abbreviated MDC, is a special kind of semi-Markov decision process (SMDP).

Definition 5-5 A *continuous-time Markov decision chain* (*MDC*) is an SMDP (cf. Definitions 5-2 and 5-4) in which there are numbers $\{\lambda(s, a) : (s, a) \in \mathscr{C}\}$ such that

$$Q_{sj}^a(t) = p_{sj}^a(1 - e^{-\lambda(s, a)t}) \qquad t \geq 0 \qquad (s, a) \in \mathscr{C} \qquad j \in \mathcal{S}$$

Therefore, the times between the transitions of an MDC are exponential r.v.'s. It follows from the form of $Q_{sj}^a(t)$ that these r.v.'s do not depend on the "next" state:

$$P\{\tau_n > x \mid s_n = s,\ a_n = a,\ s_{n+1} = j\} = e^{-\lambda(s, a)x}$$

does not depend on j.

The MDC is important because (1) it is the kind of SMDP most frequently encountered in the operations research literature and (2) it is equivalent to an MDP. Reason 2 simplifies the qualitative analysis of certain models, as shown in Chapters 7 and 8. This section explains the sense in which an MDC is equivalent to an MDP. To facilitate applications in later chapters, the set of states $\mathcal{S}$ may be a countably infinite set in this section.

Sections 8-2 and 8-3 in Volume I develop the structure of a continuous-time Markov chain. The principal parameters there are the transition probabilities and the Q-matrix (called "generator" or "infinitesimal generator" by some authors).

It is convenient to require that the MDC be *conservative*;† i.e., for all policies, finite $t \geq 0$, and initial state s,

$$\sum_{j \in S} P\{s_{N(t)} = j \mid s_1 = s\} = 1$$

All MDCs encountered in the operations research literature are conservative (so far as the authors are aware).

Just as a stationary policy in an SMDP induces a semi-Markov process, a stationary policy δ in an MDC induces a continuous-time Markov chain $\{X^\delta(t); t \geq 0\}$. The assumption of a conservative MDC implies that M_δ‡, the Q-matrix of $\{X^\delta(t); t \geq 0\}$, has elements

$$\left.\begin{aligned} m_{sj}^\delta &= p_{sj}^a \lambda(s, a) & s \neq j \\ m_{ss}^\delta &= \lambda(s, a) = \sum_{j \neq s} m_{sj}^\delta & \end{aligned}\right\} \quad a = \delta(s)$$

The following definition extends terminology in Section 8-7 of Volume 1.

Definition 5-6 An MDC is *uniformizable* if it is conservative and there is a (finite) positive number Λ such that

$$(1 - p_{ss}^a)\lambda(s, a) \leq \Lambda \qquad (s, a) \in \mathscr{C} \tag{5-51}$$

It follows from Section 8-7 of Volume I that a unformizable MDC has the following property. For every δ and distribution of s_1, $\{X^\delta(t); t \geq 0\}$ has the same finite dimensional distributions as a continuous-time Markov chain $\{Y^\delta(t); t \geq 0\}$ constructed as follows. The transitions of $\{Y^\delta(t); t \geq 0\}$ are embedded in the jump epochs of a Poisson process with intensity Λ. The transitions from state to state occur according to a Markov chain with the following transition probabilities $\{\rho_{sj}^\delta\}$:

$$\rho_{sj}^\delta = \begin{cases} p_{sj}^a \dfrac{\lambda(s, a)}{\Lambda} & s \neq j \\ 1 - \dfrac{\lambda(s, a)(1 - p_{ss}^a)}{\Lambda} & s = j \end{cases} \qquad \text{where } a = \delta(s)$$

$$= \begin{cases} \dfrac{m_{sj}^\delta}{\Lambda} & s \neq j \\ 1 - \dfrac{m_{ss}^\delta}{\Lambda} & s = j \end{cases} \tag{5-52}$$

† See Chapter 8 in Volume I, particularly Section 3.

‡ We use M for the Q-matrix because Q has already been employed for the transition function. It is unfortunate that the literature uses Q as the standard notation for both of these objects.

The essential feature of the Y^{δ} process is that all the exponential distributions (i.e., the distributions of all $\tau_n \mid s_n = s, a_n = a$) have the same mean Λ^{-1}.

Let $\mathscr{D}$ denote a uniformizable MDC with the expected-present-value criterion and with data $\{r(s, a)\}$, α, $\{\lambda(s, a)\}$, and $\{p_{sj}^a\}$. We shall specify another MDC, $\hat{\mathscr{D}}$, which is equivalent to $\mathscr{D}$ in a certain useful way. Quantities which pertain to $\hat{\mathscr{D}}$ are denoted $\hat{r}(s, a)$, $\hat{p}_{sj}^a$, etc.

Let

$$\hat{p}_{sj}^a = \begin{cases} p_{sj}^a \dfrac{\lambda(s, a)}{\Lambda} & s \neq j \\ 1 - \dfrac{\lambda(s, a)(1 - p_{ss}^a)}{\Lambda} & s = j \end{cases} \tag{5-53}$$

and

$$\hat{\lambda}(s, a) = \Lambda \qquad (s, a) \in \mathscr{C} \tag{5-54}$$

From the discussion between (5-51) and (5-52), the application of a stationary policy to (5-53) and (5-54) induces a continuous-time Markov chain with the same finite-dimensional distributions as would be induced in $\mathscr{D}$.

Exploiting the Uniformizable Property

This presentation follows Serfozo (1979).

Theorem 5-6 For a uniformizable MDC $\mathscr{D}$, let $\hat{\mathscr{D}}$ denote the uniformizable MDC with data (5-53), (5-54), continuous-time interest rate α, and expected discounted rewards

$$\hat{r}(s, a) = r(s, a)\, \frac{\alpha + \lambda(s, a)}{\alpha + \Lambda} \qquad (s, a) \in \mathscr{C} \tag{5-55}$$

Let $\delta \in \Delta$ and suppose that $\mathbf{v}_\delta$ (the vector of expected present values in $\mathscr{D}$) exists. ($\mathscr{C}$ is no longer necessarily a finite set.) Let $\hat{\mathbf{v}}_\delta$ denote the vector of expected present values in $\hat{\mathscr{D}}$ of the stationary policy δ^∞. Then $\hat{\mathbf{v}}_\delta$ exists and

$$\mathbf{v}_\delta = \hat{\mathbf{v}}_\delta \qquad \delta \in \Delta \tag{5-56}$$

Proof For each δ, $\mathscr{D}$ and $\hat{\mathscr{D}}$ have the same Q-matrix; so, for each distribution of the initial state s_1,

$$P\{s_{N(t)} = j\} = P\{\hat{s}_{N(t)} = j\} \tag{5-57}$$

for all t and j (where $\hat{s}_n$ is the nth state occupied in $\hat{\mathscr{D}}$). Let a *sojourn* in a state denote the elapsed time between entry and exit from that state. For example, if $s_n = s_{n+1} = s$, $s_{n-1} \neq s$, and $s_{n+2} \neq s$, the sojourn in state s is $\tau_n + \tau_{n+1}$. From (5-57), $\mathbf{v}_\delta = E[\mathbf{B}(\delta)]$, and definition (5-15) of $B_s(\delta)$, it is sufficient to show that the expected discounted rewards during sojourns are the same in $\mathscr{D}$ and $\hat{\mathscr{D}}$.

For given δ and s, the number of repeated visits to state s during sojourns in state s are independent geometric r.v.'s distributed in $\mathscr{D}$ and $\hat{\mathscr{D}}$ as the r.v.'s Z and $\hat{Z}$ with distributions

$$P\{Z = l\} = [1 - p_{ss}^{\delta(s)}][p_{ss}^{\delta(s)}]^l \qquad l \in I$$

and

$$P\{\hat{Z} = l\} = [1 - \hat{p}_{ss}^{\delta(s)}][\hat{p}_{ss}^{\delta(s)}]^l$$

respectively. Therefore, the expected discounted reward in $\mathscr{D}$ during a sojourn is

$$r[s, \delta(s)]E\left[\sum_{l=0}^{Z} \exp(-\alpha t_l)\right]$$

where t_l is the sum of l i.i.d. exponential r.v.'s with parameter $\lambda[s, \delta(s)]$. Therefore, t_l has a gamma distribution with parameters $\lambda[s, \delta(s)]$ and l. Let r, p, and λ denote $r[s, \delta(s)]$, $p_{ss}^{\delta(s)}$, and $\lambda[s, \delta(s)]$, respectively, and let $\xi \triangleq \lambda/(\alpha + \lambda)$. Then

$$E\left(\sum_{l=0}^{Z} e^{-\alpha t_l}\right) = \sum_{l=0}^{\infty}(1 - p)p^l \sum_{i=0}^{l} E[\exp(-\alpha t_i) \mid l \geq i]$$

$$= \sum_{l=0}^{\infty}(1 - p)p^l \sum_{i=0}^{l} \xi^i = (1 - p\xi)^{-1}$$

Similarly,

$$E\left[\sum_{l=0}^{Z} \exp(-\alpha t_l)\right] = (1 - \hat{p}\hat{\xi})^{-1}$$

where $\hat{\xi} = \hat{\lambda}/(\hat{\alpha} + \hat{\lambda}) = \Lambda/(\alpha + \Lambda)$. Therefore, the expected discounted rewards in $\mathscr{D}$ and $\hat{\mathscr{D}}$ during a sojourn are

$$\frac{r(\alpha + \lambda)}{\alpha + (1 - p)\lambda}$$

and, from (5-53), (5-54), and (5-55),

$$\frac{\hat{r}(\hat{\alpha} + \hat{\lambda})}{\hat{\alpha} + (1 - \hat{p})\hat{\lambda}} = \frac{[r(\alpha + \lambda)/(\alpha + \Lambda)](\alpha + \Lambda)}{\alpha + \Lambda\lambda(1 - p)/\Lambda} = \frac{r(\alpha + \lambda)}{\alpha + (1 - p)\lambda}$$

which are the same. □

Corollary 5-6*a* Under the assumptions of Theorem 5-6, let an MDP (discrete-time) have transition probabilities and rewards given by (5-53) and (5-55), respectively, and single-period discount factor $\beta = \Lambda/(\alpha + \Lambda)$. Let $\delta \in \Delta$ such that $\mathbf{v}_\delta$ is assumed to exist in Theorem 5-6. Let $\mathbf{v}_\delta^*$ be the MDP's vector of expected present values associated with δ. Then $\mathbf{v}_\delta^*$ exists and

$$\mathbf{v}_\delta^* = \mathbf{v}_\delta = \hat{\mathbf{v}}_\delta$$

PROOF Let $\hat{P}_\delta = (\hat{p}_{sj}^{\delta(s)})$ and $\hat{\mathbf{r}}_\delta = \{\hat{r}[s, \delta(s)]\}$. Then (5-19) and (4-80) yield

$$\mathbf{v}_\delta^* = (I - \beta\hat{P}_\delta)^{-1}\hat{\mathbf{r}}_\delta \qquad \text{and} \qquad \mathbf{v}_\delta = \hat{\mathbf{v}}_\delta = (I - \hat{M})^{-1}\hat{\mathbf{r}}_\delta$$

where $\hat{M}_\delta$ has as (s, j)th element

$$\begin{aligned}
&\hat{p}_{sj}^{\delta(s)} E(e^{-\alpha\tau_1} \mid s_1 = s, a_1 = a) \\
&= \hat{p}_{sj}^{\delta(s)} \int_0^\infty \Lambda e^{-(\alpha+\Lambda)x}\, dx = \hat{p}_{sj}^{\delta(s)} \frac{\Lambda}{\alpha + \Lambda} \\
&= \beta \hat{p}_{sj}^{\delta(s)}
\end{aligned}$$

Therefore, $\hat{M}_\delta = \beta \hat{P}_\delta$. □

It follows from Corollary 5-6*a* that a uniformizable MDC can be replaced by a discrete-time MDP if the criterion is expected present value. This conclusion simplifies the qualitative analysis of models and makes it unnecessary to develop separate computer software for algorithms for MDCs. The conclusion is valid too if the criterion is average reward per unit time.

Corollary 5-6*b* For a uniformizable MDC $\mathscr{D}$ and $\delta \in \Delta$, let $\hat{\mathscr{D}}$ be given by (5-53), (5-54), and (5-55) and suppose that $\mathbf{v}_\delta$ exists for all $\alpha > 0$. Suppose also that the vector $\mathbf{h}_\delta$ of average rewards per unit time exists. Let $\mathbf{g}_\delta$ be the vector of average rewards per period of the discrete-time MDP with transition probabilities and rewards given by (5-53) and (5-55), respectively. Suppose in the MDP that $\mathbf{v}_\delta^*$ exists for all discount factors $\beta < 1$. Then $\hat{\mathbf{h}}_\delta$ and $\hat{\mathbf{g}}_\delta$ exist and

$$\mathbf{h}_\delta = \hat{\mathbf{h}}_\delta = \mathbf{g}_\delta \Lambda$$

PROOF Let $\mathbf{v}_{\delta,\alpha}$ and $\hat{\mathbf{v}}_{\delta,\alpha}$ make explicit the dependence of the expected present values on α, the continuous-time interest rate. From Proposition 4-7 and Theorem 5-6,

$$\mathbf{h}_\delta = \lim_{\alpha \downarrow 0} \alpha \mathbf{v}_{\delta,\alpha} = \lim_{\alpha \downarrow 0} \alpha \hat{\mathbf{v}}_{\delta,\alpha} = \hat{\mathbf{h}}_\delta$$

In the discrete-time MDP given by (5-53) and (5-55), let $\mathbf{v}_{\delta,\beta}^*$ make explicit the dependence of the expected present value on β, the single-period discount factor. Let $\hat{r}_n$ denote the r.v. $\hat{r}[\hat{s}_n, \delta(\hat{s}_n)]$. Then $\beta = \Lambda/(\alpha + \Lambda)$; so $\alpha = \Lambda(1 - \beta)/\beta$ yields

$$\begin{aligned}
h_\delta(s) &= \lim_{\alpha \downarrow 0} \alpha v_{\delta,\alpha}(s) = \lim_{\alpha \downarrow 0} \alpha E\left[\sum_{n=1}^{\infty} \left(\frac{\Lambda}{\Lambda + \alpha}\right)^{n-1} \hat{r}_n \,\Big|\, \hat{s}_1 = s\right] \\
&= \Lambda \lim_{\beta \uparrow 1} \frac{1 - \beta}{\beta} E\left[\sum_{n=1}^{\infty} \beta^{n-1} \hat{r}_n \,\Big|\, s_1 = s\right] \\
&= \Lambda \lim_{\beta \uparrow 1} (1 - \beta) v_{\delta,\beta}^*(s) = \Lambda g_\delta(s)
\end{aligned}$$

□

An Example

Example 5-5: Continuation of Example 5-4 The SMDP in Examples 5-2, 5-3, and 5-4 is not an MDC because $Q_{0j}^0(\cdot)$ and $Q_{1j}^0(\cdot)$, $j \in \mathbb{S}$, do not satisfy

Table 5-9 $\lambda(s, a)$ **in Example 5-5**

$s \backslash a$	0	1
0	1.1	—
1	1.4	10
2	—	2

Definition 5-5. We shall construct an MDC so that first moments are unchanged, that is,

$$\lambda(s, a)^{-1} = \sum_{j=0}^{3} \int_0^\infty x \, dQ_{sj}^a(x)$$

where the right-side refers to the SMDP model in Example 5-4. Therefore,

$$\lambda(0, 0)^{-1} = 0.7(1) + 0.2(1) + 0.1(0.5) = 0.95$$

$$\lambda(1, 0)^{-1} = 0.4(1) + 0.6(0.5) = 0.70$$

In Example 5-4, action 1 in state 1 causes an instantaneous transition to state 2. Therefore, we assign a "large" value to $\lambda(1, 1)$, say $\lambda(1, 1) = 10$. Finally, $\lambda(2, 1) = 2$ is already an element of Example 5-4. Table 5-9 summarizes the specification of $\{\lambda(s, a)\}$, and Tables 5-10 and 5-11 give $\{r(s, a)\}$ and $\{p_{sj}^a\}$ from Example 5-4.

In Example 5-4, $\alpha = 0.7$ (which is reflected in the numerical values in Table 5-9). Inequality (5-51) is satisfied by $\Lambda = 10$; so $\beta = \Lambda/(\alpha + \Lambda) = 0.9346$. The parameters of $\hat{\mathscr{D}}$ are given by (5-53) and (5-55) in Tables 5-12 and 5-13.

Let δ be the decision rule $\delta(0) = 0$, $\delta(1) = 1$, and $\delta(2) = 1$. Viewing this MDC model as an SMDP, the vector $\mathbf{v}_\delta$ of expected present values satisfies (5-18*a*) and (5-18*b*), which is $\mathbf{v}_\delta = \mathbf{r}_\delta + M_\delta v_\delta$. Here, the transpose of $\mathbf{r}_\delta$ is (6.8737, 0, 0). The generic component of M_δ is $\beta[s, \delta(s)]p_{sj}^{\delta(s)}$, where $\beta(s, a)$ is given by (5-11*a*). In an MDC,

$$\beta(s, a) = \sum_{j \in S} \int_0^\infty e^{-\alpha x} \, dQ_{sj}^a(x) = \frac{\lambda(s, a)}{\alpha + \lambda(s, a)}$$

Table 5-10 $r(s, a)$ **in Example 5-4**

$s \backslash a$	0	1
0	6.8737	—
1	2.6417	0
2	—	0

Table 5-11 p_{sj}^a **in Example 5-4**

s \ j	0	1	2
0	0.7	0.2	0.1
1	0	0.4	0.6

p_{sj}^0

s \ j	0	1	2
1	0	0	1
2	0.8	0.2	0

p_{sj}^1

Therefore, the (s, j)th element of M_δ is

$$\frac{\lambda[s, \delta(s)]}{\alpha + \lambda[s, \delta(s)]} p_{sj}^{\delta(s)}:$$

$$M_\delta = \begin{pmatrix} 0.4117 & 0.1176 & 0.0588 \\ 0 & 0 & 0.9346 \\ 0.5926 & 0.1481 & 0 \end{pmatrix}$$

Solving $\mathbf{v}_\delta = \mathbf{r}_\delta + M_\delta \mathbf{v}_\delta$ for $\mathbf{v}_\delta$ yields

$$v_\delta(0) = 14.5 \qquad v_\delta(1) = 9.4 \qquad v_\delta(2) = 10.0$$

The equivalent MDP has

$$P = \begin{pmatrix} 0.97 & 0.02 & 0.01 \\ 0 & 0 & 1 \\ 0.16 & 0.04 & 0.8 \end{pmatrix}$$

and the transpose of $\hat{r}_\delta$ is (1.0921, 0, 0).

The solution of $\mathbf{v}_\delta^* = \hat{r}_\delta + \beta \hat{P}_\delta \mathbf{v}_\delta^*$ is

$$v_\delta^*(0) = 14.5 \qquad v_\delta^*(1) = 9.4 \qquad v_\delta^*(2) = 10.0$$

so $\mathbf{v}_\delta^* = \mathbf{v}_\delta$.

In order to verify that $\hat{\mathbf{v}}_\delta = \mathbf{v}_\delta^*$, observe that

$$\hat{\beta}(s, a) = \frac{\hat{\lambda}(s, a)}{\alpha + \hat{\lambda}(s, a)} = \frac{\Lambda}{\alpha + \Lambda} = \beta$$

so

$$\hat{\beta}(s, a)\hat{p}_{sj}^a = \beta \hat{p}_{sj}^a$$

and $\hat{M}_\delta = \beta \hat{P}_\delta$. □

Table 5-12 $\hat{r}(s, a)$ **in Example 5-5**

s \ a	0	1
0	1.0921	—
1	0.5185	0
2	—	0

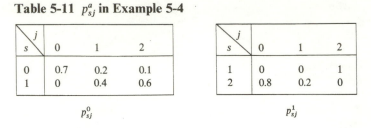

Table 5-13 $\hat{p}^a_{sj}$ in Example 5-5

s \ j	0	1	2
0	0.97	0.02	0.01
1	0	0.916	0.084

$\hat{p}^0_{sj}$

s \ j	0	1	2
1	0	0	1
2	0.16	0.04	0.8

$\hat{p}^1_{sj}$

Time-Continuous Control Models

The SMDP, including the MDC, is a continuous-time model in which control is exerted at a denumerable sequence of epochs. For example, in an inventory model, suppose unit demands occur according to a renewal process and ordered goods are delivered immediately. An SMDP model admits replenishment decisions only at epochs at which demands occur. One can imagine a continuous-time inventory model in which a replenishment decision can be made at any epoch, not only at epochs at which demands occur.

Some authors have written about stochastic control models in which actions can be varied at any epoch. Such models possess rich and interesting opportunities for mathematical analysis. See the last section of this chapter's Bibliographic Guide for references to the literature.

EXERCISES

5-29 See Exercise 5-13 for a model of the amount of equipment stocked by a firm that rents equipment to customers. The inventory of rental equipment rises when items are returned and falls when demands occur. We concentrate on one product line and make the following assumptions: the demands and returns form independent Poisson processes with means λ and μ, respectively. If a demand occurs when no items are on hand, the firm purchases an item which is delivered immediately. In order to keep the inventory level from getting too high, items in inventory can be returned to a central warehouse.

Make the following assumptions about costs:

(*a*) The cost of returning n items to the warehouse is $R + nr$.

(*b*) The cost of purchasing an item is b.

(*c*) There is a holding-cost rate of h per item per unit time applied to each item in inventory.

Formulate an MDC with the expected-present-value criterion to determine when to return items and, when items are returned, the appropriate quantities.

5-30 (Continuation) Suppose that the firm may purchase new items, at a unit cost of b, even if the inventory level is positive. Suppose also that it may purchase more than one item at a time.

(*a*) Explain why an optimal policy will specify purchases if, and only if, a demand occurs when the inventory is zero and then one item will be purchased.

(*b*) Prove that there is a finite state i such that a stationary policy insists on making a return in state i.

(*c*) Use this result to prove that there is an optimal stationary policy of the following form: There are two numbers c and d such that the firm should return $j - d$ items whenever the inventory is $j \geq c$.

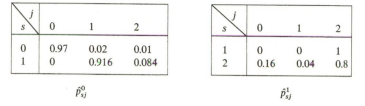

5-4* CONTRACTION MAPPINGS AND MDPs

The first portion of Section 4-3 conjectures a connection between the existence of an optimal policy and the existence of a unique solution to equation (4-39) which is

$$f(s) = \max\ \{r(s, a) + \beta E\{f[M(s, a, \xi_{s,a})]\}: a \in A_s\} \qquad s \in \mathcal{S} \qquad (5\text{-}58)$$

Equation (4-85) defines the following mapping **L** from $\mathbb{R}^{\# \mathcal{S}}$ to $\mathbb{R}^{\# \mathcal{S}}$. Let $\mathbf{v} \in \mathbb{R}^{\# \mathcal{S}}$; then

$$\mathbf{L}\mathbf{v}(s) = \max\ \{r(s, a) + \beta \sum_{j \in \mathcal{S}} p_{sj}^a v_j: \quad a \in A_s\}$$

where v_j is the jth coordinate of $\mathbf{v}$. Theorem 4-8 asserts that an MDP has a unique solution $\mathbf{f}$ to $\mathbf{Lf} = \mathbf{f}$ and $\mathbf{L}^n\mathbf{f}_0 \to \mathbf{f}$ for all $\mathbf{f}_0$ under the following conditions: finitely many states, finitely many actions in each state, and discount factor $\beta \in [0, 1)$. The equation $\mathbf{Lf} = \mathbf{f}$ is essentially (5-58). Moreover, it follows from (4-88) that a solution $\mathbf{f}$ to (5-58) yields the following optimal policy δ: for each $s \in \mathcal{S}$, let $\delta(s)$ be an a which attains the maximum in (5-58).

The results we have reviewed are sufficient to justify computational procedures which search for a solution **f** to (5-58). Numerical problems solved on a digital computer will have only finitely many states and actions.† However, the qualitative features of optimal policies often are as important as the numerical details of the policies. When models are being posed prior to qualitative analysis, sometimes it seems more natural to assume that the sets of states $\mathcal{S}$ and actions A_s, $s \in \mathcal{S}$, are continuous (i.e., contain intervals of positive width) or are countably infinite rather than finite. This section presents sufficient conditions for the results we reviewed to be applicable to MDPs whose states and actions are more than finitely numerous.

There are two other reasons to present this section's material. First, it encompasses models that are not MDPs; so it comprises more than a generalization of Section 4-5. In Chapter 9, we use some results from this section to deduce properties of a sequential game model. Second, the contraction-mapping approach, on which some of this section's results are based, is particularly elegant. This approach was initiated in Shapley (1953) and brought to maturity in Denardo (1967).

The remainder of this section dispenses with boldface notation for vectors. Entities which are vectors in other sections of the book can have a continuum of "coordinates" in this section.

Fixed-Point Theorem for Contraction Mappings

Let V be a set and $d(\cdot, \cdot)$ a real-valued function on $V \times V$. Recall from Section 4-5 [below (4-84)] that $d(\cdot, \cdot)$ is a metric on V if, for all u, v, and w in V, $d(u, v) \geq 0$, $d(u, v) = 0 \Leftrightarrow u = v$, and $d(u, v) + d(v, w) \geq d(u, w)$.

† It is conceivable that MDP problems with a continuum of states and actions may be solved on analog computers. We know of no attempts to do so.

Definition 5-7 If $d(\cdot, \cdot)$ is a metric for V, then (V, d) is a *metric space*. A metric space is *complete* if, for all Cauchy sequences $v_1, v_2, \ldots$ in V [i.e., for all $\epsilon > 0$, there exists an integer N such that $m > N$ and $n > N \Rightarrow d(v_m, v_n) < \epsilon$], there exists $v \in V$ such that $\lim_{n\to\infty} d(v_n, v) = 0$.

Definition 5-8 Let V be a metric space with metric $d(\cdot, \cdot)$ and W a function from V to V. If there exists c, $0 \le c < 1$, such that

$$d(Wu, Wv) \le cd(u, v) \qquad \text{for all } u \in V, v \in V \tag{5-59}$$

then W is a *contraction mapping on* V. Any value of c which satisfies (5-59) and $0 \le c < 1$ is a *modulus* [of W with metric $d(\cdot, \cdot)$].

For a function W from a metric space V to itself, let $W^1 = W$ and $W^{n+1} = WW^n$. Then (5-59) implies

$$\begin{aligned} d(W^n u, W^n v) &\le cd(W^{n-1}u, W^{n-1}v) \\ &\le c^2 d(W^{n-2}u, W^{n-2}v) \end{aligned} \tag{5-60}$$

In words, $0 \le c < 1$ causes the images of u and v under W to get closer and closer [in the sense of $d(\cdot, \cdot)$] to each other, as W iterates on itself. In other words, $W^n u$ and $W^n v$ are converging to the same element of V.

Proposition 5-3: S. Banach's Fixed-Point Theorem for Contraction Mappings† Suppose V is a complete metric space and there exists $N \in I_+$ such that W^N is a contraction mapping on V. Then there is a unique element $v^* \in V$ such that $Wv^* = v^*$. Moreover, for all $v \in V$, $\lim_{n\to\infty} d(W^n v, v^*) = 0$.

Contraction-Mapping Approach

Let $\mathcal{S}$ be a nonempty set and V be the set of all bounded real-valued functions on $\mathcal{S}$. For each $s \in \mathcal{S}$, let A_s be a nonempty set, let

$$\mathscr{C} = \{(s, a): \quad a \in A_s, \quad s \in \mathcal{S}\}$$

and let $\Delta = \underset{s \in \mathcal{S}}{\times} A_s$. For each $u \in V$ and $v \in V$, let

$$d(u, v) = \sup \{|u(s) - v(s)|: \quad s \in \mathcal{S}\}$$

Exercise 5-31 asks you to prove the following lemma.

Lemma 5-2 V with $d(\cdot, \cdot)$ is a complete metric space.

The most elementary function in the contraction-mapping approach is a real-valued income function $h(s, a, v)$ defined for each $(s, a) \in \mathscr{C}$ and $v \in V$. In

† J. Dugundji, *Topology*, Allyn and Bacon, Boston (1966).

Section 4-5, where $\mathcal{S}$ is countable (and finite),

$$h(s, a, v) = r(s, a) + \beta \sum_{j \in \mathcal{S}} p_{sj}^a v_j \tag{5-61}$$

where $v = (v_j, j \in \mathcal{S})$. More generally, consider an MDP where $\mathcal{S} \subset \mathbb{R}$ but $\mathcal{S}$ is not necessarily a discrete set. Let

$$G(x \mid s, a) = P\{s_{n+1} \le x \mid s_n = s, a_n = a\}$$

Then (5-61) would become

$$h(s, a, v) = r(s, a) + \beta \int_{-\infty}^{\infty} v(x)\, dG(x \mid s, a) \tag{5-62}$$

if the integral exists (i.e., if the integral does not assume the form $\infty - \infty$). Later in this section, the issue of the existence of such integrals is finessed by the contraction-mapping approach. In fact, $\mathcal{S}$ could be a subset of $\mathbb{R}^n$ or more general spaces in the following development.

It is useful to keep (5-61) and (5-62) in mind in order to interpret the following results. However, we emphasize that $h(\cdot, \cdot, \cdot)$ is the most elementary object henceforth; it is not necessarily a derived object as in (5-61) and (5-62).

For each $\delta \in \Delta$ and $v \in V$, let $H_\delta v$ be the mapping on $\mathcal{S}$ which assigns to $s \in \mathcal{S}$ the value $h[s, \delta(s), v]$.

Assumption 5-5: Contraction There exists c, $0 \le c < 1$, such that

$$| h(s, a, u) - h(s, a, v) | \le c d(u, v) \tag{5-63}$$

for all $u \in V$, $v \in V$, and $(s, a) \in \mathscr{C}$.

Assumption 5-6: Boundedness For all $v \in V$ there exists $m < \infty$ such that $| h(s, a, v) | \le m$, $(s, a) \in \mathscr{C}$.

Lemma 5-3 Suppose the boundedness assumption is valid. Then the contraction assumption is valid $\Leftrightarrow$ for all $\delta \in \Delta$, H_δ is a contraction mapping on V.

PROOF Left as Exercise 5-32. □

Example 5-6 In the case of an MDP with a countable number of states, if u and v are bounded functions on $\mathcal{S}$, (5-61) yields

$$\begin{aligned} | h(s, a, u) - h(s, a, v) | &= \beta \left| \sum_{j \in \mathcal{S}} p_{sj}^a (u_j - v_j) \right| \\ &\le \beta \sum_{j \in \mathcal{S}} p_{sj}^a | u_j - v_j | \\ &\le \beta \sup \{ | u_j - v_j | : j \in \mathcal{S} \} \sum_{j \in \mathcal{S}} p_{sj}^a \\ &= \beta\, d(u, v) \end{aligned}$$

which is (5-63) and verifies the contraction assumption.

Some restrictions are necessary to satisfy the boundedness assumption. For example, suppose $\mathcal{S} = I$, $\beta = 0$, and $r(s, \cdot) \equiv s$ for all $s \in I$. Then $H_\delta v(s) = s$ for all $s \in I$; so $H_\delta v$ is not a bounded function on $\mathcal{S}$. To guarantee $H_\delta v \in V$ for all $\delta \in \Delta$ and $v \in V$, it is sufficient that $r(\cdot, \cdot)$ be uniformly absolutely bounded on $\mathscr{C}$. That is, suppose

$$\sup \{ |r(s, a)| : (s, a) \in \mathscr{C} \} = m < \infty \tag{5-64}$$

for some $m \in \mathbb{R}_+$. Then, if $v \in V$,

$$\begin{aligned} |h(s, a, v)| &= \left| r(s, a) + \beta \sum_{j \in \mathcal{S}} p_{sj}^a v_j \right| \\ &\le |r(s, a)| + \left| \beta \sum_{j \in \mathcal{S}} p_{sj}^a v_j \right| \\ &\le m + \beta \sup \{ |v_j| : j \in \mathcal{S} \} < \infty \end{aligned}$$ □

Condition (5-64) is necessary as well as sufficient.

Proposition 5-4 Suppose $h(\cdot, \cdot, \cdot)$ is specified by (5-61) for an MDP with a discrete set of states. Then the boundedness assumption is satisfied $\Leftrightarrow$ (5-64) is satisfied.

PROOF Left as Exercise 5-34. □

Theorem 5-7 The boundedness and contraction assumptions imply for each $\delta \in \Delta$ that H_δ has a unique fixed point $v_\delta \in V$, that is,

$$v_\delta(s) = h[s, \delta(s), v_\delta]$$

has a unique solution $v_\delta \in V$. Moreover,

$$d(v_\delta, v) \le \frac{d(H_\delta v, v)}{1 - c} \qquad v \in V$$

PROOF The first assertion is implied by Lemma 5-3 and the fixed-point theorem. For the second assertion, $d(\cdot, \cdot)$ is a metric on V; so as in (5-60), for each $n \in I_+$,

$$\begin{aligned} d(v_\delta, v) &\le d(v_\delta, H_\delta^n v) + d(H_\delta^n v, v) \\ &\le d(v_\delta, H_\delta^n v) + \sum_{k=1}^{n} d(H_\delta^i v, H_\delta^{i-1} v) \\ &\le d(v_\delta, H_\delta^n v) + \sum_{k=1}^{n} c^{k-1}\, d(H_\delta v, v) \\ &\le d(v_\delta, H_\delta^n v) + \frac{d(H_\delta v, v)}{1 - c} \end{aligned}$$

The fixed-point theorem asserts $d(v_\delta, H_\delta^n v) \to 0$ as $n \to \infty$. □

Let L be the following mapping on V:

$$Lv(s) = \sup \{h(s, a, v): \quad a \in A_s\} \qquad s \in S \tag{5-65}$$

Lemma 5-4 The boundedness assumption implies $Lv \in V$ for all $v \in V$.

PROOF Left as Exercise 5-40. □

Theorem 5-8 The contraction and boundedness assumptions imply that L is a contraction mapping on V, that is,

$$d(Lu, Lv) \leq cd(u, v) \qquad u \in V, v \in V \tag{5-66}$$

PROOF Choose arbitrary u, v, and s. Label u and v so that $Lu(s) \geq Lv(s)$; let $x = Lu(s) - Lv(s) \geq 0$. We shall prove

$$0 \leq Lu(s) - Lv(s) \leq cd(u, v)$$

which implies (5-66). By the boundedness assumption and Lemma 5-4, for each $n \in I_+$ there exists $a_n \in A_s$ such that

$$Lu(s) \leq h(s, a_n, u) + \frac{x}{n}$$

Also,

$$\begin{aligned} Lu(s) - \frac{x}{n} &= Lv(s) + \frac{x(n-1)}{n} \\ &\geq Lv(s) \geq h(s, a_n, v) \end{aligned}$$

Therefore,

$$\begin{aligned} 0 &\leq Lu(s) - Lv(s) - \frac{x}{n} \\ &\leq h(s, a_n, u) - h(s, a_n, v) \leq cd(u, v) \end{aligned}$$

The last inequality is implied by the contraction assumption. □

The following corollary is an immediate consequence of Theorem 5-8 and the fixed-point theorem.

Corollary 5-8*a* The contraction and boundedness assumptions imply that L has a unique fixed point F; that is, there exists a unique solution $F \in V$ to $LF = F$.

Theorem 5-8 and its corollary comprise a generalization of Theorem 4-8. However, they do not assert that $F(s)$ is the highest possible "return" associated with a fixed point. In symbols, let

$$f(s) = \sup \{v_\delta(s): \quad \delta \in \Delta\} \qquad s \in S \tag{5-67}$$

where v_δ denotes the fixed point in V of H_δ.

Corollary 5-8*b* The contraction and boundedness assumptions imply for all $\gamma > 0$ there exists $\delta \in \Delta$ such that $d(H_\delta F, F) \le \gamma(1-c)$ and $d(v_\delta, F) \le \gamma$. Hence, $F(s) \le f(s)$ for all $s \in \mathcal{S}$.

PROOF By definition (5-65) of L and Corollary 5-8*a*, for all $\gamma > 0$ there is $\delta \in \Delta$ such that

$$F(s) = LF(s) \le H_\delta F(s) + \gamma(1-c) \qquad s \in \mathcal{S}$$

so $d(H_\delta F, F) \le \gamma(1-c)$. From Theorem 5-7, $d(v_\delta, F) \le d(H_\delta F, F)/(1-c)$. Therefore, $d(v_\delta, F) \le \gamma$. □

Not only may $F(s) < f(s)$ occur but there may not exist any $\delta \in \Delta$ such that $F = v_\delta$.

Example 5-7 Let $\mathcal{S} = \{0\}$, $A_0 = [0, 1]$, $h(0, a, v) = a$ if $0 \le a < 1$, and $h(0, 1, v) = 0$. Exercise 5-33 asks you to prove that this example satisfies the contraction and boundedness assumptions. If $0 \le a < 1$, then $H_a v = a$ so $v_a = a$. If $a = 1$, $H_a v = 0$ so $v_1 = 0$. However, for all v, $Lv = \sup\{h(0, a, v): 0 \le a \le 1\} = 1$. Therefore, $F = 1 > v_\delta$ for all δ. □

The anomaly of the preceding example is precluded by this result.

Corollary 5-8*c* Under the contraction and boundedness assumptions, if, for each $s \in \mathcal{S}$ $h(s, \cdot, F)$ is continuous on A_s and A_s is compact,† there exists $\delta \in \Delta$ such that $v_\delta = F$.

PROOF A continuous function attains its maximum on a compact set; so for each $s \in \mathcal{S}$, there exists $\delta(s) \in A_s$ such that

$$h[s, \delta(s), F] = \sup\{h(s, a, F): \quad a \in A_s\} \qquad s \in \mathcal{S} \tag{5-68}$$

Let $\delta \in \Delta$ denote the function which assigns $\delta(s)$ to $s \in \mathcal{S}$. Then (5-68) is

$$H_\delta F = LF$$

From Corollary 5-8*a*, $LF = F$. Therefore,

$$H_\delta F = F$$

But v_δ is the unique fixed point of H_δ (Theorem 5-7); so $F = v_\delta$. □

We use the notation $u \ge v$ if $u(s) \ge v(s)$ for all $s \in \mathcal{S}$.

Assumption 5-7: Monotonicity For all $\delta \in \Delta$, $u \in V$, and $v \in V$, if $u \ge v$, then $H_\delta u \ge H_\delta v$.

† In settings more general than $\mathbb{R}$, the requirement is continuity in a topology for which A_s is compact.

Monotonicity is essentially the same as (4-51) in Lemma 4-2: for every $\delta \in \Delta$, T_δ is an isotone function on $(\gtrsim, Y)$. At the end of Section 4-4, we discuss an axiomatic approach to Sections 4-3 and 4-4 which includes isotonicity as an axiom.

A broad class of income functions $h(\cdot, \cdot, \cdot)$ satisfies the monotonicity assumption. One that does not satisfy it but lacks theoretical or practical interest is

$$h(s, a, v) = r(s, a) - \beta \sum_{j \in \mathcal{S}} p_{sj}^a v_j$$

in which $\#\mathcal{C} < \infty$ is assumed. If $0 \le \beta < 1$, the contraction and boundedness assumptions are satisfied.

Monotonicity causes $F = f$.

Theorem 5-9 The contraction, boundedness, and monotonicity assumptions imply $F = f$; so f is the unique solution to $Lf = f$.

Proof $F \le f$ from Corollary 5-8*b*. Also, from (5-65) and Corollary 5-8*a*, for each $\delta \in \Delta$,

$$H_\delta F \le LF = F \tag{5-69}$$

Therefore, $H_\delta^2 F \le H_\delta F$ from Assumption 5-7 (monotonicity) and $H_\delta F \le F$ from (5-69). Recursive use of this argument yields $H_\delta^n F \le F$ for all $n \in I_+$. But Assumption 5-5 (contraction) implies $d(H_\delta^n F, v_\delta) \to 0$ as $n \to \infty$ so $v_\delta \le F$. Therefore, the definition (5-67) of f implies $f \le F$, hence $f = F$. Corollary 5-8*a* yields f as the unique solution to $Lf = f$. □

Corollary 5-8*b* and Theorem 5-9 lead immediately to the next corollary.

Corollary 5-9 The contraction, boundedness, and monotonicity assumptions imply for all $\gamma > 0$ there exists $\delta \in \Delta$ such that $d(v_\delta, f) \le \gamma$.

A policy is called *γ-optimal* if it satisfies $d(v_\delta, f) \le \gamma$.

The monotonicity assumption causes monotone convergence.

Theorem 5-10 Suppose the contraction, boundedness, and monotonicity assumptions are satisfied. Then

(*a*) $u \ge v \Rightarrow Lu \ge Lv$
(*b*) $H_\delta v \ge v \Rightarrow v \le H_\delta v \le H_\delta^2 v \le \ldots$
(*c*) $Lv \ge v \Rightarrow v \le Lv \le L^2 v \le \ldots$

Proof Definition (5-65) of L, monotonicity, and $u \ge v$ imply, for each δ,

$$H_\delta v \le H_\delta u \le Lu$$

so $Lv \le Lu$. If $H_\delta v \ge v$, repeated use of monotonicity yields (*b*). If $v \le Lv, \ldots, L^{n-2} v \le L^{n-1} v$ for some $n \ge 2$, monotonicity implies $H_\delta L^{n-2} v \le H_\delta L^{n-1} v$

for each δ; so $L^n v \geq H_\delta L^{n-2} v$; hence $L^n v \geq L^{n-1}$. Since $v \leq Lv$, induction establishes (*c*). □

It is convenient to review the consequences of the theorems for a finite MDP. There,

$$h(s, a, v) = r(s, a) + \beta \sum_{j \in \mathcal{S}} p_{sj}^a$$

with $\#\mathcal{C} < \infty$. The contraction, boundedness, and monotonicity assumptions are satisfied; so Theorems 5-8 and 5-9, Proposition 3, and Corollary 5-8*c* yield Theorem 4-8 as a special case. That theorem states $Lf = f$ and $L^n v \to f$ for all v.

MDPs with Uncountably Many States

In (5-62) we define $h(\cdot, \cdot, \cdot)$ if $\mathcal{S} \subset \mathbb{R}$:

$$h(s, a, v) = r(a, a) + \beta \int_{-\infty}^{\infty} v(x)\, dG(x \mid s, a) \tag{5-62}$$

where $$G(x \mid s, a) = P\{s_{n+1} \leq x \mid s_n = s, \quad a_n = a\}$$

The specification in (5-62) is contingent upon existence of the integral. However, the following construction avoids the issue of integrability.†

We define

$$K(v \mid s, a) = \{u: u \in V \text{ and } u(s) \geq v(s) \text{ for all } s \in \mathcal{S} \text{ and there exists } \int_{-\infty}^{\infty} u(x)\, dG(x \mid s, a)\}$$

and $$\psi(v \mid s, a) = \inf \left\{ \int_{-\infty}^{\infty} u(x)\, dG(x \mid s, a) : u \in K(v \mid s, a) \right\}$$

Then $K(v \mid s, a)$ is the set of integrable functions u which are everywhere as big as v; and $\psi(v \mid s, a)$ is the pointwise closest approximation, from above, of the "integral" in (5-62). If the integral in (5-62) actually exists, $r(s, a) + \beta\psi(v \mid s, a)$ is the same as (5-62). If it does not exist, $r(s, a) + \beta\psi(v \mid s, a)$ is a reasonable substitute for (5-62). Therefore, we use

$$h(s, a, v) = r(s, a) + \beta\psi(v \mid s, a) \tag{5-70}$$

instead of (5-62) if an MDP has uncountably many states.‡

† It also avoids the issue of measurability of $v(\cdot)$.

‡ The same approach succeeds if $\mathcal{S} \subset \mathbb{R}^n$ with $n > 1$. Partly for this reason, in Chapters 7 and 8 we generally ignore the issue of the existence of expectations.

The proof of Proposition 5-4 (Exercise 5-34), essentially unchanged, suffices for

Proposition 5-5 $h(\cdot, \cdot, \cdot)$ in (5-70) satisfies the boundedness assumption if and only if property (5-64) is satisfied, that is,

$$\sup \{|r(s, a)| : (s, a) \in \mathscr{C}\} < \infty$$

It is simple to verify that (5-70) satisfies the monotonicity assumption. Fix s and a. For the contraction assumption, we first show that $\psi(\cdot \mid s, a)$ is subadditive, that is, $\psi(u + v) \leq \psi(u) + \psi(v)$, where s, a has been suppressed in the notation; that is, $\psi(\cdot) \triangleq \psi(\cdot \mid s, a)$. Note that

$$\begin{aligned} K(u + v \mid s, a) = \{w: w \in V,\ & w(s) \geq u(s) + v(s) \\ & \text{for all } s \in \mathcal{S} \text{ and there exists} \\ & \int_{-\infty}^{\infty} w(x)\, dG(x \mid s, a)\} \\ \supset \Big\{ y + z: y(s) \geq u(s) \quad & \text{and} \quad z(s) \geq v(s) \\ & \text{for all } s \in \mathcal{S} \text{ and there exists} \\ & \int_{-\infty}^{\infty} z(x)\, dG(x \mid s, a) \quad \text{and} \quad \int_{-\infty}^{\infty} y(x)\, dG(x \mid s, a) \Big\}. \end{aligned}$$

Therefore,

$$\begin{aligned} \psi(u + v) &\leq \inf \left\{ \int_{-\infty}^{\infty} [y(x) + z(x)]\, dG(x \mid s, a): y \in K(u \mid s, a),\ z \in K(v \mid s, a) \right\} \\ &= \psi(u) + \psi(v) \end{aligned}$$

so
$$\psi(u) = \psi[(u - v) + v] \leq \psi(u - v) + \psi(v)$$

and $\psi(u) - \psi(v) \leq \psi(u - v)$. Also, by definition of ψ, for every $v \in V$, $|\psi(v)| \leq \sup \{|v(s)| : s \in \mathcal{S}\}$. Therefore,

$$|\psi(u) - \psi(v)| \leq \max \{|\psi(u - v)|, |\psi(v - u)|\} \leq d(u, v)$$

$$|h(s, a, v) - h(s, a, u)| = \beta |\psi(u) - \psi(v)| \leq \beta\, d(u, v)$$

which verifies the contraction assumption.

EXERCISES

5-31 Prove Lemma 5-2.

5-32 Prove Lemma 5-3.

5-33 Verify that Example 5-7 satisfies the contraction and boundedness assumptions.

5-34 Prove Proposition 5-4.

5-35 Suppose an SMDP (semi-Markov decision process) has continuous-time interest rate $\alpha > 0$, $\#\mathscr{C} < \infty$, and satisfies Assumption 5-1 [see (5-14)]. Show that the SMDP satisfies the contraction, boundedness, and monotonicity assumptions.

5-36 A function W from V to V is a monotone contraction mapping with modulus c if $0 \le c < 1$, condition (5-59) is satisfied, and

$$u \le v \Rightarrow Wu \le Wv \qquad \text{for all} \qquad u \in V, v \in V$$

Prove that W is a monotone contraction with modulus c if, and only if, for all $u \in V$, $v \in V$, and numbers $z \ge 0$,

$$u \le v + \mathbf{z} \Rightarrow Wu \le Wv + c\mathbf{z} \tag{5-71}$$

where $\mathbf{z} = (z, z, z, \ldots) \in V$.

5-37 (Continuation) Let W be a monotone contraction mapping with modulus c, $k \in I$, $v \in V$, and z a nonnegative number. Let w denote the fixed point in V of W and $\mathbf{z} = (z, z, \ldots) \in V$. Prove (5-72) through (5-75).

$$Wv \le v + \mathbf{z} \Rightarrow w \le W^k v + \frac{c^k}{1-c}\mathbf{z} \tag{5-72}$$

$$v \le Wv + \mathbf{z} \Rightarrow W^k v \le w + \frac{c^k}{1-c}\mathbf{z} \tag{5-73}$$

$$v \le Wv + \mathbf{z} \Rightarrow W^k v \le W^{k+1} v + c^k \mathbf{z} \tag{5-74}$$

$$v \le Wv + \mathbf{z} \Rightarrow Wv \le W^k v + \sum_{l=1}^{k-1} c^l \mathbf{z} \tag{5-75}$$

5-38 The results in this section depend on Banach's fixed-point theorem (Proposition 5-3). There is a different fixed-point approach in Shapiro (1975). It uses Brouwer's fixed-point theorem (see Proposition 9-3 and Theorem 9-2 for another application of Brouwer's theorem), which can be stated as follows: Let W be a mapping with domain and range V. Suppose that V is a compact and convex (see Appendix B) set and W is continuous on V. Then there exists $v^* \in V$ such that $Wv^* = v^*$. Notice that Proposition 5-3 makes the stronger assertion that v^* is unique.

Suppose that $H(\cdot, \cdot, \cdot)$ is specified by (5-61) for an MDP with a discrete set of states, (5-64) is satisfied, and $0 \le \beta < 1$. Use Brouwer's theorem to prove for each $\delta \in \Delta$ that H_δ has a fixed point $v_\delta \in V$. [Hint: (*a*) Restrict V to $|v(s)| \le m/(1-\beta)$, $s \in \mathbb{S}$. (*b*) If $v \in V$, then $H_\delta v \in V$ for all $\delta \in \Delta$. (*c*) H_δ is continuous on V.]

5-39 (Continuation) Under the assumptions of Exercise 5-38, suppose also that $\#A_s < \infty$ for each $s \in \mathbb{S}$. Use Brouwer's theorem to prove that L [defined in (5-65)] has a fixed point† in V.

5-40 Prove Lemma 5-4.

BIBLIOGRAPHIC GUIDE

The conference proceedings cited in Chapter 4's Bibliographic Guide contain results pertinent to this chapter. So also do some of the other citations in Chapter 4.

Markov decision processes continues to be an active research area, and new

† The assumptions in Exercises 5-38 and 5-39 are made to ensure continuity and compactness. These properties are valid under less restrictive assumptions than are made in the exercises [cf. Shapiro (1975)].

results are frequently published. There are many outlets, but the following journals are particularly worth checking: *Journal of Applied Probability*, *Mathematics of Operations Research*, and *Operations Research*.

The first three subsections here correspond to the chapter's sections. The fourth subsection lists references on two closely related topics. These are (1) the connections between alternative specifications of the average reward per unit time (or per transition) criterion, and (2) the connections between average-reward criteria and the expected-present-value criterion as the (discrete-time) discount factor tends to unity. The fifth subsection lists references on controlled diffusion processes. The underlying model is a controlled continuous-time Markov process with a continuum of states. The sixth subsection lists some references on MDPs with incomplete information. We regret not having room in this volume to present some of the elegant research results on these topics.

Semi-Markov decision processes

deCani, J. S.: "A Dynamic Programming Algorithm for Embedded Markov Chains When the Planning Horizon Is at Infinity," *Manage. Sci.* **10**: 716–733 (1964).

Denardo, E. V.: "Markov Renewal Programs with Small Interest Rates," *Ann. Math. Stat.* **42**: 477–496 (1971).

——— : "Computing Bias-Optimal Policies in Discrete and Continuous Markov Decision Problems," *Oper. Res.* **18**: 279–289 (1970).

Federgruen, A., and D. Spreen: "A New Specification of the Multichain Policy Iteration Algorithm in Undiscounted Markov Renewal Programs," *Manage. Sci.* **26**: 1211–1217 (1980).

Fox, B. L.: "Markov Renewal Programming by Linear Fractional Programming," *SIAM J. Appl. Math.* **14**: 1418–1422 (1966).

Howard, R. A.: "Semi-Markovian Decision Processes," *Proc. Int. Stat. Inst. (34th Sess.)*, Ottawa, Canada (1963).

Jewell, W. S.: "Markov-Renewal Programming I and II," *Oper. Res.* **11**: 938–971 (1963).

Schweitzer, P.: *Perturbation Theory and Markovian Decision Processes*, Ph.D. dissertation, Massachusetts Institute of Technology (1965).

Veinott, A. F., Jr.: "Discrete Dynamic Programming with Sensitive Discount Optimality Criteria," *Ann. Math. Stat.* **40**: 1635–1660 (1969).

Continuous-time Markov decision chains

Chitgopekar, S. S.: "Continuous Time Markovian Sequential Control Processes," *SIAM J. Control* **7**: 367–389 (1969).

Howard, R. A.: *Dynamic Programming and Markov Processes*, Wiley, New York (1960).

Kakumanu, P.: "Continuously Discounted Markov Decision Model with Countable State and Action Space," *Ann. Math. Stat.* **42**: 919–926 (1971).

Lippman, S.: "Applying a New Device in the Optimization of Exponential Queueing Systems," *Oper. Res.* **23**: 687–710 (1975).

Miller, B. L.: "Finite State Continuous Time Markov Decision Processes with a Finite Planning Horizon," *SIAM J. Control* **6**: 266–280 (1968).

——— : "Finite State Continuous Time Markov Decision Processes with an Infinite Planning Horizon," *J. Math. Anal. Appl.* **22**: 552–569 (1968).

Serfozo, R. F.: "An Equivalence between Continuous and Discrete Time Markov Decision Processes," *Oper. Res.* **27**: 616–620 (1979).

Contraction mappings and MDPs

Blackwell, D.: "Discounted Dynamic Programming," *Ann. Math. Stat.* **36**: 226–235 (1965).

Denardo, E. V.: "Contraction Mappings in the Theory Underlying Dynamic Programming," *SIAM Rev.* **9**: 165–177 (1967).

Federgruen, A., and P. J. Schweitzer: "Discounted and Undiscounted Value-Iteration in Markov Decision Problems: A Survey," in *Dynamic Programming and Its Applications*, edited by M. L. Puterman, Academic Press, New York (1978), pp. 23–52.

Harrison, J. M.: "Discrete Dynamic Programming with Unbounded Rewards," *Ann. Math. Stat.* **43**: 636–644 (1972).

van Nunen, J. A. E. E.: *Contracting Markov Decision Processes*, Tract 71, Mathematisch Centrum, Amsterdam (1976).

——— and J. Wessels: *The Generation of Successive Approximations for Markov Decision Processes by Using Stopping Times*, Tract 93, Mathematisch Centrum, Amsterdam (1977).

Shapiro, J. F.: "Brouwer's Fixed Point Theorem and Finite State Markovian Decision Theory," *J. Math. Anal. Appl.* **49**: 710–712 (1975).

Shapley, L. S.: "Stochastic Games," *Proc. Nat. Acad. Sci. U.S.A.* **39**: 1095–1100 (1953).

Wessels, J.: "Markov Programming by Successive Approximations with Respect to Weighted Supremum Norms," *J. Math. Anal. Appl.* **58**: 326–335 (1977).

Average-reward and sensitive-discount criteria

Blackwell, D.: "Discrete Dynamic Programming," *Ann. Math. Stat.* **33**: 719–726 (1962).

Denardo, E. V., and B. L. Miller: "An Optimality Condition for Discrete Dynamic Programming with No Discounting," *Ann. Math. Stat.* **39**: 1220–1227 (1968).

Derman, C.: "Denumerable State Markovian Decision Processes—Average Cost Criterion," *Ann. Math. Stat.* **37**: 582–585 (1966).

——— and A. F. Veinott, Jr.: "A Solution to a Countable System of Equations Arising in Markovian Decision Process," *Ann. Math. Stat.* **38**: 582–584 (1967).

Federgruen, A., and P. J. Schweitzer: "A Survey of Asymptotic Value-Iteration for Undiscounted Markovian Decision Processes," in *Recent Developments in Markov Decision Processes*, edited by R. Hartley, L. C. Thomas, and D. J. White, Academic Press, New York (1980), pp. 73–110.

Flynn, J.: "Conditions for the Equivalence of Optimality Criteria in Dynamic Programming," *Ann. Stat.* **4**: 936–953 (1976).

Hordijk, A.: *Dynamic Programming and Markov Potential Theory*, Tract 51, Mathematisch Centrum, Amsterdam (1974).

Lippman, S.: "Criterion Equivalence in Discrete Dynamic Programming," *Oper. Res.* **17**: 920–923 (1969).

Miller, B. L., and A. F. Veinott, Jr.: "Discrete Dynamic Programming with a Small Interest Rate," *Ann. Math. Stat.* **40**: 366–370 (1969).

Ross, S.: "Non-Discounted Denumerable Markovian Decision Models," *Ann. Math. Stat.* **39**: 412–432 (1968).

Sladky, K.: "On the Set of Optimal Controls for Markov Chains with Rewards," *Kybernetica*, **10**: 350–367 (1974).

Taylor, H. M.: "Markovian Sequential Replacement Process," *Ann. Math. Stat.* **36**: 1677–1694 (1965).

Veinott, A. F., Jr.: "On Finding Optimal Policies in Discrete Dynamic Programming with No Discounting," *Ann. Math. Stat.* **37**: 1284–1294 (1966).

———: "Discrete Dynamic Programming with Sensitive Discount Optimality Criteria," *Ann. Math. Stat.* **40**: 1635–1660 (1969).

van der Wal, J.: *Stochastic Dynamic Programming*, doctoral dissertation, Eindhoven Technical University (1980).

Wijngaard, J.: *Stationary Markovian Decision Problems*, doctoral dissertation. Eindhoven Technical University (1975).

Controlled diffusion processes

Bather, J.: "Diffusion Models in Stochastic Control Theory," *J. Roy. Stat. Soc. (A)* **132**: 335–345 (1969).

Benes, V.: "Existence of Optimal Stochastic Control Laws," *SIAM J. Control* **9**: 446–472 (1971).

Bensoussan, A., and J. L. Lions: "Nouvelle Methodes en Controle Impulsionnel," *Appl. Optimization Quart.* **1**: 289–312 (1975).

Davis, M. H. A., and P. Varaiya: "Dynamic Programming Conditions for Partially Observable Stochastic Systems," *SIAM J. Control* **11**: 226–261 (1973).

Doshi, B. T.: "Two-Mode Control of Brownian Motion with Quadratic Loss and Switching Costs," *Stochastic Processes and Their Applications* **6**: 277–289 (1978).

——— : "Optimal Control of a Diffusion Process with Reflecting Boundaries and Both Continuous and Lump Costs," in *Dynamic Programming and Its Applications*, edited by M. L. Puterman, Academic Press, New York (1979) pp. 269–288.

Fleming, W.: "Optimal Cost-Parameter Stochastic Control," *SIAM Rev.* **11**: 470–509 (1969).

——— and R. Rischel: *Deterministic and Stochastic Optimal Control*, Springer-Verlag, New York (1975).

Gihman, I. I., and A. V. Skorohod: *Controlled Stochastic Processes*, Springer-Verlag, New York, (1979).

Harrison, J. M., and A. J. Taylor: "Optimal Control of a Brownian Storage System," *Stochastic Processes and Their Applications*, **6**: 179–194 (1977).

Krylov, N. V.: "Optimal Stopping of Controlled Diffusion Processes," *Proc. Third Japan-USSR Symposium on Probability Theory*, Lecture Notes in Mathematics no. 550, Springer-Verlag, New York (1979).

Kushner, H. J.: "Optimality Conditions for the Average Cost per Unit Time Problem with a Diffusion Model," *SIAM J. Control Optimization* **16**: 330–346 (1978).

Mandl, P.: *Analytic Treatment of One-Dimensional Markov Processes*, Academia, Prague, and Springer-Verlag, New York (1968).

Morton, R.: "On the Optimal Control of Stationary Diffusion Processes with Inaccessible Boundaries and No Discounting," *J. Appl. Probability* **8**: 551–560 (1971).

——— : "On the Optimal Control of Stationary Diffusion Processes with Natural Boundaries and Discounted Cost," *J. Appl. Probability* **8**: 561–572 (1971).

Pliska, S. R.: "Single Person Controlled Diffusions with Discounted Costs," *J. Optimization Theory Appl.* **12**: 248–255 (1973).

——— : "A Diffusion Model for the Optimal Operation of a Reservoir System," *J. Appl. Probability* **12**: 859–863 (1975).

Puterman, M. L.: "Sensitive Discount Optimality in Controlled One-Dimensional Diffusions," *Ann. Probability* **2**: 408–419 (1974).

Richard, S. F.: "Optimal Impulse Control of a Diffusion Process with Both Fixed and Proportional Costs of Control," *SIAM J. Control Optimization* **15**: 79–91 (1975).

MDPs with incomplete information

van Hee, K. M.: *Bayesian Control of Markov Chains*, Tract 95, Mathematisch Centrum, Amsterdam (1978).

Martin, J. J.: *Bayesian Decision Problems and Markov Chains*, Wiley, New York (1967).

Monahan, G.: "A Survey of Partially Observable Markov Decision Processes: Theory, Models and Algorithms," *Manage. Sci.* **28**: 1–16 (1982).

Platzman, L.: "A Method to Determine Simple Suboptimal Strategies for Infinite Horizon Partially Observed Markov Decision Problems," *Oper. Res.*, to appear.

Silver, E.: "Markovian Decision Processes with Uncertain Transition Probabilities or Rewards," *Tech. Rept.* 1, Operations Research Center, Massachusetts Institute of Technology (1963).

Smallwood, R., and E. Sondik: "The Optimal Control of Partially Observable Markov Processes over a Finite Horizon," *Oper. Res.* **21**: 1071–1088 (1973).

Sondik, E.: "The Optimal Control of Partially Observable Markov Processes over the Infinite Horizon: Discounted Costs," *Oper. Res.* **26**: 282–304 (1978).

CHAPTER

SIX

SOME COMPUTATIONAL CONSIDERATIONS FOR MDPs

Most of this chapter concerns algorithms and computations. Section 6-1 continues the analysis of linear programs presented in Chapters 4 and 5 for expected-present-value and average-reward criteria. In the expected-present-value case, the dual variables are shown to possess a probabilistic interpretation analogous to the dual variables in the average-reward case (cf. Section 4-6). The linear program for the average-reward criterion is developed in Section 4-6 under a "unichain assumption." Section 6-1 shows that a linear-programming solution is very useful even if the unichain assumption is not valid.

Section 6-2 explains various ideas for accelerating the convergence of algorithms for the expected-present-value criterion. The ideas there include kth-order algorithms, successive overrelaxation, bounds for the elimination of suboptimal actions, reordering of states, and extrapolation.

Most applications of operations research involve several criteria of evaluation. The application of multiple-criteria optimization techniques to practical problems has become more common in recent years. Section 6-3 concerns algorithms and theory for the optimization of MDPs with several criteria.

Some real phenomena invite MDP models of an olympian size. This deterrent to the use of MDP models is sometimes called the *curse of dimensionality*. Section 6-4 discusses the use of aggregation techniques to exorcise the curse. The basic idea is to replace the original MDP with a much smaller approximating MDP model.

6-1* FURTHER RESULTS FOR LINEAR PROGRAMS

Section 4-6 concerns MDPs with the average-reward criterion. The algorithms in that section are predicated on the following "unichain assumption": every stationary policy induces a Markov chain with one ergodic class of states and, perhaps, some transient states. The validity of this assumption is both hard to check numerically and violated by many models. What, then, is the appropriate interpretation of solutions to the linear program (4-129*a*) to (4-129*d*):

$$\text{Maximize} \sum_{(s,\,a) \in \mathscr{C}} r(s, a)x_{sa}$$

$$\text{subject to } x_{sa} \geq 0 \qquad (s, a) \in \mathscr{C}$$

$$\sum_{(s,\,a) \in \mathscr{C}} x_{sa} = 1$$

$$\sum_{a \in A_j} x_{ja} = \sum_{(s,\,a) \in \mathscr{C}} p_{sj}^a x_{sa} \qquad j \in \mathcal{S}$$

In Section 5-2, we recognize that this linear program is a special case of the following average-reward linear program [(5-46*a*) to (5-46*d*)] for SMDPs (semi-Markov decision processes):

$$\text{Maximize} \sum_{(s,\,a) \in \mathscr{C}} r(s, a)x_{sa} \tag{6-1a}$$

$$\text{subject to } x_{sa} \geq 0 \qquad (s, a) \in \mathscr{C} \tag{6-1b}$$

$$\sum_{(s,\,a) \in \mathscr{C}} v(s, a)x_{sa} = 1 \tag{6-1c}$$

$$\sum_{a \in A_j} x_{ja} = \sum_{(s,\,a) \in \mathscr{C}} p_{sj}^a x_{sa} \tag{6-1d}$$

The same question of interpretation can be asked of solutions to (6-1*a*) to (6-1*d*). The third part of this section gives a unified answer to these questions. It turns out that solutions to (6-1*a*) to (6-1*d*) and its MDP counterpart have an optimality property that does not depend on a unichain assumption.

It follows from the material between (4-116) and (4-123) on pages 180 and 181 that the variables $\{x_{sa}\}$ in (6-1*a*) to (6-1*d*) have a useful probabilistic interpretation (under the unichain assumption):

$$x_{sa} = \lim_{n \to \infty} P\{s_n = s, \quad a_n = a\}$$

That is, x_{sa} is the stationary joint probability of being in state s and taking action a. Therefore, the stationary state probabilities can be computed via

$$\lim_{n \to \infty} p\{s_n = s\} = \sum_{a \in A_s} x_{sa}$$

In Section 4-5, there is no comparable probabilistic interpretation of the variables in the linear program for the MDP with the expected-present-value criterion. That linear program, namely, (4-91), is

$$\text{Minimize } \sum_{s \in \mathcal{S}} \alpha_s w_s$$

$$\text{subject to} \tag{6-2}$$

$$w_s - \sum_{j \in \mathcal{S}} q_{sj}^a w_j \geq r(s, a) \qquad (s, a) \in \mathcal{C}$$

where $q_{sj}^a = \beta p_{sj}^a$ and $\alpha_s > 0$, $s \in \mathcal{S}$, is arbitrary. If q_{sj}^a is assigned the value $\beta(s, a) p_{sj}^a$ in (5-23), then (6-2) is the same as (5-23), the linear program for SMDPs with expected-present-value criterion. The first part of this section provides a probabilistic interpretation of the dual of (6-2). The presentation is unified in the sense that it applies to SMDPs and MDPs.

Probabilistic Interpretation for Expected-Present-Value Linear Program

Theorems 4-9 and 4-11 imply that $\mathbf{w}$ is an optimal solution of (6-2), i.e., (4-91), if, and only if, there is an optimal stationary policy δ^∞ such that $\mathbf{w} = \mathbf{v}(\delta)$. The assertion requires $\alpha_s > 0$ for each $s \in \mathcal{S}$. This subsection justifies and applies a probabilistic interpretation for the variables in the dual of (6-2).

It is convenient to interpret α_s as $P\{s_1 = s\}$ and require

$$\alpha_s \geq 0 \qquad s \in \mathcal{S} \qquad \text{and} \qquad \sum_{s \in \mathcal{S}} \alpha_s = 1 \tag{6-3}$$

rather than insist on $\alpha_s > 0$ for all $s \in \mathcal{S}$. In the notation of (4-116) and (4-117) let

$$D_{sa} = P\{a_n = a \mid s_n = s\} \qquad n \in I_+ \tag{6-4}$$

and $\mathbf{D} = (D_{sa}, (s, a) \in \mathcal{C})$. Then $\delta \in \Delta$ is a special case of $\mathbf{D}$ having only 0's and 1's. In game and decision-theory terminology, $\mathbf{D}$ is a *randomized* stationary policy and δ is a *pure* stationary policy.

For given $\boldsymbol{\alpha}$ and $\mathbf{D}$, let

$$y_{sa} = \sum_{n=1}^{\infty} \beta^{n-1} P\{s_n = s, a_n = a\} \qquad \{s, a) \in \mathcal{C} \tag{6-5}$$

which is the expected discounted number of occurrences of the joint event $\{s_n = s, a_n = a\}$. This variable arises in the linear program which is dual to (6-2). The probabilities in (6-5) are determined by $\boldsymbol{\alpha}$ and $\mathbf{D}$. For fixed $\boldsymbol{\alpha}$, let $\mathbf{y}^{\mathbf{D}}$ be the vector (components in no particular order) of variables y_{sa} induced by $\mathbf{D}$ and let

$$Y \triangleq \left\{ \mathbf{y}^{\mathbf{D}} : \mathbf{D} \geq \mathbf{0} \text{ and } \sum_{a \in A_s} D_{sa} = 1, \qquad s \in \mathcal{S} \right\}$$

Finally, let X denote the set of nonnegative solutions of

$$\sum_{a \in A_s} x_{sa} = \alpha_s + \sum_{(j, k) \in \mathcal{C}} q_{js}^k x_{jk} \qquad s \in \mathcal{S} \tag{6-6}$$

The main result is that, if $\alpha_s > 0$ for all $s \in \mathcal{S}$, the set X is equivalent to the set of randomized stationary policies. This assertion is part (*b*) in the following theorem, which is proved at the end of this subsection.

Theorem 6-1 (*a*) $Y \subset X$. (*b*) If $\alpha_s > 0$ for all $s \in \mathcal{S}$, then $Y = X$.

The objective of the usual infinite-horizon discounted MDP is

$$\text{Maximize } E\left[\sum_{n=1}^{\infty} \beta^{n-1} r(s_n, a_n)\right] \tag{6-7}$$

From (6-5),

$$E\left[\sum_{n=1}^{\infty} \beta^{n-1} r(s_n, a_n)\right] = \sum_{n=1}^{\infty} \beta^{n-1} \sum_{(s, a) \in \mathscr{C}} r(s, a) P\{s_n = s, a_n = a\}$$
$$= \sum_{(s, a) \in \mathscr{C}} r(s, a) y_{sa}$$

(Why is the interchange of summations valid?). Therefore, if $\alpha_s > 0$ for all $s \in \mathcal{S}$, Theorem 6-1 implies that the usual problem is equivalent to the linear program

$$\begin{aligned} &\text{Maximize } \sum_{(s, a) \in \mathscr{C}} r(s, a) x_{sa} \\ &\text{subject to (6-6) and } x_{sa} \geq 0 \qquad (s, a) \in \mathscr{C} \end{aligned} \tag{6-8}$$

The pair (6-2) and (6-8) are dual linear programs. Moreover, $\alpha_s > 0$ for all $s \in \mathcal{S}$ permits interpretation of x_{sa} as y_{sa} in (6-5). Recall the notation $S = \#\mathcal{S}$.

If $\mathbf{x}$ is an *extreme-point* optimum of (6-8), there are at most S variables x_{sa} that are positive.† However, $\alpha_s > 0$ in (6-6) requires $\sum_a x_{sa} > 0$; so there is at least one $a \in A_s$ with $x_{sa} > 0$. This is true for each of the S states s; so $\mathbf{x}$ must have at least S positive variables. Therefore, for each s there is exactly one a for which $x_{sa} > 0$ and all other variables are zero. The proof of Theorem 6-1 implies that $\mathbf{x}$ is induced by the pure policy $\mathbf{D}$ (δ) that assigns $D_{sa} = 1$ $[\delta(s) = a]$ if $x_{sa} > 0$ and $D_{sa} = 0$ if $x_{sa} = 0$.

The above argument fails if $\alpha_j = 0$ for some state j. Then it is possible that

$$0 = \sum_{n=1}^{\infty} \beta^{n-1} P\{s_n = j\} = \sum_{a \in A_s} y_{sa}$$

In this case, D_{ja}, $a \in A_j$, does not affect the objective (6-7); so values can be assigned arbitrarily and we may let $D_{ja} = 1$ for some $a \in A_j$ and $D_{jk} = 0$ if $k \neq a$.

An application of Theorem 6-1 in Section 6-3 concerns multiple criteria in MDPs.

Solutions obtained via policy improvement, successive approximations, linear complementarity, and linear programming are equivalent (Theorems 4-8, 4-9, 4-11, and 4-12). Therefore, Theorem 6-1 and (6-5) permit probabilistic statements regardless of the algorithm actually used.

† Read about *basic* optimal solutions in a linear-programming text [e.g., G. B. Dantzig, *Linear Programming and Extensions*, Princeton University Press, Princeton, N.J. (1963)].

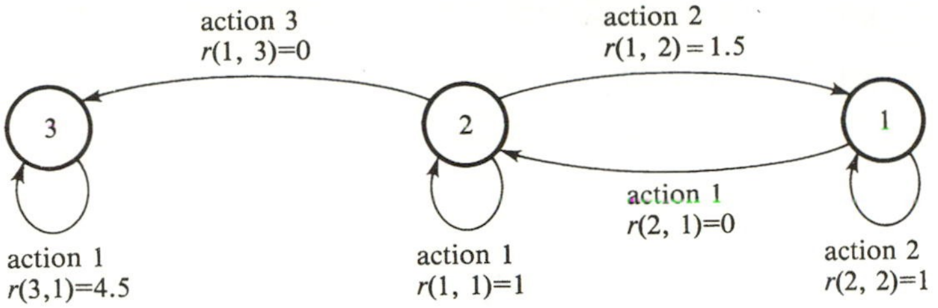

Figure 6-1 A deterministic MDP.

Example 6-1: Continuation of Examples 4-9, 4-10, and 4-17 The basic data of this deterministic MDP are summarized by $\beta = 0.5$ and Figure 4-1 which is repeated as Figure 6-1. The network that corresponds to an optimal policy is given by Figure 6-2. By direct computation

$$
\begin{aligned}
y_{21} &= \alpha_2(1 + 0\beta + 0\beta^2 + \cdots) = \alpha_2 \\
y_{13} &= \alpha_1(1 + 0\beta + 0\beta^2 + \cdots) + \alpha_2(0 + \beta + 0\beta^2 + \cdots) = \alpha_1 + \beta\alpha_2 \\
y_{31} &= \alpha_3(1 + \beta + \beta^2 + \cdots) + \alpha_1(0 + \beta + \beta^2 + \cdots) \\
&\quad + \alpha_2(0 + 0 + \beta^2 + \beta^3 + \cdots) \\
&= \frac{\alpha_3 + \beta\alpha_1 + \beta^2\alpha_2}{1 - \beta}
\end{aligned}
$$

in which $\beta = 0.5$ could have been substituted. Now we seek $\mathbf{x} \in X$ such that $x_{sa} = 0$ unless $(s, a) \in \{(1, 3), (2, 1), (3, 1)\}$. Then (6-6) becomes

$$
\begin{aligned}
x_{21} &= \alpha_2 + 0 = \alpha_2 \\
x_{13} &= \alpha_1 + \beta x_{21} = \alpha_1 + \beta\alpha_2 \\
x_{31} &= \alpha_3 + \beta(x_{13} + x_{31}) = \frac{\alpha_3 + \beta\alpha_1 + \beta^2\alpha_2}{1 - \beta}
\end{aligned}
$$

so $\mathbf{x}$ and $\mathbf{y}$ are the same. Notice that the y_{sa}'s sum to $(1 - \beta)^{-1}$ so $(1 - \beta)y_{sa}$ is the fractional expected discounted number of times that $s_n = s$ and $a_n = a$. For example, if $\alpha_1 = \alpha_2 = \alpha_3 = 1/3$, then $(1 - \beta)y_{31} = 7/12$. □

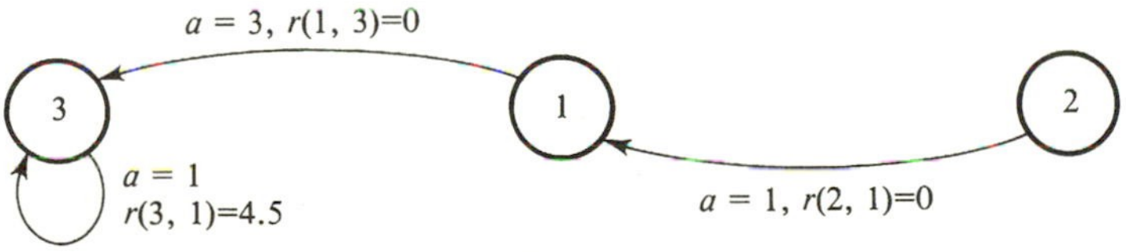

Figure 6-2 Network of an optimal policy.

Proof of Theorem 6-1

Theorem 6-1 Let X be the set of nonnegative solutions of

$$\sum_{a \in A_s} x_{sa} = \alpha_s + \sum_{(j,\, k) \in \mathscr{C}} q_{js}^k x_{jk} \qquad s \in \mathcal{S} \tag{6-9}$$

and let

$$Y \triangleq \{\mathbf{y^D}\colon \mathbf{D} \geq \mathbf{0}, \sum_{a \in A_s} D_{sa} = 1 \text{ for all } s \in \mathcal{S}\}$$

where the components

$$y_{sa} = \sum_{n=1}^{\infty} \beta^{n-1} P\{s_n = s, a_n = a\} \qquad (s, a) \in \mathscr{C} \tag{6-10}$$

of $\mathbf{y^D}$ are induced by $\mathbf{D}$. Then (*a*) $Y \subset X$. (*b*) If $\alpha_s > 0$ for all $s \in \mathcal{S}$, then $Y = X$.

Proof Sum both sides of (6-9) over all $s \in \mathcal{S}$ and on the right, interchange the order of summation to yield

$$\sum_{(s,\, a) \in \mathscr{C}} x_{sa} = (1 - \beta)^{-1}$$

From $D_{sa} = P\{a_n = a \mid s_n = s\}$ and the definition of y_{sa} in (6-5),

$$y_{sa} = \sum_{n=1}^{\infty} \beta^{n-1} P\{s_n = s, a_n = a\} = D_{sa} \sum_{n=1}^{\infty} \beta^{n-1} P\{s_n = s\}$$

$$= D_{sa}\left(\alpha_s + \beta \sum_{n=1}^{\infty} \beta^{n-1} P\{s_{n+1} = s\}\right)$$

However,

$$P\{s_{n+1} = s\} = \sum_{j \in \mathcal{S}} P\{s_{n+1} = s \mid s_n = j\} P\{s_n = j\}$$

$$= \sum_{(j,\, k) \in \mathscr{C}} P\{s_n = j\} D_{jk} p_{js}^k$$

so

$$y_{sa} = D_{sa}\left(\alpha_s + \beta \sum_{n=1}^{\infty} \beta^{n-1} \sum_{(j,\, k) \in \mathscr{C}} P\{s_n = j\} D_{jk} p_{js}^k\right)$$

and thus

$$y_{sa} = D_{sa}\left(\alpha_s + \beta \sum_{(j,\, k) \in \mathscr{C}} p_{js}^k y_{jk}\right) \tag{6-11}$$

because

$$\sum_{n=1}^{\infty} \beta^{n-1} D_{jk} P\{s_n = j\} = \sum_{n=1}^{\infty} \beta^{n-1} P\{s_n = j, a_n = k\} = y_{jk}$$

Sum both sides of (6-11) over all $a \in A_s$ and use $\sum_{a \in A_s} D_{sa} = 1$ to obtain

$$\sum_{a \in A_s} y_{sa} = \alpha_s + \beta \sum_{(j,\, k) \in \mathscr{C}} p_{js}^k y_{jk} \qquad s \in \mathcal{S} \tag{6-12}$$

which is (6-9). Also, $y_{sa} \geq 0$ because (6-10) is a nonnegative linear combination of probabilities. Therefore, $\mathbf{y^D} \in X$ for each $\mathbf{D}$ so $Y \subset X$.

Suppose $\alpha_s > 0$ for all $s \in \mathcal{S}$ and let $\mathbf{x} \in X$. To prove $Y \supset X$, hence $Y = X$, observe that $\alpha_s > 0$ in (6-9) implies $\sum_{a \in A} x_{sa} > 0$. Let $\mathbf{D}$ be specified by

$$D_{sa} = \frac{x_{sa}}{\sum_{k \in A_s} x_{sk}} \qquad (s, a) \in \mathscr{C} \tag{6-13}$$

It will be shown that $\mathbf{D}$ induces $x_{sa} = y_{sa}^{\mathbf{D}}$. From (6-13) and (6-10), as in the derivation of (6-11),

$$y_{sa}^{\mathbf{D}} = \frac{x_{sa}}{\sum_{i \in A_s} x_{si}} \left(\alpha_s + \beta \sum_{(j, k) \in \mathscr{C}} p_{js}^k \frac{x_{jk}}{\sum_{i \in A_j} x_{ji}} \sum_{n=1}^{\infty} \beta^{n-1} P\{s_n = j\} \right) \tag{6-14}$$

From (6-14) and $\alpha_s > 0$, $y_{sa}^{\mathbf{D}} > 0$ if, and only if, $x_{sa} > 0$. Therefore, $y_{sa}^{\mathbf{D}} = \mu_s x_{sa}$ where

$$\mu_s = \frac{\alpha_s + \beta \sum_{(j, k) \in \mathscr{C}} p_{js}^k x_{jk} z_j}{\sum_{a \in A_s} x_{sa}} \tag{6-15}$$

and

$$z_j = \frac{\sum_{n=1}^{\infty} \beta^{n-1} P\{s_n = j\}}{\sum_{i \in A_j} x_{ji}} = \frac{\sum_{i \in A_j} y_{ji}^{\mathbf{D}}}{\sum_{i \in A_j} x_{ji}} \tag{6-16}$$

Substitution of $y_{ji}^{\mathbf{D}} = \mu_j x_{ji}$ in (6-14) yields $z_j = \mu_j$. In (6-15), with (6-9), this yields

$$z_s \left(\alpha_s + \beta \sum_{(j, k) \in \mathscr{C}} p_{js}^k x_{jk} \right) = \alpha_s + \beta \sum_{(j, k) \in \mathscr{C}} p_{js}^k x_{jk} z_j$$

or

$$z_s \left(\alpha_s + \beta \sum_{j \neq s} \sum_{k \in A_j} p_{js}^k x_{jk} \right) - \beta \sum_{j \neq s} z_j \sum_{k \in A_j} p_{js}^k x_{jk} = \alpha_s \qquad s \in \mathcal{S} \tag{6-17}$$

Viewed as a system of linear equations, (6-17) has a dominant diagonal; so its solution, if any, is unique. But $z_j = 1$ for all j solves (6-17); so $y_{sa}^{\mathbf{D}} = \mu_s x_{sa} = z_s x_{sa} = x_{sa}$ for all $(s, a) \in \mathscr{C}$. Hence $X \subset Y$. □

Expected-Present-Value and Generalized Programs

The linear program (6-8) is

$$\begin{aligned} &\text{Maximize} \sum_{(s, a) \in \mathscr{C}} r(s, a) x_{sa} \\ &\text{subject to } x_{sa} \geq 0 \qquad (s, a) \in \mathscr{C} \\ &\sum_{a \in A_s} x_{sa} = \alpha_s + \sum_{(j, k)} q_{js}^k x_{jk} \qquad s \in \mathcal{S} \end{aligned} \tag{6-18}$$

We now restate this problem as a *generalized program*. This allows special linear-programming algorithms, *decomposition* in particular, to be applied to MDPs.

Unfortunately, decomposition seems computationally less effective than was once thought likely.

We know that (6-18) has a solution and, for any linear program, there is an extreme-point solution if there is any solution at all. We know also that an extreme point of (6-18) has the property for each s that $x_{sa} = 0$ for all but at most one $a \in A_s$. Therefore, an extreme point corresponds to a collection of S actions, one for each state $s \in \mathcal{S}$. Let $a(s)$ denote the action for state s in this collection and let $z_s = x_{sa(s)}$. Corresponding to z_s, there is (1) a row $\mathbf{q}_j$ of transition probabilities in (6-18), where $q_{js}^{a(j)}$ is the sth component of $\mathbf{q}_j$, and (2) a coefficient $r_j(\mathbf{q}_j)$ in the objective, where $r_j(\mathbf{q}_j) \triangleq r[j, a(j)]$.

The set of candidate row vectors $\mathbf{q}_j$ is $W_j \triangleq \{(q_{js}^a, s \in \mathcal{S}): a \in A_j\}$. Therefore, (6-18) can be described as selecting $q_1 \in W_1, \mathbf{q}_2 \in W_2, \ldots, \mathbf{q}_S \in W_S$ so that

$$\sum_{j \in \mathcal{S}} r_j(\mathbf{q}_j) z_j \text{ is maximized for all } \mathbf{z} \text{ satisfying} \tag{6-19a}$$

$$\mathbf{z} \geq \mathbf{0} \qquad \text{and} \qquad \mathbf{z} = \boldsymbol{\alpha} + \sum_{j \in \mathcal{S}} \mathbf{q}_j z_j \tag{6-19b}$$

Problem (6-19a) and (6-19b) is a *generalized program.* Let q_j^a, $a \in A_j$, denote a generic element of W_j. Then $\mathbf{q}_j$ can be represented as a weighted average

$$\mathbf{q}_j = \sum_{a \in A_j} \lambda_j^a \mathbf{q}_j^a \qquad \boldsymbol{\lambda} \geq 0 \qquad \sum_{a \in A_j} \lambda_j^a = 1 \tag{6-20}$$

From the preceding discussion, there is an optimum with all the λ's being zero or unity. Substitution of (6-20) in (6-19a) and (6-19b) gives

$$\text{Maximize} \sum_{(j,a) \in \mathscr{C}} r_j(\mathbf{q}_j^a) \lambda_j^a z_j \tag{6-21a}$$

$$\text{subject to } \boldsymbol{\lambda} \geq \mathbf{0} \qquad \mathbf{z} \geq \mathbf{0} \tag{6-21b}$$

$$\sum_{a \in A_j} \lambda_j^a = 1 \qquad j \in \mathcal{S} \tag{6-21c}$$

$$\mathbf{z} = \boldsymbol{\alpha} + \sum_{(j,a) \in \mathscr{C}} \lambda_j^a \mathbf{q}_j^a z_j \tag{6-21d}$$

Let $Z_{ja} = \lambda_j^a z_j$ so (6-21b) and (6-21c) is equivalent to

$$\mathbf{Z} \geq \mathbf{0} \qquad \text{and} \qquad \sum_{a \in A_j} Z_{ja} = z_j \qquad j \in \mathcal{S}$$

Therefore, (6-21a) through (6-21d) is equivalent to

$$\begin{aligned} &\text{Maximize} \sum_{(j,a) \in \mathscr{C}} Z_{ja} r_j(q_j^a) \\ &\text{subject to } \mathbf{Z} \geq \mathbf{0} \\ &\sum_{a \in A_s} Z_{sa} = \alpha_s + \sum_{(j,k) \in \mathscr{C}} q_{js}^k Z_{jk} \qquad s \in \mathcal{S} \end{aligned} \tag{6-22}$$

We have come full circle; (6-18) and (6-22) are the same with $r_j(\mathbf{q}_j^a) \equiv r(j, a)$ and $Z_{ja} \equiv x_{ja}$.

MDPs and Leontief Matrices

The focus thus far is the use of linear programs to solve MDPs. However, MDP algorithms are effective in solving some kinds of linear programs. A real matrix $M = (m_{ij})$ is *pre-Leontief* if each row has at most one positive element; it is *totally Leontief* if† (1) it has exactly one positive element per row, (2) there is $\mathbf{x} = (\mathbf{x}_i) \geq \mathbf{0}$ such that $\sum_i m_{ij} x_i > 0$ for all j. Exercise 6-2 asks you to show that the constraint matrix of linear program (6-2) is totally Leontief.

Suppose M is totally Leontief in the linear program‡ of choosing $\mathbf{y} = (y_j)$ to

$$\begin{gathered}\text{Minimize} \sum_j b_j y_j \\ \text{subject to } M\mathbf{y} \geq \mathbf{c}\end{gathered} \tag{6-23}$$

where $\mathbf{c}$ is a column vector with as many rows as M and $\mathbf{b} = (b_j) \geq \mathbf{0}$. We assume that M and $\mathbf{c}$ are scaled so that the positive elements of M are at most unity. For column j let $A_j = \{i: m_{ij} > 0\}$ and let $\Delta = \times_j A_i$. For $\delta \in \Delta$ let M_δ denote the submatrix of M which includes row $\delta(j)$ for each j and suppose $Q_\delta = \mathbf{I} - M_\delta \geq \mathbf{0}$ with the magnitude of every eigenvalue of Q_δ less than unity for each $\delta \in \Delta$. We assume that (6-23) has a bounded objective.

Then (6-23) is equivalent to optimization of the following MDP. Since M is Leontief, $A_j \neq \phi$ for each j. Let $q_{ij}^i = 1 - m_{ij}$ for each $i \in A_j$ and each j. Since M is pre-Leontief, $A_s \cap A_j = \phi$ for all $s \neq j$; so let $k(i)$ denote the j such that $i \in A_j$, that is, $i \in A_{k(i)}$. If $m_{ij} < 0$, let $-m_{ij} = q_{k(i),j}^i$. Let all other q_{sj}^a be zero. Finally, for each i let $r[k(i), i] = c_i$ and let $\alpha_j = b_j / \sum_s b_s$ for each j. The equivalence with (6-2) is complete except that the sum $\sum_j q_{sj}^a$ may vary with a. Also, $Q_\delta^n \to \mathbf{0}$ for every δ because all eigenvalues of Q_δ have magnitude less than 1. Therefore, the linear program (6-23) can be solved with algorithms such as successive approximations and policy improvement (or, more generally, with kth-order algorithms described in Section 6-2).

Recent numerical experiments by Koehler (1976) on linear programs such as (6-23) suggest that policy improvement is more efficient than the standard revised simplex algorithm (with product form of the inverse). Therefore, the kth-order algorithms for MDPs (described in Section 6-2) may be more efficient than (6-2), i.e., solution of an MDP as a linear program.

Average-Reward Linear Program without the Unichain Assumption

The following linear program is (5-46), which corresponds to the average-reward criterion for SMDPs which satisfy the unichain assumption:

$$\text{Maximize} \sum_{(s,a) \in \mathscr{C}} r(s, a) x_{sa} \tag{6-24a}$$

† Other authors sometimes use the transpose of M; so in (1) they would replace "row" with "column."

‡ See Koehler, Winston, and Wright (1975) and Veinott (1969) for examples of phenomena whose optimization models are linear programs with totally Leontief coefficient matrices.

$$\text{subject to } x_{sa} \geq 0 \qquad (s, a) \in \mathscr{C} \tag{6-24b}$$

$$\sum_{(s,a) \in \mathscr{C}} v(s, a)x_{sa} = 1 \tag{6-24c}$$

$$\sum_{a \in A_j} x_{ja} = \sum_{(s,a) \in \mathscr{C}} p_{sj}^a x_{sa} \qquad j \in \mathcal{S} \tag{6-24d}$$

The MDP linear program is the special case in which $v(\cdot, \cdot) \equiv 1$ in (6-24c). We merely assume that $v(s, a) \geq 0$ for all $(s, a) \in \mathscr{C}$ and $v(s, a) > 0$ for at least one $(s, a) \in \mathscr{C}$. For MDPs where the unichain assumption is not valid or is too laborious to validate we shall see that a solution to (6-24a) through (6-24d) is nevertheless useful. Our approach follows Denardo (1970).

Let $\mathbf{x} = (x_{sa})$ be a feasible solution of (6-24a) through (6-24d) and define

$$J = \left\{s: \sum_{a \in A_s} x_{sa} > 0\right\} \qquad \text{and} \qquad H = \{(s, a): x_{sa} > 0\} \tag{6-25}$$

Theorem 6-2 If $\mathbf{x}$ is a basic feasible solution of (6-24b) through (6-24d), then:

(i) $\#J = \#H$

(ii) $\{p_{sj}^a : (s, a) \in H, j \in J\}$ comprises an irreducible transition matrix of a semi-Markov process with state space J

(iii) The average reward per unit time of the semi-Markov process in (ii) is

$$\sum_{(s,a) \in \mathscr{C}} r(s, a)x_{sa} = \sum_{(s,a) \in H} r(s, a)x_{sa} \tag{6-26}$$

PROOF Deferred to page 255. □

Before proving Theorem 6-2, we state its consequence. It follows from part (i) of the theorem that, for each $s \in J$, there is exactly one member of A_s, say $a_s \in A_s$, such that $(s, a_s) \in H$. Let δ be any single-stage rule that specifies $\delta(s) = a_s$ if $s \in J$. From parts (ii) and (iii) of the theorem, if $s_1 \in J$, then (6-26) is the average reward per unit time associated with δ^∞. The following definition will help state a corollary.

Definition 6-1 Let $H \subset \mathscr{C}$ such that

$$\#\{a: (s, a) \in H\} \leq 1 \qquad \text{for each } s \in \mathcal{S}$$

and let

$$J = \{s: (s, a) \in H \qquad \text{for some } a \in A_s\}$$

A single-stage decision rule δ is *consistent* with H if

$$[s, \delta(s)] \in H \qquad s \in J$$

In the definition, if $s \in J$, let a_s label the element of A_s such that $(s, a_s) \in H$. Then a single-stage decision rule is consistent with H if $\delta(s) = a_s$ for all $s \in J$.

Therefore, if $\mathbf{x}$ is a basic feasible solution of (6-24b) to (6-24d), J and H are specified by (6-25), δ is consistent with H, and $s_1 \in J$, then (6-26) is the average reward per unit time associated with δ^∞.

Corollary 6-2 Suppose $\mathbf{x}^*$ is an optimal basic solution of (6-24a) through (6-24d),

$$H = \{(s, a): x_{sa}^* > 0\}$$

and δ_* is consistent with H. Then (6-26) (with $\mathbf{x} = \mathbf{x}^*$) is the maximal average reward per unit time which can be achieved by any communicating class under any policy. Also, (6-26) is the average reward per unit time associated with δ_*^∞ from every initial state in

$$J = \left\{s : \sum_{a \in A_s} x_{sa}^* > 0\right\}$$

PROOF OF THE COROLLARY Let G be a subset of states which is a communicating class under some policy. Then there is an $\mathbf{x}$ feasible in (6-24a) through (6-24d) with $G = J$ and (6-26) as the value of the objective. If any linear-program optimum exists, there is a basic optimal solution $\mathbf{x}^*$. But $v(\cdot, \cdot) \geq 0$ and $v(\cdot, \cdot) \not\equiv 0$ imply that the set of feasible solutions is nonempty and bounded so an optimum exists. By Theorem 6-2, $\mathbf{x}^*$ corresponds to an irreducible semi-Markov process with state space J. By optimality of $\mathbf{x}^*$ and part (iii) of Theorem 6-2, the average reward per unit time of the process associated with $\mathbf{x}^*$ is at least as great as that of the process associated with $\mathbf{x}$.

□

Let $\mathbf{x}^*$ be an optimal basic solution of (6-24a) through (6-24d). Let δ^∞ be any policy which specifies $\delta(s) = a$ if $(s, a) \in H$. Unless $J = S$, this procedure does not completely specify δ. If the initial state is in J, an arbitrary completion of δ causes the average reward per unit time to be maximal (due to Corollary 6-2). However, if $s_1 \notin J$, some completions of δ may induce average rewards per unit time which are inferior to those induced by other completions. How should $\delta(s)$ be specified if $s_1 \notin J$?

First, consider any state outside J which can be made transient owing to absorption in J with probability 1. The average reward per unit time from such an initial state will then be the maximal rate, the one associated with J. Therefore, if δ can be completed so that J remains the only ergodic class, complete it that way.

This leaves unspecified what action to take in states from which it is not possible to guarantee absorption in J. Denardo (1970), Derman (1970), and Hordijk and Kallenberg (1979) present algorithms which complete the specification of δ in such a way that the average reward per unit time is maximized from each initial state.

Two Examples

Example 6-2 Let $\mathcal{S} = \{1, 2, 3\}$, $A_1 = \{1\}$, and $A_2 = A_3 = \{1, 2\}$ in an MDP. Suppose states 1 and 3 are absorbing states and $p_{21}^1 = p_{22}^2 = 1$. Let $r(3, 1) = 2$, $r(2, 2) = 1$, and $r(1, 1) = r(2, 1) = r(3, 2) = 0$. Figure 6-3 portrays this model. Linear program (6-24*a*) to (6-24*d*) for this example is

$$\text{Maximize } 2x_{31} + x_{22}$$

$$\text{subject to } x_{11}, x_{21}, x_{22}, x_{31}, x_{32} \geq 0$$

$$x_{11} + x_{21} + x_{22} + x_{31} + x_{32} = 1$$

$$x_{11} = x_{11} + x_{21}$$

$$x_{21} + x_{22} = x_{22}$$

$$x_{31} + x_{32} = x_{31} + x_{32}$$

The unique optimal solution is $x_{21} = 1$ and $x_{11} = x_{21} = x_{22} = x_{32} = 0$, and the corresponding value of the objective is 2. This solution corresponds to the fact that $\{3\}$ is the communicating class with the maximal average reward per unit time (per transition) which is attained only if $\delta(3) = 1$. The linear-program solution correctly corresponds to $\delta(3) = 1$, but it does not specify $\delta(1)$ or $\delta(2)$. Of course, δ^∞ with $\delta(2) = 1$ maximizes the vector of average rewards per unit time. □

Example 6-3: Continuation of Example 6-2 Suppose Figure 6-3 is altered so $r(2, 1) = r(2, 2) = 0$, $p_{21}^1 = p_{23}^2 = 0.4$ and $p_{23}^1 = p_{21}^2 = 0.6$. Linear program (6-24*a*) to (6-24*d*) becomes

$$\text{Maximize } 2x_{31}$$

$$\text{subject to } x_{11}, x_{21}, x_{22}, x_{31}, x_{32} \geq 0$$

$$x_{11} + x_{21} + x_{22} + x_{31} + x_{32} = 1$$

$$x_{11} = 0.4x_{21} + 0.6x_{22}$$

$$x_{21} + x_{22} = 0$$

$$x_{31} + x_{32} = 0.6x_{21} + 0.4x_{22} + x_{31} + x_{22}$$

The unique optimal solution is $x_{31} = 1$ and $x_{11} = x_{21} = x_{22} = x_{32} = 0$.

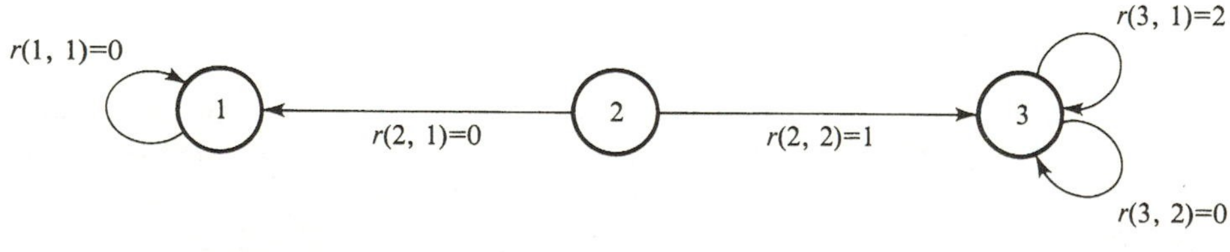

Figure 6-3 The MDP in Example 6-2.

If $s = 2$, both $a = 1$ and $a = 2$ have positive probabilities of causing absorption in $\{3\}$, the communicating class with the maximal average reward per unit time. However, if $s_1 = 2$, the average reward per unit time is not the same regardless of whether $a = 1$or $a = 2$. Let $\delta(1) = 1$, $\delta(2) = 1$, and $\delta(3) = 1$ and $\gamma(1) = 1$, $\gamma(2) = 2$, and $\gamma(3) = 1$. If $s_1 = 2$ and δ^∞ is used, the average reward per unit time is $0.4(0) + 0.6(2) = 1.2$. If $s_1 = 2$ and γ^∞ is used, the same measure of effectiveness is $0.4(2) + 0.6(0) = 0.8$. Therefore, even if (i) a state s ($s = 2$) is transient, (ii) there is a state j ($j = 3$) in the communicating class with the maximal reward rate (rate = 2), and (iii) there is an action $a \in A_s$ such that $p_{sj}^a > 0$ ($a = 2$), it is not necessarily optimal to use a policy which takes action a in state s. □

In practice, it is usually sufficient to find any policy which is consistent with the communicating class of states having the maximal average reward per unit time. For example, suppose an inventory model which satisfies the myopia assumptions in Section 3-3 is formulated as an MDP and linear program (6-24*a*) through (6-24*d*) is solved. From Sections 3-1 and 3-2, an optimal policy induces a Markov chain in which all the states above some base stock level, say a^*, are transient; that is, s is transient if $s > a^*$. If δ is consistent with an optimal solution to (6-24*a*) through (6-24*d*), then $\delta(s)$ is not specified for $s > a^*$. In practice, it is usually sufficient to know that a^* is the highest recurrent stock level and to know how much stock to order when the stock level is lower than a^*. Therefore, it is usually sufficient to stop with an optimal solution to (6-24*a*) through (6-24*d*) rather than use the previously mentioned lengthier algorithms to determine a policy which maximizes the average reward per unit time from every initial state.

PROOF OF THEOREM 6-2 Recall that the linear program (6-24*a*) through (6-24*d*) is

$$\text{Maximize} \sum_{(s,\,a) \in \mathscr{C}} r(s, a)x_{sa}$$

$$\text{subject to } x_{sa} \geq 0 \qquad (s, a)\mathscr{C}$$

$$\sum_{(s,\,a) \in \mathscr{C}} v(s, a)x_{sa} = 1$$

$$\sum_{a \in A_j} x_{ja} - \sum_{(s,\,a) \in \mathscr{C}} p_{sj}^a x_{sa} = 0 \qquad j \in \mathbb{S} \tag{6-27}$$

For a basic feasible solution $\mathbf{x}$, let

$$J = \left\{s: \sum_{a \in A_s} x_{sa} > 0\right\} \qquad \text{and} \qquad H = \{(s, a) : x_{sa} > 0\}$$

(i) $\# J = \# H$:

From (6-27), if $j \notin J$, then

$$0 = \sum_{a \in A_j} x_{ja} = \sum_{(s,\,a) \in \mathscr{C}} p_{sj}^a x_{sa} = \sum_{(s,\,a) \in H} p_{sj}^a x_{sa}$$

Since $p_{sj}^a \geq 0$ and $x_{sa} > 0$ if $(s, a) \in H$, it follows that $p_{sj}^a = 0$. That is, $p_{sj}^a = 0$ if $(s, a) \in H$ and $j \notin J$.

Let $\mathbf{c}_{sa}$ denote the vector of coefficients of x_{sa} in the linear program and let $\mathbf{w}$ denote the "right-hand side" vector whose transpose is $(1, 0, \ldots, 0)$. Then a feasible solution $\mathbf{x}$ satisfies

$$\sum_{(s, a) \in H} \mathbf{c}_{sa} x_{sa} = \mathbf{w} \tag{6-28}$$

Let H_1 be the subset of H such that

$$\#\{a: (s, a) \in H\} = 1$$

Hence, if $(s, a) \in H_1$, there is a unique $a_s \in A_s$ such that $(s, a_s) \in H$. Let J_1 denote the subset of J whose members are found in H_1, that is,

$$J_1 = \{s: s \in J \text{ and } (s, a_s) \in H_1 \text{ for some } a_s \in A_s\}$$

The transition probabilities $\{p_{sj}^a: (s, a) \in H_1, j \in J\}$ describe at least one communicating class. Let J_2 label the set of states in such a communicating class and let

$$H_2 = \{(s, a): (s, a) \in H_1 \text{ and } s \in J_2\}$$

In this communicating class, let

$$\xi_s = \lim_{T \to \infty} \frac{1}{T} \int_0^T P\{X(t) = s\}\, dt \tag{6-29}$$

and let

$$z_{sa} = \begin{cases} \dfrac{\xi_s}{v(s, a)} & \text{if } v(s, a) > 0 \\ 0 & \text{if } v(s, a) = 0 \end{cases} \qquad (s, a) \in H_2 \tag{6-30}$$

Since J_2 is a communicating class, $\{z_{sa}\}$ is the unique solution to a system of equations "$\mathbf{e}\pi = 1$ and $\pi = \pi P$" (cf. Theorem 7-10 on page 242 of Volume I), namely,

$$\sum_{(s, a) \in H_2} \mathbf{c}_{sa} z_{sa} = \mathbf{w} \tag{6-31}$$

Subtracting (6-31) from (6-28),

$$\sum_{(s, a) \in H - H_2} \mathbf{c}_{sa} x_{sa} + \sum_{(s, a) \in H_2} \mathbf{c}_{sa}(x_{sa} - z_{sa}) = 0 \tag{6-32}$$

By hypothesis, $\mathbf{x}$ is a basic feasible solution so $\{\mathbf{c}_{sa}: (s, a) \in H\}$ is a linearly independent set of vectors. Therefore, the coefficients of these vectors in (6-32) must all be zero so $H = H_2$, which implies $\#J = \#H$, and $x_{sa} = z_{sa}$ if $(s, a) \in H_2$.

(ii) $\{p_{sj}^a: (s, a) \in H, j \in J\}$ is the transition matrix of exactly one communicating class:

The result $H = H_2$ implies $J = J_2$.

(iii) The long-run average reward per unit time of the communicating class is the following expression:

$$\sum_{(s,\,a)\,\in\,\mathscr{C}} r(s, a)x_{sa} = \sum_{(s,\,a)\,\in\,H} r(s, a)x_{sa}$$

We have $x_{sa} = z_{sa}$ if $(s, a) \in H_2$ and $H = H_2$. In the embedded Markov chain with the communicating class J, let π_s denote the stationary probability of being in state s. Then

$$\xi_s = \frac{\pi_s \nu(s, a_s)}{\sum_{j \in J} \pi_j \nu(j, a_j)}$$

so (6-30) yields

$$\sum_{(s,\,a)\,\in\,H} r(s, a)x_{sa} = \sum_{(s,\,a)\,\in\,H} r(s, a)z_{sa} = \frac{\sum_{s \in J} r(s, a_s)\pi_s}{\sum_{j \in J} \nu(j, a_j)\pi_j} \tag{6-33}$$

The numerator of (6-33) is the average reward per transition and the denominator is the average duration of a transition. From (5-32), this ratio is the long-run average reward per unit time. □

6-2* ACCELERATION OF COMPUTATIONS

Section 4-5 presents several kinds of algorithms which solve an infinite-horizon discounted finite MDP. When any one of them is used, it is possible to accelerate computations. This section improves the algorithms' computational effectiveness. As in Section 4-5, we assume $0 \le r(s, a) \le u < \infty$ for all $(s, a) \in \mathscr{C}$, $\#\mathscr{C} < \infty$, and $0 \le \beta < 1$.

kth-Order Algorithms

Policy improvement and successive approximations are two extremes in a family of algorithms. We describe this family, called *k*th-order algorithms,† in order to unify the discussion of acceleration methods. This subsection describes the family, verifies that its constituent algorithms converge to an optimum, and then presents methods to accelerate computations. We follow Puterman and Brumelle (1979) and Puterman and Shin (1979).

The following restatement of the policy-improvement algorithm (from the proof of Theorem 4-5) uses Theorem 4-7. In the notation below, v_s^i is the sth component of the S-vector $\mathbf{v}^i$. Also, write $\mathbf{r}_i$ and Q_i instead of $\mathbf{r}_{\delta_i}$ and βP_{δ_i}, respectively, and q_{sj}^a instead of βp_{sj}^a.

(*a*) Let $i = 1$ and select $\delta_1 \in \Delta$.

(*b*) Solve the following equation for $\mathbf{v}^i$:

$$\mathbf{v}^i = (I - Q_i)^{-1}\mathbf{r}_i \tag{6-34}$$

† Some authors use the label *modified policy iteration* algorithm.

(*c*) For each $s \in \mathcal{S}$ compute

$$\psi_s^i \triangleq \max \left\{ r(s, a) + \sum_{j \in \mathcal{S}} q_{sj}^a v_j^i : a \in A_s \right\} - v_s^i \tag{6-35}$$

If $\psi_s^i > 0$, let $\delta_{i+1}(s)$ be any $a \in A_s$ that causes

$$r(s, a) + \sum_{j \in \mathcal{S}} q_{sj}^a v_j^i - v_s^i > 0$$

If $\psi_s^i = 0$, let $\delta_{i+1}(s) = \delta_i(s)$

(*d*) If $\delta_{i+1} = \delta_i$, stop and δ_i^∞ is optimal. Otherwise replace i with $i + 1$ and return to (*b*).

The lion's share of the computation is in (6-34), which is called *value determination.* Several schemes have been suggested to accelerate or modify (6-34). The simplest scheme stems from the observation that $(\mathbf{I} - Q_i)$ differs from $(\mathbf{I} - Q_{i-1})$ only in columns that correspond to states s for which $\delta_i(s) \neq \delta_{i-1}(s)$. Therefore, if there are relatively few such states, it will be much easier to "pivot" from $(\mathbf{I} - Q_{i-1})^{-1}$ to $(\mathbf{I} - Q_i)^{-1}$ than to compute $(I - Q_i)^{-1}$ from scratch.†

Step (*c*) during the *i*th iteration of the algorithm is to find $\delta_{i+1} \in \Delta$ such that

$$\mathbf{r}_{i+1} + Q_{i+1}\mathbf{v}_i = \mathbf{L}\mathbf{v}^i = \text{vmax}\,\{\mathbf{r}_\delta + Q_\delta \mathbf{v}^i : \delta \in \Delta\} \tag{6-36}$$

Then step (*b*) during the $(i + 1)$th iteration is to compute

$$\mathbf{v}^{i+1} = (I - Q_{i+1})^{-1}\mathbf{r}_{i+1} = \sum_{l=0}^{\infty} Q_{i+1}^l \mathbf{r}_{i+1} \tag{6-37}$$

The successive-approximations recursion (4-84) is similar to (*a*) through (*d*). It can be constructed‡ as follows:

(*A*) Let $i = 1$ and $\mathbf{v}^0 \in \mathbb{R}^S$ be the return vector associated with some policy π_0.

(*B*) Solve (6-36) for δ_{i+1}.

(*C*) Evaluate $\mathbf{v}^{i+1}$ from

$$\mathbf{v}^{i+1} = \mathbf{r}_{i+1} + Q_{i+1}\mathbf{v}^i = \mathbf{L}\mathbf{v}^i \tag{6-38}$$

(*D*) Replace i with $i + 1$ and return to (B).

Recall the notation T_δ^π from Section 4-3. If π is a Markov policy and δ is a single-stage decision rule, $T_\delta \pi$ denotes the Markov policy where δ is used for the first period and the one-period deferral of π is used thereafter. Let $T_\delta^1 \pi$ denote $T_\delta \pi$ and $T_\delta^{k+1}\pi = T_\delta T_\delta^k \pi$ for $k \in I$. An important connection between the two

† Generally, the policy-improvement algorithm is equivalent to "block pivoting" in the linear program (4-91).

‡ It is not apparent in (4-84) that $f_0(\cdot) \equiv 0$ would be the return vector associated with some policy π_0. Under assumption (4-78), $0 \le r(\cdot, \cdot) < u$, the original MDP is equivalent to one in which a "dummy" action labeled ∞ augments A_s for each $s \in \mathcal{S}$. Assign $p_{ss}^\infty = 1$ and $r(s, \infty) = 0$ for all $s \in \mathcal{S}$. Let π_0 be the stationary policy that always takes action ∞ in every state. Then $\pi_0 \lesssim \pi$ for all π and $\mathbf{v}(\pi_0) = \mathbf{0}$.

algorithms is based on the following sequences v^0, v^1, ... which depend on an integer parameter k:

$$\mathbf{v}^{i+1} = \sum_{l=0}^{k-1} Q_{i+1}^l \mathbf{r}_{i+1} + Q_{i+1}^k \mathbf{L}\mathbf{v}^i \qquad i \in I_+ \tag{6-39}$$

$$\pi_{i+1}^k = T_{\delta_{i+1}}^k \pi_i^k \tag{6-40}$$

where δ_{i+1} satisfies (6-36). Then (6-37) and (6-38) are special cases of (6-39) with $k \to \infty$ and $k = 0$, respectively (we use the convention $\sum_{k=i}^{j} x_k \triangleq 0$ if $j < i$). By varying k and using (6-39) instead of (6-38), (*A*) through (*D*) generates a large family of algorithms. Call k the order of the algorithm. Then (*a*) through (*d*) is the ∞-order algorithm and (*A*) through (*D*) is the 0-order algorithm.

The trade-off between (*a*) through (*d*) and (*A*) through (*D*) [with (6-38)] is that the former needs fewer iterations (finite vs. infinite), but each iteration requires more work. The attractiveness of (6-39) is that the total computation may be less for some $1 \le k < \infty$ than for either of the two extremes. There are two issues: does (6-39) converge to the value vector of an optimal policy, and how does k influence the amount of computation?

Example 6-4: Continuation of Examples 4-9, 4-10, 4-17 and 4-19 The basic data for this deterministic MDP are $\beta = 0.5$ and Figure 6-4. In Example 4-9, a $k = 0$-order algorithm (successive approximations) is used for six iterations. In Example 4-19, a $k = \infty$-order algorithm (policy improvement) is used until an optimum is reached. The optimal policy is $\delta(1) = 3$, $\delta(2) = 1$, and $\delta(3) = 1$ with associated return vector $\mathbf{v} = (4.5, 2.25, 9.0)$. Here we use a 3-order algorithm.

Example 4-9 has a zero-salvage-value vector; so here let $\mathbf{v}^0 = (0, 0, 0)$. The mechanics of the policy-improvement step, namely, (*B*) above or (6-36), do not depend on k; so $\delta_1 = (2, 2, 1)$ is the same here as in Example 4-9 (Table 4-11, $n = 1$) where $\mathbf{L}\mathbf{v}^0 = (1.5, 1, 5.5)$. Then (*C*) with (6-39) instead of (6-38) and $i = 1$ is

$$v_1^1 = 1.5 + 1(\beta + \beta^2) + \beta^3 \cdot 1 = 2.375$$

$$v_2^1 = 1(1 + \beta + \beta^2) + \beta^3 \cdot 1 = 1.875$$

$$v_3^1 = 4.5(1 + \beta + \beta^2 + \beta^3) = 8.4375$$

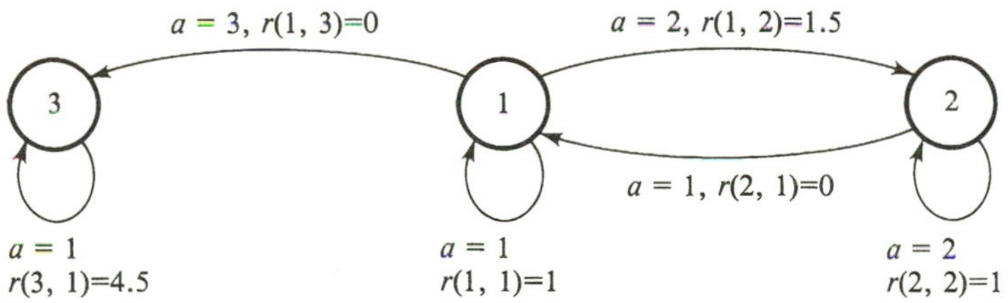

Figure 6-4 A deterministic MDP.

The next step determines δ_2 via (6-36) with $\mathbf{v}^1 = (2.375, 1.875, 8.4375)$ and $i = 1$. Then $\delta_2(3) = 1$ and

$$\begin{aligned}
\mathbf{L}\mathbf{v}^1(1) &= \max\{r(1,1) + \beta v_1^1, r(1,2) + \beta v_2^1, r(1,3) + \beta v_3^1\} \\
&= \max\left\{1 + \frac{2.375}{2}, 1.5 + \frac{1.875}{2}, \mathbf{0} + \frac{\mathbf{8.4375}}{\mathbf{2}}\right\} \\
&= 4.2188 \Rightarrow \delta_2(1) = 3 \\
\mathbf{L}\mathbf{v}^1(2) &= \max\{r(2,1) + \beta v_1^1, r(2,2) + \beta v_2^1 \\
&= \max\left\{0 + \frac{2.375}{2}, \mathbf{1} + \frac{\mathbf{1.875}}{\mathbf{2}}\right\} \\
&= 1.9375 \Rightarrow \delta_2(2) = 2 \\
\mathbf{L}\mathbf{v}'(3) &= r(3,1) + \beta v_3^3 = 4.5 + \frac{8.4375}{2} = 8.7188
\end{aligned}$$

Therefore,

$$\begin{aligned}
v_1^2 &= 0 + 4.5(\beta + \beta^2) + \beta^3(8.7188) = 4.4649 \\
v_2^2 &= 1(1 + \beta + \beta^2) + \beta^3(1.9375) = 1.9922 \\
v_3^2 &= 4.5(1 + \beta + \beta^2) + \beta^3(8.7188) = 8.9649
\end{aligned}$$

so $\mathbf{v}^2 = (4.4649, 1.9922, 8.9649)$. The next step determines δ_3 via (6-36) with $i = 2$. Then $\delta_3(3) = 1$ and

$$\begin{aligned}
\mathbf{L}\mathbf{v}^2(1) &= \max\{r(1,1) + \beta v_1^2, r(1,2) + \beta v_2^2, r(1,3) + \beta v_3^2\} \\
&= \max\left\{1 + \frac{4.4649}{2}, 1.5 + \frac{1.9922}{2}, \mathbf{0} + \frac{\mathbf{8.9649}}{\mathbf{2}}\right\} \\
&= 4.4825 \Rightarrow \delta_3(1) = 3 \\
\mathbf{L}\mathbf{v}^2(2) &= \max\{r(2,1) + \beta v_1^2, r(2,2) + \beta v_2^2\} \\
&= \max\left\{\mathbf{0} + \frac{\mathbf{4.4649}}{\mathbf{2}}, 1 + \frac{1.9922}{2}\right\} = 2.2325 \Rightarrow \delta_3(2) = 1 \\
\mathbf{L}\mathbf{v}^2(3) &= r(3,1) + \beta v_3^2 = 4.5 + \frac{8.9649}{2} = 8.9825
\end{aligned}$$

Therefore,

$$\begin{aligned}
v_1^3 &= 0 + 4.5(\beta + \beta^2) + \beta^3(8.9825) = 4.4978 \\
v_2^3 &= 0 + 0 + 4.5\beta^2 + \beta^3(8.9825) = 2.2478 \\
v_3^3 &= 4.5(1 + \beta + \beta^2) + \beta^3(8.9825) = 8.9978
\end{aligned}$$

Table 6-1 summarizes the results and compares them with Table 4-11 for the 0-order algorithm (successive approximations).

Table 6-1 $k = 0$ **and** $k = 3$ **for the** k**th-order algorithm**

	$\delta_i(1)$		$\delta_i(2)$		v_1^i		v_2^i		v_3^i	
i	$k = 0$	$k = 3$	$k = 0$	$k = 3$	$k = 0$	$k = 3$	$k = 0$	$k = 3$	$k = 0$	$k = 3$
1	2	2	2	2	1.5	2.38	1	1.88	4.5	8.44
2	3	3	2	2	2.25	4.46	1.5	1.99	6.75	8.96
3	3	3	2	1	3.38	4.50	1.75	2.25	7.88	9.00
4	3		2		3.94		1.88		8.44	
5	3		1		4.22		1.97		8.72	
6	3		1		4.36		2.11		8.86	

With $k = 3$, the algorithm converges much more rapidly than with $k = 0$. With $k = 3$, after three iterations, each component of $\mathbf{v}^3$ is within 1 percent of its asymptotic value. With $k = 0$, the comparable figure is 25 percent; after six iterations, v_2^6 is still 17 percent distant from its asymptote. □

Let π_i denote the policy which yields return vector $\mathbf{v}^i$. It follows from (6-40) that $\mathbf{v}^{i+1}$ is the return vector from $T_{\delta_{i+1}}^k \pi$ if δ_{i+1} is a best rule to delay π_i for one period. In the notation of (4-71) and (4-72), δ_{i+1} maximizes $T_\delta \pi_i$ over $\delta \in G_1(\pi_i)$. From Proposition 4-5, either $G_1(\pi_i)$ is empty and π_i is optimal or $\pi_{i+1} > \pi_i$. Let $\mathbf{f}$ denote the vector with components $f(s)$. We say that the sequence $\{\pi_i\}$ *converges* to π^* if $\{\mathbf{v}^i\}$ converges† to $\mathbf{f}$ and $\mathbf{f} = \mathbf{v}(\pi^*)$. Does $\{\pi_i\}$ converge, say to π^*, and is π^* optimal?

For an S-vector $\mathbf{u}$ let

$$\|\mathbf{u}\| \triangleq \max\{|u_s| : s \in \mathcal{S}\}$$

This repeats the notation of Section 4-5 and is a special case of the notation $\|\mathbf{u}\|$ in Section 5-4.

Theorem 6-3 Suppose $\#\mathscr{C} < \infty$, $\mathbf{v}^0 \in \mathbb{R}^S$, and $k \in I \cup \{\infty\}$. Generate $\mathbf{v}^0$, $\mathbf{v}^1$, $\mathbf{v}^2, \ldots$ with (A) through (D) and (6-39) replacing (6-38).

(*a*) Then

$$\mathbf{f} = \lim_{i \to \infty} \mathbf{v}^i$$

(*b*) If $\mathbf{L}\mathbf{v}^0 \geq \mathbf{v}^0$, then

$$\mathbf{v}^0 \leq \mathbf{v}^1 \leq \mathbf{v}^2 \leq \cdots \to \mathbf{f} \tag{6-41}$$

and there is $K < \infty$ such that

$$\|\mathbf{f} - \mathbf{v}^i\| \leq K\beta^i \tag{6-42}$$

Part (*a*) asserts convergence of $\mathbf{v}^i$ to the optimum, namely, $\mathbf{f}$. Part (*b*) asserts

† We say that a sequence $\mathbf{x}^1$, $\mathbf{x}^2$, ..., of S-vectors converges to $x^* \in \mathbb{R}^S$ if the sequence of sth components $x_s^1, x_s^2, \ldots$, converges to x_s^*, the sth component of x^*, for $s = 1, 2, \ldots, S$.

that convergence is monotone and geometrically fast if $\mathbf{Lv}^0 \geq \mathbf{v}^0$. This condition is easily satisfied by letting $\mathbf{v}^0 = \mathbf{0}$. Then $r(\cdot, \cdot) \geq 0$ implies $\mathbf{L0} \geq \mathbf{0}$.

The proof of the theorem is presented at the end of this section (page 273). We sketch the proof here in order to identify other properties of the sequence $\mathbf{v}^0$, $\mathbf{v}^1, \mathbf{v}^2, \ldots$. The proof of part (*a*) follows from establishing

$$\frac{-\beta^{i(k+1)}}{1-\beta}\mathbf{y} \leq \mathbf{f} - \mathbf{v}^i \leq \frac{(i+1)\beta^i}{1-\beta}\mathbf{x}$$

where $\mathbf{x}$ and $\mathbf{y}$ are specified as follows. Let

$$x = \max\{[\mathbf{Lv}^0(s) - v_s^0]^+ : s \in \mathcal{S}\}$$

$$y = \max\{[v_s^0 - \mathbf{Lv}^0(s)]^+ : s \in \mathcal{S}\}$$

Then $\mathbf{x} \triangleq x\mathbf{e}$ and $\mathbf{y} \triangleq y\mathbf{e}$.

The proof of (*b*) uses the following mappings from $\mathbb{R}^S$ to $\mathbb{R}^S$. For $\delta \in \Delta$ and $\mathbf{v} \in \mathbb{R}^S$,

$$\mathbf{G}_\delta \mathbf{v} \triangleq \sum_{l=0}^{k-1} Q_\delta^l r_\delta + Q_\delta^k \mathbf{Lv}$$

$$\mathbf{Jv} \triangleq \text{vmax}\{\mathbf{G}_\delta \mathbf{v} : \delta \in \Delta\}$$

Suppose $\gamma \in \Delta$ is maximal for $\mathbf{Lv}$, that is,

$$\mathbf{r}_\gamma + Q_\gamma \mathbf{v} = \text{vmax}\{\mathbf{r}_\gamma + Q_\delta \mathbf{v} : \delta \in \Delta\} \triangleq \mathbf{Lv}$$

Then $\mathbf{Hv} \triangleq \mathbf{G}_\gamma \mathbf{v}$. In this notation, $\mathbf{v}^{i+1} = \mathbf{Hv}^i$. Let $\mathbf{H}^1 = \mathbf{H}$, $\mathbf{J}^1 = \mathbf{J}$, $\mathbf{H}^{i+1} = \mathbf{HH}^i$, and $\mathbf{J}^{i+1} = \mathbf{JJ}^i$. Then $\mathbf{v}^i = \mathbf{H}^i\mathbf{v}^0$. Let $\mathbf{V}_k^i \equiv \mathbf{v}^i$ make the effect of k explicit and let $\mathbf{U}_k^i = \mathbf{J}^i\mathbf{v}^0$.

The proof of (*b*) has three parts. First,

$$\|\mathbf{f} - \mathbf{V}_0^i\| \leq K_1\beta^i \qquad \text{and} \qquad \|\mathbf{f} - \mathbf{U}_k^i\| \leq K_2\,\beta^{i(k+1)} \tag{6-43}$$

Theorem 4-7 already asserts geometric convergence for $\mathbf{V}_0$. Second, the monotonicity in (6-41) is implied by

$$\mathbf{v}^0 \leq \mathbf{Lv}^0 \Rightarrow \mathbf{V}_k^i \leq \mathbf{V}_k^{i+1} \qquad \text{for all } i \tag{6-44}$$

which is verified by showing that $\mathbf{v} \leq \mathbf{Lv}$ implies $\mathbf{v} \leq \mathbf{Hv} \leq \mathbf{LHv}$. Last,

$$\mathbf{v}^0 \leq \mathbf{Lv}^0 \Rightarrow \mathbf{V}_0^i \leq \mathbf{V}_k^i \leq \mathbf{U}_k^i \qquad \text{for all } i \tag{6-45}$$

so $\mathbf{V}_k^i$, that is, $\mathbf{v}^i$, is squeezed between two sequences that are each converging to $\mathbf{f}$ geometrically fast. It follows that $\mathbf{v}^i$ converges to $\mathbf{f}$ geometrically fast.

You might wonder if $\mathbf{V}_k^i$ is monotone in k in the following sense. If $\mathbf{V}_k^0 = \mathbf{V}_{k+1}^0 \leq \mathbf{LV}_k^0$, is it true for all i that $\mathbf{V}_k^i \leq \mathbf{V}_{k+1}^i$? The answer is "no" (unless $k = 0$) [cf. van der Wal and van Nunen (1977) or Puterman and Shin (1979) for a counterexample].

Acceleration of kth-Order Algorithms

Every kth-order algorithm uses (6-36) to identify δ_{i+1} during the ith iteration. Then k affects the computation of $\mathbf{v}^{i+1}$ according to (6-39). The following modification of (6-36) seeks a "better" δ_{i+1} for a given $\mathbf{v}^i$. Suppose that elements of $\mathcal{S}$ are labeled 1, 2, ..., S. A restatement of (6-36) is

(i) Let $s = 1$.
(ii) Compute

$$\mathbf{Lv}^i(s) = \max \{r(s, a) + \sum_{j \in \mathcal{S}} q_{sj}^a v_j^i : a \in A_s\} \tag{6-46}$$

(iii) Assign to $\delta_{i+1}(s)$ any $a \in A_s$ that attains the maximum in (6-46).
(iv) If $s < S$, replace s with $s + 1$ and return to (ii). If $s = S$, stop.

The basic idea of the following modification of (6-46) is that $\mathbf{Lv}^i$ is a more appealing estimate of $\mathbf{f}$ than is $\mathbf{v}^i$. Therefore, whenever possible, replace components of $\mathbf{v}^i$ with components of $\mathbf{Lv}^i$. In (6-46) this means that v_j^i should be replaced by $\mathbf{Lv}^i(j)$ if $j < s$ because $\mathbf{Lv}^i(j)$ has already been computed. However,

$$Lv^i(s) \neq \max \{r(s, a) + \sum_{j<s} q_{sj}^a \, Lv^i(j) + \sum_{j \geqslant s} q_{sj}^a v_j^i : a \in A_s\}$$

so different notation is needed. Also, why leave the term $q_{ss}^a v_s^i$ in the maximand when the entire maximization is aimed at replacement of v_s^i? This suggests replacing the right side of (6-46) with

$$\max \left\{ \frac{r(s, a) + \sum_{j<s} q_{sj}^a \mathbf{Lv}^i(j) + \sum_{j>s} q_{sj}^a v_j^i}{1 - q_{ss}^a} : a \in A_s \right\}$$

whose denominator is not zero because $0 \leq q_{ss}^a \leq \beta < 1$.

These remarks lead to the following modification of (6-46), (ii), and (iii): let

$$w_s = \max \left\{ \frac{r(s, a) + \sum_{j<s} q_{sj}^a w_j + \sum_{j>s} q_{sj}^a v_j^i}{1 - q_{ss}^a} : a \in A_s \right\} \tag{6-47}$$

and assign to $\delta_{i+1}(s)$ any $a \in A_s$ that causes the maximum in (6-47) to be attained. The modification is analogous to Gauss-Seidel iterations in the solution of a system of equations. Gauss-Seidel iterations comprise a version of *successive overrelaxation* (SOR) methods.† Exercise 6-1 asks you to prove that the SORs modification above converges to an optimum, but we cannot prove that it accelerates convergence if $k > 0$. However, many numerical experiments confirm that it is dramatically effective if the states are shrewdly ordered when the labels 1, 2, ..., S are assigned.

† Here is the SOR generalization of Gauss-Seidel. Let $x_i(s, a)$ denote the maximand in (6-47) and let $0 < \xi < 2$. Then

$$v_s^{i+1} = \xi \max \{x_i(s, a): a \in A_s\} + (1 - \xi) v_s^i$$

Example 6-5: Continuation of Example 6-4 The basic data are $\beta = 0.5$ and Figure 6-4. First we use SOR with $k = 0$ and $v^0 = \mathbf{0}$. Then (6-47) in place of (6-46) with $i = 0$ is

$$w_1 = \max\left\{\frac{r(1, 1)}{1-\beta}, r(1, 2), r(1, 3)\right\}$$

$$= \max\{\mathbf{2}, 1.5, 0\} = 2 \Rightarrow \delta_1(1) = 1$$

$$w_2 = \max\left\{r(2, 1) + \beta w_1, \frac{r(2, 2)}{1-\beta}\right\}$$

$$= \max\{1, \mathbf{2}\} = 2 \Rightarrow \delta_1(2) = 2$$

$$w_3 = \frac{r(3, 1)}{1-\beta} = 9.0 \qquad [\delta_i(3) = 1 \text{ for all } i]$$

so $\mathbf{v}^1 = (2, 2, 9.0)$. Note that $w_3 = f(3)$ after one iteration!

For $i = 1$, (6-47) yields

$$w_1 = \max\left\{\frac{1}{1-0.5}, 1.5 + 2(0.5), \mathbf{0 + 9(0.5)}\right\}$$

$$= 4.5 \Rightarrow \delta_2(1) = 3$$

$$w_2 = \max\left\{r(2, 1) + \beta w_1, \frac{r(2, 2)}{1-\beta}\right\}$$

$$= \max\{\mathbf{2.25}, 2\} = 2.25 \Rightarrow \delta_2(2) = 1$$

$$w_3 = \frac{r(3, 1)}{1-\beta} = 9.0$$

Therefore, $\mathbf{w} = \mathbf{f}$ and δ_2^∞ is optimal after two iterations!

Permuting the labels of states can further accelerate the SOR modified algorithm. Suppose $s = 1$ becomes $s = 2$, $s = 2$ becomes $s = 3$, and $s = 3$ becomes $s = 1$. Then Figure 6-4 becomes Figure 6-5. The actions are relabeled too; so taking action $a \in A_s$ causes the next state to be a.

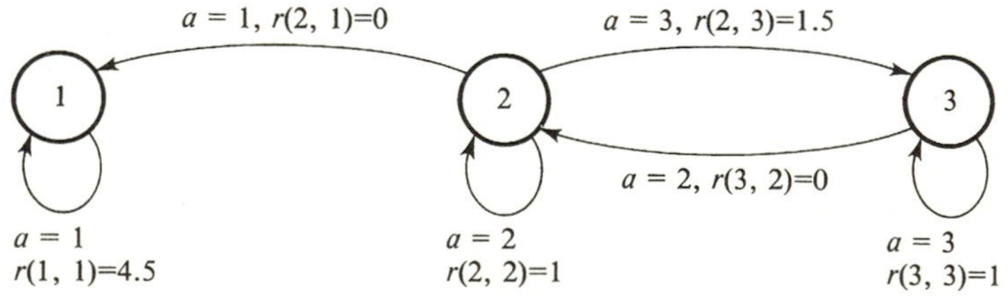

Figure 6-5 Relabeled states in a deterministic MDP.

Table 6-2 Optimal actions in Example 6-5

	Unmodified algorithm		SOR		SOR and relabeled states	
i	$\delta_i(1)$	$\delta_i(2)$	$\delta_i(1)$	$\delta_i(2)$	$\delta_i(1)$	$\delta_i(2)$
1	2	2	1	2	3	1
2	3	2	3	1		
3	3	2				
4	3	2				
5	3	1				

Starting with $\mathbf{v}^0 = 0$, SOR yields

$$w_1 = \frac{r(1, 1)}{1 - \beta} = 9.0$$

$$w_2 = \max\left\{r(2, 1) + \beta w_1, \frac{r(2, 2)}{1 - \beta}, r(2, 3) + \beta \cdot 0\right\}$$

$$= \max\{\mathbf{4.5}, 2, 1.5\} = 4.5 \Rightarrow \delta_1(2) = 1$$

$$w_3 = \max\left\{\mathbf{0 + 4.5\boldsymbol{\beta}}, \frac{1}{1 - \beta}\right\} = 2.25 \Rightarrow \delta_1(3) = 2$$

Therefore, $\mathbf{w} = \mathbf{f}$ and δ_1^∞ is optimal after one iteration! You should verify that $i = 2$ in (6-47) leads to the same $\mathbf{w}$.

Tables 6-2 and 6-3 display the results from the unmodified algorithm (Table 4-11), SOR, and SOR applied to relabeled states. *For* SOR *with relabeled states*, the results are given in terms of the original labels of states and actions. This simplifies comparisons.

SOR switches to action 1 (from action 2) in state 2 when $i = 2$, whereas

Table 6-3 v^i in Example 6-5

	Unmodified algorithm			SOR			SOR and relabeled states		
i	$v^i(1)$	$v^i(2)$	$v^i(3)$	$v^i(1)$	$v^i(2)$	$v^i(3)$	$v^i(1)$	$v^i(2)$	$v^i(3)$
1	1.5	1	4.5	2	2	9	4.5	2.25	9
2	2.25	1.5	6.75	4.5	2.25	9			
3	3.38	1.75	7.88						
4	3.94	1.88	8.44						
5	4.22	1.97	8.72						

the switch occurs when $i = 5$ in the unmodified algorithm. The switch occurs when $i = 1$ when SOR is used with relabeled states. □

The effectiveness of an algorithm may depend on the scale of the problems to which it is applied and on the features of the specific problem. Therefore, we do not claim that the comparisons in the preceding example lead to accurate forecasts of effectiveness on large MDPs.

Elimination of Suboptimal Actions

Another way to accelerate computations is to ignore in (6-46) [or (6-47)] necessarily suboptimal elements of A_s. The elimination of suboptimal actions depends on having bounds $\mathbf{v}^- \leq \mathbf{f} \leq \mathbf{v}^+$.

Theorem 6-4 If there are bounds $\mathbf{v}^-$ and $\mathbf{v}^+$ such that

$$\mathbf{v}^- \leq \mathbf{f} \leq \mathbf{v}^+ \tag{6-48}$$

and

$$r(s, a) + \sum_{j \in \mathcal{S}} q_{sj}^a v_j^+ < v_s^- \tag{6-49}$$

then δ^∞ is not optimal if $\delta(s) = a$.

Proof Inequalities (6-48) and (6-49) imply

$$r(s, a) + \sum_{j \in \mathcal{S}} q_{sj}^a f(j) \leq r(s, a) + \sum_{j \in \mathcal{S}} q_{sj}^a v_j^+ < v_s^- \leq f_s$$

so $a = \delta(s)$ fails to satisfy the optimality test of the policy improvement algorithm, i.e., (4-90). Then Theorem 4-9 implies δ^∞ is not optimal if $\delta(s) = a$. □

The effectiveness of the suboptimality test (6-49) depends on the stringency of the bounds (6-48). From part (*a*) of Theorem 4-8, if $\mathbf{v}^0, \mathbf{v}^1, \ldots$ is generated by a 0-order algorithm, i.e., the successive-approximations recursion (4-84), then (6-48) is satisfied by

$$v_s^- = v^i(s) - \frac{\| \mathbf{L}\mathbf{v}^i - \mathbf{v}^i \|}{1 - \beta} \qquad v_s^+ = v^i(s) + \frac{\| \mathbf{L}\mathbf{v}^i - \mathbf{v}^i \|}{1 - \beta} \tag{6-50}$$

where $\mathbf{v}^{i+1} = \mathbf{L}\mathbf{v}^i$ and $\| w \| = \max \{ | w_s |: s \in \mathcal{S}\}$. However, these bounds can be tightened. For any algorithm that uses the policy-improvement step (6-36), the ith iteration begins with some vector $\mathbf{v}^i \in \mathbb{R}^S$. Then δ_{i+1} is determined via policy improvement, that is, $\mathbf{r}_{i+1} + Q_{i+1}\mathbf{v}^i = \mathbf{L}\mathbf{v}^i$; so $\mathbf{L}\mathbf{v}^i$ is computed. Of course, in kth-order algorithms with $k > 0$, $\mathbf{v}^{i+1} \not\equiv \mathbf{L}\mathbf{v}^i$. The bounds in Theorem 6-5 below can be used with any algorithm that computes $\mathbf{L}\mathbf{v}^i$ for a sequence $\mathbf{v}^0, \mathbf{v}^1, \mathbf{v}^2, \ldots$.

For $\mathbf{w} \in \mathbb{R}^S$, let

$$(\mathbf{w})_m = \min \{w_s: s \in \mathcal{S}\} \qquad \text{and} \qquad (\mathbf{w})^M = \max \{w_s: s \in \mathcal{S}\}$$

Then the tightening of (6-50) depends on the observation that

$$-\|\mathbf{w}\| \le (\mathbf{w})_m \le w_s \le (\mathbf{w})^M \le \|\mathbf{w}\| \qquad s \in \mathcal{S}$$

and the following strengthening of (4-87). For $\mathbf{v} \in \mathbb{R}^S$ and $\mathbf{w} \in \mathbb{R}^S$,

$$\beta(\mathbf{v} - \mathbf{w})_m \le (\mathbf{Lv} - \mathbf{Lw})_m \le (\mathbf{Lv} - \mathbf{Lw})^M \le \beta(\mathbf{v} - \mathbf{w})^M \tag{6-51}$$

The proof of (6-51) uses the following basic inequality.

Lemma 6-1 For $\mathbf{v} \in \mathbb{R}^S$, and $\mathbf{w} \in \mathbb{R}^S$, suppose γ attains $\mathbf{Lv}$, that is,

$$\mathbf{r}_\gamma + Q_\gamma \mathbf{v} = \text{vmax}\,\{\mathbf{r}_\delta + Q_\delta \mathbf{v} \colon \delta \in \Delta\} \triangleq \mathbf{Lv}$$

Then

$$\mathbf{Lw} - \mathbf{Lv} \ge Q_\gamma(\mathbf{w} - \mathbf{v})$$

PROOF By definition of **Lw**,

$$\begin{aligned}\mathbf{Lw} - \mathbf{Lv} &= v\max\,\{\mathbf{r}_\delta + Q_\delta \mathbf{w} \colon \delta \in \Delta\} - \mathbf{r}_\gamma - Q_\gamma \mathbf{v} \\ &\ge (\mathbf{r}_\gamma + Q_\gamma \mathbf{w}) - \mathbf{r}_\gamma - Q_\gamma \mathbf{v} \\ &= Q_\gamma(\mathbf{w} - \mathbf{v})\end{aligned}$$

□

Suppose in (6-51) that $\lambda \in \Delta$ attains **Lw**. Then the lemma yields

$$Q_\lambda(\mathbf{v} - \mathbf{w}) \le \mathbf{Lv} - \mathbf{Lw} \le Q_\gamma(\mathbf{v} - \mathbf{w})$$

which implies (6-51) because the rows of Q_γ and Q_λ sum to β.

Recall the notation $\mathbf{e} = (1, 1, \ldots, 1) \in \mathbb{R}^S$ and for any $\delta \in \Delta$ and $k \in I$,

$$\mathbf{G}_\delta \gamma = \sum_{l=0}^{k-1} Q_\delta^l \mathbf{r}_\delta + Q_\delta^k \mathbf{Lv}$$

Suppose $\gamma \in \Delta$ satisfies $\mathbf{r}_\gamma + Q_\gamma \mathbf{v} = \mathbf{Lv}$ and recall the notation $\mathbf{Hv} = \mathbf{G}_\gamma \mathbf{v}$. This notation is used below to provide upper and lower bounds on the optimal value function **f** of a discounted finite MDP.

Theorem 6-5† For all $\mathbf{v} \in \mathbb{R}^S$,

$$\begin{aligned}\mathbf{v} + \frac{\mathbf{e}(\mathbf{Lv} - \mathbf{v})_m}{1 - \beta} &\le \mathbf{Lv} + \frac{\beta}{1 - \beta}\,\mathbf{e}(\mathbf{Lv} - \mathbf{v})^M \\ &\le \mathbf{Hv} + \frac{\beta^{k+1}}{1 - \beta}\,\mathbf{e}(\mathbf{Lv} - \mathbf{v})_m \\ &\le \mathbf{f} \le \mathbf{Lv} + \frac{\beta}{1 - \beta}\,\mathbf{e}(\mathbf{Lv} - \mathbf{v})^M\end{aligned} \tag{6-52}$$

† In Theorem 6-5 and its proof, interpret **e** as a vector with the same orientation (row or column) as **v**, **Lv**, **Hv**, and **f**.

PROOF Suppose Q^* is induced by a policy in Δ which attains **Lf**. Then Lemma 6-1 implies

$$(\mathbf{Lv} - \mathbf{Lf}) \geq Q^*(\mathbf{v} - \mathbf{f})$$

so
$$(\mathbf{Lv} - \mathbf{v}) - (\mathbf{Lf} - \mathbf{f}) \geq (Q^* - I)(\mathbf{v} - \mathbf{f})$$

Thus $\mathbf{Lf} - \mathbf{f} = \mathbf{0}$ yields

$$\mathbf{v} - \mathbf{Lv} \leq (I - Q_*)(\mathbf{v} - \mathbf{f}) \tag{6-52$'$}$$

Therefore,

$$\begin{aligned}
\mathbf{f} &\leq \mathbf{v} + (I - Q_*)^{-1}(\mathbf{Lv} - \mathbf{v}) \\
&= \mathbf{Lv} + (I - Q_*)^{-1}Q_*(\mathbf{Lv} - \mathbf{v}) \\
&\quad -(\mathbf{Lv} - \mathbf{v}) + (I - Q_*)^{-1}[(\mathbf{Lv} - \mathbf{v}) - Q_*(\mathbf{Lv} - \mathbf{v})] \\
&= \mathbf{Lv} + (I - Q_*)^{-1}Q_*(\mathbf{Lv} - \mathbf{v}) \\
&\quad -[I - (I - Q_*)^{-1}(I - Q_*)](\mathbf{Lv} - \mathbf{v}) \\
&= \mathbf{Lv} + (I - Q_*)^{-1}Q_*(\mathbf{Lv} - \mathbf{v}) \\
&\leq \mathbf{Lv} + (I - Q_*)^{-1}Q_*\,\mathbf{e}(\mathbf{Lv} - \mathbf{v})^M \\
&= \mathbf{Lv} + (I - Q_*)^{-1}\mathbf{e}\beta(\mathbf{Lv} - \mathbf{v})^M \\
&= \mathbf{Lv} + \frac{\beta}{1 - \beta}\,\mathbf{e}(\mathbf{Lv} - \mathbf{v})^M
\end{aligned}$$

because the row sums of Q_* are β.

For the lower bounds on **f**, (6-52′) yields

$$\mathbf{Lv} - \mathbf{v} \geq (Q_* - I)(\mathbf{v} - \mathbf{f})$$

so $\quad -(I - Q_*)^{-1}(\mathbf{Lv} - \mathbf{v}) \geq \mathbf{v} - \mathbf{f}$

Thus

$$\begin{aligned}
\mathbf{f} &\geq \mathbf{v} + (I - Q_\gamma)^{-1}(\mathbf{Lv} - \mathbf{v}) \\
&= \mathbf{v} + \sum_{l=1}^{\infty} Q_\gamma^l(\mathbf{Lv} - \mathbf{v}) \\
&= \mathbf{v} + \sum_{l=0}^{k} Q_\gamma^l(\mathbf{Lv} - \mathbf{v}) + \sum_{l=k+1}^{\infty} Q_\gamma^l(\mathbf{Lv} - \mathbf{v}) \\
&= \mathbf{Lv} + \sum_{l=1}^{k} Q_\gamma^l(\mathbf{Lv} - \mathbf{v}) + \sum_{l=k+1}^{\infty} Q_\gamma^l(\mathbf{Lv} - \mathbf{v}) && \text{(6-53}a\text{)} \\
&\geq \mathbf{Lv} + \sum_{l=1}^{l} Q_\gamma^l(\mathbf{Lv} - \mathbf{v}) + \frac{\beta^{k+1}}{1 - \beta}\,\mathbf{e}(\mathbf{Lv} - \mathbf{v})_m && \text{(6-53}b\text{)}
\end{aligned}$$

so the first lower bound on **f** in (6-52) is implied by the following representation.

Lemma 6-2 For all $\mathbf{v} \in \mathbb{R}^S$,

$$\mathbf{Hv} = \mathbf{Lv} + \sum_{l=1}^{k} Q_\gamma^l(\mathbf{Lv} - \mathbf{v}) \tag{6-54}$$

PROOF By definition of **H**,

$$\mathbf{Hv} = \mathbf{G}_\gamma \mathbf{v} = \sum_{l=0}^{k-1} Q_\gamma^l \mathbf{r}_\gamma + Q_\gamma^k \mathbf{Lv}$$

$$= \sum_{l=0}^{k-1} Q_\gamma^l \mathbf{r}_\gamma + \mathbf{Lv} + \sum_{l=1}^{k} Q_\gamma^l \mathbf{Lv} - \sum_{l=0}^{k-1} Q_\gamma^l \mathbf{Lv}$$

so

$$= \mathbf{Lv} + \sum_{l=0}^{k-1} Q_\gamma^l(\mathbf{r}_\gamma + Q_\gamma \mathbf{Lv} - \mathbf{Lv})$$

However, $\mathbf{Lv} = \mathbf{r}_\gamma + Q_\gamma \mathbf{v}$ implies

$$\mathbf{r}_\gamma + Q_\gamma \mathbf{Lv} - \mathbf{Lv} = Q_\gamma(\mathbf{Lv} - \mathbf{v})$$

$$\mathbf{Hv} = \mathbf{Lv} + \sum_{l=1}^{k} Q_\gamma^l(\mathbf{Lv} - \mathbf{v}) \qquad \square$$

Substitution of (6-54) in (6-53*b*) yields the first lower bound on **f** in (6-52). Continuing with (6-53),

$$\mathbf{f} \geq \mathbf{Lv} + \sum_{l=1}^{\infty} Q_\gamma^l(\mathbf{Lv} - \mathbf{v})$$

$$\geq \mathbf{Lv} + \mathbf{e}\beta(\mathbf{Lv} - \mathbf{v})_m$$

Also,

$$\mathbf{f} \geq \mathbf{Lv} + \sum_{l=1}^{\infty} Q_\gamma^l(\mathbf{Lv} - \mathbf{v})$$

$$= \mathbf{v} + \sum_{l=0}^{\infty} Q_\gamma^l(\mathbf{Lv} - \mathbf{v})$$

$$\geq \mathbf{v} + \mathbf{e}(\mathbf{Lv} - \mathbf{v})_m$$

This establishes (6-52) if

$$\mathbf{Lv} + \frac{\beta}{1-\beta}\mathbf{e}(\mathbf{Lv} - \mathbf{v})_m \geq \mathbf{v} + \mathbf{e}(\mathbf{Lv} - \mathbf{v})_m$$

However, $\mathbf{w} - \mathbf{e}(\mathbf{w})_m \geq \mathbf{0}$ for all $\mathbf{w} \in \mathbb{R}^S$ implies

$$\mathbf{Lv} + \frac{\beta}{1-\beta}\mathbf{e}(\mathbf{Lv}-\mathbf{v}) - \mathbf{v} - \mathbf{e}(\mathbf{Lv}-\mathbf{v})_m = \mathbf{Lv} - \mathbf{v} - \mathbf{e}(\mathbf{Lv}-\mathbf{v})_m \geq \mathbf{0} \quad \square$$

The following bounds use (6-52) and are stronger than (6-50) even for successive approximations. For *any* $\mathbf{v} \in \mathbb{R}^S$,

$$\begin{aligned} \mathbf{v}^- &= \mathbf{Hv} + \frac{\beta^{k+1}}{1-\beta}\mathbf{e}(\mathbf{Lv}-\mathbf{v})_m \\ \mathbf{v}^+ &= \mathbf{Lv} + \frac{\beta}{1-\beta}\mathbf{e}(\mathbf{Lv}-\mathbf{v})^M \end{aligned} \tag{6-55}$$

which would ordinarily be used with $\mathbf{v} = \mathbf{v}^{i-1}$. Substitution of (6-55) in (6-49) yields the following test. Discard $a \in A_s$ if

$$\mathbf{Hv}(s) - \sum_{j \in \mathcal{S}} q_{sj}^a \mathbf{Lv}(j) - r(s, a) > \frac{\beta[\beta(\mathbf{Lv}-\mathbf{v})^M - \beta^k(\mathbf{Lv}-\mathbf{v})_m]}{1-\beta} \tag{6-56}$$

where $\mathbf{Hv}(s)$ and $\mathbf{Lv}(j)$ denote the sth and jth coordinates of $\mathbf{Hv}$ and $\mathbf{Lv}$. If $\mathbf{v} \equiv \mathbf{v}^{i-1}$ in a kth-order algorithm, $\mathbf{Hv}(s) = v_s^i$. If also $k = 0$, then $\mathbf{Lv}(j) = v_j^i$. In practice, (6-56) would be used only every several iterations on all $(s, a) \in \mathscr{C}$ which have not already been eliminated.

Label the coordinates of $\mathbb{R}^S$ from 1 to S and let $\mathbf{K}$ denote the mapping from $\mathbb{R}^S$ to $\mathbb{R}^S$ given by SOR, i.e., (6-47). For $\mathbf{v} \in \mathbb{R}^S$, the sth component of $\mathbf{Kv}$ is

$$\mathbf{Kv}(s) = \max\left\{\frac{r(s, a) + \sum_{j<s} q_{sj}^a \mathbf{Kv}(j) + \sum_{j>s} q_{sj}^a v_j}{1 - q_{ss}^a} : a \in A_s\right\}$$

Example 6-5 and many other numerical experiments suggest that $\mathbf{K}^i\mathbf{v}$ converges faster to $\mathbf{f}$ than does $\mathbf{L}^i\mathbf{v}$. Therefore, you might conjecture that it is valid to modify (6-55) by replacing $\mathbf{Lv}^i$ with $\mathbf{Kv}^i$. However, the validity of the replacement is an open question.

Example 6-6: Continuation of Examples 6-4 and 6-5 First we use action elimination with unmodified successive approximations, i.e., a 0-order algorithm. Also, we use $k = 0$ in the suboptimality test (6-56). Since $\beta = 0.5$, (6-56) asserts that a should be discarded from A_s if

$$v_s^i = \sum_{j=1}^{3} q_{sj}^a v_j^i - r(s, a) > \frac{(\mathbf{v}^i - \mathbf{v}^{i-1})^M}{2} - (\mathbf{v}^i - \mathbf{v}^{i-1})_m$$

Table 6-4 repeats part of Table 4-11 and presents $(\mathbf{v}^i - \mathbf{v}^{i-1})^M$, $(\mathbf{v}^i - \mathbf{v}^{i-1})_m$, and $(\mathbf{v}^i - \mathbf{v}^{i-1})^M/2 - (\mathbf{v}^i - \mathbf{v}^{i-1})_m$.

Table 6-4 Value functions and bounds

i	v_1^i	v_2^i	v_3^i	$(\mathbf{v}^i - \mathbf{v}^{i-1})^M$	$(\mathbf{v}^i - \mathbf{v}^{i-1})_m$	$(\mathbf{v}^i - \mathbf{v}^{i-1})^M/2 - (\mathbf{v}^i - \mathbf{v}^{i-1})_m$
0	0	0	0			
1	1.5	1	4.5	4.5	1	1.25
2	2.25	1.5	6.75	2.25	0.5	0.625
3	3.375	1.75	7.875	1.125	0.25	0.3125
4	3.9375	1.875	8.4375	0.5625	0.125	0.1613
5	4.2188	1.9688	8.7188	0.2813	0.0938	0.0469
6	4.3594	2.1094	8.8594	0.1406	0.1406	−0.0703
7	4.4297	2.1797	8.9297	0.0703	0.0703	−0.0357

The test first causes $a = 1$ to be deleted from A_1 when $i = 4$:

$$i = 1:\ v_1^0 - \frac{v_1^0}{2} - r(1, 1) = -1 \le 1.25$$

$$i = 2:\ \frac{v_1^1}{2} - r(1, 1) = -0.25 \le 0.625$$

$$i = 3:\ \frac{v_2^2}{2} - r(1, 1) = 0.125 \le 0.3125$$

$$i = 4:\ \frac{v_1^3}{2} - r(1, 1) = 0.6875 > 0.1613$$

For $s = 1$, the test deletes $a = 2$ from A_1 when $i = 4$:

$$i = 1:\ v_1^0 - \frac{v_2^0}{2} - r(1, 2) = -1.5 \le 1.25$$

$$i = 2:\ v_1^1 - \frac{v_2^1}{2} - r(1, 2) = -0.5 \le 0.625$$

$$i = 3:\ v_1^2 - \frac{v_2^2}{2} - r(1, 2) = 0 \le 0.3125$$

$$i = 4:\ v_1^3 - \frac{v_2^3}{2} - r(1, 2) = 1 > 0.1613$$

For $s = 2$, the test deletes $a = 2$ from A_2 when $i = 6$:

$$i = 5:\ v_2^4 - \frac{v_2^4}{2} - r(2, 2) = -0.0625 \le 0.0469$$

$$i = 6:\ \frac{v_2^5}{2} - r(2, 2) = -0.0156 > -0.0703$$

Table 6-5 Calculations for SOR, a 3-order algorithm, and action elimination

i	v_1^i	v_2^i	v_3^i	$\mathbf{Lv}^i(1)$	$\mathbf{Lv}^i(2)$	$\mathbf{Lv}^i(3)$	$(\mathbf{Lv}^i - \mathbf{v}^i)^M$	$(\mathbf{Lv}^i - \mathbf{v}^i)_m$
0	0	0	0	1.5	1	4.5	4.5	1
1	2.2375	1.875	8.4375	4.2188	1.9375	8.7188	1.9813	0.0625
2	4.4649	1.9922	8.9648	4.4825	2.2325	8.9825	0.2403	0.0177
3	4.4978	2.2478	8.9978	4.4989	2.2489	8.9989	0.0011	0.0011

After $i = 6$, exactly one element remains in each set of actions; so it must be the optimal action in that state.

Now we combine SOR, a 3-order algorithm, and action elimination. Table 6-5 contains calculations which will be useful in the action-elimination tests.

All the entries in Table 6-5 are calculated in Example 6-4 except for $\mathbf{Lv}^3$ and the two rightmost columns.

We shall use $k = 3$ in the suboptimality test (6-56), which is equivalent to deleting a from A_s if

$$v_s^i - \sum_{j=1}^{3} q_{sj}^a \mathbf{Lv}^{i-1}(j) - r(s, a) > \frac{(\mathbf{Lv}^{i-1} - \mathbf{v}^{i-1})^M}{2} - \frac{(\mathbf{Lv}^{i-1} - \mathbf{v}^{i-1})_m}{8}$$

If $i = 3$, the right side is $0.2403/2 - 0.0177/8 = 0.1179$, which yields the following conclusions:

$$s = 1,\ a = 1:\ v_1^3 - \frac{\mathbf{Lv}^2(1)}{2} - r(1, 1) = 1.2536$$

$$> 0.1179 \Rightarrow \text{delete } a = 1 \text{ from } A_1$$

$$s = 1,\ a = 2:\ v_1^3 - \frac{\mathbf{Lv}^2(2)}{2} = r(1, 2) = 1.8816$$

$$> 0.1179 \Rightarrow \text{delete } a = 2 \text{ from } A_1$$

$$s = 2,\ a = 2:\ v_2^3 - \frac{\mathbf{Lv}^2(2)}{2} - r(2, 2) = 0.1316$$

$$> 0.1179 \Rightarrow \text{delete } a = 2 \text{ from } A_2$$

At this point, the reduced set of feasible policies has exactly one element; so it is optimal. □

Extrapolation

We have discussed how bounds on **f**, such as those in Theorem 6-5, can be employed to eliminate suboptimal actions. Extrapolation is another use for bounds. For example, suppose $\mathbf{v}^0$, $\mathbf{v}^1$, ... have been generated by a kth-order algorithm and the upper and lower bounds $\mathbf{v}^+$ and $\mathbf{v}^-$ in (6-55) have been computed with $\mathbf{v} = \mathbf{v}^i$:

$$\mathbf{v}^{i+} = \mathbf{L}\mathbf{v}^i + \frac{\beta}{1-\beta}\,\mathbf{e}(\mathbf{L}\mathbf{v}^i - \mathbf{v}^i)^M$$

$$\mathbf{v}^{i-} = \mathbf{H}\mathbf{v}^i + \frac{\beta^{k+1}}{1-\beta}\,\mathbf{e}(\mathbf{L}\mathbf{v}^i - \mathbf{v}^i)_m$$

In a kth-order algorithm, $\mathbf{v}^{i+1} = \mathbf{H}\mathbf{v}^i$. However, the average of the bounds, $(\mathbf{v}^{i+} + \mathbf{v}^{i-})/2$, may converge to $\mathbf{f}$ more quickly than $\mathbf{H}\mathbf{v}^i$.

The following rule is commonly used to decide when to halt computations. Choose the parameter $\epsilon > 0$ and stop computations at the smallest i for which $\| \mathbf{v}^{i+1} - \mathbf{v}^i \| \le \epsilon$. Then the issue is whether the algorithm where $\mathbf{v}^{i+1} = \mathbf{H}\mathbf{v}^i$ involves less computation, until stopping (given ϵ and $\mathbf{v}^0$), than the algorithm where $\mathbf{v}^{i+1} = (\mathbf{v}^{i+} + \mathbf{v}^{i-})/2$.

The kth-order algorithm has the form

$$\mathbf{v}^{i+1} = \mathbf{H}\mathbf{v}^i + \mathbf{c}^i$$

with $\mathbf{c}^i = \mathbf{0}$ for each i. The algorithm where $\mathbf{v}^{i+1} = (\mathbf{v}^{i+} + \mathbf{v}^{i-})/2$ has the same form with

$$\mathbf{c}^i = \frac{\mathbf{L}\mathbf{v}^i - \mathbf{H}\mathbf{v}^i}{2} + \frac{\mathbf{e}\beta^{k+1}(\mathbf{L}\mathbf{v}^i - \mathbf{v}^i)_m + \mathbf{e}\beta^2(\mathbf{L}\mathbf{v}^i - \mathbf{v}^i)^M}{2(1-\beta)}$$

The correction vector $\mathbf{c}^i$ is called an *extrapolation* of $\mathbf{H}\mathbf{v}^i$. There are many conceivable ways in which to choose an extrapolation. Recent research concerns (1) how extrapolations should be chosen to reduce the total computational effort, and (2) which extrapolations yield properly convergent algorithms in the sense that $\| \mathbf{f} - \mathbf{v}^i \| \to 0$. See Porteus (1971), Porteus (1980*b*), Schweitzer and Federgruen (1979), and Shin (1980) for some recent research results and further references concerning extrapolations.

Proof† of Theorem 6-3

Theorem 6-3 Suppose $\#\mathscr{C} < \infty$, $v^0 \in \mathbb{R}^S$, and $k \in I \cup \{\infty\}$. Generate $\mathbf{v}^0, \mathbf{v}^1, \mathbf{v}^2, \ldots$ with (*A*) through (*D*) and

$$\mathbf{v}^{i+1} = \sum_{l=0}^{k-1} Q_{i+1}^l \mathbf{r}_{i+1} + Q_{i+1}^k \mathbf{L}\mathbf{v}^i \qquad i \in I_+ \tag{6-57}$$

replacing (6-38).

(*a*) Then

$$\mathbf{f} = \lim_{i\to\infty} \mathbf{v}^i \tag{6-58}$$

† This proof follows unpublished material by U. Rothblum which is summarized in Rothblum (1979).

(*b*) If $\mathbf{Lv}^0 \geq \mathbf{v}^0$, then

$$\mathbf{v}^0 \leq \mathbf{v}^1 \leq \mathbf{v}^2 \leq \cdots \to \mathbf{f} \tag{6-59}$$

and there is $K < \infty$ such that

$$\| \mathbf{f} - \mathbf{v}^i \| \leq \beta^i K \tag{6-60}$$

The proof of part (*a*) follows from letting $i \to \infty$ in†

$$-\frac{\beta^{i(k+1)}}{1-\beta}\mathbf{y} \leq \mathbf{f} - \mathbf{v}^i \leq \frac{(i+1)\beta^i}{1-\beta}\mathbf{x} \tag{6-61}$$

where $\mathbf{x}$ and $\mathbf{y}$ remain to be specified and inequality (6-61) has yet to be established. Let

$$x = \max \{[\mathbf{Lv}^0(s) - v_s^0]^+ : s \in \mathcal{S}\}$$

$$y = \max \{[v_s^0 - \mathbf{Lv}^0(s)]^+ : s \in \mathcal{S}\}$$

and $\mathbf{x} = (x, x, \ldots, x) \in \mathbb{R}^S$ and $\mathbf{y} = (y, y, \ldots, y) \in \mathbb{R}^S$. Then $\mathbf{x} \geq \mathbf{0}$, $\mathbf{y} \geq \mathbf{0}$, and

$$\mathbf{Lv}^0 \leq \mathbf{v}^0 + \mathbf{x} \qquad \mathbf{v}^0 \leq \mathbf{Lv}^0 + \mathbf{y} \tag{6-62}$$

In order to prove (6-61), we use the following lemma.

Lemma 6-3

$$\mathbf{Lv}^{i+1} - \frac{1-\beta^k}{1-\beta}\beta^{(k+1)i-k}\mathbf{y} \leq \mathbf{v}^i \leq \mathbf{Lv}^i + \beta^{(k+1)i}\mathbf{y} \tag{6-63}$$

with the right side valid for all $i \in I$ and the left side valid for all integers $i \geq 2$.

PROOF We initiate an inductive proof of the upper bound with $i = 0$, in which case the inequality is trivially true. Suppose the inequality is valid for some $i \in I_+$.

Since $\#\mathscr{C} < \infty$, there exists δ_{i+1} such that

$$\mathbf{Lv}^i = \mathbf{r}_{i+1} + Q_{i+1}\mathbf{v}^i$$

Let $\mathbf{W}$ denote the mapping from $\mathbb{R}^S$ to $\mathbb{R}^S$, which specifies

$$\mathbf{Wv} = \mathbf{r}_{i+1} + Q_{i+1}\mathbf{v}$$

Then

$$\mathbf{v}^{i+1} = \mathbf{H}v^i = \mathbf{W}^{k+1}\mathbf{v}^i$$

† The upper bound in (6-61) tends to zero as $i \to \infty$ because $i\beta^i \to 0$ due to $(i+1)\beta^{i+1}/(i\beta^i) < 1$ if $i > \beta/(1-\beta)$.

In the terminology of Section 6-3, **W** is a monotone contraction mapping on $\mathbb{R}^S$. The inductive hypothesis satisfies the assumption of (5-74)† with $v = \mathbf{v}^i$, $c = \beta$, and $z = \beta^{(k+1)i}y$ Therefore,

$$\begin{aligned}\mathbf{v}^{i+1} &= \mathbf{W}^{k+1}\mathbf{v}^i \le \mathbf{W}^{k+2}\mathbf{v}^i + \beta^{k+1}\beta^{(k+1)i}\mathbf{y} \\ &= \mathbf{W}\mathbf{W}^{k+1}\mathbf{v}^i + \beta^{(k+1)(i+1)}\mathbf{y} = \mathbf{W}\mathbf{v}^{i+1} + \beta^{(k+1)(i+1)}\mathbf{y} \\ &\le \mathbf{L}\mathbf{v}^{i+1} + \beta^{(k+1)(i+1)}\mathbf{y}\end{aligned}$$

which establishes the right inequality in (6-63) for $i + 1$, hence for all i.

In order to prove the left inequality, observe that the right inequality satisfies the assumption of (5-75)‡ with $v = \mathbf{v}^i$, $c = \beta$, and $z = \beta^{(k+1)i}\mathbf{y}$. Therefore,

$$\begin{aligned}\mathbf{v}^{i+1} &= \mathbf{W}^{k+1}\mathbf{v}^i \ge \mathbf{W}\mathbf{v}^i - \sum_{l=1}^{k} \beta^l \beta^{(k+1)i}\mathbf{y} \\ &= \mathbf{L}\mathbf{v}^i - \frac{1-\beta^k}{1-\beta}\beta^{(k+1)i+1}\mathbf{y}\end{aligned}$$

which proves the left inequality in (6-63). □

In order to prove $\mathbf{v}^i \to \mathbf{f}$, we establish the inequality (6-61). The upper bound in (6-63) satisfies the hypothesis of (5-73)§ with $v = \mathbf{v}^{i+1}$, $\mathbf{W} = \mathbf{L}$, and $z = \beta^{(k+1)i}\mathbf{y}$. Therefore, $k = 0$ in (5-73) yields

$$\mathbf{v}^i \le \mathbf{f} + \frac{1}{1-\beta}\beta^{(k+1)i}\mathbf{y}$$

which results in the lower bound in (6-61).

For the upper bound in (6-61), observe that the lower bound in (6-63) implies that the hypothesis of (5-71)‖ is satisfied with $u = \mathbf{L}\mathbf{v}^{m+j}$, $v = \mathbf{v}^{m+j+1}$, $c = \beta^{i-j-1}$, $z = [\beta^{(m+j)(k+1)+1}/(1-\beta)]x$, and $\mathbf{W} = \mathbf{L}^{i-j-1}$ for any $i \in I_+$ and $j \in I$ such that $j < i$. Therefore,

$$\begin{aligned}\mathbf{L}^{i-j}\mathbf{v}^{m+j} &\le \mathbf{L}^{i-j-1}\mathbf{v}^{m+j+1} + \beta^{i-j-1}\frac{\beta^{(m+j)(k+1)+1}}{1-\beta}\mathbf{x} \\ &= \mathbf{L}^{i-j-1}\mathbf{v}^{m+j+1} + \frac{\beta^{(m+j)(k+1)+i-j}}{1-\beta}\mathbf{x} \\ &\le \mathbf{L}^{i-j-1}\mathbf{v}^{m+j+1} + \frac{\beta^{m(k+1)+i}}{1-\beta}\mathbf{x}\end{aligned}$$

† Property (5-74) asserts $\mathbf{v} \le \mathbf{W}\mathbf{v} + z\mathbf{e} \Rightarrow \mathbf{W}^k\mathbf{v} \le \mathbf{W}^{k+1}\mathbf{v} + c^k z\mathbf{e}$ where **W** is a monotone contraction mapping with modulus c and $z \ge 0$.

‡ Property (5-75) asserts $\mathbf{v} \le \mathbf{W}\mathbf{v} + z\mathbf{e} \Rightarrow \mathbf{W}\mathbf{v} \le \mathbf{W}^k\mathbf{v} + \sum_{l=1}^{k-1} c^l z\mathbf{e}$ where **W** is a monotone contraction mapping with modulus c and $z \ge 0$.

§ Property (5-73) asserts $\mathbf{v} \le \mathbf{W}\mathbf{v} + z\mathbf{e} \Rightarrow \mathbf{W}^k\mathbf{v} \le \mathbf{w} + [c^k/(1-c)]z\mathbf{e}$ where **w** is the fixed point of the monotone contraction mapping **W** with modulus c and $z \ge 0$.

‖ Property (5-71) asserts $\mathbf{u} \le \mathbf{v} + z\mathbf{e} \Rightarrow \mathbf{W}\mathbf{u} \le \mathbf{W}\mathbf{v} + cz\mathbf{e}$ where **W** is a monotone contraction mapping with modulus c and $z \ge 0$.

Therefore

$$\sum_{j=0}^{i-1}(\mathbf{L}^{i-j}\mathbf{v}^{m+j} - \mathbf{L}^{i-j-1}\mathbf{v}^{m+j+1}) \leq \sum_{j=0}^{i-1} \frac{\beta^{m(k+1)+i}}{1-\beta}\,\mathbf{x}$$

so

$$\mathbf{L}^i\mathbf{v}^m \leq \mathbf{v}^{i+m} + \frac{i\beta^{m(k+1)+i}}{1-\beta}\,\mathbf{x} \tag{6-64}$$

From (5-72)† if $\mathbf{L}\mathbf{v}^m \leq \mathbf{v}^m + \mathbf{z}$, then

$$\mathbf{f} \leq \mathbf{L}^i\mathbf{v}^m + \frac{\beta^i}{1-\beta}\,\mathbf{z}$$

which implies, with (6-64), that

$$\mathbf{f} \leq \mathbf{v}^{m+i} + \frac{i\beta^{m(k+1)+i}}{1-\beta}\,\mathbf{x} + \frac{\beta^i}{1-\beta}\,\mathbf{z}$$

Use $\mathbf{L}\mathbf{v}^0 \leq \mathbf{v}^0 + \mathbf{x}$ to substitute $\mathbf{x}$ for $\mathbf{z}$ and let $m = 0$ to obtain

$$\mathbf{f} - \mathbf{v}^i \leq \frac{(i+1)\beta^i}{1-\beta}\,\mathbf{x}$$

which is the upper bound in (6-61).

The hypothesis of part (*b*) of the theorem is $\mathbf{L}\mathbf{v}^0 \geq \mathbf{v}^0$, and the consequences to be proved are (6-59) and (6-60) repeated here:

$$\mathbf{v}_0 \leq \mathbf{v}^1 \leq \mathbf{v}^2 \leq \cdots \tag{6-59}$$

$$\|\mathbf{f} - \mathbf{v}^i\| \leq \beta^i K \qquad \text{for some } K < \infty \tag{6-60}$$

Recall the notation

$$\mathbf{G}_\delta \mathbf{v} \triangleq \sum_{l=0}^{k-1} Q_\delta^l \mathbf{r}_\delta + Q_\delta^k \mathbf{v}$$

$$\mathbf{J}\mathbf{v} \triangleq \text{vmax}\,\{\mathbf{G}_\delta \mathbf{v}\colon \delta \in \Delta\}$$

If $\gamma \in \Delta$ is maximal for $\mathbf{L}\mathbf{v}$, that is,

$$\mathbf{r}_\gamma + Q_\gamma \mathbf{v} = \text{vmax}\,\{\mathbf{r}_\delta + Q_\delta \mathbf{v}\colon \delta \in \Delta\} \triangleq \mathbf{L}\mathbf{v} \tag{6-65}$$

then $\mathbf{H}\mathbf{v} = \mathbf{G}_\gamma \mathbf{v}$. Let $\mathbf{L}^1 = \mathbf{L}$, $\mathbf{H}^1 = \mathbf{H}$, $\mathbf{J}^1 = \mathbf{J}$, $\mathbf{H}^{i+1} = \mathbf{H}\mathbf{H}^i$, $\mathbf{J}^{i+1} = \mathbf{J}\mathbf{J}^i$, $\mathbf{L}^{i+1} = \mathbf{L}\mathbf{L}^i$, $\mathbf{V}_k^i = \mathbf{H}^i\mathbf{v}^0$, $\mathbf{U}_k^i = \mathbf{J}^i\mathbf{v}^0$, $\mathbf{V}_0^i = \mathbf{L}^i\mathbf{v}^0$, and $\theta = \{\mathbf{v}\colon \mathbf{L}\mathbf{v} \geq \mathbf{v},\ \mathbf{v} \in \mathbb{R}^S\}$. The major steps in the proof are (6-43) through (6-45):

$$\|\mathbf{f} - \mathbf{V}_0^i\| \leq K_1\beta^i \qquad \text{and} \qquad \|\mathbf{f} - U_k^i\| \leq K_2\,\beta^{i(k+1)} \tag{6-43}$$

$$\mathbf{v}^0 \leq \mathbf{L}\mathbf{v}^0 \Rightarrow \mathbf{V}_k^i \leq \mathbf{V}_k^{i+1} \qquad \text{for all } i \tag{6-44}$$

$$\mathbf{v}^0 \leq \mathbf{L}\mathbf{v}^0 \Rightarrow \mathbf{V}_0^i \leq \mathbf{V}_k^i \leq \mathbf{U}_k^i \qquad \text{for all } i \tag{6-45}$$

† Property (5-72) asserts $\mathbf{W}\mathbf{v} \leq \mathbf{v} + z\mathbf{e} \Rightarrow \mathbf{w} \leq \mathbf{W}^k\mathbf{v} + [c^k/(1-c)]z\mathbf{e}$ where $\mathbf{w}$ is the fixed point of the monotone contraction mapping $\mathbf{W}$ with modulus c and $z \geq 0$.

The first part of (6-43), geometric convergence of v_0^n, is asserted by Theorem 4-8 with $K_1 = \| \mathbf{Lv}^0 - \mathbf{v}^0 \| / (1 - \beta)$. The analogous statement for $\mathbf{U}_k^n$ depends on the following generalization of (4-87) whose proof (see Exercise 6-3) and consequences are similar to those of (4-87).

Lemma 6-4 (*a*) $\| \mathbf{Jv} - \mathbf{Jw} \| \le \beta^{k+1} \| \mathbf{v} - \mathbf{w} \|$.

(*b*) **f** is the unique fixed point of **J**; that is, **f** is the unique solution of $\mathbf{Jf} = \mathbf{f}$. Therefore, (6-43) is valid with $K_2 = \| J\mathbf{v}^0 - \mathbf{v}^0 \| / (1 - \beta^{k+1})$.

The proof of monotonicity, i.e., (6-44), has two parts. The first shows $\mathbf{Hv} \ge \mathbf{v}$ if $\mathbf{v} \in \theta$ and the second shows $\mathbf{Hv} \in \theta$ if $\mathbf{v} \in \theta$. These parts permit inductive proof of (6-44) starting with $\mathbf{Lv}^0 \ge \mathbf{v}^0$, that is, $\mathbf{v}^0 \in \theta$.

From Lemma 6-2,

$$\mathbf{Hv} = \mathbf{Lv} + \sum_{l=1}^{k} Q_\gamma^l (\mathbf{Lv} - \mathbf{v}) \tag{6-66}$$

Therefore, $\mathbf{Hv} \ge \mathbf{Lv} \ge \mathbf{0}$ if $\mathbf{v} \in \theta$.

The following property completes the proof of (6-44).

Lemma 6-5 $\mathbf{v} \in \theta \Rightarrow \mathbf{Hv} \in \theta$.

Proof For any $\mathbf{v} \in \mathbb{R}^S$ and $\mathbf{w} \in \mathbb{R}^S$, using the notation of (6-65), Lemma 6-1 asserts

$$\mathbf{Lw} - \mathbf{Lv} \ge Q_\gamma(\mathbf{w} - \mathbf{v})$$

so

$$\mathbf{Lw} - \mathbf{w} \ge \mathbf{Lv} - \mathbf{v} + (Q_\gamma - I)(\mathbf{w} - \mathbf{v})$$

Replace **w** with **Hv** and substitute (6-66) to obtain

$$\begin{aligned}
\mathbf{LHv} - \mathbf{Hv} &\ge \mathbf{Lv} - \mathbf{v} + (Q_\gamma - I)\left[\mathbf{Lv} + \sum_{i=1}^{k} Q_\gamma^l(\mathbf{Lv} - \mathbf{v}) - \mathbf{v}\right] \\
&= \left[I + (Q_\gamma - I)\left(I + \sum_{l=1}^{k+1} Q_\gamma^l\right)\right](\mathbf{Lv} - \mathbf{v}) \\
&= \left(I + \sum_{l=1}^{k+1} Q_\gamma^l - I - \sum_{l=1}^{k} Q_\gamma^l\right)(\mathbf{Lv} - \mathbf{v}) \\
&= Q_\gamma^{k+1}(\mathbf{Lv} - \mathbf{v})
\end{aligned}$$

Therefore, $\mathbf{v} \in \theta$ and $Q_\gamma \ge 0$ imply $\mathbf{LHv} - \mathbf{Hv} \ge \mathbf{0}$ so $\mathbf{Hv} \in W$. □

As a consequence of (6-66) and Lemma 6-5, $\mathbf{v}^0 \in \theta$ implies $\mathbf{H}^i\mathbf{v}^0 \in \theta$ for all i; so $\mathbf{H}^{i+1}\mathbf{v}^0 = \mathbf{HH}^i\mathbf{v}^0 \ge \mathbf{LH}^i\mathbf{v}^0 \ge \mathbf{L}^{i+1}\mathbf{v}^0$, which is the lower inequality in (6-45).

From (6-65) and the definitions of **J** and **H**, if $\mathbf{v} \le \mathbf{w}$, then

$$\begin{aligned}\mathbf{Jw} &= \text{vmax}\,\{G_\delta \mathbf{w}: \delta \in \Delta\}\\ &\ge \text{vmax}\,\{\mathbf{G}_\delta \mathbf{v}: \delta \in \Delta\} \ge \mathbf{G}_\gamma \mathbf{v} = \mathbf{Hv}\end{aligned}$$

Then $\mathbf{V}_k^0 = \mathbf{U}_k^0$ initiates an inductive proof of the upper inequality in (6-45).

The proof is completed by using (6-43) and (6-45) to establish (6-42). From (6-45), $\|\mathbf{U}_k^i - \mathbf{V}_k^i\| \le \|\mathbf{U}_k^i - \mathbf{V}_0^i\|$ so

$$\begin{aligned}\|\mathbf{f} - \mathbf{v}^i\| &= \|\mathbf{f} - \mathbf{V}_k^i\| \le \|\mathbf{f} - \mathbf{U}_k^i\| + \|\mathbf{U}_k^i - \mathbf{V}_k^i\|\\ &\le \|\mathbf{f} - \mathbf{U}_k^i\| + \|\mathbf{U}_k^i - \mathbf{f}\| + \|\mathbf{f} - \mathbf{V}_k^i\|\\ &\le K_1\beta^i + 2K_2\,\beta^{i(k+1)} = \beta^i(K_1 + 2K_2\,\beta^{ik})\\ &\le \beta^i(K_1 + 2K_2)\end{aligned}$$

□

EXERCISES

6-1 Let $\delta_1, \delta_2, \ldots$ be a sequence of elements of Δ, let $\pi = (\delta_1, \delta_2, \ldots)$, and extend (6-38) and (6-39) to

$$\mathbf{v}(\pi) = \sum_{i=1}^{\infty} \beta^{i-1}\left(\prod_{j=0}^{i-1} P_j\right)\mathbf{r}_j$$

where $P_0 \triangleq \mathbf{I}$. For $\gamma \in \Delta$,

$$\mathbf{v}(T_\gamma \pi) = \mathbf{r}_\gamma + \beta P_\gamma \mathbf{v}(\pi)$$

Suppose $\mathbf{v}(\pi) \le \mathbf{v}(T_\gamma \pi)$ [that is, $v_s(\pi) \le v_s(T_\gamma \pi)$ for all $s \in \mathcal{S}$]. Prove $\mathbf{v}(\pi) \le \mathbf{v}(T_\gamma^k \pi) \le \mathbf{v}(\gamma)$ for all $k \in I_+$.

6-2 In notation similar to (4-86), let $\mathbf{L}'$ be the mapping from $\mathbb{R}^S$ to $\mathbb{R}^S$ such that $\mathbf{w} \in \mathbb{R}^S$ causes $\mathbf{L}'\mathbf{w}$ to have (6-47) as the sth coordinate. Let $\|\mathbf{w}\|$ denote max $\{|w_s|: s \in \mathcal{S}\}$. Prove $\|\mathbf{L}'\mathbf{v} - \mathbf{Lw}\| \le \beta\|\mathbf{v} - \mathbf{w}\|$ for all $\mathbf{v} \in \mathbb{R}^S$ and $\mathbf{w} \in \mathbb{R}^S$.

6-3 Prove Lemma 6-4 (in the proof of Theorem 6-3).

6-3* MULTIPLE CRITERIA

In the professional practice of operations research, a phenomenon which merits a model usually is evaluated with several criteria. Criteria often are termed *measures of effectiveness*, or MOEs. The prospect of analyzing the trade-offs among the MOEs is a major motivation to construct a model. In inventory models, for example, we usually wish to analyze the trade-offs among ordering costs, holding costs, and stockout frequency.

One approach to the analysis of trade-offs among MOEs is to attach a cost or benefit to *all* MOEs (the major ones) and then to optimize the sum of the benefit MOEs minus the cost MOEs. The difficulty with this approach is that different MOEs may be measured in incommensurate units. For example, in an inventory model we might include a stockout cost in the objective and forgo a constraint related to stockout frequency. But what is the cost per unit of unsatisfied demand? It might include both lost unit profit and the adverse impact on subsequent profit due to a reduction in patronage by disappointed customers. How does one measure the latter effect (in monetary units)? One resolution of this difficulty is to analyze the sensitivity of the solution and its cost to variations in the parameters which appear in the optimization objective.

Another approach to the analysis of such trade-offs is to choose a linear combination of some MOEs as the optimization criterion and to constrain other MOEs. In an inventory model, for example, we might minimize the sum of expected ordering and holding costs subject to an upper bound on stockout probability. We would then analyze the sensitivity of the solution and its cost to variations in cost parameters and the upper bound.

A third approach to the analysis of trade-offs among MOEs is to "optimize" several MOEs simultaneously. "Optimize" is used in the following sense. For a function $\mathbf{f}$ from some domain Z to $\mathbb{R}^m$ let $\mathbf{f}(z) = [f_i(z); i = 1, \ldots, m]$.

Definition 6-2 Let $\mathbf{f}$ be a function from Z to $\mathbb{R}^m$ $(m < \infty)$. Then $z^* \in Z$ is a *Pareto optimum* (also called *admissible, undominated, or efficient*), and written PO, if, for all $z \in Z$, either

(*a*) $f_i(z^*) \geq f_i(z)$, $i = 1, \ldots, m$; or

(*b*) $\mathbf{f}(z^*) \not\geq \mathbf{f}(z)$ but $f_i(z^*) > f_i(z)$ for some i.

If z^* is PO, then for each $z \in Z$, if there is an i with $f_i(z) > f_i(z^*)$, there is a j with $f_j(z^*) > f_j(z)$. Hence, if z^* is PO, then $\mathbf{f}(z^*)$ cannot be bettered along some dimension (the ith) without being damaged along some other dimension (the jth). The third approach to the analysis of trade-offs among several MOEs is to form the vector-valued criterion $\mathbf{f}(\cdot)$, where $f_i(\cdot)$ is the numerical value of the ith MOE. Then one characterizes the set of Pareto optima. We shall see that the second and third approaches are equivalent.

This section focuses on the second and third approaches when the underlying model is an MDP. We consider discounted MDPs in which "discounted constraints" are appended to the usual problem. Then we embed the search for POs in the ordinal framework discussed on pages 150 to 152.

Measures of Effectiveness as Constraints

Let

$$\alpha_s > 0 \qquad s \in \mathcal{S} \qquad \text{and } \sum_{s \in \mathcal{S}} \alpha_s = 1 \tag{6-67}$$

$$\delta_{sj} = \begin{cases} 1 & \text{if } s = j \\ 0 & \text{if } s \neq j \end{cases} \tag{6-68}$$

and let Z denote the set of solutions $(\mathbf{z}, \mathbf{c})$ to

$$z_{sa} \geq 0 \qquad (s, a) \in \mathscr{C}$$

$$\sum_{(j,\,a)\in\mathscr{C}} (\delta_{sj} - \beta p_{js}^{a}) z_{ja} = \alpha_s \qquad s \in \mathcal{S} \tag{6-69}$$

$$\sum_{(s,\,a)\in\mathscr{C}} r_q(s, a) z_{sa} \geq c_q \qquad q = 1, \ldots, Q$$

The set Z can be interpreted in several ways. The problem of maximizing

$$E\left[\sum_{n=1}^{\infty} \beta^{n-1} r_k(s_n, a_n)\right] \tag{6-70}$$

for some fixed $k \in \mathscr{Q} \triangleq \{1, \ldots, Q\}$, is equivalent to the linear program

$$\text{Maximize } c_k \tag{6-71}$$

$$\text{subject to } (\mathbf{z}, \mathbf{c}) \in Z \tag{6-72}$$

The remainder of this subsection leads to additional interpretations of Z.

From Definition 6-2 of PO (Pareto optimality), $(\mathbf{z}^*, \mathbf{c}^*) \in Z$ is PO if there is not any $(\mathbf{z}, \mathbf{c}) \in Z$ with $\mathbf{c} \geq \mathbf{c}^*$ and $\mathbf{c} \neq \mathbf{c}^*$. The following result is an immediate consequence of this equivalent definition of PO.

Theorem 6-6 Consider the linear program

$$\begin{aligned} &\text{Maximize } c_k \\ &\text{subject to } (\mathbf{z}, \mathbf{c}) \in Z \\ &\text{and } c_q \geq c_q^0 \qquad q \in \mathscr{Q} - \{k\} \end{aligned} \tag{6-73}$$

where $k \in \mathscr{Q}$ is fixed and c_q^0 are numbers, $q \neq k$. Suppose (i) $(\mathbf{z}^*, \mathbf{c}^*)$ is optimal for this linear program and (ii) all $Q - 1$ constraints $c_q \geq c_q^0$, $q \neq k$, are binding. Then $(\mathbf{z}^*, \mathbf{c}^*)$ is PO.

From (6-67) and Theorem 6-1, the set of randomized stationary policies is one-to-one with the set $\{\mathbf{z}\colon (\mathbf{z}, \mathbf{c}) \in Z \text{ for some } \mathbf{c}\}$. If $(\mathbf{z}, \mathbf{c}) \in Z$, the construction which specifies the randomized stationary policy corresponding to $\mathbf{z}$ is

$$D_{sa} \triangleq P\{a_n = a \mid s_n = s\} = \frac{z_{sa}}{\sum_{k\in A_s} z_{sk}} \qquad (s, a) \in \mathscr{C} \tag{6-74}$$

The following result is an immediate consequence of the one-to-one relation.

Theorem 6-7 If $(\mathbf{z}, \mathbf{c})$ is PO in Z, then (6-74) yields a policy which is PO among randomized stationary policies.

Let $(\mathbf{z}, c_k')$ be optimal in (6-73) for $\mathbf{c}^0 = (c_q^0, q \neq k)$ with all components of $\mathbf{c}^0$ being associated with binding constraints. Then $\mathbf{z}$ would be part of an optimal solution too for all $Q - 1$ linear programs

$$\begin{aligned}&\text{Maximize } c_i\\&\text{subject to } (\mathbf{z}, \mathbf{c}) \in Z \text{ and}\\&c_q \geq c_q^0 \qquad q \neq k, i\\&c_k \geq c_k'\end{aligned}$$

where $i \in \mathscr{Q} \sim \{k\}$ is fixed.

It would seem possible to iterate this argument to conclude that $\mathbf{z}$ would be part of an optimal solution to

$$\begin{aligned}&\text{Maximize } \sum_{q \neq k} \lambda_q c_q\\&\text{subject to } (\mathbf{z}, \mathbf{c}) \in Z \text{ and}\\&c_k \geq c_k'.\end{aligned}$$

for some $\lambda_1 > 0, \ldots, \lambda_Q > 0$. A stronger result is a corollary of the following property of multiple-objective linear programs which we state without proof.†

Theorem 6-8 Let A, B, and $\mathbf{d}$ be given $m \times n$, $Q \times 1$, and $m \times 1$ matrices, respectively. Then $(\mathbf{z}^*, \mathbf{c}^*)$ is PO in the set of solutions of

$$A\mathbf{z} + B\mathbf{c} \geq \mathbf{d}$$

if, and only if, there is a row vector $\boldsymbol{\lambda} = (\lambda_1, \ldots, \lambda_Q)$ with

$$\lambda_q > 0 \text{ for } q = 1, \ldots, Q \qquad \text{and} \qquad \sum_{q=1}^{Q} \lambda_q = 1 \tag{6-75}$$

such that $(\mathbf{z}^*, \mathbf{c}^*)$ is an optimal solution of the linear program

$$\begin{aligned}&\text{Maximize } \boldsymbol{\lambda}\mathbf{c}\\&\text{subject to } A\mathbf{z} + B\mathbf{c} \geq \mathbf{d}\end{aligned}$$

The next result is an immediate consequence of Theorem 6-8.

Corollary 6-8 $\{D_{sa}: (s, a) \in \mathscr{C}\}$ is PO among randomized stationary policies if, and only if, (6-74) is valid for a $\mathbf{z}$ which is part of an optimal solution of

$$\begin{aligned}&\text{Maximize } \boldsymbol{\lambda}\mathbf{c}\\&\text{subject to } (\mathbf{z}, \mathbf{c}) \in Z\end{aligned} \tag{6-76}$$

where $\boldsymbol{\lambda}$ satisfies (6-75).

† J. P. Evans and R. E. Steur, "A Revised Simplex Method for Linear Multiple Objective Programming," *Math. Programming* **5**: 54–72 (1973). H. Isermann, "Proper Efficiency and the Linear Vector Maximum Problem," *Oper. Res.* **22**: 189–191 (1974).

We now have two procedures to generate PO policies. In order to exploit Theorem 6-6, first use any algorithm in Sections 4-5 or 6-1 to maximize

$$E\left[\sum_{n=1}^{\infty} \beta^{n-1} r_k(s_n, a_n)\right]$$

for any specific $k \in \mathscr{Q}$. An optimal policy is associated with an optimal $\mathbf{z}$ in linear program (6-73) when

$$c_q = \sum_{(s,a) \in \mathscr{C}} r_q(s, a) z_{sa} \qquad \text{for } q = 1, \ldots, Q$$

$$c_q^0 = c_q \qquad q \neq k$$

Now increase each c_q^0 until the qth constraint is binding.

Another procedure to obtain a PO policy is to choose any $\boldsymbol{\lambda}$ which satisfies (6-75) and then solve (6-76). Here too, we emphasize that the procedure can be divided into two stages where any algorithm can be used for the first stage. First, substitute $\mathbf{e}_k$ (the kth unit vector) for $\boldsymbol{\lambda}$ in (6-76) [this violates (6-75)], which is then equivalent to maximizing (6-77), which may be accomplished, for example, with policy improvement. Identify the feasible $\mathbf{z}$ in (6-76) associated with an optimal solution and thereafter pivot to an $(\mathbf{z}, \mathbf{c})$ which is optimal in (6-76) for given $\boldsymbol{\lambda}$. This may be accomplished, for example, with standard parametric programming procedures.

The linear program (6-76) has $S + Q$ constraints [see (6-69) for the constraints]. An optimal basic solution has at most $S + Q$ positive variables, and for each s there is at least one $a \in A_s$ with $z_{sa} > 0$ [see the discussion below (6-8)]. However, we may assume $r_q(s, a) > 0$ for all $q \in \mathscr{Q}$ and $(s, a) \in \mathscr{C}$ without loss of generality. Therefore, if $(\mathbf{z}, \mathbf{c})$ is PO for Z, $c_q > 0$ for all $q \in \mathscr{Q}$ so $c_1, \ldots, c_Q$ are among the positive variables in an optimal basic solution of (6-76). As a result, every optimal basic solution $(\mathbf{z}, \mathbf{c})$ of (6-76) with $\boldsymbol{\lambda}$ satisfying (6-75) has $\mathbf{z}$ corresponding [in the sense of (6-74)] to an unrandomized policy. The simplex algorithm and its variants examine only basic solutions.

Example 6-7: Continuation of Examples 4-8, 4-20, 4-22, and 4-23 The derived form of this inventory model has $\beta = 0.9$, single-period expected costs, and transition probabilities in Tables 4-7 and 4-8 repeated below as Tables 6-6 and 6-7, $\mathcal{S} = \{0, 1, \ldots, 4\}$, and $A_s = \{s, s + 1, \ldots, 4\}$ for each

Table 6-6 Single-period costs, $r(s, a)$

	s \ a (Inventory after ordering)	0	1	2	3	4
Inventory before ordering	0	0	0	0	1	3
	1	—	−1	0	1	3
	2	—	—	−1	1	3
	3	—	—	—	0	3
	4	—	—	—	—	2

Table 6-7 β × transition probabilities, βp_{sj}^a

Subsequent inventory level

a \ j	0	1	2	3	4
0	0.9	0	0	0	0
1	0.675	0.225	0	0	0
2	0.45	0.225	0.225	0	0
3	0.225	0.225	0.225	0.225	0
4	0	0.225	0.225	0.225	0.225

$s \in \mathcal{S}$. Here, $s \in \mathcal{S}$ is the inventory level at the beginning of a period before additional goods are ordered, and $a \in A_s$ is the inventory level after ordered goods, if any, are received. The data in Table 6-7 are based on a "lost sales" assumption, that is, $s_{n+1} = (a_n - D_n)^+$, where D_n denotes the quantity demanded in the nth period, and $P\{D_n = i\} = 0.25$ for $i = 0$, 1, 2, and 3. We assume that $D_1, D_2, \ldots$ are independent and identically distributed random variables.

Table 4-16 presents the system of constraints which is dual to the equality constraints in (6-69).

We shall let $Q = 2$ where $r_1(s, a)$ denotes the single-period costs in Table 6-6. The second criterion is the expected discounted frequency demand exceeds supply:

$$\begin{aligned}\sum_{n=1}^{\infty} \beta^{n-1} P\{D_n > a_n\} &= \sum_{n=1}^{\infty} \beta^{n-1} \sum_{i=0}^{4} P\{D_n > s_n \mid s_n = i\}\; P\{s_n = i\} \\ &= \sum_{n=1}^{\infty} \beta^{n-1} \sum_{i=0}^{4} P\{D_n > i\}\; P\{s_n = i\} \\ &= \sum_{i=0}^{4} P\{D_1 > i\} \sum_{n=1}^{\infty} \beta^{n-1}\; P\{s_n = i\}\end{aligned}$$

because $D_1, D_2, \ldots, D_n, \ldots$ all have the same distribution. Theorem 6-1 implies that a feasible $\{z_{sa}\}$ in (6-67) has the property

$$\sum_{n=1}^{\infty} \beta^{n-1} P\{s_n = i\} = \sum_{a \in A_i} z_{ia}$$

if $\alpha_s > 0$ for all $s \in S$ and $\sum_{s \in \mathcal{S}} \alpha_s = 1$. Assuming so,

$$\sum_{n=1}^{\infty} \beta^{n-1} P\{D_n > s_n\} = \sum_{i=0}^{4} P\{D_1 > i\} \sum_{a \in A_i} z_{ia}$$

Here, $P\{D_1 > 0\} = 0.75$, $P\{D_1 > 1\} = 0.5$, $P\{D_1 > 2\} = 0.25$, and $P\{D_1 > 3\} = P\{D_1 > 4\} = 0$; so

$$\sum_{n=1}^{\infty} \beta^{n-1} P\{D_n > s_n\} = 0.75 \sum_{a=0}^{4} z_{0a} + 0.5 \sum_{a=1}^{4} z_{1a} + 0.25 \sum_{a=2}^{4} z_{2a}$$

Table 6-8 Constraints for linear program (6-76) in Example 6-7

Variable																	
z_{00}	z_{01}	z_{02}	z_{03}	z_{04}	z_{11}	z_{12}	z_{13}	z_{14}	z_{22}	z_{23}	z_{24}	z_{33}	z_{34}	z_{44}	c_1	c_2	
0.1	0.325	0.55	0.775	1	−0.675	−0.45	−0.225		−0.45	−0.225							$=0.2$
	−0.255	−0.225	−0.225	−0.225	0.775	0.775	0.775	0.775	−0.225	−0.225	−0.225	−0.225	−0.225	−0.225			$=0.2$
		−0.225	−0.225	−0.225	−0.225	−0.225	−0.225	−0.225	0.775	0.775	0.775	−0.225	−0.225	−0.225			$=0.2$
			−0.225	−0.225			−0.225	−0.225		−0.225	−0.225	−0.225	0.725	−0.225			$=0.2$
				−0.225				−0.225			−0.225	0.775	−0.225	0.775			$=0.2$
			1	3	−1		1	3	−1	1	3		3	2	−1		≤ -10
0.75	0.5	0.25			0.5	0.25			0.25							−1	≤ 0

All variables nonnegative

Therefore, $r_2(0, a) = 0.75$, $r_2(1, a) = 0.5$, $r_2(2, a) = 0.25$, and $r_2(3, a) = r_2(4, a) = 0$.

Table 6-8 shows the constraints of linear program (6-76) for this problem. Note that the last two constraints in (6-76) have their inequalities reversed in Table 6-8 because both criteria measure undesirable characteristics (cost and excess demand). Also, c_1 in Table 6-8 represents c_1 in (6-76) plus 10 to ensure that $c_1 \geq 0$ in Table 6-8 is not binding. Ten is a sufficiently large constant because

$$\sum_{(s,a)\in\mathscr{C}} r_1(s, a)z_{sa} \geq \sum_{(s,a)\in\mathscr{C}} z_{sa} \min\{r_1(s, a): (s, a) \in \mathscr{C}\}$$
$$= -\sum_{(s,a)\in\mathscr{C}} z_{sa} = -(1-\beta)^{-1} = -10$$

Table 6-9 summarizes the basic optimal solutions of the linear program whose constraints are given in Table 6-8 and whose objective is to minimize $[\lambda c_1 + (1 - \lambda)c_2]$.

Of course, a linear-programming algorithm need not be used for most of the calculations. For example, let $\lambda = 0$ and use policy improvement to optimize the ordinary MDP with criterion $r_2(\cdot, \cdot)$. This yields the alternative optimal policies $a = 3$ or $a = 4$ from $s = 0, 1, 2, 3$ and $a = 4$ from $s = 4$. The optimal value of

$$E\left[\sum_{n=1}^{\infty} \beta^{n-1} r_2(s_n, a_n)\right]$$

is zero. Letting $\lambda = 0$ in the linear program objective yields only $a = 3$ from $s = 0, 1, 2, 3$ as optimal. This policy is PO, whereas, for example, $a = 4$ from $s = 0, 1, 2, 3$ is not PO. Also, the other two policies in Table 6-9 are PO.

□

Henig (1982) has shown that Corollary 6-8 is valid for a more general class of models than finite MDPs. Other discussions of algorithms for PO policies include Shin (1980), Viswanathan, Aggarwal, and Nair (1977), and White and Kim (1980).

Table 6-9 Parametric solution of the linear program in Example 6-7

	Optimal action‡				
Range of λ†	$s = 0$	$s = 1$	$s = 2$	$s = 3$	$s = 4$
$0 < \lambda \leq 0.12$	3	3	3	3	3
$0.12 < \lambda \leq 0.24$	3	3	2	3	4
$0.24 < \lambda < 1$	2	1	2	3	4

† Values of λ are shown to only two decimal places.

‡ The table does not show two changes of optimal basis which occur as λ traverses (0.243900, 0.243903).

Stationary Policies

The procedures thus far are confined to stationary policies. Now we show that the restriction is without loss of Pareto optimality. First, recall some binary relations terminology from Sections 2-4 and 4-3. Let D be a nonempty set and $B \subset D \times D$. Then (B, D) is *transitive* if $(a, b) \in B$ and $(b, c) \in B$ implies $(a, c) \in B$; (B, D) is *reflexive* if $(b, b) \in B$ for all $b \in D$; and B is a *preorder* and (B, D) is a *preordered set* if B is transitive and reflexive. A mapping f from D to D is *isotone* if $(a, b) \in B$ implies $[f(a), f(b)] \in B$.

We introduce a new definition and briefly review Sections 4-3 and 4-4.

Definition 6-3 If D is a nonempty set and $B \subset D \times D$, then $b \in D$ is B-maximal if $(b, c) \in B$ for all $c \in D$.

Thus B-maximality of b has two requirements. First, b must be comparable via B with all elements in D. Second, b must be at least as favorable via B as any element in D.

Let $\mathcal{S}$ be a countable or finite set of *states* and A_s a nonempty set of *actions* for each $s \in \mathcal{S}$. Let

$$\mathscr{C} = \{(s, a): a \in A_s, s \in \mathcal{S}\} \qquad \Delta = \mathop{\times}_{s \in \mathcal{S}} A_s \qquad \text{and} \qquad Y = \mathop{\times}_{n=1}^{\infty} \Delta \tag{6-77}$$

Elements of Δ are *single-stage decision rules* and elements of Y are *Markov policies.*

Let W be the set of probability distributions on $\mathcal{S}$ so $\mathbf{w} \in W \Leftrightarrow \mathbf{w} \geq \mathbf{0}$ and $\sum_{j \in \mathcal{S}} w(j) = 1$. For each $\mathbf{w} \in W$, $\delta \in \Delta$, and $j \in \mathcal{S}$ let

$$M(\mathbf{w}, \delta)(j) = \sum_{i \in \mathcal{S}} w(i) p_{ij}^{\delta(i)}$$

denote the conditional probability of being in state j next period if $\mathbf{w}$ is the distribution of the state this period and we use decision rule δ. Let $\mathbf{w}_n$ denote the probability distribution of the state in period n and suppose that Markov policy $(\delta_1, \delta_2, \ldots)$ is used. Then $\mathbf{w}_{n+1} = M(\mathbf{w}_n, \delta_n)$.

A *posterity* is a feasible sequence $(\mathbf{w}_1, \delta_1, \mathbf{w}_2, \delta_2, \ldots)$ of successive distributions and decision rules. The set $\Phi_{\mathbf{w}}$ of all posterities with the initial distribution $\mathbf{w}$ is

$$\Phi_{\mathbf{w}} = \{\mathbf{w}_1, \delta_1, \mathbf{w}_2, \delta_2, \ldots): \mathbf{w}_1 = \mathbf{w}, \delta_n \in \Delta, \text{and } \mathbf{w}_{n+1} = M(\mathbf{w}_n, \delta_n) \text{ for all } n\} \tag{6-78}$$

A Markov policy $\pi = (\delta_1, \delta_2, \ldots)$ is a *stationary policy* if $\delta = \delta_1 = \delta_2 = \ldots$ for some $\delta \in \Delta$. Then the policy π is written as δ^∞. For each $\gamma \in \Delta$, T_γ is the mapping from Y to Y given by $T_\gamma \pi = (\gamma, \delta_1, \delta_2, \ldots)$ where $\pi = (\delta_1, \delta_2, \ldots)$.

For a distribution $\mathbf{w} \in W$ and decision rule $\delta \in \Delta$ let

$$\rho(\mathbf{w}, \delta) = \sum_{j \in \mathcal{S}} w(j) r[j, \delta(j)]$$

denote the expected single-stage reward if the state is distributed according to $\mathbf{w}$ and δ is used to choose the action. If the initial state is distributed according to $\mathbf{w}$ and Markov policy $\pi = (\delta_1, \delta_2, \ldots)$ is employed, the expected present value is

$$v_{\mathbf{w}}(\pi) = \sum_{n=1}^{\infty} \beta^{n-1} \rho(\mathbf{w}_n, \delta_n) \qquad (0 \leq \beta < 1) \tag{6-79a}$$

where $\mathbf{w}_1 = \mathbf{w}$ and $\mathbf{w}_{n+1} = M(\mathbf{w}_n, \delta_n)$ for all n. If $\mathbf{w}$ is degenerate, i.e., if $w(s) = 1$ and $w(j) = 0$ for all $j \neq s$ so s is the initial state for certain, then we write $v_s(\pi)$ instead of $v_{\mathbf{w}}(\pi)$.

For Markov policies π and π' write

$$\pi \gtrsim \pi' \Leftrightarrow v_{\mathbf{w}}(\pi) \geq v_{\mathbf{w}}(\pi') \qquad \text{for all } \mathbf{w} \in W \tag{6-79b}$$

In the proof of Corollary 4-4a we show that $\pi \gtrsim \pi' \Leftrightarrow v_s(\pi) \geq v_s(\pi')$ for all $s \in \mathcal{S}$.

Lemma 4-2 asserts that if $0 \leq \beta < 1$ and $\mathcal{S}$ is at most countably infinite, then:

$(\gtrsim, Y)$ is reflexive and transitive (6-80)

For each $\gamma \in \Delta$, T_γ is isotone on $(\gtrsim, Y)$ (6-81)

For each $\pi = (\delta_1, \delta_2, \ldots) \in Y$ and $\xi \in Y$, (6-82)

$T_{\delta_1} T_{\delta_2} \ldots T_{\delta_{k+1}} \xi \lesssim T_{\delta_1} T_{\delta_2} \ldots T_{\delta_k} \xi$ for all $k \in I \Rightarrow \pi \lesssim \xi$

and

$T_{\delta_1} T_{\delta_2} \ldots T_{\delta_{k+1}} \xi \gtrsim T_{\delta_1} T_{\delta_2} \ldots T_{\delta_k} \xi$ for all $k \in I \Rightarrow \pi \gtrsim \xi$

As we observe on pages 150–152, the proofs of several of the principal results in Sections 4-3 and 4-4 depend on (6-80), (6-81), and (6-82) but not on (6-79a) and (6-79b). Therefore, any other model or procedure to induce a preference ordering which satisfies (6-80), (6-81), and (6-82) will share the aforementioned principal results. This observation suggests that we regard (6-80), (6-81), and (6-82) as axioms and then list their consequences. We shortly find that a model related to multiple criteria satisfies these axioms, hence shares their consequences.

Proposition 6-1. Suppose $\mathcal{S}$ is at most countably infinite and axioms (6-80), (6-81), and (6-82) are valid. Then:

$T_\delta \gtrsim \pi \Rightarrow \delta^\infty \gtrsim \pi$ (Theorem 4-2) (6-83)

If there is a $\gtrsim$-maximal $\pi \in Y$, there is $\delta \in \Delta$ such that δ^∞ is $\gtrsim$-maximal (Theorem 4-3) (6-84)

$\pi \in Y$ is $\gtrsim$-maximal $\Leftrightarrow \pi \gtrsim T_\delta \pi$ for all $\delta \in \Delta$ (Theorems 4-4 and 4-5) (6-85)

If $\#\mathscr{C} < \infty$, there is a $\gtrsim$-maximal δ^∞ (Theorem 4-6) (6-86)

For every $\gamma \in \Delta$, either γ^∞ is $\gtrsim$-maximal or there is another $\delta \in \Delta$ such that $\delta^\infty \gtrsim \gamma^\infty$ and $\gamma^\infty \not\gtrsim \delta^\infty$ (6-87)

Property (6-87) is Theorem 4-5, which is proved with the policy-improvement algorithm. By embedding multiple-criteria MDPs in the present ordinal framework, we shall gain the general existence theorem (6-84) and a policy-improvement algorithm. However, a major obstacle noted below is that $\gtrsim$ for PO is not generally transitive. Here are the details of a partial remedy.

Consider an MDP with $\mathscr{C}$, Δ, and Y defined in (6-77), distribution-to-distribution mapping M, and definition (6-78) of the set $\Phi_{\mathbf{w}}$ of posterities emanating from the initial state $\mathbf{w}$. Let $\mathscr{Q}$ denote a nonempty set of MOEs. For each $\mathbf{w} \in W$ and $q \in \mathscr{Q}$ let $\theta_{\mathbf{w}}^q \subset \Phi_{\mathbf{w}} \times \Phi_{\mathbf{w}}$ indicate preferences with respect to the qth MOE if $\mathbf{w}_1 = \mathbf{w}$. We interpret $(\tau, \tau') \in \theta_{\mathbf{w}}^q$ as "posterity τ is at least as desirable as posterity τ', according to the qth MOE, if $\mathbf{w}_1 = \mathbf{w}$." For each $\mathbf{w} \in W$ and $\pi = (\delta_1, \delta_2, \ldots) \in Y$, let $\tau_{\mathbf{w}}(\pi)$ denote the posterity generated by the Markov policy π from the initial state $\mathbf{w}$:

$$\tau_{\mathbf{w}}(\pi) = [\mathbf{w}, \delta_1, M(\mathbf{w}, \delta_1), \delta_2, \ldots]$$

The following definition specifies a binary relation $\gtrsim_p$ with the property that a $\gtrsim_p$-maximal policy is a Pareto optimum.

Definition 6-4 Let π and ξ be Markov policies. We define $\gtrsim_p$ by

$$\pi \gtrsim_p \xi \Leftrightarrow \text{either } [\tau_{\mathbf{w}}(\pi), \tau_{\mathbf{w}}(\xi)] \in \theta_{\mathbf{w}}^q \text{ for all } \mathbf{w} \in W \text{ and } q \in \mathscr{Q},$$

or there are $\mathbf{w}$ and $\mathbf{y}$ in W and q and u in $\mathscr{Q}$ such that

$$[\tau_{\mathbf{w}}(\pi), \tau_{\mathbf{w}}(\xi)] \notin \theta_{\mathbf{w}}^q \qquad \text{and} \qquad [\tau_{\mathbf{y}}(\xi), \tau_{\mathbf{y}}(\pi)] \notin \theta_{\mathbf{y}}^u \tag{6-88}$$

Definition 6-5 A Markov policy π is a *Pareto optimum*, written *PO*, if π is $\gtrsim_p$-optimal.

The following example shows that $\gtrsim_p$ is not necessarily transitive.

Example 6-8 Consider an MDP with two states, two actions available in each state, and two MOEs. Let (i, j) denote the unrandomized policy of taking action i in state 1 and action j in state 2. The entries in the ith row and jth column of the following table are

$$\left\{E\left[\sum_{n=1}^{\infty} \beta^{n-1} r_1(s_n, a_n) \mid s_1 = 1\right], E\left[\sum_{n=1}^{\infty} \beta^{n-1} r_2(s_n, a_n) \mid s_1 = 1\right]\right\}$$

Action in state 2

		1	2
Action in state 1	1	0, 1	3, 0
	2	2, 2	0, 0

Here, (1, 1) $\gtrsim_p$ (1, 2) because the second MOE is 1 at the policy (1, 1) but only 0 at the policy (1, 2). Also (1, 2) $\gtrsim_p$ (2, 1) because the first MOE is 3 at the policy (1, 2) but only 2 at the policy (2, 1). However, (1, 1) $\not\gtrsim_p$ (2, 1) because both MOEs are lower at (1, 1) than at (2, 1). Therefore, $\gtrsim_p$ is not transitive. □

Solutions in Stationary Policies

Some additional notation is needed for a partial resolution of the intransitivity of $(\gtrsim_p, Y)$.

Definition 6-6 Let D be a nonempty set and $B \subset D \times D$. An *inconsistency cycle* connects x and y under B if there is a finite sequence $x_1, \ldots, x_n$ such that $x_1 = x_n = x$, $x_k = y$ for some $1 < k < n$, $(x_i, x_{i+1}) \in B$ for all $i < n$, and $(x_{i+1}, x_i) \notin B$ for some i. The *completion* of a binary relation (B, D), written (B', D) is

$$B' = B \cup \{x, y): \in D \times D, (x, y) \notin B, \text{ and } (y, x) \notin B\}$$

The *transitive completion* of a binary relation (B, D), written (B_c, D), is

$$B_c = B' \cup \{(x, y) : \text{there is an inconsistency cycle which connects } x \text{ and } y \text{ under } B'\}$$

The completion of a binary relation causes pairs of elements to become equivalent if they are at first not comparable. The transitive completion causes all elements of an inconsistency cycle to become equivalent.

Example 6-9 In Example 6-8, $\gtrsim_p$ yields the following set of ordered pairs:

$$\begin{aligned} B = \{&[(1, 1), (1, 1)], [(1, 1), (1, 2)], [(1, 1), (2, 2)], \\ &[(1, 2), (1, 1)], [(1, 2), (1, 2)], [(1, 2), (2, 1)], [(1, 2), (2, 2)], \\ &[(2, 1) (1, 1)], [(2, 1), (1, 2)], [(2, 1), (2, 1)], [(2, 1), (2, 2)]\} \end{aligned}$$

Here, $B' = B$ because $[(2, 1), (1, 1)] \in B$ although $[(1, 1), (2, 1)] \notin B$, and $[(i, j), (2, 2)] \in B$, although $[(2, 2), (i, j)] \notin B$ for all $(i, j) \neq (2, 2)$.

The only inconsistency cycle under $B' = B$ is (1, 1), (1, 2), (2, 1), and (1, 1). Therefore, the transitive completion of B is

$$B_c = B \cup \{[(1, 1), (2, 1)]\}$$ □

You are asked to prove the following properties of transitive completions in Exercise 6-4.

Lemma 6-6 If (B, D) is a binary relation with D nonempty, then:

$$(B_c, D) \text{ is complete and transitive} \tag{6-89}$$

$$B \subset B_c \tag{6-90}$$

$$x \in D \text{ is } B\text{-maximal} \Rightarrow x \text{ is } B_c\text{-maximal} \tag{6-91}$$

Thus B_c is "coarser" than B but preserves B-maximality.

Let $(\geq, Y)$ be any binary relation on the set Y of Markov policies. Two examples of $\geq$ are $\gtrsim$ defined by (6-79*a*) and (6-79*b*) and $\gtrsim_p$ defined by (6-88). The relation $(\geq, Y)$ is not assumed to be complete, reflexive, or transitive. Let $(\geq_c, Y)$ denote the transitive completion of $(\geq, Y)$.

Theorem 6-9 Suppose $\mathcal{S}$ is at most countably infinite and (6-81) and (6-82) are valid when $\geq$ replaces $\gtrsim$, that is,

$$\text{For each } \gamma \in \Delta,\ T_\gamma \text{ is isotone on } (\geq, Y) \tag{6-92}$$

For each $\pi = (\delta_1, \delta_2, \ldots) \in Y$ and $\xi \in Y$,

$$T_{\delta_1} T_{\delta_2} \ldots T_{\delta_{k+1}} \xi \leq T_{\delta_1} T_{\delta_2} \ldots T_{\delta_k} \xi \text{ for all } k \in I \Rightarrow \pi \leq \xi$$

$$\text{and } T_{\delta_1} T_{\delta_2} \ldots T_{\delta_{k+1}} \xi \geq T_{\delta_1} T_{\delta_2} \ldots T_{\delta_k} \xi \text{ for all } k \in I \Rightarrow \pi \geq \xi \tag{6-93}$$

Then:

$$T_\delta \pi \geq_c \pi \Rightarrow \delta^\infty \geq_c \pi \tag{6-94}$$

If $\pi \in Y$ is $\geq$-maximal, π is $\geq_c$-maximal and there is a $\geq_c$-maximal δ^∞ (6-95)

If $\pi \in Y$ is $\geq$-maximal, then $\pi \geq_c T_\delta \pi$ for all $\delta \in \Delta$ (6-96)

If $\#\mathscr{C} < \infty$, then there is a $\geq_c$-maximal δ^∞ (6-97)

For every $\gamma \in \Delta$, either γ^∞ is $\geq_c$-maximal or there is another $\delta \in \Delta$ such that $\delta^\infty \geq_c \gamma^\infty$ (6-98)

PROOF Here is a sketch; Exercise 6-5 asks you to fill in the details. Part (6-90) of Lemma 6-6 implies that $\geq_c$ inherits (6-92) and (6-93) from $\geq$. Parts (6-94) through (6-98) depend on (6-89) in Lemma 6-6 and $\geq_c$ inheriting (6-92) and (6-93) from $\geq$. Parts (6-94) through (6-98) depend on (6-91) of Lemma 6-6. Each part of this theorem depends on its corresponding part of Proposition 6-1. □

The objective now is to determine conditions under which $\gtrsim_p$, the binary relation for a Pareto optimum, satisfies (6-92) and (6-93).

Definition 6-7 (*a*) $\{\theta_{\mathbf{w}}^q : \mathbf{w} \in W, q \in \mathscr{Q}\}$ has *consistent choice* (also called *temporal persistence*) if $\mathbf{w} \in W$, $\delta \in \Delta$, $M(\mathbf{w}, \delta) = \mathbf{y}$, and $(\tau, \tau') \in \theta_{\mathbf{y}}^q$ implies $[(\mathbf{w}, \delta, \tau), (\mathbf{w}, \delta, \tau')] \in \theta_{\mathbf{w}}^q$.

(*b*) Let $\tau^0, \tau^1, \tau^2, \ldots$ be a sequence of posterities with $\tau^j = (\mathbf{w}, \delta_1^j, \mathbf{w}_2^j, \delta_2^j, \ldots) \in \Phi_{\mathbf{w}}$ such that τ^j matches τ^0 up to $\mathbf{w}_j^j$ and δ_j^j (hence $\mathbf{w}_{j+1}^j$), that is, $\mathbf{w}_n^0 = \mathbf{w}_n^j$ and $\delta_n^0 = \delta_n^j$ for all $n \leq j$. Then $\{\theta_{\mathbf{w}}^q : \mathbf{w} \in W, q \in \mathscr{Q}\}$ is *continuous* (also called *countably transitive*) if both

$$(\tau^{j+1}, \tau^j) \in \theta^q_{\mathbf{w}}$$

$$\text{for all } j \geq 1 \Rightarrow (\tau^0, \tau^j) \in \theta^q_{\mathbf{w}} \qquad \text{for all } j \geq 1 \tag{6-99a}$$

and

$$(\tau^j, \tau^{j+1}) \in \theta^q_{\mathbf{w}}$$

$$\text{for all } j \geq 1 \Rightarrow (\tau^j, \tau^0) \in \theta^q_{\mathbf{w}} \qquad \text{for all } j \geq 1 \tag{6-99b}$$

Consistent choice and continuity are discussed following Lemma 4-1. In particular, continuity is akin to the use of discounted criteria.

Definition 6-8 A distribution $\mathbf{w} \in W$ is *reachable* if there exists $\mathbf{y} \in W$ and $\delta \in \Delta$ such that $\mathbf{w} = M(\mathbf{y}, \delta)$. A *subset* G of W is *reachable* if $\mathbf{w}$ is reachable for all $\mathbf{w} \in G$.

Theorem 6-10 If $\mathcal{S}$ is at most countably infinite, W is reachable, and $\{\theta^q_{\mathbf{w}}: \mathbf{w} \in W, q \in \mathcal{Q}\}$ has consistent choice and is continuous, then $(\gtrsim_p, Y)$ satisfies (6-92) and (6-93).

PROOF Continuity of $\{\theta^q_{\mathbf{w}}\}$ implies (6-93) for $\gtrsim_p$. Suppose W is reachable and define $\pi > \xi \Leftrightarrow [\tau_{\mathbf{w}}(\pi), \tau_{\mathbf{w}}(\xi)] \in \theta^q_{\mathbf{w}}$ for all $\mathbf{w}$ and q. Then $(>, Y)$ satisfies (6-92) because $\{\theta^q_{\mathbf{w}}\}$ has consistent choice. Now there are two cases. If $\pi \gtrsim_p \xi$ and $\pi > \xi$, then (6-92) for $(>, Y)$ implies $T_\delta \pi \gtrsim_p T_\delta \xi$. If $\pi \gtrsim_p \xi$ but not $\pi > \xi$, then necessarily $\xi \gtrsim_p \pi$. Therefore, by Definition 6-4 of $\gtrsim_p$, there exist $\mathbf{w}$ and $\mathbf{y}$ in W and q and u in $\mathcal{Q}$ such that $[\tau_{\mathbf{w}}(\xi), \tau_{\mathbf{w}}(\pi)] \notin \theta^q_{\mathbf{w}}$ and $[\tau_{\mathbf{y}}(\pi), \tau_{\mathbf{y}}(\xi)] \notin \theta^u_{\mathbf{y}}$. As a result, continuity of $\{\theta^q_{\mathbf{w}}\}$ and the reachability of $\mathbf{w}$ and $\mathbf{y}$ imply existence of $\mathbf{x}$ and $\mathbf{z}$ in W, and δ and γ in Δ that $\mathbf{w} = M(\mathbf{x}, \delta)$, $\mathbf{y} = M(\mathbf{z}, \gamma)$, and

$$\{[\mathbf{x}, \delta, \tau_{\mathbf{w}}(\xi)], [\mathbf{x}, \delta, \tau_{\mathbf{w}}(\pi)]\} \notin \theta^q_{\mathbf{x}}$$

and

$$\{[\mathbf{z}, \gamma, \tau_{\mathbf{y}}(\pi)], [\mathbf{z}, \gamma, \tau_{\mathbf{y}}(\xi)]\} \notin \theta^u_{\mathbf{z}}$$

Therefore, $T_\delta \pi \gtrsim_p T_\delta \xi$ (and $T_\delta \xi \gtrsim_p T_\delta \pi$) so $\{\theta^q_{\mathbf{w}}\}$ has consistent choice.

□

Let $\gtrsim_c$ denote the transitive completion of $\gtrsim_p$. The next result states properties of $\gtrsim_c$; its proof is an immediate consequence of Theorems 6-9 and 6-10.

Corollary 6-10 If $\mathcal{S}$ is at most countably infinite, W is reachable, and $\{\theta^q_{\mathbf{w}}\}$ has consistent choice and is continuous, then $\gtrsim_p$ and $\gtrsim_c$ have properties (6-94) through (6-98) (with $\gtrsim_p$ replacing $\geq$ and $\gtrsim_c$ replacing $\geq_c$).

In particular, if there is a Pareto optimum, (6-95) implies that there is a stationary policy which is Pareto optimal with respect to the transitive completion $\gtrsim_c$.

EXERCISES

6-4 Prove Lemma 6-6.

6-5 Prove Theorem 6-9 in detail.

6-6 Exercise 4-30 observes that several results in Sections 4-3 and 4-4 are valid for deterministic MDPs without restricting $\mathcal{S}$ to be a countable (or finite) set. Verify that Theorems 6-9 and 6-10 and Corollary 6-10 are valid for deterministic MDPs if we assume merely that $\mathcal{S}$ is a nonempty set and A_s is a nonempty set for each $s \in \mathcal{S}$.

6-4* AGGREGATION TECHNIQUES

The "curse of dimensionality" refers to the prodigious number of states S in some realistic models. This section discusses the use of aggregation techniques† to exorcise the "curse." The basic idea is to replace the original MDP model with a smaller approximate MDP model. The approximate model may be relatively small but absolutely large. Hence it may be important to use acceleration techniques (in Section 6-2) to solve the approximate model.

> **Example 6-10** Consider a managerial model of a commercial fishery with 10 interacting species (a realistic number in some places). Details for a single-species fishery are given in Section 8-3. Under various biological assumptions, successive years' vectors of quantities of the 10 species constitute a sequence of states in an MDP. Thus a state is a vector with 10 components. Suppose the range of possible quantities for each species is discretized into 50 categories. Then $S = 50^{10}$, which is approximately 10^{17}. Such a model would have to be reduced in size by a factor of at least a billion before any presently known algorithm could be used. □

Let $\mathscr{C}$, $\{p_{sj}^a\}$, $\{r(s, a)\}$, and β be the data of a finite MDP. We assume $0 \le \beta < 1$ and consider the objective of maximizing the expected present value over an infinite planning horizon. Let f_s denote the maximal expected present value if s is the initial state. Presumably, $\#\mathscr{C}$ is a large number; so we wish to replace the given MDP with a much smaller one. Suppose this is done and the smaller MDP is solved. Our goals are to use information embedded in the solution of the smaller problem in order to:

1. Specify a feasible policy δ in the original MDP; we hope δ^∞ is a "good" policy. Let $v_s(\delta)$ denote the expected present value of policy δ^∞ if $s_1 = s$. For each s, of course, $v_s(\delta) \le f_s$, but the numerical value of f_s and the identity of an optimal policy are both unknown. Also, the model may be too large to compute $\mathbf{v}(\delta)$.
2. Compute an upper bound on the difference $f_s - v_s(\delta)$.

† Some of this section's material is based on unpublished work by M. J. Sobel and P. Zipkin. Professor Zipkin has graciously permitted the work to be published in this volume.

There are several conceivable ways in which one could build a smaller MDP as an approximation to a given MDP. Here, we consider only aggregation techniques. The general idea is to partition an MDP's data and to replace each partitioned subset by a single number.

Let $\mathcal{S}_1, \mathcal{S}_2, \ldots, \mathcal{S}_K$ partition $\mathcal{S}$, that is, $\bigcup_{k=1}^{K} \mathcal{S}_k = \mathcal{S}$ and $\mathcal{S}_k \cap \mathcal{S}_l = \phi$ if $k \neq l$. Let n_k denote $\#\mathcal{S}_k$ and $\mathcal{K} = \{1, \ldots, K\}$, which will be the set of states in the approximating MDP. For each $k \in \mathcal{K}$, let $\mathcal{A}_k$ be a nonempty subset of $\bigcap_{s \in \mathcal{S}_k} A_s$. We assume that the labels of elements of $\{A_s\}$ and the partition of $\mathcal{S}$ make it possible to choose $\mathcal{A}_k$ nonempty for each k. The set $\mathcal{A}_k$ is the actions available from state k in the approximating MDP.

For $k \in \mathcal{K}$, $l \in \mathcal{K}$, and $a \in \mathcal{A}_k$, let q_{kl}^a and ρ_k^a denote a transition probability and single-stage reward, respectively, in the approximating MDP. The numbers q_{kl}^a and ρ_k^a represent the sets $\{p_{sj}^a: s \in \mathcal{S}_k \text{ and } j \in \mathcal{S}_l\}$ and $\{r(s, a): s \in \mathcal{S}_k\}$, respectively, in the original MDP. If the numbers in each of these sets are very close to one another, it may be possible to replace them with their representatives q_{kl}^a and ρ_k^a without significantly degrading the quality of a solution. However, when all such sets are replaced by their representatives, the resulting problem is equivalent to an MDP with state space $\mathcal{K}$.

Fixed-Weight Aggregation

For each $k \in \mathcal{K}$ let $\lambda_s^k \geq 0$, $s \in \mathcal{S}_k$, with $\sum_{s \in \mathcal{S}_k} \lambda_s^k = 1$. These numbers can be used to take weighted averages as follows:

$$q_{kl}^a = \sum_{s \in \mathcal{S}_k} \lambda_s^k \sum_{j \in \mathcal{S}_l} p_{sj}^a \qquad \rho_k^a = \sum_{s \in \mathcal{S}_k} \lambda_s^k r(s, a) \tag{6-100}$$

The $\{q_{kl}^a\}$ behave as transition probabilities in an MDP:

$$\begin{aligned}\sum_{l \in \mathcal{K}} q_{kl}^a &= \sum_{l \in \mathcal{K}} \sum_{s \in \mathcal{S}_k} \lambda_s^k \sum_{j \in \mathcal{S}_l} p_{sj}^a \\ &= \sum_{s \in \mathcal{S}_k} \lambda_s^k \sum_{l \in \mathcal{K}} \sum_{j \in \mathcal{S}_l} p_{sj}^a = \sum_{s \in \mathcal{S}_k} \lambda_s^k \sum_{j \in \mathcal{S}} p_{sj}^a \\ &= \sum_{s \in \mathcal{S}_k} \lambda_s^k = 1\end{aligned}$$

In (6-100), $\sum_{j \in \mathcal{S}_l} p_{sj}^a$ is the probability of moving from state s to subset $\mathcal{S}_l$ if action a is taken. Then q_{kl}^a is the weighted average of these probabilities where s ranges over $\mathcal{S}_k$ and $\{\lambda_s^k: s \in \mathcal{S}_k\}$ provides the weights. Similarly, ρ_k^a is the weighted average of $r(s, a)$ as s ranges over $\mathcal{S}_k$.

Aggregation with Minima

Another aggregation method replaces the numbers in each set by their minima as follows:

$$q_{kl}^a = n_l \min \{p_{sj}^a: s \in \mathcal{S}_k, j \in \mathcal{S}_l\} \qquad \rho_k^a = \min \{r(s, a): s \in \mathcal{S}_k\} \tag{6-101}$$

This aggregation yields substochastic transition probabilities, that is,

$$\sum_{l \in \mathscr{K}} q_{kl}^a \leq 1 \tag{6-102}$$

with the inequality typically strict. However, as we mention in Section 4-5, (6-102) is sufficient for the principal properties of all the algorithms to remain valid.

An Aggregate MDP

Suppose that either (6-100) or (6-101) has been used to specify $\{q_{kl}^a\}$ and $\{\rho_k^a\}$. Consider the following alteration of the original MDP:

For each $k \in \mathscr{K}$, for each $s \in \mathcal{S}_k$ replace A_s with $\mathscr{A}_k$ (6-103)

For each $s \in \mathcal{S}$ and $j \in \mathcal{S}$, if $s \in \mathcal{S}_k$ and $j \in \mathcal{S}_l$, (6-104)

$$\text{replace } p_{sj}^a \text{ with } \frac{q_{kl}^a}{n_k} \text{ for each } a \in \mathscr{A}_k$$

For each $s \in \mathcal{S}$, if $s \in \mathcal{S}_k$, replace $r(s, a)$ with $\dfrac{\rho_k^a}{n_k}$ for each $a \in \mathscr{A}_k$. (6-105)

The primal linear program for the unaltered original problem is

$$\text{Minimize } \sum_{s \in \mathcal{S}} \alpha_s w_s \tag{6-106a}$$

subject to

$$w_s \geq r(s, a) + \beta \sum_{j \in \mathcal{S}} p_{sj}^a w_j \qquad (s, a) \in \mathscr{C} \tag{6-106b}$$

where $\alpha_s > 0$ for all $s \in \mathcal{S}$. If replacements (6-104) and (6-105) are made in a constraint in (6-106*b*), the result, for $s \in \mathcal{S}_k$ and $a \in \mathscr{A}_k$, is

$$\begin{aligned} w_s &\geq \frac{\rho_k^a}{n_k} + \beta \sum_{l \in \mathscr{K}} \sum_{j \in \mathcal{S}_l} \frac{w_j q_{kl}^a}{n_k} \\ &= \frac{\rho_k^a}{n_k} + \frac{\beta}{n_k} \sum_{l \in \mathscr{K}} q_{kl}^a \sum_{j \in \mathcal{S}_l} w_j \end{aligned} \tag{6-107}$$

Let

$$y_l = \sum_{j \in \mathcal{S}_l} w_j$$

Then (6-107) is

$$w_s \geq \frac{\rho_k^a}{n_k} + \frac{\beta}{n_k} \sum_{l \in \mathscr{K}} q_{kl}^a y_l \qquad a \in \mathscr{A}_k,\ s \in \mathcal{S}_k \tag{6-108}$$

whose right side is the same for all $s \in \mathcal{S}_k$.

From Theorem 4-11, the solution to (6-106*a*) and (6-106*b*) is invariant with respect to strictly positive vectors $\boldsymbol{\alpha}$ in (6-106*a*). Hence there is no loss of optimality in the assumption

$$\alpha_s = \alpha_j \qquad \text{if } s \in \mathbb{S}_k \text{ and } j \in \mathbb{S}_k \text{ for some } k \in \mathscr{K} \tag{6-109}$$

Henceforth making this assumption, there is a solution of (6-106*a*) and (6-108) if, and only if, there is a solution with

$$w_s = \frac{y_k}{n_k} \qquad s \in \mathbb{S}_k,\ k \in \mathscr{K} \tag{6-110}$$

Substitution of (6-110) in (6-108) yields

$$\frac{y_k}{n_k} \geq \frac{\rho_k^a}{n_k} + \frac{\beta}{n_k} \sum_{l \in \mathscr{K}} q_{kl}^a y_l \qquad a \in \mathscr{A}_k,\ s \in \mathbb{S}$$

so

$$y_k \geq \rho_k^a + \beta \sum_{l \in \mathscr{K}} q_{kl}^a y_l \qquad a \in \mathscr{A}_k,\ s \in \mathbb{S}_k$$

which is a system of n_k identical inequalities for each $a \in \mathscr{A}_k$. Deleting $n_k - 1$ of these inequalities leaves

$$y_k \geq \rho_k^a + \beta \sum_{l \in \mathscr{K}} q_{kl}^a y_l \qquad a \in \mathscr{A}_k,\ k \in \mathscr{K}$$

Under assumption (6-109), let

$$0 < \lambda_k = \sum_{s \in \mathbb{S}_k} \alpha_s \qquad \text{so } \alpha_s = \frac{\lambda_k}{n_k} \text{ if } s \in \mathbb{S}_k \tag{6-111}$$

Then (6-106*a*) and (6-106*b*) becomes

$$\text{Minimize } \sum_{k \in \mathscr{K}} \lambda_k y_k \tag{6-112}$$

$$\text{subject to } y_k \geq \rho_k^a + \beta \sum_{l \in \mathscr{K}} q_{kl}^a y_l \qquad a \in \mathscr{A}_k,\ k \in \mathscr{K}$$

This linear program corresponds to an MDP with state space K. There are K variables and $\sum_{k=1}^{K} (\#\mathscr{A}_k)$ constraints in (6-112). In the linear program for the original MDP, (6-106*a*) and (6-106*b*), there are S variables and $\#\mathscr{C} = \sum_{s \in \mathbb{S}} (\# A_s)$ constraints.

The exposition thus far relates the MDP with data $\{q_{kl}^a\}$ and $\{\rho_k^a\}$ to the original MDP and compares the sizes of the two problems via their associated linear programs. It is important to appreciate that the *aggregate* MDP, the one with data $\{q_{kl}^a\}$ and $\{\rho_k^a\}$, may be solved with *any* MDP algorithm.

Bounds

Recall that f_s denotes the expected present value of an optimal policy in the original MDP if s is the initial state. Also in the original MDP, let v_s denote $v_s(\delta)$ for any specific policy δ^∞. Let $\mathbf{f}$ and $\mathbf{v}$ indicate the vectors with components f_s and v_s, respectively. Our first goal is to obtain an upper bound on $\mathbf{0} \leq \mathbf{f} - \mathbf{v}$ and then to apply the bound to the aggregate MDP.

The linear program which is dual to (6-106a) and (6-106b) is

$$\text{Maximize} \sum_{(s,a)\in\mathscr{C}} r(s,a)x_{sa}$$

$$\text{subject to } x_{sa} \geq 0 \qquad (s,a)\in\mathscr{C}, \tag{6-113}$$

and

$$\sum_{(j,a)\in\mathscr{C}} (\delta_{sj} - \beta p_{js}^a)x_{ja} = \alpha_s \qquad s \in \mathcal{S}$$

where

$$\delta_{sj} = \begin{cases} 1 & \text{if } s = j \\ 0 & \text{if } s \neq j \end{cases}$$

Let $\mathbf{x}$ be optimal in (6-113). The values of the objectives of (6-106a) and (6-106b) and (6-113) must be the same at optimal solutions; so for any S-vector $\mathbf{v}$,

$$\boldsymbol{\alpha}\cdot(\mathbf{f}-\mathbf{v}) = \sum_{(j,a)\in\mathscr{C}} r(j,a)x_{ja} - \sum_{s\in\mathcal{S}} \alpha_s v_s$$

From the constraints of (6-113),

$$\alpha_s = \sum_{(j,a)\in\mathscr{C}} (\delta_{sj} - \beta p_{js}^a)x_{ja}$$

so

$$\begin{aligned}
\boldsymbol{\alpha}\cdot(\mathbf{f}-\mathbf{v}) &= \sum_{(j,a)\in\mathscr{C}} r(s,a)x_{ja} - \sum_{s\in\mathcal{S}} v_s \sum_{(j,a)\in\mathscr{C}} (\delta_{sj} - \beta p_{js}^a)x_{ja} \\
&= \sum_{(j,a)\in\mathscr{C}} x_{ja}\left[r(j,a) - \sum_{s\in\mathcal{S}} v_s(\delta_{sj} - \beta p_{js}^a)\right] \\
&\leq \sum_{j\in\mathcal{S}}\left[\sum_{a\in A_j} x_{ja} \max\left\{r(j,a) - \sum_{s\in\mathcal{S}} v_s(\delta_{sj} - \beta p_{js}^a) \colon a \in A_j\right\}\right] \qquad (6\text{-}114) \\
&\leq \sum_{(j,a)\in\mathscr{C}} x_{ja} \max\left\{r(j,a) + \beta\sum_{s\in\mathcal{S}} p_{js}^a v_s - v_j \colon (j,a)\in\mathscr{C}\right\} \\
&= (1-\beta)^{-1} \max\left\{r(j,a) + \beta\sum_{s\in\mathcal{S}} p_{js}^a v_s - v_j \colon (j,a)\in\mathscr{C}\right\} \qquad (6\text{-}115)
\end{aligned}$$

The usefulness of (6-114) as a bound depends on (1) being able to compute or approximate $\mathbf{v} = \mathbf{v}(\delta)$ for a "good" policy δ^∞, and (2) being able to bound $\sum_{a\in A_j} x_{ja}$ for each $j \in \mathcal{S}$. For the second task, at least

$$\begin{aligned}
\sum_{a\in A_j} x_{ja} &= \sum_{n=1}^{\infty} \beta^{n-1} P\{s_n = j\} \\
&\leq P\{s_1 = j\} + \sum_{n=2}^{\infty} \beta^{n-1}
\end{aligned}$$

Assigning $\alpha_j = P\{s_1 = j\}$ for each j yields

$$\sum_{a\in A_j} x_{ja} \leq \alpha_j + \frac{\beta}{1-\beta} \qquad j\in\mathcal{S} \tag{6-116}$$

This bound can be tightened in many cases. Substituting (6-115) and (6-116) in (6-114) yields the following result.

Theorem 6-11 For any S-vectors $\mathbf{v}$ and $\boldsymbol{\alpha}$ such that $P\{s_1 = j\} = \alpha_j, j \in \mathcal{S}$,

$$\boldsymbol{\alpha} \cdot (\mathbf{f} - \mathbf{v}) \leq (1 - \beta)^{-1} \max \left\{ r(j, a) + \beta \sum_{s \in \mathcal{S}} p_{js}^a v_s - v_j : (j, a) \in \mathscr{C} \right\} \tag{6-117a}$$

$$\boldsymbol{\alpha} \cdot (\mathbf{f} - \mathbf{v}) \leq \sum_{j \in \mathcal{S}} \left(\alpha_j + \frac{\beta}{1 - \beta} \right) \max \left\{ r(j, a) + \beta \sum_{s \in \mathcal{S}} p_{js}^a v_s - v_j : a \in A_j \right\} \tag{6-117b}$$

Now we turn to the task of finding a "good" policy δ^∞ for which $\mathbf{v}(\delta)$ can be approximated with an easily computable $\mathbf{v}$. Specifically, we use aggregation with minima to specify the data in (6-112). Suppose (6-112) is solved (using policy improvement, linear programming, or any other algorithm) and that $\mathbf{y}$ is optimal in (6-112) with γ^∞ an optimal policy in the aggregate MDP. Thus $\gamma(k) \in \mathscr{A}_k$ for each $k \in \mathscr{K}$. A single-stage decision rule δ in the original MDP is

$$\delta(s) = a \qquad \text{if } s \in \mathcal{S}_k \text{ and } \gamma(k) = a \tag{6-118a}$$

Let

$$v_s = y_k \qquad \text{if } s \in \mathcal{S}_k \tag{6-118b}$$

Theorem 6-12 Suppose $r(s, a) \geq 0$ for all $(s, a) \in \mathscr{C}$, there are numbers $\lambda_k > 0$, $k \in \mathscr{K}$, with $\alpha_s = \lambda_k/n_k$ if $s \in \mathcal{S}_k$, $\mathbf{y}$ is optimal in (6-112), γ^∞ is optimal in the (substochastic) MDP corresponding to (6-112), and (6-118a) and (6-118b) specify $\boldsymbol{\delta}$ and $\mathbf{v}$. Then

$$\mathbf{v} \leq \mathbf{v}(\delta) \leq \mathbf{f} \tag{6-119}$$

Comments (1) The right inequality is true for any policy, hence for δ^∞. (2) Using (6-118b) it is trivial to compute $\mathbf{v}$; so it can be used in (6-116), the bound in Theorem 6-11. (3) From (6-119), $\mathbf{f} - \mathbf{v}(\delta) \leq \mathbf{f} - \mathbf{v}$; so

$$0 \leq \alpha \cdot [\mathbf{f} - \mathbf{v}(\delta)] \leq \alpha \cdot (\mathbf{f} - \mathbf{v})$$

and (6-117a) and (6-117b), the bounds in Theorem 6-11, are valid for $\alpha \cdot [\mathbf{f} - \mathbf{v}(\delta)]$.

Corollary 6-12*a* Under the assumptions of Theorem 6-12, if $\sum_{k \in \mathscr{K}} \lambda_k = 1$, then

$$0 \leq \boldsymbol{\alpha} \cdot [\mathbf{f} - \mathbf{v}(\delta)] \leq \text{each of the following expressions:}$$

$$(1 - \beta)^{-1} \max \left\{ r(j, a) + \beta \sum_{s \in \mathcal{S}} p_{js}^a v_s - v_j : (j, a) \in \mathscr{C} \right\} \tag{6-120a}$$

$$\sum_{j \in \mathcal{S}} \left(\alpha_j + \frac{\beta}{1 - \beta} \right) \max \left\{ r(j, a) + \beta \sum_{s \in \mathcal{S}} p_{js}^a v_s - v_j : a \in A_j \right\} \tag{6-120b}$$

PROOF OF THEOREM 6-12 From (6-101), (6-118b), and the exposition between (6-103) and (6-112), $\mathbf{v}$ is optimal in the following alteration of (6-106a) and (6-106b). For each $s \in \mathcal{S}$, suppose $s \in \mathcal{S}_k$; delete all rows corresponding to $a \notin \mathcal{A}_k$. For each $s \in \mathcal{S}$ and $j \in \mathcal{S}$, suppose $s \in \mathcal{S}_k$ and $j \in \mathcal{S}_l$; for each $a \in \mathcal{A}_k$ replace p_{sj}^a with

$$\min \{p_{sj}^a: s \in \mathcal{S}_k, j \in \mathcal{S}_l\}$$

If $s \in \mathcal{S}_k$ and $a \in \mathcal{A}_k$, replace $r(s, a)$, with $\min \{r(j, a): j \in \mathcal{S}_k\}$. This series of alterations of (6-106a) and (6-106b) replaces numbers with smaller nonnegative numbers [deletion of a row is the same as replacing $r(s, a)$ and p_{sj}^a with zero, which yields $w_s \geq 0$ in (6-106b), which is not binding when $r(s, a) \geq 0$ for all $(s, a) \in \mathscr{C}$]. Therefore, (6-119) follows from the next lemma.

Lemma 6-7 Suppose $\mathbf{r}$ and $\mathbf{r}'$ are S-vectors, P and P' are $S \times S$ substochastic matrices, $0 \leq \beta < 1$, and

$$\mathbf{0} \leq \mathbf{r} \leq \mathbf{r}' \qquad \text{and} \qquad P \leq P'$$

(the last inequality means $p_{sj} \leq p'_{sj}$ for all s and j). Then

$$\mathbf{0} \leq (I - \beta P)^{-1}\mathbf{r} \leq (I - \beta P')^{-1}\mathbf{r}'$$

PROOF From Theorem 7-3 on page 223 of Volume I† and the hypothesis,

$$\mathbf{0} \leq \sum_{n=0}^{\infty} \beta^n P^n \mathbf{r} = (I - \beta P)^{-1}\mathbf{r}$$

$$(I - \beta P')^{-1}\mathbf{r}' = \sum_{n=0}^{\infty} \beta^n (P')^n \mathbf{r} \leq \sum_{n=0}^{\infty} (P)^n \mathbf{r} = (I - \beta P)^{-1}\mathbf{r} \qquad \square$$

Without loss of generality, suppose $\mathbf{v} > \mathbf{0}$. Then it is comparatively easy to compute an upper bound on the relative error attributable to $\mathbf{v}(\delta)$ in (6-119). The following corollary is an immediate consequence of Corollary 6-12a.

Corollary 6-12b Under the assumptions of Corollary 6-12a, if $v_s > 0$ for all $s \in \mathcal{S}$, then

$$0 \leq \sum_{j \in \mathcal{S}} \frac{f_j - v_j(\delta)}{f_j} \leq \text{each of the following expressions:}$$

$$\frac{\max \left\{ r(j, a) + \beta \sum_{s \in \mathcal{S}} p_{js}^a v_s - v_j : (j, a) \in \mathscr{C} \right\}}{(1 - \beta) \min \{v_j : j \in \mathcal{S}\}} \tag{6-121a}$$

† $(\beta P)^n \to 0$ implies $(I - \beta P)^{-1}$ exists and is given by

$$(I - \beta P)^{-1} = \sum_{n=0}^{\infty} \beta^n P^n$$

$$\sum_{j \in \mathcal{S}} \max \left\{ r(j, a) + \beta \sum_{s \in \mathcal{S}} p_{js}^a v_s - v_j \colon a \in A_j \right\} \left(\alpha_j + \frac{\beta}{1 - \beta} \right) v_j \qquad \text{(6-121}b\text{)}$$

Example 6-11 As in Section 3-1, suppose a single-product inventory model has immediate delivery of purchased goods and excess demand is lost. Suppose also that there is storage space for at most 999 items. Let r, c, and h denote linear price, purchase cost, and holding cost, respectively. Suppose that demands in successive periods are i.i.d. r.v.'s $D_1, D_2, \ldots$, with the uniform distribution

$$P\{D_1 = j\} = 0.001 \qquad j = 0, 1, \ldots, 999$$

Let $\beta = 0.99$, $r = \$20$, $c = \$5$, and $h = \$5$. Let $Q(\cdot)$ denote the distribution function of demand so

$$Q(x) = 0.001(x + 1) \qquad x = 0, 1, \ldots, 999$$

From Section 3-1, an optimal policy consists of ordering $(a^* - s)^+$ units each period, where s denotes the inventory on hand at the start of the period and

$$a^* = Q^{-1}\left[\frac{r - c}{r + h - \beta c}\right] = 748 \qquad \text{(6-122)}$$

which minimizes

$$\begin{aligned} G(a) &= (r + h - \beta c)E[(a - D_1)^+] - (r - c)a \\ &= -15a + 0.010025a(a + 1) \end{aligned} \qquad \text{(6-123)}$$

Also,

$$\begin{aligned} f(s) &= cs - \frac{G(a^*)}{1 - \beta} \\ &= 5s + 560{,}347.1225 \qquad 0 \le s \le a^* = 748 \end{aligned} \qquad \text{(6-124)}$$

The numerical results (6-122) and (6-124) are benchmarks against which the aggregation results will be measured.

Suppose that this problem is analyzed without exploiting its myopic structure. Let s and a, respectively, denote the inventory levels before and immediately after ordered goods are delivered at the start of a period. Then

$$r(s, a) = -ca + r[a - E(a - D_1)^+] - hE[(a - D_1)^+] \qquad \text{(6-125}a\text{)}$$

The transition probabilities are

$$p_{sj}^a = \begin{cases} 1 - 0.001a & \text{if } j = 0 \\ 0.001 & \text{if } 0 < j \le a \\ 0 & \text{if } j > a \end{cases} \qquad \text{(6-125}b\text{)}$$

with s, j, and a integers between 0 and 999.

Now we consider a succession of aggregations with minima applied to

(6-125*a*) and (6-125*b*), $\mathcal{S} = \{0, 1, \ldots, 999\}$, and $A_s = \{0, 1, \ldots, 999 - s\}$, $s \in \mathcal{S}$. We index these aggregations with b (for "batch size"). When $b = 2$, $\mathcal{S}$ is reduced to $\{0, 2, 4, \ldots, 998\}$ and, for states in this smaller set, $A_s = \{0, 2, \ldots, 998 - s\}$. Generally, the index b refers to an aggregation in which approximately $100(b - 1)/b$ percent of the original states and actions have been deleted. The numerical results below concern $b \in \{2, 4, 8, 16, 32, 64, 100, 128, 140, 150, 160\}$. For example, when $b = 150$, $\mathcal{S}$ is replaced by $\{0, 150, 300, 450, 600, 750\}$. Then $\mathscr{K} = \{1, 2, \ldots, 6\}$, $n_k = 150$ for each $k \in \mathscr{K}$, and $\mathcal{S}_1 = \{0, 1, \ldots, 149\}$. The aggregate action i denotes $a = ib$.

Generally, $\mathscr{K} = \{1, \ldots, K\}$ with† $K = \lfloor 999/b \rfloor$. Then

$$\begin{aligned} \mathcal{S}_k &= [(k-1)b,\, kb - 1] \cap I \\ \mathscr{A}_k &= [k, K] \cap I \end{aligned} \qquad (6\text{-}126)$$

because, if $i \in \mathscr{A}_k$, then $ib = a \le 999$ is ensured via

$$i \le \left\lfloor \tfrac{999}{b} \right\rfloor = K$$

For example, if $b = 150$, then $\mathscr{A}_3 = \{3, 4, 5, 6\}$.

Aggregation with minima involves the application of (6-101) to (6-125*a*) and (6-125*b*). Then [Exercise 6-4], (6-101), (6-125*a*), and (6-125*b*) yield

$$\rho_k^i = 15a - 0.0125a(a+1) + 5b(k-1) \qquad (6\text{-}127a)$$

for the aggregate version of the single-stage expected cost and

$$q_{kl}^i = \begin{cases} 0.001b & \text{if } 1 \le l \le i \\ 0 & \text{if } l > i \end{cases} \qquad (6\text{-}127b)$$

for the aggregate version of the transition subprobabilities.

Fortunately, the aggregate MDP with structure (6-127*a*) and (6-127*b*) satisfies the assumptions in Section 3-3; so it can be solved myopically. Let $G_b(i)$ indicate the dependence on b of the objective in the myopic problem equivalent to (6-127*a*) and (6-127*b*). Then (Exercise 6-4)

$$G_b(i) = 0.010025(ib)^2 + 0.002475ib^2 - 14.9875ib \qquad (6\text{-}128)$$

Let i_b denote the value of i which minimizes $G_b(\cdot)$ on $\{0, b, 2b, \ldots, \lfloor 999/b \rfloor b\}$. Table 6-10 shows i_b, bi_b, $G_b(i_b)$, $G(bi_b)$, and, using (6-122) and (6-123), $100[G(bi_b) - G(a^*)]/G(a^*)$. The disaggregate decision rule δ_b which corresponds to (6-118*a*) is to order $(bi_b - s)^+$ items. The expected present value of δ_b^∞ is

$$v_s(\delta_b) = cs - \frac{G(bi_b)}{1 - \beta} \qquad s \le bi_b$$

so $100[G(bi_b) - G(a^*)]/|G(a^*)|$ measures the percentage relative error from the initial state $s = 0$.

† The notation $\lfloor x \rfloor$ denotes the largest integer not exceeding x.

Table 6-10 Aggregation results in Example 6-11

Batch size b (1)	i_b (2)	Disaggregate policy parameter bi_b (3)	Error $G_b(i_b) - G(bi_b)$ (4)	Relative error (%) $\frac{100[G(bi_b) - G(a^*)]}{-G(a^*)}$ (5)
2	374	748	0	0
4	187	748	0	0
8	93	744	16.7	0.002
16	47	752	31.6	0.003
32	23	736	60.1	0.024
64	11	704	113.3	0.340
128	6	768	245.2	0.074
256	3	768	488.5	0.074

Column (5) in Table 6-10 shows that the relative error introduced by aggregation in this example is negligible. The reason is that the single-stage rewards (6-125*a*) and transition probabilities (6-125*b*) are smooth functions of *s*, *j*, and *a*. Many problems encountered in professional practice have this desirable feature.

How well do the bounds in (6-120*a*) match column (5) in Table 6-10? For (6-120*a*), one must compute

$$\max \left\{r(j, a) + \beta \sum_{s \in \mathcal{S}} p_{js}^{a} v_s - v_j : j \le a \le 999;\ 0 \le j \le 999\right\} \tag{6-129}$$

where
$$v_j = cj - \frac{G_b(i_b)}{1-\beta} = 50j - 100G_b(i_b) \qquad j \le bi_b \tag{6-130}$$

and (6-125*a*) and (6-125*b*) specify $r(j, a)$ and p_{js}^{a}. Exercise 6-8 asks you to

Table 6-11 Error bounds in Example 6-11

b (1)	Actual Error $100[G(bi_b) - G(a^*)]$ (2)	Bound on Actual Error $100[5{,}603.47 + G_b(i_b)]$ (3)	Relative Error (%) $\frac{100[G(bi_b) - G(a^*)]}{-G(a^*)}$ (4)	Bound on Relative Error (%) $\frac{100[G_b(i_b) - G(a^*)]}{-G_b(i_b)}$ (5)
2	0	555.1	0	0.091
4	0	925.4	0	0.165
8	0.13	1,684.2	0.002	0.301
16	0.19	3,164.8	0.003	0.568
32	1.35	6,146.5	0.024	1.109
64	19.08	13,233.5	0.340	2.419
128	4.16	24,935.9	0.074	4.657
256	4.16	49,266.2	0.074	9.640

verify that the maximum in (6-129) is attained at $a = 748$ and $j = 0$, where (6-129) has the value

$$5{,}603.471225 + G_b(i_b) \tag{6-131}$$

Here, $(1 - \beta)^{-1} = 100$; so column (3) in Table 6-11 tabulates the bound in (6-120*a*) as 100 times the value in (6-131). The bound (6-121*a*) on relative error, as a percentage, is shown in column (5) of Table 6-11. Column (4) in Table 6-11 and Column (5) in Table 6-10 are the same. □

EXERCISES

6-7 Verify (6-127*a*), (6-127*b*), and (6-128).

6-8 Verify that the maximum in (6-130) is achived at $a = 748$ and $j = 0$.

BIBLIOGRAPHIC GUIDE

The conference proceedings cited in the Bibliographic Guide in Chapter 4 contain results pertinent to this chapter.

Further results for linear programs

Denardo, E. V.: "On Linear Programming in a Markov Decision Problem," *Manage. Sci.* **16**: 281–288 (1970).

——— and B. L. Fox: "Multichain Markov Renewal Programs," *SIAM J. Appl. Math.* **16**: 468–487 (1968).

Derman, C.: *Finite State Markovian Decision Processes*, Academic Press, New York (1970).

———, and A. F. Veinott, Jr.: "Constrained Markov Decision Chains," *Manage. Sci.* **19**: 389–390 (1972).

Heilman, W. R.: "Solving Stochastic Dynamic Programming Problems by Linear Programming: An Annotated Bibliography," *Z. Oper. Res.* **22**: 43–53 (1978).

Hordijk, A., and L. C. M. Kallenberg: "Linear Programming and Markov Decision Chains," *Manage. Sci.* **25**: 352–362 (1979).

Kallenberg, L. C. M.: *Linear Programming and Finite Markovian Control Problems*, Mathematisch Centrum, Amsterdam (1980).

Koehler, G. J., A. B. Whinston, and G. P. Wright: *Optimization over Leontief Substitution Systems*, North-Holland, Amsterdam (1975).

Kushner, H.: *Introduction to Stochastic Control*, Holt, New York (1971).

Mine, H., and S. Osaki: *Markovian Decision Processes*, American Elsevier, New York (1970).

Veinott, A. F., Jr.: "Minimum Concave-Cost Solution of Leontief Substitution Models of Multi-Facility Inventory Systems," *Oper. Res.* **17**: 262–291 (1969).

Wessels, J., and J. A. E. E. van Nunen: "Discounted Semi-Markov Decision Processes: Linear Programming and Policy Iteration," *Statistica Neerlandica*, **29**: 1–7 (1975).

Wolfe, P., and G. B. Dantzig: "Linear Programming in a Markov Chain," *Oper. Res.* **10**: 702–710 (1962).

Acceleration of computations

Denardo, E. V.: "Contraction Mappings in the Theory Underlying Dynamic Programming," *SIAM Rev.* **9**: 165–177 (1967).

———, and B. L. Fox: "Shortest-Route Methods; 1 and 2," *Oper. Res.* **27**: 161–186, 548–466 (1979).

Fox, B. L.: "Coupled Successive Approximation for Markov Programs," *Oper. Res.* **30**: 400–403 (1982).

Grinold, R.: "Elimination of Suboptimal Actions in Markov Decision Problems," *Oper. Res.* **21**: 848–851 (1973).

Hastings, N. A. J., and J. M. C. Mello: "Tests for Suboptimal Actions in Discounted Markov Programming," *Manage. Sci.* **19**: 1019–1022 (1973).

Hinderer, K.: "Estimates for Finite-Stage Dynamic Programs," *J. Math. Anal. Appl.* **55**: 207–238 (1976).

———: "On Approximate Solutions of Finite-Stage Dynamic Programs," in *Dynamic Programming and Its Applications*, edited by M. L. Puterman, Academic Press, New York (1978), pp. 289–319.

Kushner, H. J., and A. J. Kleinman: "Accelerated Procedures for the Solution of Discrete Markov Control Problems," *IEEE Trans. Auto. Control* **AC-16**: 147–152 (1971).

MacQueen, J.: "A Test for Suboptimal Actions in Markov Decision Problems," *Oper. Res.* **15**: 559–561 (1967).

Morin, T. L.: "Computational Advances and Reduction of Dimensionality in Dynamic Programming: A Survey," in *Dynamic Programming and Its Applications*, edited by M. L. Puterman, Academic Press, New York (1979).

Morton, T.: "Undiscounted Markov Renewal Programming via Modified Successive Approximations," *Oper. Res.* **19**: 1081–1089 (1971).

van Nunen, J. A. E. E.: "A Set of Successive Approximation Methods for Discounted Markovian Decision Problems," *Z. Oper. Res.* **20**: 203–208 (1979).

———: *Contracting Markov Decision Processes*, Mathematical Centre Tract 71, Mathematisch Centrum, Amsterdam (1976).

Porteus, E. L.: "Some Bounds for Discounted Sequential Decision Processes," *Manage. Sci.* **18**: 7–11 (1971).

———: "Bounds and Transformation for Discounted Finite Markov Decision Chains," *Oper. Res.* **23**: 761–784 (1975).

———: "Improved Iterative Computation of the Expected Discounted Return in Markov and Semi-Markov Chains," *Z. Oper. Res.* **24**: 155–170 (1980*a*).

———: "Overview of Iterative Methods for Discounted Finite Markov and Semi-Markov Decision Chains," in *Recent Developments in Markov Decision Processes*, edited by R. Hartley, L. C. Thomas, and D. J. White, Academic Press, New York, pp. 1–20 (1980*b*).

———: "Computing the Discounted Return in Markov and Semi-Markov Chains," *Nav. Res. Log. Q.* **28**: 567–578 (1981).

———, and J. C. Totten: "Accelerated Computation of the Expected Discounted Return in a Markov Chain," *Oper. Res.* **26**: 350–358 (1978).

———, and ———: "Optimal Extrapolations for Discounted Finite Markov Reward Chains," Research Paper 315, Graduate School of Business, Stanford University (1976).

Puterman, M. L., and S. L. Brumelle: "Policy Iteration in Stationary Dynamic Programming," *Math. Oper. Res.* **4**: 60–69 (1979).

———, and M. C. Shin: "Modified Policy Iteration Algorithms for Discounted Markov Decision Problems," *Manage. Sci.* **24**: 1127–1137 (1979).

———, and ———: "Action Elimination Procedures for Modified Policy Iteration Algorithms," *Oper. Res.* **30**: 301–318 (1982).

Reetz, D.: "Solution of a Markovian Decision Problem by Overrelaxation," *Z. Oper. Res.* **17**: 29–32 (1973).

Rothblum, U. G.: "Iterated Successive Approximation for Sequential Decision Processes," *Stochastic Control and Optimization*, report of an April 1979 International Conference at the Vrije Universiteit in Amsterdam, edited by J. W. B. van Overhagan and H. C. Tijms: 30–32 (1979).

Schellhaas, H.: "Zur Extrapolation in Markoffschen Entscheidungsmodellen mit Diskontierung," *Z. Oper. Res.* **18**: 91–104 (1974).

Schweitzer, P. J., and A. Federgruen: "Geometric Convergence of Value-Iteration in Multichain Markov Renewal Programming," *Adv. Appl. Probability* **11**: 188–217 (1979).

Shin, M. C.: *Computational Methods for Markov Decision Problems*, Ph.D. dissertation, University of British Columbia (1980).

van der Wal, J., and J. A. E. E. van Nunen: "A Note on the Convergence of the Value Oriented Successive Approximation Method," COSOR Note R77-05, Department of Mathematics, Eindhoven University of Technology (1977).

White, D. J.: "Elimination of Non-Optimal Actions in Markov Decision Processes," in *Dynamic Programming and Its Applications*, edited by M. L. Puterman, Academic Press, New York (1978), pp. 131–160.

Multiple criteria

Brown, T. A., and R. E. Strauch: "Dynamic Programming in Multiplicative Lattices," *J. Math. Anal. Appl.* **12**: 364–370 (1965).

Cochrane, J. L., and M. Zeleny, eds.: *Multiple Criteria Decision-Making*, University of South Carolina Press, Columbia, S.C. (1973).

Furukawa, N.: "Vector-Valued Markovian Decision Processes with Countable State Space," in *Recent Developments in Markov Decision Processes*, edited by R. Hartley, L. C. Thomas, and D. J. White, Academic Press, New York (1980), pp. 205–224.

———: "Characterization of Optimal Policies in Vector-Valued Markovian Decision Processes," *Math. Oper. Res.* **5**: 271–279 (1980).

Henig, M. I.: *Multicriterion Dynamic Programming*, Ph.D. dissertation, Yale University (1978).

———: "Vector-Valued Dynamic Programming," *SIAM J. Control Optimization* (1982).

Jeroslow, R. L.: "Linear Multiple-Objective Programs," unpublished notes for Management Science 6410, Georgia Institute of Technology, Atlanta (1980).

Mendelssohn, R.: "A Systematic Approach to Determining Mean-Variance Tradeoffs When Managing Randomly Varying Populations," *Math. Biosci.* **50**:75–84 (1980).

Mitten, L. G.: "Preference Order Dynamic Programming," *Manage. Sci.* **21**: 43–46 (1974).

Schmee, J., E. Hannan, and M. Mirabile: "An Examination of Patient Referral and Discharge Policies using a Multiple Objective Semi-Markov Decision Process," *J. Oper. Res. Soc.* **30**: 121–129 (1979).

Shin, M. C.: *Computational Methods for Markov Decision Problems*, Ph.D. dissertation, University of British Columbia (1980).

Sobel, M. J.: "Ordinal Sequential Games," *Economies et Sociétés*, **XIV**: 1571–1582 (1980).

Starr, M. K., and M. Zeleny, eds., *Multiple Criteria Decision Making*, vol. 6, TIMS Studies in the Management Sciences, North-Holland, Amsterdam (1977).

Viswanathan, B., V. V. Aggarwal, and K. P. K. Nair: "Multiple Criteria Markov Decision Processes," 263–272 in Starr and Zeleny (1977).

White, C. C., and K. W. Kim: "Solution Procedures for Vector Criterion Markov Decision Processes," *Large Scale Systems*, **1**: 129–140 (1980).

Zeleny, M.: *Multiple Criteria Decision Making*, McGraw-Hill, New York (1982).

Zionts, S. ed.: *Multiple Criteria Problem Solving*, Springer-Verlag, Berlin (1977).

Approximation techniques

Bertsekas, D. P.: "Convergence of Discretization Procedures in Dynamic Programming," *IEEE Trans. Auto. Control* **20**: 415–419 (1975).

Federgruen, A., and P. Zipkin: "Approximations of Dynamic, Multilocation Production and Inventory Problems," Research Working Paper 381A, Graduate School of Business, Columbia University, New York (1980).

Fox, B. L.: "Finite-State Approximation to Denumerable-State Dynamic Programs," *J. Math. Anal. Appl.* **34**: 665–670 (1971).

Hahnewalk-Busch, A., and V. Nollau: "An Approximation Procedure for Stochastic Dynamic Programming in Countable State Space," *Math. Operationsforsch. Stat. Ser. Optimization* **9**: 109–117 (1978).

Hinderer, K.: "On Approximate Solutions of Finite-Stage Dynamic Programs," in *Dynamic Program-*

ming and Its Applications, edited by M. L. Puterman, Academic Press, New York (1979), pp. 289–318.

Langen, H.-J.: "Convergence of Dynamic Programming Models," *Math. Oper. Res.* **6**: 493–512 (1981).

Mendelssohn, R.: "An Iterative Aggregation Procedure for Markov Decision Processes," *Oper. Res.* **30**: 62–73 (1982).

———: "Improved Bounds for Aggregated Linear Programs," *Oper. Res.* **28**: 1450–1453 (1980).

Schweitzer, P. J.: "State Aggregation in Finite Markov Chains," Research Report RC 6200 (26635), Mathematics, Research Division, IBM, Yorktown Heights, N.Y. (1976).

———, M. Puterman, and K. W. Kindle: "Iterative Aggregation-Disaggregation for Solving Discounted Semi-Markovian Reward Processes," Working Paper 8123 (revised), Graduate School of Management, University of Rochester (1982).

Thomas, A.: *Models for Optimal Capacity Expansion*, Ph.D. dissertation, Yale University (1977).

White, C. C. III, and K. Schlussel: "Suboptimal Design for Large Scale, Multimodule Systems," *Oper. Res.* **29**: 865–875 (1981).

White, D. J.: "Finite State Approximations for Denumerable State Infinite Horizon Discounted Markov Decision Processes," in *Recent Developments in Markov Decision Processes*, edited by R. Hartley, L. C. Thomas, and D. J. White, Academic Press, New York (1980), pp. 23–34.

Whitt, W.: "Approximations of Dynamic Programs, I," *Math. Oper. Res.* **3**: 231–243 (1978).

———: "Approximations of Dynamic Programs, II," *Math. Oper. Res.* **4**: 179–185 (1979).

———: "A Priori Bounds for Approximations of Markov Programs," *J. Math. Anal. Appl.* **71**: 297–302 (1979).

Zipkin, P.: "Bounds on the Effect of Aggregating Variables in Linear Programs," *Oper. Res.* **28**: 403–418 (1980).

———: "Bounds for Row-Aggregation in Linear Programming," *Oper. Res.* **28**: 903–916 (1980).

CHAPTER

SEVEN

OTHER STRUCTURAL PROPERTIES

The three preceding chapters examine quite general models used in stochastic optimization. This chapter concerns several specific structural properties which are important in sequential decision processes and which can unify their analyses.

The first two sections analyze the classical inventory problem with setup costs. The third section analyzes a continuous-time production model which is a natural extension of the inventory model. Special cases of the production model include maintenance and queueing models with start-up and shut-down costs.

The inventory problem with setup costs in Section 7-1 is an example of an MDP that is *separable*. Section 7-4 defines a separable MDP and shows that often it may be transformed to a smaller MDP.

Some MDPs have the property that (1) replacement of random variables with their expected values and (2) optimization of the resulting deterministic model is equivalent to optimization of the original MDP. Moreover, step (2) is much simpler than optimization of the original MDP. Section 7-5 defines such an equivalence and presents sufficient conditions for an MDP to possess this *certainty equivalence* property.

7-1 INVENTORY MODELS WITH SETUP COSTS

Actual inventory systems often have the following characteristic. The total cost of reordering goods exhibits economies of scale. In Section 2-1 we explain why these economies occur and observe that they can be modeled in several ways. The setup cost is the most common method used to model them in practice.

This section analyzes a discrete-time dynamic single-product inventory model with setup costs. In such a model, it is foolish to order goods frequently and in small amounts, because you fail to exploit the economy of scale. If the setup cost is $K \geq 0$ and an order quantity is $z > 0$, the setup cost per unit ordered, K/z, is reduced by increasing z. Of course, higher inventory costs are incurred as z increases; so a compromise must be struck.

In such a model, it might appear that a policy of the following simple form is optimal. Do not order any goods at all unless the inventory level is sufficiently low, say lower than σ. Whenever the inventory is lower than σ, order enough to balance the setup cost with the subsequent inventory holding costs, say order q. In fact, this kind of policy is *not* optimal in the model.†

Suppose in period 1 that the inventory level is at σ; so no ordering occurs. Then suppose an unusually large demand for $q + 1$ units drives the inventory down to $\sigma - q - 1$ in period 2. The policy above results in an inventory level of $\sigma - 1$ if ordered goods are delivered instantaneously. Therefore, another order will surely be necessary in period 3, and meanwhile there are heightened risks of a stockout with attendant costs. You might conjecture now that it would have been better in period 2 to order $2q + 1$ units than merely q. Let $\Sigma = \sigma + q$. This new policy orders goods only if the inventory level is below σ, and then orders an amount sufficient to raise inventory to Σ. This is called a (σ, Σ) policy.‡

Scarf (1960) presented the first proof that (σ, Σ) policies are optimal under reasonably general conditions. His proof, a major achievement in inventory theory, depends on a geometric notion which we use in our first proof that (σ, Σ) policies are optimal. Section 7-2 contains an entirely analytical proof which is due to Veinott (1966*a*). It extends the class of problems for which there is an optimal (σ, Σ) policy.§

This section focuses on proving that some (σ, Σ) policy is optimal. At the end of the section we comment on the problem of computing an optimal (σ, Σ) policy. The computational problem is discussed further in Sections 7-3, 8-5, and 8-6.

† The problem of finding a best (σ, q) policy has practical value but it is ill defined. Different (σ, q) policies may be best from different initial states. In other words, an optimal stationary policy may lie outside the class of (σ, q) policies.

‡ The literature on inventory theory uses the term (s, S) policy where we use (σ, Σ) policy. However, the symbols s and S are already used extensively in this book. We choose σ and Σ because they are mnemonically close to s and S.

§ Schäl (1976) has a unified proof with conditions that subsume those of Scarf and Veinott. However, we believe that Scarf's geometric notion and Veinott's analytical approach should first be read separately to appreciate their basic ideas.

A Back-Order Inventory Model

The following formulation is similar to those in Sections 3-1 and 3-2. Suppose that the planning horizon is N periods long and let s_n denote the inventory level at the beginning of period n. At the beginning of each period, s_n is reviewed, additional goods (if any) are ordered and received immediately, demand occurs, and costs are incurred. We saw in Section 3-2 that, if excess demand is backlogged, instantaneous delivery is equivalent to a fixed positive delivery delay. Exercise 7-7 asks you to verify that the equivalence is valid here too. Let a_n denote the inventory level after ordered goods (if any) are delivered. Therefore, $a_n - s_n$ is the quantity ordered and $a_n - s_n \geq 0$ or

$$s_n \leq a_n \qquad n = 1, \ldots, N \tag{7-1}$$

is a constraint.

Let $D_1, D_2, \ldots, D_N$ be independent nonnegative r.v.'s that are the demands in successive periods. Suppose that excess demand is backlogged so

$$s_{n+1} = a_n - D_n \qquad n = 1, \ldots, N \tag{7-2}$$

We interpret s_{N+1} as the terminal inventory level to which we shortly assign a salvage value. If $s_{N+1} < 0$, then $-s_{N+1}$ units are owed to consumers. Assumption (7-2) is relaxed in Section 7-2; then excess demand may be lost as well as backlogged.

The cost structure is a nonstationary version of the model in Section 3-2 augmented by setup costs. Suppose $a_n = a$ and $D_n = d$. Then $g_n(a, d)$ denotes the present value at the beginning of period 1 of the inventory or back-order costs in period n minus that period's revenue. For example, if β is the discount factor per period, h_1 and h_2 are the respective unit costs of inventories and back orders, and r is the price, then

$$g_n(a, d) = \beta^{n-1}h_1(a - d)^+ + \beta^{n-1}h_2(d - a)^+ - \beta^{n-1}rd \tag{7-3}$$

if all revenue is received at the time that goods are requested.

Let $\delta(u) = 1$ if $u > 0$ and $\delta(u) = 0$ if $u \leq 0$. If $s_n = s$ and $a_n = a$, the cost of ordered goods in period n is

$$K_n \delta(a - s) + c_n \cdot (a - s)$$

Finally, let c_{N+1} denote the salvage value per unit of s_{N+1}. If $s_{N+1} < 0$, then c_{N+1} is the cost per unit of back-ordered demand. For reasons briefly mentioned in Section 3-1, in practice the unit cost of back-ordered demand is likely to be higher than the unit salvage value. We assume that all costs are nonnegative. The discount factor for period n, β^{n-1}, is incorporated in K_n and c_n for notational convenience. The assumption of a linear salvage-value function is relaxed later in the section.

Let B denote the present value, at the beginning of period 1, of all costs that are incurred during periods $1, \ldots, N + 1$. Then

$$B = \sum_{n=1}^{N} [K_n \delta(a_n - s_n) + c_n(a_n - s_n) + g_n(a_n, D_n)] - c_{N+1}s_{N+1}$$

Substitution of $s_n = a_{n-1} - D_{n-1}$ for $n > 1$ yields

$$B = \sum_{n=1}^{N} [K_n \delta(a_n - s_n) + g_n(a_n, D_n) + (c_n - c_{n+1})a_n] - c_1 s_1 + \sum_{n=1}^{N} c_{n+1} D_n$$

Let

$$G_n(a) \triangleq (c_n - c_{n+1})a + E[g_n(a, D_n)] \qquad a \in \mathbb{R} \tag{7-4}$$

Then

$$E(B) = \sum_{n=1}^{N} E[K_n \delta(a_n - s_n) + G_n(a_n)] - c_1 s_1 + \sum_{n=1}^{N} c_{n+1} E(D_n) \tag{7-5}$$

The second and third terms in (7-5) are not affected by the policy used to choose $a_1, a_2, \ldots, a_N$; so there is no loss of optimality in using the first term in (7-5) as the criterion of minimization.

Exercise 7-1 asks you to verify that the minimization of

$$\sum_{n=1}^{N} E[K_n \delta(a_n - s_n) + G_n(a_n)]$$

is equivalent to solving the following dynamic program:

$$f_n(s) = \inf \{K_n \delta(a - s) + J_n(a) \colon a \geq s\} \qquad s \in \mathbb{R} \tag{7-6a}$$

$$J_n(a) = G_n(a) + E[f_{n+1}(a - D_n)] \qquad a \in \mathbb{R} \tag{7-6b}$$

for $n = 1, 2, \ldots, N$, with $f_{N+1}(\cdot) \equiv 0$.

(σ, Σ) Policies and Scarf's Notion of *K*-Convexity

Here is the formal definition of a (σ, Σ) policy.

Definition 7-1 A (σ, Σ) *policy* consists of parameters $\sigma_n \leq \Sigma_n$, $n = 1, 2, \ldots, N$ such that

$$a_n = \begin{cases} \Sigma_n & \text{if} \quad s_n < \sigma_n \\ s_n & \text{if} \quad s_n \geq \sigma_n \end{cases} \tag{7-7}$$

Figure 7-1 shows how such a policy might not be optimal. The proof of Theorem 7-1 below precludes such graphs. In Figure 7-1, Σ_n denotes the global minimum of $J_n(\cdot)$. It is optimal to order up to Σ_n, that is, $f_n(s) = K_n + J_n(\Sigma_n)$, for any $s < \Sigma_n$ such that $J_n(s) > K_n + J_n(\Sigma_n)$. This claim follows from rewriting (7-6*a*) as

$$f_n(s) = \min \{J_n(s), K_n + \inf \{J_n(a) \colon a > s\}\}$$

Therefore, in Figure 7-1, an optimal policy is

$$a_n = \begin{cases} \Sigma_n & \text{if } s_n < w_1 \\ s_n & \text{if } w_1 \leq s_n \leq w_2 \\ \Sigma_n & \text{if } w_2 < s_n < w_3 \\ s_n & \text{if } w_3 \leq s_n \leq w_4 \\ w_6 & \text{if } w_4 < s_n < w_5 \\ s_n & \text{if } w_5 \leq s_n \end{cases}$$

This specification requires seven parameters versus only two needed in (7-6).

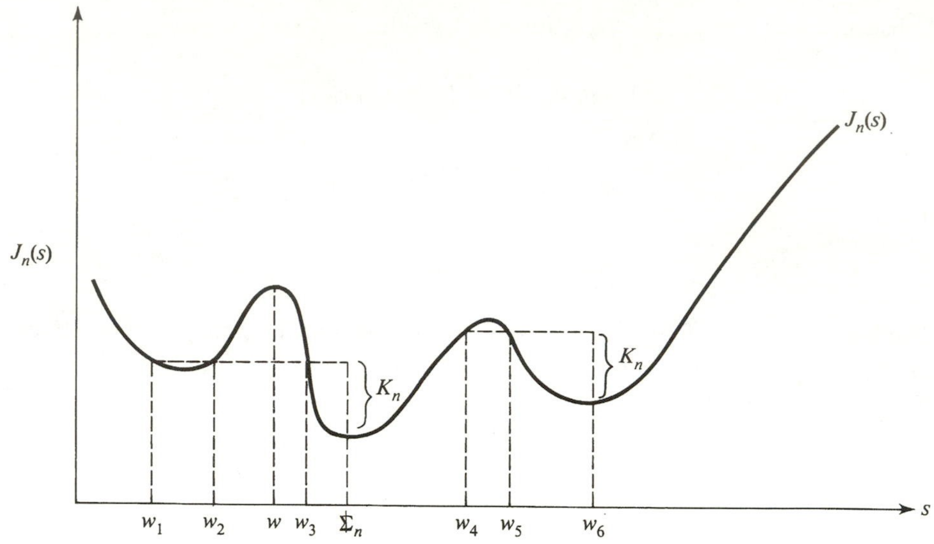

Figure 7-1 Hypothetical case lacking optimal (σ_n, Σ_n) policy.

Definition 7-2 A real-valued function $f(\cdot)$ on $\mathbb{R}$ is *K-convex* ($K \geq 0$) if

$$K + f(x + \gamma) \geq f(x) + \frac{\gamma}{\theta}[f(x) - f(x - \theta)] \tag{7-8}$$

for all $x \in \mathbb{R}$, $\gamma \geq 0$, and $\theta > 0$.

Inequality (7-8) asserts that K plus the value of $f(\cdot)$ at $x + \gamma$ is at least as great as the projection at $x + \gamma$ of the line segment that connects $f(x)$ and $f(x - \theta)$.

For any real-valued function $f(\cdot)$ on $\mathbb{R}$, $\gamma \geq 0$ and $\theta > 0$, we define

$$\Delta_f(x, \gamma, \theta) = K + f(x + \gamma) - f(x) - \frac{\gamma}{\theta}[f(x) - f(x - \theta)]$$

Then K-convexity is the same as $\Delta_f(\cdot, \cdot, \cdot) \geq 0$.

In order to relate K-convexity to ordinary convexity, in (7-8) let $x_1 = x - \theta$, $x_2 = x + \gamma$, and $\lambda = \gamma/(\gamma + \theta)$ so $x = \lambda x_1 + (1 - \lambda)x_2$. Then $\gamma/\theta = \lambda/(1 - \lambda)$ so (7-8) is

$$K + f(x_2) \geq \frac{1}{\lambda} f[\lambda x_1 + (1 - \lambda)x_2] - \frac{\lambda}{1 - \lambda} f(x_2)$$

which becomes

$$f[\lambda x_1 + (1 - \lambda)x_2] \leq \lambda f(x_1) + (1 - \lambda)f(x_2) + (1 - \lambda)K$$

which is ordinary convexity if $K = 0$.

Exercise 7-1 asks you to explain why the hypothetical $J_n(\cdot)$ in Figure 7-1 is not K_n-convex.

Most of the following properties (which Exercise 7-2 asks you to verify) of K-convex functions are used in the optimality proof.

Lemma 7-1

(a) $H(\cdot)$ is 0-convex $\Leftrightarrow H(\cdot)$ is convex on $\mathbb{R}$.
(b) $H(\cdot)$ is K-convex $\Rightarrow H(\cdot + u)$ is K-convex for all $\mu \in \mathbb{R}$.
(c) $H_i(\cdot)$ is K-convex, $i = 1, 2 \Rightarrow \alpha_1 H_1(\cdot) + \alpha_2 H_2(\cdot)$ is $\alpha_1 K_1 + \alpha_2 K_2$-convex ($\alpha_1 > 0, \alpha_2 > 0$).
(d) $H(\cdot)$ is K-convex $\Rightarrow H(\cdot)$ is V-convex for all $V \geq K$.
(e) $H(\cdot)$ is K-convex $\Rightarrow H(\cdot)$ is continuous on $\mathbb{R}$.
(f) $H(\cdot)$ is K-convex $\Rightarrow H(\cdot)$ is differentiable on $\mathbb{R}$ except for at most countably many points.

Optimality of (σ, Σ) Policies

Recall the notation $L(s_{N+1})$ for the salvage-value function and $s_{n+1} = v_n(a_n, D_n)$ for the dependence of next period's starting inventory on this period's supply and demand.

Theorem 7-1 Suppose $L(s) = c_{N+1}s$ and for each $n = 1, 2, \ldots, N$,
(a) $E[g_n(\cdot, D_n)]$ is a convex function on $\mathbb{R}$,
(b) $G_n(a) \to \infty$ as $|a| \to \infty$,
(c) $K_1 \geq K_2 \geq \cdots \geq K_N \geq 0$
(d) $v_n(a, d) = a - d$.

Then there is an optimal (σ, Σ) policy.

Assumptions (a) and (b) are relaxed in Theorem 7-4. Nevertheless, convexity of $E[g_n(\cdot, D_n)]$ with (7-4) is a weaker assumption than convexity of $g_n(\cdot, d)$ for each d. For example, suppose

$$g_n(a, d) = h_1(a - d)^+ + h_2(d - a)^+ + h_3\,\delta(d - a)$$

where h_3 is a fixed charge if a stockout occurs. Then $E[g_n(\cdot, D_n)]$ is convex if $h_3 = 0$ or if $h_3 > 0$ and D_n has a concave distribution function. In the latter case, $g_n(\cdot, d)$ is not convex for each d.† In Theorems 7-2, 7-3, and 7-4 in Section 7-2, we relax assumption (d) and admit models in which excess demand is lost.

PROOF OF THEOREM 7-1 To initiate an induction, $f_{N+1}(\cdot) \equiv 0$ in (7-6) yields $J_N(a) = G_N(a)$, which is convex due to (a), hence 0-convex due to part (a) of Lemma 7-1, hence K_N-convex due to part (d) of Lemma 7-1. For any n, suppose $J_n(\cdot)$ is K_n-convex. Then $f_n(\cdot)$ is K_n-convex and (σ_n, Σ_n) is optimal due to the next lemma.

† The reader who wishes to skip the proof should move to page 314, on which computational issues are discussed.

Lemma 7-2 Suppose $J(\cdot)$ is K-convex, attains its global minimum at Σ, and there is a smallest number $\sigma \le \Sigma$ such that

$$J(\sigma) \le V + J(\Sigma) \tag{7-9}$$

where $K \le V$. Then $f(\cdot)$ is V-convex where

$$f(s) = \inf \{V\delta(a - s) + J(a) \colon a \ge s\} \qquad s \in \mathbb{R} \tag{7-10}$$

Proof The first step is to prove optimality of (σ, Σ) in (7-10), that is,

$$f(s) = \begin{cases} V + J(\Sigma) & s < \sigma \\ J(s) & s \ge \sigma \end{cases} \tag{7-11}$$

By definition of σ and Σ, $J(s) > V + J(\Sigma)$ for all $s < \sigma$; so $f(s) = V + J(\Sigma)$. Therefore, if (7-11) is false, there is some $u \ge \sigma$ such that $J(u) > V + \inf \{J(a) \colon a > u\}$. If $\sigma < u < \Sigma$ as in Figure 7-2, then $J(u) > V + J(\Sigma) \ge J(\sigma)$ and $\sigma < \Sigma$; so

$$\begin{aligned} \Delta_J(u, \Sigma - u, u - \sigma) &= K + J(\Sigma) - J(u) - \frac{\Sigma - u}{u - \sigma}[J(u) - J(\sigma)] \\ &< V + J(\Sigma) - J(\sigma) - \frac{\Sigma - u}{u - \sigma}[J(u) - J(\sigma)] \\ &\le -\frac{\Sigma - u}{u - \sigma}[J(u) - J(\sigma)] < 0 \end{aligned}$$

which is contrary to K-convexity of $J(\cdot)$. Instead of $u \in (\sigma, \Sigma)$, suppose $\Sigma < u$

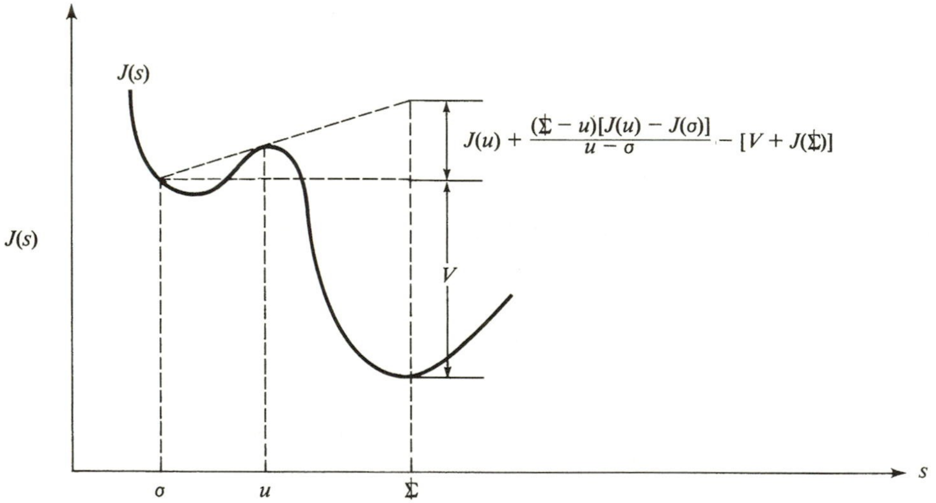

Figure 7-2 Suppose (σ, Σ) is suboptimal at u, $\sigma < u < \Sigma$.

as in Figure 7-3. Then there is some $\gamma > 0$ such that $J(u) > V + J(u + \gamma)$. Using an argument similar to the case $\sigma < u < \Sigma$, $\Delta_J(u, \gamma, u - \Sigma) < 0$, which is contrary to K-convexity $J(\cdot)$. Therefore, (7-11) is valid.

We use four cases to prove V-convexity of $f(\cdot)$, that is,

$$0 \le \Delta_f(s, \gamma, \theta) = V + f(s + \gamma) - f(s) - \frac{\gamma}{\theta}\,[f(s) - f(s - \theta)]$$

If $s - \theta \ge \sigma$, then

$$\Delta_f(s, \gamma, \theta) = V + J(s + \gamma) - J(s) - \frac{\gamma}{\theta}\,[J(s) - J(s - \theta)]$$

$$\ge K + J(s + \gamma) - J(s) - \frac{\gamma}{\theta}\,[J(s) - J(s - \theta)]$$

which is nonnegative because $J(\cdot)$ is K-convex.

If $s + \gamma < \sigma$, then $\Delta_f(s, \gamma, \theta) = V \ge 0$.

If $s < \sigma < s + \gamma$, then

$$\Delta_f(s, \gamma, \theta) = V + J(s + \gamma) - [V + J(\Sigma)] - \gamma \cdot \frac{0}{\theta} = J(s + \gamma) - J(\Sigma) \ge 0$$

If $s - \theta < \sigma \le s$, then

$$\Delta_f(s, \gamma, \theta) = V + J(s + \gamma) - J(s) - \frac{\gamma}{\theta}\,[J(s) - V - J(\Sigma)]$$

$$\ge -\frac{\gamma}{\theta}\,[J(s) - V - J(\Sigma)] \tag{7-12}$$

due to the following property (which Exercise 7-5 asks you to prove).

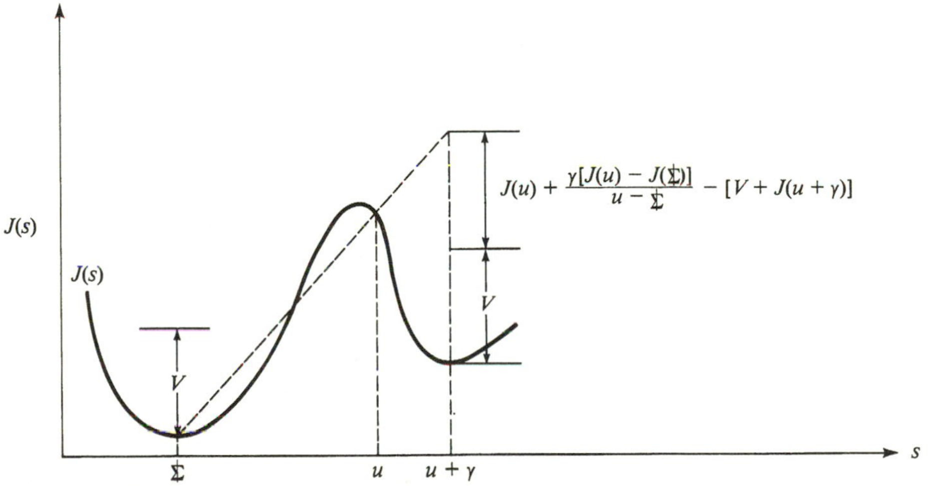

Figure 7-3 Suppose (σ, Σ) is suboptimal at u, $\Sigma < u$.

Lemma 7-3 For a real-valued function $J(\cdot)$ on $\mathbb{R}$ and $V \geq 0$, let

$$f(s) = \inf \{V\delta(a - s) + J(a): a \geq s\} \qquad s \in \mathbb{R}$$

For all $s \in \mathbb{R}$ and $\gamma \geq 0$,

$$f(s) \leq V + f(s + \gamma)$$

From (7-12), $\Delta_f(s, \gamma, \theta) \geq 0$ if $J(s) \leq V + J(\Sigma)$. Suppose $J(s) > V + J(\Sigma)$. Then $V \geq K$ and the definition of σ imply $\sigma < s$ because $J(\sigma) \leq K + J(\Sigma)$. Then $s - \theta < \sigma < s$, or $0 < s - \sigma < \theta$, implies

$$\Delta_f(s, \gamma, \theta) \geq V + J(s + \gamma) - J(s) - \frac{\gamma}{s - \sigma} [J(s) - V - J(\Sigma)]$$

$$\geq K + J(s + \gamma) - J(s) - \frac{\gamma}{s - \sigma} [J(s) - J(\sigma)]$$

which is nonnegative because $J(\cdot)$ is K-convex [in (7-8) let $L(\cdot) = J(\cdot)$, $x = s$, and $\theta = s - \sigma$]. This completes the proof of Lemma 7-2. □

Assumption (*b*) implies that $J_n(a) \to \infty$ as $|a| \to \infty$. Then continuity of $J_n(\cdot)$ [part (*e*) of Lemma 7-1] due to K_n-convexity implies that the global minimum of $J_n(\cdot)$ is attained, say at Σ_n. Also, σ_n, the smallest number x such that $x \leq \Sigma_n$ and $J_n(x) \leq K_n + J_n(\Sigma_n)$, is well defined.

To complete the theorem's proof, it must be shown that K_n-convexity of $f_n(\cdot)$ implies K_{n-1}-convexity of

$$J_{n-1}(a) = G_{n-1}(a) + E[f_n(a - D_{n-1})]$$

Part (*c*) of Lemma 7-1 implies that a weighted average of K_n-convex functions is itself K_n-convex. A limiting argument implies $E[f_n(a - D_{n-1})]$ is K_n-convex. Assumption (*a*) and part (*a*) of Lemma 7-1 yield 0-convexity of $G_{n-1}(\cdot)$; so part (*c*) of Lemma 7-1 now yields K_n-convexity of the sum $J_{n-1}(\cdot)$. Finally, part (*d*) of Lemma 7-1 and $K_{n-1} \geq K_n$ due to assumption (*c*) imply K_{n-1}-convexity of $J_{n-1}(\cdot)$. □

Computing an Optimal (σ, Σ) Policy

Sections 8-5 and 8-6 establish that infinite-horizon versions of the inventory models in this section and Section 7-2 possess optimal (σ, Σ) policies. There are several approaches to the computation of an optimal (σ, Σ) policy in such models. One approach is explored in Section 7-3. Two other approaches exploit the structure of the stationary Markov process induced by a stationary (σ, Σ) policy. Veinott and Wagner (1965) brought these two approaches to fruition and presented the details of the algorithm. The algorithm depends on formulas which are presented in Examples 6-3 and 6-8 on pages 176 and 190 of Volume I. Other approaches, refinements, and approximations are cited in the Bibliographic Guide at the end of this chapter.

EXERCISES

7-1 The hypothetical $J_n(\cdot)$ in Figure 7-1 is not K_n-convex. Why?

7-2 Prove parts (*a*) to (*e*) of Lemma 7-1.

7-3 Prove part (*f*) of Lemma 7-1.

7-4 Under the hypotheses of Theorem 7-1, prove that $J_n(\cdot)$ cannot possess a strict relative minimum at u if $u < \sigma_n$.

7-5 Prove Lemma 7-3.

7-6 Lemma 7-2 can be strengthened if $J(\cdot)$ is differentiable (one-sided derivatives suffice). Suppose $J(\cdot)$ is differentiable, K-convex, attains its global minimum at Σ, and there is a smallest number σ such that $J(\sigma) \le V + J(\Sigma)$ where $0 \le V \le K$. Define $f(\cdot)$ with (7-10). Prove that $f(\cdot)$ is K-convex. [Hint: A differentiable function $J(\cdot)$ is K-convex if, and only if,

$$0 \le K + J(s + \gamma) - J(s) - \gamma J'(s)$$

for all $s \in \mathbb{R}$ and $\gamma \ge 0$.] We do not know if this result is valid without a differentiability assumption.

7-7 Suppose the model on which (7-6*a*) and (7-6*b*) is based is altered so that delivered goods are received after a delay of v periods, where $v \in I$. Assume that excess demand is back-ordered and that payment (setup cost and linear purchase cost) for ordered goods occurs at the time that the order is placed. Prove that the minimization of $E(B)$ is equivalent to (7-6*a*) and (7-6*b*), where s denotes the sum of the amounts of goods on hand and on order. (The wording is deliberately vague. You must specify B and the sense of the equivalence. Hint: Section 3-2.)

It follows that the assumptions of Theorem 7-1 imply the existence of an optimal (σ, Σ) policy.

7-8 How is the answer to Exercise 7-7 affected if the payment for ordered goods takes place at the time they are delivered?

7-9 (From E. L. Porteus) Prove that a real-valued function $L(\cdot)$ on $\mathbb{R}$ is K-convex if, and only if,

$$L[\lambda x_1 + (1 - \lambda)x_2] \le \lambda L(x_1) + (1 - \lambda)[L(x_2) + K]$$

for all $x_1 < x_2$ and $0 \le \lambda \le 1$.

7-10 A real-valued function $L(\cdot)$ on $-\infty \le a < b \le \infty$ is *non-K-decreasing* if

$$L(x) \le L(y) + K$$

for $a \le x \le y \le b$. It is *quasi-K-convex* if there is a number x^* such that $L(\cdot)$ is nonincreasing on $(-\infty, x^*]$ and non-K-decreasing on $[x^*, \infty)$. Prove that $L(\cdot)$ is quasi-K-convex if, and only if,

$$L[\lambda x_1 + (1 - \lambda)x_2] \le \max\{L(x_1), L(x_2) + K\}$$

for all $x_1 < x_2$ and $0 \le \lambda \le 1$.

7-11: Cash management at a retail bank branch†. Let $D_1, D_2, \ldots$ be independent r.v.'s which denote the amounts of cash withdrawn on successive business days at a bank branch. We treat cash deposits as negative contributions to D_n so $D_n < 0$ has positive probability. The cost of having cash in the branch at day's end is the interest h which would otherwise be earned per dollar invested. The other significant cost is the armored car's fee for delivering or picking up cash. This fee is K if the delivery or pickup is scheduled early in the morning, but it is $L (L > K)$ if it is an emergency delivery caused by D_n exceeding cash on hand. Let s_n and a_n denote the cash on hand at the beginning of the day before and after the ordered cash is delivered, respectively. Then we model the single-period holding-penalty cost function as

$$g(a_n, D_n) = h \cdot (a_n - D_n)^+ + L\delta(D_n - a_n)$$

† See Section 8-4 for a different cash-management model.

The fee on day n for the scheduled delivery or pickup can be written $K\delta(|a_n - s_n|)$ so the minimization of expected discounted costs yields the dynamic program

$$f_n(s) = \inf\{K\delta(|a - s|) + E\{g(a, D_n) + \beta f_{n-1}[(a - D_n)^+]\}: \quad a \geq 0\}$$

with $f_0(\cdot) \equiv 0$. It is implicit in this formulation that an emergency delivery, if one is necessary, provides only $D_n - a$ dollars. This unrealistic assumption simplifies the analysis, as does the assumption, made now, that $D_1, D_2, \ldots$ are identically distributed. Let $F(\cdot)$ denote their distribution function and assume that $F(\cdot)$ is continuous. Under these assumptions, prove that there are parameters $\sigma_n^L \leq \Sigma_n \leq \sigma_n^U$ for each n such that

$$a_n = \begin{cases} \Sigma_n & \text{if } s_n < \sigma^L \\ s_n & \text{if } \sigma_n^L \leq s_n \leq \sigma_n^U \\ \Sigma_n & \text{if } \sigma_n^U < s_n \end{cases}$$

is an optimal policy.

7-12 The primary shortcoming of the model in Exercise 7-11 is the assumption that successive amounts of cash withdrawn, $D_1, D_2, \ldots$ are i.i.d. r.v.'s. For example, we observe smaller amounts withdrawn on Mondays than on Tuesdays, Wednesdays, and Thursdays. Suppose $D_1, D_2, \ldots$ are independent but not necessarily identically distributed. What additional assumption is sufficient for the kind of policy described in Exercise 7-11 to remain optimal? Justify your answer.

7-13 Suppose that the back-order inventory model in this section is altered so that $D_1, D_2, \ldots$ comprise a Markov chain with state space I and transition probabilities $\{p_{ij}\}$. Then the dynamic program (7-6a) and (7-6b) is altered to

$$f_n(s, i) = \inf\{K_n\delta(a - s) + J_n(a, i): a \geq s\}$$

$$J_n(a, i) = G_n(a, i) + \sum_{j \in I} p_{ij} f_{n+1}(a - j, j)$$

for $1 \leq n \leq N$, where the argument i in $f_n(s, i)$ indicates that $D_{n-1} = i$. Let $f_{N+1}(\cdot, \cdot) \equiv 0$. The following generalization of a (σ, Σ) policy is optimal under certain assumptions. For each $i \in I$ and $n = 1, \ldots, N$, there are parameters $\sigma_{ni} \leq \Sigma_{ni}$ such that

$$a_n = \begin{cases} \Sigma_{ni} & \text{if } s_n < \sigma_{ni} \text{ and } D_{n-1} = i \\ s_n & \text{if } s_n \geq \sigma_{ni} \text{ and } D_{n-1} = i \end{cases}$$

Specify sufficient conditions for such a policy to be optimal and justify your answer.

7-2* FURTHER RESULTS FOR INVENTORY MODELS WITH SETUP COSTS

This section has two purposes. The first is to show that a (σ, Σ) policy is optimal for inventory models in which excess demand is lost (instead of being back-ordered as Theorem 7-1 assumes). The second purpose is to use an optimality proof which is entirely analytical so it is dramatically different from the geometric basis of the proof of Theorem 7-1. We show that a general inventory model has an optimal (σ, Σ) policy. The model includes lost sales and back ordering as special cases.

Lost Sales

Recall the notation $s_{n+1} = v_n(a_n, D_n)$ for the dependence of next period's starting inventory on this period's supply and demand. Assumption (d) in Theorem 7-1 is that all excess demand is back-ordered, that is, $v_n(a, d) = a - d$ for all n, a, and d. Now we extend Theorem 7-1 to the lost-sales case, that is, $v_n(a, d) = (a - d)^+$.

The substitution of

$$s_n = (a_{n-1} - D_{n-1})^+ = a_{n-1} - D_{n-1} + (D_{n-1} - a_{n-1})^+$$

for $n > 1$ in the present-value expression

$$B = \sum_{n=1}^{N} [K_n \delta(a_n - s_n) + c_n \cdot (a_n - s_n) + g_n(a_n, D_n)] - c_{N+1}s_{N+1}$$

yields

$$B = \sum_{n=1}^{N} [K_n \delta(a_n - s_n) + g_n(a_n, D_n) + (c_n - c_{n+1})a_n - c_{n+1}(D_n - a_n)^+] - c_1 s_1 + \sum_{n=1}^{N} c_{n+1}D_n$$

Let

$$G_n(a) \triangleq (c_n - c_{n+1})a + E[g_n(a, D_n) - c_{n+1}(D_n - a)^+] \qquad a \geq 0 \qquad (7\text{-}13)$$

Then

$$E(B) = \sum_{n=1}^{N} E[K_n \delta(a_n - s_n) + G_n(a_n)] - c_1 s_1 + \sum_{n=1}^{N} c_{n+1}E(D_n) \qquad (7\text{-}14)$$

It is instructive to compare this expression for the expected present value with (7-5), which is the analogous expression in the backlog case. The only difference is the term $-c_{n+1}E(D_n - a)^+$, which appears in (7-13) but not in (7-4). Veinott and Wagner (1965), who obtained the result in this subsection, interpret the term as follows. The quantity $c_{n+1}E(D_n - a_n)^+$ is the expected purchase cost† in period $n + 1$, which would be present in a back-order model but not in a lost-sales model. Hence, $G(a)$ in the two models should differ by $c_{n+1}E(D_n - a_n)^+$.

The dynamic program for (7-14) (neglecting the "nuisance terms") is

$$f_n(s) = \inf \{K_n \delta(a - s) + J_n(a) \colon a \geq s\} \qquad s \geq 0 \qquad (7\text{-}15)$$

$$J_n(a) = G_n(a) + E\{f_{n+1}[(a - D_n)^+]\} \qquad a \geq 0 \qquad (7\text{-}16)$$

for $n = 1, \ldots, N$ with $f_{N+1}(\cdot) \equiv 0$. Expanding (7-16),

$$J_n(a) = G_n(a) + f_{n+1}(0)P\{D_n > a\} + E[f_{n+1}(a - D_n) \mid D_n \leq a]P\{D_n \leq a\} \quad (7\text{-}17)$$

We wish to extend the domains of $f_n(\cdot)$ and $J_n(\cdot)$ from $\mathbb{R}_+$ to $\mathbb{R}$ in such a way that (7-15) remains valid on $\mathbb{R}_+$ but

$$J_n(a) = G_n(a) + E[f_{n+1}(a - D_n)] \qquad a \in \mathbb{R} \qquad (7\text{-}18)$$

† This interpretation depends on the assumption that $\sigma_{n+1} > 0$ in the back-order model; so $s_{n+1} < 0$ implies that the order quantity in period $n + 1$ is at least s_{n+1} units.

instead of (7-16). A comparison of (7-17) with (7-18) shows that a sufficient condition for the extension to preserve (7-15) is $f_{n+1}(s) = f_{n+1}(0)$ for all $s < 0$. The extended function has this property if it has an optimal (σ, Σ) policy with $\sigma_{n+1} > 0$. Then

$$f_{n+1}(s) = K + J_n(\Sigma_n) \qquad \text{for all } s \le 0$$

The extension of $G_n(\cdot)$ from $\mathbb{R}_+$ to $\mathbb{R}$ is usually easy. From (7-13), $g_n(\cdot, d)$ is the only component of $G_n(\cdot)$ which might not at first be defined on $\mathbb{R}$. However, the most common inventory cost functions are already defined on $\mathbb{R}$. A good example is (7-3), which stipulates linear holding costs and linear penalty costs.

Suppose that $G_n(\cdot)$ in (7-13) has been extended to $\mathbb{R}$ and (7-15) with (7-18) has an optimal (σ, Σ) policy with $\sigma_n > 0$ for $n = 1, 2, \ldots, N$. It follows from the preceding discussion that the same policy is optimal for the original lost-sales model. Therefore, Theorem 7-1 has the following corollary whose proof is immediate.

Theorem 7-2 Suppose $L(s) = c_{N+1}s$, $G(\cdot)$ is specified by (7-13) and extended to $\mathbb{R}$, and for each $n = 1, 2, \ldots, N$,
(*a*) $E[g_n(\cdot, D_n) - c_{n+1}(D_n - \cdot)^+]$ is a convex function on $\mathbb{R}$.
(*b*) $G_n(a) \to \infty$ as $|a| \to \infty$.
(*c*) $v_n(a, d) = (a - d)^+$.
(*d*) $K_1 \ge K_2 \ge \cdots \ge K_N \ge 0$.

Then there is an optimal (σ, Σ) policy for (7-15) with (7-18), and if $\sigma_n > 0$ for $n = 1, 2, \ldots, N$, this policy is optimal for (7-15) with (7-16).

The term $-c_{n+1}E[(D_n - a)^+]$ is not a convex function of a if $c_{n+1} > 0$. Nevertheless, assumption (*a*) is often valid. For example, if there are linear holding and penalty costs, $g_n(a, d) - c_{n+1}(d - a)^+$ has the form

$$h_{1,n} \cdot (a - d)^+ + h_{2,n} \cdot (d - a)^+ - c_{n+1}(d - a)^+$$

If $h_{2,n} \ge c_{n+1}$, this expression is a convex function of a for each value of d; so assumption (*a*) is satisfied. In practice, the unit penalty cost of excess demand is greater than the unit purchase cost.

The assumption $\sigma_n > 0$ can be verified in many cases before solving for an optimal (σ, Σ) policy. Below, in (7-29*a*) to (7-29*c*), (7-30), and Theorem 7-4, there are sufficient conditions for $\sigma_n \ge \underline{\sigma}_n$ where $\underline{\sigma}_n$ is an easily determined lower bound. Hence $\underline{\sigma}_n > 0$ is an adequate condition for $\sigma_n > 0$. Finally, $\sigma_n > 0$ in most practical problems.

Lost or Back-Ordered Excess Demand

Recall the notation $s_{n+1} = v_n(a_n, D_n)$ for the dependence of next period's starting inventory on this period's supply and demand. Assumption (*d*) in Theorem 7-1 is $v_n(a, d) = a - d$ for all n, a, and d; that is, all excess demand is back-ordered. Now we extend Theorem 7-2 to the case

$$v_n(a, d) = \xi_{1,n}(a - d)^+ - \xi_{2,n} \cdot (d - a)^+$$

where† $0 \le \xi_{2,n} \le \xi_{1,n}$.

In particular, if $\xi_{2,n} = 0$ and $\xi_{1,n} = 1$, all excess demand is lost ("lost-sales" case), and if $\xi_{2,n} = \xi_{1,n} = 1$, all excess demand is back-ordered. If $\xi_{1,n} < 1$, we interpret $1 - \xi_{1,n}$ as the fraction of inventoried goods which "perish" in storage each period. Goods may "perish" because of obsolescence, theft, fragility, and physical deterioration. If $0 < \xi_{2,n} < 1$, we interpret $\xi_{2,n}$ as the fraction of excess demand which is back-ordered.

Instead of the back-ordering assumption, we write merely $s_{n+1} = v_n(a_n, D_n)$ for each n and assume

$$g_n(a, d) = h_n(a, d) - r_n \cdot [a - v_n(a, d)] \tag{7-19}$$

The term $h_n(a, d)$ denotes the inventory and penalty costs not offset by revenue. The second term in (7-19) is the revenue represented as the product of the price r_n and the amount of goods sold in period n, that is,

$$a_n - s_{n+1} = a_n - v_n(a_n, D_n)$$

This interpretation is valid only if goods are not "perishable."

The substitution of (7-19) and $s_n = v_{n-1}(a_{n-1}, D_{n-1})$ for $n > 1$ in the expression

$$B = \sum_{n=1}^{N} [K_n \delta(a_n - s_n) + c_n(a_n - s_n) + g_n(a_n, D_n)] - c_{N+1}s_{N+1}$$

yields

$$B = \sum_{n=1}^{N} [K_n \delta(a_n - s_n) - (r_n - c_n)a_n + h_n(a_n, D_n) + (r_n - c_{n+1})v_n(a_n, D_n)] - c_1 s_1$$

The *re*definition

$$G_n(a) = -(r_n - c_n)a + E[h_n(a, D_n) + (r_n - c_{n+1})v_n(a, D_n)] \tag{7-20}$$

yields

$$E(B) = \sum_{n=1}^{N} E[K_n \delta(a_n - s_n) + G_n(a_n)] - c_1 s_1$$

By neglecting the "nuisance term" $-c_1 s_1$, the resulting dynamic problem is

$$f_n(s) = \inf \{K_n \delta(a - s) + J_n(a): a \ge s\} \qquad s \in \mathbb{R} \tag{7-21a}$$

$$J_n(a) = G_n(a) + E\{f_{n+1}[v_n(a, D_n)]\} \qquad a \in \mathbb{R} \tag{7-21b}$$

for $n = 1, 2, \ldots, N$ with $f_{N+1}(\cdot) \equiv 0$.

† The restriction $0 \le \xi_{2,n} \le \xi_{1,n}$ implies for each n and d that $v_n(\cdot, d)$ is nondecreasing and convex on $\mathbb{R}$, as we shall assume in Theorem 7-3.

Suppose that $G_n(\cdot)$ in (7-20) is convex for each n, we make assumptions (*b*) and (*c*) in Theorem 7-1 and try to apply the proof of Theorem 7-1 to (7-21*a*) and (7-21*b*). No obstacle arises until the last paragraph of the proof, which tries to verify that K_n-convexity of $f_n(\cdot)$ implies K_{n-1}-convexity of

$$J_{n-1}(a) = G_{n-1}(a) + E\{f_n[v_{n-1}(a, D_{n-1})]\} \tag{7-22}$$

The obstacle is that it is not apparent that $E\{f_n[v_{n-1}(a, D_{n-1})]\}$ is K_n-convex. Heretofore, part (*c*) of Lemma 7-1 implied K_n-convexity of $E[f_n(a - D_{n-1})]$.

Theorem 7-3 Suppose $L(s) = c_{N+1}s$ and for each $n = 1, 2, \ldots, N$,
(*a*) $r_n \geq c_{n+1}$ and $E[h_n(\cdot, D_n)]$ is a convex function on $\mathbb{R}$.
(*b*) $v_n(\cdot, d)$ is a nondecreasing and convex function on $\mathbb{R}$ for each d.
(*c*) $G_n(a) \to \infty$ as $|a| \to \infty$.
(*d*) $K_1 \geq K_2 \geq \cdots \geq K_N \geq 0$.

Then there is an optimal (σ, Σ) policy.

The assumptions are rather mild. If costs and revenues are discounted so $r_n = \beta^{n-1}r$ and $c_{n+1} = \beta^n c$, then $r_n \geq c_{n+1}$ if $r \geq \beta c$, which is satisfied if $r \geq c$, which is valid in all commercially viable inventory systems (price $\geq$ unit purchase cost). If $v_n(a, d) = \xi_1 \cdot (a - d)^+ - \xi_2 \cdot (d - a)^+$, assumption (*b*) is satisfied if $0 \leq \xi_2 \leq \xi_1$. We have already observed that many systems can be modeled in this manner.

PROOF Assumptions (*a*) and (*b*) and (7-20) imply for each n that $G_n(\cdot)$ is convex. As we discuss above, it remains to prove that K_n-convexity of $f_n(\cdot)$ implies K_n-convexity of $E\{f_n[v_{n-1}(a, D_{n-1})]\}$. Then assumption (*d*) and part (*d*) of Lemma 7-1 imply that $J_{n-1}(\cdot)$ is K_{n-1}-convex. The proof depends on the following lemma from Denardo (1982).

Lemma 7-4 Let $v(\cdot)$ be a convex and nondecreasing function on $\mathbb{R}$, let $f(\cdot)$ be K-convex, and

$$f(s) \leq K + f(s + \gamma) \tag{7-23}$$

for all $s \in \mathbb{R}$ and $\gamma \geq 0$. Then $f[v(\cdot)]$ is K-convex.

PROOF Let $h(\cdot) = f[v(\cdot)]$. From the remarks prior to Lemma 7-1, $h(\cdot)$ is K-convex if, and only if, $\mu_b \geq 0$ for all $x_1 \leq x_2$ and $0 \leq \lambda \leq 1$ where

$$\mu_n \triangleq (1 - \lambda)K + \lambda h(x_1) + (1 - \lambda)h(x_2) - h(x) \tag{7-24}$$

and $x = \lambda x_1 + (1 - \lambda)x_2$. By assumption, $v(\cdot)$ is nondecreasing so

$$v(x_1) \leq v(x) \leq v(x_2)$$

and there must exist $0 \leq \alpha \leq 1$ which satisfies

$$v(x) = \alpha v(x_1) + (1 - \alpha)v(x_2) \tag{7-25}$$

Hence, K-convexity of $f(\cdot)$ implies $\mu_f \geq 0$ or

$$0 \leq (1-\alpha)K + \alpha f[v(x_1)] + (1-\alpha)f[v(x_2)] - f[\alpha v(x_1) + (1-\alpha)v(x_2)]$$

Rearrangement of terms produces

$$h(x) \leq \alpha h(x_1) + (1-\alpha)h(x_2) + (1-\alpha)K$$

Subtracting this inequality from (7-24) yields

$$\mu_h \geq (\alpha - \lambda)[K + h(x_2) - h(x_1)] \tag{7-26}$$

In (7-23), let $s = v(x_1)$ and $\gamma = v(x_2) - v(x_1) \geq 0$ because $x_1 \leq x_2$ and $v(\cdot)$ is nondecreasing. Therefore, (7-23) implies

$$0 \leq K + f[v(x_s)] - f[v(x_1)] = K + h(x_2) = h(x_1)$$

so (7-26) implies $\mu_h \geq 0$ if $\alpha \geq \lambda$

Since $v(\cdot)$ is convex,

$$v(x) \leq \lambda v(x_1) + (1-\lambda)v(x_2)$$

whose subtraction from (7-25) gives

$$0 \geq (\alpha - \lambda)[v(x_1) - v(x_2)]$$

or

$$0 \leq (\alpha - \lambda)[v(x_2) - v(x_1)]$$

But $x_1 \leq x_2$ so $v(x_2) - v(x_1) \geq 0$, which implies $\lambda \leq \alpha$ and completes the lemma's proof. □

In order to complete the proof of the theorem, it is sufficient to verify that the hypothesis of Lemma 7-4 is valid with $f = f_n$ and $K = K_n$, that is,

$$f_n(s) \leq K_n + f_n(s + \gamma)$$

for all $\gamma > 0$ and s. But this inequality is an immediate consequence of (7-21*a*) and Lemma 7-3. Therefore, $f_n[v_{n-1}(\cdot, d)]$ is K_n-convex for each d. Hence, part (*c*) of Lemma 7-1 and a limiting argument yields K_n-convexity of $E\{f_n[v_{n-1}(\cdot, D_{n-1})]\}$. □

A General Model

In the back-order model in Section 7-1, suppose that $K_N = 0.5$, $c_N = c_{N+1}$, and $G_N(a) = 1 \wedge |a|$ as in Figure 7-4. This function violates conditions (*a*) and (*b*) of Theorem 7-1, but a (σ_N, Σ_N) policy is optimal with $\sigma_N = -0.5$ and $\Sigma_N = 0$.

Theorem 7-4 below replaces convexity of $G_n(\cdot)$ [assumption (*a*)] with *quasiconvexity* of $G_n(\cdot)$. A function $G(\cdot)$ is said to be quasiconvex† if $-G(\cdot)$ is unimodal. A quasiconvex function has at most one strict local minimum. It cannot go

† Formally, let $G: X \to \mathbb{R}$ with X a convex set. Then G is *quasiconvex* if

$$f(v) \vee f(y) \geq f[\lambda x + (1-\lambda)y] \qquad x \in X,\ y \in X,\ \lambda \in [0, 1]$$

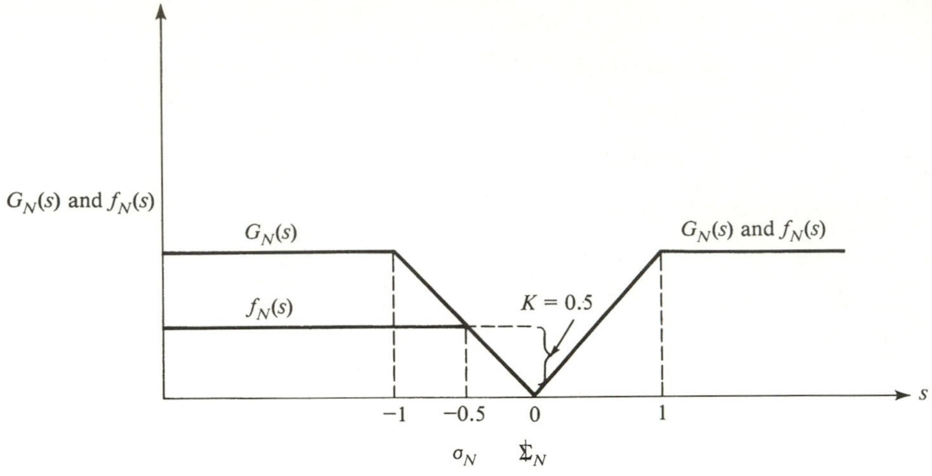

Figure 7-4 A (σ_N, Σ_N) policy is optimal although $G_N(\cdot)$ is not convex.

up and then go down. In Figure 7-4, both $G_N(\cdot)$ and $f_N(\cdot)$ are quasiconvex but not convex. Also, $f_N(\cdot)$ is not K_N-convex.

The most important difference between the first model and the present one is that the single-period expected cost is not necessarily convex. However, Veinott's method of proof, which differs dramatically from Scarf's, is the most significant feature of this subsection.

Discount factors in the first model were implicit in the notation K_n, g_n, and c_n with (7-3) an example. Here, the discount factors are explicit and non-stationary. Let β_n be the factor that reduces one cost unit at the beginning of period $n + 1$ to a cost unit at the beginning of period n. Let α_n denote the factor that reduces one cost unit at the start of period n to a cost unit at the start of period 1. Then $\alpha_1 = 1$ and

$$\alpha_n = \prod_{i=1}^{n-1} \beta_i \qquad \text{for } n = 2, \ldots, N + 1$$

If $\beta_i = \beta$ for all i, $\alpha_n = \beta^{n-1}$ as usual.

Review the proof of Theorem 7-1 and observe that the assumption $D_n \geq 0$ is not used. Here, we merely assume for each n that the sample space of D_n is a half line $[\theta_n, \infty)$:

$$P\{D_n \geq \theta_n\} = 1$$

If $\theta_n < 0$, this assumption admits at most $-\theta_n$ "returns" in period n.

Let B denote the present value of all costs minus the discounted salvage value of s_{N+1}. Then

$$B = \sum_{n=1}^{N} \alpha_n[K_n \delta(a_n - s_n) + c_n(a_n - s_n) + g_n(a_n, D_n)] - \alpha_{N+1}c_{N+1}s_{N+1}$$

Substitution of $\alpha_{n+1} = \beta_n \alpha_n$ and $s_n = v_{n-1}(a_{n-1}, D_{n-1})$ for $n > 1$ yields

$$B = \sum_{n=1}^{N} \alpha_n [K_n \delta(a_n - s_n) + g_n(a_n, D_n) + c_n a_n - \beta_n c_{n+1} v_n(a_n, D_n)] - c_1 s_1$$

Instead of (7-4), (7-13), and (7-20), let

$$G_n(a) \triangleq c_n a + E[g_n(a, D_n) - \beta_n c_{n+1} v_n(a, D_n)]$$

Then

$$E(B) = \sum_{n=1}^{N} \alpha_n E[K_n \delta(a_n - s_n) + G_n(a_n)] - c_1 s_1$$

The term $-c_1 s_1$ is constant with respect to the policy used to choose $a_1, \ldots, a_N$; so it can be deleted from the criterion of optimality. The resulting dynamic program is the same as (7-21*a*) and (7-21*b*), namely,

$$f_n(s) = \inf \{K_n \delta(a - s) + J_n(a) \colon a \geq s\} \qquad s \in \mathbb{R} \tag{7-27a}$$

$$J_n(a) = G_n(a) + \beta_n E\{f_{n+1}[v_n(a, D_n)]\} \qquad a \in \mathbb{R} \tag{7-27b}$$

for $n = 1, 2, \ldots, N$, with $f_{N+1}(\cdot) \equiv 0$.

Veinott's Conditions

The following assumptions ensure the existence of an optimal (σ, Σ) policy, i.e., (7-7), for $n = 1, \ldots, N$. More modestly, they imply that $G_n(\cdot)$ attains its global minimum on $\mathbb{R}$, say at Σ_n. Suppose for each $n = 1, \ldots, N$ [$n = 1, \ldots, N - 1$ for (7-28*c*), (7-28*f*), and (7-28*g*)]:

$$G_n(\cdot) \text{ and } v_n(\cdot, d) \text{ are continuous on } \mathbb{R} \text{ for each } d \geq \theta_n \tag{7-28a}$$

$$G_n(\cdot) \text{ is quasiconvex on } \mathbb{R} \tag{7-28b}$$

$$G_n(\Sigma_n) + \beta_n K_{n+1} < \lim_{a \to \infty} G_n(a) \tag{7-28c}$$

$$\lim_{a \to -\infty} G_n(a) > G_n(\Sigma_n) + K_n \tag{7-28d}$$

For each $d \geq \theta_n$, $v_n(\cdot, d)$ is nondecreasing on $\mathbb{R}$; (7-28*e*)
for each $a \in \mathbb{R}$, $v_n(a, \cdot)$ is bounded above on $[\theta_n, \infty)$

$$K_n \geq \beta_n K_{n+1} \tag{7-28f}$$

$$v_n(\Sigma_n, d) \leq \Sigma_{n+1} \qquad \text{for all } d \geq \theta_n \tag{7-28g}$$

By comparison with the conditions of Theorem 7-1, (7-28*a*) and (7-28*b*) replace convexity of $G(\cdot)$, (7-28*c*) and (7-28*d*) replace the assumption that $G(a) \to \infty$ as $|a| \to \infty$, and (7-28*f*) is essentially the same as $K_1 \geq K_2 \geq \cdots \geq K_n \geq 0$. Although (7-28*a*) to (7-28*g*) includes cases excluded from Theorem 7-1, (7-28*a*) to (7-28*g*) does not encompass the assumptions of Theorem 7-1 as a special case. As an example, suppose the model is stationary, that is, $D_1, \ldots, D_N$ are i.i.d. r.v.'s, and for $n = 1, \ldots, N$, $c_n = c$, $g_n(\cdot, \cdot) = g(\cdot, \cdot)$, $v_n(a, d) = a - d$, $K_n = K$, and $\theta_n = 0$. If $c_{N+1} = c$, then $G_n(\cdot) = G(\cdot)$ so $\Sigma_n = \Sigma$ for all $n = 1, \ldots, N$. Suppose $G(\cdot)$ is convex on $\mathbb{R}$. If instead $c_{N+1} = 0$, then $G_N(a) = G(a) + \beta c[a - E(D_N)]$ so, if $\beta c > 0$, $\Sigma_N \leq \Sigma$ with the inequality strict if $G'(\cdot)$ is continuously differentiable.

In (7-28*c*) and (7-28*d*), the limits are interpreted as $+\infty$ if $G_n(\cdot)$ is unbounded. It follows from (7-28*a*), (7-28*c*), and (7-28*d*) that there are numbers $\underline{\sigma}_n$ and $\bar{\Sigma}_n$ such that $\underline{\sigma}_n \leq \underline{\Sigma}_n \leq \bar{\Sigma}_n$ and

$$G_n(\bar{\Sigma}_n) = G_n(\underline{\Sigma}_n) + \beta_n K_{n+1} \tag{7-29a}$$

$$G_n(\underline{\sigma}_n) = G_n(\bar{\Sigma}_n) + K_n \tag{7-29b}$$

Then (7-28*f*) yields existence of a number $\bar{\sigma}_n$ such that $\underline{\sigma}_n \leq \bar{\sigma}_n \leq \underline{\Sigma}_n$ and

$$G_n(\bar{\sigma}_n) = G_n(\underline{\Sigma}_n) + K_n - \beta_n K_{n+1} \tag{7-29c}$$

The numbers $\underline{\sigma}_n$, $\bar{\sigma}_n$, $\underline{\Sigma}_n$, and $\bar{\Sigma}_n$ are more than technical artifacts in a proof. The parameters (σ_n, Σ_n) of an optimal policy, it turns out, satisfy

$$\underline{\sigma}_n \leq \sigma_n \leq \bar{\sigma}_n \leq \underline{\Sigma}_n \leq \Sigma_n \leq \bar{\Sigma}_n \tag{7-30}$$

$$G_n(\sigma_n) \geq G_n(\Sigma_n) + K_n - \beta_n K_{n+1} \tag{7-31}$$

for all n as shown in Figure 7-5.

Proof of Optimality

Theorem 7-4 Assumptions (7-28*a*) to (7-28*g*) for $n = 1, \ldots, N$ [$n = 1, \ldots, N - 1$ for (7-28*c*), (7-28*f*), and (7-28*g*)] imply existence of an optimal (σ_n, Σ_n) policy whose parameters satisfy (7-30) and (7-31).

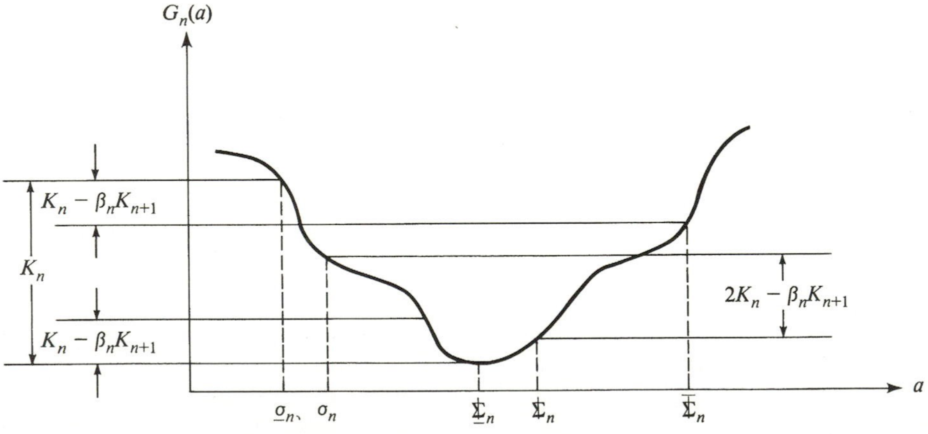

Figure 7-5 Bounds on an optimal (σ_n, Σ_n) policy.

PROOF Assumption (7-28a) and $f_{N+1}(\cdot) \equiv 0$ yield $J_N(\cdot) = G_N(\cdot)$ and continuity of $J_N(\cdot)$. The inductive assumption is continuity of $J_n(\cdot)$. The proof consists of the following steps:

(i) $\min \{J_n(a): a \in \mathbb{R}\} = \min \{J_n(a): \underline{\Sigma}_n \le a \le \overline{\Sigma}_n\}$

$$= J_n(\underline{\Sigma}_n)$$

(ii) There exists σ_n satisfying (7-30), (7-31), and

$$J_n(\sigma_n) = K_n + J_n(\underline{\Sigma}_n) \tag{7-32}$$

(iii) The infimum in (7-27a) is attained by

$$a = \begin{cases} \underline{\Sigma}_n & \text{if } s < \sigma_n \\ s & \text{if } s \ge \sigma_n \end{cases}$$

(iv) $J_{n-1}(\cdot)$ is continuous on $\mathbb{R}$

Step (iii) verifies optimality of a $(\sigma_n, \underline{\Sigma}_n)$ policy, steps (i) and (ii) establish the bounds in (7-30) and (7-31), and step (iv) completes the induction.

Step (i) consists of showing that $J_n(\cdot)$ is nonincreasing on $(-\infty, \underline{\Sigma}_n)$ and $a > \overline{\Sigma}_n$ implies $J_n(a) \ge J_n(\underline{\Sigma}_n)$. Therefore, the global minimum of $J_n(\cdot)$, if any, is attained on $[\underline{\Sigma}_n, \overline{\Sigma}_n]$. Continuity of $J_n(\cdot)$ implies that the minimum is attained.

Lemma 7-5 Assumption (7-28e) and $a \le a'$ imply

$$J_n(a') - J_n(a) \ge G_n(a') - G_n(a) - \beta_n K_{n+1} \qquad n = 1, \ldots, N \tag{7-33}$$

PROOF Assumption (7-28e) [$v_n(\cdot, d)$ is nondecreasing and $v_n(a, \cdot)$ is bounded above] and $a \le a'$ imply $v_n(a, d) \le v_n(a', d)$. Hence, (7-27a) and (7-27b) [specification of $f_n(s)$ and $J_n(a)$] and Lemma 7-3 [if $f(s) = \inf \{V\delta(a - s) + J(a): a \ge s\}$, then $f(s) \le V + f(s + \gamma)$ for $\gamma > 0$] yield

$$\begin{aligned} J_n(a') - J_n(a) &= G_n(a') - G_n(a) \\ &\quad + \beta_n E\{f_{n+1}[v_n(a', D_n)] - f_{n+1}[a, D_n]\} \qquad (7\text{-}34) \\ &\ge G_n(a') - G_n(a) - \beta_n K_{n+1} \qquad \square \end{aligned}$$

Lemma 7-5, quasiconvexity of $G_n(\cdot)$ [(7-28b)], the definition (7-29a) of $\overline{\Sigma}_n$, and $a > \overline{\Sigma}_n \ge \underline{\Sigma}_n$ imply

$$\begin{aligned} J_n(a) - J_n(\underline{\Sigma}_n) &\ge G_n(a) - G_n(\underline{\Sigma}_n) - \beta_n K_{n+1} \\ &\ge G_n(\overline{\Sigma}_n) - G_n(\underline{\Sigma}_n) - \beta_n K_{n+1} = 0 \end{aligned}$$

Step (i) is completed by showing that $J_n(\cdot)$ is nonincreasing on $(-\infty, \underline{\Sigma}_n]$. That property is implied by (7-28b), (7-28g), and the following lemma.

Lemma 7-6 Suppose for $n = 1, \ldots, N$ that (7-28e) [$v_n(\cdot, d)$ is nondecreasing and $v_n(a, \cdot)$ is bounded above] is valid, $\{y_n\}$ is a sequence for which $v_n(y_n, d) \le y_{n+1}$ if $d \ge \theta_n$, and $G_n(\cdot)$ is nonincreasing on $(-\infty, y_n]$. Then for $n = 1, \ldots, N$,

$$J_n(a') - J_n(a) \le G_n(a') - G_n(a) \le 0 \qquad a \le a' \le y_n \tag{7-35}$$

$$f_n(s') - f_n(s) \le 0 \qquad s \le s' \le y_n \tag{7-36}$$

PROOF Suppose (7-35) and (7-36) are valid at $n + 1$. Then monotonicity of $v_n(\cdot, d)$ [(7-28e)] and $a \le a' \le y_n$ imply

$$v_n(a, d) \le v_n(a', d) \le y_{n+1} \qquad d \ge \theta_n$$

Hence, (7-36) at $n + 1$ implies

$$E\{f_{n+1}[v_n(a', D_n)] - f_{n+1}[v_n(a, D_n)]\} \le 0$$

Then (7-34) and the assumption that $G_n(\cdot)$ is nonincreasing on $(-\infty, y_n]$ imply (7-35) at n. For (7-36), if $s \le s' \le y_n$, then (7-35) yields

$$\begin{aligned} f_n(s) &= \min \{J_n(s'), K_n + \inf \{J_n(a)\colon a \ge s\}\} \\ &\ge \min \{J_n(s'), K_n + \inf \{J_n(a)\colon a > s'\}\} = f_n(s') \end{aligned}$$

which is (7-36). To initiate the induction, $f_{N+1}(\cdot) \equiv 0$ and $G_N(\cdot)$ nonincreasing on $(-\infty, y_N)$ imply $J_N(a') - J_N(a) = G_N(a') - G_N(a) \le 0$ if $a \le a' \le y_N$. Then (7-36) for $n = N$ is established with the same argument as above. □

Step (ii) of the proof must show that there exists σ_n that satisfies (7-30), (7-31), and (7-32). Now step (i) established existence of $\Sigma_n \in [\underline{\Sigma}_n, \bar{\Sigma}_n]$ at which $J_n(\cdot)$ attains its global minimum. Therefore, Lemma 7-6 and the definitions of $\underline{\Sigma}_n$ and $\underline{\sigma}_n$ [in (7-29b)] imply

$$\begin{aligned} J_n(\Sigma_n) + K_n - J_n(\underline{\sigma}_n) &\le J_n(\underline{\Sigma}_n) + K_n - J_n(\underline{\sigma}_n) \\ &\le G_n(\underline{\Sigma}_n) + K_n - J_n(\underline{\sigma}_n) = 0 \end{aligned} \tag{7-37}$$

On the other hand, Lemma 7-5 and the definitions of $\bar{\Sigma}_n$ and $\bar{\sigma}_n$ [in (7-29c)] imply

$$\begin{aligned} J_n(\Sigma_n) + K_n - J_n(\bar{\sigma}_n) &\ge G_n(\Sigma_n) + K_n - G_n(\bar{\sigma}_n) - \beta_n K_{n+1} \\ &\ge G_n(\bar{\Sigma}_n) - G_n(\bar{\sigma}_n) + K_n - \beta_n K_{n+1} = 0 \end{aligned} \tag{7-38}$$

Continuity of $J_n(\cdot)$, (7-37), and (7-38) imply that there exists σ_n that satisfies (7-30) and (7-32). From (7-32) and Lemma 7-5,

$$\begin{aligned} 0 &= J_n(\Sigma_n) + K_n - J_n(\sigma_n) \\ &\ge G_n(\Sigma_n) - G_n(\sigma_n) + K_n - \beta_n K_{n+1} \end{aligned}$$

which verifies (7-31).

Step (iii) of the proof must show that the infimum in (7-27a) [$f_n(s) = \inf \{K_n \delta(a - s) + J_n(a)\colon a \ge s\}$] is attained by $a = \Sigma_n$ if $s < \sigma_n$, and by $a = s$ if

$s \geq \sigma_n$. Suppose $s < \sigma_n$. Steps (i) and (ii) show that $\sigma_n \leq \bar{\sigma}_n \leq \Sigma_n$ and $J_n(\cdot)$ is nonincreasing on $(-\infty, \Sigma_n]$; so (7-32) yields

$$J_n(s) \geq J_n(\sigma_n) = K_n + J_n(\Sigma_n) = K_n + \min\{J_a(a)\colon a \in \mathbb{R}\}$$

This inequality with

$$f_n(s) = \min\{K_n + \inf\{J_n(a)\colon a > s\}, J_n(s)\} \tag{7-39}$$

shows that $a = \Sigma_n$ attains the infimum in (7-27*a*) if $s < \sigma_n$.

Suppose $\sigma_n \leq s \leq \Sigma_n$. Then $J_n(\cdot)$ is nonincreasing on $(-\infty, \Sigma_n]$ so

$$J_n(s) \leq J_n(\sigma_n) = K_n + \min\{J_n(a)\colon a \in \mathbb{R}\}$$

which, with (7-39), implies $a = s$ attains the infimum in (7-27*a*).

Suppose $\Sigma_n < s$. If $s < a$, then

$$J_n(a) + K_n - J_n(s) \geq G_n(a) - G_n(s) + K_n - \beta_n K_{n+1} \geq 0 \tag{7-40}$$

with the first inequality due to Lemma 7-5 and the second due to quasiconvexity of $G_n(\cdot)$ and $K_n \geq \beta_n K_{n+1}$ [assumptions (7-28*b*) and (7-28*f*)]. Inequality (7-40) in (7-39) implies $a = s$ attains the infimum in (7-27*a*).

The final step of the proof must show that $J_{n-1}(\cdot)$ is continuous on $\mathbb{R}$. To initiate an induction, the continuity of $G_N(\cdot)$ [(7-28*a*)] and $f_{N+1}(\cdot) \equiv 0$ imply $J_N(\cdot) = G_N(\cdot)$ is continuous on $\mathbb{R}$. If $J_{n+1}(\cdot)$ is continuous on $\mathbb{R}$, step (iii) of the proof establishes

$$f_{n+1}(s) = \begin{cases} K_{n+1} + J_{n+1}(\Sigma_n) & \text{if } s < \sigma_n \\ J_{n+1}(s) & \text{if } s \geq \sigma_n \end{cases}$$

Then $J_{n+1}(\sigma_{n+1}) = K_{n+1} + J_{n+1}(\Sigma_{n+1})$ and continuity of $J_{n+1}(\cdot)$ imply continuity of $f_{n+1}(\cdot)$. Now $G_n(\cdot)$ is continuous [(7-28*a*)]; so, from (7-27*b*) [definition of $J_n(\cdot)$] $J_n(\cdot)$ is continuous if continuity of $f_{n+1}(\cdot)$ implies continuity of $\zeta(\cdot)$ where

$$\zeta(a) \triangleq E\{f_{n+1}[v_n(a, D_n)]\}$$

For each $d \geq \theta_n$, $v_n(\cdot, d)$ is continuous [(7-28*a*)]; so $f_{n+1}[v_n(\cdot, d)]$ is continuous. The dominated convergence theorem (Proposition A-6) asserts that $\zeta(\cdot)$ is continuous on every interval $[x, z]$ on which $f_{n+1}[v_n(\cdot, d)]$ is uniformly bounded for all $d \geq \theta_n$. We prove there exists a number w such that

$$J_{n+1}(\Sigma_{n+1}) \leq f_{n+1}[v_n(a, d)] \leq f_{n+1}(w) + K_{n+1} \tag{7-41}$$

for all $d \geq \theta_n$ and $x \leq a \leq z$. The left inequality of (7-41) is implied by (7-39) and $J_{n+1}(\Sigma_{n+1}) = \inf\{J_{n+1}(a)\colon a \in \mathbb{R}\}$. Assumption (7-28*e*) and $x \leq a \leq z$ imply

$$v_n(a, d) \leq v_n(z, d) \leq w \qquad d \geq \theta_n$$

for some number w. Hence, Lemma 7-3 [if $f(s) = \inf\{V\delta(a - s) + J(a)\colon a \geq s\}$, then $f(s) \leq V + f(s + \gamma)$ for $\gamma > 0$] establishes the right inequality in (7-41). □

EXERCISE

7-14 (*a*) Under the assumptions of Theorem 7-4, prove for each $n = 1, \ldots, N$ that $J_n(\cdot)$ is quasi-K-convex. You may invoke the results in the proof of Theorem 7-4 only up to and including Lemma 7-5 (but not results after Lemma 7-5). (*b*) Explain how to use your proof of part (*a*) as part of an alternative proof of Theorem 7-4.

7-3* PRODUCTION MODEL WITH START-UP AND SHUT-DOWN COSTS

The models in this section are motivated by the chronology associated with a setup cost in a production context. Suppose that a dormant production activity is activated and production continues awhile and eventually halts. The setup cost is the sum of the costs of starting and stopping production. If production continues for any length of time, the cost of stopping should be discounted back to the point in time that production starts. If the production rate is much higher than the demand rate, we may model production as an instantaneous phenomenon. However, in many production settings, an appreciable demand may occur while production continues. Then the length of the production "run" may depend on the demand that occurs during the run.

The preceding paragraph argues that a continuous time scale is sometimes appropriate. What then does a (σ, Σ) policy signify? We might construe it to mean that production starts when inventory drops below σ and production continues until the inventory level is up to Σ. The quantity produced during a run is Σ minus the triggering inventory level (less than σ) plus the demand that occurs while production is occurring. Therefore, a stochastic demand process causes randomness in the length of the run, in both time and quantity produced.

The model in Section 7-1 does not capture the internal dynamics of a period; so sometimes it is inappropriate for managing inventories in conjunction with a production process. Nevertheless, a continuous-time model preserves the optimality of a (σ, Σ) policy (with the interpretation above). This section confirms the optimality by analyzing the sample paths induced by a large class of stationary policies. Eventually, these sample paths are identical to those which a two-parameter policy would induce.

Sample-Path Approach—The Process

Suppose that the cumulative demand process $\{D(t);\, t \geq 0\}$ is a renewal process with interarrival times $U_1, U_2, \ldots$. Let $U \equiv U_1$. Let $V_1, V_2, \ldots$ denote the i.i.d. times to manufacture successive units with $V \equiv V_1$. We assume

$$P\{U > 0\} = P\{V > 0\} = 1 \tag{7-42a}$$

$$P\{U < V\} > 0 \tag{7-42b}$$

$$E(V) < E(U) < \infty \tag{7-42c}$$

Assumption (7-42*a*) insists that demands arrive one at a time and that production takes a positive length of time. Then (7-42*b*) asserts that demands can arrive faster than goods are produced, but (7-42*c*) says that, on average, production outstrips demand. Observe that (7-42*b*) and (7-42*c*) exclude models where U and V are constants.

Let

$$T_n = \sum_{i=1}^{n} U_i$$

which is the epoch at which the nth unit is demanded. Let $N(t)$ denote the number of units produced up to and including time t. Observe that $\sum_{i=1}^{n} V_i \leq t$ does not imply $N(t) \geq n$ unless production occurs continuously during $[0, t]$. Let $X(t)$ denote the inventory level at time t, so

$$X(t) = x_0 + N(t) - D(t) \qquad t \geq 0 \tag{7-43}$$

if excess demand is backlogged. We make a backlogging assumption because its consequences are *more* troublesome than those of a lost-sales assumption.

Let $\Lambda(t)$ indicate whether or not production is occurring at time t so that $\Lambda(t) = 0$ if production is not occurring and $\Lambda(t) = 1$ if production is occurring. We assume that production of a unit is not interrupted once it starts, and define $\Lambda(\cdot)$ to be left-continuous; therefore, if $N(t^-) + 1 = N(t) = n$, then $\Lambda(t') = 1$ for all $t' \in (t - V_n, t]$. In other words, the production process may be halted only when some unit has just been made. The other restriction on $\Lambda(\cdot)$ is that a dormant production process may be revived only at epochs when demand arrives, i.e., only at $0, T_1, T_2, \ldots$.

If $\Lambda(\cdot)$ obeys both restrictions, the set of epochs t which are candidates for $\Lambda(t) \neq \Lambda(t^+)$ is

$$\theta \triangleq \{t: \Lambda(t) = 1, N(t) = N(t^-) + 1\} \cup \{0, T_1, T_2, T_3, \ldots\}$$

The first set contains the epochs at which production completions occur, and the second contains the epochs at which demands arrive. Let the ordered sequence of elements in θ be $0 = \tau_0 < \tau_1 < \tau_2 < \ldots$. Notice that $\tau_{n+1} - \tau_n > 0$ for all n by definition of θ. Let $X_n = X(\tau_n)$, $\Lambda_n = \Lambda(\tau_n)$, and $s_n = (X_n, \Lambda_n)$. The set of all possible values of s_n is

$$\mathcal{S} \triangleq \{\ldots, -2, -1, 0, 1, 2, \ldots\} \times \{0, 1\}$$

Observe that $s_0, s_1, s_2, \ldots$ cannot be a Markov chain if U is not exponentially distributed, i.e., if demand does not comprise a Poisson process.

Sample-Path Approach—Stationary Policies

We consider only stationary policies on the state space $\mathcal{S}$. The set of all such policies, label it Π^*, is the set of mappings from $\mathcal{S}$ to $\{0, 1\}$. If $\pi \in \Pi^*$ determines $\Lambda(\cdot)$, then $\Lambda_{n+1} = \pi(s_n)$ for all $n \in I$. The (σ, Σ) policies form a proper subset of

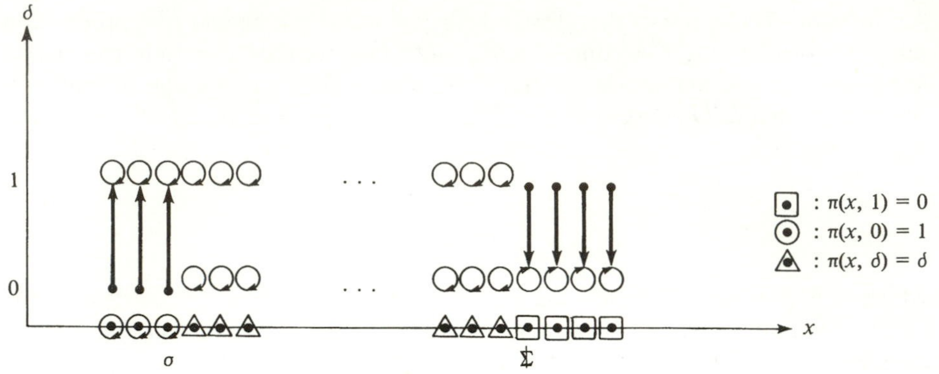

Figure 7-6 Actions taken with a $(\sigma, \bar{\Sigma})$ policy.

Π^*, label it Π^{**}. Each element of Π^{**}, namely, a $(\sigma, \bar{\Sigma})$ policy, has the form

$$\pi(x, \delta) = \begin{cases} 1 & \text{if } x \le \sigma \\ \delta & \text{if } \delta < x < \bar{\Sigma} \\ 0 & \text{if } x \ge \bar{\Sigma} \end{cases} \qquad \text{for all } (x, \delta) \in \mathcal{S} \tag{7-44}$$

Figure 7-6 portrays a $(\sigma, \bar{\Sigma})$ policy. If $\sigma < x < \bar{\Sigma}$, then $\pi(x, 0) = 0$ and $\pi(x, 1) = 1$. The appearance of Figure 7-6 suggests the label "hysteresis policy."

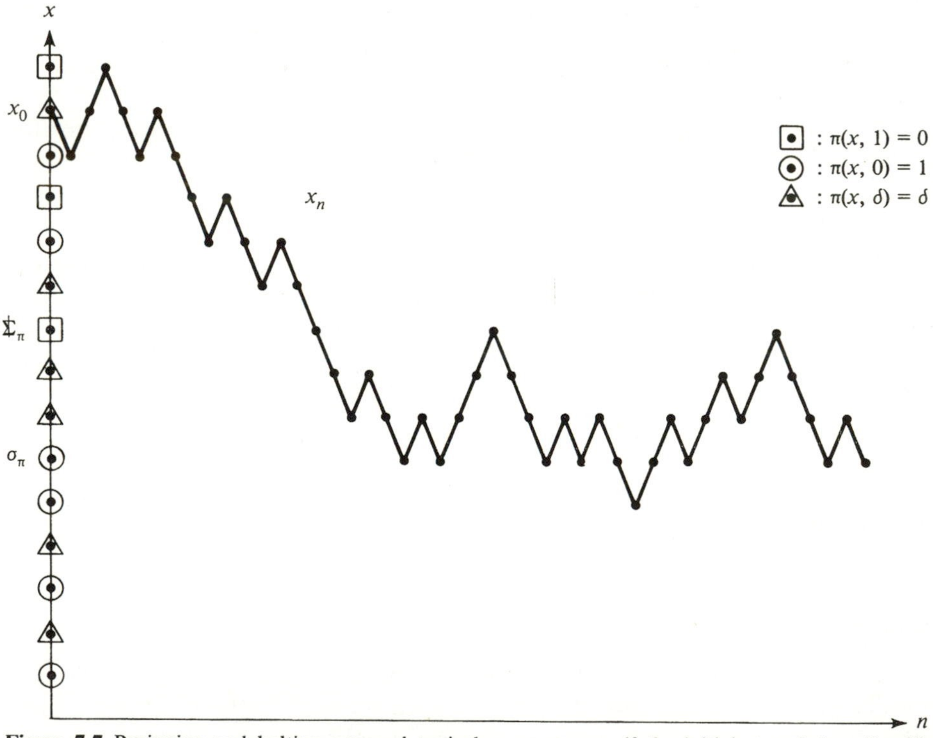

Figure 7-7 Beginning and halting sets and typical $x_0, x_1, x_2, \ldots$ if the initial state is $(x_0, 1)$ with $x_0 > \bar{\Sigma}_\pi$.

The basic idea of the sample-path approach is to define a random epoch ξ for each $\pi \in \Pi^*$ with the following property. The sequence of "states" after epoch ξ is identical to the sequence of states generated by some (σ, Σ) policy. Theorem 7-5 below proves that $E(\xi) < \infty$.

Each $\pi \in \Pi^*$ is completely determined by the sets of x's where it switches from on to off and from off to on. Let B_π and H_π denote these "beginning" and "halting" sets:

$$B_\pi = \{x\colon \pi(x, 0) = 1\} \qquad \text{and} \qquad H_\pi = \{x\colon \pi(x, 1) = 0\} \tag{7-45}$$

Let Σ_π denote the lowest halting element and σ_π denote the largest beginning element that is less than Σ_π:

$$\begin{aligned} \Sigma_\pi &= \inf \{x\colon x \in H_\pi\} \\ \sigma_\pi &= \sup \{x\colon x < \Sigma_\pi,\ x \in B_\pi\} \end{aligned} \tag{7-46}$$

Figure 7-7 portrays some of the elements of a policy π's halting and beginning sets. It shows a typical sequence $x_0, x_1, x_2, \ldots$ starting from an initial state $(x_0, 1)$ with $x_0 > \Sigma_\pi$.

Heuristic Proof that $E(\xi) < \infty$

We assume that $-\infty < \Sigma_\pi$ and that $\{\ldots, \Sigma_\pi - 3, \Sigma_\pi - 2, \Sigma_\pi - 1\}$ contains infinitely many elements of B_π. Then even if the initial inventory level x_0 is lower than Σ_π, policy π eventually causes production to begin. Figure 7-8 illustrates this case. Production continues until a halting element is reached, and that event

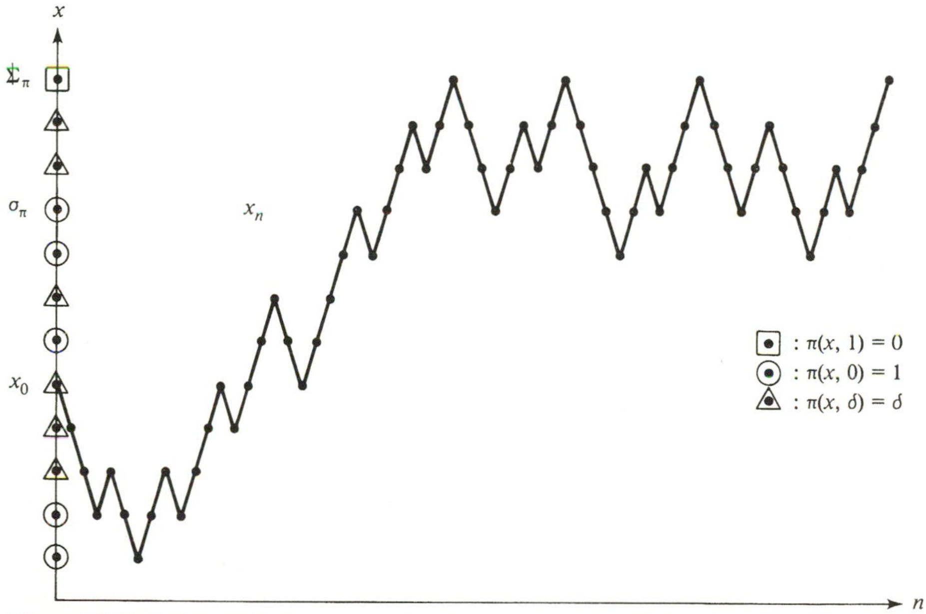

Figure 7-8 Beginning and halting sets and typical $x_0, x_1, x_2, \ldots$ if the initial state is $(x_0, 0)$ and $x_0 < \sigma_\pi$.

first occurs when inventory reaches $\ddagger_\pi$. This event occurs with probability 1 because of assumption (7-42*c*). Then production halts and remains dormant until the inventory drops as low as a beginning element. It first drops that low when inventory reaches σ_π.

On the other hand, suppose that the initial inventory level x_0 is above $\ddagger_\pi$. Figure 7-7 illustrates this case. Suppose also that $\{\ddagger_\pi, \ddagger_\pi + 1, \ddagger_\pi + 2, \ldots\}$ contains infinitely many elements of H_π. Then even if the production is "on" initially, that is, $\delta_0 = 1$, production eventually halts for awhile. Indeed, our argument above shows that inventory will never climb higher than the first element of the halting set that it encounters. How low may inventory drop? It is least likely to drop low if $[\ddagger_\pi, x_0] \cap H_\pi = \phi$, that is, if there are no halting elements between $\ddagger_\pi$ and x_0. Suppose so. Then assumption (7-42*b*), namely, $P\{U < V\} > 0$, ensures that demand will (with probability 1) occasionally cause arbitrarily large downward fluctuations in the inventory level. In particular, inventory will drop as low as $\ddagger_\pi$. Thereafter, inventory will never exceed $\ddagger_\pi$ because $\ddagger_\pi \in H_\pi$.

Formal Proof

Define ξ as the earliest time t at which $[X(t), \Lambda(t)] = (\ddagger_\pi, 1)$. For any rule $\pi \in \Pi^*$, let

$$L_\pi(t) = \sup\{n\colon \tau_n \le t\} \qquad t \ge 0$$

Then $L_\pi(t) \ge D(t)$ (because $\tau_n \le T_n$) and $U_n > 0$ and $V_n > 0$ for all n [from (7-42*a*)] imply $L_\pi(t) \to \infty$ as $t \to \infty$. Let

$$\xi_\pi = \inf\{t\colon s_{L_\pi}(t) = (\ddagger_\pi, 1),\ t \ge 0\} \qquad \pi \in \Pi^* \tag{7-47}$$

The distribution of ξ_π depends on the policy π and on the initial state $s_0 = (x_0, \delta_0)$. Theorem 7-5 below gives sufficient conditions for $E(\xi_\pi) < \infty$, which implies ξ_π is a nondefective r.v. for all $s_0 \in \mathcal{S}$. Its lengthy proof demonstrates the intimate bonds between queueing and production models. These bonds are described frequently in Volume I, particularly in Section 2-6 starting on page 32 and Example 11-16 on page 444.

Theorem 7-5 For $\pi \in \Pi^*$ suppose (7-42*a*) to (7-42*c*) are valid and that

$$\text{for all integers } x \text{ there are } x'' \in H_\pi \text{ and } x' \in B_\pi \text{ such that} \\ -\infty < x' < x < x'' < \infty \tag{7-48}$$

and

$$-\infty < \ddagger_\pi \tag{7-49}$$

Then $E(\xi_\pi) < \infty$ for all $s_0 \in \mathcal{S}$.

Proof Suppose first that $x_0 \le \ddagger_\pi$. Then $E[\xi_\pi \mid s_0 = (x_0, 0)] \ge E(\xi_\pi \mid s_0 = (x_0, 1)]$; so suppose $\delta_0 = 0$. Let $w = \min\{x\colon x < x_0, x \in B_\pi\}$ with $w > -\infty$ because of (7-48). Then ξ_π is the length of time until inventory drops to w and production begins plus the additional time until production drives the inventory up to $\ddagger_\pi$. Let Q_0 denote the second quantity. The first is $T_{x_0 - w}$; so $Q_0 = \xi_\pi - T_{x_0 - w}$. The distribution of Q_0 has the same distribution as the time until first emptiness of a $GI/G/1$ queue which has $\ddagger_\pi - w$ customers in the system at epoch 0, generic service time V, and generic interarrival time U. Then (7-42b) and (7-42c) imply $E(Q_0) < \infty$.† Also, $E(T_{x_0 - w}) = (x_0 - w)E(U) < \infty$ from (7-42c); so $E(\xi_\pi) < \infty$.

If $x_0 > \Sigma_\pi$, then $E[\xi_\pi \,|\, s_0 = (x_0, 0)] \le E[\xi_\pi \,|\, s_0 = (x_0, 1)]$; so suppose $\delta_0 = 1$. Let $w = \min\{x\colon x < x_0, x \in H_\pi\}$ with $-\infty < \Sigma < x_0 < w < \infty$ from (7-48). Let γ denote the (σ, Σ) policy where $\Sigma = w$ and $\sigma = w - 1$. Let ω denote a sample path and let $N'(t, \omega)$ and $X'(t, \omega)$ denote the cumulative quantity produced by time t and the inventory level at time t under policy γ. Then $N'(t, \omega) \ge N(t, \omega)$; so

$$X(t, \omega) = x_0 + N(t, \omega) - D(t, \omega) \le x_0 + N'(t, \omega) - D(t, \omega) = X'(t, \omega)$$

Hence $\xi_\pi(\omega) \le \xi_\gamma(\omega)$; we shall prove $E(\xi_\gamma) < \infty$, hence $E(\xi_\pi) < \infty$.

The r.v. ξ_γ is distributed as the length of time until the first occurrence of $w - \Sigma_\pi$ customers in a $GI/G/1$ queue which has $w - x_0$ customers at epoch 0, generic service time V, and generic interarrival time U. The queue-inventory correspondence is that the number of customers drops from $w - x_0$ to 0 and eventually gets as high as $w - \Sigma_\pi$, as the inventory level rises from x_0 to w and eventually drops as low as Σ_π.

In the queueing model, let Q_0 denote the first epoch when the system is empty and a customer arrives. Let $Q_1, Q_2, \ldots$ denote the i.i.d. lengths of subsequent busy cycles and let $Q \equiv Q_1$. Let $\mathscr{E}_0$ be the event that $\xi_\gamma \le Q_0$ and let $\mathscr{E}$ denote the event that as many as $w - \Sigma_\pi$ customers are waiting at some epoch during a busy cycle. Let $p_0 = P\{\mathscr{E}_0\}$ and $p = P\{\mathscr{E}\}$. Let $\mathscr{E}'_0$ and $\mathscr{E}'$ denote the complements of $\mathscr{E}_0$ and $\mathscr{E}$ so their probabilities are $q_0 = 1 - p_0$ and $q = 1 - p$.

From (7-42b) and (7-42c),‡ $E(Q_0) < \infty$, and $E(Q) < \infty$. Also, (7-42b) implies $p_0 > 0$ and $p > 1$ while (7-42c) implies $q_0 > 0$ and $q > 0$. Therefore, $E(Q_0 \,|\, \mathscr{E}_0) < \infty$, $E(Q_0 \,|\, \mathscr{E}'_0) < \infty$, $E(Q \,|\, \mathscr{E}) < \infty$, and $E(Q \,|\, \mathscr{E}') < \infty$; so

$$\begin{aligned} E(\xi_\pi) &\le E(\xi_\gamma) \\ &\le p_0 E(Q_0 \,|\, \mathscr{E}_0) + q_0\big(E(Q_0 \,|\, \mathscr{E}'_0) + pE(Q \,|\, \mathscr{E}) \\ &\qquad + q\{E(Q \,|\, \mathscr{E}') + pE(Q \,|\, \mathscr{E}) \\ &\qquad + q[E(Q \,|\, \mathscr{E}') + pE(Q \,|\, \mathscr{E}) + \ldots]\}\big) \\ &= p_0 E(Q_0 \,|\, \mathscr{E}_0) + q_0\left[E(Q_0 \,|\, \mathscr{E}'_0) + \frac{qE(Q \,|\, \mathscr{E}')}{p} + E(Q \,|\, \mathscr{E})\right] < \infty \end{aligned}$$

□

† Let Q be the length of a busy period of a $GI/G/1$ queue. Then (7-42c) implies $E(Q) < \infty$ from Example 6-5 on page 183 of Volume I. Let M denote the maximum number in the queue during the first busy period. Assumption (7-42b) implies $P\{M = n\} > 0$ for all $n \in I$; so (7-42c) yields

$$\infty > E(Q) = \sum_{n=0}^{\infty} E(Q \,|\, M = n)P\{M = n\}$$

Thus $E(Q \,|\, M = n) < \infty$ for every $n \in I$. Lastly, $E(Q_0) \le E(Q \,|\, M = \Sigma_\pi - w) < \infty$. An alternative proof which does not invoke (7-42b) is obtained by answering Exercise 11-9 on page 391 of Volume I using results in that volume's Sections 11-1 and 9-4.

‡ Paraphrase the argument in the previous footnote.

If a policy π satisfies (7-48) and (7-49), then Theorem 7-5 and the theorems in Section 6-4 of Volume I assert that the long-run behavior of $\{[X(t), \Lambda(t)]; t \geq 0\}$ would be generated also by the $(\sigma, \not\Sigma)$ policy in which $\sigma = \sigma_\pi$ and $\not\Sigma = \not\Sigma_\pi$. Therefore, the long-run average cost per unit time associated with π should be the same as would be generated by the $(\sigma_\pi, \not\Sigma_\pi)$ policy. Now we examine this assertion carefully.

Let $C_s(t)$ be the cumulative cost using π during the interval $[0, t]$ if $s_0 = s$. The long-run average-cost criterion is

$$\phi(\pi \mid s) \triangleq \liminf_{t \to \infty} \frac{E_\pi[C_s(t)]}{t} \qquad \pi \in \Pi^* \qquad s \in \mathcal{S}$$

The next result gives a sufficient condition for the long-run average cost of a policy π to be the same as the long-run average cost of the policy $(\sigma_\pi, \not\Sigma_\pi)$ starting from the initial state $(\not\Sigma_\pi, 1)$.

Corollary 7-5 If (7-42*a*) to (7-42*c*), (7-48), and (7-49) are valid and

$$E_\pi[C_s(\xi_\pi)] < \infty \tag{7-50}$$

then

$$\phi(\pi \mid s) = \phi[(\sigma_\pi, \not\Sigma_\pi) \mid (\not\Sigma_\pi, 1)] \tag{7-51}$$

PROOF Manifestly,

$$C_s(t) = [C_s(t) - C_s(\xi_\pi)] + C_s(\xi_\pi)$$

Assumption (7-50) implies $E_\pi[C_s(\xi_\pi)]/t \to 0$ as $t \to \infty$. Also,

$$\frac{1}{t} = \frac{(t - \xi_\pi)/t}{t - \xi_\pi}$$

and $(t - \xi_\pi)/t \to 1$; so

$$\begin{aligned} &\liminf_{t \to \infty} \frac{E_\pi[C_s(t) - C_s(\xi_\pi)]}{t} \\ &= \liminf_{t \to \infty} \frac{E_\pi[C_s(t) - C_s(\xi_\pi)]}{t - \xi_\pi} \\ &= \phi[(\sigma_\pi, \not\Sigma_\pi) \mid (\not\Sigma_\pi, 1)] \end{aligned}$$

with probability 1. □

Assumption (7-50) would fail to hold if ξ_π has an infinite variance and $C_s(t) \sim t^2$. This type of behavior does not seem to arise in practice. Conditions which ensure (7-50) [cf. Sobel (1969)] seem to be satisfied by realistic production models.

Poisson Demand Process—Further Results

Suppose that demand $\{D(t);\ t \geq 0\}$ is a Poisson process, there are start-up and shut-down costs, there is a holding-cost rate which is proportional to the inventory level, and there is a back-order cost rate which is proportional to the amount of backlogged demand. Then the model is an SMDP (semi-Markov decision process; cf. Sections 5-1 and 5-2). Exercise 5-15 asks you to specify the canonical elements of the SMDP.

The criterion of expected present value does not seem amenable to the arguments employed to prove Theorem 7-5 and Corollary 7-5. Instead, Heyman (1968) and Bell (1971) exploit the features of the SMDP model to prove that some (σ, Σ) policy is optimal for the criterion of expected present value.†

Suppose that the SMDP production model is altered to permit several production lines to be working simultaneously. Suppose also that excess demand is lost and there are start-up and shut-down costs when dormant lines become active or active lines become dormant. The multiple server queueing control model in Section 8-4 can be interpreted as such a production model. The results in Section 8-4 include the optimality of a generalization of a (σ, Σ) policy.

Now we discuss the optimization of the SMDP induced by the following costs. Let K be the sum of the start-up and shut-down costs, c be the cost per unit time incurred while production is active, h be the holding cost per unit time per unit in stock, and L be the back-order cost per unit time per unit out of stock. Let λ be the intensity of the Poisson demand process and μ^{-1} the expected value of the time it takes to produce an item. We assume

$$0 < \lambda < \mu < \infty$$

which implies (7-42*c*).

Recall that $X(t)$ denotes the inventory level at time t. Under the assumption above, the use of a (σ, Σ) policy with $-\infty < \sigma < \Sigma < \infty$ implies the existence of limiting probabilities

$$p_i = \lim_{t \to \infty} P\{X(t) = i \mid X(0) = x_0,\ \Lambda(0) = \delta_0\}$$

for all i, x_0, and δ_0.

Let a *cycle* be the first passage time from $(\Sigma, 0)$ to $(\Sigma, 0)$. It includes a dormant period as the inventory drops from Σ to σ and the subsequent period while production is active in order to raise the inventory to Σ. Let a *busy period* be the length of time that production is active during a cycle. From Exercise 11-49 on page 449 of Volume I, the expected value of a busy period depends on Σ and σ only via $\Sigma - \sigma$. Let $Q \triangleq \Sigma - \sigma$, z_Q be the expected value of the busy period, and $z \triangleq z_1$ in the same model with a $(\sigma, \sigma + 1)$ policy. Then $z_Q = Qz$ from equation (11-118) on page 446 of Volume I. A cycle consists of a busy period plus the time needed for inventory to drop from Σ to σ. The latter is the sum of Q demand interarrival times; so the expected value of a cycle is $Qz + Q/\lambda$.

† Heyman and Bell analyze a cost model of an $M/G/1$ queue with a removable server. Their proofs can be extended to a production model with start-up and shut-down costs.

The model, because of the preceding argument, has the following long-run average cost per unit time, say b:

$$b = \frac{K}{Qz + Q/\lambda} + \frac{cQz}{Qz + Q/\lambda} + h\sum_{i=1}^{\ddagger} ip_i - L\sum_{i=-\infty}^{-1} ip_i \tag{7-52}$$

By paraphrasing the proof of Theorem 11-19 in page 445 of Volume I,

$$\frac{\lambda}{\mu} = \lim_{t\to\infty} P\{\Lambda(t) = 1\}$$

$$= \lim_{t\to\infty} P\{\text{production is active at } t\} = \frac{z}{z + 1/\lambda}$$

so $z = (\mu - \lambda)^{-1}$. Therefore, (7-52) is

$$b = \frac{K/Q + cz}{1/(\mu - \lambda) + 1/\lambda} + h\sum_{i=1}^{\ddagger} ip_i - L\sum_{i=-\infty}^{-1} ip_i \tag{7-53}$$

We wish to optimize (7-53) with respect to σ and $\ddagger$ or, equivalently, with respect to $Q = \ddagger - \sigma$ and $\ddagger$. Unfortunately, the two sums on the right side of (7-53) lead to cumbersome expressions. It is possible to complete the specification of (7-53), but it involves a tedious derivation of messy generating functions. Instead, we turn to a model closely related to the SMDP production model but which can be optimized more easily.

M/*G*/1 Queue with a Removable Server

Consider a service facility, that is, a queueing model, with a Poisson arrival process, unlimited waiting room, and a single server whose mean service time is μ^{-1}. Let λ be the intensity of the Poisson arrival process. We assume $0 < \lambda < \mu < \infty$. Suppose that start-up and shut-down costs are incurred each time the server resumes or halts service, respectively. Let K be the sum of the start-up and shut-down costs, c the operating cost per unit time when the server is active, and h the holding cost per unit time per customer in the facility.

Let $X(t)$ denote the number of people in the system at time t and let $\Lambda(t)$ be 1 or 0 depending on whether or not the server is active at time t. By analogy with the production model, we restrict start-ups to epochs when a customer arrives at the service facility (and the server is inactive). We restrict shut-downs to epochs when a customer has just completed being served (so the server is active). Let $\tau_0 < \tau_1 < \ldots$ be the ordered sequence of times which are candidates for starting up or shutting down and let $(X_n, \Lambda_n) = [X(\tau_n), \Lambda(\tau_n)]$.

The presence of the start-up and shut-down costs suggests the use of the following kind of policy, say π. For a parameter Q,†

$$\pi(x, \delta) = \begin{cases} 1 & x \geq Q \\ \delta & 1 \leq x < Q \\ 0 & x = 0 \end{cases} \tag{7-54}$$

Thus a busy period begins when Q people have accumulated. It ends when the facility next becomes empty. The ensuing idle period is the length of time for Q people to arrive, that is, Q interarrival times.

By analogy with the discussion of the production model with a Poisson demand process, the long-run average cost per unit time, say b, is similar to (7-52). It is

$$b = \frac{K}{Qz + Q/\lambda} + \frac{cQz}{Qz + Q/\lambda} + h\sum_{i=0}^{\infty} ip_i \tag{7-55}$$

where z is the expected value of the busy period of an ordinary $M/G/1$ queue (that is, when $Q = 1$), and p_i is the long-run probability of having i customers in the service facility. The sum in (7-55) is much easier to evaluate than the two sums in (7-52) and (7-53).

We have already observed that $z = (\mu - \lambda)^{-1}$. The sum on the right side of (7-55) is the long-run average number of customers in the service facility, say L. From Section 11-3 in Volume I, $L = \lambda W$, where W is the long-run average customer waiting time and, from Theorem 11-19 on page 445 of Volume I,

$$W = \frac{Q-1}{2\lambda} + \frac{1}{\mu} + \frac{\lambda v_2}{2(1-\rho)}$$

where $\rho = \lambda/\mu$ and v_2 is the second moment of the service time. Making the substitutions for z in the sum in (7-55) yields

$$b = \frac{K/Q + cz}{1/(\mu - \lambda) + 1/\lambda} + h\left[\frac{Q-1}{2} + \rho + \frac{\lambda^2 v_2}{2(1-\rho)}\right] \tag{7-56}$$

Formula (7-56) is convex in Q. Treating Q as a continuous variable, differentiating (7-56) with respect to Q, and setting $db/dQ = 0$ yields a minimum of b at

$$Q = \sqrt{\frac{2K\lambda(1-\rho)}{h}} \tag{7-57}$$

Therefore, the optimal value of Q is one of the two integer neighbors to the (generally) fractional solution provided by (7-57). The two neighbors can be compared easily by substituting each of them for Q in (7-56). That comparison requires specification of v_2 in (7-56), whereas (7-57) involves the distribution of service time only via its mean μ^{-1}.

† In Section 11-6 of Volume I, we call this an N-policy.

EXERCISES

7-15 Suppose that the production model to which Theorem 7-5 refers is altered so that excess demand is lost instead of backlogged. Explain (in detail) how to weaken the assumptions of Theorem 7-5 and alter its proof so that $E(\xi_\pi) < \infty$ for all $s_0 \in \mathcal{S}$ is still valid.

7-16 Dishon and Weiss† study the replacement of communications satellites which have fallen out of orbit. They consider (σ, Σ) policies because there is a start-up cost to initiate launching activities. They assume that different satellites remain in orbit for lengths of time which are independent and identical exponential r.v.'s with mean, say, θ^{-1}. As a result, $\lambda_i \triangleq i\theta$ is the "rate" at which satellites fall from orbit when there are i of them in orbit. This feature generalizes Exercise 7-15 by introducing a state-dependent demand rate (a unit "demand" corresponds to one satellite falling out of orbit). Do *not* assume $\lambda_i = i\theta$ but suppose that demands have interarrival times which are independent exponential r.v.'s whose means depend on the stock level. Let λ_i^{-1} be the mean interarrival time when the stock level is i. Suppose $\lambda_0 = 0$ so excess demand is lost as in Exercise 7-15. Prove $E(\xi_\pi) < \infty$, as in Theorem 7-5, if (7-42*a*), (7-42*b*), and (7-48) are valid.

7-4* SEPARABLE MDPs

In this section we identify a class of structured MDPs which are equivalent to smaller MDPs. This class, called separable MDPs, is similar to the myopic MDPs in Chapter 3. We illustrate the similarities and differences with a (σ, Σ) inventory model.

A Special Case

Here is a variant of the models for which a (σ, Σ) policy is optimal in Sections 7-1 and 7-2. Suppose that the model is stationary (i.e., time homogeneous), excess demand is lost, demand is a discrete r.v., ordered goods are delivered instantaneously, order quantities are integers, and the initial inventory level is an integer. Let

$$\phi_i = P\{D_1 = i\} \qquad i \in I$$

specify the distribution of demand. As in Section 7-1, let s_n and a_n denote the respective amounts of stock on hand immediately before and after goods are ordered (and delivered). It follows from (7-15) and (7-16) that this model is equivalent to the following MDP with discount factor β, $0 \le \beta < 1$:

$$r(s, a) = K\delta(a - s) + G(a) \tag{7-58a}$$

$$p_{sj}^a = P\{(a - D_1)^+ = j\} = \begin{cases} \phi_{a-j} & \text{if } j > 0 \\ \sum_{i=1}^{\infty} \phi_i & \text{if } j = 0 \end{cases} \tag{7-58b}$$

In (7-58*a*), we interpret $r(s, a)$ as a single-stage cost rather than a reward.

Suppose that there is room to store at most u items ($u < \infty$) and that the

† M. Dishon and G. H. Weiss, "A Communications Satellite Replenishment Policy," *Technometrics*, **8:** 399–410 (1966).

"order-up-to" quantity must be a nonnegative integer.† The second assumption, that $a \in I$, is without loss of generality if $s_1 \in I$. It could not be optimal to order up to a fractional amount, because demand is integer-valued.

It follows from the assumptions that the state space is

$$\mathcal{S} = \{0, 1, \dots, u\} \tag{7-58c}$$

and the sets of "order-up-to" actions are

$$A_s = \{s, s+1, \dots, u\} \qquad s \in \mathcal{S} \tag{7-58d}$$

so $\#\mathscr{C} < \infty$.

How does (7-58*a*) to (7-58*d*) differ from the myopic MDP described in Section 3-3? The first two assumptions of a myopic MDP are‡

$$r(s, a) = X(a) + L(s) \qquad (s, a) \in \mathscr{C} \tag{7-59a}$$

$$p_{sj}^a = \rho_j^a \qquad a \in A \triangleq \bigcup_{s \in \mathcal{S}} A_s \qquad j \in \mathcal{S} \tag{7-59b}$$

where $X(\cdot)$ and $L(\cdot)$ are real-valued functions (on A and $\mathcal{S}$, respectively) and $\{\rho_j^a\}$ are nonnegative numbers with

$$\sum_{j \in \mathcal{S}} \rho_j^a = 1 \qquad a \in A$$

Observe that (7-58*b*) is a special case of (7-59*b*) but (7-58*a*) does not satisfy (7-59*a*).

Specifically, $K\delta(a - s) + G(a)$ on the domain $\{(s, a): s \le a\}$ cannot be represented as the sum of a function of a and a function of s (if $K > 0$ and $G(\cdot) \not\equiv 0$). However, $K\delta(a - s) + G(a)$ trivially has this property on the subset $\{(s, a): s < a\}$. That is, if $s < a$,

$$K\delta(a - s) + G(a) = K + G(a) = X(a) + L(s)$$

with $L(\cdot) \equiv K$ and $X(a) = G(a)$.

If $0 \le s < u$, the substitution of (7-58*a*) and (7-58*b*) in the optimality equation (4-89*a*) yields

$$\begin{aligned} f_s &= \min\left\{r(s, a) + \beta \sum_{j \in \mathcal{S}} p_{sj}^a f_j : a \in A_s\right\} \\ &= \min\left\{K\delta(a - s) + G(a) + \beta \sum_{j=1}^{a} \phi_{a-j} f_j + \beta f_0 \sum_{j=a}^{\infty} \phi_j : s \le a \le u\right\} \\ &= \min\left\{G(s) + \beta \sum_{j=1}^{s} \phi_{s-j} f_j + \beta f_0 \sum_{j=s}^{\infty} \phi_j,\right. \\ &\qquad \left. K + \min\left\{G(a) + \beta \sum_{j=1}^{a} \phi_{a-j} f_j + \beta f_0 \sum_{j=a}^{\infty} \phi_j : s + 1 \le a \le u\right\}\right\} \end{aligned} \tag{7-60a}$$

† The assumption $u < \infty$ is made to ensure $\#\mathscr{C} < \infty$; so equation (7-60) below can be invoked. It can be shown that the assumption is essentially unnecessary. The argument in Section 8-5 is one of several methods which could be used to justify (7-60) without the assumption.

‡ The counterpart of (7-59*a*) in Section 3-3 uses the notation $K(a)$ where (7-59*a*) uses $X(a)$. The symbol K is usually used in inventory theory to represent a setup cost.

Recall that we allow only integer values of a. The third equality results from splitting the minimum over $a \in \{s, s+1, \ldots, u\}$ into the minimum of $a = s$ and the best choice of $a \in \{s+1, \ldots, u\}$.

Let

$$J_s \triangleq G(s) + \beta \sum_{j=1}^{s} \phi_{s-j} f_j + \beta f_0 \sum_{j=s}^{\infty} \phi_j \tag{7-60b}$$

and

$$y_s \triangleq \min \{J_a : s+1 \le a \le u\} \tag{7-60c}$$

Since y_s depends on s only via the constraint $s + 1 \le a$,

$$y_s = \min \{y_{s+1}, J_{s+1}\} \tag{7-61a}$$

The substitution of (7-60b) and (7-60c) in (7-60a) yields

$$f_s = \min \{J_s, K + y_s\} \tag{7-61b}$$

Now we use (7-61a) to write (7-61b) as

$$f_s = \min \{J_s, K + J_{s+1}, K + y_{s+1}\} \tag{7-61c}$$

The network in Figure 7-9 depicts (7-61a) and (7-61b). There is a node for each variable and an arc for each "action" which is available in (7-61a) and (7-61b). In comparison with Figure 7-9, the network in Figure 7-10 depicts (7-60a). The average number of arcs per node is higher in Figure 7-10 than in Figure 7-9, but Figure 7-9 has more nodes than Figure 7-10. Therefore, it is not apparent from the figures that (7-61a) and (7-61b) is a smaller problem, in some sense, than (-60a). We see below that the value of u determines which formulation is more efficient.

The minimization of (7-60a) involves $u - s + 1$ comparisons while (7-61a)

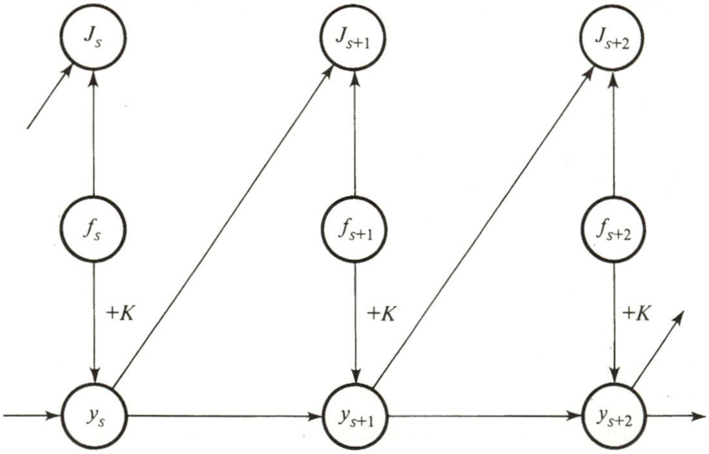

Figure 7-9 Part of a network which corresponds to (7-61a) and (7-61b).

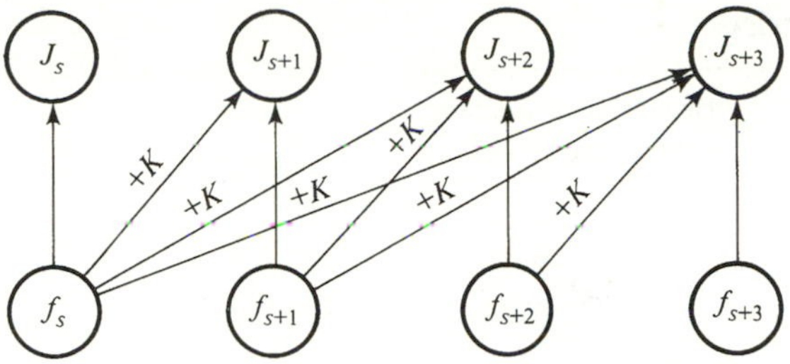

Figure 7-10 Part of a network which corresponds to (7-60*a*).

and (7-61*b*) requires only four comparisons. If $s = u$, then

$$f_u = J_u \tag{7-61d}$$

In order to appreciate the advantage of (7-61*a*) and (7-61*b*) versus (7-59), we write (7-61*a*), (7-61*b*), and (7-61*d*) as the following linear program in which $\{\lambda_s\}$ are arbitrary positive numbers:

$$\text{Maximize} \sum_{s=0}^{u} \lambda_s(y_s + f_s) \tag{7-62a}$$

subject to

$$\left.\begin{aligned} &f_s \le G(s) + \beta \sum_{j=1}^{s} \phi_{s-j} f_j + \beta f_0 \sum_{j=s}^{\infty} \phi_j && \text{(7-62b)} \\ &f_s \le K + y_s && \text{(7-62c)} \end{aligned}\right\} \quad 0 \le s \le u - 1$$

$$\left.\begin{aligned} &y_s \le y_{s+1} && \text{(7-62d)} \\ &y_s \le G(s+1) + \beta \sum_{j=1}^{s} \phi_{s-j} f_j + \beta f_0 \sum_{j=s}^{\infty} \phi_j && \text{(7-62e)} \end{aligned}\right\} \quad 0 \le s \le u - 1$$

$$f_u = G(u) + \beta \sum_{j=1}^{u} \phi_{u-j} f_j + \beta f_0 \sum_{j=u}^{\infty} \phi_j \tag{7-62f}$$

Exercise 7-17 asks you to state and prove the sense in which (7-62*a*) to (7-62*f*) is equivalent to (7-61*a*) and (7-61*b*). The linear program has $2(u + 1)$ variables in $4u + 1$ constraints.

Instead of (7-62*a*) to (7-62*f*), if

$$f_s = \min \left\{ r(s, a) + \beta \sum_{j \in S} p_{sj}^a f_j \colon s \le a \le u \right\} \qquad 0 \le s \le u$$

were straightforwardly replaced by a linear program, there would be $u + 1$ variables in $(u + 1)(u + 2)/2$ constraints. Table 7-1 compares the sizes of the resulting matrices of coefficients. For (7-62*a*) to (7-62*d*), the size (the number of elements in the coefficient matrix) is $2(u + 1)(4u + 1)$. For the straightforward linear program, the size is $u(u + 1)(u + 2)/2 - u$.

For larger problem sizes, (7-62*a*) to (7-62*f*) is smaller than the usual linear program. Why? If $0 \le s < u - 1$, we may rewrite (7-60*a*) and (7-61*a*) and (7-61*b*)

Table 7-1 Sizes of coefficient matrices

u	Linear program (7-62a) to (7-62d) $2(u+1)(4u+1)$	Usual linear program $(u+1)^2(u+2)/2$
5	252	126
10	902	726
20	3,402	4,851
50	20,502	67,626

as

$$f_s = \min \{J_s, K + J_{s+1}, K + \min \{J_a: a > s+1\}\} \tag{7-63a}$$

$$f_{s+1} = \min \{J_{s+1}, K + \min \{J_a: a > s+1\}\} \tag{7-63b}$$

Most of the labor in (7-63a) and (7-63b) consists of the searches in the identical third and second terms, respectively. The recursion (7-61a) and (7-61b) and its associated linear program (7-62a) to (7-62f) avoid the repetition of identical searches.

Exercises 7-18, 7-19, 7-20, and 7-21 show that it is possible to reduce further the size of (7-61a) and (7-61b) and (7-62a) to (7-62f).

The preceding inventory model is an example of a separable MDP. The computational advantages of (7-61a) and (7-61b) and (7-62a) to (7-62f) exemplify the efficiencies which can be obtained by exploiting a separable structure. Now we define a separable MDP and show how to exploit its structure computationally.

Definition of a Separable MDP

The model in (7-58a) to (7-58d) is a special case of a finite MDP which satisfies the following definition. Let $\mathcal{S} = \{1, \ldots, S\}$.

Definition 7-3 A *separable MDP* is an MDP with a discrete state space for which each A_s contains a (possibly empty) subset H_s such that, for each s,

$$H_s \supset H_{s+1} \qquad (s < S) \tag{7-64a}$$

$$r(s, a) = X(a) + L(s) \qquad a \in H_s \tag{7-64b}$$

$$p_{sj}^a = \rho_j^a \qquad a \in H_s \qquad j \in \mathcal{S} \tag{7-64c}$$

In words, if $a \in H_s$, then $r(s, a)$ is additively separable and the transition probabilities do not depend on s. The model in (7-58a) to (7-58d) satisfies the definition† with $X(a) = G(a)$, $L(s) = K$, $\rho_j^a = \phi_{a-j}$, and

$$H_s = \begin{cases} \{s+1, \ldots, u\} & \text{if } 0 \le s < u \\ \phi & \text{if } s = u \end{cases}$$

where we use the symbol ϕ to denote the empty set.

† The model in (7-58a) to (7-58d) has $\mathcal{S} = \{0, \ldots, u\}$ instead of $\{1, \ldots, S\}$.

Exploiting the Separable Structure

Let $M_s = H_s - H_{s+1}$, $W_s = A_s - H_s$,

$$x_s = \max\left\{r(s, a) + \beta \sum_{j\in S} p_{sj}^a f_j : a \in W_s\right\}$$
$$y_s = \max\left\{X(a) + \beta \sum_{j\in S} \rho_j^a f_j : a \in H_s\right\} \tag{7-65}$$
$$z_s = \max\left\{X(a) + \beta \sum_{j\in S} \rho_j^a f_j : a \in M_s\right\} \tag{7-66}$$

Substitution of (7-64*b*) and (7-64*c*) in the optimality equation (4-89*a*) yields

$$\begin{aligned} f_s &= \max\left\{r(s, a) + \beta \sum_{j\in S} p_{sj}^a f_j : a \in A_s\right\} \\ &= \max\left\{x_s, \max\left\{X(a) + L(s) + \beta \sum_{j\in S} \rho_j^a f_j : a \in H_s\right\}\right\} \\ &= \max\{x_s, L(s) + y_s\} \end{aligned} \tag{7-67}$$

However, $H_s = H_{s+1} \cup M_s$; so

$$y_s = \max\{y_{s+1}, z_s\} \tag{7-68}$$

Figure 7-11 shows part of the network which corresponds to (7-65) through (7-68). Notice that the network has $\#W_s + \#M_s + 2$ arcs and two nodes for each s. The node marked f_s would be present even if we did not exploit the separable structure. The second node, marked y_s, corresponds to (7-68). It denotes the optimum among those actions with separable structure which are not feasible at any state s' with $s' > s$.

The network for the straightforward dynamic program of an MDP has $\#A_s$ arcs and one node for each s. For each s, the reduction in the number of arcs is

$$\#A_s - (\#W_s + \#M_s + 2) = \#H_{s+1} - 2$$

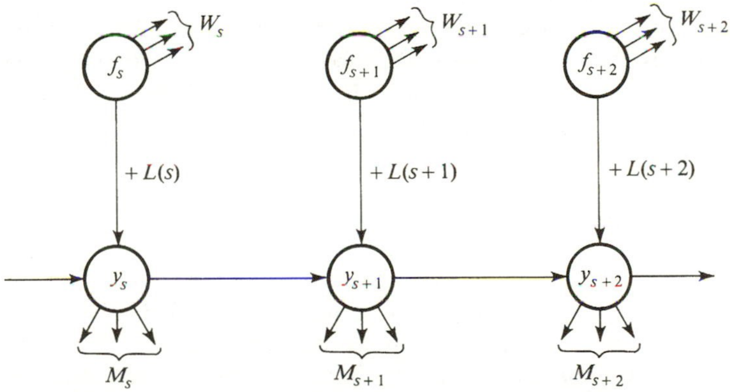

Figure 7-11 Part of a network which corresponds to (7-65) through (7-68).

but the number of nodes is doubled. Hence, it is profitable to exploit the separable structure via (7-65) through (7-68) only if $\# H_{s+1}$ is sufficiently large for sufficiently many states s.

Let $\{\lambda_s\}$ be arbitrary positive numbers and let b denote the largest s for which H_s is nonempty. Then a linear program which is equivalent to (7-65) through (7-68) is

$$\text{Minimize } \sum_{s \in \mathcal{S}} \lambda_s(f_s + y_s) \tag{7-69a}$$

subject to

$$f_s \geq r(s, a) + \beta \sum_{j \in \mathcal{S}} p_{sj}^a f_j \qquad a \in W_s \qquad s \in \mathcal{S} \tag{7-69b}$$

$$f_s \geq L(a) + y_s \qquad s \leq b \tag{7-69c}$$

$$y_s \geq y_{s+1} \qquad s \leq b \tag{7-69d}$$

$$y_s \geq X(a) + \beta \sum_{j \in \mathcal{S}} \rho_j^a f_j \qquad a \in M_s \qquad s \leq b \tag{7-69e}$$

Observe that (7-62a) to (7-62f) is a special case of (7-69a) to (7-69e) with each lettered part of (7-62a) to (7-62e) corresponding to the same lettered part of (7-69a) to (7-69e).

Some authors call (7-69a) to (7-69e) the *reduced* or *streamlined* linear program. What is the computational advantage of the streamlined program? Let

$$c_1 = \sum_{s=1}^{S} \# A_s \qquad c_2 = \sum_{s=2}^{b} \# H_s$$

Then (7-69a) to (7-69e) has $S + b + 1$ variables in $2b + c_1 - c_2$ constraints. The usual linear program (4-91) has S variables in c_1 constraints. The difference in the size of the coefficient matrices is the following expression:

$$\begin{aligned} \text{Usual} - \text{streamlined} &= Sc_1 - (S + b + 1)(2b + c_1 - c_2) \\ &= (c_2 - 2b)(S + b + 1) - c_1(b + 1) \end{aligned} \tag{7-70}$$

The quantity (7-70) is positive, i.e., favorable for the streamlined problem, if c_2 is close to c_1. This was true for the (σ, Σ) model at the beginning of the section. There $\# A_s = \# H_s + 1$ so $c_1 - c_2 = 2u + 1$. On the other hand, (7-70) would be negative if $S = 10$ and for all s, $\# A_s = 10$ and $\# H_s = 10 - s$. Then $b = 9$, $c_1 = 100$, and $c_2 = 36$; so (7-70) would take the value -640.

A Machine-Replacement Model

Example 7-1 Suppose that a machine ages as it is used and, if it fails, must be replaced. At the beginning of each period, if it has not failed during the previous period, it may be either retained or replaced. Let $u - 1$ be the maximum acceptable age of a machine; a machine must be replaced at age u. Thus $\mathcal{S} = \{0, 1, \ldots, u\}$, where the state represents the machine's age.

Let ϕ_a be the probability that an age a machine does not fail (during a period) and let c_a be the expected single-period operating cost of an age a machine. If an age s machine is replaced by an age a machine, the expected single-period costs are $c_a + b_a - d_s$, where b_a is the purchase cost of an age a machine and d_s is the salvage value of an age s machine. If a machine fails in use, it is categorized as an age u machine.

It would be natural to let $A_s = \{0, 1, \ldots, u-1\}$, where we would interpret $a \neq s$ as replacement of the current machine with an age a machine and $a = s$ as retention of the current machine. This formulation would preclude replacement of an age s machine with another age s machine. Such an action would be suboptimal if $b_s > d_s$, i.e., if purchase costs are higher than salvage values. This inequality is valid in practice, and we assume that it holds for all $s < u$.

Let $A' = \{0, 1, \ldots, u-1\}$. Instead of letting $A_s = A'$ for all s, let

$$A_s = \begin{cases} A' \cup \{s^*\} & \text{if } s < u \\ A' & \text{if } s = u \end{cases}$$

We interpret action s^* as retention of an age s machine and action s as replacement of an age s machine with another age s machine. Therefore,

$$r(s, a) = \begin{cases} c_a + b_a - d_s & \text{if } a \neq s^* \\ c_s & \text{if } a = s^* \end{cases}$$

Let $\phi_{s*} = \phi_s$ for each $s < u$. Then

$$p^a_{s,\,a+1} = \phi_a \qquad p^a_{su} = 1 - \phi_a \qquad a \neq s^*$$

$$p^{s*}_{s,\,s+1} = \phi_s \qquad p^{s*}_{su} = 1 - \phi_s$$

There is no loss of optimality in including the spurious option of replacing a machine with another one of the same age. An optimal policy would not specify taking such an action.

This model is a separable MDP with $H_s = A'$ for all $s < u$, $H_u = \phi$, $X(a) = c_a + b_a$, $L(s) = -d_s$, $M_s = \phi$ if $s < u - 1$, $M_{u-1} = A'$, $W_s = \{s^*\}$ if $s < u$, and $W_u = A'$. Hence, (7-69*a*) to (7-69*e*) takes the form

$$\text{Maximize } \sum_{s=0}^{u-1} \lambda_s(f_s + y_s) + \lambda_u f_u \tag{7-71a}$$

subject to

$$f_s \leq c_s + \beta[\phi_s f_{s+1} + (1 - \phi_s)f_u] \qquad 0 \leq s \leq u-1 \tag{7-71b}$$

$$f_u \leq c_a + b_a + \beta[\phi_a f_{a+1} + (1 - \phi_a)f_u] \qquad 0 \leq a \leq u-1 \tag{7-71c}$$

$$f_s \leq -d_s + y_s \qquad 0 \leq s \leq u-1 \tag{7-71d}$$

$$y_s \leq y_{s+1} \qquad 0 \leq s \leq u-2 \tag{7-71e}$$

$$y_{u-1} \leq c_a + b_a + \beta[\phi_a f_{a+1} + (1 - \phi_a)f_u] \qquad 0 \leq a \leq u-1 \tag{7-71f}$$

This streamlined problem has $2(u + 1)$ variables in $5u$ constraints.

Problem (7-71*a*) to (7-71*f*) can be further reduced by observing that (7-71*c*) and (7-71*f*) imply $f_u = y_{u-1}$ and (7-71*e*) implies $y_s = y_{u-1}$ for all s. These observations permit the elimination of $u - 1$ variables and $2u - 1$ constraints. In this further streamlined version, there are $u + 1$ variables in $3u$ constraints. Figure 7-12 shows the network which corresponds to the further streamlined version.

The usual linear program for the replacement model is

$$\text{Maximize } \sum_{s=0}^{\infty} \lambda_s f_s$$

subject to

$$\begin{aligned}
&f_s \leq c_s + \beta[\rho_s f_{s+1} + (1 - \rho_s) f_u] && 0 \leq s \leq u - 1 \\
&f_s \leq c_a + b_a - d_a \\
&\qquad + \beta[\rho_a f_{a+1} + (1 - \rho_a) f_u] && 0 \leq a \leq u - 1,\ a \neq s,\ 0 \leq s \leq u - 1 \\
&f_u \leq c_a + b_a - d_u \\
&\qquad + \beta[\rho_a f_{a+1} + (1 - \rho_a) f_u] && 0 \leq a \leq u - 1
\end{aligned} \tag{7-72}$$

This linear program has $u + 1$ variables in $u(u + 1)$ constraints. The difference in the sizes of the coefficient matrices of (7-71) and (7-72) is

$$\text{Size of (7-72)} - \text{size of (7-71)} = u(u + 1)^2 - 10u(u + 1) = u(u + 1)(u - 9)$$

At $u = 5$, 10, and 20, this difference is -120, 110, and 4,620, respectively. If $u \geq 10$, the reduction in the number of constraints [in (7-71) compared with (7-72)] more than compensates for the increase in the number of variables. Incidentally, the coefficient matrix of (7-71) is less dense than that of (7-72). Streamlined problems generally have this advantage (cf. Exercise 7-27). □

A Generalization

The computational efficiency of a streamlined problem stems from (7-64*a*), that is,

$$\ldots \supset H_s \supset H_{s+1} \supset \ldots$$

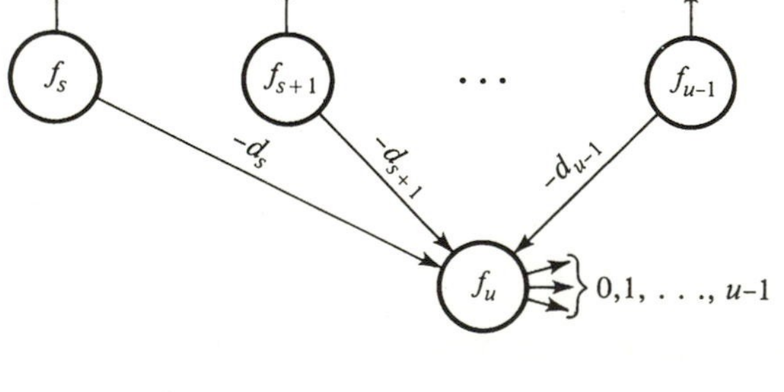

Figure 7-12 Part of a network for the further streamlined version of (7-71*a*) to (7-71*f*).

which permits the recursion (7-68), that is,

$$y_s = \max \{y_{s+1}, z_s\}$$

Some MDPs fail to satisfy (7-64*a*) but possess subsets of states with the telescoping feature (7-64*a*). We use two examples to illustrate the fact that computations in such MDPs still can be streamlined.

Inventory Model with Dependent Demands

Example 7-2 Suppose the inventory model at the beginning of this section is altered by permitting successive demands $D_1, D_2, \ldots$ to comprise a Markov chain with a state space which is a subset of the integers $\{0, 1, \ldots, m\}$. Let $G_i(a)$ denote the expected "generalized" inventory costs in period n if $D_{n-1} = i$ and a is the order-up-to decision in period n, and let

$$\phi(i, j) = P\{D_n = j \mid D_{n-1} = i\}$$

denote the transition probabilities. The state in this MDP is now a 2-vector, $s = (z, i)$, where z denotes the inventory on hand at the beginning of a period and i denotes the previous period's demand. Instead of (7-60), the optimality equation becomes

$$f_{zi} = \min \Big\{K\delta(a - x) + G_i(a) + \beta \sum_{j=1}^{a} \phi(i, a - j) f_{j,\, a-j}$$

$$+ \beta \sum_{j=a}^{m} \phi(i, j) f_{0j}\colon z < a < u\Big\}$$

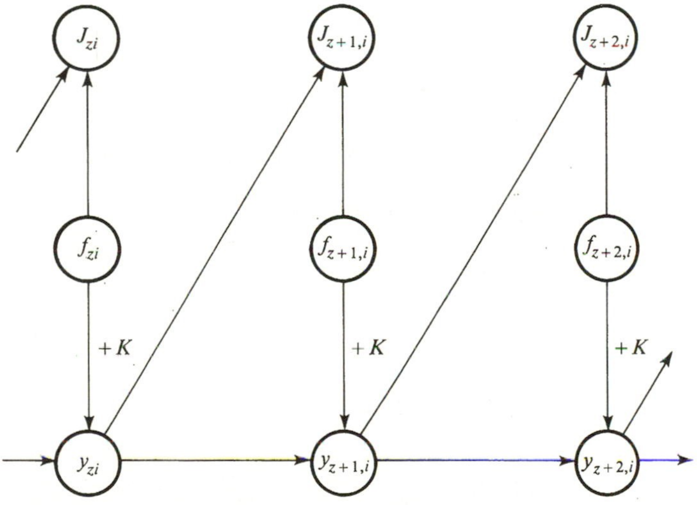

Figure 7-13 Part of a subnetwork for Example 7-2.

Reasoning as in the transition from (7-60) to (7-61a) yields

$$f_{zi} = \min\{J_{zi}, K + y_{zi}\} \qquad y_{zi} = \min\{y_{z+1,i}, J_{z+1,i}\}$$

$$J_{zi} \triangleq G_i(z) + \beta \sum_{j=1}^{z} \phi(i, z-j) f_{j,z-j} + \beta \sum_{j=z}^{m} \phi(i,j) f_{0j}$$

$$y_{zi} \triangleq \min\{J_{ai}: z+1 \le a \le u\}$$

Figure 7-13 depicts part of the concomitant network. It is a collection of subnetworks, and each has the same structure as the network in Figure 7-9. Let $\{\lambda_{zi}\}$ be arbitrary positive numbers. The linear program, instead of (7-62a) to (7-62d), is

$$\text{Maximize} \sum_{i=0}^{m} \sum_{z=0}^{u} \lambda_{zi}(y_{zi} + f_{zi}) \tag{7-73a}$$

subject to

$$\left.\begin{aligned} f_{zi} &\le G_i(z) + \beta \sum_{j=1}^{z} \phi(i, z-j) f_{j,z-j} \\ &\quad + \beta \sum_{j=z}^{m} \phi(i,j) f_{0j} \qquad (7\text{-}73b) \\ f_{zi} &\le K + y_{zi} \qquad (7\text{-}73c) \end{aligned}\right\} 0 \le z \le u-1,\ 0 \le i \le m$$

$$\left.\begin{aligned} y_{zi} &\le y_{z+1,i} \qquad (7\text{-}73d) \\ y_{zi} &\le G_i(z+1) + \beta \sum_{j=1}^{z+1} \phi(i, z+1-j) f_{j,z+1-j} \\ &\quad + \beta \sum_{j=z+1}^{m} \phi(i,j) f_{0j} \qquad (7\text{-}73e) \end{aligned}\right\} 0 \le z \le u-1,\ 0 \le i \le m$$

$$f_{ui} = G_i(u) + \beta \sum_{j=1}^{u} \phi(i, u-j) + \beta \sum_{j=u}^{m} \phi(i,j) f_{0j} \qquad 0 \le i \le m \tag{7-73f}$$

This linear program has $(m+1)(u+1)$ variables in $(m+1)(4u+1)$ constraints.

The usual linear program for this MDP is

$$\text{Maximize} \sum_{i=0}^{m} \sum_{z=0}^{u} \lambda_{zi} f_{zi} \tag{7-74a}$$

subject to

$$\begin{aligned} f_{zi} &\le K\delta(a-z) + G_i(a) + \beta \sum_{j=1}^{a} \phi(i, a-j) f_{j,a-j} && 0 \le z \le u-1 \\ & && 0 \le i \le m \\ &\quad + \beta \sum_{j=a}^{m} \phi(i,j) f_{0j} && z \le a \le u \end{aligned} \tag{7-74b}$$

$$f_{ui} = G_i(u) + \beta \sum_{j=1}^{u} \phi(i, u-j) f_{j, u-j} + \beta \sum_{j=u}^{m} \phi(i, j) f_{0j} \qquad 0 \le i \le m \qquad \text{(7-74c)}$$

This problem has $(m + 1)(u + 1)$ variables in $(m + 1)(u + 1)(u + 2)/2$ constraints.

The difference in the number of elements in the coefficient matrices of (7-73) and (7-74), after some algebra, is

$$\text{Size of (7-74)} - \text{size of (7-73)} = \frac{(u-5)(u+1)u(m+1)^2}{2} \qquad \text{(7-75)}$$

Regardless of the value of m, this difference is positive, i.e., favorable for (7-73), if $u \ge 6$. □

Short-Run Changes in Production with a Fixed Employment Level

Example 7-3 In a factory with a given employment level, often there are opportunities to make temporary adjustments to that level via overtime, the use of subcontractors, or a curtailed work week.† Suppose that w units are produced each period in which no adjustment is made. If an adjustment is made, suppose that there is a cost proportional to the size of the adjustment. Assume that excess demand is lost, successive demands $D_1, D_2, \ldots$ are i.i.d. nonnegative integer-valued r.v.'s with

$$\phi_i = P\{D_1 = i\} \qquad i \in I$$

and goods are produced quickly enough to satisfy demand during the same period.‡ Let s_n and a_n denote the respective amounts of stock on hand in period n immediately before and after goods are produced during period n. Hence, $a_n - s_n$ is the quantity produced during period n. Let u denote the storage capacity.

Suppose that the production cost in period n is $c \cdot (a_n - s_n)$ and the adjustment cost of $a_n - s_n$ relative to the output level w is

$$\begin{array}{ll} \gamma \cdot (a_n - s_n - w) & \text{if } a_n - s_n \ge w \\ \rho \cdot [w - (a_n - s_n)] & \text{if } a_n - s_n < w \end{array}$$

where γ and ρ are nonnegative constants. Let $g(a_n, D_n)$ represent the inventory holding and shortage costs in period n and let β be the single-period discount factor with $0 \le \beta < 1$.

† Section 8-4 has a production model in which permanent adjustments can be made.

‡ Exercise 7-26 alters this model by assuming that excess demand is back-ordered and ordered goods are delivered several periods later.

The present value of the sequence of costs is

$$B = \sum_{n=1}^{\infty} \beta^{n-1}[c(a_n - s_n) + g(a_n, D_n) + \gamma(a_n - s_n - w)^+ + \rho(w + s_n - a_n)^+] \tag{7-76}$$

It is convenient to rewrite (7-76) using the identity

$$\gamma \cdot (u - v)^+ + \rho \cdot (v - u)^+ = d\,|u - v| + h \cdot (u - v)$$

where

$$d = \frac{\gamma + \rho}{2} \qquad \text{and} \qquad h = \frac{\gamma - \rho}{2} \tag{7-77}$$

Therefore,

$$\begin{aligned} B &= \sum_{n=1}^{\infty} \beta^{n-1}[c \cdot (a_n - s_n) + g(a_n, D_n) + d\,|a_n - s_n - w| + h \cdot (a_n - s_n - w)] \\ &= \sum_{n=1}^{\infty} \beta^{n-1}[(c + h)(a_n - s_n) + g(a_n, D_n) + d\,|a_n - s_n - w|] - \frac{hw}{1 - \beta} \end{aligned}$$

Since excess demand is lost, substitution of

$$s_n = (a_{n-1} - D_{n-1})^+ \qquad n > 1$$

in the expression for B yields

$$B = \sum_{n=1}^{\infty} \beta^{n-1}\{(c + h)[a_n - (a_{n-1} - D_{n-1})^+] + g(a_n, D_n) + d\,|a_n - s_n - w|\} - \frac{hw}{1 - \beta} - (c + h)s_1$$

Let

$$G(a) = (c + h)\{a - \beta E[(a - D_1)^+]\} + E[g(a, D_1)] \tag{7-78}$$

so

$$E(B) = E\left\{\sum_{n=1}^{\infty} \beta^{n-1}[G(a_n) + d\,|a_n - s_n - w|]\right\} - \left[\frac{hw}{1 - \beta} + (c + h)s_1\right] \tag{7-79a}$$

The second term on the right side of (7-79*a*) is not affected by $a_1, a_2, \ldots$; so a policy minimizes $E(B)$ if, and only if, it minimizes

$$E\left\{\sum_{n=1}^{\infty} \beta^{n-1}[G(a_n) + d\,|a_n - s_n - w|]\right\} \tag{7-79b}$$

Exercise 7-25 asks you to explain the sense in which minimization of (7-79*b*) is equivalent to the dynamic program

$$f_s = \min \{d\,|a - s - w| + J_a\colon s \le a \le u\} \qquad 0 \le s \le u \qquad \text{(7-80}a\text{)}$$

$$J_a = G(a) + \beta \sum_{j=1}^{a} \phi_{a-j} f_j + \beta f_0 \sum_{j=a}^{\infty} \phi_j \qquad 0 \le s \le u \qquad \text{(7-80}b\text{)}$$

where the minimization in (7-80*a*) considers only integer values of *a*.

The representation (7-80*a*) and (7-80*b*) satisfies Definition 7-3 of a separable MDP in two alternative ways. For the first way, let

$$H_s^+ = \{a\colon s + w < a \le u,\ a \in I\}$$

where $H_s^+ = \phi$ if $s \ge u - w$. Then $H_s^+ \subset H_{s+1}^+$ as required by (7-64*a*). Also,

$$\begin{aligned} r(s, a) &= d\,|a - s - w| + G(a) \\ &= da + G(a) - d\cdot(s + w) \qquad a \in H_s^+ \end{aligned}$$

which satisfies (7-64*b*), namely,

$$r(s, a) = X(a) + L(s) \qquad a \in H_s \qquad \text{(7-81)}$$

with $X(a) = X^+(a) \triangleq da + G(a)$ and $L(s) = L^+(s) \triangleq d(s + w)$. Last, (7-64*c*) is

$$p_{sj}^a = \begin{cases} \phi_{a-j} & \text{if } j > 0 \\ \sum_{i=a}^{\infty} \phi_i & \text{if } j = 0 \end{cases} \qquad \text{(7-82)}$$

which does not depend on *s*.

A second way in which (7-80*a*) and (7-80*b*) satisfies the definition of a separable MDP uses†

$$H_s^- = \{a\colon 0 \le a < (s + w) \wedge (u + 1),\ a \in I\}$$

Here, $H_s^- \supset H_{s+1}^-$, so a formal relabeling of states would yield $H_s^- \subset H_{s+1}^-$; we shall *not* relabel the states. Also,

$$\begin{aligned} r(s, a) &= d\,|a - s - w| + G(a) \\ &= -da + G(a) + d\cdot(s + w) \qquad a \in H_s^- \end{aligned}$$

which satisfies (7-81), hence (7-64*b*), with $X(a) = X^-(a) \triangleq -da + G(a)$ and $L(s) = L^-(s) \triangleq d\cdot(s + w)$. Last, (7-64*c*) is satisfied because (7-82) still is valid.

It is unnecessary to exploit separability based on only one of the two structures, i.e., either $\{H_s^+\}$ or $\{H_s^-\}$ but not both. Instead, the following analysis exploits both structures. We mimic the argument and notation between (7-65) and (7-68) on page 343.

† Recall our notation $a \wedge b = \text{minimum } \{a, b\}$.

Let

$$M_s^+ = H_s^+ - H_{s+1}^+ = \{s + w + 1\} \qquad M_s^- = H_s^- - H_{s-1}^- = \{s + w - 1\}$$

$$W_s = A_s - H_s^+ - H_s^- = \begin{cases} \{s + w\} & \text{if } s + w \le u \\ \phi & \text{if } s + w > u \end{cases}$$

Let

$$x_s = J_{s+w}$$

$$z_s^+ = d(s + w + 1) + J_{s+w+1}$$

$$= \min \left\{ X^+(a) + \beta \sum_{j \in \mathcal{S}} \rho_j^a f_j : \quad a \in M_s^+ \right\}$$

$$z_s^- = -d(s + w - 1) + J_{s+w-1}$$

$$= \min \left\{ X^-(a) + \beta \sum_{j \in \mathcal{S}} \rho_j^a f_j : \quad a \in M_s^- \right\}$$

$$y_s^+ = \min \{da + J_a : s + w + 1 \le a \le u\}$$

$$y_s^- = \min \{-da + J_a : s \le a \le s + w - 1\}$$

Since† $A_s = W_s \cup H_s^+ \cup H_s^-$ and these sets are mutually exclusive,

$$f_s = \min \begin{Bmatrix} J_{s+w} \\ \min \{d(a - s - w) + J_a : s + w + 1 \le a \le u\} \\ \min \{d(s + w - a) + J_a : s \le a \le s + w - 1\} \end{Bmatrix}$$

$$= \min \{x_s, y_s^+ - d(s + w), y_s^- + d(s + w)\} \tag{7-83a}$$

with $\quad y_s^+ = \min \{y_{s+1}^+, z_s^+\} \quad$ and $\quad y_s^- = \min \{y_{s-1}^-, z_s\} \qquad$ (7-83b)

Figure 7-14 displays the network which corresponds to (7-83*a*) and (7-83*b*).

Let $\lambda_0, \ldots, \lambda_u$ be arbitrary positive constants. A linear program which corresponds to (7-83*a*) and (7-83*b*) is

$$\text{Minimize} \sum_{s=0}^{u-w-1} \lambda_s(f_s + y_s^+ + y_s^-) + \sum_{s=u-w}^{u} \lambda_s(f_s + y_s^-)$$

subject to

$$f_s \le G(s + w) + \beta \sum_{j=1}^{s+w} \phi_{s+w-j} f_j + \beta f_0 \sum_{j=s+w}^{\infty} \phi_j \qquad 0 \le s \le u - w$$

$$f_s \le y_s^+ - d(s + w) \qquad 0 \le s \le u - w - 1$$

$$f_s \le y_s^- + d(s + w) \qquad 0 \le s \le u$$

$$y_s^+ \le -d(s + w + 1) + G(s + w + 1)$$

$$+ \beta \sum_{j=1}^{s+w+1} \phi_{s+w+1-j} f_j + \beta f_0 \sum_{j=s+w+1}^{\infty} \phi_j \qquad 0 \le s \le u - w - 1$$

† We adopt the usual convention that a minimum (maximum) over the empty set is assigned the value $+\infty(-\infty)$.

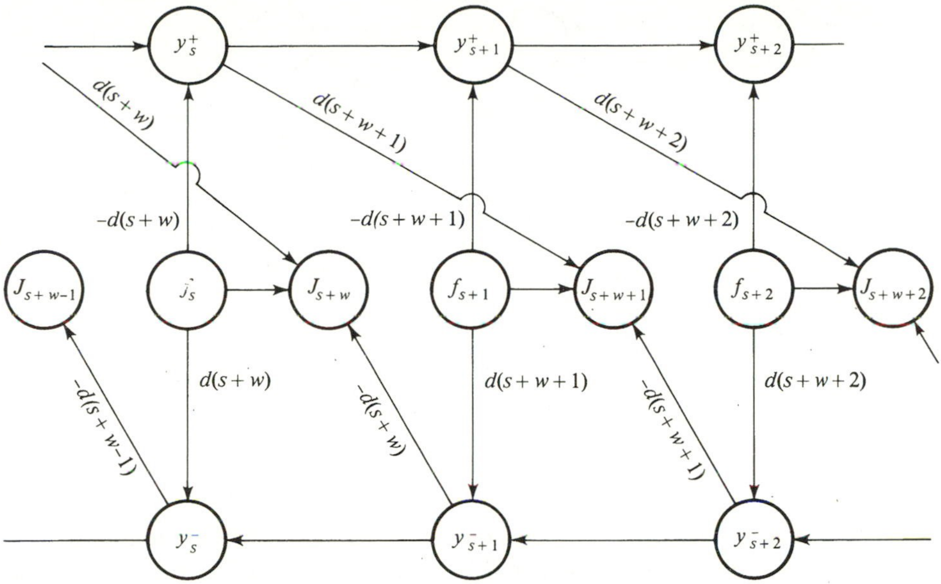

Figure 7-14 Part of the network associated with (7-83*a*) and (7-83*b*).

$$y_s^+ \le y_{s+1}^+ \qquad 0 \le s \le u - w - 2$$

$$y_s^- \le d(s+w-1) + G(s+w-1)$$

$$+ \beta \sum_{j=1}^{s+w-1} \phi_{s+w-1-j} f_j + \beta f_0 \sum_{j=s+w-1}^{\infty} \phi_j \qquad 0 \le s \le u - w + 1$$

$$y_s^- \le y_{s-1}^- \qquad 1 \le s \le u \tag{7-84}$$

This linear program has $3u - w + 2$ variables in $7u - 5w + 3$ constraints. The straightforward linear-program version of (7-80*a*) and (7-80*b*) has $u + 1$ variables in $(u + 1)(u + 2)/2$ constraints. Therefore, (7-84) is superior only for certain values of u and w.

Exercise 7-24 asks you to prove that (7-83*a*) and (7-83*b*) can be reduced to

$$f_s = \min \{x_s, d + f_{s+1}, d + f_{s-1}\} \tag{7-85}$$

Figure 7-15 shows the network associated with (7-85). The linear program which corresponds to (7-85) is

Minimize $\sum_{s=0}^{u} \lambda_s f_s$

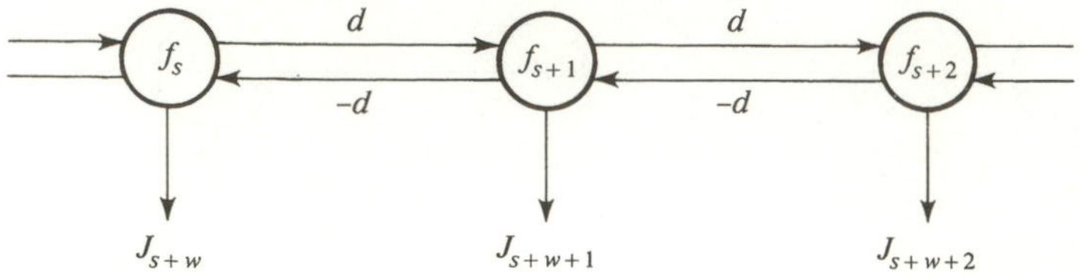

Figure 7-15 The network associated with (7-85).

subject to

$$f_s \le G(s+w) + \beta \sum_{j=1}^{s+w} \phi_{s+w-j} f_j + \beta f_0 \sum_{j=s+w}^{\infty} \phi_j \qquad 0 \le s \le u - w$$

$$f_s \le d + f_{s+1} \qquad 0 \le s \le u-1$$

$$f_s \le d + f_{s-1} \qquad 1 \le s \le u \qquad (7\text{-}86)$$

This linear program has $u + 1$ variables in $3u - w + 1$ constraints. For all values of u and w, it is smaller than both (7-84) and the straightforward linear-program version of (7-80*a*) and (7-80*b*). □

EXERCISES

7-17 In what sense is (7-62*a*) to (7-62*f*) equivalent to (7-61*a*), (7-61*b*), and (7-61*d*)? (Hint: Theorem 4-11.)

7-18 Prove that (7-61*a*) to (7-61*c*) can be reduced to

$$f_s = \min \{x_s, K + x_{s+1}, K + f_{s+1}\} \qquad (7\text{-}87)$$

if $0 \le s < u$. [Hint: Write (7-61*b*) for s and $s + 1$ and consider the four cases which are possible.] Draw a network diagram for (7-87) which mimics the correspondence between Figure 7-9 and (7-61*a*) and (7-61*b*).

7-19 (Continuation) Write the linear program for (7-87) and (7-61*d*) which is analogous to (7-62*a*) to (7-62*f*) written for (7-61*a*), (7-61*b*), and (7-61*d*). How many rows and columns does your linear program have? Compute the resulting size of the constraint matrix for $u = 5$, 10, 20, and 50, and compare your results with the entries in Table 7-1.

7-20† Suppose in (7-60*a*) and (7-61*a*), (7-61*b*), and (7-61*d*) that a (σ, Σ) policy is optimal; that is, there are integers σ and Σ with $0 \le \sigma \le \Sigma$ and

$$f_s = \begin{cases} K + x_{\Sigma} & \text{if } 0 \le s < \sigma \\ x_s & \text{if } \sigma \le s \le u \end{cases}$$

Suppose that one solves (with a linear program, for example)

$$z_s = \min \left\{ G(s) + \beta \sum_{j=1}^{s} \phi_{s-j} z_j + \beta z_0 \sum_{j=s+1}^{\infty} \phi_j, K + y \right\} \qquad (7\text{-}88a)$$

† This exercise uses ideas in Denardo (1968) and Johnson (1967).

$$y = \min \left\{ G(a) + \beta \sum_{j=1}^{a} \phi_{a-j} z_j + \beta z \sum_{j=a+1}^{\infty} \phi_j : \quad 0 \le a \le u \right\} \tag{7-88b}$$

for the variables $y, z_0, \ldots, z_u$. Let Λ satisfy

$$y = G(\Lambda) + \beta \sum_{j=1}^{\Lambda} \phi_{\Lambda-j} z_j + \beta z_0 \sum_{j=\Lambda+1}^{\infty} \phi_j$$

and let λ be the largest integer less than or equal to Λ such that $z_{\lambda-1} = K + y$. Prove that $f_s = z_s$ if $s \le \Lambda$ and that (λ, Λ) is an optimal (σ, Σ) policy.

7-21 (Continuation) Under the assumptions of Exercise 7-20 repeat Exercise 7-19 for the linear program (7-88*a*) and (7-88*b*) [instead of (7-87)], which is analogous to (7-62*a*) to (7-62*f*) written for (7-61*a*), (7-61*b*), and (7-61*d*).

7-22 Suppose the inventory model at the start of this section is altered so that excess demand is back-ordered rather than lost and we assume there is an integer $m < \infty$ such that

$$P\{D_1 = i\} = \phi_i = 0 \text{ if } i > m$$

Then instead of (7-58*b*), (7-58*c*), and (7-58*d*), there is

$$p_{sj}^a = \phi_{a-j}$$

$$\mathcal{S} = \{-m, -m+1, \ldots, 0, \ldots, u\}$$

$$A_s = \begin{cases} \{0, 1, \ldots, u\} & \text{if } -m \le s < 0 \\ \{s, s+1, \ldots, u\} & \text{if } 0 \le s \le u \end{cases}$$

(*a*) For $0 \le s < u$, what equation and identity replace (7-60*a*) and (7-60*b*) under these assumptions?

(*b*) Let $y_{-1} = \min\{x_0, y_0\}$. If $-m \le u < 0$, explain why $f_s = K + y_{-1}$.

(*c*) Use your answers to (*a*) and (*b*) to write a linear program analogous to (7-62*a*) to (7-62*f*).

(*d*) For this altered model, compute the entries in a table similar to Table 7-1.

(*e*) Why is the claim in Exercise 7-18 valid for this altered model?

(*f*) State and prove a version of the claim in Exercise 7-20 which is valid for the altered model.

7-23 (You may need a computer to solve this problem.) Exercise 3-16 describes a hypothetical model of a small commercial catfish farming pond. Harvests occur annually, the harvest takes only a few days, the farmer can accurately estimate the amount (tons) of fish in the pond before selecting the quantity to be harvested, the annual discount factor is 0.9, the net profit is \$500 per ton harvested, and the young and optimistic farmer uses an infinitely long planning horizon with the expected-present-value criterion. Suppose that the pond cannot hold more than 5 tons of fish and that the following probabilities pertain (instead of the ones in Exercise 3-16):

Tons of catfish remaining after harvest this year	Tons of catfish at harvest time next year					
	0	1	2	3	4	5
0	1	0	0	0	0	0
1	0	0.9	0.1	0	0	0
2	0	0.1	0.7	0.1	0.1	0
3	0	0.05	0.05	0.7	0.1	0.1
4	0	0	0.05	0.05	0.6	0.3
5	0	0	0	0.1	0.1	0.8

What policy is optimal? What is the farmer's maximal expected present value of profit if there are 3 tons in the pond and harvesting is about to start?

7-24 Prove that the solution common to (7-80*a*) and (7-80*b*) and (7-83*a*) and (7-83*b*) satisfies (7-85). [Hint: Write (7-85) for $s - 1$, s, and $s + 1$. Then there are $3^3 = 27$ cases to consider! Can you devise a shorter proof?]

7-25 In what sense is the minimization of (7-79*b*) equivalent to the dynamic program (7-80*a*) and (7-80*b*)?

7-26 (*a*) Suppose that Example 7-3 is altered as follows. Excess demand is back-ordered rather than lost and, for each n, goods ordered in period n are received at the beginning of period $\lambda + n$ rather than delivered immediately. What is the dynamic program which replaces (7-80*a*) and (7-80*b*)? (Hint: Section 3-2.)

(*b*) Under the assumptions of part (*a*), derive equations which replace (7-83*a*) and (7-83*b*) and (7-85).

7-27 (*a*) At several points in this section, we compare the size of a streamlined problem with that of an ordinary MDP formulation. The criterion in each case is the number of elements in the matrix of coefficients of a linear program [for example, Table 7-1 and expressions (7-70) and (7-75)]. One of the other criteria which are pertinent to such a comparison is the number of nonzero elements in the coefficient matrix. Another criterion is the fraction of elements which are nonzero (called *density*; 1 − density = *sparseness*). These two additional criteria are important because they are closely related to storage requirements when a linear program is being solved on a computer. Compare (7-71) and (7-72) with respect to both of the additional criteria.

(*b*) Use the two additional criteria in part (*a*) to compare (7-69*a*) to (7-69*e*) with the usual linear program (4-91).

7-5 CERTAINTY EQUIVALENCE

This section concerns MDPs with the following structural features: states and actions are vectors, actions are unconstrained by the state, the transition function depends linearly on the state and action, and the single-stage reward function is the sum of quadratic functions of the state and action. We call an MDP with these features a *linear-quadratic model* (Definition 7-4 below is more precise). Under certain assumptions, these features yield explicit formulas for the optimal policy and the optimal value function. The formula for the optimal policy depends on random variables in the model only through their expected values. In this sense, a linear-quadratic model is equivalent to an MDP in which random variables are replaced by their expected values. The equivalence justifies the term *certainty equivalent.*

We begin with an example of a linear-quadratic model. Another example suggests that it may be easy to optimize such models. Then we formally define a linear-quadratic model and optimize it.

Example 7-4 Consider the following model of a production process with severe bottleneck costs of production and storage. If a_n is the quantity produced in period n, the production cost is $c_1 a_n + c_2 a_n^2$. If s_n is the inventory level at the beginning of period n, the storage cost is $h_1 s_n + h_2 s_n^2$. Let successive inventory levels be *modeled* as

$$s_{n+1} = s_n + a_n - D_n \tag{7-89}$$

where $D_1, D_2, \ldots$ are independent random variables representing demands. We assume that the demands do not depend on the inventory levels or the order quantities. Although (7-89) is consistent with back-ordering excess demand, the following cost structure does not penalize negative inventory levels. Therefore, the model may be unreasonable.

The objective is to minimize

$$E\left[\sum_{n=1}^{N} \beta^{n-1}(c_1 a_n + c_2 a_n^2 + h_1 s_n + h_2 s_n^2) + \beta^N(h_1 s_{N+1} + h_2 s_{N+1}^2)\right] \quad (7\text{-}90)$$

which is the expected present value of costs during an N-period planning horizon subject to the constraints in (7-89). We place no other constraints on $a_1, a_2, \ldots$ or, implicitly, on $s_1, s_2, \ldots$. Since (7-89) admits back-ordering but (7-90) does not penalize it, the model is unreasonable unless it has the following properties:

(i) Optimization yields a policy such that $a_n \geq 0$ for every n; that is, production quantities are nonnegative.

(ii) An optimal policy induces a negligibly small probability that any inventory level will be negative; that is, $s_{n+1} < 0$ in (7-89).

Some numerical versions of this model will satisfy (i) and (ii). □

The next example suggests that linear-quadratic models may possess a simple explicit optimum.

Example 7-5 Consider the problem of choosing a number a in order to maximize

$$E[gs + ha + \xi)^2]v \quad (7\text{-}91)$$

where s, g, h, and v are numbers and ξ is a random variable. Let $\mu = E(\xi)$ and $\sigma^2 = \text{Var}(\xi)$. We assume $|\mu| < \infty$ and $\sigma^2 < \infty$.

The objective (7-91) is

$$vh^2a^2 + 2vh(gs + \mu)a + v(g^2s^2 + \sigma^2 + \mu^2 + 2gs\mu)$$

whose first and second derivatives with respect to a are

$$2vh^2a + 2vh(gs + \mu)$$

and $2vh^2$, respectively. If $v < 0$ and $h \neq 0$, therefore, the maximum of (7-91) is achieved uniquely at

$$a = -\frac{gs + \mu}{h} = -\frac{\mu}{h} - \frac{g}{h}s \quad (7\text{-}92)$$

When (7-92) is substituted in (7-91), the value of the objective is $v\sigma^2$.

There are three notable features of the solution: the optimal value of a in (7-92) depends on the random variable ξ only via $\mu = E(\xi)$, the value of a in (7-92) depends only linearly on s, and the optimal value of the objective does not depend on s. These features persist below when we analyze a dynamic model in which s and a are vectors rather than numbers, and ξ is a random vector instead of a scalar random variable. □

The Linear-Quadratic Model

In the following definition, we use the notation Z' for the transpose of a matrix Z. All matrices and vectors are assumed to have the dimensions needed to justify the indicated matrix addition and multiplication. We permit states and actions to be finite-dimensional vectors; so to be specific, suppose that they are column vectors.

Definition 7-4 A *linear-quadratic model* is a nonstationary† MDP with an N-period planning horizon such that

$$\mathcal{S} = \mathbb{R}^m \qquad \text{and} \qquad A_{\mathbf{s}} = \mathbb{R}^\theta \qquad \text{for all } \mathbf{s} \in \mathcal{S}$$

where $m < \infty$ and $\theta < \infty$, and

$$L(s) = \mathbf{s}' U_{N+1} \mathbf{s} \tag{7-93a}$$

$$r_n(\mathbf{s}, \mathbf{a}) = \mathbf{s}' U_n \mathbf{s} + \mathbf{a}' V_n \mathbf{a} + \mathbf{u}'_n \mathbf{s} + \mathbf{v}'_n \mathbf{a} \qquad n = 1, \ldots, N \tag{7-93b}$$

$$\mathbf{s}_{n+1} = G_n \mathbf{s}_n + H_n \mathbf{a}_n + \boldsymbol{\xi}_n \qquad n = 1, \ldots, N \tag{7-93c}$$

where $\boldsymbol{\xi}_1, \boldsymbol{\xi}_2, \ldots, \boldsymbol{\xi}_N$ are independent random vectors such that

$$\begin{array}{ccc} |E(\xi_{ni})| < \infty & \text{and} & |E(\xi_{ni}\xi_{nj})| < \infty \\ i, j = 1, \ldots, M, & & n = 1, \ldots, N \end{array} \tag{7-93d}$$

In (7-93d), ξ_{ni} and ξ_{nj} denote the ith and jth components of $\boldsymbol{\xi}_n$, respectively. Now we verify that Example 7-4 is indeed a linear-quadratic model.

Example 7-6: Continuation of Example 7-4 In this inventory model, $s_{n+1} = s_n + a_n - D_n \in \mathbb{R}$; so $m = \theta = 1$, $G_n = H_n = 1$, and $\xi_n = -D_n$. From (7-90),

$$r_n(s, a) = -\beta^{n-1}(c_1 a + c_2 a^2 + h_1 s + h_2 s^2)$$

and
$$L(s) = \beta^N (c_1 s + c_2 s^2)$$

with the minus sign due to cost minimization. Therefore, $U_n = -h_2$, $u_n = -h_1$, $V_n = -c_2$, and $v_n = -c_1$. □

In Section 8-4 we formulate a linear-quadratic model in which $m > 1$ and $\theta > 1$. That formulation, known as the HMMS model,‡ is the most widely known linear-quadratic model in operations research.

† A nonstationary MDP has a structure which varies from period. See Exercises 3-20 and 4-12.

‡ The HMMS model is named for C. C. Holt, F. Modigliani, J. F. Muth, and H. A. Simon and is presented in detail in their book, *Planning Production, Inventories, and Work Force*, Prentice-Hall, Englewood Cliffs, N.J. (1960).

Optimization

The recursive optimality equations (4-18) for the linear-quadratic model are

$$f_{N+1}(\mathbf{s}) = \mathbf{s}'U_{N+1}\mathbf{s} \tag{7-94a}$$

$$f_n(\mathbf{s}) = \max\{J_n(\mathbf{s}, \mathbf{a}): \mathbf{a} \in \mathbb{R}^m\} \tag{7-94b}$$

$$J_n(\mathbf{s}, \mathbf{a}) = \mathbf{s}'U_n\mathbf{s} + \mathbf{a}'V_n\mathbf{a} + \mathbf{u}_n'\mathbf{s} + \mathbf{v}_n'\mathbf{a} + E[f_{n+1}(G_n\mathbf{s} + H_n\mathbf{a} + \xi_n)] \tag{7-94c}$$

where (7-94*b*) and (7-94*c*) apply to $n = 1, \ldots, N$.

Shortly, we state and prove the principal result. To motivate the assumptions, consider (7-94*b*) and (7-94*c*) when $E(\xi_n) = \mathbf{0}$, $\mathbf{u}_n = \mathbf{0}$, $\mathbf{v}_n = \mathbf{0}$, and $f_{n+1}(\mathbf{s}) = z + \mathbf{s}'Z\mathbf{s}$, where Z is an $m \times m$ symmetric matrix. Then (7-94*c*) is

$$\begin{aligned} J_n(\mathbf{s}, \mathbf{a}) &= \mathbf{s}'U_n\mathbf{s} + \mathbf{a}'V_n\mathbf{a} + E[z + (G_n\mathbf{s} + H_n\mathbf{a} + \xi_n)'Z(G_n\mathbf{s} + H_n\mathbf{a} + \xi_n)] \\ &= \mathbf{s}'U_n\mathbf{s} + \mathbf{a}'V_n\mathbf{a} + z + \mathbf{s}'G_n'ZG_n\mathbf{s} + \mathbf{a}'H_n'ZH_n\mathbf{a} + 2\mathbf{s}'G_n'ZH_n\mathbf{a} \\ &\quad + E(\xi_n'Z\xi_n) \\ &= \mathbf{s}'X\mathbf{s} + \mathbf{a}'Y\mathbf{a} + 2\mathbf{w}'\mathbf{a} + z + E(\xi_n'Z\xi_n) \end{aligned} \tag{7-95}$$

where $\quad X = U_n + G_n'ZG_n \qquad Y = V_n + H_n'ZH_n \qquad \mathbf{w} = H_n'ZG_n\mathbf{s} \qquad$ (7-96)

From Proposition B-18, (7-95) is uniquely maximized with respect to **a** by

$$\mathbf{a} = -Y^{-1}\mathbf{w} = -Y^{-1}H_n'ZG_n s$$

if Y is a negative definite matrix.† From (7-96) and Proposition B-17, Y is negative definite when V_n is negative definite and Z is negative semidefinite.

Theorem 7-6 In the linear-quadratic model, suppose $E(\xi_n) = \mathbf{0}$, $\mathbf{u}_n = \mathbf{0}$, and $\mathbf{v}_n = \mathbf{0}$ for all n, $U_1, \ldots, U_{N+1}$ are negative definite and $V_1, \ldots, V_N$ are negative semidefinite. Then

$$f_n(\mathbf{s}) = z_n + \mathbf{s}'Z_n\mathbf{s} \qquad n = 1, \ldots, N+1 \tag{7-97}$$

where Z_n is negative semidefinite, and

$$\mathbf{a} = W_n\mathbf{s} \qquad n = 1, \ldots, N$$

attains $f_n(\mathbf{s})$ in (7-94*b*). The constants z_n and matrices Z_n and W_n are specified below in (7-100*b*) and (7-100*a*) and (7-100*b*).

Proof We initiate an inductive proof by observing that (7-94*a*) implies (7-97) for $n = N + 1$ with $z_{N+1} = 0$ and $Z_{N+1} = U_{N+1}$, which is negative semidefinite. Suppose (7-97) is valid at $n + 1$, that is,

$$f_{n+1}(\mathbf{s}) = z_{n+1} + \mathbf{s}'Z_{n+1}\mathbf{s} \qquad \mathbf{s} \in \mathbb{R}^m$$

† From Definition B-4, negative definite and negative semidefinite matrices are necessarily symmetric. From the comment following that definition, the symmetry assumption is without loss of generality.

where Z_{n+1} is negative semidefinite. Then the argument leading to (7-95) yields

$$J_n(\mathbf{s}, \mathbf{a}) = \mathbf{s}'X_n\mathbf{s} + \mathbf{a}'Y_n\mathbf{a} + 2\mathbf{w}_n'\mathbf{a} + z_{n+1} + E(\xi_n' Z_{n+1}\xi_n) \tag{7-98}$$

where

$$X_n = U_n + G_n' Z_{n+1} G_n \qquad Y_n = V_n + H_n' Z_{n+1} H_n \qquad \mathbf{w}_n = H_n' Z_{n+1} G_n \mathbf{s} \tag{7-99}$$

The matrix Y_n is negative semidefinite because V_n has that property and Z_{n+1} is negative semidefinite by the inductive assumption; so $J_n(\mathbf{s}, \cdot)$ is maximized on $\mathbb{R}^{\theta}$ by

$$\mathbf{a} = -Y_n^{-1}\mathbf{w}_n = W_n\mathbf{s} \tag{7-100a}$$

$$W_n \triangleq -Y_n^{-1}H_n' Z_{N+1} G_n \tag{7-100b}$$

The substitution of (7-100*a*) and (7-100*b*) in (7-97) yields

$$\begin{aligned} f_n(\mathbf{s}) &= J_n(\mathbf{s}, W_n\mathbf{s}) = \mathbf{s}'X_n\mathbf{s} + \mathbf{s}'W_n' Y_n W_n\mathbf{s} + 2\mathbf{w}_n' W_n\mathbf{s} + z_{n+1} + E(\xi_n' Z_{n+1}\xi_n) \\ &= \mathbf{s}'X_n\mathbf{s} + \mathbf{s}'G_n' Z_{n+1} H_n Y_n^{-1} H_n' Z_{n+1} G_n\mathbf{s} \\ &\quad - 2\mathbf{s}'G_n' Z_{n+1} H_n Y_n^{-1} H_n' Z_{n+1} G_n\mathbf{s} + z_{n+1} + E(\xi_n' Z_{n+1}\xi_n) \\ &= \mathbf{s}'Z_n\mathbf{s} + z_n \end{aligned}$$

where, from (7-93),

$$z_n = z_{n+1} + E(\xi_n' Z_{n+1}\xi_n) \tag{7-101a}$$

$$\begin{aligned} Z_n &= X_n - G_n' Z_{n+1} H_n Y_n^{-1} H_n' Z_{n+1} G_n \\ &= U_n + G_n' Z_{n+1}(I - H_n Y_n^{-1} H_n' Z_{n+1}) G_n \end{aligned} \tag{7-101b}$$

The symmetry of Z_n in (7-101*b*) follows from the symmetry of U_n, Y_n, and Z_{n+1}. The following justification of the negative semidefinite property of Z_n is taken from Bertsekas (1976). For any $\mathbf{s} \in \mathbb{R}^m$,

$$\mathbf{s}'Z_n\mathbf{s} = \min\{\mathbf{s}'U_n\mathbf{s} + \mathbf{a}'V_n\mathbf{a} + (G_n\mathbf{s} + H_n\mathbf{a})'Z_{n+1}(G_n\mathbf{s} + H_n\mathbf{a}) : \mathbf{a} \in \mathbb{R}^{m'}\}$$

However, U_n, V_n, and Z_{n+1} are negative semidefinite; so each term of the minimand is nonpositive. Hence, $\mathbf{s}'Z_n\mathbf{s} \leq 0$; so Z_n is negative semidefinite and the induction continues. □

It follows from (7-99) and (7-100*a*) that the optimal value of $\mathbf{a}_n$ is

$$\mathbf{a}_n = -(V_n + H_n' Z_{n+1} H_n)^{-1} H_n' Z_{n+1} G_n \mathbf{s}_n \tag{7-102}$$

From (7-93*c*), if $n > 1$,

$$\mathbf{s}_n = G_{n-1}\mathbf{s}_{n-1} + H_{n-1}\mathbf{a}_{n-1} + \xi_{n-1}$$

which, upon substitution in (7-102), shows that the action vector $\mathbf{a}_n$ *does* depend on the random vector ξ_{n-1}. However, if $n = 1$, then $\mathbf{s}_n = \mathbf{s}_1$ is presumed known; so

$$\mathbf{a}_1 = -(V_1 + H_1' Z_2 H_1)^{-1} H_1 Z_2 G_1 \mathbf{s}_1 \tag{7-103}$$

does *not* depend on the distributions of $\boldsymbol{\xi}_1, \ldots, \boldsymbol{\xi}_N$. This argument justifies the following claim.

Corollary 7-6: Certainty Equivalence Under the assumptions of Theorem 7-6, if the linear-quadratic model is altered by replacing $\boldsymbol{\xi}_n$ with $\mathbf{0}$ for each $n = 1, \ldots, N$, then (7-103) still attains $f_1(\mathbf{s})$ in (7-94*b*).

In comparison, if $n > 1$, we interpret $\mathbf{a}_n$ in (7-98) as a random vector.

For each n, let $\boldsymbol{\mu}_n = E(\boldsymbol{\xi}_n)$ and let $Y_n = V_n + H_n' Z_{n+1} H_n$ as in (7-99). The following generalization of Theorem 7-6 abandons the restriction $\boldsymbol{\mu}_n = \mathbf{u}_n = \mathbf{0}$ and $\mathbf{v}_n = \mathbf{0}$ for all n. The major qualitative conclusions in Theorem 7-6 remain valid, but the formulas are messier.

Theorem 7-7 In the linear-quadratic model, if $U_1, \ldots, U_{N+1}$ are negative definite and $V_1, \ldots, V_N$ are negative semidefinite, then

$$f_n(\mathbf{s}) = z_n + \mathbf{s}' Z_n \mathbf{s} + \mathbf{b}_n' \mathbf{s} \qquad \mathbf{s} \in \mathbb{R}^m \tag{7-104a}$$

where $Z_{N+1} = U_{N+1}$, $z_{N+1} = 0$, $\mathbf{b}_{N+1} = \mathbf{u}_{N+1}$, Z_n is given by (7-101*b*) for $n \le N$,

$$z_n = z_{n+1} + E(\boldsymbol{\xi}_n' Z_{n+1} \boldsymbol{\xi}_n) + b_{n+1}' \boldsymbol{\mu}_n$$

$$- (H_n' \mathbf{b}_{n+1} + 2H_n' Z_{n+1} \boldsymbol{\mu}_n + \mathbf{v}_n)' Y_n^{-1} \left(H_n' Z_{n+1} \boldsymbol{\mu}_n + \frac{H_n' \mathbf{b}_{n+1}}{2} + \frac{\mathbf{v}_n}{2}\right) \tag{7-104b}$$

and

$$\mathbf{b}_n = \mathbf{u}_n + G_n'[\mathbf{b}_{n+1} + 2Z_{n+1}\boldsymbol{\mu}_n - Z_{n+1} H_n Y_n^{-1} (H_n' \mathbf{b}_{n+1} + \mathbf{v}_n + 2H_n' Z_{n+1} \boldsymbol{\mu}_n)] \tag{7-104c}$$

In (7-94*b*), $f_n(s)$ is attained by

$$\mathbf{a} = -Y_n^{-1} \left[H_n' \left(Z_{n+1} G_n \mathbf{s} + Z_{n+1} \boldsymbol{\mu}_n + \frac{\mathbf{b}_{n+1}}{2} \right) + \frac{\mathbf{v}_n}{2} \right] \tag{7-105}$$

PROOF Left as Exercise 7-28. □

If $n = 1$, (7-105) has the form

$$\mathbf{a}_1 = M \mathbf{s}_1 + \mathbf{c}$$

where neither the matrix M nor the vector $\mathbf{c}$ depends on the vectors $\xi_1, \ldots, \xi_N$ except through their expected values $\boldsymbol{\mu}_1, \ldots, \boldsymbol{\mu}_N$. Hence the following obvious analog of Corollary 7-6 is valid.

Corollary 7-7: Certainty Equivalence Under the assumptions of Theorem 7-7, if the linear-quadratic model is altered by replacing ξ_n with μ_n for each $n = 1, \ldots, N$, then (7-105) still attains $f_n(s)$.

The computational effort to apply the formulas in Theorems 7-6 and 7-7 is centered on (7-101*b*), which is repeated below:

$$Z_n = U_n + G_n' Z_{n+1}[I - H_n(V_n + H_n' Z_{n+1} H_n)^{-1} H_n' Z_{n+1}]G_n \qquad (7\text{-}106a)$$

Suppose the linear-quadratic model has a stationary structure; i.e., the model's data are invariant with respect to n. Then (7-106*a*) is

$$Z_k = U + G'Z_{k-1}[I - H(V + H'Z_{k-1}H)^{-1}H'Z_{k-1}]G \qquad (7\text{-}106b)$$

in which the following change of variable has been made in the subscripts: $N + 1 \to 0$, $N \to 1, \ldots, 1 \to N$. Under the assumptions of Theorem 7-7, it can be shown that the recursion (7-106*b*) has a limit Z which satisfies

$$Z = U + G'Z[I - H(V + H'ZH)^{-1}H'Z]G \qquad (7\text{-}107)$$

which is known as a *Ricatti equation.* See Bertsekas (1976) for details.

A Note of Caution

The primary virtue of the linear-quadratic model is the power and relative simplicity of the formulas in Theorems 7-6 and 7-7. A major deficiency of the model is its absence of constraints on actions. The absurdity in some stochastic problems of being unable to restrict A_s to a proper subset of $\mathbb{R}^\theta$ is exemplified by the following version of Example 7-4.

Example 7-7: Continuation of Example 7-4 In Example 7-4, let $c_1 = 10$, $c_2 = 1$, $h_1 = 5$, $h_2 = 1$, $\beta = 0.9$, and for each n let D_n be uniformly distributed on [5, 15] so $\mu_n = 10$. Therefore, $G_n = H_n = 1$, $U_n = -1$, $u_n = -5$, $V_n = -1$, and $v_n = -10$. Also,

$$f_{N+1}(s) = 10s + s^2$$

so $z_{N+1} = 0$, $Z_{N+1} = 1$, and $b_{N+1} = 10$. Substitution in (7-101*b*) and (7-105) for $n = N$ yields

$$a_N = 9s_N - 95 \qquad (7\text{-}108a)$$

Since a_N represents an order quantity, it should be nonnegative. From (7-108*a*), $0 \leq a_N$ only if $s_N \geq 95/9$. In practice we might replace (7-108*a*) with

$$a_N = (9s_N - 95)^+ \qquad (7\text{-}108b)$$

except that $f_N(s)$ would no longer be given by (7-104*a*) to (7-104*c*) with $n = N$. As a consequence, (7-105) with $n = N - 1$ would *not* be an optimal selection of a_{N-1}.

The difficulty vanishes if s_{N-1} is sufficiently large that the sum of s_{N-1}

plus a_{N-1} [specified by (7-105) with $n = N - 1$] is at least $95/9 + 15$. Then

$$P\{D_{N-1} \leq 15\} = 1 \Rightarrow P\{s_N \geq 95/9\} = 1 \qquad \square$$

The example suggests that the linear-quadratic model is less apt for highly stochastic phenomena than for those which are nearly deterministic.

EXERCISES

7-28 Prove Theorem 7-7.

7-29 In Example 7-7:
(*a*) Verify that (7-108*a*) results from (7-105) with $n = N$.
(*b*) Specify $f_N(s)$.
(*c*) Compute (7-105) with $n = N - 1$.
(*d*) Solve for Z in (7-107).

BIBLIOGRAPHIC GUIDE

Arrow, Harris, and Marschack (1951) is the seminal paper in dynamic inventory models. Dvoretzky, Kiefer, and Wolfowitz (1952) contributed to the mathematical foundations, and the book Arrow, Karlin, and Scarf (1958) contains important contributions. Scarf (1960) introduces the notion of K-convexity to establish the optimality of (σ, Σ) policies under reasonably general conditions [cf. Zabel (1962)]. Veinott (1966*a*) establishes optimality under general conditions which overlap Scarf's, but the proof does not use K-convexity. Schäl (1976) has an optimality proof which subsumes the conditions in Scarf (1960) and Veinott (1966*a*). Johnson (1967) and Kalin (1980) establish the optimality of a multidimensional (σ, Σ) policy for a multi-item inventory model. Porteus (1971) shows that a generalization of a (σ, Σ) policy is optimal for a single-item inventory model whose setup cost is replaced by a piecewise linear concave function.

Wagner (1962) analyzes Markov chains induced by stationary (σ, Σ) policies. Veinott and Wagner (1965) exploit renewal theory and "stationary analysis" (discrete-time Markov chains) to develop an algorithm for computing an optimal policy. In Volume I, Examples 5-10, 6-3, and 6-8 on pages 134, 176, and 190 use the Veinott-Wagner approach. Johnson (1968) has (σ, Σ) optimality conditions which overlap those of Scarf (1960) and Veinott (1966*a*) and presents a constructive proof of optimality using policy improvement. Bell (1970) uses stopping-rule theory to accelerate the Veinott-Wagner algorithm. Roberts (1962), Wagner, O'Hagan, and Lundh (1965), Sivazlian (1971), Hordijk and Tijms (1974), Snyder (1974), Nahmias (1979), Freeland and Porteus (1980), and Ehrhardt and Wagner (1982) discuss the computation and implementation of approximately optimal policies. Surveys of the optimization of dynamic inventory models include Scarf (1963), Veinott (1966*b*), Iglehart (1967), Clark (1972), and Aggarwal (1974).

Models with setup costs have been proposed for many phenomena other than commercial inventories. For example, Spulber (1982) and its references discuss the use of (σ, Σ) policies to manage fisheries and wildlife populations.

An on-off model of production is essentially equivalent to a single-server queueing model with a removable server. The optimization of such a model first appeared in Yadin and Naor (1963) with refinements in Heyman (1968), Sobel (1969), Bell (1971), Balachandran (1973), and Heyman (1977). Many authors have written about models which are more general or more elaborate than the one in Yadin and Naor (1963). Recent work includes algorithms; see Tijms (1980) and its references. Surveys and bibliographies concerning the optimization of queueing models include Prabhu and Stidham (1974), Sobel (1974), and Crabill, Gross, and Magazine (1977).

The notion of a separable MDP occurs independently in Kaufmann and Cruon (1967), de Ghellinck and Eppen (1967), and in a restricted form in Johnson (1967). Denardo (1968) uses direct methods to unify and extend the results in Kaufmann and Cruon (1967) and de Ghellinck and Eppen (1967). The exposition in Section 7-3 is significantly influenced by Denardo (1968).

It has long been appreciated that linear functions maximize concave quadratic functions. Theil (1954), which has the first rigorous statement of such a result, concerns a static model. Simon (1956) discusses a deterministic and dynamic linear-quadratic model. Theil (1957) appends uncertainty to the model in Simon (1956) and observes that the optimum possesses the certainty equivalence property. Our proof of Theorem 4-1 follows the exposition in Bertsekas (1976), which cites many references to linear-quadratic models in the control-theory literature. See the special issue of *IEEE Transactions Automatic Control* (1971) for linear-quadratic models from the control-theory perspective. Dreyfus and Law (1977) present a detailed analysis of a linear-quadratic model, with the state and action both being scalar variables (rather than vector variables as in Section 7-4). Holt, Modigliani, Muth, and Simon (1960) is a detailed assessment of linear-quadratic models for the management of production and inventories.

The optimality of a linear decision rule is a striking characteristic of a linear-quadratic model. Linear decision rules have been advocated as heuristic devices in some models which lack the linear-quadratic features; chance-constrained programming is an example. Charnes and Cooper (1963) is a basic paper in this area, and Stancu-Minasian and Wets (1976) is a research bibliography in stochastic programming, the larger area of mathematical programming of which chance-constrained programming is a part. Many water-resource models have been heuristically optimized with linear decision rules. See ReVelle and Gundelach (1975) for an example and for references to other examples of this kind.

Inventory models with setup costs

Aggarwal, S. C.: "A Review of Current Inventory Theory and Its Applications," *Int. J. Prod. Res.* **12**: 443–482 (1974).

Arrow, K. J., T. Harris, and J. Marschack: "Optimal Inventory Policy," *Econometrica* **19**: 250–272 (1951).

Arrow, K. J., S. Karlin, and H. Scarf: *Studies in the Mathematical Theory of Inventory and Production*, Stanford University Press, Stanford, Calif. (1958).

Bell, C. E.: "Improved Algorithms for Inventory and Replacement-Stocking Problems," *SIAM J.* **18**: 558–566 (1970).

Boylan, E. S.: "Multiple (s, S) Policies and the n-Period Inventory Problem," *Manage. Sci.* **14**: 196–204 (1967).

Clark, A. J.: "An Informal Survey of Multi-Echelon Inventory Theory," *Nav. Res. Logistics Quart.* **19**: 621–650 (1972).

Denardo, E. V.: *Dynamic Programming*, Prentice-Hall, Englewood Cliffs, N.J. (1982).

Dvoretzky, A., J. Kiefer, and J. Wolfowitz: "The Inventory Problem," *Econometrica* **20**: 187–222, 450–466 (1952).

Ehrhardt, R.: "The Power Approximation for Computing (s, S) Inventory Policies," *Manage. Sci.* **25**: 777–786 (1979).

———: "Analytic Approximation for (s, S) Inventory Policy Operating Characteristics," *Nav. Res. Logistics Quart.* **28**: 255–266 (1981).

——— and H. M. Wagner: "Inventory Models and Practice," in *Advanced Techniques in the Practice of Operations Research*, edited by H. G. Greenberg, F. H. Murphy, and S. H. Shaw, American Elsevier, New York (1982).

Federgruen, A., and P. Zipkin: "An Efficient Algorithm for Computing Optimal (*s*, *S*) Policies," Research Working Paper 458A, Graduate School of Business, Columbia University, New York (1981).

Freeland, J. R., and E. L. Porteus: "Evaluating the Effectiveness of a New Method for Computing Approximately Optimal (*s*, *S*) Inventory Policies," *Oper. Res.* **28**: 353–364 (1980).

Hordijk, A., and H. C. Tijms: "Convergence Results and Approximations for Optimal (*s*, *S*) Policies," *Manage. Sci.* **20**: 1432–1438 (1974).

Iglehart, D. L.: "Recent Results in Inventory Theory," *J. Ind. Eng.* **18**: 48–51 (1967).

Johnson, E. L.: "Optimality and Computation of (σ, *S*) Policies in the Multi-Item Infinite Horizon Inventory Problem," *Manage. Sci.* **13**: 475–491 (1967).

———: "On (*s*, *S*) Policies," *Manage. Sci.* **15**: 80–101 (1968).

Kalin, D.: "On the Optimality of (σ, *S*) Policies," *Math. Oper. Res.* **5**: 293–307 (1980).

Lippman, S. A.: "Optimal Inventory Policy with Subadditive Ordering Costs and Stochastic Demands," *SIAM J. Appl. Math.* **17**: 543–559 (1969).

Nahmias, S.: "Simple Approximations for a Variety of Dynamic Leadtime Lost-Sales Inventory Models," *Oper. Res.* **27**: 904–924 (1979).

Porteus, E. L.: "On the Optimality of Generalized (s, S) Policies," *Manage. Sci.* **17**: 411–462 (1971).

Roberts, D. M.: "Approximations to Optimal Policies in a Dynamic Inventory Model," in *Studies in Applied Probability and Management Science*, edited by K. J. Arrow, S. Karlin, and H. Scarf, Stanford University Press, Stanford, Calif. (1962).

Scarf, H.: "The Optimality of (s, S) Policies in the Dynamic Inventory Problem," in *Mathematical Methods in the Social Sciences 1959*, Stanford University Press, Stanford, Calif. (1960).

———: "A Survey of Analytic Techniques in Inventory Theory," chap. 7 in *Multistage Inventory Models and Techniques*, edited by H. Scarf, D. M. Gilford, and M. W. Shelly, Stanford University Press, Stanford, Calif. (1963).

Schäl, M.: "On the Optimality of (s, S) Policies in Dynamic Inventory Models with Finite Horizon," *SIAM J. Appl. Math.* **30**: 528–537 (1976).

Sivazlian, B. D.: "Dimensional and Computational Analysis in Stationary (s, S) Inventory Problems with Gamma Distributed Demand," *Manage. Sci.* **17**: B307–B311 (1971).

Snyder, R. D.: "Computation of (S, s) Ordering Policy Parameters," *Manage. Sci.* **21**: 223–229 (1974).

Spulber, D. F.: "Adaptive Harvesting of a Renewable Resource and Stable Equilibrium," chap. 6 in *Essays in the Economics of Renewable Resources*, edited by L. J. Mirman and D. F. Spulber, North Holland, Amsterdam (1982).

Tijms, H. C.: *Analysis of (s, S) Inventory Models*, Tract 40, Mathematisch Centrum, Amsterdam (1972).

Veinott, A. F., Jr.: "On the Optimality of (s, S) Inventory Policies: New Conditions and a New Proof," *SIAM J.* **14**: 1067–1083 (1966*a*).

———: "The Status of Mathematical Inventory Theory," *Manage. Sci.* **12**: 745–777 (1966*b*).

——— and H. M. Wagner: "Computing Optimal (s, S) Inventory Policies," *Manage. Sci.* **11**: 525–552 (1965).

Wagner, H. M.: *Statistical Management of Inventory Systems*, Wiley, New York (1962).

Wagner, H. M., M. O'Hagan, and B. Lundh: "An Empirical Study of Exactly and Approximately Optimal Inventory Policies," *Manage. Sci.* **11**: 690–723 (1965).

Zabel, E.: "A Note on the Optimality of (s, S) Policies in Inventory Theory," *Manage. Sci.* **9**: 123–125 (1962).

Production model with start-up and shut-down costs

Balachandran, K. R.: "Control Policies for a Single Server System," *Manage. Sci.* **19**: 1013–1018 (1973).

Bell, C. E.: "Characterization and Computation of Optimal Policies for Operating an M/G/1 Queueing System with Removable Server," *Oper. Res.* **19**: 208–218 (1971).

Crabill, T. B., D. Gross, and M. J. Magazine: "A Classified Bibliography of Research on Optimal Design and Control of Queues," *Oper. Res.* **25**: 219–232 (1977).

Heyman, D. P.: "Optimal Operating Policies for M/G/1 Queueing Systems," *Oper. Res.* **16**: 363–383 (1968).

———: "The T-Policy for the M/G/1 Queue," *Manage. Sci.* **23**: 775–778 (1977).

Prabhu, N. U., and S. Stidham: "Optimal Control of Queueing Systems," in *Mathematical Methods in Queueing Theory*, edited by A. B. Clarke, Lecture Notes in Economics and Mathematical Systems 98, Springer-Verlag, Berlin (1974).

Sobel, M. J.: "Optimal Average-Cost Policy for a Queue with Start-Up and Shut-Down Costs," *Oper. Res.* **17**: 145–162 (1969).

———: Optimal Operation of Queues,"in *Mathematical Methods in Queueing Theory*, edited by A. B. Clarke, Lecture Notes in Economics and Mathematical Systems 98, Springer-Verlag, Berlin (1974).

Tijms, H. C.: "An Algorithm for Average Cost Denumerable State Semi-Markov Decision Problems with Applications to Controlled Production and Queueing Systems," in *Recent Developments in Markov Decision Processes*, edited by R. Hartley, L. C. Thomas, and D. J. White (editors): Academic Press, New York (1980).

Yadin, M., and P. Naor: "Queueing Systems with a Removable Service Station," *Oper. Res. Quart.* **14**: 393–405 (1963).

Separable MDPs

Denardo, E. V.: "Separable Markovian Decision Problems," *Manage. Sci.* **14**: 451–462 (1968).

de Ghellinck, G. D., and G. D. Eppen: "Linear Programming Solutions for Separable Markovian Decision Problems," *Manage. Sci.* **13**: 371–394 (1967).

Johnson, E. L.: "Computation and Structure of Optimum Reset Policies," *J. Am. Stat. Assoc.* **62**: 1462–1487 (1967).

Kaufmann, A., and R. Cruon: *Dynamic Programming*, Academic Press, New York (1967) [originally published in French as *La Programmation Dynamique*, Dunod, Paris (1965)].

Sobel, M. J.: "Production Smoothing with Stochastic Demand II: Infinite Horizon Case," *Manage. Sci.* **17**: 724–735 (1971).

———: "Making Short-Run Changes in Production When the Employment Level Is Fixed," *Oper. Res.* **18**: 35–51 (1970).

Certainty equivalence

Bertsekas, D. P.: *Dynamic Programming and Stochastic Control*, Academic Press, New York (1976).

Charnes, A., and W. W. Cooper: "Deterministic Equivalents for Optimizing and Satisficing under Chance Constraints," *Oper. Res.* **11**: 18–19 (1963).

Dreyfus, S. E., and A. M. Law: *The Art and Theory of Dynamic Programming*, Academic Press, New York (1977).

Holt, C. C., F. Modigliani, J. F. Muth, and H. A. Simon: *Planning, Production, Inventories, and Work Force*, Prentice-Hall, Englewood Cliffs, N.J. (1960).

IEEE Trans. Auto. Control, Special Issue on the Linear-Quadratic Gaussian Problem, **AC-16** (1971).

Malinvaud, E.: "First Order Certainty Equivalence," *Econometrica* **37**: 706–718 (1969).

ReVelle, C., and J. Gundelach: "Linear Decision Rule in Reservoir Management and Design 4: A Rule That Minimizes Output Variance," *Water Resour. Res.* **11**: 197–203 (1975).

Simon, H. A.: "Dynamic Programming under Uncertainty with a Quadratic Criterion Function," *Econometrica* **24**: 74–81 (1956).

Stancu-Minasian, I. M., and M. J. Wets: "A Research Bibliography in Stochastic Programming 1955–1975," *Oper. Res.* **24**: 1078–1119 (1976).

Theil, H.: "A Note on Certainty Equivalence in Dynamic Planning," *Econometrica* **25**: 346–349 (1957).

CHAPTER

EIGHT

MONOTONE OPTIMAL POLICIES

A policy for a Markov decision process is a contingency plan that specifies what action to take as the process unfolds. An optimal policy, if one exists, is at least as good as any other policy. But what kind of policy is optimal? Suppose an optimal policy is a stationary policy. Is there a qualitative property that describes how the action should depend on the state? For example, in an inventory model, does the optimal order quantity decrease if the inventory level rises? In other words, is the optimal order quantity a monotone function of the inventory level? Such a qualitative insight is informative and may accelerate the computation of an optimal policy.

Section 8-1 presents sufficient conditions for quite general models to possess monotone optimal policies. The conditions are stated in terms of submodular functions on lattices. Succeeding sections apply the general theory to models from numerous contexts.

Section 8-2 concerns MDPs in which there are at most two actions available at each state. The phenomena whose models frequently have the binary decision feature include equipment replacement, maintenance, and production control.

Section 8-3 presents a general class of resource-management models. Motivating phenomena include fisheries, reservoirs, oil wells, and capital asset portfolios. The timing of consumption is a central feature of these phenomena.

Many managed systems have the property that significant costs are incurred when there is a change in the rate at which activities occur. Section 8-4 analyzes several models with such costs. Motivating phenomena include corporate short-run cash accounts, production control, and scheduling the number of open tellers' windows in a retail bank.

The analyses in Sections 8-2 through 8-4 concern finite-horizon models. Sections 8-5 and 8-6 show that infinite-horizon models inherit the qualitative properties of their finite-horizon counterparts. This conclusion sometimes accelerates the computation of an optimal stationary policy.

The analyses in this chapter frequently exploit properties of convex sets and convex functions. You should read those parts of Appendix B whose contents are new to you.

8-1* LATTICE PROGRAMMING

Suppose an MDP has states and actions which are real variables and an optimal decision rule $a_n(\cdot)$ as a function of the state when n periods remain to the end of the planning horizon. If s is a state, so $s \in \mathbb{S} \subset \mathbb{R}$, then $a_n(\cdot)$ is monotone (nondecreasing) if $a_n(s) \leq a_n(s + \gamma)$ for each $\gamma > 0$ and s. This section presents reasonably general conditions for $a_n(\cdot)$to be monotone for each n.

What is the crux of the argument that $a_n(s) \leq a_n(s + \gamma)$? First, there is a feasibility issue. If s rises to $s + \gamma$, could $a_n(s + \gamma)$ be constrained to be less than $a_n(s)$? The answer is "no" if the smallest feasible action, that is, the least element of A_s, is a nondecreasing function of s.

Second, there is an optimality issue. Let $J_n(s, a)$ be the objective which is maximized by $a = a_n(s)$. That is, if the current state and action are s and a, respectively, then $J_n(s, a)$ is the minimal expected present value of the single-stage rewards in n periods and the salvage value. Let† $J_n^{(2)}(s, a) = \partial J_n(s, a)/\partial a$ and suppose that $J_n(s, \cdot)$ and $J_n(s + \gamma, \cdot)$ are concave functions. Then $J_n^{(2)}(s, \cdot)$ and $J_n^{(2)}(s + \gamma, \cdot)$ are nonincreasing functions; suppose that both functions are continuous too. By concavity, either $a_n(s)$ is at a boundary point of A_s, or $J_n^{(2)}[s, a_n(s)] = 0$. The same is true for $a_n(s + \gamma)$. Suppose that neither $a_n(s)$ nor $a_n(s + \gamma)$ is at a boundary so

$$0 = J_n^{(2)}[s, a_n(s)] = J_n^{(2)}[s + \gamma, a_n(s + \gamma)]$$

Suppose also that

$$J_n^{(2)}(s, a) \leq J_n^{(2)}(s + \gamma, a) \qquad \text{for all } a \tag{8-1}$$

† We emphasize that $J_n^{(2)}(s, a)$ does not indicate a second derivative. It is the partial derivative (from the left) with respect to the second argument of the function $J_n(\cdot, \cdot)$.

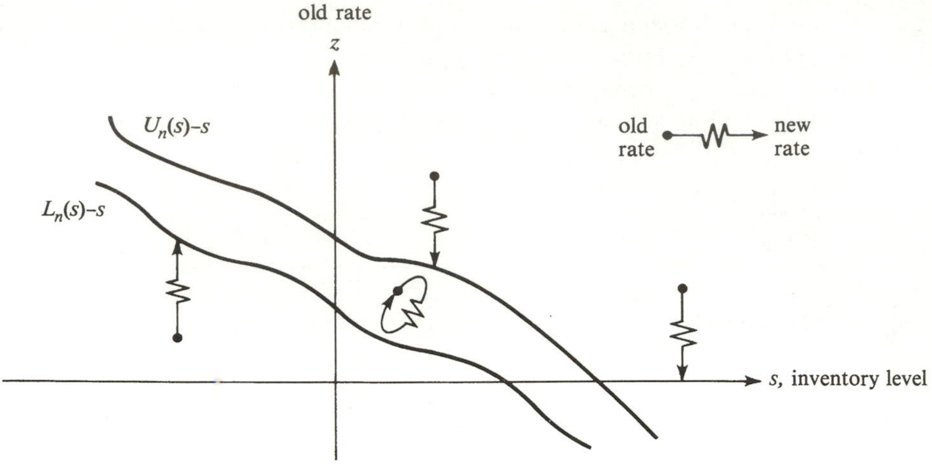

Figure 8-1 Optimal adjustment of production rate.

Then $\gamma > 0$ and (8-1) imply

$$J_n^{(2)}[s + \gamma, a_n(s + \gamma)] = 0 = J_n^{(2)}[s, a_n(s)] \leq J_n^{(2)}[s + \gamma, a_n(s)]$$

If the inequality is weak, then $a_n(s) = a_n(s + \gamma)$. If $J_n^{(2)}[s + \gamma, a_n(s)] > 0$, the nonincreasing property of $J_n^{(2)}(s + \gamma, \cdot)$ implies that $a > a_n(s)$ if $J_n^{(2)}(s + \gamma, a) = 0$. Therefore, $a_n(s) \leq a_n(s + \gamma)$. Figure 8-1 corresponds to our assumptions. In summary, concavity and (8-1) imply monotonicity of $a_n(\cdot)$.

The left side of (8-1) is the limit, as $\epsilon \downarrow 0$, of

$$\frac{J_n(s + \gamma, a + \epsilon) - J_n(s + \gamma, a) - J_n(s, a + \epsilon) + J_n(s, a)}{\epsilon}$$

If this quantity is nonnegative for all $\epsilon > 0$ and $\gamma > 0$ for which J_n is defined, the limit as $\epsilon \downarrow 0$ certainly is nonnegative. The functions considered in this section possess this requisite property.

The feasibility and optimality issues are resolved in this section in the general setting of ordered solutions to optimization problems. The general results are applied to particular models in Sections 8-2 through 8-6. The phenomena modeled in Section 8-2 include equipment-replacement and fruit-cannery operations. Section 8-3 includes models of fisheries, capital accumulation, and oil and natural gas well exploitation.

Partially Ordered Sets and Lattices

It is natural to embed an analysis of ordered solutions in a context of ordered sets. Let Ω be any nonempty set and B be a collection of ordered pairs of elements of Ω, that is, $B \subset \Omega \times \Omega$. Recall from Sections 2-4 and 4-3 that (B, Ω) is said to be a *partially ordered set* if it is reflexive, antisymmetric, and transitive. If

$(x, y) \in B$, then $x \leq y$ and $y \geq x$ are written interchangeably. When the ordering $\leq$, hence B, is obvious from the context, Ω is said to be a partially ordered set.

A partially ordered set Ω is a *chain* if $x \leq y$ or $y \leq x$ for all x and y in Ω. In other words, all pairs of elements are comparable in a chain. This terminology is apt because transitivity then implies that every finite collection of elements of Ω can be ordered from "least" element to "greatest" element. In still other words, take n elements in a chain Ω, $x_1, x_2, \ldots, x_n$. The subscripts $1, 2, \ldots, n$ can be assigned so that $x_1 \leq x_2$, $x_2 \leq x_3, \ldots, x_{n-1} \leq x_n$. We write $x_1 \leq x_2 \leq x_3 \leq \ldots \leq x_{n-1} \leq x_n$ and call x_1 and x_n least and greatest elements, respectively.

Consider a collection $\{\Omega_k : k \in K\}$ of partially ordered sets and let $\leq_k$ denote the ordering relation for Ω_k. The product set $\Omega = \times_{k \in K} \Omega_k$ is partially ordered by $x = (x_k) \leq y = (y_k)$ if $x_k \leq_k y_k$ for all $k \in K$. With this relation we call Ω the *direct product* of Ω_k, $k \in K$.

For the remainder of this section, Ω is a partially ordered set. For x, y, and z' in Ω, we say z' is an *upper bound* of x and y if $z' \geq x$ and $z' \geq y$. A *lower bound* w' satisfies $w' \leq x$ and $w' \leq y$. Consider the sets $L_{x,y}$ and $U_{x,y}$ of lower and upper bounds in Ω of x and y. Then $w \in L_{x,y}$ and $z \in U_{x,y}$ are a *greatest lower bound* and *least upper bound*, respectively, if $w \geq w'$ for all $w' \in L_{x,y}$ and $z \leq z'$ for all $z' \in U_{x,y}$. We write† w as $x \wedge y$ (sometimes called the *meet*) and z as $x \vee y$ (sometimes called the *join*). In this sense,

$$
\begin{aligned}
x \wedge y &= \sup \{w' : w' \leq x,\ w' \leq y,\ w' \in \Omega\} \\
x \vee y &= \inf \{z' : z' \geq x,\ z' \geq y,\ z' \in \Omega\}
\end{aligned}
\tag{8-2}
$$

Definition 8-1 A partially ordered set Ω is a *lattice* if $x \wedge y \in \Omega$ and $x \vee y \in \Omega$ for all x and y in Ω.

Example 8-1 For two real numbers a and b, let Ω be the interval $[a, b]$. Then Ω with the usual ordering of real numbers is a lattice where $x \wedge y = \text{minimum}\ \{x, y\}$ and $x \vee y = \text{maximum}\ \{x, y\}$. Also, Ω is a chain. Similarly, $(-\infty, a]$, $(-\infty, a)$, (a, ∞), and $[a, \infty)$ are lattices. Therefore, the sets of feasible decisions in Sections 3-1, 3-4, and 3-5 are lattices. □

Example 8-2 Let Ω be the line segment that connects (0, 1) and (1, 0) (inclusive) in the plane. For‡ $x = (x_i)$ and $y = (y_i)$ in Ω we write $x \leq y$ if both $x_1 \leq y_1$ and $x_2 \leq y_2$. Then $x \neq y$ implies the sets of upper bounds and lower bounds are empty so Ω with $\leq$ is *not* a lattice. The difficulty is that we want to write $x \wedge y = (\min \{x_1, y_1\}, \min \{x_2, y_a\})$ and $x \wedge y = (\max \{x_1, y_1\}, \max \{x_2, y_2\})$, but neither of these points lies in Ω. If Ω is augmented with the points (0, 0) and (1, 1) and x and y are on the line between (1, 0) and

† This notation is the same as "min" and "max," but it should not be ambiguous. The minimum (maximum) of two real numbers is their greatest lower bound (least upper bound).

‡ In this section we dispense with our usual practice of writing vectors in boldface notation because some of the results are valid for sets other than vector spaces.

(0, 1), then $x \wedge y = (0, 0)$, $x \vee y = (1, 1)$, $(0, 0) \wedge y = (0, 0)$, $(0, 0) \vee y = y$, $(1, 1) \wedge y = y$, and $(1, 1) \vee y = (1, 1)$; so Ω *is* a lattice. □

Example 8-3 The previous example shows that a nonlattice subset of a partially ordered set can sometimes become a lattice by enlarging the subset. Sometimes, the subset can become a lattice by changing the ordering. Again, let Ω denote the line segment that connects (1, 0) and (0, 1), inclusive, but alter $\leq$ to *lexicographic* ordering (cf. Example 4-13); i.e., if x and y are in Ω, then $x \leq y$ if either $x_1 < y_1$ or both $x_1 = y_1$ and $x_2 \leq y_2$. Then $x \vee y = (\min\{x_1, y_1\}, 1 - \min\{x_1, y_1\})$ and $x \vee y = (\max\{x_1, y_1\}, 1 - \max\{x_1, y_1\})$; so $x \wedge y \in \Omega$ and $x \vee y \in \Omega$ and Ω is a lattice. In fact, Ω is a chain. □

Example 8-4 Suppose Ω is $\mathbb{R}^n$ with $x \leq y$ if $x_i \leq y_i$, $i = 1, \ldots, n$. Then $x \wedge y$ has $\min\{x_i, y_i\}$ as its ith component, $i = 1, \ldots, n$. Resurrecting the notation $a \wedge b = \min\{a, b\}$ for $a, b \in \mathbb{R}$, we write $x \wedge y = (x_i \wedge y_i) \in \mathbb{R}^n$. Similarly $x \vee y = (x_i \vee y_i) \in \mathbb{R}^n$; so Ω is a lattice. However, it is not a chain if $n > 1$. For example, if $n = 2$, $x = (0, 1)$, and $y = (1, 0)$, then $x \nleq y$ and $y \nleq x$. □

Example 8-4 is a special case of the general fact that a direct product of lattices is again a lattice.

Definition 8-2 Let Ω be a lattice and $\Gamma \subset \Omega$. Then Γ is a *sublattice* of Ω if $x \wedge y \in \Gamma$ and $x \vee y \in \Gamma$ for all x and y in Γ.

An essential feature is that Γ contains $x \wedge y$ and $x \vee y$. If $\Gamma \subset \Omega$ is not a sublattice, we can assert only that Ω contains $x \wedge y$ and $x \vee y$. A less obvious feature is that $x \wedge y$ and $x \vee y$ are defined, as in (8-2), with respect to Ω rather than Γ.

Example 8-5 Consider Example 8-4 when $n = 2$ and let Γ be the line segment between (0, 1) and (1, 0) (as in Example 8-2). Then Γ does not contain $(1, 0) \wedge (0, 1) = (0, 0)$ or $(1, 0) \vee (0, 1) = (1, 1)$; so it is not a sublattice. However, the (closed) unit square is a sublattice. □

The following consequence of the definitions is useful later in this section.

Lemma 8-1 Let $\mathcal{S}$ be a nonempty set and A_s be a nonempty set for each $s \in \mathcal{S}$. Let

$$\mathscr{C} = \{(s, a) : a \in A_s, s \in \mathcal{S}\}$$

and suppose $\mathscr{C}$ is a lattice.

(i) Then $\mathcal{S}$ is a lattice and A_s is a lattice for each $s \in \mathcal{S}$.

(ii) If $(s_i, a_i) \in \mathscr{C}$, $i = 1, 2$, then

$$(s_1, a_1) \wedge (s_2, a_2) = (s_1 \wedge s_2, a_1 \wedge a_2)$$
$$(s_1, a_1) \vee (s_2, a_2) = (s_1 \vee s_2, a_1 \vee a_2)$$

PROOF Left as Exercise 8-2. □

Recall the following definition from Sections 2-4 and 4-3.

Definition 8-3 Let f be a function whose domain Ω and range Δ are partially ordered sets. Then f is *isotone* (*antitone*) if $f(x) \le f(y)$ $[f(x) \ge f(y)]$ for all x and y in Ω with $x \le y$.

Remember that $x \le y$ refers to the partial ordering of Ω and $f(x) \le f(y)$ refers to the partial ordering of Δ.

For any lattice Ω let $L(\Omega)$ be the collection of nonempty sublattices of Ω and $P(\Omega)$ be the set of nonempty subsets of Ω. Let $\lesssim$ denote the binary relation on $P(\Omega)$ where $X \lesssim Y$ if X and Y are subsets of Ω for which $x \in X$ and $y \in Y$ implies $x \wedge y \in X$ and $x \vee y \in Y$. It can be shown (Exercise 8-1) that $L(\Omega)$ is partially ordered by $\lesssim$.

Definition 8-4 Let Ω be a lattice, Γ a partially ordered set, and suppose $\lesssim$ is the partial order on $L(\Omega)$. An isotone (antitone) function from Γ into $L(\Omega)$ is called *ascending* (*descending*) on Γ.

In the definition, if x and y are in Γ, then $f(x)$ and $f(y)$ are in $L(\Omega)$; that is, they are sublattices of Ω. Suppose $x \le y$ ($\le$ is the binary relation on Γ) and $a \in f(x) \subset \Omega$ and $b \in f(y) \subset \Omega$. Then f is ascending if necessarily $a \wedge b \in f(x)$ and $a \vee b \in f(y)$ (here $a \wedge b$ and $a \vee b$ are defined via the binary relation on Ω).

Example 8-6 Let $\Gamma = \mathbb{R}$ and $\Omega = \mathbb{R}$, both with the usual ordering of scalars. Define f with $f(x) = [x, \infty)$, which is a sublattice for each x. Then $a \in [x, \infty)$ and $b \in [y, \infty)$ with $x \le y$ implies $a \wedge b \in [x, \infty)$ and $a \vee b \in [y, \infty)$. Therefore, $\{[x, \infty)\colon x \in \mathbb{R}\}$ is ascending on $\mathbb{R}$. Similarly, $\{(-\infty, x]\colon x \in \mathbb{R}\}$ is ascending on $\mathbb{R}$. □

Example 8-7 More generally, for any lattice Ω, $\{\{u\colon u \in \Omega, u \le x\}\colon x \in \Omega\}$ and $\{\{u\colon u \in \Omega, u \ge x\}\colon x \in \Omega\}$ are ascending on Ω. □

Example 8-8 $\{[0, 1/x]\colon 0 < x < 1\}$ is descending on $(0, 1)$. □

In Example 8-8 we could have written "$[0, 1/x]$ is descending in x on $(0, 1)$."

Submodular Functions

In Appendix B, convex and concave functions are the kinds of functions which are allied with convex sets. The analogous functions for lattices are defined now.

Definition 8-5 Let f be a real-valued function whose domain is a lattice Ω. Then f is *submodular* if

$$f(x \wedge y) + f(x \vee y) \leq f(x) + f(y) \tag{8-3}$$

for all x and y in Ω. If $-f$ is submodular, then f is *supermodular.*

Example 8-9 Let Ω be any subset of $\mathbb{R}$ and let f be any real-valued function on Ω. Then f is both submodular and supermodular. Therefore, the notion of submodularity is uninteresting for real-valued functions of a real variable. □

Example 8-10 Let Ω be $\mathbb{R}^2$ with $x = (x_1, x_2) \leq y = (y_1, y_2)$ if $x_1 \leq y_1$ and $x_2 \leq y_2$. Then $f(x) = x_1 x_2$ is supermodular, which we verify with two cases because of the symmetry of x and y in (8-3). If $x \leq y$, then $x \wedge y = x$ and $x \vee y = y$; so (8-3) is trivial.

Suppose neither $x \leq y$ nor $x \geq y$, say $x_1 < y_1$ and $x_2 > y_2$. Then

$$f(x) + f(y) - f(x \wedge y) - f(x \vee y) = x_1 x_2 + y_1 y_2 - x_1 y_2 - x_2 y_1 = (x_1 - y_1)(x_2 - y_2) < 0 \quad \square$$

Example 8-11 Let $\Omega = \{(x_1, x_2): x_1 > 0, x_2 > 0\}$ with the same ordering as in Example 8-10. Let $f(x) = x_1/x_2$. Then f is submodular. For example, in the case $x_1 \geq y_1$ and $x_2 \leq y_2$,

$$f(x) + f(y) - f(x \wedge y) - f(x \vee y) = \frac{x_1}{x_2} + \frac{y_1}{y_2} - \frac{y_1}{x_2} - \frac{x_1}{y_2} = (x_1 - y_1)\left(\frac{1}{x_2} - \frac{1}{y_2}\right) \geq 0 \quad \square$$

The material in this section is likely to be completely new to most readers. Therefore, it is useful to mention that some results in the theory of convex sets and convex functions are similar to the propositions in this section. The following table indicates the real-valued counterparts to some of the items in this section.

Item in this section	Real-valued counterpart
Lattice	Convex set
Submodular function	Convex function
Supermodular function	Concave function
Isotone function	Nondecreasing function on $\mathbb{R}$
Antitone function	Nonincreasing function on $\mathbb{R}$

In Appendix B, Propositions B-5, B-6, and B-10 through B-14 provide useful characterizations of convex functions and tests of convexity. The following definition is similarly useful for submodular functions.

Let Ω_1 and Ω_2 be partially ordered sets and f be a real-valued function on $\Gamma \subset \Omega_1 \times \Omega_2$. Let $\Gamma_u = \{x: (x, u) \in \Gamma\}$, $u \in \Omega_2$. For each u, Γ_u is partially ordered and so also is $\Gamma_u \cap \Gamma_v$ when u and v are in Ω_2.

Definition 8-6 Suppose $u \leq v$. If $f(\cdot, v) - f(\cdot, u)$ is an isotone (antitone) function on $\Gamma_u \cap \Gamma_v$, then f has *isotone differences* (*antitone differences*) on Γ.

Observe that the apparent asymmetry in the definition is an illusion: suppose $x \leq y$ with x and y in $\Gamma_u \cap \Gamma_v$. Then

$$f(x, v) - f(x, u) \leq f(y, v) - f(y, u)$$

if, and only if,

$$f(y, u) - f(x, u) \leq f(y, v) - f(x, v)$$

More generally, let $\Omega_1, \Omega_2, \ldots, \Omega_n$ be partially ordered sets and f be a real-valued function on $\Gamma \subset \times_{i=1}^{n} \Omega_i$. Suppose $x = (x_1, x_2, \ldots, x_n) \in \Gamma$, let j and k be integers between 1 and n with $j \neq k$, and fix all x_i with $i \neq j$, $i \neq k$. If f has isotone (antitone) differences in (x_j, x_k), and if this is true for all x, j, and k, then f is said to have *isotone* (*antitone*) *differences* on Γ.

Theorem 8-1 (*a*) Let $\Omega_1, \ldots, \Omega_n$ be lattices and f a submodular (supermodular) function on $\Omega \in L(\times_{i=1}^{n} \Omega_i)$. Then f has antitone (isotone) differences on Ω.

(*b*) If $\Omega_1, \ldots, \Omega_n$ are chains and f has antitone (isotone) differences on $\times_{i=1}^{n} \Omega_i$, then f is submodular (supermodular) on $\times_{i=1}^{n} \Omega_i$.

PROOF Part (*a*) is an immediate consequence of the definitions as we verify for $n = 2$. The same argument suffices for $n > 2$ because the definition of antitone difference uses only pairwise differences. Suppose $x \leq y$, and $u \leq v$ and let $a = (x, v)$ and $b = (y, u)$. Then submodularity implies [cf. (8-3)]

$$\begin{aligned} 0 &\leq f(a) + f(b) - f(a \wedge b) - f(a \vee b) \\ &= f(x, v) + f(y, u) - f(x, u) - f(y, v) \end{aligned}$$

so

$$f(y, v) - f(y, u) \leq f(x, v) - f(x, u)$$

which is antitonicity of $f(\cdot, v) - f(\cdot, u)$ when $v \geq u$.

For (*b*), pick $x = (x_i)$ and $y = (y_i)$ in $\times_{i=1}^{n} \Omega_i$. If $x \leq y$ or $y \leq x$, then (8-3) is trivial; so suppose $x \not\leq y$ and $y \not\leq x$. Because each Ω_i is a chain for each i, either $x_i \leq y_i$ or $y_i \leq x_i$. Therefore, the components of x and y can be arranged so $x_i \leq y_i$ for $i = 1, 2, \ldots, k$ and $y_i \leq x_i$, $i = k + 1, \ldots, n$. Also, $0 < k < n$ because x and y are unordered. Now $x \wedge y = (x_1, x_2, \ldots, x_k, y_{k+1}, \ldots, y_n)$ and $x \vee y = (y_1, \ldots, y_k, x_{k+1}, \ldots, x_n)$. For $0 \leq i \leq j \leq n$ define $z_{ij} =$

$(y_1, \ldots, y_i, x_{i+1}, \ldots, x_j, y_{j+1}, \ldots, y_n)$ so $x \wedge y = z_{0k}$, $x \vee y = z_{kn}$, $x = z_{0n}$, and $y = z_{mm}$ for each m. In particular, $y = z_{kk}$. But antitone differences implies $f(z_{ij}) - f(z_{i,j+1}) \le f(z_{i+1,j}) - f(z_{i+1,j+1})$ when $0 \le i \le k \le j \le n$. Therefore, letting i progress from 0 to $k-1$,

$$f(z_{0k}) - f(z_{0n}) = \sum_{j=k}^{n-1} [f(z_{0j}) - f(z_{0,j+1})]$$

$$\le \sum_{j=k}^{n-1} [f(z_{kj}) - f(z_{k,j+1})] = f(z_{kk}) - f(z_{kn}) \qquad \square$$

The most important consequence of the theorem is in $\mathbb{R}^n$. Let e_i denote the ith unit vector in $\mathbb{R}^n$.

Corollary 8-1 Suppose $\Omega_i \subset \mathbb{R}$, $i = 1, \ldots, n$, with the usual ordering on $\mathbb{R}$, and $\times_{i=1}^{n} \Omega_i$ is the domain of f.

(*a*) f is submodular if, and only if, $f(x + \gamma e_i) - f(x)$ is antitone in x_j for each $j \ne i$, $\gamma > 0$, and x.

(*b*) If f is twice differentiable on its domain and its domain is a convex set, it is submodular if, and only if,

$$\frac{\partial^2 f(x)}{\partial x_i \, \partial x_j} \le 0 \tag{8-4}$$

for all $i \ne j$ and x.

PROOF (*a*) is an immediate consequence of part (*b*) of Theorem 8-1. But antitonicity of $f(x + \gamma e_i) - f(x)$ in x_j for each $j \ne i$, $\gamma > 0$, and x is equivalent, for each $\lambda > 0$, to

$$f(x + \gamma e_i) - f(x) \ge f(x + \gamma e_i + \lambda e_j) - j(x + \lambda e_j)$$

or $$[f(x + \gamma e_i + \lambda e_j) - f(x + \gamma e_i)] - [f(x + \lambda e_j) - f(x)] \le 0$$

This implies (8-4) if f is twice differentiable and the domain contains the points over which the limit is taken. The latter is true because of the convexity assumption in (*b*). $\square$

As usual the corollary remains true if you change "submodular" to "supermodular," "antitone" to "isotone," and reverse the inequality in (8-4).

Example 8-12 Let $c(\cdot)$ be convex and twice differentiable on $\mathbb{R}$ and $g(x, y) = c(y - x)$, $(x, y) \in \mathbb{R}^2$. Then $\partial^2 g(x, y)/\partial x \partial y = -c''(y - x) \le 0$; so g is submodular on $\mathbb{R}^2$. Similarly, if $c(\cdot)$ is concave and twice differentiable on $\mathbb{R}^n$, then $g(x) = c(\sum_{i=1}^{n} x_i)$ is submodular on $\mathbb{R}^n$ [where $x = (x_i) \in \mathbb{R}^n$]. $\square$

Example 8-13 You saw in Example 8-10 that $f(x) = x_1 x_2$ is supermodular on $\mathbb{R}^2$. Observe that $\partial^2 f(x)/\partial x_1 \partial x_2 = 1 > 0$ for all $x \in \mathbb{R}^2$. $\square$

Economic Interpretation

Theorem 8-1 and its corollary yield an economic interpretation of submodularity. Suppose that f specifies production costs in an economic system with n kinds of inputs. If f is submodular on $\mathbb{R}^n_+$, then

$$f(x + \gamma e_i) - f(x) \geq f(x + \lambda e_j + \gamma e_i) - f(x + \lambda e_j)$$

for $x \in \mathbb{R}^n_+$, $\gamma > 0$, $\lambda > 0$, and $i \neq j$. This inequality states that the extra cost of using γ extra units of input i is not raised by having λ extra units of input j. In other words, having more of input j cannot reduce the effectiveness of using more of input i. In this sense, inputs i and j complement each other's effectiveness. In economics, antitonicity, hence submodularity, is equivalent to commodities being *complementary*. Similarly, isotonicity, hence supermodularity, is equivalent to commodities being *substitutes*.

Notions of "substitute" and "complementary" products arise in microeconomics in theories of production and consumer choice. Two products are regarded as complements (substitutes) if having more of one does not induce you to choose less (more) of the other. Shoes and shoelaces are complementary, whereas spaghetti and macaroni are substitutes. Observe that the word definition of "complementary" ("substitute") is essentially the stipulation that a utility function have antitone (isotone) differences, hence that it be submodular (supermodular).

Optimization

We wish now to prove analogs of part (*a*) of Proposition B-3 and Proposition B-4. These assert (*a*) that the set of points where a convex function attains its minimum is a convex set, and (*b*) convexity with respect to a parameter is preserved by minimization.

Theorem 8-2 Let f be a submodular (supermodular) function on a lattice X. Then the set X^* where f attains its minimum (maximum) on X is a sublattice of X.

PROOF For any x and y in X^*, submodularity of f and optimality of x and y yield

$$0 \leq f(x \vee y) - f(x) \leq f(y) - f(x \wedge y) \leq 0$$

so $x \vee y$ and $x \wedge y$ are in X^*. □

The following analog of Proposition B-4 shows that economic complementarity is preserved in optimization.

Theorem 8-3 Let $\mathcal{S}$ be a nonempty set and A_s nonempty for each $s \in \mathcal{S}$. Let

$$\mathscr{C} = \{(s, a): a \in A_s, s \in \mathcal{S}\} \tag{8-5}$$

Let J be a real-valued function on $\mathscr{C}$ and define

$$f(s) = \inf \{J(s, a): a \in A_s\} \qquad \text{for } s \in \mathbb{S} \tag{8-6}$$

If $\mathscr{C}$ is a lattice, J is submodular and finite on $\mathscr{C}$, and f is finite on $\mathbb{S}$, then f is submodular on $\mathbb{S}$.

Proof $\mathbb{S}$ is a lattice due to Lemma 8-1. Pick x_1 and x_2 in $\mathbb{S}$ so, for all $\gamma > 0$, there are $y_i \in A_{x_i}$ with $(x_i, y_i) \in \mathscr{C}$ and $f(x_i) > J(x_i, y_i) - \gamma/2$, $i = 1, 2$. Also $\mathscr{C}$ is a lattice so $(x_1, y_1) \vee (x_2, y_2) \in \mathscr{C}$ and $(x_1, y_1) \wedge (x_2, y_2) \in \mathscr{C}$. Therefore, submodularity of J, Lemma 8-1, and (8-6) imply

$$\begin{aligned} f(x_1) + f(x_2) &\geq J(x_1, y_1) + J(x_2, y_2) - \gamma \\ &\geq J[(x_1, y_1) \wedge (x_2, y_2)] + J[(x_1, y_1) \vee (x_2, y_2)] - \gamma \\ &\geq f(x_1 \wedge x_2) + f(x_1 \vee x_2) - \gamma \end{aligned}$$

Let $\gamma \to 0$ to complete the proof. □

If $\mathbb{S} \subset \mathbb{R}$, then f is trivially submodular (recall Example 8-9) without appeal to Theorem 8-3. Hence the theorem is applied only to models whose states are vectors.

The following results are the primary reason for our inclusion of a section on lattices and submodular functions. They concern monotonicity of decision variables or, more generally, the ascendancy of a set of optimal decisions.

Let $f(\cdot)$ be defined by (8-6), let $Y(s) \triangleq \{a: a \in A_s, J(s, a) = f(s)\}$ and let $\mathbb{S}' = \{s: s \in \mathbb{S}, Y(s) \text{ is nonempty}\}$. Recall Definition 8-4 of the terms *ascending* and *descending*.

Theorem 8-4 Let $\mathbb{S}$ be a nonempty partially ordered set, A a lattice, $A_s \in L(A)$ for each $s \in \mathbb{S}$, and A_s an ascending function on $\mathbb{S}$. If $J(\cdot, \cdot)$ is a real-valued function on $\mathscr{C}$ [defined by (8-5)] that satisfies

$$J(s_1, a_1 \wedge a_2) + J(s_2, a_1 \vee a_2) \leq J(s_1, a_1) + J(s_2, a_2) \tag{8-7}$$

for all s_1 and s_2 in $\mathbb{S}$ with $s_1 \leq s_2$ and $a_i \in A_{x_i}$ for $i = 1$ and 2, then $Y(\cdot)$ is an ascending function on $\mathbb{S}'$ (that is, $\{Y(s): s \in \mathbb{S}'\}$ is ascending).

If $J(\cdot, \cdot)$ is submodular, then it must satisfy (8-7) but not conversely. In most of our applications, $J(\cdot, a)$ will *not* be submodular on $\mathbb{S}$ for each a, as submodularity on $\mathscr{C}$ would imply. We say that $J(\cdot, \cdot)$ is *weakly submodular* when (8-7) is satisfied and *weakly supermodular* if $-J(\cdot, \cdot)$ satisfies (8-7). Exercise 8-7 asks you to prove that submodularity and weak submodularity are equivalent when $\mathscr{C} \subset \mathbb{R}^2$.

Proof Pick s_1 and s_2 in $\mathbb{S}'$ with $s_1 \leq s_2$ and pick $a_i \in Y(s_i)$, $i = 1, 2$. By hypothesis, $a_1 \wedge a_2 \in A_{s_1}$ and $a_1 \vee a_2 \in A_{s_2}$ so $(s_1, a_1 \wedge a_2) \in \mathscr{C}$ and $(s_2,$

$a_1 \vee a_2) \in \mathscr{C}$. Then (8-7) and $f(s_i) = J(s_i, a_i)$ yield $0 \leq J(s_2, a_1 \vee a_2) - J(s_2, a_2) \leq J(s_1, a_1) - J(s_1, a_1 \wedge a_2) \leq 0$ so $a_1 \vee a_2 \in Y(s_2)$ and $a_1 \wedge a_2 \in Y(s_1)$.

□

How can Theorem 8-4 be used to assert the existence of a monotone optimal solution? Suppose, for example, there is a unique optimal solution to (8-6) for each $s \in \mathcal{S}$. Then $Y(s)$ has a single element, call it $a^*(s)$; so the theorem gives conditions under which $s_1 \leq s_2$ imply $a^*(s_1) \leq a_2^*(s_2)$. The assumption of uniqueness is unnecessary to conclude that *some* selection $a(s)$ of an element in $Y(s)$, for each $s \in \mathcal{S}'$, yields $a(s_1) \leq a(s_2)$ if $s_1 \leq s_2$. On the other hand, some care is needed.

Example 8-14 Suppose $\mathcal{S} = \mathbb{R}_+$, $Y(s) = [s, \ s + 1]$, for each $s \in \mathbb{R}_+$. Let $\lambda(\cdot)$ be differentiable on $\mathbb{R}$ with $0 \leq \lambda(s) \leq 1$ for $s \in \mathbb{R}_+$ and consider $a(s) = \lambda(s)s + [1 - \lambda(s)](s + 1)$, which is a weighted average of the end points of $Y(s)$. Then $a(s) = s + 1 - \lambda(s)$ and $da(s)/ds = 1 - d\lambda(s)/ds$ so $da(s)/ds < 0$ if $d\lambda(s)/ds > 1$. For example, let $\lambda(s) = 1 - e^{-3s}$. Then $d\lambda(s)/ds > 1$ so $a(\cdot)$ is decreasing when $0 < s < 1/3$. □

You may properly observe that Example 8-14 had to "reach" for a counterexample. The upper and lower end points of $Y(s)$ are monotone; so the selections $a(s) = s$ and $a(s) = s + 1$ would have yielded two monotone optimal policies. What ensures existence of greatest and least elements, $a^*(s)$ and $a_*(s)$, respectively, in $Y(s)$? Corollary 8-2 below addresses the case where f is a real-valued function on a subset of $\mathbb{R}^n$.

Here are some standard notions. A set $\mathcal{S} \subset \mathbb{R}^n$ is *compact* if it is closed and bounded. Also, we use the notation $N_\delta(\mathbf{x})$, the *δ-neighborhood* of $\mathbf{x} \in \mathbb{R}^n$, for the set of all $\mathbf{y} \in \mathbb{R}^n$ such that $|\mathbf{x} - \mathbf{y}| < \delta$ [where $|\mathbf{u}| \triangleq (\sum_{i=1}^n u_i^2)^{1/2}$ for $\mathbf{u} = (u_1, \ldots, u_n)$]. Then a real-valued function $h(\cdot)$ on $\mathcal{S}$ is *lower semicontinuous* at $\mathbf{x} \in \mathbb{R}^n$ if, for all $\varepsilon > 0$, there is $\delta > 0$ such that $h(\mathbf{x}) \leq h(\mathbf{y}) + \varepsilon$ for all $\mathbf{y}$ with $|\mathbf{x} - \mathbf{y}| < \delta$ [equivalently, $h(\mathbf{x}) < \liminf_{y \to x} h(\mathbf{x})$]. If $-h(\cdot)$ is lower semicontinuous, then $h(\cdot)$ is said to be *upper semicontinuous*. In fact, $h(\cdot)$ is continuous if, and only if, it is both upper and lower semicontinuous. The following corollaries are straightforward consequences of the fact that a lower (upper) semicontinuous function on a compact set achieves its minimum (maximum).

Corollary 8-2 Let $h(\cdot)$ be a submodular (supermodular) function on a lattice X. If X is a compact subset of $\mathbb{R}^n$ and $h(\cdot)$ is lower semicontinuous (upper semicontinuous) on X, the set X^* of points where $h(\cdot)$ attains its minimum (maximum) on X is a nonempty compact sublattice of X that has a greatest and a least element.

This corollary and Theorem 8-4 yield the following "punch line" to this subsection.

Corollary 8-4 Under the hypotheses of Theorem 8-4, suppose $A \subset \mathbb{R}^n$ with A_s a compact set and $J(s, \cdot)$ lower semicontinuous on A_s for all $s \in \mathcal{S}$. Then for each $s \in \mathcal{S}$, $Y(s)$ has a greatest element $a^*(s)$ and a least element $a_*(s)$ and both are isotone on $\mathcal{S}$.

Example 8-15 Consider the optimization problem

$$f(s) = \inf \{J(s, a): s \le a \le s + u\} \qquad s \in \mathbb{R} \qquad (u > 0)$$

$$J(s, a) = g(a) + b(a - s) \qquad (s, a) \in \mathscr{C}$$

where $\mathscr{C} = \{(s, a): s \in \mathbb{R}, s \le a \le s + u\}$, b is convex and continuous on $\mathbb{R}_+$, and g is lower semicontinuous on $\mathbb{R}$. Let $\mathcal{S} = \mathbb{R}$. The inventory problem in Section 3-1 has this form if $n = 1$ and $A_s = \{s, s + u]$ is replaced by $[s, \infty)$, $s \in \mathcal{S}$. You may interpret the optimization problem as a single-period inventory model in which the present inventory level is s and previous backlogging may have caused $s < 0$ so $s \in \mathbb{R}$ rather than $s \in \mathbb{R}_+$. The order quantity is $a - s$; so $b(a - s)$ is the ordering cost. Then $g(a)$ is the expected holding and penalty costs if a is the quantity of goods available to satisfy demand.

Under these assumptions, $\mathcal{S}$ is partially ordered (with the usual ordering on $\mathbb{R}$), $A_s = [s, s + u] \subset \mathbb{R}$ is compact and ascending on $\mathbb{R}$, and $J(s, \cdot)$ is lower semicontinuous on A_s for each $s \in \mathcal{S}$. Last, (8-7) is trivially satisfied by $g(\cdot)$ because $g(a)$ does not depend on s. To see that $b(a - s)$ satisfies (8-7), we use Corollary 8-1 and the argument implicit in Example 8-12 (without assuming differentiability). Let $h(s, a) \triangleq b(a - s)$ and write

$$\Delta = h(s, a + \lambda) - h(s, a) - h(s + \gamma, a + \gamma) + h(s + \gamma, a)$$

for $\lambda > 0$, $\gamma > 0$ such that all arguments are in $\mathscr{C}$. If $\Delta \ge 0$, then h is submodular (Corollary 8-1), hence weakly submodular. Now

$$\begin{aligned} \Delta &= b(a + \lambda - s) - b(a - s) - b(a + \lambda - s - \gamma) + b(a - s - \gamma) \\ &= [b(a + \lambda - s) - b(a + \lambda - s - \gamma)] - [b(a - s) - b(a - s - \gamma)] \end{aligned}$$

which is nonnegative because b is a convex function. Therefore, $a_*(s)$ and $a^*(s)$ exist for each $s \in \mathbb{R}$ and are nondecreasing on $\mathbb{R}$. In Section 7-1, we assume that $G(x) \to \infty$ as $|x| \to \infty$, which is equivalent to the compactification assumed here with $A_s = [s, s + u]$. □

A General Dynamic Programming Result

Most of our uses of lattice programming depend on Theorem 8-5 below and its corollaries. For the remainder of this section we use a minimization criterion; so we use $c(s, a)$ to denote a single-stage cost [rather than $-r(s, a)$].

Consider the recursion

$$f_n(s) = \inf \{J_n(s, a): a \in A_s\} \qquad s \in \mathcal{S} \subset \mathbb{R}, \qquad (8\text{-}8a)$$

$$J_n(s, a) = c(s, a) + \beta E(f_{n-1}[\xi(s, a)]) \qquad (s, a) \in \mathscr{C} \subset \mathbb{R}^2 \qquad (8\text{-}8b)$$

for $n \in I_+$, where $f_0(\cdot)$ is given, $\mathbb{R}^2 \supset \mathscr{C} = \{(s, a): a \in A_s\}$, $\xi(s, a)$ is an r.v. for each $(s, a) \in \mathscr{C}$ that takes values in $\mathbb{S}$, and $c(\cdot, \cdot)$ is a real-valued function on $\mathscr{C}$. What are sufficient conditions for (8-8*a*) and (8-8*b*) to have a monotone optimal policy? It is convenient to use the notation

$$\gamma_x(s, a) \triangleq P\{\xi(s, a) > x\}$$

for the complement of the distribution function of $\xi(s, a)$.

Definition 8-7 The collection of sets $\{A_s: s \in \mathbb{S}\}$ is called *contracting* if $s \leq s'$ implies $A_s \supset A_{s'}$.

Theorem 8-5 Suppose for each $s \in \mathbb{S}$ that A_s is compact, $J_n(s, \cdot)$ is lower semicontinuous for each $n \in I_+$, $\mathscr{C}$ is a lattice, $f_0(\cdot)$ is nondecreasing and bounded below on $\mathbb{S}$, the infimum in (8-8*a*) is attained at each $s \in \mathbb{S}$, and

$$c(\cdot, a) \text{ is nondecreasing for each } a \tag{8-9a}$$

$$c(\cdot, \cdot) \text{ is submodular and bounded below} \tag{8-9b}$$

$$\gamma_x(\cdot, \cdot) \text{ is submodular on } \mathscr{C} \text{ for each } x \tag{8-9c}$$

$$\gamma_x(\cdot, a) \text{ is nondecreasing for each } x \text{ and } a \tag{8-9d}$$

$$\{A_s: s \in \mathbb{S}\} \text{ is contracting and ascending} \tag{8-9e}$$

Then for each n there exists $a_n^*(\cdot)$ nondecreasing on $\mathbb{S}$ such that

$$f_n(s) = J_n[s, a_n^*(s)] \qquad n \in I_+ \qquad s \in \mathbb{S}$$

To begin an informal proof, observe that Corollary 8-4 implies that weak submodularity of $J_n(\cdot, \cdot)$ for each n (with other assumptions) would imply that (8-8) has a monotone optimal policy. But Exercise 8-7 asks you to verify that $\mathscr{C} \subset \mathbb{R}^2$ implies that weak submodularity is equivalent to submodularity. Suppose $c(\cdot, \cdot)$ is submodular on $\mathscr{C}$; what would cause

$$E(f_{n-1}[\xi(s, a)]) \tag{8-10}$$

to be submodular too?

Suppose $F(\cdot \mid s, a)$ is the distribution function of $\xi(s, a)$. Then (8-10) is submodular if, and only if (Corollary 8-1),

$$\int_{-\infty}^{\infty} f_{n-1}(r)\, d[F(r \mid s', a) - F(r \mid s'a')]$$

$$\geq \int_{-\infty}^{\infty} f_{n-1}(r)\, d[F(r \mid s, a) - F(r \mid s, a')] \tag{8-11}$$

for all $s \leq s'$ and $a \leq a'$ such that (s, a), (s', a), (s, a'), and (s', a') are all in $\mathscr{C}$. First, (8-11) would be trivial if $\xi(s, a)$ depends only on a and not on s. This case occurs throughout Chapter 3. Second, suppose $f(\cdot)$ is nondecreasing and recall the notation $\gamma_x(s, a) \triangleq P\{\xi(s, a) \geq x\}$. Then (8-11) would be implied by submodularity of γ_x for each x and the following important result.

Lemma 8-2 Let $b(\cdot)$ and $q(\cdot)$ be integrable functions on $\mathbb{R}$. Then

$$\int_{-\infty}^{\infty} f(r)\, db(r) \leq \int_{-\infty}^{\infty} f(r)\, dq(r) \tag{8-12}$$

for all nondecreasing $f(\cdot)$ that have $f(-\infty) \triangleq \lim_{r \to -\infty} f(r) = 0$ if, and only if,

$$\int_{x}^{\infty} db(z) \leq \int_{x}^{\infty} dq(z) \qquad \text{for all } x \in \mathbb{R} \tag{8-13}$$

PROOF To obtain (8-13) from (8-12), choose $f(\cdot)$ as the indicator of the set $[x, \infty)$, that is, $f(r) = 0$ if $r < x$ and $f(r) = 1$ if $r \geq x$. To obtain (8-12) from (8-13), observe that $df(r) \geq 0$ for all r [because $f(\cdot)$ is nondecreasing], $f(-\infty) = 0$, and (8-13) imply

$$\int_{t}^{\infty} d[q(r) - b(r)] \geq 0. \qquad \square$$

The restriction $f(-\infty) = 0$ can be dropped if we add the assumption $q(\infty) \geq b(\infty)$. Of course, if $b(\cdot)$ and $q(\cdot)$ are distribution functions, $q(\infty) - b(\infty) = 1 - 1 = 0$

Compare (8-11) with (8-12); then (8-13) becomes

$$\gamma_x(s', a) - \gamma_x(s', a') \geq \gamma_x(s, a) - \gamma_x(s, a')$$

which is equivalent to submodularity. Therefore, $E(f_{n-1}[\xi(s, a)])$ is submodular if $f_{n-1}(\cdot)$ is nondecreasing and bounded below and γ_x is submodular for each x. But what causes $f_{n-1}(\cdot)$ to be nondecreasing and bounded below? Return to (8-8) and consider the effect of s, the state, on feasibility and optimality. Suppose $J_n(\cdot, a)$ is nondecreasing [on $\{s\colon a \in A_s, s \in \mathcal{S}\}$]. Then f_n is nondecreasing if $A_s \supset A_{s'}$ when $s \leq s'$, that is, if $\{A_s\colon s \in \mathcal{S}\}$ is contracting. Can $\{A_s\colon s \in \mathcal{S}\}$ be contracting and ascending, as would be needed to invoke Corollary 8-4? A case in which this occurs is $A_s = \{m(s), 1\}$, where† $m(s) = \delta(s - a^*)$, which is unity (zero) if $s > (\leq) a^*$. But what causes $J_n(\cdot, a)$ to be nondecreasing? Suppose $r(\cdot, a)$ in (8-8*b*) is nondecreasing. Again, the issue is the second term in (8-8*b*), that is, (8-10).

Let $s \leq s'$ and evaluate

$$E(f_{n-1}[\xi(s', a)]) - E(f_{n-1}[\xi(s, a)]) =$$

$$\int_{-\infty}^{\infty} f_{n-1}(r)\, d[F(r \mid s', a) - F(r \mid s, a)] \tag{8-14}$$

† We use the notation $\delta(u) = 1$ if $u > 0$ and $\delta(u) = 0$ if $u \leq 0$.

If f_{n-1} is nondecreasing, it follows from Lemma 8-2 that (8-14) is nonnegative if

$$\int_x^\infty dF(r \mid s', a) \geq \int_x^\infty dF(r \mid s, a)$$

that is, if $\gamma_x(\cdot, a)$ is nondecreasing for each x. Similarly, if $c(\cdot, \cdot)$ is bounded below, so also is f_{n-1} for each n.

The preceding exposition can be converted to a formal inductive proof of Theorem 8-5. □

The lower semicontinuity assumption for $J_n(s, \cdot)$ is established in various ways that depend upon further details. For example, if A_s is a denumerable set for each s or if $J_n(s, \cdot)$ is convex and continuous at the boundaries of A_s for each s, the assumption is satisfied.

The collection $\{A_s: s \in \mathcal{S}\}$ is said to be *expanding* if $s \leq s'$ implies $A_s \subset A_{s'}$. It is straightforward to prove the following results.

Corollary 8-5*a* Suppose for each $s \in \mathcal{S}$ that A_s is compact,

$$f_n(s) = \sup \{J_n(s, a): a \in A_s\} \qquad s \in \mathcal{S} \subset \mathbb{R}$$

$$J_n(s, a) = r(s, a) + \beta \mathrm{E}(f_{n-1}[\xi(s, a)]) \qquad (s, a) \in \mathscr{C} \subset \mathbb{R}$$

$J_n(s, \cdot)$ is lower semicontinuous for each $n \in I_+$, $\mathscr{C}$ is a lattice, $f_0(\cdot)$ is nondecreasing and bounded below on $\mathcal{S}$, the supremum is attained at each $s \in \mathcal{S}$, and

$$r(\cdot, a) \text{ is nondecreasing for each } a \tag{8-9a'}$$

$$r(\cdot, \cdot) \text{ is supermodular and bounded below} \tag{8-9b'}$$

$$\gamma_x(\cdot, \cdot) \text{ is supermodular on } \mathscr{C} \text{ for each } x \tag{8-9c'}$$

$$\gamma_x(\cdot, a) \text{ is nondecreasing for each } x \text{ and } a \tag{8-9d}$$

$$\{A_s: s \in \mathcal{S}\} \text{ is expanding and ascending} \tag{8-9e'}$$

Then for each n there exists $a_n^*(\cdot)$ nondecreasing on S such that

$$f_n(s) = J_n[s, a_n^*(s)] \qquad n \in \mathrm{I}_+ \qquad s \in \mathcal{S}$$

Corollary 8-5*b* The conclusion of Theorem 8-5 [Corollary 8-5*a*] is valid if (8-9*a*) and (8-9*c*) [(8-9*a'*) and (8-9*c'*)] are deleted and (8-9*d*) and (8-9*e*) [(8-9*d*) and (8-9*e'*)] are altered to

$$\text{For each } a, \ \xi(s, a) \text{ is the same for all } s \tag{8-9d'}$$

$$\{A_s: s \in \mathcal{S}\} \text{ is ascending} \tag{8-9e''}$$

Theorem 8-5 and its corollaries are applied in Section 8-2 to binary decision models that arise in numerous contexts and in Sections 8-3, 8-4, 8-5, and 8-6 to models with larger sets of actions.

EXERCISES

8-1 Suppose Ω is a lattice. Prove that $L(\Omega)$ is partially ordered by $\lesssim$. Hint: First show that $\lesssim$ is antisymmetric and transitive on $P(\Omega)$.

8-2 Prove Lemma 8-1.

8-3 Suppose Ω is a lattice on which the real-valued function $g(\cdot)$ is defined and $f(\cdot)$ is a real-valued function on $\mathbb{R}$. Let $g(\cdot)$ be *monotone*; that is, g is either isotone or antitone. Then prove that the composite function $f(g(\cdot))$ on Ω is supermodular (submodular) if f is a convex (concave) function and either (*a*) $f(\cdot)$ is isotone and $g(\cdot)$ is supermodular or (*b*) $f(\cdot)$ is antitone and $g(\cdot)$ is submodular. Hint: Use the following identity for all x and y in S:

$$\begin{aligned} &f(g(x)) + f(g(y)) - f(g(x \vee y)) - f(g(x \wedge y)) \\ &\quad = [f(g(x)) - f(g(x \vee y) + g(x \wedge y) - g(y))] \\ &\quad\quad + [f(g(x \vee y) + g(x \wedge y) - g(y) - f(g(x \vee y)) - f(g(x \wedge y)) + f(g(y))] \end{aligned}$$

8-4 Prove that Theorem 8-4 is valid when "ascending" is replaced by "descending."

8-5 Lemma 8-2 can be embellished in many interesting ways. For example, prove that (8-12) and (8-13) are equivalent to the following property when $b(\cdot)$ and $q(\cdot)$ are distribution functions. There exist r.v.'s X and Y on the same sample space such that $X(\omega) \leq Y(\omega)$ for all ω in the sample space,

$$P\{X \leq x\} = \int_{-\infty}^{x} dq(r) \qquad \text{and} \qquad P\{Y \leq y\} = \int_{-\infty}^{y} db(r)$$

8-6 Suppose $J(\cdot, \cdot)$ is a strictly positive function on $\mathbb{R}_+^2$, $J(x, \cdot)$ is nonincreasing for each x, $J(\cdot, y)$ is nondecreasing for each y, and $\log J(\cdot, \cdot)$ is submodular. Prove that $J(\cdot, \cdot)$ is submodular.

8-7 In Theorem 8-4, suppose $\mathscr{C} \subset \mathbb{R}^2$. Prove that $J(\cdot, \cdot)$ is submodular if it is weakly submodular.

8-8 Prove Theorem 8-5 formally and in detail.

8-2* EQUIPMENT REPLACEMENT, STOPPING, AND OTHER BINARY DECISION PROBLEMS

Many firms and public agencies operate fleets of vehicles. As a vehicle's age and accumulated usage increase, so do its operating and maintenance costs. Meanwhile its resale value, i.e., its "salvage value," diminishes. At what age and condition is it appropriate to dispose of the vehicle and replace it? The decision of whether or not to replace equipment, such as a vehicle, is an important example of a *binary decision renewal problem.* Examples 4-1 and 4-2, concerning gambler's ruin and machine maintenance, are examples of binary decision renewal problems. Both examples are closely related to replacement problems.

As individuals we face equipment-replacement problems but lack the data generated by a fleet of similar vehicles. Also, we lack the capital of a large firm; a single vehicle represents a significant portion of our total wealth and thereby induces a nonlinear utility function when we evaluate alternatives (cf. Section 1-4). Therefore, the replacement models in this section are likely to be useful only in a large firm or agency.

This section formulates the binary decision renewal model and then sketches several examples of the general model. Equipment replacement is the first example. Then the results in Section 8-1 are used to deduce conditions under which there is a monotone optimal policy. When at most two actions are available at each state, such a policy is sometimes called a control limit.

Definitions of a Binary Decision Process and Examples

A *binary decision process* is a Markov decision process with a denumerable (or finite) set of states $\mathcal{S} = \{0, 1, \ldots\}$ and $A_s = \{0, 1\}$ for each $s \in \mathcal{S}$. In this section, c_s^1 and c_s^0 denote single-stage costs [which we write as $c(s, 1)$ and $c(s, 0)$ in other sections] and, as usual, p_{sj}^a denotes a transition probability. With the criterion of expected discounted cost and β the discount factor, the dynamic programming recursion for this model is

$$f_n(s) = \min \left\{ c_s^0 + \beta \sum_{j \in \mathcal{S}} p_{sj}^0 f_{n-1}(j),\ c_s^1 + \beta \sum_{j \in \mathcal{S}} p_{sj}^1 f_{n-1}(j) \right\} \qquad (8\text{-}15)$$

for $s \in \mathcal{S}$ and $n \in I_+$. Usually $f_0(\cdot) \equiv 0$.

Example 8-16: Equipment replacement In the simplest case, s is the age of the equipment, c_s^0 is the expected cost to operate a piece of equipment for one period if its age is s, and c_s^1 is the cost to buy a new piece of equipment and operate it for one period minus the salvage value of a piece of equipment at age s. Here, we interpret the actions 1 and 0 as "replace" and "don't replace," respectively. As a result, $p_{s0}^1 = p_{s,s+1}^0 = 1$. This simplest case is usually too simple because it does not include breakdowns which force replacement. To include breakdowns, let b_s be the probability that an age s machine will break down and necessarily be replaced. Now $p_{s,s+1}^0 = 1 - b_s$ and $p_{s,0}^0 = b_s$. Similarly, $c_s^0 = b_s K_s' + (1 - b_s)K_s$, where K_s' is the expected replacement cost of a broken-down item at age s and K_s is the expected operating cost of an item at age s that does not break down. □

Often, you know more about a vehicle than merely its age. For example, you may know its aggregate repair cost thus far. On that basis you may categorize the vehicle as being a "lemon" or not. In practice there is a challenge to keep $\mathcal{S}$ small but informative.

Example 8-17: Maintenance Nuclear reactors are complex devices which are inspected periodically and then either shut down for maintenance or operated normally. Large jet engines, computers, railroad bridges, and earth dams also have this feature. A jet engine, for example, is necessarily overhauled at intervals separated by a maximum number of hours of usage. During the interim, it is inspected to see if an overhaul is needed even sooner. The incentive to overhaul sooner is that an obligatory unanticipated overhaul (i.e., a breakdown) is more expensive and may be more dangerous than a planned repair.

In this maintenance context, stage s is the condition of the device based on the most recent inspection (and past history), action 0 denotes continued operation, and action 1 denotes planned maintenance. Suppose the state labels are assigned so that high values of s correspond to a relatively poor condition. The model here is the same as that for Example 8-16 except that there may be numerous j for which $p_{sj}^0 > 0$. □

Example 8-18: Canning fruit Some canneries have available two rates at which they are able to process and can fruit. During a harvest season, a cannery might run two shifts daily at the processing rate selected at the beginning of the day. The advantage of the higher rate is that newly arrived harvested fruit will be stored for less time before it is processed. The fruit deteriorates in storage and thereby obliges labeling of the cans with a lower and less profitable grade. The attraction of the lower rate is that it is less expensive to operate. Let D_n be the amount of fruit delivered to the cannery on day n.

Here, the state s is the amount of unprocessed fruit in storage, action 1 denotes the higher rate, and action 0 denotes the lower rate. Let μ_1 and μ_0 denote the higher and lower rates, respectively, suppose $D_1, D_2, \ldots$ are i.i.d. r.v.'s, and let a_n denote the processing rate used on day n. Then $s_{n+1} = s_n + D_n - \mu_{a_n}$ so $p_{sj}^k = P\{D_1 = j - s + \mu_k\}$ $(s \geq \mu_k)$. Let w_k denote the production cost when μ_k is the rate and let $h_k(s)$ denote the expected deterioration loss during a day t when $s_n = s$ and $a_n = k$. Then $c_s^k = w_k + h_k(s)$ with $w_1 \geq w_0$ and $h_1(s) \leq h_0(s)$. □

Example 8-19: Stopping The generic "stopping problem" [cf. Breiman (1964) or Chow, Robbins, and Siegmund (1971)] is the special case of (8-15) with $c_0^1 = c_0^0 = 0$ and $p_{s0}^0 = p_{00}^1 = 1$ for all s. In words, actions 0 and 1 are "stop" and "continue." State 0 denotes the event of having stopped already. □

We shall wish to claim for (8-15) that there is a *control limit* V_n such that the action $k = 0$ is optimal if, and only if, the state $s_n \geq V_n$. Every Markov policy partitions the set of states $\mathcal{S}$ into two subsets S_n^0 and S_n^1, where S_n^k is the set of states s_n where $a_n = k$. But a control-limit policy causes $S_n^1 = \{0, 1, \ldots, V_n - 1\}$ and $S_n^0 = \{V_n, V_n + 1, \ldots\}$. The existence of an optimal control-limit policy is a special case of the following application of Theorem 8-5.

An Application of Lattice Programming

A restatement of Theorem 8-5 vis-à-vis

$$f_n(s) = \min \left\{ c_s^a + \beta \sum_{j \in \mathcal{S}} p_{sj}^a f_{n-1}(j) \colon a \in \{0, 1\} \right\} \tag{8-16}$$

will be applied to Examples 8-16 through 8-19. Let

$$\gamma_i(s, a) = \sum_{j \geq i} p_{sj}^a \qquad s \in \mathcal{S},\ i \in \mathcal{S},\ a \in \{0, 1\} \tag{8-17}$$

Proposition 8-1 Suppose

$$0 \le c_{s+1}^1 - c_s^1 \le c_{s+1}^0 - c_s^0 \qquad s \in \mathcal{S} \tag{8-18}$$

$$\text{For each } i,\ \gamma_i(\cdot, 0) - \gamma_i(\cdot, 1) \text{ is nondecreasing} \tag{8-19}$$

$$\text{For each } i,\ \gamma_i(\cdot, 0) \text{ and } \gamma_i(\cdot, 1) \text{ are nondecreasing} \tag{8-20}$$

Then for each n there is a number V_n such that an optimal policy is given by

$$a_n = \begin{cases} 0 & \text{if } s_n < V_n \\ 1 & \text{if } s_n \ge V_n \end{cases} \tag{8-21}$$

PROOF The minimum in (8-16) is attained for every n and s (why?), and (8-9a) through (8-9d) are implied by (8-17) through (8-20). Since $A_s = \{0, 1\}$ for all s, assumption (8-9e) is trivially satisfied. Therefore, (8-21) is implied by Theorem 8-5. □

Condition (8-18) of the proposition states that the single-period cost increases as s increases but that the increment in cost is greater if $a = 0$ than if $a = 1$. Condition (8-20) states that large values of s_n tend to be followed by large values of s_{n+1}; condition (8-19) states that the tendency is greater if $a = 0$ than if $a = 1$. Thus these conditions make it plausible that if $a_n = 1$ is optimal at $s_n = s$, then $a_n = 1$ is optimal at $s_n = s'$ for all $s' \ge s$.

Examples Revisited

Example 8-20: Equipment replacement Recall in Example 8-16 that $a = 1$ denotes "replacement." In the simplest case, s is the age of the equipment so $p_{s,s+1}^0 = p_{s0}^1 = 1$. Therefore, $\gamma_i(s, 0) = 0$ for $s \le i - 1$ and $\gamma_i(s, 0) = 1$ for $s > i - 1$, $\gamma_0(s, 1) = 1$ for all s, and $\gamma_i(s, 1) = 0$ for all $i \ge 1$. It is easily seen that (8-19) and (8-20) are satisfied. Recall that c_s^0 is the operating cost at age s. Suppose c_s^1 has the form $r - L_s$, where r is the cost of a new piece of equipment plus the cost to operate it one period and L_s is the salvage value of equipment at age s. Then (8-18) requires

$$0 \le L_s - L_{s+1} \le c_{s+1}^0 - c_s^0 \tag{8-22}$$

that is, operating cost increases with age while salvage value decreases. The effect of (8-18) is that operating costs are assumed to rise with age at a rate at least as great as the reduction in salvage value.

The altered replacement example included probabilistic failure with b_s as the probability that an age s machine breaks down. Then $p_{s,s+1}^0 = 1 - b_s$ and $p_{s0}^0 = b_s$ so

$$\gamma_i(s, 0) = \begin{cases} 0 & \text{if } s + 1 < i \\ 1 - b_s & \text{if } 0 < i \le s + 1 \\ 1 & \text{if } i = 0 \end{cases}$$

which is not necessarily nondecreasing in s as required by (8-20) if $b_s < b_{s+1}$.

Frequently, $b_s \le b_{s+1}$ ("increasing failure rate") is a valid assumption. Fortunately, a relabeling of states saves the day!

Let ∞ denote the "highest" state, one from which replacement is mandatory so $A_\infty = \{1\}$. Delete state 0 (you would not *choose* action 0 in that state) and for all s (including $s = \infty$) let $p^1_{s\infty} = b_0$ denote the probability that a new item breaks down. Then $p^0_{s\infty} = b_s$ and $p^0_{s,s+1} = 1 - b_s$ for $1 \le s, s \ne \infty$; so $\gamma_i(s, 0) = 1$ if $i \le s + 1$ and $\gamma_i(s, 0) = b_s$ if $i > s + 1$. Therefore, $\gamma_i(\cdot, 0)$ is nondecreasing if $b_s \le b_{s+1}$ for all $s \ne \infty$ ($k = 0$ is not feasible at $s = \infty$). Similarly, $\gamma_i(s, 1)$ is a constant with respect to s; so (8-19) and (8-20) are satisfied. Let c^1_s remain unaltered if $s \ne \infty$ and $c^1_\infty = r - L_\infty$. However, expected costs under action 0 must be incremented to include the expected replacement cost, namely, $c^0_s = b_s[r(1 - \beta) - K'_s] + (1 - b_s)K_s$. Here, K'_s is the salvage value of a broken-down item and K_s is the operating cost of one that has not broken down. Exercise 8-9 asks you to verify that $\{c^k_s\}$ is nondecreasing in s if

$$\begin{gathered} L_{s+1} \le L_s,\ b_s \le b_{s+1},\ (1 - b_s)K_s \le (1 - b_{s+1})K_{s+1} \\ b_s K'_s \ge b_{s+1}K'_{s+1} \qquad \text{and} \qquad L_\infty \le L_s \end{gathered} \tag{8-23}$$

for all s. Last, (8-18) is implied by (8-22). □

Example 8-21: Continuation of Example 8-17, Maintenance The model is the same as in Example 8-20 except that $p^0_{sj} > 0$ is possible for many j (not only $j = s + 1$ and ∞). Therefore, for (8-19) and (8-20) to hold, it is necessary that $\sum_{j \geq i} p^0_{sj}$ be nondecreasing in s for each i. If $c^1_s = r - L_s$, then (8-22) implies property (8-18). □

Example 8-22: Canning fruit Recall in Example 8-18 that $s_{n+1} = s_n + D_n - \mu_{a_n}$ so $\gamma_x(s, k) = P\{D_1 \ge x - s + \mu_k\} = 1 - F(x - s + \mu_k)$, where $F(b) = P\{D_1 < b\}$ (note the strict inequality). As s increases, $F(x - s + \mu_k)$ is nonincreasing; so $1 - F(x - s + \mu_k)$ is nondecreasing and condition (8-20) is satisfied. For (8-19) to be valid [$\gamma_i(\cdot, 0) - \gamma_i(\cdot, 1)$ nondecreasing for each i]

$$\gamma_x(s, 0) - \gamma_x(s, 1) = F(x + \mu_1 - s) - F(x + \mu_0 - s) \tag{8-24}$$

must be nondecreasing in s. Recall $\mu_1 \ge \mu_0$ and suppose, for the moment, that F has a density ϕ. Then the following argument shows that (8-24) is nondecreasing in s (for all x and $\mu_1 \ge \mu_0$) if $\phi(\cdot)$ is nondecreasing. Differentiate the right side of (8-24) with respect to s to obtain $\phi(x + \mu_0 - s) - \phi(x + \mu_1 - s)$, which is nonnegative if $\phi(\cdot)$ is nonincreasing. Also, (8-24) can be nonincreasing in s in cases when D is a discrete r.v.

The costs in this problem are a production cost w_k on a day when the fruit-processing rate is μ_k and an expected deterioration cost $h_k(s)$ if $a_n = k$ and $s_n = s$. Therefore, $c^k_s = w_k + h_k(s)$; so $c^k_s \le c^k_{s+1}$ if $h_k(\cdot)$ is nondecreasing. Condition (8-18) in Proposition 8-1 [submodularity of $\{c^k_s\}$] is equivalent to submodularity of $\{h_k(s)\}$:

$$h_0(s + 1) - h_0(s) \ge h_1(s + 1) - h_1(s) \tag{8-25}$$

Exercise 8-11 asks you to verify that this condition is satisfied if $h_k(s)$ is the expected value of a convex function of the average of beginning and ending storage levels, that is,

$$h_k(s) = E\left[H\left(s - \frac{\mu_k - D_1}{2}\right)\right]$$

with H convex on $\mathbb{R}_+$. □

Example 8-23: Stopping Recall in Example 8-19 that $c_0^0 = c_0^1 = 0$ and $p_{s0}^0 = p_{00}^1 = 1$ for all s. Therefore, $\gamma_i(s, 0)$ does not vary with s; so conditions (8-19) and (8-20) are satisfied if $\sum_{j \geq i} p_{sj}^1$ is the same for all s. This is the case in "house hunting" and some other classic search problems such as the "secretary problem." Suppose that the ratings of candidates for a secretarial position are i.i.d. r.v.'s and that candidates are interviewed and tested sequentially. Then p_{sj}^1 does not depend on s. If (8-18) is valid too, Proposition 8-1 is applicable and (8-16) is said to be a *monotone stopping problem.*

In general, even if (8-17) through (8-20) do not hold, stopping problems are simpler than suggested by the form of (8-16). Here

$$f_n(s) = \min \left\{c_s^0,\, c_s^1 + \beta \sum_j p_{sj} f_{n-1}(j)\right\} \tag{8-26}$$

where $p_{sj} \equiv p_{sj}^1$. Continuing with (8-26),

$$\begin{aligned} f_n(s) - c_s^0 &= \min \left\{0,\, c_s^1 - c_s^0 + \beta \sum_j p_{sj} f_{n-1}(j)\right\} \\ &= \min \left\{0,\, c_s^1 - c_s^0 + \beta \sum_j p_{sj}[f_{n-1}(j) - c_j^0] + \beta \sum_j p_{sj} c_j^0\right\} \end{aligned}$$

Let $g_n(s) = f_n(s) - c_s^0$ for $n > 0$, $g_0(s) = -c_s^0$, and

$$h_s = c_s^1 - c_s^0 + \beta \sum_j p_{sj} c_j^0$$

Then

$$g_n(s) = \min \left\{0,\, h_s + \beta \sum_j p_{sj} g_{n-1}(j)\right\} \tag{8-27}$$

Sometimes the terminology *entry fee* and *exit fee* is used for c_s^1 and c_s^0. In this sense, the general stopping problem with entry and exit fees can be reduced to an equivalent stopping problem, namely, (8-27), with only entry fees. This is called an *entry-fee stopping problem.* Incidentally, monotonicity of $f_n(\cdot)$ does not cause $g_n(\cdot)$ to be monotone.

Some entry-fee stopping problems possess myopic optimal policies. In (8-27), suppose $S^* \triangleq \{s\colon h_s \geq 0\}$ is an absorbing set under $\{p_{ij}\}$, that is,

$$\sum_{j \in S^*} p_{ij} = 1 \qquad \text{for all } i \in S^* \tag{8-28}$$

Then it would be foolish to "continue," i.e., take action $k = 0$, at any state

$s \in S^*$, because $h_j \geq 0$ (versus the exit fee 0) for all states j that will (with probability 1) ever be reached. On the other hand, $h_s < 0$ at any $s \in S - S^*$; so it would be foolish to stop at such a state and forgo $h_s < 0$ in favor of 0 thereafter. This heuristic argument could be made precise except for the nuisance of $g_0(s) = -c_s^1$. Exercise 8-12 asks you to verify that if $g_0(s)$ is altered to 0, it can be made precise. For large enough values of n, of course, the effect on $g_n(\cdot)$ of $g_0(s) = -c_s^0$ versus $g_0(\cdot) \equiv 0$ is negligible. Stopping problems that satisfy (8-28) are said to be *absolutely monotone*. □

EXERCISES

8-9 In the version of Example 8-16 with probabilistic failure, suppose (8-23) is valid. Verify that $\{c_s^k\}$ is nondecreasing in s, for $k = 0$ or 1.

8-10 Prove that (8-24) is nondecreasing in s if:

(*a*)
$$F(r) = \begin{cases} 0, & r < 0 \\ Lr, & 0 \leq r \leq b \\ Lb + (1 - Lb)\{1 - \exp[-\lambda(r - b)]\} & b < r \end{cases}$$

Here, D has a density consisting of a uniform portion followed by an exponential tail.

(*b*) $\mu_1 - \mu_0 \geq 1$ and D is a geometric r.v. Why is it necessary to assume $\mu_1 - \mu_0 \geq 1$?

8-11 Verify (8-25) if $h_k(s) = E[H(s - \mu_k/2 + D_1/2)]$ and $H(\cdot)$ is a convex function on $\mathbb{R}_+$.

8-12 Consider (8-27) with $g_0(\cdot) \equiv 0$. Prove that

$$a_n = \begin{cases} 0 & \text{if } s_n \notin S^* \\ 1 & \text{if } s_n \in S^* \end{cases}$$

is optimal for all n.

8-13 In Example 8-18, suppose the amounts of fruit bought on successive days, $D_1, D_2, \ldots$ is a Markov chain with transition probabilities $\{q_{mr}\}$. Then the deterioration cost and distribution of the current input are conditioned by the most recent input. Let $h_k(s, m)$ denote the expected deterioration cost on day n if $a_n = \mu_k$, $s_n = s$, and $D_{n-1} = m$. Then the state is (s_n, D_{n-1}) rather than s_n alone and (8-15) becomes

$$f_n(s, m) = \min\{J_n(s, m, k): k \in \{0, 1\}\}$$

$$J_n(s, m, k) = w_k + h_k(s, m) + \beta \sum_j q_{mj} f_{n-1}(s + j - \mu_k, j)$$

Under what conditions are there numbers $\{V_n(m)\}$ such that an optimal policy is given by

$$a_n = \begin{cases} 0 & \text{if } s_n < V_n(D_{n-1}) \\ 1 & \text{if } s_n \geq V_n(D_{n-1}) \end{cases}$$

Prove that your conditions are sufficient.

8-14 (*a*) In Example 8-18, in reality $s_n \geq 0$ for all n. Suppose you alter the model to $s_{n+1} = (s_n + D_n - \mu_{a_n})^+$. Under what conditions are (8-19) and (8-20) valid?

(*b*) In Example 8-18, let X_n denote the amount of fruit that could be processed on day n without regard to the constraint $s_n \geq 0$. Suppose X_n and D_n are independent r.v.'s with the distribution of X_n depending only on μ_{a_n}. Let $r_j(k) = P\{X_n = j \mid a_n = k\}$. Specify p_{sj}^k in this altered model and determine sufficient conditions for (8-19) and (8-20) to be valid.

8-15 Suppose in Example 8-17 (Maintenance) that there is a delay of T periods each time an item is overhauled. In each case below, specify the form taken by the dynamic program (8-15) and deduce sufficient conditions for the conclusion of Proposition 8-1 to be valid.

(*a*) T is a (constant) positive integer.

(*b*) The delays $T_1, T_2, \ldots$ following successive overhauls are i.i.d. geometric r.v.'s with $P\{T_i = k - 1\} = (1 - q)q^k$ for $k \in I_+$.

(*c*) The delays $T_1, T_2, \ldots$ following successive overhauls are i.i.d. r.v.'s with $P\{T_i = k\} = q_k$ and $\sum_{k=0}^{m} q_k = 1$ for some $m < \infty$.

8-16 A "Dutch auction" consists of a sale, over time, with periodic price reductions. Consider a sale of a single item. Each day, if the item has not yet been sold, either it will remain at its present price or the price will be reduced by \$10. Both outcomes are equally likely. The initial price is \$40. If the item is unsold by the time the price drops to \$10, the price remains at \$10 until the item is purchased. Suppose that you want to obtain this item but you know that each day you choose not to buy it, the probability is 0.1 that it will be sold before you return next day. It costs you \$1 in carfare each day that you return. The item is worth \$40 to you. You wish to maximize the expected total value, which is

$$\begin{aligned} &\$40 - \text{purchase price} - \text{number of trips} && \text{if you buy it} \\ &\qquad\qquad\qquad\qquad\; - \text{number of trips} && \text{if you do not.} \end{aligned}$$

This is an item you wish to buy; so you do not consider such actions as ignoring the Dutch auction; you return each day until either you buy it or it is gone. Therefore, each day until the item is sold, the two actions open to you are action 1, buy it, and action 0, wait another day. This problem has a discounted objective with $\beta = 1 - 0.1 = 0.9$ because a discount factor can be interpreted as a (conditional) probability of survival until tomorrow if you have survived until today. What policy is optimal in the Dutch auction that you confront? Justify your answer.

8-3* RESOURCE MANAGEMENT

In many resource-management contexts, the timing of consumption is subject to some degree of choice. In a fishery, you may catch a fish now or defer its withdrawal from the water while it gains weight and breeds to produce more fish. If you bring a huge amount to market in a short period of time, the market price may drop. But some fraction of the fish whose harvest you postpone will die naturally in the meanwhile. Also, other things being equal, you prefer catching a fish now and receiving its revenue immediately rather than waiting until next year to catch it and be paid for it. Therefore, you choose to catch fish up to the point where your discounted expected benefit from forgoing a fish would be just offset by the immediate additional benefit of catching it now.

This argument can (and will) be made rigorous, but it is laced with assumptions of concavity, attitude to risk, and intertemporal preference. It exemplifies the consumption-investment trade-off frequently found in resource-management contexts. Also, it is typical of the management of monetary investments. The exposition here is in terms of fishery management, but there will be intermittent references to other applications of the model. The pertinent fishery decision is the size of the season's catch when the amount of fish in the sea, i.e., the state, is given. The petroleum counterpart is the amount of oil to pump up. The capital-management decision is the amount of your investment portfolio that you should liquidate. In lumbering, it is the amount of trees that you should cut. In reservoir regulation, it is the amount of water that you should discharge.

The decision problem in each of these contexts is more complex than has been suggested. In petroleum, for example, you would coordinate your pumping and exploration policies. In a fishery, you would coordinate your policy for the amount of fish you caught with your policy for the minimum size at which fish may be caught (the reproductive potential of a ton of fish depends on the ages and sizes of the fish). The latter decision is implicit in the choice of a minimum net mesh size. There are several reasons why the basic model in this section is devoid of such complexity. First, some technological decisions can be added to the model without destroying the validity of the conclusions drawn from the basic model. Indeed, those conclusions often facilitate the analysis of more complicated models. Second, in practice it is often necessary to use highly aggregated models to obtain insights quickly enough to act effectively. Third, some of these other decisions are static rather than dynamic. In a fishery, the amount that can be caught is a regulatory decision that typically varies from year to year. But minimum net mesh size is almost statutory in the difficulty that an agency would encounter in trying to change it. Nets represent large investments (a purse seine used by a large tuna boat may cost \$100,000) and boat owners would block frequent changes in mesh size. We regard mesh size as a static decision problem in most fisheries, whereas catch size is a dynamic problem. The minimum size in a lobster or crab fishery, however, could well be a dynamic problem. Fourth, the admittedly aggregated model here permits more detail than the models that are actually used in many resource-management procedures. Last, aggregated models are simpler to explain than their more complicated counterparts. You enhance the likelihood that your analytical insights will alter managerial practice if you can communicate the intuition behind those insights.

Fisheries have natural seasonal features, and it is typical for a management agency to promulgate a catch limit for each "fishing year." At the beginning of the nth fishing year, $n \in I_+$, let s_n denote the biomass (i.e., weight of living organisms) of the fish species at issue. Typically, the length of a fishing season is determined by fish schooling behavior and weather conditions; let a_n denote the biomass at the end of the season. Suppose $z_n = s_n - a_n$ is the amount caught. A more detailed model would reflect the fact that some fish die naturally during the fishing season while the survivors increase their biomass and breed to some extent. These details become more important when z_n/s_n may be a relatively large fraction, say for anchovies, than if the fraction is lower, say for longer-lived species such as cod and salmon. To encompass such details, the constraint

$$a_n \in A_{s_n} \qquad n \in I_+ \tag{8-29}$$

is assumed. Typically, $A_s = [0, s]$, but the 0 lower bound is not taken seriously.

Suppose that successive biomasses are connected by

$$s_{n+1} = M(a_n, D_n) \tag{8-30}$$

where $D_1, D_2, \ldots$ are i.i.d. r.v.'s, D is a generic D_i, and M is a real-valued function on the cartesian product of $\mathbb{R}_+$ with the sample space of D. If a_n is the basis from which breeding and growth lead to a_{n+1}, the model $M(a, b) = m(a)b$ is likely to be appropriate. If the species migrates in and out of the fishery to some extent, D may be a 2-vector and $M(a, b) = ab' + b''$, where $b = (b', b'')$.

In a fishery, $s_n \in \mathbb{R}_+$ is reasonable but other contexts will suggest tighter constraints. Therefore, $s_n \in \mathcal{S} \subset \mathbb{R}$ is assumed with $\mathcal{S}$ a convex set. Let

$$\mathscr{C} = \{(s, a): a \in A_s, s \in \mathcal{S}\}$$

Let $G(s, a)$ denote the expected benefit (net of fishing expense) in any fishing year, where $s_n = s$ and $a_n = a$, $(s, a) \in \mathscr{C}$. In a fishery, a detailed model would show that $G(s, a)$ is the sum of separate functions of $s - a$, that is, the catch size, and of a (or s). The revenue stems from $s - a$, but the cost of catching $s - a$ depends on the residual density of fish in the water, namely, on a itself. We say "expected benefits" because the revenue depends on price level, which is affected by phenomena external to the fishery. Other contexts lead to different forms for G. Finally, let β denote the single-period discount factor.

It follows from Theorem 4-1 that the existence of an optimal policy is equivalent to the supremum always being attained in this recursion:

$$f_n(s) = \sup \{J_n(s, a): a \in A_s\} \qquad s \in \mathcal{S} \tag{8-31a}$$

$$J_n(s, a) = G(s, a) + \beta E(f_{n-1}[M(a, D)]) \qquad (s, a) \in \mathscr{C} \tag{8-31b}$$

for $n \in I_+$, where $f_0(\cdot) \equiv 0$. We interpret $f_n(s)$ as the maximal sum of expected discounted benefits during n years if the initial biomass is s. The appropriate catch size in the first of the n years is $s - a$, where a is any element of A_s that attains the supremum in (8-31a).

The Model in Other Contexts

Example 8-24 Consider the task of planning economic development in a region with limited "proven" reserves of an exhaustible resource such as coal or petroleum. Economists have analyzed versions of (8-31) augmented by an additional decision x_n (each period), which is the magnitude of the exploratory effort to discover additional deposits of the resource. In the augmented model, the immediate expected net benefit $G(s, a, x)$ is adversely influenced by the cost of exploration (and development), and next period's proven reserves, s_{n+1}, is a random function $M(a_n, x_n, D_n)$ of the form $a_n + \phi(x_n, D_n)$. Here, $A_s = [0, s]$ (although the 0 lower bound is not taken seriously). □

Example 8-25 Suppose a reservoir experiences i.i.d. inflows $D_1, D_2, \ldots$ in successive periods. Let s_n denote the *freeboard* at the start of period n (the freeboard is the volume of space in the reservoir). If K is the reservoir capacity, $K - s_n$ is the amount of water in storage. It follows that s_{n+1} is given by (8-30), where

$$M(a, b) = (a - b)^+ \wedge K$$

If $D_n = b > a_n = a$, the reservoir overflows; so freeboard is zero at $n + 1$, that is, $s_{n+1} = 0$. If $b = D_n < 0$ because of evaporation and leakage, $a - b > K$ is possible; in that case the reservoir is dry; that is, $s_{n+1} = K$. The expected net benefit, again, is likely to be the sum of a function of the amount discharged, $s - a$, and the residual level, $K - a$. Usually, $A_s = [s, K]$. □

Example 8-26 A less developed country faces a painful choice as it programs its economic development. It must compromise between the production of consumption goods, typically in short supply, and the expansion of its capital base in order to be able to produce and consume more goods in the future. Greater consumption now would postpone the date by which substantially greater consumption would be possible. A comparable problem would be faced by an individual with a savings program except for the important effects of bequest motives and tax structures. If the individual reinvests a_n, having consumed $s_n - a_n$, the subsequent level of wealth is s_{n+1}, a random function (interest, dividends, etc.) of a_n. The benefits of consuming $s_n - a_n$, by either the country or the individual, are given by $G(s_n, a_n)$. In general, G may be a function of a_n (or of s_n) as well as of $s_n - a_n$ because of the effects on preferences of different levels of wealth. □

The generic problem, for either an individual or a country, is called *capital accumulation*. Many economists and financial theorists have analyzed versions of (8-31) but for different reasons than in the fishery context. "The" capital accumulation problem is not, we believe, a useful basis for deciding how much of your investment portfolio to liquidate each year or the amount of consumption goods to produce in a growing economy. Instead, capital-accumulation models are analyzed because they may be reasonably good qualitative descriptions of behavior. Economists and financial theorists use the models to enhance their understanding of consumption-investment behavior rather than to guide it. In the fishery context and in other managed biological systems, versions of (8-31) are useful guides, and they influence managerial decisions in practice. Also, (8-31) is useful for reservoir regulation when the model is augmented to include dependence among the inflows.

Economists call a fishery a *renewable resource*, whereas they label a coal mine an *exhaustible resource*. Renewable resources whose management can be assisted by versions of (8-31) include livestock ranches, poultry farms, and timber holdings.

Example 8-27 Many inventory and production models are special cases of (8-31). The inventory model in Section 3-1, for example, has, $\mathcal{S} = \mathbb{R}_+$, $A_s = [s, \infty)$,

$$G(s, a) = -c(a - s) - G(a)$$

and

$$M(a, D) = (a - D)^+$$

□

The remainder of this section develops properties of policies that are optimal in (8-31). Assumptions are listed, and then (8-31) is analyzed for $n < \infty$. The infinite-horizon case is considered in Sections 8-5 and 8-6.

Assumptions

$\mathscr{C}$ is a convex set and G is concave and continuous on $\mathscr{C}$ (8-32a)

For each a, $G(\cdot, a)$ is nondecreasing on $\{s: (s, a) \in \mathscr{C}\}$ (8-32b)

For each b, $M(\cdot, b)$ is continuous and concave on

$$A \triangleq \bigcup_{s \in \mathcal{S}} A_s \tag{8-32c}$$

G is supermodular and continuous on $\mathscr{C}$, which is a lattice (8-33)

$\{A_s: s \in \mathcal{S}\}$ is expanding (i.e., if $s' > s$, then $A_{s'} \supset A_s$) and ascending and A_s is a nonempty compact set for each $s \in \mathcal{S}$ (8-34)

These assumptions are neither always in force nor, when some are in force, are they always made together. Each statement of a result will list the assumptions being made.

An important case of (8-32), (8-33), and (8-34) is $\mathcal{S} = \mathbb{R}_+$, $A_s = [0, s]$, $G(s, a) \equiv g(s - a)$ with g concave and nondecreasing on $\mathbb{R}_+$, and either $M(a, b) \equiv ab$ or $\equiv a + b$. Then $\mathscr{C}$ is the lower triangle in the nonnegative orthant, which is a convex set. If a fishery model had this structure, concavity of $g(\cdot)$ would correspond to diminishing market price and increasing unit production cost as the quantity of fish brought to market increases. If net profit is a nondecreasing function of the quantity of fish brought to market, then $g(\cdot)$ is nondecreasing. And if $M(a, b) \equiv ab$, the residual fish population a is the entire base from which next year's catch can be taken; next year there will be b units of fish for each residual unit this year.

Optimal Policies

Theorem 8-6 Assumptions (8-32) and (8-34) imply for each n that f_n is continuous, concave, and nondecreasing on $\mathcal{S}$.

Proof To begin an inductive proof of concavity, $f_0(\cdot) \equiv 0$ is trivially concave and nondecreasing. For $n \geq 1$ suppose f_{n-1} is concave and nondecreasing; then (8-32a) and (8-32c) imply concavity of

$$G(s, a) + \beta f_{n-1}[M(a, b)]$$

for each outcome b of D_n. Therefore, Proposition B-2 in Appendix B yields concavity of J_n on $\mathscr{C}$ for each n and Proposition B-4 asserts concavity of f_n on $\mathcal{S}$. The fact that $f_n(s) > -\infty$ for all $s \in \mathcal{S}$ follows from continuity of $J_n(s, \cdot)$ on the compact set A_s [assumed in (8-34)].

Exercise 8-17 asks you to verify that f_n is necessarily continuous. Finally, it is nondecreasing: let $s \in \mathcal{S}$ and $\gamma > 0$ so

$$\begin{aligned} f_n(s+\gamma) &= \sup \{J_n(s+\gamma, y): a \in A_{s+\gamma}\} \\ &\geq \sup \{J_n(s+\gamma, a): a \in A_s\} \\ &\geq \sup \{J_n(s, a): a \in A_s\} = f_n(s) \end{aligned}$$

The first inequality is due to (8-34) and the second to (8-32b) and the definition $J_n(s, a) = G(s, a) + \beta E(f_{n-1}[M(a, D)])$. □

For $u \in \mathbb{R}$ and $K \subset \mathbb{R}$ let $u - K = \{z: u - z \in K\}$. Suppose s labels the amount of fish in the fishery and A_s is the set of feasible amounts of fish remaining in the fishery at the end of the season. Then $s - A_s$ is the set of feasible seasonal catches. In fact, $z \in s - A_s \Leftrightarrow s - z \in A_s$. Note that (8-33) and (8-34) do not imply that $s - A_s$ is ascending† on $\mathcal{S}$. For example, if $A_s = [0, 2s]$ and $\mathcal{S} = \mathbb{R}_+$, then $s - A_s = [-s, s]$, which is not ascending. However, if $A_s = [0, s]$, then $s - A_s = [0, s]$ too.

Theorem 8-7 (a) Assumptions (8-32), (8-33), and (8-34) imply for each $n \in I_+$ and $s \in \mathcal{S}$ that there is $a_n(s) \in A_s$ that is optimal in (8-31) and

$$a_n(s) \leq a_n(s') \qquad \text{if } s \leq s' \qquad s \in \mathcal{S} \text{ and } s' \in \mathcal{S} \tag{8-35}$$

(b) Let

$$\{s - A_s : s \in \mathcal{S}\} \text{ be ascending} \tag{8-36a}$$

and

$$G(s, a) \equiv Q(s - a) + L(s) \tag{8-36b}$$

with Q and L concave on $\mathbb{R}_+$ and $\mathcal{S}$, respectively. Then for all $s \in \mathcal{S}$ and $\gamma > 0$ with $s + \gamma \in \mathcal{S}$,

$$0 \leq a_n(s+\gamma) - a_n(s) \leq \gamma \tag{8-37}$$

Proof To apply Theorem 8-5, observe that (8-33) implies $\{A_s: s \in \mathcal{S}\}$ is ascending (the section of a lattice is necessarily ascending). Also $f_{n-1}(\cdot)$ and $E(f_{n-1}[M(\cdot, D)])$ are trivially supermodular because they are functions of one real variable. Therefore, with (8-33), J_n is supermodular on $\mathscr{C}$; so part (b) of Corollary 8-5 asserts (a). For (b), i.e., the right inequality in (8-37), make the change of variable $z = x - y$ and rewrite (8-31) as

$$f_n(s) = \sup \{H_n(s, z): z \in s - A_s\} \tag{8-38a}$$

$$H_n(s, z) = Q(z) + L(s) + \beta E(f_{n-1}[M(s - z, D)]) \tag{8-38b}$$

If H_n is supermodular, then part (a) of Corollary 8-5 asserts existence of $z_n(s)$

† Recall that A_s is said to be ascending on $\mathcal{S}$ if $s_1 \leq s_2$ and $b_j \in A_{s_j} \Rightarrow b_1 \wedge b_2 \in A_{s_1}$ and $b_1 \vee b_2 \in A_{s_2}$.

optimal in (8-38a) such that $z_n(s) \le z_n(s + \gamma)$. But $s - a_n(s)$ is optimal in (8-38), i.e., $z_n(s) = s - a_n(s)$; so monotonicity of $z_n(\cdot)$ implies the right inequality in (8-37). To verify that H_n is indeed supermodular, use concavity and monotonicity of $f_{n-1}(\cdot)$ and concavity of $M(\cdot, b)$ for each b to establish that H_n has isotone differences. Therefore, part (b) of Theorem 8-1 implies supermodularity of H_n. □

Theorem 8-7 confines the set of possible optimal policies, but a much stronger result is valid if $G(s, a) = \rho \cdot (s - a)$, where ρ is the unit profit. In this case, a single parameter characterizes $a_n(\cdot)$ for each n. The essential feature of $G(s, a) = \rho \cdot (s - a)$ is additive separability, which was exploited in Chapter 3.

Theorem 8-8 Suppose $\mathscr{C}$ is a convex set, L and K are nondecreasing continuous functions that are concave on $\mathcal{S}$ and convex on A, respectively, assumption (8-32c) [for each b, $M(\cdot, b)$ is continuous and concave on A] is valid, $\mathcal{S} \subset \mathbb{R}_+$, m is a nonnegative, nondecreasing, and concave function on $\mathcal{S}$,

$$G(s, a) = L(s) - K(a) \qquad (s, a) \in \mathscr{C} \quad \text{and} \quad A_s = [0, m(s)] \qquad s \in \mathcal{S} \qquad (8\text{-}39)$$

and $$K'(a) < \beta L'(0) E[M^{(1)}(0, D)] \qquad \text{for sufficiently large } a \qquad (8\text{-}40)$$

Then for each n there exists a_n^*, specified below, such that an optimal policy is $a_1(\cdot) \equiv 0$ and

$$a_n(s) = m(s) \wedge a_n^* \qquad s \in \mathcal{S} \qquad n > 1 \qquad (8\text{-}41)$$

Proof Let B denote the sum of discounted benefits. Then (8-39) and $s_{n+1} = M(a_n, D_n)$ yield

$$\begin{aligned} B &= \sum_{n=1}^{N} \beta^{n-1} G(s_n, a_n) = \sum_{n=1}^{N} \beta^{n-1}[L(s_n) - K(a_n)] \\ &= L(s_1) - K(a_1) + \sum_{n=2}^{N} \beta^{n-1}\{L[M(a_{n-1}, D_{n-1})] - K(a_n)\} \\ &= \sum_{n=1}^{N-1} \beta^{n-1}\{\beta L[M(a_n, D_n)] - K(a_n)\} + L(s_1) - \beta^{N-1}K(a_N) \end{aligned}$$

In the one-period problem ($N = 1$), $B = L(s_N) - K(a_N)$; so $a_N \equiv 0$ because $K(\cdot)$ is nondecreasing and 0 is the least element of A_{s_N}. Therefore, an optimal policy induces

$$E(B) = L(s_1) - \beta^{N-1}K(0) + E\left[\sum_{n=1}^{N-1} \beta^{n-1} w(a_n)\right]$$

where $$w(a) = \beta E(L[M(a, D)]) - K(a) \qquad a \ge 0$$

Let $v_0(\cdot) \equiv 0$ and, for $n \in I_+$,

$$v_n(s) = \sup\{u_n(a) : 0 \le a \le m(s)\} \qquad s \in \mathcal{S} \qquad (8\text{-}42a)$$

$$u_n(a) = w(a) + \beta E(v_{n-1}[M(a, D)] \qquad a \geq 0 \tag{8-42b}$$

From the discussion above, a policy $\{a_n(\cdot)\}$ maximizes $E(B)$ if, and only if, $a_1(\cdot) \equiv 0$ and $a_{n+1}(\cdot)$ attains the supremum in (8-42*a*), that is, $v_n(s) = u_n[a_{n+1}(s)]$.

Let a_n^* maximize $u_n(\cdot)$ on $\mathbb{R}_+$. The maximum is attained because Theorem 8-6 implies that $u_n(\cdot)$ is concave and continuous on $\mathbb{R}_+$. Also, Exercise 8-22 asks you to verify that (8-40) implies $w'(a) \to -\infty$ as $a \to \infty$ and $u_n'(a) \to -\infty$ as $a \to \infty$. Therefore, sup $\{u_n(a)\colon a \geq 0\}$ is attained. Finally, concavity of $u_n(\cdot)$ implies (cf. Exercise B-18) that (8-41) is the form of an optimal policy.

□

You should appreciate that (8-31*a*) and (8-31*b*) and (8-42*a*) and (8-42*b*) possess different numerical values. However, (8-42) is equivalent to the original problem in the sense that a policy optimal in (8-42) for $n > 1$ is optimal in the original problem.

Effects of the Planning Horizon

How does the length of the planning horizon influence an optimal policy and the "value" of an optimal policy? In our notation, how does n affect $a_n(\cdot)$ and $f_n(\cdot)$ in (8-31), and what happens as $n \to \infty$? These questions are slightly different, and for the first, it is convenient to assume

$$G(s, a) \geq 0 \qquad (s, a) \in \mathscr{C} \tag{8-43}$$

However, this is equivalent (cf. Exercise 4-13) to the assumption that G is uniformly bounded below on $\mathscr{C}$. The latter is reasonable in practice.

Theorem 8-9 Suppose (8-43) is valid. Also assume that

$$\text{for each } a,\ G(\cdot, a) \text{ is nondecreasing on } \{s\colon (s, a) \in \mathscr{C}\} \tag{8-32b}$$

$$G \text{ is supermodular and continuous on } \mathscr{C}\text{, which is a lattice} \tag{8-33}$$

$$M(\cdot, b) \text{ is nondecreasing for all } b \tag{8-44}$$

and

$$A_s \text{ is a nonempty compact set for each } s \in \mathbb{S} \tag{8-45}$$

Then

$$f_n(s) \leq f_{n+1}(s) \qquad s \in \mathbb{S} \tag{8-46}$$

$$a_n(s) \leq a_{n+1}(s) \qquad s \in \mathbb{S} \tag{8-47}$$

$$J_n(s, a + \lambda) - J_n(s, a) \leq J_n(s, a + \lambda) - J_n(s, a)$$

$$(s, a) \in \mathscr{C} \qquad \lambda > 0 \qquad (s, a + \lambda) \in C \tag{8-48}$$

$$f_n(s + \lambda) - f_n(s) \leq f_{n+1}(s + \lambda) - f_{n+1}(s)$$

$$s \in \mathbb{S} \qquad \lambda > 0 \qquad s + \lambda \in \mathbb{S} \tag{8-49}$$

PROOF For (8-46), only (8-43) is needed because

$$f_0(s) = 0 \le \sup \{G(s, a): a \in A_s\} = f_1(s) \qquad s \in \mathcal{S}$$

For any n, if $f_n(s) \ge f_{n-1}(s)$, $s \in \mathcal{S}$, then

$$f_{n+1}(s) = \sup \{G(s, a) + \beta E(f_n[M(a, D)]): a \in A_s\}$$
$$\ge \sup \{G(s, a) + \beta E(f_{n-1}[M(a, D)]): a \in A_s\} = f_n(s)$$

so (8-46) is valid for all n.

Observe that (8-48) and (8-49) assert that $J_n(s, a)$ is supermodular in (a, n) for each s, and $f_n(s)$ is supermodular in (s, n). *If* $J_k(s, a)$ is supermodular in (a, k) for each s and $k \le n + 1$, then part (*b*) of Corollary 8-5 implies (8-47). If $f_k(s)$ is supermodular in (s, k) for all $k \le n$, then (8-31*b*) and (8-44) imply $J_k(s, a)$ is supermodular in (a, k) for all $k \le n + 1$. Therefore, (8-47) and (8-48) rest on (8-49).

Assumption (8-32*b*) yields $f_1(s + \lambda) - f_1(s) \ge 0$ if $\lambda > 0$ (as in Theorem 8-6); so $f_0(\cdot) \equiv 0$ implies (8-49) is valid for $n = 0$. As an inductive assumption, if (8-49) is valid for all $n \le k - 1$, then (8-44) implies that (8-48) is valid for all $n \le k$. Therefore, Theorem 8-3 and (8-45) yield (8-49) for all $n \le k$. □

Some of the theorem's assertions are less intuitive than others. The first, $f_n(s) \le f_{n+1}(s)$, is a trivial consequence of the fact that $f_n(s)$ is the value of the maximum of n nonnegative summands. This monotonicity in n yields a limit of $f_n(s)$ as $n \to \infty$ (see page 422). The property $a_n(s) \le a_{n+1}(s)$ also is intuitive. It asserts that the optimal catch size $s - a_n(s)$ diminishes or stays constant as the horizon lengthens. Equivalently, a longer horizon warrants a greater "investment" $a_n(s)$ because more time is available to collect the proceeds. Indeed, $J_n^{(2)}(s, a) \le J_{n+1}^{(2)}(s, a)$ [(8-47) wherever $J_n^{(2)}$ and $J_{n+1}^{(2)}$ exist] asserts that the marginal net benefit increases with the horizon length for all a [not only $a = a_n(s)$ vs. $a = a_{n+1}(s)$]. Lastly, $f'_n(s) \le f'_{n+1}(s)$ [(8-48) wherever f'_n and f'_{n+1} exist] states that the returns to scale (s) diminish more slowly as the horizon length increases. That they diminish at all is equivalent to concavity of f_n and f_{n+1}.

EXERCISES

8-17 Establish continuity of f_n in Theorem 8-6. [Hint: use induction and a contrapositive proof and show that discontinuity of f_n at s implies existence of $a \in A_s$ with J_n discontinuous at (s, a).]

8-18 The analog of (8-31) for the exhaustible-resource model in Example 8-24 is

$$f_n(s) = \sup \{J_n(s, a, x) : a \in A_s,\ x \in H\} \qquad s \ge 0$$

$$J_n(s, a, x) = G(s, a, x) + \beta E(f_{n-1}[a + \phi(x - D)])$$

for $n \in I_+$, where $f_0(\cdot) \equiv 0$. The set of feasible exploration efforts H is usually of the form $[0, h]$ or $\mathbb{R}_+$. Suppose H is a compact convex set on which $G(s, a, \cdot)$ is nonincreasing for each $(s, a) \in \mathscr{C}$ and on which $\phi(\cdot, b)$ is nondecreasing and concave for each b.

(*a*) Prove Theorem 8-6 here [using modifications of (8-32) and (8-34)].

(*b*) Prove for each n and s that there is $x_n(s) \in H$ optimal such that $x_n(\cdot)$ is nonincreasing on $\mathcal{S}$ [use modifications of (8-32), (8-33), and (8-34) if necessary].

(*c*) Prove a version of Theorem 8-7 here [use modifications of (8-32), (8-33), and (8-34) where necessary).

8-19 Under the assumptions of part (*b*) of Theorem 8-7:

(*a*) Suppose $0 < a_n(s) < s$ for some $s \in \mathcal{S}$. Verify that $0 < a_n(s') < s'$ for all $s' \in \mathcal{S}$, $s' \geq s$.

(*b*) Let S° denote any open subset of $\mathcal{S}$ and let $a_n'(s)$ denote the right-hand derivative. Verify that $a_n'(s)$ exists and

$$0 \leq a_n'(s) \leq 1 \qquad s \in S^\circ$$

8-20 For the model of Theorem 8-8, suppose $M(a, b) = b\phi(a)$ and $P\{\phi(a^*)D \geq a^*\} = 1$, where a^* maximizes $w(\cdot)$ on A. Show that $a_n^* = a^*$ for all n.

8-21 Abandon all concavity and convexity assumptions for the model in Theorem 8-8, but suppose for all n there exists a_n^* which maximizes $u_n(\cdot)$ on $\mathbb{R}_+$. This generalization is important in some applications. For each $n > 1$, prove that there is an optimal policy $a_n(\cdot)$ which specifies $a_n(s) = a_n^*$ if $m(s) \geq a_n^*$.

8-22 In the proof of Theorem 8-8 there are definitions of w and for each n of u_n, both from $\mathbb{R}_+$ to $\mathbb{R}$. Prove $w'(a) \to -\infty$ and $u_n'(a) \to -\infty$ as $a \to \infty$ under the assumptions of Theorem 8-8. Hints: (1) Concavity and monotonicity assumptions imply $w'(a) \geq \beta L'(0)E[M^{(1)}(0, D)]$. Use (8-40). (2) If $u_n'(a) \to -\infty$, then

$$v_n(s) = \begin{cases} u_n[m(s)] & \text{if } m(s) < a_n^* \\ u_n(s_n^*) & \text{if } m(s) \geq a_n^* \end{cases}$$

which implies $u_{n+1}'(a) \to -\infty$.

8-4* CASH MANAGEMENT AND OTHER SMOOTHING PROBLEMS

Most firms, large and small, actively manage the balances of their "cash accounts," i.e., checking accounts. In some instances, the primary problem is to secure credit at appropriate times. In others, receipts are transferred to short-term investments such as U.S. Treasury notes. The following model concerns firms with established lines of credit and short-term riskless investment opportunities. It applies primarily to large firms because we assume that the costs of lending and borrowing are proportional to the amounts lent and borrowed. Smaller firms would encounter "fixed" or "setup" costs too. This complexity is investigated in Section 7-1; in particular, see Exercises 7-12 and 7-13.

Large firms review their cash positions at the end of each business day; so the length of a period in the following model is likely to be 1 day. Let s_n be the amount in the checking account at the end of day $n - 1$. At that point in time, an amount z_n is borrowed (if $z_n \geq 0$) or $-z_n$ is lent (if $z_n < 0$) and the transaction is completed immediately, i.e., before the start of the next business day. Let

$$a_n = s_n + z_n \tag{8-50}$$

denote the consequent altered cash position.

The dynamics are

$$s_{n+1} = a_n - D_n \tag{8-51}$$

where D_n is the sum of disbursements minus receipts during day n. Suppose D_1,

$D_2, \ldots$ are i.i.d. r.v.'s. This assumption is based on the realistic supposition that weekly, semiweekly, and monthly payrolls and other large disbursements known in advance are offset by deposits of the requisite amounts made for that sole purpose. Neither the z_n's nor the D_n's include these amounts. See Heyman (1973) for a model where large deterministic disbursements are considered explicitly.

The costs considered here are of two kinds, namely, transfer costs and expected costs associated with the cash position. Suppose that the unit costs of borrowing and lending are c^+ and c^-. The costs c^+ and c^- represent fees which a bank charges the firm for transferring funds between the checking account and forms of short-term investments. You might regard c^+ and c^- as brokerage fees.

When $s_{n+1} > 0$, the firm is losing income in the form of interest or other returns that the funds, in amount s_{n+1}, could have been earning if they had been invested. This forgone income is called an *opportunity cost*. If $s_{n+1} < 0$, the firm incurs an out-of-pocket expense in the form of interest it must pay the bank. Risk here is due to the fact that s_{n+1} itself is not chosen, only a_n is selected, and then $s_{n+1} = a_n - D_n$. Therefore, the actual cost at the end of day $n + 1$, discounted back to the end of day n, is some r.v. $g(a_n, D_n)$ that depends on a_n and D_n.

Recall the notation $(x)^-$ for $-x \wedge 0$. There is a useful identity

$$u \cdot (x)^+ + v \cdot (x)^- = |x| \frac{(u + v)}{2} + x \frac{(u - v)}{2} \tag{8-52}$$

where u, x, and v are real numbers. Then the total cost attributable to a_n, with $c = (c^+ + c^-)/2$, is

$$c^+(a_n - s_n)^+ + c^-(a_n - s_n)^- + g(a_n, D_n)$$

$$= |a_n - s_n| c + (a_n - s_n) \frac{c^+ - c^-}{2} + g(a_n, D_n)$$

The total discounted cost is

$$B \triangleq \sum_{n=1}^{\infty} \beta^{n-1} \left[|a_n - s_n| c + g(a_n, D_n) + (a_n - s_n) \frac{c^+ - c^-}{2} \right]$$

Use (8-51) to obtain

$$\sum_{n=1}^{\infty} \beta^{n-1}(a_n - s_n) = a_1 - s_1 + \sum_{n=2}^{\infty} \beta^{n-1}(a_n - a_{n-1} + D_{n-1})$$

$$= -s_1 + \sum_{n=1}^{\infty} \beta^n D_n + \sum_{n=1}^{\infty} \beta^{n-1}(1 - \beta) a_n$$

Therefore,

$$B = \sum_{n=1}^{\infty} \beta^{n-1} \left[|a_n - s_n| c + g(a_n, D_n) + a_n(1 - \beta) \frac{(c^+ - c^-)}{2} \right]$$

$$- \left(s_1 + \sum_{n=1}^{\infty} \beta^n D_n \right) \frac{c^+ - c^-}{2}$$

whose two last terms can be neglected for purposes of optimization.

Let

$$G(a) = a(1-\beta)\frac{c^+ - c^-}{2} + E[g(a, D)] \qquad a \in \mathbb{R} \tag{8-53}$$

so the objective, $E[B + (s_1 - \sum_{n=1}^{\infty} \beta^n D_n)(c^+ - c^-)/2]$, is

$$\sum_{n=1}^{\infty} \beta^{n-1} E[c|a_n - D_n| + G(a_n)]$$

This form induces the dynamic program†

$$f_n(s) = \inf\{J_n(a) + c|a - s| : a \in \mathbb{R}\} \qquad s \in \mathbb{R} \tag{8-54a}$$

$$J_n(a) = G(a) + \beta E[f_{n-1}(a - D)] \qquad a \in \mathbb{R} \tag{8-54b}$$

where $n \in I_+$ and $f_0(\cdot) \equiv 0$.

Usually, there are unit costs of opportunity and interest, r^+ and r^-, respectively, and $g(a, b) = r^+ \cdot (a - b)^+ + r^- \cdot (b - a)^-$. In this case, $E[g(\cdot, D)]$ is a convex function; so $G(\cdot)$ is convex too. Also, $c|a - s|$ is a convex function of (s, a) on $\mathbb{R}^2$; so inductive use of Proposition B-4, as in Section 8-3, yields convexity of $f_n(\cdot)$ and $J_n(\cdot)$ on $\mathbb{R}$, for all $n \in I_+$. Convexity leads to a simple form of optimal policy.

Theorem 8-10 If $c \geq 0$, $G(\cdot)$ is convex on $\mathbb{R}$, and $G(a) \to \infty$ as $|a| \to \infty$, then for each $n \in I_+$ there are numbers (or $\pm\infty$) $L_n \leq U_n$ such that an optimal policy in (8-54) is

$$a = \begin{cases} L_n & \text{if } s < L_n \\ s & \text{if } L_n \leq s < U_n \\ U_n & \text{if } U_n \leq s \end{cases} \tag{8-55}$$

Proof Convexity of G implies convexity of J_n, hence continuity, and $G(a) \to \infty$ as $|a| \to \infty$ implies the same property for J_n. Therefore, the infimum in (8-54a) is attained and there exists an optimal policy. Convexity implies (Propositions B-5 and B-6) existence of the one-sided derivative $J_n'(a)$, say from the left for specificity.

Let $\mathscr{D}_x^n(a)$ denote the left-hand (partial) derivative of $c|a - s| + J_n(a)$ with respect to a:

$$\begin{aligned} \mathscr{D}_s^n(a) &= -c + J_n'(a) \qquad a \leq s \\ &= c + J_n'(a) \qquad a > s \end{aligned} \tag{8-56}$$

Convexity and Proposition B-7 imply that a' is optimal in (8-54a) if, and only if,

$$\mathscr{D}_s^n(a)\left\{\begin{matrix}\leq\\ \geq\end{matrix}\right\}0 \qquad \text{if } a\left\{\begin{matrix}\leq\\ >\end{matrix}\right\}a' \tag{8-57}$$

† Recall from Section 8-1 that throughout this chapter we assume all expectations are finite.

Notice in (8-54a) and (8-54b) that n denotes the number of remaining decisions so $n \to n - 1$. In (8-51), by comparison, n denotes "clock time" so $n \to n + 1$.

From (8-56), if $a \le s$, then

$$\mathscr{D}_s^n(a)\{\lesseqgtr\}0 \qquad \text{if } J_n'(a)\{\lesseqgtr\}c$$

Define

$$U_n = \sup\{a\colon J_n'(a) \le c\} \tag{8-58a}$$

If $U_n \le s$, then (8-57) is valid with $a' = U_n$ so $a = U_n$ is optimal. Parallel reasoning, with

$$L_n = \sup\{a\colon J_n'(a) \le -c\} \tag{8-58b}$$

implies $a = L_n$ is optimal if $s < L_n$.

If $L_n \le s < U_n$, then (8-58) yields

$$-c \le J_n'(s) \le c$$

Convexity implies (Proposition B-11) that $J_n'(\cdot)$ is nondecreasing; so $a \le s$ yields

$$\mathscr{D}_s^n(a) = J_n'(a) - c \le J_n'(s) - c \le 0$$

while $a > s$ causes

$$\mathscr{D}_s^n(a) = J_n'(a) + c \ge J_n'(s) + c \ge 0$$

Therefore, $a' = s$ satisfies (8-57) and $L_n \le s < U_n$ implies $a = s$ is optimal in (8-54). Monotonicity of $J_n'(\cdot)$ and (8-58a) and (8-58b) imply $L_n \le U_n$. □

The combination of (8-50) and (8-55) shows that the cash balance is incremented by an amount $(L_n - s)^+$ and it is decremented by an amount $(s - U_n)^+$. At most one of these amounts can be positive because $L_n \le U_n$.

The unit transfer costs of changing the cash level are a principal feature of the cash-management model. These costs induce an optimal policy that "adjusts" the process, that is, $z_n \ne 0$, only if s_n is either "much" too low ($s_n < L_n$) or "much" too high ($s_n > U_n$). If s_n lies in an "acceptable" range ($L_n \le s_n \le U_n$), no adjustment occurs, that is, $z_n = 0$. Therefore, the transfer costs make it relatively unlikely that transfers will occur frequently. For this reason, these costs are termed *smoothing costs* and (8-54) is called a *smoothing problem.*

Other contexts have costs that are incurred when the rate of an activity is altered. The remainder of this section examines the effects of smoothing costs in congestion models and production models.

Adjusting the Number of Servers for a Queue

Consider a bank branch with w tellers and teller windows. If fewer than w retail customers are inside the branch, it might be prudent to close some windows and transfer their tellers to other activities. Later, if there are more customers than tellers, some windows may be reopened. The disadvantage of too frequent closings and openings is the excessive amount of wasted time while tellers set up and shut down their cash drawers. The opportunity costs associated with the wasted time are smoothing costs.

Suppose the number of customers in the branch is frequently compared with the number of tellers with open windows, say at times 1, 2, 3, ... Let s_n† and z_{n-1} denote the number of retail customers in the bank and the number of open teller windows, respectively, at the instant of the nth periodic review. At this instant, z_n is chosen and, we assume, a cost

$$c^+ \cdot (z_n - z_{n-1})^+ + c^-(z_{n-1} - z_n)^+$$

is incurred. Here, c^+ and c^- are the unit opportunity costs associated with opening and closing a window, respectively. From (8-52), this cost is

$$c|z_n - z_{n-1}| + (z_n - z_{n-1})\frac{c^+ - c^-}{2} \tag{8-59}$$

where $c = (c^+ + c^-)/2$.

Let D_n denote the number of customers who arrive during $[n, n+1)$ and suppose $D_1, D_2, \ldots$ are i.i.d. integer-valued r.v.'s. Exercise 8-29 asks you to verify that continuous-time processes that are compatible with this assumption include cumulative processes (cf. page 169 in Volume I) generated by either a Poisson process or a sequence of identical constants of length $1/k$ for some $k \in I_+$. We assume that service times are constant, each one unit in length, and that customer services are initiated only at times 0, 1, 2, ... Exercise 8-35 removes this restriction and permits service times to be exponentially distributed. Therefore, $s_n \wedge z_n$ is the number of customers served during $[n, n+1]$; so the dynamics are given by

$$s_{n+1} = s_n + D_n - s_n \wedge z_n \tag{8-60}$$

The costs during $[n, n+1)$ include the smoothing costs, an expense $r(z_n)$ due to the number of tellers with open windows, and an imputed expense $h(s_n, z_n)$ caused by the delays experienced by waiting customers. The total discounted cost is

$$B \triangleq \sum_{n=1}^{\infty} \beta^{n-1}\left[c|z_n - z_{n-1}| + (z_n - z_{n-1})\frac{c^+ - c^-}{2} + r(z_n) + h(s_n, z_n)\right] \tag{8-61}$$

The identity

$$\sum_{n=1}^{\infty} \beta^{n-1}(z_n - z_{n-1}) = -z_0 + \sum_{n=1}^{\infty} \beta^{n-1}(1-\beta)z_n$$

yields

$$B = -z_0 + \sum_{n=1}^{\infty} \beta^{n-1}[c|z_n - z_{n-1}| + G(s_n, z_n)] \tag{8-62}$$

† The notation in this subsection is analogous to (but different from) the notation for the cash-balance model. Here, s_n is the number of customers and there it is the cash balance; here z_n is the action, namely, the number of open windows, and there it is the increment to the cash balance; etc.

where† $$G(s, z) = (1 - \beta)z \frac{c^+ - c^-}{2} + r(z) + h(s, z) \tag{8-63}$$

The domain of G is the nonconvex set $I \times A$, where $A = \{0, 1, \ldots, w\}$. In order to exploit the properties of convex functions, we adopt the following definition: *A function $G(\cdot)$ on a nonconvex domain X is said to be a convex function if there is a convex function $G^*(\cdot)$ on the convex hull X^* of X such that $G^*(x) = G(x)$ for all $x \in X$.* With this definition, Exercise 8-34 asks you to show that $G(\cdot, \cdot)$ on $I \times A$ is a convex function if it has the following properties:

(*a*) $G(s + 1, z) - G(s, z)$ is a nondecreasing function of $s \in I$ (for each $z \in A$).

(*b*) $G(s, z + 1) - G(s, z)$ is a nondecreasing function of $z \in \{0, 1, \ldots, w - 1\}$ (for each $s \in I$).

(*c*) $G(s + 1, z) - G(s, z)$ is a nondecreasing function of $z \in A$ (for each $s \in I$).

Condition (*a*) causes $G(\cdot, z)$ not to lie above the chord connecting any two of its points (where defined), (*b*) causes $G(s, \cdot)$ not to lie above any of its chords, and (*c*) causes $G(s + 1, z + 1)$ not to lie below the plane that connects $G(s, z)$, $G(s + 1, z)$, and $G(s, z + 1)$.

Underlying Cost Models

In (8-63), usually $r(z)$ is of the form $r \cdot z$ so that the convexity of G rests on properties of $h(\cdot, \cdot)$. As you saw in Theorem 8-10, convexity can be used to deduce the structure of an optimal policy. What underlying cost models yield convexity of $h(\cdot, \cdot)$? Suppose $\rho(s)$ is the cost per unit time when s customers are inside the bank. Let γ be the instantaneous discount rate that yields a periodic rate of β, that is, $\beta = e^{-\gamma \cdot 1}$ so $\gamma = \log \beta^{-1}$. Then

$$h(s, z) = E\left\{\int_0^1 \rho[s(1 + x)]e^{-\gamma x}\, dx \,|\, s(1) = s,\ z_1 = z\right\} \tag{8-64}$$

where $s(n)$ denotes the number of customers in the bank at epoch n; $s(n) = s_n$ if $n \in I_+$. Note that (8-64) does not depend on z; so convexity of $\rho(\cdot)$ implies convexity of $h(\cdot, \cdot)$ in this case.

In practice, instead of (8-64) one of the following approximations might be used for $h(s, z)$:

$$\rho(s)\int_0^1 e^{-\gamma x}\, dx$$

or
$$\rho(s)\int_0^{1/2} e^{-\gamma x}\, dx + \int_{1/2}^1 e^{-\gamma x}\, dx E[\rho(s + 1 - s \wedge z)] \tag{8-65}$$

If $\rho(\cdot)$ is nondecreasing and convex, then both these expressions are convex in (s, z). Expression (8-65) approximates the right side of (8-64) with‡ $s(n) = s_{\lfloor n \rfloor}$ if $\lfloor n \rfloor \le n < \lfloor n \rfloor + 1/2$ and $s(n) = s_{\lfloor n \rfloor + 1}$ if $\lfloor n \rfloor + 1/2 \le n < \lfloor n \rfloor + 1$. Exercise 8-35

† The functions $G(a)$ in (8-53) and $G(s, z)$ in (8-63) are different.

‡ Recall our notation $\lfloor u \rfloor$ for the largest whole number which does not exceed u.

asks you to show that the substitution of (8-65) for $h(s, z)$ in (8-61) leads to a simple expression for B similar to (8-62).

Return to the General Case of Adjusting the Number of Servers

Return to the general case and consider the minimization of $E(B)$ without the nuisance term $-z_0$ in (8-62). Recall $A = \{0, 1 \ldots, w\}$ and note that $s - s \wedge z = (s - z)^+$. Then the dynamic program that corresponds to (8-60) and (8-62) is†

$$f_n(s, z) = \min \{J_n(s, a) + c|a - z| : a \in A\} \qquad (s, z) \in I \times A \qquad \text{(8-66}a\text{)}$$

$$J_n(s, a) = G(s, a) + \beta E\{f_{n-1}[D + (s - a)^+, a]\} \qquad (s, a) \in I \times A \quad \text{(8-66}b\text{)}$$

for $n \in I_+$, where $f_0(\cdot) \equiv 0$. Observe in (8-66*a*) and (8-66*b*) that (s_n, z_{n-1}) is the nth state of this MDP. The costs during $[n, n + 1)$ depend on z_{n-1} via $c|z_n - z_{n-1}|$ and on s_n via $G(s_n, z_n)$. Also, the transition probabilities depend on s_n via (8-60).

It is instructive to compare (8-54) and (8-66). In the former, the state is a single variable s and J_n depends only on the action taken, a. In the latter, the state is a pair (s, z) and J_n depends on s as well as a. As a result, a policy optimal for (8-66) is more complicated than (8-55) in Theorem 8-10 in the sense that L_n and U_n become functions of s.

Theorem 8-11 Suppose $c \geq 0$, G is convex on $I \times A$, and $G(\cdot, a)$ is nondecreasing on I for each $a \in A$. Then for each $n \in I_+$ and $s \in I$ there are numbers (or $\pm\infty$) $L_n(s) \leq U_n(s)$ such that an optimal policy in (8-66) is

$$y = \begin{cases} L_n(s) & \text{if } \quad z < L_n(s) \\ z & \text{if } L_n(s) \leq z < U_n(s) \\ U_n(s) & \text{if } U_n(s) \leq z \end{cases} \qquad \text{(8-67)}$$

PROOF The assumption that $G(\cdot, a)$ is nondecreasing for each a implies, by induction starting with f_1, that $f_n(\cdot, z)$ is nondecreasing for all n and z. Exercise 8-24 asks you to prove that the further assumption that G is convex implies convexity of J_n and f_n for all n. Let

$$J_n^{(2)}(s, a) = J_n(s, a) - J_n(s, a - 1)$$

so

$$\mathscr{D}_{sz}^n(a) = J_n^{(2)}(s, a) + c|a - z| - c|a - 1 - z|$$

for $s \in I$ and $a \in \{1, \ldots, w\}$. Then

$$\mathscr{D}_{sz}^n(a) = c + J_n^{(2)}(s, a) \qquad a > z, a > 0$$

$$= -c + J_n^{(2)}(s, a) \qquad 0 < a \leq z$$

† In (8-60) and (8-62), n denotes the chronologically ordered point in time. In (8-66*a*) and (8-66*b*) n denotes the remaining number of points in time at which decisions can be made. Thus, $n \to n + 1$ in (8-60) and (8-62), whereas $n \to n - 1$ in (8-66*a*) and (8-66*b*).

Therefore, if $a \leq z$,

$$\mathscr{D}^n_{sz}(a)\left\{\begin{smallmatrix}\leq\\ \geq\end{smallmatrix}\right\}0 \qquad \text{if } J_n(s, a) - J_n(s, a-1)\left\{\begin{smallmatrix}\leq\\ \geq\end{smallmatrix}\right\}c$$

Recall the convention that the supremum of an empty set of numbers is $-\infty$ and let

$$U_n(s) = \sup\,\{a\colon J_n(s, a) - J_n(s, a-1) \leq c,\ a \in A - \{0\}\} \tag{8-68a}$$

Then $U_n(s) \leq z$ implies (8-57) [with $\mathscr{D}^n_{sz}(a)$ in place of $\mathscr{D}^n_s(a)$] is valid with $a' = U_n(s)$ so $a = U_n(s)$ is optimal. Parallel reasoning with

$$L_n(s) = \sup\,\{a\colon J_n(s, a) - J_n(s, a-1) \leq c,\ a \in A - \{0\}\} \tag{8-68b}$$

implies $a = L_n(s)$ is optimal if $z < L_n(s)$. If $L_n(s) \leq z < U_n(s)$, then (8-68) and monotonicity in a of $J_n(s, a) - J_n(s, a-1)$ (due to convexity of J_n) yield, as in the proof of Theorem 8-10, $\mathscr{D}^n_{sz}(a) \leq 0$ if $a \leq s$ and $\mathscr{D}^n_{sz}(a) > 0$ if $a > s$ so $a = s$ is optimal. Convexity of J_n and $c \geq 0$ implies $L_n(s) \leq U_n(s)$. □

Let $a_n(s, z)$ denote the value of y in (8-67). From the form of the policy which is optimal, namely, (8-67), $a_n(s, \cdot)$ is nondecreasing for each n and s. Hence, an optimal policy specifies a nondecreasing number of open teller windows as a function of the number of windows open already. The following result shows that the number of open teller windows should be nondecreasing too as a function of the number of customers present.

Corollary 8-11 If G is submodular on its domain, the assumptions of Theorem 8-11 imply for each n that $L_n(\cdot)$ and $U_n(\cdot)$ are nondecreasing on I. Therefore, $a_n(\cdot, z)$ and $a_n(s, \cdot)$ are nondecreasing for each n, s, and z.

Proof Exercise 8-25 asks you to prove that the assumptions imply for each n that f_n and J_n are submodular on $I \times A$ and that the minimand in (8-66) is submodular in (s, a, z) on $I \times A \times A$. Consider the class of problems

$$\min\,\{J_n(s, a) - ba\colon a \in A\} \qquad b \in \mathbb{R} \tag{8-69}$$

If the minimum is attained, let $m(s, b)$ denote the largest value of $a \in A$ at which the minimum is attained; let $m(s, b) = -\infty$ otherwise. From (8-68), $L_n(s) = m(s, -c)$ and $U_n(s) = m(s, c)$. However, each term of the minimand in (8-69) is submodular in $(s, a) \in I \times A$; so Corollary 8-4 implies $m(\cdot, b)$, hence $L_n(\cdot)$ and $U_n(\cdot)$ for each n, is nondecreasing on I. Then the monotonicity of $a_n(\cdot, z)$ and $a_n(s, \cdot)$ follows from (8-67). □

Incidentally, the submodularity assumption for G is not redundant because of convexity assumed in Theorem 8-11 (cf. Exercise 8-27).

Production Smoothing

The inventory models in Sections 3-1, 3-2, and 7-1 ignore some impacts of a policy on the manufacturing process. In particular, consider the aggregate production as a function of time. Abrupt changes in production rate entail costs of overtime, extra shifts, hiring and training, or subcontracting if the rate increases. If the production rate decreases, there are costs due to severance pay, higher premiums for unemployment insurance, and exit interviews. In general, these are *smoothing costs*. See page 13 for an expanded discussion of the origins of these costs.

Consider a production process whose inventory of completed units is reviewed periodically, say at epochs 1, 2, 3, ... Let s_n denote the inventory level at the nth review and D_n the demand during $[n, n+1)$. Excess demand is assumed to be backlogged; so

$$s_{n+1} = a_n - D_n \tag{8-70}$$

where a_n is the total number of units available to satisfy demand during $[n, n+1)$. Suppose that $z_n \geq 0$ is the amount produced during $[n, n+1)$ and that all of z_n is available to satisfy D_n. Then

$$a_n = s_n + z_n \tag{8-71}$$

Let $b(z_n)$ denote the direct cost of labor and materials to produce z_n units during $[n, n+1)$. We assume that $b(\cdot)$ is convex; the linear case is typical. Let $g(s_n, a_n, D_n)$ denote the costs of shortage plus inventory in $[n, n+1)$ with the most frequently used model being

$$h \cdot (a_n - D_n)^+ + \pi \cdot (D_n - a_n)^+$$

which does not depend on s_n.

Let c^+ and c^- denote the unit costs of increasing and decreasing production and let $c = (c^+ + c^-)/2$. Then the total discounted cost is

$$B \triangleq \sum_{n=1}^{\infty} \beta^{n-1}[b(z_n) + g(s_n, a_n, D_n) + c^+ \cdot (z_n - z_{n-1})^+ + c^- \cdot (z_{n-1} - z_n)^+]$$

From (8-70) and (8-71),

$$\begin{aligned}
\sum_{n=1}^{\infty} \beta^{n-1}(z_n - z_{n-1}) = \sum_{n=1}^{\infty} \beta^{n-1}(a_n - s_n - z_{n-1}) &= a_1 - s_1 - z_0 \\
&\quad + \sum_{n=2}^{\infty} \beta^{n-1}(a_n - a_{n-1} + D_{n-1} - a_{n-1} + s_{n-1}) \\
&= -z_0 + a_1(1-2\beta) - s_1(1-\beta) + \beta D_1 + \beta a_2 \\
&\quad + \sum_{n=3}^{\infty} \beta^{n-1}(a_n - 2a_{n-1} + D_{n-1} + a_{n-2} - D_{n-2}) \\
&= \sum_{n=1}^{\infty} \beta^{n-1}(1-\beta)^2 a_n - s_1(1-\beta) \\
&\quad + \sum_{n=1}^{\infty} \beta^n (1-\beta) D_n - z_0
\end{aligned}$$

Therefore, the identity (8-52) yields

$$B = \sum_{n=1}^{\infty} \beta^{n-1}\left[b(a_n - s_n) + g(s_n, a_n, D_n) + c\,|z_n - z_{n-1}| + (1-\beta)^2 \frac{c^+ - c^-}{2} a_n\right]$$
$$- \left[s_1(1-\beta) - \sum_{n=1}^{\infty} \beta^n (1-\beta) D_n - z_0\right] \frac{c^+ - c^-}{2}$$

Suppose that $D_1, D_2, \ldots$ are i.i.d. r.v.'s, that D has the same distribution as D_1, and $Eg(\cdot, \cdot, D)$ is convex on $\mathscr{C} \triangleq \{(s, a)\colon s \in \mathbb{R}, a \geq s\}$. Let

$$G(s, a) = E[g(s, a, D)] + b(a - s) + (1-\beta)^2 \frac{(c^+ - c^-)}{2} a \qquad \text{(8-72}a\text{)}$$

so
$$E(B) = \sum_{n=1}^{\infty} \beta^{n-1} E[c\,|a_n - s_n - z_{n-1}| + G(s_n, a_n)] + \text{constants} \qquad \text{(8-72}b\text{)}$$

The constants are uninfluenced by the a_n's or z_n's, $n \in I_+$.

The constraint $a_n \geq s_n$, or $0 \leq a_n - s_n$, results from $0 \leq z_n = a_n - s_n$. The dynamic program that corresponds to (8-72) is

$$f_n(s, z) = \inf\{c\,|a - s - z| + J_n(s, a)\colon a \geq s\} \qquad s \in \mathbb{R} \qquad z \geq 0 \qquad \text{(8-73}a\text{)}$$

$$J_n(s, a) = G(s, a) + \beta E[f_{n-1}(a - D, a - s)] \qquad (s, a) \in \mathscr{C} \qquad n \in I_+ \qquad \text{(8-73}b\text{)}$$

where $f_0(\cdot, \cdot) \equiv 0$. The argument $a - D$ in $f_{n-1}(a - D, a - s)$ results from $s_{n+1} = a_n - D_n$, and the argument $a - s$ results from $z_n = a_n - s_n$. Comparison of (8-73*a*) with the dynamic programs for the cash management and queueing models, namely

$$f_n(s) = \inf\{c\,|a - s| + J_n(a)\colon a \in \mathbb{R}\} \qquad s \in \mathbb{R} \qquad \text{(8-54}a\text{)}$$

$$f_n(s, z) = \min\{c\,|a - z| + J_n(s, a)\colon a \in A\} \qquad s \in I \qquad z \in A \quad \text{(8-66}a\text{)}$$

shows that the smoothing cost $c\,|a - s - z|$ in (8-73*a*) contrasts a with $s + z$, whereas a is compared only with s in (8-54) and only with z in (8-66). This feature of (8-73) complicates its analysis. Nevertheless, the optimal policy is the same as (8-67) except that $L_n(s)$ and $U_n(s)$ are compared with $s + z$ instead of z.

Theorem 8-12 Suppose $c \geq 0$ and G is convex on $\mathscr{C}$. Then for each $n \in I_+$ and $s \in \mathbb{R}$ there are numbers (or $\pm\infty$) $L_n(s) \leq U_n(s)$ such that an optimal policy in (8-73) is

$$a = \begin{cases} L_n(s) & \text{if } s + z < L_n(s) \\ s + z & \text{if } L_n(s) \leq s + z < U_n(s) \\ U_n(s) & \text{if } s \leq U(s) \leq s + z \\ s & \text{if } U_n(s) < s \end{cases} \qquad \text{(8-74)}$$

PROOF Convexity of G implies convexity of f_n and J_n for each n (Exercise 8-28). With the earlier notation, the partial derivative (from the left) of the minimand in (8-73*a*) is

$$\mathscr{D}^n_{sz}(a) = c + J^{(2)}_n(s, a) \qquad a > s + z$$
$$= -c + J^{(2)}_n(s, a) \qquad a \leq s + z$$

Let

$$L_n(s) = \sup \{a: J^{(2)}_n(s, a) \leq -c\} \qquad s \in \mathbb{R} \tag{8-75a}$$
$$U_n(s) = \sup \{a: J^{(2)}_n(s, a) \leq c\} \qquad s \in \mathbb{R} \tag{8-75b}$$

By analogy with the proof of Theorem 8-11, if $U_n(s) \leq s + z$, then $\mathscr{D}^n_{sz}(a) \leq 0$ if $a \leq U_n(s)$ and $\mathscr{D}^n_{sz}(a) > 0$ if $a > U_n(s)$; so $a = U_n(s)$ is optimal if feasible. Here, feasibility is $a \geq s$; so $a = U_n(s)$ is optimal if $s \leq U_n(s) \leq s + z$. If $U_n(s) < s$, then $z \geq 0$ implies $U_n(s) < s + z$; so $a = U_n(s)$ would be optimal if it were feasible. However, $a = s$ is then the feasible value closest to $U_n(s)$; so Exercise B-18 implies $a = s$ is optimal.

If $s + z < L_n(s)$, then $L_n(s) - s > z \geq 0$; so $a = L_n(s)$ is feasible. Optimality is verified as it was for $U_n(s)$. An argument similar to the case $L_n(s) \leq z < U_n(s)$ in the proof of Theorem 8-11 shows that $a = s + z$ is optimal if $L_n(s) \leq s + z < U_n(s)$. □

Figure 8-1 on page 370 displays the form of an optimal policy specified by (8-74). If $U_n(s) - s$ and $L_n(s) - s$ are nonincreasing, as in Figure 8-1, $U_n(s)$ and $L_n(s)$ cannot increase faster than s; that is, $s \leq s'$ implies $\max \{U_n(s') - U_n(s), L_n(s') - L_n(s)\} \leq s' - s$. Indeed this is true sometimes because a version of Corollary 8-11 is valid for the production-smoothing problem (cf. Exercise 8-30).

It follows from (8-74) that $z_n = z_{n-1}$ is optimal if $L(s_n) - s_n \leq z_{n-1} \leq U(s_n) - s_n$. Thus a typical sample path $s_1, z_1, s_2, z_2, \ldots$ will exhibit "clumps" of successive periods during which the production rate remains the same. This property is caused by the lack of differentiability of the smoothing cost function $c|a - s - z|$, as a function of a, at $a = s + z$. A quadratic smoothing cost function *would* be differentiable so it would yield an optimal policy which would change the production rate every period.

The *HMMS model*† is a linear-quadratic (see Section 7-5) production smoothing cost model. It is the most widely known linear-quadratic model in operations research. Let s_n and z_n retain their respective definitions as the beginning inventory level and the quantity produced during period n. Let w_n be the size of the work force during period n and let qw_n be the "normal" full-time output in period n. Then z_n may differ from qw_n due to overtime, undertime, or other reasons. Let $1 - x$ be a period's "normal" fractional reduction of the work force size due to retirement and resignation. Smoothing costs are incurred if w_n differs from xw_{n-1}.

The HMMS model has a finite horizon, has nonstationary demand distri-

† The HMMS model is named for C. C. Holt, F. Modigliani, J. F. Muth, and H. A. Simon and is presented in detail in their book *Planning Production, Inventories, and Work Force*, Prentice-Hall, Englewood Cliffs, N.J., 1960.

butions, and all costs are convex and quadratic. Two decisions are made each period, namely z_n and w_n. In the HMMS model, these variables may take negative values. In comparison with (8-72*b*), the objective is to minimize

$$\sum_{n=1}^{N} E[c(w_n - xw_{n-1}) + h(s_n + z_n - D_n) + b(z_n) + d(qw_n - z_n)]$$

where $c(\cdot)$, $h(\cdot)$, $b(\cdot)$, and $d(\cdot)$ are convex quadratic functions. The respective costs of work-force smoothing, holding inventory, and overtime/undertime are given by $c(\cdot)$, $h(\cdot)$, and $d(\cdot)$. Notice that negative inventory levels are not penalized. As in (8-54*a*), (8-66*a*), and (8-73*a*), the state variable is a vector, namely (s_n, w_{n-1}).

See Section 7-5 for the major properties of linear-quadratic models and a discussion of their strengths and weaknesses.

EXERCISES

8-23 Prove that L_n and U_n in (8-58) satisfy $L \le L_n \le U_n \le U$, where

$$L = \sup\{a\colon G'(a) \le -c(1+\beta)\},\ U = \sup\{y\colon G'(a) \le c(1+\beta)\}$$

Hint: Prove $f'_{n-1}(s) \in [-c, c]$ for all n and s so $J'_n(a) - G'(a) \in [-\beta c, \beta c]$ for all n and a.

8-24 In (8-66*a*), why is the minimum necessarily attained for each n, s, and z?

8-25 Prove, in detail, that f_n in (8-66*a*) is convex for all n if $f_n(\cdot, z)$ is nondecreasing for all n and z and G is convex.

8-26 In Corollary 8-11, prove submodularity of f_n and J_n on $I \times A$ and of $c|a - z| + J_n(s, a)$ in $(s, a, z) \in I \times A \times A$. (Hint: Use convexity of f_n and J_n established in Exercise 8-25.)

8-27 Let G be a real-valued function on $\mathbb{R}^2$. Give an example of G that is convex but not submodular and another example that is submodular but not convex.

8-28 Prove, in detail, that f_n in (8-73*a*) is convex on $\mathbb{R} \times \mathbb{R}_+$ for each n. Assume that G is convex on $\mathscr{C}$.

8-29 Let $\{X(n);\ n \ge 0\}$ be a cumulative process† and let $D_n = X(n + 1^-) - X(n)$. Prove that D_1, $D_2, \ldots$ are i.i.d. integer-valued r.v.'s in each of the following cases. (This assumption is made in the model of a queue with an adjustable number of servers.)

(*a*) The cumulative process is generated by a Poisson process.

(*b*) The cumulative process is generated by a sequence of identical constants of length $1/k$ for some $k \in I_+$.

8-30 For the production-smoothing problem in (8-73), suppose G is submodular and convex on $\mathscr{C}$. Suppose also $|L_n(s)| < \infty$ and $|U_n(s)| < \infty$, $s \in \mathbb{R}$ and $n \in I_n^+$, where $L_n(\cdot)$ and $U_n(\cdot)$ are given by (8-75). Prove that $L_n(\cdot)$ and $U_n(\cdot)$ satisfy

$$0 \le L_n(s') - L_n(s) \le s' - s$$

$$0 \le U_n(s') - U_n(s) \le s' - s$$

for all $s \le s'$.

8-31 In the production-smoothing problem, suppose that the cost of labor and materials $b(z)$ is proportional to z; so $b(z) \equiv b \cdot z$. Suppose also that $g(s, a, d)$ does not depend on s.

† See p. 169 in Volume I for a definition.

(*a*) Verify that $G(s_n, a_n)$ in (8-72*a*) and $G(s, a)$ in (8-73*b*) can be replaced by $G(a_n)$ and $G(a)$, respectively.

(*b*) Suppose that $G(\cdot)$ is convex and nonnegative on $\mathbb{R}$ and let

$$L = \sup\{y: G'(a) \le -c(1+\beta)^2\}$$

$$U = \sup\{y: G'(a) \le c(1+\beta)^2\}$$

$$V = \sup\{y: G'(a) \le c[(1+\beta)^2 - 2]\}$$

$$W = \sup\{y: G'(a) \le -c[(1+\beta)^2 - 2]\}$$

Assume $-\infty < L$ and $U < \infty$. Prove that $L \le L_n(s) \le V$ and $W \le U_n(s) \le U$ for all $s \in \mathbb{R}$ and $n \in I_+$. Hint: Use an inductive proof starting with $n = 1$. For each n, obtain $x_n(\cdot)$ and $h_n(\cdot)$ such that $x_n(a) \le J_n^{(2)}(s, a) \le h_n(a)$ for all $s \in \mathbb{R}$. Then define $l_n = \sup\{a: h_n(a) \le -c\}$, $u_n = \sup\{a: x_n(a) \le c\}$, $v_n = \sup\{a: x_n(a) \le -c\}$, and $w_n = \sup\{a: h_n(a) \le c\}$. Hence, $l_n \le L_n(s) \le v_n$ and $w_n \le U_n(s) \le u_n$. Show $L \le l_n$, $u_n \le U$, $v_n \le V$, and $W \le w_n$ for all n.

8-32 In the production-smoothing problem, under the assumptions of Exercise 8-31, suppose $g(a, d) = h\cdot(a-d)^+ + \pi\cdot(d-a)^+ (h > 0,\ \pi > 0)$. Let F be the distribution function of demand and let $F^{-1}(p)$ be the pth fractile of F, that is, $F^{-1}(p) = \inf\{y: F(y) \ge p\}$.

(*a*) Prove that L and U in Exercise 8-31 are given by

$$L = F^{-1}\left[\frac{\pi - c(1+\beta)^2 - b(1-\beta) - (c^+ - c^-)(1-\beta)^2/2}{h+\pi}\right]$$

$$U = F^{-1}\left[\frac{\pi + c(1+\beta)^2 - b(1-\beta) - (c^+ - c^-)(1-\beta)^2/2}{h+\pi}\right]$$

(*b*) Establish that $-\infty < L$ and $U < \infty$ if, and only if,

$$-h + c(1+\beta)^2 < b(1-\beta) + \frac{(c^+ - c^-)(1-\beta)^2}{2} \le \pi - c(1+\beta)^2$$

8-33 Consider a fishery in which s_n is the biomass of commercial species at the beginning of the nth fishing year. Let $z_n \ge 0$ denote the amount caught in the nth year and $a_n = s_n - z_n$. Suppose that u is a basic minimum catch quantity and that an implicit cost $c^-\cdot(u - z_n)$ is assessed if $z_n < u$. Let $g(s_n, a_n)$ denote the expected costs in year n of fishing minus the revenue from the amount caught. Then the total discounted costs, net of revenues, is

$$B = \sum_{n=1}^{\infty} \beta^{n-1}[g(s_n, a_n) + c^-(u - z_n)^+]$$

Suppose that the dynamics are given, as in Section 8-3, by $s_{n+1} = M(a_n, D_n)$ with $D_1, D_2, \ldots$ being i.i.d. r.v.'s.

(*a*) Show that

$$E(B) = \sum_{n=1}^{\infty} \beta^{n-1} E[G(s_n, a_n) + c\,|u - z_n|] + c\left(\frac{u}{1-\beta} - s_1\right)$$

where $c = c^-/2$ and

$$G(s, a) = g(s, a) - cE[\beta M(a, D) - a]$$

(*b*) The corresponding dynamic program is

$$f_n(s) = \inf\{c\,|a - s - u| + J_n(s, a): 0 \le a \le s\} \qquad s \ge 0$$

$$J_n(s, a) = G(s, a) + \beta E f_{n-1}[M(a, D)] \qquad 0 \le a \le s$$

Deduce the form of an optimal policy [as in (8-55), (8-67), and (8-74)]. Make your assumptions concerning $M(\cdot, \cdot)$ and $g(\cdot, \cdot)$ explicit.

(c) Suppose $g(\cdot, \cdot)$ takes the form

$$g(s, a) = b \cdot (s - a) + b_1(a)$$

with $b_1(\cdot)$ convex on $\mathbb{R}_+$. Simplify your results in (b).

8-34 On page 405 we define convexity for a function on a nonconvex domain. Then we make three assumptions for $G(\cdot, \cdot)$ defined by (8-63) on the domain $I \times A$. Prove that $G(\cdot, \cdot)$ is a convex function if it satisfies the three assumptions.

8-35 Show that the substitution of (8-65) for $h(s, z)$ in (8-61) leads to

$$B = -z_0 \frac{c^+ - c^-}{2} - \rho^{\#}(s_1)\beta^{-1/2} + \sum_{n=1}^{\infty} \beta^{n-1}[c\,|z_n - z_{n-1}| + G(s_n, z_n)]$$

where
$$\rho^{\#}(s) = \frac{(1 - e^{-\gamma/2})\rho(s)}{\gamma}$$

$$G(s, z) = (1 - \beta)z \frac{c^+ - c^-}{2} + r(z) + \rho^{\#}(s)(1 + \beta^{-1/2})$$

and G is convex if r and ρ are convex.

8-5* OPTIMAL INFINITE-HORIZON POLICIES

This section and the next one are motivated by questions of the following kinds. Let $a_n(s)$ be an optimal action in an MDP with an n-period planning horizon and initial state s. Does $a_n(s)$ converge as $n \to \infty$? If $a_n(\cdot)$ is monotone for each n (as in the models in Sections 8-2 and 8-3) and if $a_n(s)$ converges to $a(s)$ as $n \to \infty$, does $a(\cdot)$ inherit monotonicity? If convergence occurs, is $a(\cdot)$ an optimal policy for the infinite-horizon MDP? Is there a (σ, Σ) policy which is optimal for the infinite-horizon setup-cost inventory problem? When does an infinite-horizon binary decision problem have a control-limit policy which is optimal?

There is a close connection between optimal infinite-horizon policies and the limit of the return functions $f_n(\cdot)$ as $n \to \infty$. Therefore, this section obtains sufficient conditions for the existence of a limit function and for the limit function to solve the functional equation of dynamic programming. Then Section 8-6 uses convergence of return functions to analyze convergence of policies. The last part of Section 8-6 presents algorithms to compute a monotone optimal policy (when one exists). Throughout both sections we use the models in Sections 8-2, 8-3, 8-4, and 7-1 to illustrate the results.

Suppose an MDP is finite and has discount factor $\beta < 1$. Then it satisfies the contraction, boundedness, and monotonicity assumptions in Section 5-4 and the maximand in (8-76) below is (trivially) continuous as a function of a. Therefore, Theorems 5-8, 5-9, and 5-10, Proposition 5-3, and Corollary 5-8c have the following consequences. In a finite MDP, let $f_0(s)$, $s \in \mathcal{S}$, be arbitrary and

$$f_n(s) = \max \left\{ r(s, a) + \beta \sum_{j \in \mathcal{S}} p_{sj}^a f_{n-1}(j) : a \in A_s \right\} \qquad s \in \mathcal{S},\ n \in I_+ \qquad (8\text{-}76)$$

Proposition 8-2 Suppose an MDP is finite and has discount factor $0 \le \beta < 1$.

(*a*) For each $s \in \mathcal{S}$, there exists

$$f(s) \triangleq \lim_{n \to \infty} f_n(s) \tag{8-77}$$

and the limit does not depend on $f_0(\cdot)$.
(*b*) Moreover, $f(\cdot)$ is the unique solution of

$$f(s) = \max \left\{ r(s, a) + \beta \sum_{j \in \mathcal{S}} p_{sj}^a f(j) \colon a \in A_s \right\} \qquad s \in \mathcal{S} \tag{8-78}$$

(*c*) For each s let $\delta(s)$ be an $a \in A_s$ which attains the maximum on the right side of (8-78). Then δ^∞ is an optimal stationary policy and $f(\cdot)$ is its return function.

Part (*a*) states that a finite discounted finite-horizon MDP has a return function which converges to a limit which does not depend on the salvage-value function. Part (*b*) states that the limit function is the only solution of (8-78). Equation (8-78) is called the *functional equation of dynamic programming*. Part (*c*) explains how to use a solution of (8-78) in order to identify an optimal policy.

Some of the MDPs in Sections 8-2, 8-3, and 8-4 are not finite but possess the properties listed in Proposition 8-2. For example, Example 8-4 is an equipment-replacement model in which the set of states $\mathcal{S}$ is discrete but not necessarily finite. Each state has at most two actions ("replace" and "don't replace"). In this model, it is reasonable to assume that there is a finite uniform bound θ on the single-period expected cost $c(s, a)$ of taking action a in state s:

$$|c(s, a)| \le \theta < \infty \qquad \text{for all } (s, a) \in \mathscr{C}$$

This assumption will yield the conclusions in Proposition 8-2 even though the MDP is not finite.

In general, boundedness is the only assumption in Section 5-4 which is at risk when $\mathscr{C}$ is a discrete but nonfinite set. Suppose that

$$|r(s, a)| \le \theta < \infty \qquad \text{for all } (s, a) \in \mathscr{C} \tag{8-79}$$

and $\mathscr{C}$ is a discrete set (but not necessarily a finite set). Assumption (8-79) implies $|r(s_n, a_n)| \le \theta$ in every period n; so

$$\left| \sum_{n=1}^{\infty} \beta^{n-1} r(s_n, a_n) \right| \le \frac{\theta}{1 - \beta}$$

for every policy, initial state, and outcome of the MDP. Hence, the boundedness assumption in Section 5-4 is satisfied and the conclusions of Proposition 8-2 remain valid.

Proposition 8-3 In an MDP, if $\mathscr{C}$ is a discrete set and (8-79) is satisfied, the conclusions of Proposition 8-2 are valid, with "max" in (*b*) replaced by "sup."

The only benefit in Proposition 8-3 of assuming that $\mathscr{C}$ is a discrete set is that it trivially yields continuity in a of the maximands in (8-76) and (8-78). Sections 8-3 and 8-4 present MDPs where $\mathscr{C}$ is not a discrete set. If $\mathcal{S}$ (hence $\mathscr{C}$) is not necessarily a discrete set, for each $(s, a) \in \mathscr{C}$ let $\xi(s, a)$ be a random variable which takes values in $\mathcal{S}$. We specify the dynamics with $s_{n+1} \sim \xi(s_n, a_n)$ and replace the sums in (8-76) and (8-78) with

$$E\{f_{n-1}[\xi(s, a)]\} \qquad \text{and} \qquad E\{f[\xi(s, a)]\} \tag{8-80}$$

respectively. It is no longer necessarily true that the maximands in (8-76) and (8-78) are continuous functions of a. Since Corollary 5-8*c* is the only result in Section 5-4 which requires continuity in a, the conclusions in Proposition 8-2 remain valid except that there may not exist a policy δ^∞ which attains $f(\cdot)$.

Proposition 8-4 If a nonfinite MDP satisfies (8-79), parts (*a*) and (*b*) of Proposition 8-2 remain valid with "max" in (*b*) replaced by "sup." If $r(s, \cdot)$ and $E\{f[\xi(s, \cdot)]\}$ are continuous on A_s and A_s is compact for each $s \in \mathcal{S}$, part (*c*) of Proposition 8-2 is valid; so there is an optimal stationary policy δ^∞ and $f(\cdot)$ is its return function.

The sufficiency of the continuity and compactness assumptions is an immediate consequence of Corollary 5-8*c*.

Some of the models in Sections 8-3 and 8-4 do not satisfy the uniform boundedness assumption (8-79).

Example 8-28 Consider a single-species fishery model in which s and a denote population biomass before and after fishing, respectively, and there is a unit profit $\rho > 0$ per unit biomass caught. Then

$$r(s, a) = \rho \cdot (s - a)$$

with $s \in \mathcal{S} = [0, \infty)$ and $a \in A_s = [0, s]$. Here, $0 \leq r(s, a) \leq \rho s$ but there is no (finite) upper bound on $r(\cdot, \cdot)$, which is uniform in s. □

Nonnegative Costs

When $r(\cdot, \cdot)$ cannot be uniformly bounded, the contraction-mapping approach in Section 5-4 cannot be used.† However, particular unbounded models often possess the properties stated in Proposition 8-2. We shall illustrate several approaches that have succeeded with numerous structured models.

First, suppose that nonnegative costs are being minimized. Most of the models in Sections 7-1 and 8-4 have this property. Recall (cf. Exercise 4-13) that nonnegativity is equivalent to the costs having a lower bound. We assume that single-stage costs satisfy

$$c(s, a) \geq 0 \qquad (s, a) \in \mathscr{C} \tag{8-81}$$

† See Harrison (1972) for a generalization of the contraction-mapping approach with unbounded rewards.

Lemma 8-3 Assumption (8-81) and $f_0(\cdot) \equiv 0$ imply $f_{n+1}(s) \geq f_n(s)$ for all n and s.

PROOF Use essentially the same argument which verifies (8-46) in the proof of Theorem 8-9. □

Structured MDPs often have simple suboptimal policies whose return functions can be computed easily. For an arbitrary stationary policy λ^∞ let $v^\lambda(s)$ denote the expected present value of using λ^∞ if s is the initial state.

Theorem 8-13 In an MDP with nonnegative single-period cost function $c(\cdot, \cdot)$ and salvage-value function $f_0(\cdot) \equiv 0$ let λ^∞ be a stationary policy for which

$$v^\lambda(s) < \infty \qquad s \in \mathcal{S} \tag{8-82}$$

Then there exists $f(s) = \lim_{n\to\infty} f_n(s)$ for all $s \in \mathcal{S}$.

PROOF The infimum over all n-period policies yields $f_n(s)$, which is a lower bound for the return function if λ^∞ is used for n periods. Therefore,

$$f_n(s) = \inf\left\{E\left[\sum_{i=1}^{n} \beta^{i-1} c(s_i, a_i) \mid s_i = s\right]\right\}$$

$$\leq E\left\{\sum_{i=1}^{\infty} \beta^{i-1} c[s_i, \lambda(s_i)] \mid s_1 = s\right\} = v^\lambda(s) < \infty$$

It follows from Lemma 8-3 that $\{f_n(s)\}$ is a bounded monotone sequence as $n \to \infty$; so its limit exists. □

The restriction to a stationary policy simplifies the notation in the proof but is otherwise unnecessary. That is, if π is any policy, let $v_\pi(s)$ denote the expected discounted cost when $s_1 = s$. If there is any policy π for which $v_\pi(s) < \infty$ for all $s \in \mathcal{S}$, then $f_n(s) \to f(s)$ as $n \to \infty$ for all $s \in \mathcal{S}$.

Applications of Theorem 8-13

Existence of the limit function $f(\cdot)$ can sometimes be used to prove existence of an optimal policy with certain features. The following example is an infinite-horizon version of a finite-horizon model on page 385 which has an optimal control-limit policy.

Example 8-29: Equipment replacement As in Example 8-16, let s be the age of the equipment, $a = 1$ and $a = 0$ denote the actions "replace" and "don't replace," respectively, $c(s, 0)$ be the expected cost to operate the equipment one period if its age is s, and $c(s, 1)$ be the cost to buy a new piece of equipment and operate it one period minus the salvage value of an age s

piece of equipment. Let b_s be the probability that an age s equipment breaks down and forces replacement; then $p^0_{s,s+1} = 1 - b_s$, $p^0_{s0} = b_s$, and $p^1_{s0} = 1$.

In any application we can find a big enough age u such that it is best to replace any equipment no later than age u. Then let $\mathcal{S} = \{0, 1, \ldots, u\}$ instead of $\mathcal{S} = I$. In this example we let $\mathcal{S} = I$ and later (in Example 8-32) show how one might assign a value to u in order to truncate $\mathcal{S}$ to $\{0, 1, \ldots, u\}$. We assume $c(s, 0) \geq 0$ and $c(s, 1) \geq 0$ for all $s \in \mathcal{S}$.

Let λ be the decision rule which replaces the equipment if its age s is positive. Then

$$v^\lambda(s) = c(s, 1) + \beta v^\lambda(0) \qquad s \in I_+ \tag{8-83a}$$

$$v^\lambda(0) = c(0, 0) + \beta[(1 - b_0)v^\lambda(1) + b_0 v^\lambda(0)] \tag{8-83b}$$

so $v^\lambda(s) < \infty$ if $v^\lambda(0) < \infty$. Substitution of (8-83a) with $s = 1$ in (8-83b) yields

$$v^\lambda(0) = \frac{c(0, 0) + \beta(1 - b_0)c(1, 1)}{1 - \beta^2(1 - b_0) - \beta b_0}$$

which is finite because the denominator is positive $[1 - \beta^2(1 - b_0) - \beta b_0 > 0 \Leftrightarrow b_0 < (1 - \beta^2)/[\beta(1 - \beta)] = (1 + \beta)/\beta > 1]$. Therefore, $f_n(s) \to f(s)$ for all s.

□

Example 8-30: Setup cost inventory model Here is an infinite-horizon version of the (σ, Σ) inventory model in Section 7-1. Suppose that $D_1, D_2, \ldots$ are independent and identically distributed nonnegative demands, excess demand is backlogged, ordered goods are delivered immediately, β is the discount factor, $g(a, d)$ is the single-period expected cost of inventories and backlogs if a units are available to satisfy demand d, c is the unit cost of ordering goods, and K is the setup cost. Let s_n and a_n denote the inventory levels in period n immediately before and after ordered goods (if any) are delivered. We assume that all costs are nonnegative. By letting $N \to \infty$ in (7-5), the expected present value is

$$E(B) = E\left\{\sum_{n=1}^{\infty} \beta^{n-1}[K\delta(a_n - s_n) + G(a_n)]\right\} - cs_1 + \sum_{n=1}^{\infty} \beta^n E(D_1)$$

where
$$G(a) \triangleq c(1 - \beta)a + E[g(a, D_1)]$$

We assume $G(a) < \infty$ for all $a \in \mathbb{R}$. The second and third terms on the right side of $E(B)$ are constants with respect to alternative policies; so we shall use only the first term as the criterion of minimization.

Consider the simple policy λ^∞ which selects $a_n = s_1$ for all n so $a_n - s_n = D_{n-1}$ for all $n > 1$. This policy is feasible because D_{n-1} and s_n are known when a_n is chosen and the order quantity $a_n - s_n$ is nonnegative for each n. The expected present value from the initial inventory level s satisfies

$$v^\lambda(s) = E\left\{\sum_{n=2}^{\infty} \beta^{n-1}[K\delta(D_{n-1}) + G(s)]\right\} + G(s) \leq \frac{K + G(s)}{1 - \beta} < \infty$$

It follows from Theorem 8-13 that $f_n(s) \to f(s)$ for all $s \in \mathcal{S}$. □

Example 8-31: Production smoothing Equations (8-70) through (8-73*a*) and (8-73*b*) describe a model for making production decisions when excess demand is backlogged and there are smoothing costs. With the simplification described in Exercise 8-31 the expected present value can be considered to be [cf. (8-72)]

$$E(B) = \sum_{n=1}^{\infty} \beta^{n-1} E[c\,|a_n - s_n - z_{n-1}| + G(a_n)]$$

where s_n and a_n are the inventory levels before and after production occurs in period n and $z_n = a_n - s_n$ is the quantity produced in period n. We assume $s_{n+1} = a_n - D_n$ for each n with the demands $D_1, D_2, \ldots$ being independent and identically distributed r.v.'s. Recall from Section 8-4 that the state variable in period n in this MDP is the vector (s_n, z_{n-1}).

Let λ be the decision rule which specifies $a_n = s_1$ for all n. Then $z_1 = 0$ and $z_n = D_{n-1}$ for all $n > 1$; so with $s = s_1$ and $z = z_0$,

$$\begin{aligned} v^{\lambda}(s, z) &= cE\left[|z_0| + |D_1| + \sum_{n=3}^{\infty} \beta^{n-1} |D_{n-1} - D_{n-2}| \right] + \sum_{n=1}^{\infty} \beta^{n-1} G(s_1) \\ &= c\left[z + E(D_1) + \frac{\beta^2 E(|D_2 - D_1|)}{1-\beta} \right] + \frac{G(s)}{1-\beta} \end{aligned}$$

which is finite if $E(D_1) < \infty$ and $E(|D_2 - D_1|) < \infty$. Under those assumptions and nonnegativity of c and $G(\cdot)$, $f_n(s, z) \to f(s, z)$ for all s and z. □

The Functional Equation of Dynamic Programming

If $\mathcal{S}$ is not necessarily discrete, (8-80) suggests

$$f_n(s) = \sup \{r(s, a) + \beta E\{f_{n-1}[\xi(s, a)]\} : a \in A_s\} \qquad s \in \mathcal{S} \tag{8-84}$$

instead of (8-76) and

$$f(s) = \sup \{r(s, a) + \beta E\{f[\xi(s, a)]\} : a \in A_s\} \qquad s \in \mathcal{S} \tag{8-85}$$

instead of (8-78). The contraction-mapping approach in Section 5-4 leads to sufficient conditions for the limit function $f(\cdot)$ to be the unique solution of (8-85). These conditions include uniform boundedness of $|r(\cdot, \cdot)|$ on $\mathscr{C}$, but boundedness often is lacking if the state space $\mathcal{S}$ is not finite (or not compact). The single-stage reward functions in Examples 8-30 and 8-31 are not uniformly bounded. Nevertheless, these examples satisfy (8-85) with only a few additional assumptions. Here is a general result and its proof. Then we apply the result to Examples 8-29, 8-30, and 8-31.

Theorem 8-14 Let (8-84) specify $f_1(\cdot), f_2(\cdot), \ldots$ with $f_0(\cdot) \equiv 0$. Suppose:

(*a*) For each $s \in \mathcal{S}$ there exists $f(s) \triangleq \lim_{n\to\infty} f_n(s)$.
(*b*) Either $r(s, a) \geq 0$ for all $(s, a) \in \mathscr{C}$ or $r(s, a) \leq 0$ for all $(s, a) \in \mathscr{C}$.

(*c*) For each $s \in \mathcal{S}$, $A_s \subset \mathbb{R}$ and A_s is a compact set.

(*d*) For each $s \in \mathcal{S}$ and $n \in I_+$, $J_n(s, \cdot)$ is continuous on A_s, where $J_n(s, a) \triangleq r(s, a) + \beta E\{f_{n-1}[\xi(s, a)]\}$ $(s, a) \in \mathscr{C}$.

Then $f(\cdot)$ satisfies (8-85), the functional equation of dynamic programming, and for each $s \in \mathcal{S}$ the supremum is attained by some $a \in A_s$.

PROOF $J_n(s, a)$ is the maximand in (8-84). Let $J(s, a)$ denote the maximand in (8-85) and suppose $r(\cdot, \cdot) \geq 0$. Then $f_0(\cdot) \equiv 0$ implies $f_n(s) \leq f_{n+1}(s)$ for all n and s and by (*a*) the limit $f(s)$ exists. Using $f_{n-1}(s) \leq f(s)$ in (8-85),

$$f_n(s) \leq \sup \{r(s, a) + \beta E\{\xi(s, a)]\}: a \in A_s\}$$

Let $n \to \infty$ to obtain

$$f(s) \leq \sup \{J(s, a): a \in A_s\} \qquad (8\text{-}86)$$

In order to derive the opposite inequality, we start with $f(s) \geq f_n(s)$ in (8-85) to obtain

$$f(s) \geq f_n(s) = \sup \{J_n(s, a): a \in A_s\} \qquad (8\text{-}87)$$

By assumption, the supremum is a bounded monotone sequence (as $n \to \infty$); so it has a limit. Therefore,

$$f(s) \geq \lim_{n \to \infty} [\sup \{J_n(s, a): a \in A_s\}] \qquad s \in \mathcal{S}$$

Assumptions (*c*) and (*d*) imply for each $s \in \mathcal{S}$ that $J_n(s, \cdot)$ converges uniformly to $J(s, \cdot)$ on A_s. Also, the monotone convergence theorem (Proposition A-9) justifies

$$\lim_{n \to \infty} E\{f_{n-1}[\xi(s, a)]\} = E\{f[\xi(s, a)]\}$$

Therefore,

$$\lim_{n \to \infty} [\sup \{J_n(s, a): a \in A_s\}] = \sup \{J(s, a): a \in A_s\} \qquad (8\text{-}87)$$

so $n \to \infty$ in (8-87) yields

$$f(s) \geq \sup \{J(s, a): a \in A_s\} \qquad s \in \mathcal{S}$$

This inequality and (8-86) imply (8-85).

If the single-state rewards are nonpositive, reverse all inequalities and the same proof is valid. □

Applications of Theorem 8-14

Example 8-32: Equipment replacement In Example 8-29 we established for this model that $f(s)$ exists for each s. Condition (*b*) in Theorem 8-14 is satisfied because all costs are nonnegative, which is equivalent to nonpositive single-stage rewards. Condition (*c*) is satisfied because $A_s = \{0, 1\}$ if $s > 0$ and $A_0 = \{0\}$, which are trivially compact. Discreteness of A_s trivially implies that condition (*d*) is met; so $f(\cdot)$ satisfies (8-85).

Now we shall use (8-85) to truncate the state space $\mathcal{S} = I$ to $\mathcal{S} = \{0, 1, \ldots, u\}$ for some $u < \infty$. For this model, (8-85) is

$$f(s) = \min \{c(s, 0) + \beta f(0), c(s, 1) + \beta b_s f(0) + \beta(1 - b_s) f(s + 1)\}$$

The first term in the braces is associated with "planned replacement" and the second term with "don't replace unless forced to do so by equipment failure." Let Δ_s denote the difference between the second and first terms. If Δ_s is nonnegative, planned replacement is optimal at age s. Using $c(s, 1) = r - L_s$ (from Example 8-16),

$$\begin{aligned} \Delta_s &= c(s, 1) - c(s, 0) + \beta b_s f(0) - \beta f(0) + \beta(1 - b_s) f(s + 1) \\ &= r - L_s - c(s, 0) + \beta(1 - b_s)[f(s + 1) - f(0)] \end{aligned}$$

It follows from Example 8-16 that $f(\cdot)$ is nondecreasing if

$$0 \le L_s - L_{s+1} \le c(s + 1, 0) - c(s, 0) \qquad s \in I$$

which we now assume. These inequalities stipulate that salvage values decrease with age, operating costs increase with age, and the rise in operating cost is at least as great as the reduction in salvage value. Also, $c(s, 0) \le c(s, 1)$ for all s if $r > c(s, 0) + L_s$, which we assume is valid for all s. Under these assumptions

$$f(s) \ge c(s, 0) + \beta f(0)$$

so

$$\begin{aligned} \Delta_s &\ge r - L_s - c(s, 0) + \beta(1 - b_s)[c(s, 0) + \beta f(0) - f(0)] \\ &= r - L_s - c(s, 0)[1 - \beta(1 - b_s)] - \beta(1 - \beta)(1 - b_s) f(0) \end{aligned}$$

It follows from Example 8-29 that

$$f(0) \le \frac{c(0, 0) + \beta(1 - b_0)(r - L_1)}{1 - \beta^2(1 - b_0) - \beta b_0}$$

so

$$\begin{aligned} \Delta_s \ge\; & r - L_s - c(s, 0)[1 - \beta(1 - b_s)] \\ & - \beta(1 - \beta)(1 - b_s) \frac{c(0, 0) + \beta(1 - b_0)(r - L_1)}{1 - \beta^2(1 - b_0) - \beta b_0} \end{aligned}$$

whose right side is nonnegative if

$$\beta(1 - \beta) \frac{c(0, 0) + \beta(1 - b_0)(r - b_1)}{1 - \beta^2(1 - b_0) - \beta b_0} \le \frac{r - L_s - c(s, 0)[1 - \beta(1 - b_s)]}{1 - b_s}$$

We already have $c(s, 0) + L_s < r$; so the right side becomes arbitrarily large as $s \to \infty$ if $b_s \to 1$. However, b_s is the probability of failure at age s; so it is quite reasonable to assume $b_s \to 1$ as $s \to \infty$. Therefore, if $b_s \to 1$ as $s \to \infty$, there exists $u < \infty$ such that $\Delta_s \geq 0$ for all $s \geq u$; so truncating $\mathcal{S}$ to $\{0, 1, \ldots, u\}$ is without loss of optimality. □

Example 8-33: Setup cost inventory model In Example 8-30 we established for this model that $f(s)$ exists for each $s \in \mathcal{S}$; so condition (*a*) in Theorem 8-14 is met. All costs are nonnegative; so (*b*) is satisfied. In this model, $A_s = [s, \infty)$, which is not compact. However, Theorem 7-4 implies that there is no loss of optimality in truncating A_s to $[s, s \vee \bar{\Sigma}]$, where $\bar{\Sigma}$ is explained in the discussion on pages 323 to 327. The truncated version of A_s is compact for each s.

Condition (*d*) of Theorem 8-14 is equivalent to continuity in a of

$$K\delta(a - s) + G(a) + \beta E[f_{n-1}(a - D_1)] \tag{8-89}$$

From Theorem 7-1, if $G(\cdot)$ is convex on $\mathbb{R}$, then $f_{n-1}(\cdot)$ is K-convex; so the second and third terms of (8-90) are continuous in a. The first term is not continuous; so Theorem 8-14 cannot be applied directly. [Exercise 8-36 asks you to prove, nevertheless, that (8-85) is valid.] □

Example 8-34: Production smoothing In Example 8-31 we assume that costs are nonnegative and verify that the limit $f(s)$ exists for each s. Exercise 8-31 asks you to establish sufficient conditions for $A_s = [s, \infty)$ to be truncated to $A_s = [s, U \vee s]$ (with $U < \infty$ for each s) without loss of optimality. Condition (*d*) of Theorem 8-14 is equivalent to continuity in a of

$$c\,|a - s - z| + G(a) + \beta E[f_{n-1}(a - D_1, a - s)] \tag{8-90}$$

However, convexity of $G(\cdot)$ (assumed in Section 8-4 and Exercise 8-31) implies convexity of $f_{n-1}(\cdot, \cdot)$ (cf. Exercise 8-28) so (why?) expression (8-90) is continuous in a. Therefore, $f(\cdot)$ satisfies (8-85). □

Example 8-35: Resource management Consider the specific case of the general model in Section 8-4 where $\mathcal{S} = [0, \infty)$, $A_s = [0, s]$, $G(s, a) = s - a$, and $M(s, \cdot) \equiv \theta s$. Theorem 8-13 cannot be invoked for existence of $f(\cdot)$ because the *rewards* here (rather than the costs) are nonnegative. Let $f_0(\cdot) \equiv 0$. Then $f_n(s)$ is at least as great as the value of the policy that delays all consumption (consumption in period n is $s_n - a_n$) until the last period. Therefore

$$f_1(s) = s, \; f_2(s) \geq G(s, s) + \beta G(\theta s, 0) = \beta\theta s$$

$f_3(s) \geq \beta^2\theta^2 s$ and $f_n(s) \geq (\beta\theta)^{n-1}s$; so $f_n(s)$ diverges to ∞ as $n \to \infty$ if $\theta > 1/\beta$. Clearly, some restrictions are needed in the general model.

Now consider a more general case of the resource-management model in Section 8-3 where we maximize

$$E\left[\sum_{n=1}^{\infty} \beta^{n-1} G(s_n, a_n)\right]$$

with $A_s = [0, s]$ for each s, $G(s, \cdot)$ is nonnegative and nonincreasing on A_s for each s, $G(\cdot, 0)$ is concave† and nondecreasing on $\{s : s \geq 0 \text{ and } s \in \mathcal{S}\}$, $s_{n+1} = M(a_n, D_n)$, and $M(\cdot, d)$ is nondecreasing for each d. Under these assumptions $G(s_n, a_n) \leq G(s, 0)$ and $s_{n+1} \leq M(s_n, D_n)$ for each n. The first inequality describes an effect of consuming everything; the second inequality is associated with consuming nothing at all in period n. Let $Y_1 \equiv s_1$ and $Y_{n+1} = M(Y_n, D_n)$ for all $n > 1$. Then

$$E\left[\sum_{n=1}^{\infty} \beta^{n-1} G(s_n, a_n) \mid s_1 = s\right] \leq E\left[\sum_{n=1}^{\infty} \beta^{n-1} G(Y_n, 0) \mid s_1 = s\right]$$

Concavity of $G(\cdot, 0)$, Proposition B-6, and inequality (B-7) yield

$$G(y, 0) \leq G(0, 0) + yG^{(1)}(0, 0)$$

for each y [where $G^{(1)}(s, a)$ denotes the right-hand partial derivative $\partial G(s, a)/\partial s$]; so

$$E\left[\sum_{n=1}^{\infty} \beta^{n-1} G(s_n, a_n) \mid s_1 = s\right]$$

$$\leq \frac{G(0, 0)}{1 - \beta} + G^{(1)}(0, 0)\, E\left(\sum_{n=1}^{\infty} \beta^{n-1} Y_n \mid Y_1 = s\right) \quad (8\text{-}91)$$

Let $\mathcal{Y}_1(s) \equiv s$ and $\mathcal{Y}_{n+1}(s) = M[\mathcal{Y}_n(s), D_n]$ for $n \in I_+$. Then $Y_n \equiv \mathcal{Y}_n(s)$; so the right side of (8-91) is finite if $G^{(1)}(0, 0) < \infty$ and

$$E\left[\sum_{n=1}^{\infty} \beta^{n-1} \mathcal{Y}_n(s)\right] < \infty \quad (8\text{-}92)$$

Condition (8-92) is satisfied in many specific cases (cf. Exercise 8-37).

Under the previously stated assumptions, for each $s \in \mathcal{S}$, $f_n(s)$ converges monotonely to $f(s)$ and $f(\cdot)$ satisfies (8-85), the functional equation of dynamic programming. □

EXERCISES

8-36 In the inventory model with setup costs in Examples 8-30 and 8-33 suppose that $G(\cdot)$ is convex on $\mathbb{R}$ and there is a number $\ddagger < \infty$ such that truncating A_s to $[s, s \vee \ddagger]$ is without loss of optimality. Prove that $f(\cdot)$ satisfies (8-85), which is

$$f(s) = \inf \{K\delta(a - s) + G(a) + \beta E[f(a - D_1)] : a \geq s\} \qquad s \in \mathbb{R}$$

Hint: This equation can be written

$$f(s) = \min \{G(s) + \beta E[f(s - D_1)], K + \inf \{G(a) + \beta E[f(a - D_1)] : a > s\}\}$$

Also,

$$f_n(s) = \min \{G(s) + \beta E[f_{n-1}(s - D_1)], K + \inf \{G(a) + \beta E[f_{n-1}(a - D_1)] : a > s\}\}$$

† In Section 8-3 we make the more restrictive assumption that $G(\cdot, \cdot)$ is concave on $\mathscr{C}$.

Argue that continuity permits this representation of $f_n(s)$ to be altered by replacing $a > s$ with $a \geq s$. Then embellish the proof of Theorem 8-14.

8-37 In Example 8-35 show that (8-92) is satisfied in the following cases:
(*a*) $M(a, d) = (a + d)^+$ with $E[(D_1)^+] < \infty$.
(*b*) $M(a, d) = ad$ with $|E(D_1)| < 1/\beta$.

8-38 Prove Lemma 8-3.

8-6* CONVERGENCE IN POLICY

Section 8-5 focuses on convergence of $f_n(s)$ as $n \to \infty$. Here we consider the closely related issue of convergence of n-period optimal policies. Suppose $a_n(s)$ is an optimal action in an MDP with an n-period planning horizon whose initial state is s. Suppose also for each n that $a_n(\cdot)$ is monotone as in Sections 8-2, 8-3, and 8-4. What are sufficient conditions for existence of

$$a(s) \triangleq \lim_{n \to \infty} a_n(s) \tag{8-93}$$

for each $s \in \mathcal{S}$ and for $a(\cdot)$ to be monotone on $\mathcal{S}$? Part (*a*) of the following result shows that $a(\cdot)$ is necessarily monotone if the limit in (8-93) exists.

Lemma 8-4 (*a*) Let $\mathcal{S}$ be a partially ordered set and for each n let $g_n(\cdot)$ be a real-valued isotone (antitone) function on $\mathcal{S}$ such that $g_n(s) \to g(s)$ as $n \to \infty$ for each $s \in \mathcal{S}$. Then $g(\cdot)$ is an isotone (antitone) function on $\mathcal{S}$.

(*b*) Let $\mathcal{S}$ be a convex set and for each n let $h_n(\cdot)$ be a convex (concave) function on $\mathcal{S}$ such that $h_n(s) \to h(s)$ as $n \to \infty$ for each $s \in \mathcal{S}$. Then $h(\cdot)$ is a convex (concave) function on $\mathcal{S}$.

PROOF (*a*) Let $\gtrsim$ be the partial order on $\mathcal{S}$ and $y \gtrsim x$. If $g_n(\cdot)$ is isotone for each n, then $g_n(y) \geq g_n(x)$ for each n; so

$$\begin{aligned} g(y) - g(x) &= \lim_{n \to \infty} g_n(y) - \lim_{n \to \infty} g_n(x) \\ &= \lim_{n \to \infty} [g_n(y) - g_n(x)] \geq 0 \end{aligned}$$

and $g(\cdot)$ is isotone. Reverse the inequalities if $g_n(\cdot)$ is antitone for each n.

(*b*) Suppose $\mathcal{S}$ is a convex set and $h_n(\cdot)$ is a convex function on $\mathcal{S}$. From the definition of a convex function, for each $x \in \mathcal{S}$, $y \in \mathcal{S}$, and $0 \leq \lambda \leq 1$,

$$0 \leq \lambda h_n(x) + (1 - \lambda)h_n(y) - h_n[\lambda x + (1 - \lambda)y]$$

Letting $n \to \infty$ yields

$$0 \leq \lambda h(x) + (1 - \lambda)h(y) - h[\lambda x + (1 - \lambda)y]$$

so $h(\cdot)$ is a convex function on $\mathcal{S}$. If $h_n(\cdot)$ is concave for each n, reverse the inequalities. □

It follows from Lemma 8-4 that Section 8-2 lists conditions under which $f(\cdot)$ in the equipment-replacement model in Examples 8-29 and 8-32 is nondecreasing. Section 8-3 states sufficient conditions for the resource-management model in Example 8-35 to be a concave function. Section 8-4 presents sufficient conditions for $f(\cdot, \cdot)$ in the production-smoothing model of Examples 8-31 and 8-34 to be a convex function. Exercise 8-39 asks you to list these conditions. Exercise 8-40 asks you to perform the analogous task for the setup-cost inventory model in Examples 8-30 and 8-33.

In the resource-management model, it is easy to prove that finite-horizon monotone optimal policies $a_n(\cdot)$ converge to an infinite-horizon monotone optimal policy $a(\cdot)$.

Example 8-36: Resource management Theorems 8-7 and 8-9 state sufficient conditions for the model in Example 8-35 to yield monotonicity of $a_n(\cdot)$ and $a_n(s) \leq a_{n+1}(s)$ for each n and s. The assumptions include compactness of A_s so $a_n(s)$ for each s is a bounded monotone sequence whose limit, say $a(s)$, must exist. Part (*a*) of Lemma 8-4 implies that $a(\cdot)$ is monotone. □

Suppose for each s that $a = a(s)$ attains the optimand in (8-85) repeated here:

$$f(s) = \sup \{r(s, a) + \beta E\{f[\xi(s, a)]\}: a \in A_s\} \qquad s \in \mathcal{S} \tag{8-94}$$

If $a_n(\cdot)$ is monotone for each n, one suspects that $a(\cdot)$ too is monotone. One way to verify the suspicion is to prove for each s that $a_n(s)$ converges as $n \to \infty$ and that the limiting value is optimal in (8-94). Then $a(\cdot)$ is monotone because of part (*a*) of Lemma 8-4. In Example 8-36 we prove for the resource-management model that $a_n(s)$ converges as $n \to \infty$, hence the limit function is monotone, but we have not yet shown that the limit is optimal in (8-94).

The convergence for each s of $a_n(s)$ as $n \to \infty$ cannot be proved for many models as easily as it is in Example 8-36. An alternative convergence proof is often valid when $f_n(\cdot)$ is concave (or convex) for each n. Let $J(s, a)$ denote the maximand in (8-94) and let $J_n(s, a)$ denote the maximand in the finite-horizon recursion (8-84) which corresponds to (8-94):

$$J(s, a) = r(s, a) + \beta E\{f[\xi(s, a)]\} \tag{8-95a}$$

$$J_n(s, a) = r(s, a) + \beta E\{f_{n-1}[\xi(s, a)]\} \tag{8-95b}$$

Suppose $J_n(s, \cdot)$ is concave on $A_s \subset \mathbb{R}$, which is a convex set. Let $J_n^{(2)}(s, a)$ and $J^{(2)}(s, a)$ denote left-hand partial derivatives with respect to a and

$$\alpha_n(s) = \sup \{a: J_n^{(2)}(s, a) \geq 0 \text{ and } a \in A_s\} \tag{8-96a}$$

$$\alpha(s) = \sup \{a: J^{(2)}(s, a) \geq 0 \text{ and } a \in A_s\} \tag{8-96b}$$

It follows from Proposition B-7 that $a = \alpha_n(s)$ maximizes $J_n(s, \cdot)$ on A_s; so $a_n(s) = \alpha_n(s)$ is valid. Suppose for each s that $f_n(s) \to f(s)$ as $n \to \infty$ and we have used part (*b*) of Lemma 8-4 to deduce that $f(\cdot)$ is concave on $\mathcal{S}$ and $J(s, \cdot)$ is concave on A_s. Then $\alpha(s)$ maximizes $J(s, \cdot)$ on A_s. Does $a_n(s) = \alpha_n(s)$ converge to $\alpha(s)$ as $n \to \infty$?

Even if $f_n(s) \to f(s)$ implies $J_n(s, a) \to J(s, a)$, it is not necessarily true that $J_n^{(2)}(s, a) \to J^{(2)}(s, a)$; so a direct appeal to (8-96*a*) and (8-96*b*) does not yield $\alpha_n(s) \to \alpha(s)$.

Example 8-37 Let c_n be a nondecreasing sequence of negative numbers with $c_n \to 0$ as $n \to \infty$. Let $g(x) = |x|$ and $g_n(x) = c_n + |x - c_n|$, $x \in \mathbb{R}$. Let $g^-(x)$ and $g_n^-(x)$ denote the left-hand derivatives of $g(\cdot)$ and $g_n(\cdot)$ at x. Then $g_n^-(0) = 1$ for all n; so $g_n^-(0) \to 1$ as $n \to \infty$ while $g^-(0) = -1$. Instead, if c_n is a nonincreasing sequence of positive numbers which converges to 0 and $g^+(\cdot)$ and $g_n^+(\cdot)$ denote right-hand derivatives, $g_n^+(0) = -1$ for all n; so $g_n^+(0) \to -1$ as $n \to \infty$ while $g^+(0) = 1$. □

In Example 8-37, the limit of the one-sided derivative is not always the one-sided derivative of the limit function. However,

$$\sup \{x: g^-(x) \le 0\}$$

and

$$\lim_{n \to \infty} [\sup \{x: g_n^-(x) \le 0\}] = \lim_{n \to \infty} c_n = 0$$

so we may hope in (8-96*a*) and (8-96*b*) that $\alpha_n(s) \to \alpha(s)$ even if $J^{(2)}(s, a)$ does not converge to $J^{(2)}(s, a)$. That hope is well met if the convergence of $f_n(s)$ to $f(s)$ is monotone. Exercise B-19 asks you to prove the following result.

Lemma 8-5 Let $g(\cdot), g_1(\cdot), g_2(\cdot), \ldots$ be convex functions on an open convex subset X of $\mathbb{R}$ such that $g_n(x) \to g(x)$ as $n \to \infty$ and $g_n(x) \le g_{n+1}(x)$ for all n and x. Let $g_n^-(x)$ and $g^-(x)$ denote derivatives from the left and $g_n^+(x)$ and $g^+(x)$ derivatives from the right. Then for all $x \in X$

$$g^-(x) \le \liminf_{n \to \infty} g_n^-(x) \le \limsup_{n \to \infty} g_n^+(x) \le g^+(x)$$

The straightforward application of Theorem 8-14 and Lemmas 8-4 and 8-5 to (8-95*a*) and (8-95*b*) and (8-96*a*) and (8-96*b*) yields the next theorem.

Theorem 8-15 For some $s \in \mathcal{S}$ suppose:

(*a*) For each $a \in A_s$ there exists

$$J(s, a) \triangleq \lim_{n \to \infty} J_n(s, a)$$

(*b*) $J_n(s, a) \le J_{n+1}(s, a)$ for each $n \in I$ and $a \in A_s$.

(*c*) A_s is a convex subset of $\mathbb{R}$.

(*d*) For each n, $J_n(s, \cdot)$ is concave and continuous on A_s.

(*e*) In (8-96*a*), $|\alpha_n(s)| < \infty$ for each n.

Then $\alpha(s)$ maximizes $J(s, \cdot)$ on A_s, and

$$\alpha(s) = \lim_{n \to \infty} \alpha_n(s)$$

If assumptions (*a*) through (*d*) are valid for all $s \in \mathcal{S}$ and

(*f*) A_s is a compact set for all $s \in \mathcal{S}$ [so (*e*) is necessarily true], then $f(\cdot)$ satisfies (8-94), the functional equation of dynamic programming. The conclusions remain valid if the inequality in (*b*) is reversed.

Theorem 8-15 concerns an MDP with a maximizing objective. The same result for a minimizing objective would reverse the inequalities in (8-96*a*) and (8-96*b*), replace "concave" with "convex" in (*d*), and replace "maximizes" with "minimizes" in the theorem's conclusion.

An MDP may have many optimal policies; so $a_n(\cdot)$ and $a(\cdot)$ may not be well defined. Convergence of $a_n(s)$ as $n \to \infty$ may depend on the manner in which the set of multiple optima is mapped into $a_n(\cdot)$ (Exercise 8-41). Equation (8-96*a*) is a mapping which yields convergence for concave MDPs; so we use it in Theorem 8-15 [rather than using $a_n(\cdot)$ and $a(\cdot)$ in the statement of the theorem].

Example 8-38: Resource management Examples 8-35 and 8-36 establish convergence of $f_n(\cdot)$ and $a_n(\cdot)$. The examples show that the limit functions $f(\cdot)$ and $a(\cdot)$ are endowed with the concavity and monotonicity properties of their finite-horizon counterparts and that $f(\cdot)$ satisfies (8-94), the functional equation of dynamic programming. It follows from Theorem 8-15 that $a(\cdot)$ attains the optimum in (8-94). □

Example 8-39: Production smoothing Examples 8-31 and 8-34 show that $f_n(s, z)$ is convergent for each (s, z) and $f(\cdot, \cdot)$ satisfies (8-94), the functional equation of dynamic programming. Under the assumptions in those examples and Theorem 8-12, it follows from Theorem 8-15 that the decision rule in (8-74) converges to a decision rule which attains the optimum in (8-94).

In the present case (8-94) can be written

$$f(s, z) = \inf\{c|a - s - z| + Q(s, a): a \geq s\} \qquad s \in \mathbb{R} \qquad z \geq 0$$

$$Q(s, a) = G(a) + \beta E[f(a - D_1, a - s)]$$

The convexity of f and the proof of Theorem 8-12 imply that the limiting decision rule has the structure

$$a = \begin{cases} L(s) & \text{if } s + z < L(s) \\ s + z & \text{if } L(s) \leq s + z < U(s) \\ U(s) & \text{if } s \leq U(s) \leq s + z \\ s & \text{if } U(s) < s \end{cases}$$

where

$$L(s) = \sup\{a: Q^{(2)}(s, a) \leq -c\} \qquad s \in \mathbb{R}$$

$$U(s) = \sup\{a: Q^{(2)}(s, a) \leq c\} \qquad s \in \mathbb{R}$$

Exercise 8-42 asks you to prove that $L(\cdot)$ and $U(\cdot)$ satisfy

$$0 \leq L(s') - L(s) \leq s' - s$$

$$0 \leq U(s') - U(s) \leq s' - s$$

under the assumptions in Exercise 8-31. □

The following argument was used in the previous example:

(i) f_n is convergent.
(ii) The limit function f inherits a property (convexity) from f_n.
(iii) f satisfies (8-94).
(iv) The inherited property of f and (8-94) identify the structure of a decision rule which attains (8-94).

The same argument sometimes succeeds when the property which f inherits is neither convexity nor concavity. By combining Theorems 8-14 and 8-5 and Corollary 8-5*a*, we obtain the next result. Let the dynamics be described as $s_{n+1} \sim \xi(s_n, a_n)$ and let $\gamma_x(s, a) \triangleq P\{\xi(s, a) > x\}$. In this notation

$$J_n(s, a) = r(s, a) + \beta E\{f_{n-1}[\xi(s, a)]\}$$

Theorem 8-16 In an MDP suppose:

(*a*) For each $s \in \mathcal{S}$, A_s is compact.
(*b*) $\mathscr{C}$ is a sublattice of $\mathbb{R}^2$.
(*c*) $f_0(\cdot)$ is nondecreasing and bounded below.
(*d*) For each n and s, $J_n(s, \cdot)$ is continuous on A_s.
(*e*) For each a, $r(\cdot, a)$ is nonincreasing on $\{s\colon (s, a) \in \mathscr{C}\}$.
(*f*) $r(\cdot, \cdot)$ is supermodular and bounded below on $\mathscr{C}$.
(*g*) $\gamma_x(\cdot, \cdot)$ is supermodular on $\mathscr{C}$ for each $x \in \mathbb{R}$.
(*h*) $\{A_s\colon s \in \mathcal{S}\}$ is expanding and ascending.†
(*i*) Either $f_n(s) \leq f_{n+1}(s)$ for all n and s or $f_n(s) \geq f_{n+1}(s)$ for all n and s.
(*j*) For each $s \in \mathcal{S}$ there exists

$$f(s) = \lim_{n \to \infty} f_n(s)$$

and

$$J(s, a) \triangleq \lim_{n \to \infty} J_n(s, a)$$

which satisfies

$$J(s, a) = r(s, a) + \beta E\{f[\xi(s, a)]\}$$

Then $f(\cdot)$ is nondecreasing and there exists $a(\cdot)$ nondecreasing on $\mathcal{S}$ such that

$$f(s) = J[s, a(s)] \qquad s \in \mathcal{S}$$

Example 8-40: Equipment replacement From Examples 8-16, 8-20, 8-29, and 8-32 the following conditions yield an optimal infinite-horizon replacement policy which is a nondecreasing function of equipment age. Let b_s be the probability that an age s equipment breaks down and L_s the salvage value if it does not break down. The cost of a planned replacement is modeled as $c(s, 1) = r - L_s$, where r is the cost of a new piece of equipment.

† From our definition on page 383 $\{A_s\colon s \in \mathcal{S}\}$ is said to be *expanding* if $s \leq s'$ implies $A_s \subset A_{s'}$.

Then the sufficient conditions for a monotone optimal policy are

$$0 \le L_s - L_{s+1} \le c(s+1, 0) - c(s, 0) \qquad \text{and} \qquad b_s \le b_{s+1}$$

for all s. These inequalities are discussed in Example 8-20. □

Algorithms for Monotone Optimal Policies

For the remainder of this section the objective is to maximize the expected present value of a finite MDP. Let $\mathcal{S} = \{1, \dots, S\}$ so $\#\mathcal{S} = S$. For $s \in \mathcal{S}$ let g_s be positive but otherwise arbitrary. Then a restatement of the linear program (4-91) is to find $w_1, \dots, w_S$ in order to

$$\begin{aligned} &\text{Minimize } \sum_{s=1}^{S} g_s w_s \\ &\text{subject to } w_s - \beta \sum_{j=1}^{S} p_{sj}^a w_j \ge r(s, a) \qquad (s, a) \in \mathscr{C} \end{aligned} \tag{8-97}$$

From Theorem 4-11 an optimal solution to (8-97) satisfies $w_s = f(s)$, $s \in \mathcal{S}$. Therefore, monotonicity of $f(\cdot)$ implies monotonicity in s of $\{w_s\}$. To be specific, suppose $f(\cdot)$ is nondecreasing, which is true under the assumptions of Theorem 8-16. Then $w_1 \le w_2 \le \cdots \le w_S$ at optimality in (8-97).

For an equivalent linear program, let

$$d_1 = w_1 \qquad \text{and} \qquad d_s = w_s - w_{s-1} \qquad \text{for } s > 1 \tag{8-98}$$

so $d_s \ge 0$ without loss of optimality. Then $w_s = \sum_{j=1}^{s} d_j$, $s \in \mathcal{S}$. Temporarily, let $\mathscr{G}_s = \sum_{j=s}^{S} g_j$, $s \in \mathcal{S}$. Then substitution of (8-98) in (8-97) yields

$$\begin{aligned} &\min \sum_{s=1}^{S} \mathscr{G}_s d_s \\ &\text{subject to } d_s \ge 0 \qquad s \in \mathcal{S} \\ &\sum_{j=1}^{s} d_j - \beta \sum_{j=1}^{S} d_j \sum_{i=j}^{S} p_{si}^a \ge r(s, a) \qquad (s, a) \in \mathscr{C} \end{aligned} \tag{8-99}$$

which is equivalent to (8-97). Since $g_s > 0$ is arbitrary, the only restriction on $\mathscr{G}_s = \sum_{j=s}^{S} g_j$ is $\mathscr{G}_1 > \mathscr{G}_2 > \cdots > \mathscr{G}_S > 0$. Thus we have established this proposition.

Proposition 8-5 Suppose a finite MDP has a nondecreasing value function $f(\cdot)$. Let $\{w_s\}$ and $\{d_s\}$ be optimal solutions to (8-97) and (8-99), respectively. Then (8-98) is valid.

Linear programs (8-97) and (8-99) have the same number of variables and the same number of constraints [not including the nonnegativity constraints in (8-

99)]. Moreover, we know that $r(\cdot, \cdot) \geq 0$ without loss of generality in a finite MDP; so $w_s \geq 0$, $s \in \mathcal{S}$ may be appended to (8-97). The two linear programs might then appear to require exactly the same numerical effort. In fact, their numerical burdens may differ.

Linear programs with inequality constraints generally converge to an optimum more rapidly if additional constraints are binding at an optimum. If $r(\cdot, \cdot) \geq 0$ so $\{w_s \geq 0\}$ is appended to (8-97), the nonnegativity constraints are superfluous and therefore not binding at an optimum. However, $\{d_s \geq 0\}$ in (8-99) generally includes binding constraints which correspond to the monotonicity of $f(\cdot)$. Therefore, (8-99) may converge to an optimum in fewer iterations than (8-97). On the other hand, its constraint matrix is denser than that of (8-97).

Example 8-41 Let $\mathcal{S} = \{1, 2\}$, $A_1 = \{1, 2, 3\}$, $A_2 = \{1, 2\}$, $r(1, 1) = r(2, 2) = 3$, $r(2, 1) = r(1, 2) = 0$, $r(1, 3) = 2.6$, $p_{11}^1 = p_{22}^2 = p_{12}^2 = p_{21}^1 = 1$, $p_{11}^3 = 7/9$, $p_{12}^3 = 2/9$,

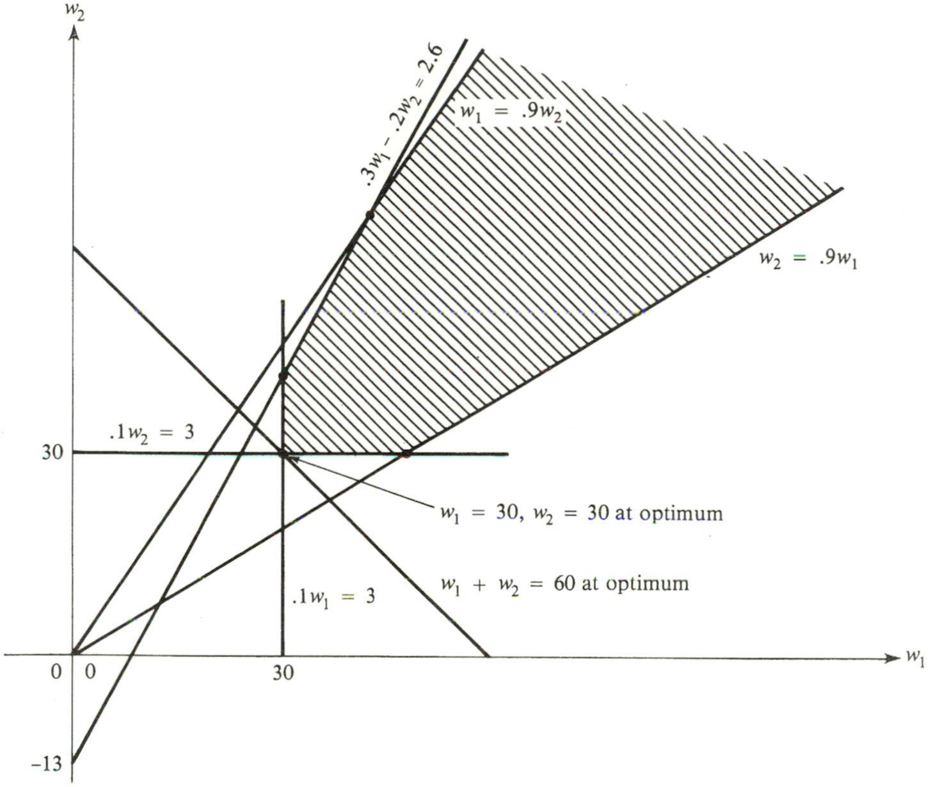

Figure 8-2 The usual linear program (8-97).

and $\beta = 0.9$. Then linear program (8-97), with $g_1 = g_2 = 1$, is

$$\begin{aligned} &\min w_1 + w_2 \\ \text{subject to } &0.1w_1 \geq 3 \\ &w_1 - 0.9w_2 \geq 0 \\ &0.3w_1 - 0.2w_2 \geq 2.6 \\ &-0.9w_1 + w_2 \geq 3 \\ &0.1w_2 \geq 3 \end{aligned}$$

This linear program is shown graphically in Figure 8-2.

Appending the constraint $w_1 \leq w_2$ to the linear program, as shown in Figure 8-3, reduces the feasible region and deletes one of the previously feasible extreme points. Linear program (8-99), with $\mathscr{G}_1 = 2$ and $\mathscr{G}_2 = 1$, is

$$\begin{aligned} &\min 2d_1 + d_2 \\ \text{subject to } &0.1d_1 \geq 3 \end{aligned}$$

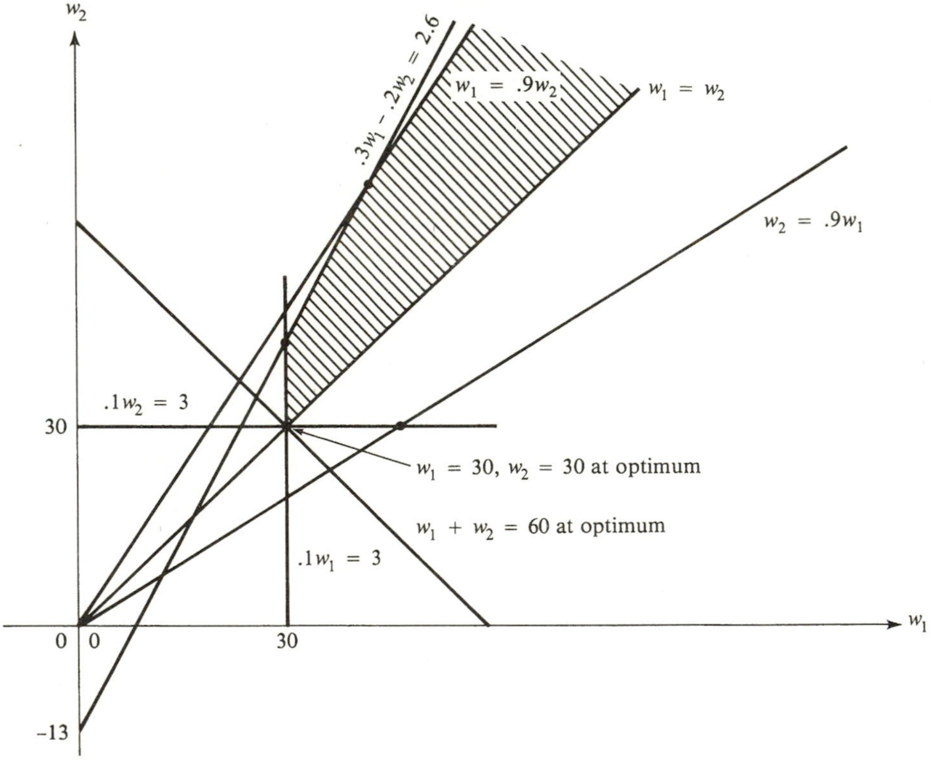

Figure 8-3 The usual linear program (8-97) augmented by $w_j \leq w_{j+1}$ for each j.

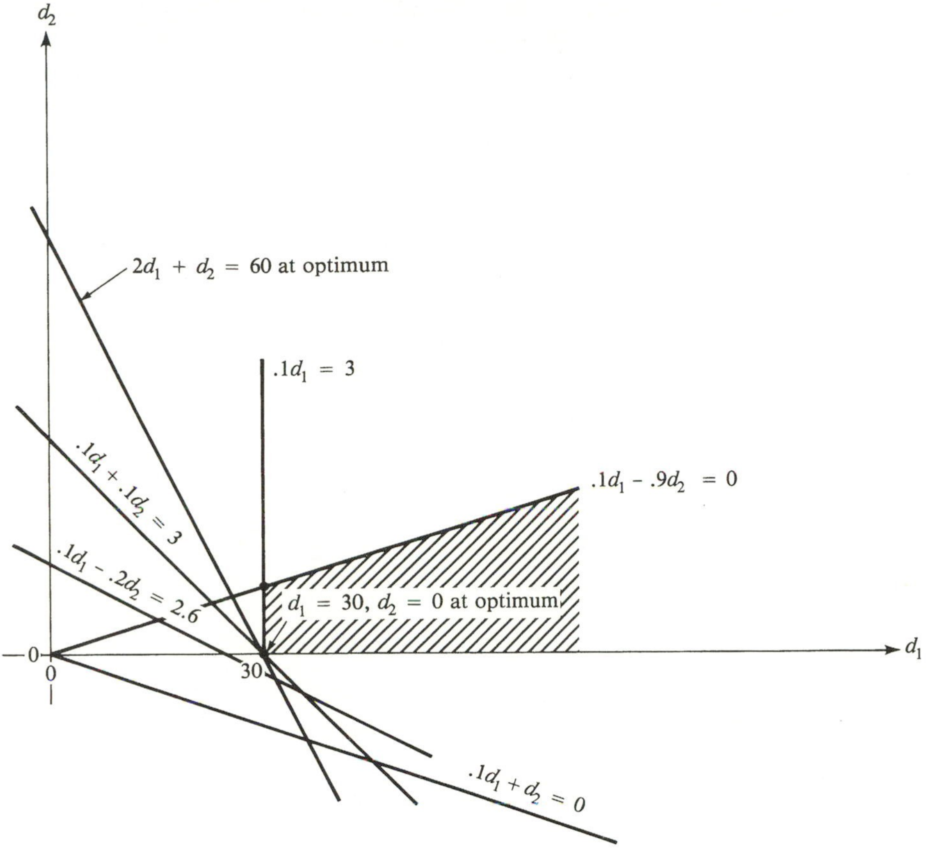

Figure 8-4 The linear program in transformed variables, (8-99).

$$
\begin{aligned}
0.1d_1 - 0.9d_2 &\geq 0 \\
0.1d_1 - 0.2d_2 &\geq 2.6 \\
0.1d_1 + \quad d_2 &\geq 0 \\
0.1d_1 + 0.1d_2 &\geq 3 \\
d_1 \qquad &\geq 0 \\
d_2 &\geq 0
\end{aligned}
$$

This linear program is shown graphically in Figure 8-4. There are fewer feasible extreme points than in Figures 8-2 or 8-3. □

Successive Approximations

Now we present a successive-approximations algorithm for finite MDPs with monotone optimal policies. To be specific, suppose an MDP with $\mathcal{S} = \{1, \ldots, S\}$

is known for each n to possess an n-period optimal decision rule such that $a_n(1) \le a_n(2) \le \cdots \le a_n(S)$. Corollary 8-5$a$ presents sufficient conditions for such monotonicity, and the conditions are satisfied by versions of the equipment-replacement and resource-management models in this section and Section 8-5.

Without loss of generality, suppose $A_s \subset \mathbb{R}$ for each s and $r(s, a) \ge 0$ for all $(s, a) \in \mathscr{C}$. Let $f_0(\cdot)$ be nondecreasing† and

$$f_n(s) = \max \left\{ r(s, a) + \beta \sum_{j=1}^{S} p_{sj}^{a} f_{n-1}(j) \colon a \in A_s \right\} \qquad s \in \mathcal{S} \tag{8-100}$$

for $n \ge 1$. Let $a_n(s)$ be the largest a that attains the maximum in (8-100):

$$a_n(s) = \max \{a\colon a \in A_s \text{ and } f_n(s) = J_n(s, a)\} \qquad s \in \mathcal{S} \tag{8-101}$$

where $J_n(s, a)$ denotes the maximand in (8-100). Since $a_n(s) \ge a_n(s-1)$ without loss of optimality, in (8-100) and (8-101) we may replace the search over $a \in A_s$ with a search over $a \in A_s \cap \{a \ge a_n(s-1)\}$. Therefore, for each n, (8-100) and (8-101) may be replaced with

$$\begin{aligned}
f_n(1) &= \max \{J_n(1, a)\colon a \in A_1\} \\
a_n(1) &= \max \{a\colon a \in A_1 \text{ and } f_n(1) = J_n(1, a)\} \\
f_n(2) &= \max \{J_n(2, a)\colon a \in A_2 \text{ and } a \ge a_n(1)\} \\
a_n(2) &= \max \{a\colon a \in A_2,\, a \ge a_n(1), \text{ and } f_n(2) = J_n(2, a)\} \\
&\vdots \\
f_n(s) &= \max \{J_n(s, a)\colon a \in A_s \text{ and } a \ge a_n(s-1)\} \\
a_n(s) &= \max \{a\colon a \in A_s,\, a \ge a_n(s-1), \text{ and } f_n(s) = J_n(s, a)\} \\
&\vdots \\
f_n(S) &= \max \{J_n(S, a)\colon a \in A_S \text{ and } a \ge a_n(S-1)\} \\
a_n(S) &= \max \{a\colon a \in A_S,\, a \ge a_n(S-1), \text{ and } f_n(S) = J_n(S, a)\}
\end{aligned} \tag{8-102}$$

Some MDPs with monotone optimal policies $a_n(\cdot)$ have the further property that $s - a_n(s)$ is monotone. That is, for all n,

$$0 \le a_n(s') - a_n(s) \le s' - s \qquad s \le s' \tag{8-103}$$

Some versions of the models in Sections 8-3 [part (b) of Theorem 8-7] and 8-4 (Exercise 8-30) have this property. If $\mathcal{S} = \{1, 2, \ldots, S\}$ and $A_s = \{l_s, l_s + 1, \ldots, u_s\}$, $s \in \mathcal{S}$, then (8-103) is equivalent to

$$a_n(s+1) \in \{a_n(s), a_n(s) + 1\} \qquad [a_n(s) < u_{s+1}]$$

which permits the following simplification of (8-102):

$$\begin{aligned}
f_n(1) &= \max \{J_n(1, a)\colon a \in A_1\} \\
a_n(1) &= \max \{a\colon a \in A_1 \text{ and } f_n(1) = J_n(1, a)\} \\
f_n(2) &= \max \{J_n(2, a)\colon a \in A_1 \text{ and } a = a_n(1) \text{ or } a = a_n(1) + 1\}
\end{aligned}$$

† We assume that Corollary 8-5a is the basis for the existence of a monotone optimal policy.

$$a_n(2) = \max \{a: a \in A_2, f_n(2) = J_n(2, a) \text{ and } a = a_n(1) \text{ or } a = a_n(1) + 1\} \tag{8-104}$$

$$\vdots$$

$$f_n(S) = \max \{J_n(S, a): a \in A_S \text{ and } a = a_n(S-1) \text{ or } a = a_n(S-1) + 1\}$$

$$a_n(S) = \max \{a: a \in A_S, f_n(S) = J_n(S, a), \text{ and } a = a_n(S-1) \text{ or } a = a_n(S-1) + 1\}$$

Example 8-42 The MDP is the same as the one in Example 8-41, and let $f_0(\cdot) \equiv 0$ so Corollary 8-5*a* is satisfied. Then

$$f_1(s) = \max \{s - a: 1 \le a \le s\} = s - 1 \qquad s = 1, 2, 3, 4$$

and $a_1(\cdot) \equiv 1$. To the nearest 0.01,

$$f_2(1) = J_2(1, 1) = 0.9[0.8(0) + 0.2(1)] = 0.18 \qquad a_2(1) = 1$$

$$f_2(2) = \max \{J_2(2, 1), J_2(2, 2)\} = \max \{1.18, 1.35\} = 1.35 \qquad a_2(2) = 2$$

$$f_2(3) = \max \{J_2(3, 2), J_2(3, 3)\} = \max \{2.35, 1.89\} = 2.35 \qquad a_2(3) = 2$$

$$f_2(4) = \max \{J_2(4, 2), J_2(4, 3), J_2(4, 4)\} = \max \{3.35, 2.89, 2.52\} = 3.35 \qquad a_2(4) = 2 \tag{8-105a}$$

$$f_3(1) = J_3(1, 1) = 0.9[0.8(0.18) + 0.2(1.35)] = 0.37 \qquad a_3(1) = 1$$

$$f_3(2) = \max \{J_3(2, 1), J_3(2, 2)\} = \max \{1.37, 1.65\} = 1.65 \qquad a_3(2) = 2$$

$$f_3(3) = \max \{J_3(3, 2), J_3(3, 3)\} = \max \{2.65, 2.21\} = 2.65 \qquad a_3(3) = 2$$

$$f_3(4) = \max \{J_3(4, 2), J_3(4, 3), J_3(4, 4)\} = \max \{3.65, 3.21, 2.84\} = 3.65 \qquad a_3(4) = 2 \tag{8-105b}$$

Had (8-100) and (8-101) been used instead of (8-102), the maximization for $f_2(3)$ would have included $J_2(3, 1)$, the one for $f_2(4)$ would have included $J_2(4, 1)$, and similarly for $f_3(3)$ and $f_3(4)$.

Exercise 8-46 asks you to prove that the MDP in this example satisfies (8-103). Therefore, (8-104) may be used instead of (8-102). If (8-104) is used, it replaces (8-105*a*) and (8-105*b*) with

$$f_2(4) = \max \{J_2(4, 2), J_2(4, 3)\} = \max \{3.35, 2.89\} = 3.35 \qquad a_2(4) = 2 \tag{8-106a}$$

$$f_3(4) = \max \{J_3(4, 2), J_3(4, 3)\} = \max \{3.65, 3.21\} = 3.65 \qquad a_3(4) = 2 \tag{8-106b}$$

which involves fewer comparisons than are needed in (8-105*a*) and (8-105*b*). Also, $J_2(4, 4)$ and $J_3(4, 4)$ are never computed when (8-104) is used instead of (8-102). □

EXERCISES

8-39 What are sufficient conditions for the following functions to have the stated properties?

(*a*) $f(\cdot)$ in Examples 8-29 and 8-32 to be nondecreasing?

(*b*) $f(\cdot)$ in Example 8-35 to be a concave function?

(*c*) $f(\cdot, \cdot)$ in Examples 8-31 and 8-34 to be a convex function?

8-40 Section 7-1 states sufficient conditions for $f_n(\cdot)$ in the model of Examples 8-30 and 8-33 to be K-convex on $\mathbb{R}$. State sufficient conditions for $f(\cdot)$ in those examples to be K-convex (prove that the conditions are sufficient).

8-41 Specify an MDP with the criterion of maximizing the expected present value, and a sequence $a_1(\cdot)$, $a_2(\cdot)$, ... of finite-horizon optimal policies such that the assumptions of Theorem 8-15 are satisfied and $a_1(s)$, $a_2(s)$, ... does not converge for any $s \in \mathcal{S}$. (Hint: An MDP with one state is sufficient.)

8-42 Formulate linear program (8-99) for the equipment-replacement model in Examples 8-29, 8-32, and 8-39. Remember that costs are being minimized in this model, whereas rewards are being maximized in the MDP which leads to (8-99).

8-43 The replacement of (8-100) and (8-101) with (8-102) can be improved if $s \le s'$ implies

$$0 \le a_n(s') - a_n(s) \quad \text{and} \quad a_n(s') - a_n(s) \le s' - s \tag{8-107}$$

Some versions of the models in Sections 8-3 and 8-4 (and in the exercises in those sections) were shown to satisfy (8-107). Without loss of generality, let $A_s = \{1, 2, \ldots, m_s\}$ and $m_s \le m_{s+1}$ for Corollary 8-5*a* to be valid. For $s > 1$ in (8-102), show that (8-107) implies

$$f_n(s) = \max \{J_n(s, a)\colon a = a_n(s-1) \text{ or } a = a_n(s-1) + 1, \text{ and } a \in A_s\}$$

$$a_n(s) = \max \{a\colon f_n(s) = J_n(s, a),\ a \in A_s, \text{ and } a = a_n(s-1) \text{ or } a = a_n(s-1) + 1\}$$

so, in effect, $\# A_s \le 2$.

8-44 Suppose that the assumptions of Exercises 8-30 and 8-31 are valid for the production-smoothing model, and let $L(\cdot)$ and $U(\cdot)$ be defined as in Example 8-42. Then Exercise 8-31 gives numbers L and U such that $L \le L(s) \le U(s) \le U$ for all s. Suppose demand is an integer-valued r.v. such that $P\{D_1 \le m\} = 1$ for some $m < \infty$. Assume $s_1 \le U$ and that the limiting decision rule in Example 8-39 is used as a stationary policy.

(*a*) Let $v = \inf \{s\colon U(s) = s\}$. Prove $s_n \le U$ for all n (that is, on every sample path).

(*b*) Recall that $z_n = a_n - s_n$ is the production quantity in period n and that the state space is $\mathcal{S} = \{(s, z)\colon s \in \mathbb{R}, z \ge 0\}$. Let $h = U - L + m$. If $s_1 \le U$, prove that $(s_n, z_{n-1}) \in \mathcal{S}^*$ for all $n \ge 2$, where $\mathcal{S}^*$ is the following finite subset of $\mathcal{S}$:

$$\mathcal{S}^* = \{(s, z)\colon L - m \le s \le U,\ 0 \le z < h, \text{ and } s \text{ and } z \text{ are integers}\}$$

Therefore, the MDP is finite without loss of optimality.

(*c*) For $(s, z) \in \mathcal{S}^*$ prove†

$$f(s, z) = \begin{cases} c(L - s - z) + f(s, L - s) & \text{if } s + z < L \\ c(z + s - U) + f(s, U - s) & \text{if } s + z > U \end{cases} \tag{8-108}$$

† It can be shown that substitution of (8-108) in $E[f(a - D_1, a - s)]$ makes it possible to reduce $\mathcal{S}^*$ for computational purposes to

$$\{(s, z)\colon L - m \le s \le U,\ (L - s)^+ \le z \le U - s, \text{ and } s \text{ and } z \text{ are integers}\}.$$

8-45 (Continuation) (*a*) Under the assumptions of Exercise 8-44 (not including the use of the stationary policy) explain why the following computation of $L_n(s)$ and $U_n(s)$ is valid [see (8-75*a*) and (8-75*b*) for definitions of $L_n(s)$ and $U_n(s)$], $L - m \leq s \leq U$. All values of a are restricted to be integers and

$$Q_n(s, a) \triangleq G(a) + \beta E[f_{n-1}(a - D_1, a - s)]$$

First, compute

$$L_n(L - m) = \max \{a: Q_n(L - m, a) - Q_n(L - m, a - 1) \leq -c, \quad L \leq a \leq U\}$$

$$U_n(L - m) = \max \{a: Q_n(L - m, a) - Q_n(L - m, a - 1) \leq c, \quad L_n(L - m) \leq a \leq U\}$$

Then in turn, for $s = L - m + 1, L - m + 2, \ldots, U$ compute

$$L_n(s) = \begin{Bmatrix} L_n(s-1) \\ L_n(s-1) + 1 \end{Bmatrix} \quad \text{if } Q_n[s, L_n(s-1) + 1] - Q_n[s, L_n(s-1)] \begin{Bmatrix} > \\ \leq \end{Bmatrix} - c$$

$$U_n(s) = \begin{Bmatrix} U_n(s-1) \\ U_n(s-1) + 1 \end{Bmatrix} \quad \text{if } Q_n[s, U_n(s-1) + 1] - Q_n[s, U_n(s-1)] \begin{Bmatrix} > \\ \leq \end{Bmatrix} c$$

(*b*) Show that the procedure in (*a*) requires at most $3(U - L) + 2m - 1$ subtractions and comparisons, whereas (8-75*a*) and (8-75*b*) would require at most $(U - L + 1)^2 + m(U - L + 1)$ such operations [assuming for each s that $L_n(s)$ is computed first and then $a \geq L_n(s)$ is included in (8-75*b*)].

8-46 For the MDP in Examples 8-41 and 8-42, prove for each n that there is an optimal decision rule $a_n(\cdot)$ which satisfies (8-103) [which is (8-107)].

BIBLIOGRAPHIC GUIDE

Many authors have shown that a particular model has a monotone optimal policy. Kalmykov (1962) and Daley (1968) contain related results for Markov processes. Topkis (1968) presents the first general framework for monotone optimal policies. Section 1 (Lattice Programming) is based on Topkis (1978). See page 511 of the important paper Whitney (1935) for an early antecedent.

The analysis of the "stopping problem" in Section 8-2 hardly scratches the surface of the large and elegant literature on this subject. See Breiman (1964) and Chow, Robbins, and Siegmund (1971) for expositions and references. The idea of monotone stopping problems is introduced in Derman and Sacks (1960) and Chow and Robbins (1961). Other examples in Section 8-2 include MDPs which model equipment replacement. See the surveys McCall (1965) and Pierskalla and Voelker (1976) and the book Jorgenson, McCall, and Radner (1967) for many more models and references. Some papers which emphasize the MDP properties in replacement and maintenance models include Derman (1963*a*, 1963*b*), Kalymon (1972), Klein (1962), Kolesar (1967), Rosenfield (1976), and Ross (1971). White (1979) has a model in which the state is only partially observable.

Section 8-3 is related to normative models in at least four literatures: capital accumulation, fisheries management, reservoir operation, and exploration and mining of nonrenewable resources. The earliest work in the modern genre is Massé (1964) (the work was completed decades earlier than the publication date). Phelps (1962) and Levhari and Srinavasan (1969) attracted attention by economists, and Brock and Mirman (1972) is a sophisticated analysis. Much of Section 8-3 follows Mendelssohn and Sobel (1980).

The cash-management model in Section 8-4 is drawn from a literature which includes Eppen and Fama (1969), Heyman (1973), Neave (1970), and Porteus (1972). See copies of the *Journal of Cash Management* (Volume 1 in 1981) for institutional details of cash management. The queueing-optimization model in Section 8-4 is a variant of Huang, Brumelle, Sawaki, and Vertinsky (1977). See the Bibliographic Guide at the end of Chapter 7 for other references to the literature on optimal operation of queues. The analysis of the production-smoothing model extends Beckman (1961).

Section 8-5 is a generalization of results obtained by many authors for specific models. Our proof of Theorem 8-14 is drawn from Iglehart (1963). Stidham and van Nunen (1981) extend Theorems 8-13, 8-14, and 8-15 in ways which are particularly appropriate for models of inventories, production, queues, and replacement.

Lattice programming

Birkhoff, G.: *Lattice Theory*, 3d ed., American Mathematics Society Colloquium Publications 25, Providence, R.I. (1967).

Daley, D. J.: "Stochastically Monotone Markov Chains," *Z. Wahrscheinlichkeitstheorie Verwandte Gebiete* **10**: 305–317 (1968).

Kalmykov, G. I.: "On the Partial Ordering of One-Dimensional Markov Processes," *Theory of Probability and Its Applications* **7**: 456–459 (1962).

Serfozo, R. F.: "Monotone Optimal Policies for Markov Decision Processes," *Mathematical Programming Study* **6**: 202–215 (1976).

Topkis, D. T.: *Ordered Optimal Solutions*, Ph.D. dissertation, Stanford University, Stanford, Calif. (1968).

———: "Minimizing a Submodular Function on a Lattice," *Oper. Res.* **26**: 305–321 (1978).

———: "Applications of Minimizing a Subadditive Function on a Lattice," unpublished manuscript (1978).

——— and A. F. Veinott, Jr.: "Isotone Solutions of Extremal Problems on a Lattice," unpublished manuscript (1972).

White, C. C. III: "The Optimality of Isotone Strategies for Markov Decision Problems with Utility Criterion," in *Recent Developments in Markov Decision Processes*, edited by R. Hartley, L. C. Thomas, and D. J. White, Academic Press, New York (1980), pp. 261–275.

———: "Monotone Control Laws for Noisy Countable-State Markov Chains," *Eur. J. Oper. Res.* **5**: 124–132 (1980).

Whitney, H.: "On the Abstract Properties of Linear Dependence," *Am. J. Math.* **57**: 509–533 (1935).

Replacement, stopping, and other binary decision models

Breiman, L.: "Stopping Rule Problems," in *Applied Combinatorial Mathematics*, edited by E. F. Beckenbach, Wiley, New York (1964), pp. 284–319.

Chow, Y. S., and H. Robbins: "A Martingale System Theorem and Applications," in *Proceedings of the Fourth Berkeley Symposium on Mathematical Statistics and Probability*, vol. 1, edited by J. Neyman, University of California Press, Berkeley (1961), pp. 93–104.

———, ———, and D. Siegmund: *Great Expectations: The Theory of Optimal Stopping*, Houghton Mifflin, Boston (1971).

Derman, C.: "On Optimal Replacement Rules When Changes of State Are Markovian," in *Mathematical Optimization Techniques*, edited by R. Bellman, University of California Press, Berkeley (1963*a*), pp. 201–210.

———: "Optimal Replacement under Markovian Deterioration with Probability Bounds on Failure," *Manage. Sci.* **9**: 478–481 (1963*b*).

——— and J. Sacks: "Replacement of Periodically Inspected Equipment (An Optimal Stopping Rule)," *Nav. Res. Logistics Quart.* **7**: 597–607 (1960).

Jorgenson, D. W., J. J. McCall, and R. Radner: *Optimal Replacement Policy*, North-Holland, Amsterdam (1967).

Kalymon, B. A.: "Machine Replacement with Stochastic Costs," *Manage. Sci.* **18**: 288–298 (1972).

Klein, M.: "Inspection-Maintenance-Replacement Schedules under Markovian Deterioration," *Manage. Sci.* **9**: 469–475 (1962).

Kolesar, P.: "Minimum Cost Replacement under Markovian Deterioration," *Manage. Sci.* **12**: 694–766 (1966).

McCall, J. J.: "Maintenance Policies for Stochastically Failing Equipment: A Survey," *Manage. Sci.* **21**: 493–525 (1965).

Miller, B. L.: "Countable State Average Cost Regenerative Stopping Problems," *J. Appl. Probability* **18**: 361–377 (1981).

Pierskalla, W. P., and J. A. Voelker: "A Survey of Maintenance Models: The Control and Surveillance of Deteriorating Systems," *Nav. Res. Logistics Quart.* **23**: 353–388 (1976).

Rosenfield, D.: "Markovian Deterioration with Uncertain Information," *Oper. Res.* **24**: 141–154 (1976).

Ross, S.: "Quality Control under Markovian Deterioration," *Manage. Sci.* **17**: 587–596 (1971).

White, C. C. III: "Optimal Control-Limit Strategies for a Partially Observed Replacement Model," *Int. J. Syst. Sci.* **10**: 321–331 (1978).

Resource management

Abrams, R., and U. S. Karmarkar: "Optimal Multiperiod Investment-Consumption Policies," *Econometrica* **48**: 333–353 (1980).

Bewley, T.: "The Permanent Income Hypothesis: A Theoretical Formulation," *J. Econ. Theory* **16**: 252–292 (1977).

Brock, W. A., and L. J. Mirman: "Optimal Economic Growth and Uncertainty: The Discounted Case," *J. Econ. Theory* **4**: 479–513 (1972).

Deshmukh, S. D., and S. R. Pliska: "Optimal Consumption and Exploration of Nonrenewable Resources under Uncertainty," *Econometrica* **48**: 177–200 (1980).

Hakansson, N.: "Optimal Investment and Consumption Strategies under Risk for a Class of Utility Functions," *Econometrica* **38**: 587–607 (1970).

Iglehart, D. L.: "Capital Accumulation and Production for the Firm: Optimal Dynamic Policies," *Manage. Sci.* **12**: 193–205 (1965).

Kleindorfer, P., and H. Kunreuther: "Stochastic Horizons for the Aggregate Planning Problem," *Manage. Sci.* **24**: 485–497 (1978).

Levhari, D., and T. Srinivasan: "Optimal Savings under Uncertainty," *Rev. Econ. Studies* **36**: 153–164 (1969).

Massé, P.: *Les Réserves et la Régulation d'Avenir dans la Vie Économique*, 2 vols., Hermann, Paris (1946).

Massy, W. F., R. C. Grinold, D. S. P. Hopkins, and A. Gerson: "Optimal Smoothing Rules for University Financial Planning," *Oper. Res.* **29**: 1121–1136 (1981).

Mendelssohn, R.: "Optimal Harvesting Strategies for Stochastic, Single Species, Multiage Class Models," *Math. Biosci.* **41**: 159–174 (1978).

———: "Managing Stochastic Multispecies Models," *Math. Biosci.* **49**: 249–262 (1980).

——— and M. J. Sobel: "Capital Accumulation and the Optimization of Renewable Resource Models," *J. Econ. Theory* **23**: 243–260 (1980).

Miller, B. L.: "Optimal Consumption with a Stochastic Income Stream," *Econometrica* **42**: 253–266 (1974).

Phelps, E.: "The Accumulation of Risky Capital," **30**: 729–743 (1962).

Cash management and other smoothing problems

Beckmann, M. J.: "Production Smoothing and Inventory Control," *Oper. Res.* **9**: 446–467 (1961).

Beja, A.: "The Optimality of Connected Policies for Markovian Systems with Two Types of Service," *Manage. Sci.* **18**: 683–686 (1972).

Crabill, T. B.: "Optimal Control of a Service Facility with Variable Exponential Service Times and Constant Arrival Rate," *Manage. Sci.* **18**: 560–566 (1972).

Eppen, G. D., and E. F. Fama: "Solutions for Cash-Balance and Simple Dynamic-Portfolio Problems," *J. Business* **41**: 94–112 (1968).

——— and ———: "Cash Balance and Simple Dynamic Portfolio Problems with Proportional Costs," *Int. Econ. Rev.* **10**: 119–113 (1969).

Gallisch, E.: "On Monotone Optimal Policies in a Queueing Model of $M/G/1$ Type with Controllable Service Time Distribution," *Adv. Appl. Probability* **11**: 870–887 (1979).

Girgis, N. M.: "Optimal Cash Balance Levels," *Manage. Sci.* **15**: 130–140 (1968).

Heyman, D. P.: "A Model for Cash Balance Management," *Manage. Sci.* **19**: 1407–1413 (1973).

Huang, C. C., S. L. Brumelle, K. Sawaki, and I. Vertinsky: "Optimal Control for Multi-Server Queueing Systems under Periodic Review," *Nav. Res. Logistics Quart.* **24**: 127–135 (1977).

Kleindorfer, P. R., and K. Glover: "Linear Convex Stochastic Optimal Control with Applications in Production Planning," *IEEE Trans. Auto. Control* **18**: 56–59 (1973).

——— and H. Kunreuther: "Stochastic Horizons for the Aggregate Planning Problem," *Manage. Sci.* **23**: 485–497 (1978).

Lehoczky, J., S. Sethi, and S. Shreve: "Optimal Consumption and Investment Policies Allowing Consumption Constraints, Bankruptcy, and Welfare," *Math. Oper. Res.*, to appear (1983).

Lippman, S. A.: "Applying a New Device in the Optimization of Exponential Queueing Systems," *Oper. Res.* **23**: 687–710 (1975).

Lu, F. V., and R. Serfozo: "M/M/1 Queueing Decision Processes with Monotone Hysteretic Optimal Policies," *Oper. Res.* **31**: to appear (1983).

Mendelssohn, R.: "The Use of Markov Decision Models and Related Techniques for Purposes Other than Simple Optimization: Analyzing the Consequences of Policy Alternatives on the Management of Salmon Runs," *Fish. Bull.* **78**: 35–50 (1980).

Neave, E. H.: "The Stochastic Cash Balance Problem with Fixed Costs for Increases and Decreases," *Manage. Sci.* **16**: 474–490 (1970).

Orr, D.: "A Random Walk Production-Inventory Policy: Rationale and Implementation," *Manage. Sci.* **9**: 108–122 (1963).

Porteus, E. L.: "Equivalent Formulations of the Stochastic Cash Balance Problem," *Manage. Sci.* **19**: 250–253 (1972).

——— and E. H. Neave: "The Stochastic Cash Balance Problem with Charges Levied against the Balance," *Manage. Sci.* **18**: 600–602 (1972).

Schassberger, R.: "A Note on Optimal Service Selection in a Single Server Queue," *Manage. Sci.* **21**: 1326–1331 (1975).

Serfozo, R.: "Optimal Control of Random Walks, Birth and Death Processes, and Queues," *Adv. Appl. Probability* **13**: 61–83 (1981).

Sobel, M. J.: "Production Smoothing with Stochastic Demand II: Infinite Horizon Case," *Manage. Sci.* **17**: 724–735 (1971).

———: "Making Short-Run Changes in Production When the Employment Level Is Fixed," *Oper. Res.* **18**: 35–51 (1970).

Optimal infinite-horizon policies

Bellman, R., I. Glicksberg, and O. Gross: "On the Optimal Inventory Equation," *Manage, Sci.* **2**: 83–104 (1955).

Blackwell, D.: "Positive Dynamic Programming," *Proceedings Fifth Berkeley Symposium Mathematical Statistics and Probability*, edited by L. M. Le Cam and J. Neyman, University of California Press, Berkeley (1967), vol. I, pp. 415–418.

Harrison, J. M.: "Discrete Dynamic Programming with Unbounded Rewards," *Ann. Math. Stat.* **43**: 636–644 (1972).

Iglehart, D. L.: "Optimality of (*s*, *S*) Policies in the Infinite Horizon Inventory Problem," *Manage. Sci.* **9**: 259–267 (1963).

———: "Dynamic Programming and Stationary Analysis of Inventory Problems," chap. 1 in *Multi-stage Inventory Models and Techniques*, edited by H. Scarf, D. Gilford, and M. Shelly, Stanford University Press, Stanford, Calif. (1963).

Morton, T. E., and W. E. Wecker: "Discounting, Ergodicity, and Convergence for Markov Decision Processes," *Manage. Sci.* **23**: 890–900 (1977).

Porteus, E. L.: "On the Optimality of Structured Policies in Countable Stage Decision Processes," *Manage. Sci.* **22**: 148–157 (1975).

Schäl, M.: "Conditions for Optimality in Dynamic Programming and for the Limit of *n*-stage Optimal Policies to Be Optimal," *Z. Wahrscheinlichkeitstheorie* **32**: 179–196 (1975).

Stidham, S., Jr., and J. van Nunen: "The Shift-Function Approach for Markov Decision Processes with Unbounded Returns," Technical Report 60, Department of Operations Research, Stanford University, Stanford, Calif. (1981).

Strauch, R. E.: "Negative Dynamic Programming," *Ann. Math. Stat.* **37**: 871–890 (1966).

CHAPTER

NINE

SEQUENTIAL GAMES

A sequential game is a multiperson decision process in which each participant makes a sequence of decisions. The participants' sequences of decisions influence the evolution of the process and affect the time streams of rewards to the participants. A sequential game is also called a *stochastic game* and a *Markov game.*

Sequential game models have been constructed of diverse phenomena in management science, biology, economics, psychology, and military affairs. Some of the phenomena are arms control and disarmament, advertising decisions of competing firms, pricing and production decisions of competing firms, interactions of biological species, harvesting decisions in a fishery, the entry and exit of firms to and from an industry, pursuit-evasion tactics for opposing submarines, duels between opposing aircraft, and various paradigms in experimental social psychology.

The dynamic games in this chapter are discrete in time. There is a largely separate literature on continuous-time sequential games called *differential games.* The Bibliographic Guide at the end of the chapter includes several general references on differential games.

A sequential game is a natural generalization of both Markov decision processes (MDPs) and "static" game theory. Therefore, this chapter leans heavily on earlier chapters. In particular, Chapter 4 influences the entire chapter.

9-1 THE MODEL

The canonical elements of a sequential game model include nonempty sets $\mathcal{Q}$ of *players*, $\mathcal{S}$ of *states*, and A_s^q of alternative *actions* for each $q \in \mathcal{Q}$ and $s \in \mathcal{S}$. The other canonical elements are a *transition function* and a *single-stage reward function*. The general idea is that (i) each player must make a sequence of decisions, and (ii) the state of affairs when the players are about to make their nth decisions (for each n) is adequately summarized by some state $s \in \mathcal{S}$.

The "adequately" in (ii) has three qualifications. First, the constraining effects of the past history on player q's set of feasible alternative actions is completely specified by s in A_s^q. Second, the subsequent state of affairs is conditionally independent of the past history given the current state and current action. Third, the immediate reward, possibly an r.v., is conditionally independent of the past history given the current state and current action. These are Markovian assumptions.

The sample path (or outcome) of a sequential game specifies the successive states and actions. Let a_n^q denote the action taken by player q in period n, let $\mathbf{a}_n = (a_n^q, q \in \mathcal{Q})$ be the vector of all players' actions in period n, and let s_n denote the state at the beginning of period n. In order to be consistent with our MDP notation, let

$$A_s = \underset{q \in \mathcal{Q}}{\times} A_s^q$$

$$\mathscr{C} = \{(s, a) \colon a \in A_s,\ s \in \mathcal{S}\}$$

Hence, A_s is the set of feasible action vectors (actions of all players) in period n if the state is $s_n = s$.

Definitions and Assumptions

Only in this subsection, we use boldface type to denote random quantities and ordinary typeface to denote the values taken by random quantities.

Definition 9-1 The *history* up to the time at which the nth action is taken is

$$H_n \triangleq (s_1, a_1, s_2, a_2, \ldots, s_{n-1}, a_{n-1}, s_n) \tag{9-1}$$

As with an MDP, at time $j < n$, it is assumed that all the players know H_j but that they do not know $\mathbf{s}_n$ and $\mathbf{a}_n$. The quantities $\mathbf{s}_n$ and $\mathbf{a}_n$ are r.v.'s because they depend on how the game evolves during periods $j + 1, j + 2, \ldots, n$.

Assumption 9-1 For any $J \subset \mathcal{S}$,

$$P\{\mathbf{s}_{n+1} \in J \mid \mathbf{H}_n, \mathbf{a}_n\} = P\{\mathbf{s}_{n+1} \in J \mid \mathbf{s}_n, \mathbf{a}_n\} \tag{9-2}$$

This Markovian assumption implies existence of a transition function $p(\cdot \mid \cdot, \cdot)$ defined as follows.

Definition 9-2 The *transition function* $p(\cdot \mid \cdot, \cdot)$ satisfies

$$p(J \mid s, a) = P\{\mathbf{s}_{n+1} \in J \mid \mathbf{s}_n = s, \mathbf{a}_n = a\} \qquad n \in I_+ \qquad (s, a) \in \mathscr{C}$$

When the set of states $\mathcal{S}$ is discrete, we write $p_{sj}(a)$ for $p(\{j\} \mid s, a)$.

Let $\mathbf{X}_{nq}$ denote the reward received by player q in period n. Here is the Markovian assumption concerning $\mathbf{X}_{nq}$.

Assumption 9-2 For each $n \in I_+$ and $q \in \mathscr{Q}$, $E(\mathbf{X}_{nq} \mid \mathbf{H}_n, \mathbf{a}_n) = E(\mathbf{X}_{nq} \mid \mathbf{s}_n, \mathbf{a}_n)$.

Definition 9-3 The *single-stage reward function* is

$$r_q(s, a) = E(\mathbf{X}_{nq} \mid \mathbf{s}_n = s, \mathbf{a}_n = a) \qquad (s, a) \in \mathscr{C} \qquad q \in \mathscr{Q}$$

The extant sequential game theory is based on the assumption that all the players have the same planning horizon. Let N denote that planning horizon (possibly infinity). We call N the *duration* of the game. If $N < \infty$, let $L_q(s)$ denote the salvage value received by player q if the ultimate state is $\mathbf{s}_{N+1} = s$. In this chapter we assume that N is not an r.v. because a sequential game with a random duration is equivalent to a sequential game whose duration is not random (see Exercise 9-1).

Definition 9-4 The *duration* of the game N is a positive integer or infinity. The *salvage-value* function is

$$L_q(s) = E(\mathbf{X}_{N+1,q} \mid \mathbf{s}_{N+1} = s) \qquad s \in \mathcal{S} \qquad q \in \mathscr{Q}$$

Definition 9-5 A *sequential game* (SG) is a model which consists of the non-empty sets $\mathscr{Q}$, $\mathcal{S}$, and A_s^q, $s \in \mathcal{S}$ and $q \in \mathscr{Q}$, the transition function, the single-stage reward function, the duration N, and the salvage-value function if $N < \infty$, and which satisfies Assumptions 9-1 and 9-2. Let Q denote the number of players, i.e., number of elements in the set $\mathscr{Q}$.

Definition 9-6 An SG is *finite* if $\mathscr{C}$ is a finite set.

It follows from the definition that a finite SG has only finitely many players, finitely many states, and each player in each state has only finitely many alternative actions.

Let β_q be player q's discount factor, $0 \leq \beta_q \leq 1$. Most solution concepts for

SGs (as is true for MDPs) concern expected values of the following r.v.'s:

$$\mathbf{V}^q(N) = \sum_{n=1}^{N} \beta_q^{n-1} r_q(\mathbf{s}_n, \mathbf{a}_n) + \beta_q^N L_q(\mathbf{s}_{N+1}) \qquad (N < \infty) \tag{9-3a}$$

$$\mathbf{V}^q = \sum_{n=1}^{\infty} \beta_q^{n-1} r_q(\mathbf{s}_n, \mathbf{a}_n) \qquad (\beta_q < 1) \tag{9-3b}$$

and

$$\mathbf{G}^q = \liminf_{M \to \infty} \frac{1}{M} \sum_{n=1}^{M} r(\mathbf{s}_n, \mathbf{a}_n) \tag{9-3c}$$

Of course, these are the sums of discounted rewards for finite and infinite durations, and the average reward per period.

In Sections 4-2 and 4-3 we discuss the notion of a policy in an MDP. A policy in an MDP is a nonanticipative and deterministic contingency plan for making feasible decisions. A *policy for player q* in an SG is exactly the same except that randomizations are permitted. Hence, a policy $\Pi_q = (\pi_{1q}, \pi_{2q}, \ldots, \pi_{Nq})$ for player q is a sequence (countably infinite if $N = \infty$) such that $\pi_{nq}(H_n)$ is a probability distribution on A_s^q if s_n, the last element of H_n, specifies $s_n = s$.

Definition 9-7 A *policy* $\Pi = (\Pi_q; q \in \mathcal{Q})$ is a Q-tuple consisting of a policy for player q, for each $q \in \mathcal{Q}$.

Our definition of policy is restricted to the *behavior strategies* of game theory. Briefly, a behavior strategy separates for each n the randomization for period $n + 1$'s decision from the randomizations for decisions in periods 1 through n. Example 9-5 below specifies a policy which is not a behavior strategy. The restriction to behavior strategies is without loss of optimality in the following sense. For the solution concepts in Section 9-2, if $\#\mathcal{C} < \infty$ and the other players are using behavior strategies, you may confine yourself to the class of behavior strategies without loss of optimality.

Definition 9-7 admits policies for player q which are not Markov policies (cf. Definition 4-11). In words, a policy for player q permits the nth decision to depend on more of the past history than merely the nth state. The results in Section 9-3 state sufficient conditions for existence of an SG solution which *is* a Markov policy.

We amend the notation in (9-3*a*), (9-3*b*), and (9-3*c*) to indicate the dependence on the players' policy: $\mathbf{V}^q(\Pi, N)$, $\mathbf{V}^q(\Pi)$, and $\mathbf{G}^q(\Pi)$. Let lowercase letters denote expected values† of these r.v.'s:

$$v_s^q(\Pi, N) = E[\mathbf{V}^q(\Pi, N) \mid \mathbf{s}_1 = s] \tag{9-4a}$$

$$v_s^q(\Pi) = E[\mathbf{V}^q(\Pi) \mid \mathbf{s}_1 = s] \tag{9-4b}$$

and

$$g_s^q(\Pi) = E[\mathbf{G}^q(\Pi) \mid \mathbf{s}_1 = s] \tag{9-4c}$$

† As with MDPs, sufficiently general models admit policies Π for which $V^q(\Pi, N)$, $V^q(\Pi)$, or $G^q(\Pi)$ may not be bona fide r.v.'s or, if they are, for which the expectations may not exist. We do not treat such general models.

Table 9-1 Movements of the nickel-tossing game

	Outcome of flip	
Jeff's choice	Heads	Tails
+	Mutt gives Jeff 1 cent	Jeff gives Mutt 1 cent
−	Jeff gives Mutt 1 cent	Mutt gives Jeff 1 cent

For the remainder of the chapter, we revert to our customary use of boldface type to denote vectors.

Example 9-1: Tossing Nickels Two gamblers, Mutt and Jeff, are going to play the following game. Mutt has two nickels. Nickel 1 is biased so that its probability of heads is 1/3; nickel 2 has probability 3/4 of heads. Mutt and Jeff each have five pennies. At each play of the game, Mutt chooses which of his nickels he will flip and, simultaneously, Jeff decides positive (+) or negative (−). Depending on whether the flipped nickel falls heads (H) or tails (T), and whether Jeff decided on + or −, one of the players gives a penny to the other according to Table 9-1. The game ends when one of the players has all 10 pennies (nickels are not convertible to pennies).

Here, $\mathcal{Q} = \{1, 2\}$; let $q = 1$ label Mutt and $q = 2$ label Jeff. Let state s label Mutt's penny holdings so Jeff holds 10-s pennies, the initial state is $s_1 = 5$, and $\mathcal{S} = \{0, \ldots, 10\}$. State $s = 0$ indicates that Mutt had no pennies when the game ended and Jeff had 10. Similarly, $s = 10$ indicates termination with Mutt holding 10 pennies. Mutt's actions are chosen from $A_s^1 = \{1, 2\}$, where 1(2) indicates nickel 1(2), if $s \in \{1, 2, \ldots, 9\}$. Similarly, Jeff's actions are chosen from $A_s^2 = \{+, -\}$, if $s \in \{1, \ldots, 9\}$. If $s \in \{0, 10\}$, let $A_s^1 = A_s^2 = \{0\}$, where "action 0" is a "dummy action."

Most of the transition probabilities are given in Table 9-2, which specifies $p_{s,s+1}(a) = 1 - p_{s,s-1}(a)$ if $s \in \{1, \ldots, 9\}$. For these values of s, $p_{sj}(a) = 0$ if $j \notin \{s+1, s-1\}$. Finally, let $p_{0,0}(0, 0) = p_{10,10}(0, 0) = 1$ so states 0 and 10 are absorbing.

The single-stage reward function specifies $r_1[10, (0, 0)] = r_2[0, (0, 0)] = 0$. In words, Mutt keeps 10 pennies if the game ends at $s = 10$, and Jeff keeps 10 pennies if the game ends at $s = 0$. The remaining values of the single-stage reward function are $r_2(s, a) = -r_1(s, a)$ with $r_1(s, a)$ given below if $s \in \{1, \ldots, 9\}$.

a	$r_1(s, a)$
(1, +)	1/3
(1, −)	−1/3
(2, +)	−1/2
(2, −)	1/2

Table 9-2 $p_{s,s+1}(a^1, a^2) = 1 - p_{s,s-1}(a^1, a^2)$ **if** $1 \le s \le 9$

	Jeff's action A^2	
Mutt's action A^1	+	−
1	2/3	1/3
2	1/4	3/4

For example, if $a = (2, +)$, Mutt wins $+1$ if the coin falls tails and he wins -1 if the coin falls heads. Since nickel 2 is flipped, $P(H) = 3/4$ so

$$r_1[s, (2, +)] = (+1)(1 - 3/4) + (-1)(3/4) = -1/2$$

We use the artifice of $N = \infty$, that is, an infinite duration, to reflect the fact that the actual length of the game is an r.v. whose distribution depends on how the players play the game. □

Example 9-2: Advertising and pricing Consider two retail stores which compete with one another primarily on the basis of price of goods sold and advertising. Suppose that wholesale costs are approximately the same for both stores and proportional to quantities sold. Let c denote the wholesale cost per unit sold. Suppose also that the stores experience negligible inventory costs because wholesalers are located nearby.

Let $\rho_{n,q}$ and $z_{n,q}$ denote store q's price and advertising expenditure in period n, respectively. Let $\mathbf{a}_{n,q} = (\rho_{n,q}, z_{n,q})$ and $\mathbf{a}_n = (\mathbf{a}_{n,1}, \mathbf{a}_{n,2})$.

As in Section 3-4, suppose that "goodwill" represents the impacts of advertising expenditures on demand. Let $s_{n,q}$ denote store q's goodwill at the beginning of period n; let $\mathbf{s}_n = (s_{n,1}, s_{n,2})$. We assume that each store's goodwill in period n depends, perhaps probabilistically, on both stores' goodwills and advertising expenditures in period $n - 1$. A simple specific model of goodwill that is similar to the model in Section 3-4 is

$$s_{n,q} = \theta_q(s_{n-1,q} + z_{n-1,q}) \qquad q = 1, 2$$

where $0 < \theta_q < 1$. The parameter θ_q represents the rate at which store q's goodwill deteriorates.

Let $D_{n,q}$ denote the number of units of store q's goods demanded in period n; let $\mathbf{D}_n = (D_{n,1}, D_{n,2})$. We assume that the random vector $\mathbf{D}_n$ has a distribution which depends only on $\mathbf{s}_n$ and $\mathbf{a}_n$.

If each firm uses the discount factor β, the sum of discounted profits for store q is

$$\sum_{n=1}^{\infty} \beta^{n-1}[(\rho_{n,q} - c)D_{n,q} - z_{n,q}] \tag{9-5}$$

Let

$$\mu_q(\mathbf{s}, \mathbf{a}) = E(D_{n,q} \mid \mathbf{s}_n = \mathbf{s}, \mathbf{a}_n = \mathbf{a})$$

$$r_q(\mathbf{s}, \mathbf{a}) = (\rho_{n,q} - c)\mu_q(\mathbf{s}, \mathbf{a}) - z_q$$

where $\mathbf{a} = [(\rho_1, z_1), (\rho_2, z_2)]$. Then the expected value of (9-5) is the same as the expected value of

$$\sum_{n=1}^{\infty} \beta^{n-1} r_q(s_n, \mathbf{a}_n)$$

This model satisfies the definition of an SG with $\mathscr{Q} = \{1, 2\}$, $\mathcal{S} = \mathbb{R}_+^2$ (goodwill is scaled to be nonnegative), and $A_s^q = \mathbb{R}_+^2$ for each $s \in \mathcal{S}$ and $q \in \mathscr{Q}$. □

Example 9-3: Competing banks Suppose that a town has three banks and that nearly all the residents maintain their checking accounts at these banks. Let $\mathscr{Q} = \{1, 2, 3\}$ and let $s_{n,q}$ denote the fraction of the residents with accounts at bank q at the beginning of month n; let $\mathbf{s}_n = (s_{n,1}, s_{n,2}, s_{n,3})$. Then $\mathbf{s}_n \in \mathcal{S} = \{(x_1, x_2, x_3)\colon x_i \ge 0 \text{ all } i \text{ and } \Sigma x_i = 1\}$.

Suppose that the banks do not engage in price competition but that they vary in the lengths of time that customers must wait at tellers' windows. Let $a_{n,q}$ denote the number of tellers employed at bank q during month n; let $\mathbf{a}_n = (a_{n,1}, a_{n,2}, a_{n,3})$. In the short run, each bank is limited by its physical structure; let m_q denote the number of tellers' windows in bank q. Then $A_{\mathbf{s}}^q = \{1, \ldots, m_q\}$ for each $\mathbf{s}$ and q. A simple model for the dependence of the fraction of accounts on quality of service is

$$s_{n+1,q} = (1 - \gamma_q)s_{n,q} + \frac{\gamma_q a_{n,q}}{a_{n,1} + a_{n,2} + a_{n,3}}$$

where $0 < \gamma_q < 1$. Hence, each bank's patronage next month is a weighted average of its patronage this month and its fraction of the total number of tellers.

Suppose that the operating profit in a month depends on the amount of funds deposited in a bank and that the amount of funds is proportional to the fraction of residents with accounts at the bank. Then a reasonable model for the expected monthly operating profit is

$$r_q(\mathbf{s}, \mathbf{a}) = w_q(s_q) - c_q(a_q)$$

where $w_q(\cdot)$ describes revenue due to invested assets and $c_q(\cdot)$ describes the cost of employing a_q tellers. □

Example 9-4: Oligopoly with pricing, production, and inventories Suppose that a collection $\mathscr{Q}$ of manufacturers are competitors. Let $s_{n,q}$ denote the amount of the qth firm's stock of finished goods at the beginning of period n, $\mathbf{s}_n = (s_{n,q}; q \in \mathscr{Q})$, and $\mathcal{S} = \mathbb{R}^{\mathscr{Q}}$. Let $z_{n,q}$ denote the quantity of goods produced by the qth firm during period n and $y_{n,q} = s_{n,q} + z_{n,q}$. The constraint $z_{n,q} \ge 0$

implies $y_{n,q} \geq s_{n,q}$. Suppose that production is sufficiently rapid (relative to a period's length) so that $y_{n,q}$ is the total amount of goods available to satisfy demand during period n. Let $\mathbf{y}_n = (y_{n,q}; q \in \mathcal{Q})$.

Let $D_{n,q}$ denote firm q's demand during period n and $\mathbf{D}_n = (D_{n,q}; q \in \mathcal{Q})$. Under the assumption that excess demand is backlogged, $s_{n+1,q} = y_{n,q} - D_{n,q}$ so

$$\mathbf{s}_{n+1} = \mathbf{y}_n - \mathbf{D}_n \tag{9-6}$$

Let $\rho_{n,q}$ denote the price charged by firm n during period n, $\mathbf{a}_{n,q} = (y_{n,q}, \rho_{n,q})$, and $\mathbf{a}_n = (a_{n,q}; q \in Q)$. Then $A_s^q = [s_q, \infty) \times [0, \infty)$. We assume that $\mathbf{D}_1$ given $\mathbf{a}_1$, $\mathbf{D}_2$ given $\mathbf{a}_2, \ldots$ is a sequence of conditionally independent and identically distributed random vectors. With (9-6), this assumption implies that the distribution of $\mathbf{s}_{n+1}$ is entirely determined by $\mathbf{a}_n$:

$$P\{\mathbf{s}_{n+1} \leq \mathbf{x} \mid \mathbf{a}_n = \mathbf{a}\} = P\{D_n \geq \mathbf{a} - \mathbf{x} \mid \mathbf{a}_n = \mathbf{a}\}$$

where $\mathbf{x} \in \mathbb{R}^Q$.

We make revenue and cost assumptions similar to those made in Section 3-2 for the monopoly case. Suppose that c_q is the qth firm's unit cost of production so $z_{n,q}$ costs $c_q z_{n,q}$. Let $g_q(\mathbf{y}, \boldsymbol{\rho}, \mathbf{d})$ denote the qth firm's revenue minus inventory and stockout costs in any period in which $\mathbf{y}_n = \mathbf{y}$, $\boldsymbol{\rho}_n = \boldsymbol{\rho}$, and $\mathbf{D}_n = \mathbf{d}$. An example is

$$g_q(\mathbf{y}, \boldsymbol{\rho}, \mathbf{d}) = -h_q \cdot (y_q - d_q)^+ - \pi_q \cdot (d_q - y_q)^+ + \rho_q d_q$$

Recall that $\mathbf{a}_{n,q} = (y_{n,q}, \rho_{n,q})$. Therefore, the specification of a model which satisfies the definition of an SG is completed with

$$r_q(\mathbf{s}, \mathbf{a}) = E[g_q(\mathbf{y}, \boldsymbol{\rho}, \mathbf{D}_1) \mid \mathbf{a}_1 = \mathbf{a}] - c_q \cdot (y_q - s_q)$$

where s_q and y_q are the qth components of $\mathbf{s}$ and $\mathbf{y}$, respectively, and $\mathbf{a}_1 = (a_{1,q}; q \in \mathcal{Q})$. □

Example 9-5 Consider a two-period MDP. This is an SG with $\mathcal{Q} = \{1\}$ and $N = 2$. Let $\mathcal{S} = \{1\}$ and $A_1^1 = \{1, 2\}$. The set of all policies, according to Definition 9-7, is the set of all triples $(\alpha_1, \alpha_{21}, \alpha_{22})$, where α_1 is the probability of taking action 1 in period 1 and, for $j = 1$ or 2, α_{2j} is the probability of taking action j in period 2 if action j was taken in period 1. In each instance, one minus α is the probability of taking action 2.

Here is a decision rule which is *not* a policy, i.e., is not admitted by Definition 9-7. With probability 1/2, take action 1 in both periods; with probability 1/2 take action 2 in both periods. This rule requires a joint randomization which is precluded by Definition 9-7. □

EXERCISES

9-1 In Example 9-1, the duration of the game N is infinite. Let $\mathbf{N}$ be the random number of plays of the game until one of the players wins, i.e., has all 10 pennies. Specify a model which is a sequential game except that the infinite duration $N = \infty$ is replaced by the random duration $\mathbf{N}$ and

$$\sum_{n=1}^{\infty} \beta_q^{n-1} r_q(\mathbf{s}_n, \mathbf{a}_n) \equiv \sum_{n=1}^{\mathbf{N}} \beta_q^{n-1} r_q'(\mathbf{s}_n', \mathbf{a}_n')$$

where primed entities refer to the model with duration $\mathbf{N}$ and unprimed entities refer to the model in Example 9-1.

9-2 SOLUTION CONCEPTS

Specific sequential game models are descriptions of interactions among several persons or institutions. What is a "solution" of such a model? It can be construed as a prediction of the behavior that would actually occur in the modeled context. Also, it can be viewed as a recommendation of how the players ought to play the game. For either purpose, one must at least anticipate the "other" players' behavior. But behavior often seems to depend on the type of setting and personality as much as on the actual rewards and dynamics. This pluralism leads to numerous concepts of solution for a game. In this section, we define several concepts of solution for sequential games.

This section begins with some definitions and existence theorems concerning static games. The definitions and theorems extend naturally to sequential games.

Equilibrium Point of a Static Game

In order to define an equilibrium point of an SG (sequential game) and prove that one exists, it is convenient first to define an equilibrium point of a single-period SG. A single-period SG is called a *static game*, *noncooperative game*, or *Nash game* (after John Nash, who first proved the existence of equilibrium points under reasonably general conditions).

Definition 9-8 A *static game* consists of a nonempty set $\mathcal{Q}$, for each $q \in \mathcal{Q}$ a nonempty set W_q, and for each $q \in \mathcal{Q}$ a real-valued function $m_q(\cdot)$ defined on $W = \times_{i \in \mathcal{Q}} W_i$. A *bimatrix game* is a static game with two players and $\# W_1 < \infty$ and $\# W_2 < \infty$.

We interpret W_q as the set of *choices* available to player q. In some applications, W_q consists of the specific actions available to player q. In others, W_q is a set of probability distributions on the set of specific actions available to player q. The latter interpretation arises if we admit randomized strategies. We regard $m_q(\cdot)$ as player q's reward function.

Suppose W_q is a set of randomized strategies and Y_q is the set of specific actions available to player q. If Y_q is a denumerable set, usually W_q is the set of all probability distributions on Y_q, that is, the set of all $[y(q, k);\ k \in Y_q] \geq \mathbf{0}$ such that

$$\sum_{k \in Y_q} y(q, k) = 1$$

If there are more than denumerably many specific actions in Y_q, then (for technical reasons) W_q is a proper subset of the set of all probability distributions on Y_q. Where this issue might arise in SGs, we shall make an assumption such as $\# Y_q < \infty$ for each $q \in \mathcal{Q}$.

Suppose for each $q \in \mathcal{Q}$ that $w_q \in W_q$ is player q's choice, and let $\mathbf{w} = (w_q; q \in \mathcal{Q})$. Sometimes it is convenient to separate a specific player's component in $\mathbf{w}$ from those of the other players. Hence, we abuse notation and for a specific $q \in \mathcal{Q}$ write $\mathbf{w} = (w_q, \mathbf{w}_{-q})$, where $\mathbf{w}_{-q}$ denotes the choices of all the players except player q. Similarly, for any $i \in \mathcal{Q}$, $q \in \mathcal{Q}$, and $\mathbf{w} \in W$, we write $m_i(\mathbf{w}) = m_i(w_q, \mathbf{w}_{-q})$.

Definition 9-9 A static game has an *equilibrium point* (EP) $\mathbf{w}^*$ *with respect to* W if

$$m_q(\mathbf{w}^*) \geq m_q(w_q, \mathbf{w}^*_{-q}) \qquad \text{for all } w_q \in W_q \text{ and } q \in \mathcal{Q} \tag{9-7}$$

We postpone behavioral interpretations of an EP until Example 9-6. However, the basic idea is that no player's lot can be unilaterally improved if the other players adhere to their portions of the EP. When W is clear from the context, we often write "EP" rather than "EP with respect to W." The qualification "with respect to W" is important. For example, in some static games no EP exists if W represents only unrandomized specific actions. In many such games an EP exists if W is expanded to include randomizations of actions. If a static game has an EP with respect to unrandomized specific actions, that multiplayer action remains an EP with respect to randomizations of specific actions.

Recall (see Appendix B) that a set X is a closed convex polyhedron if, and only if, it is the convex hull of a finite set of points. Hence, there must be only finitely many extreme points $x_1, \ldots, x_m$ such that for each $x \in X$ there are nonnegative numbers $\lambda_1, \ldots, \lambda_m$ satisfying

$$\sum_{i=1}^{m} \lambda_i = 1 \qquad \text{and} \qquad x = \sum_{i=1}^{m} \lambda_i x_i$$

Proposition 9-1: If a static game has finitely many players and, for each $q \in \mathcal{Q}$, $m_q(\cdot)$ is continuous on W and W_q is a closed, convex, and bounded polyhedron, then there is an EP with respect to W.

PROOF See J. Nash, "Equilibrium Points in n-Person Games," *Proc. Nat. Acad. Sci. USA* **36**: 48–49 (1950). □

Proposition 9-2: John Nash Suppose a static game has finitely many players Q, and for each q, W_q consists of all the probability distributions on a finite set Y_q. Then there is an EP with respect to W.

PROOF We shall apply Proposition 9-1. For each q, W_q consists of the nonnegative solutions $[y(q, k); k \in Y_q]$ to $\sum_{k \in Y_q} y(q, k) = 1$. This is a closed, bounded, and convex polyhedron.

Let $z_q(k_1, \ldots, k_Q)$ denote player q's payoff if the players take unrandomized actions $(k_1, \ldots, k_Q)$. If the players use $\mathbf{y} = [y(q, k);\ k \in Y_q,\ q \in \mathcal{Q}] \in W$, player q's expected payoff is

$$m_q(\mathbf{y}) = \sum_{k_1 \in Y_1} \cdots \sum_{k_Q \in Y_Q} \prod_{i \in \mathcal{Q}} y(i, k_i) z_q(k_1, \ldots, k_Q)$$

It follows that $m_q(\cdot)$ is continuous on W_i for every i, hence on W. Proposition 9-1 now yields existence of an EP. □

We refer to Propositions 9-1 and 9-2 together as *Nash's Theorem.*

The following example shows that several solution concepts can be interpreted as EPs. It also introduces the *Pareto optimum*, a solution concept to which we return later in the section. Pareto optima are discussed also in Section 6-3.

Example 9-6 Suppose that there are two players and that each has the two choices in $W_1 = W_2 = \{1, 2\}$. Let c_1 and c_2 label the choices of players 1 and 2, respectively. The entries in Table 9-3 specify $[m_1(c_1, c_2), m_2(c_1, c_2)]$; that is, the first entry in a cell is player 1's payoff and the second is player 2's.

Suppose $(c_1, c_2) = (2, 1)$. Then the payoffs are (5, 3). If player 2 adheres to $c_2 = 1$, how attractive to player 1 is $c_1 = 2$ vs. $c_1 = 1$? Player 1's payoff is 4 if $c_1 = 1$ vs. 5 if $c_1 = 2$. We assume "more is better"; so player 1 would prefer to use $c_1 = 2$ if player 2 adheres to $c_2 = 1$.

Similarly, if player 1 adheres to $c_1 = 2$, what should player 2 do? The comparison is a payoff of 3 via $c_2 = 1$ vs. 2 if $c_2 = 1$. Hence, player 2 would prefer to use $c_2 = 1$ if player 2 adheres to $c_1 = 2$. Therefore, $(c_1, c_2) = (2, 1)$ satisfies Definition 9-9; so it is an equilibrium point. Neither player has an incentive to deviate from his or her portion of $(c_1, c_2) = (2, 1)$ if the other player is steadfast. Each player checks, or balances, the other.

If the players use $(c_1, c_2) = (2, 1)$, their payoffs sum to $5 + 3 = 8$. However, if they use $(c_1, c_2) = (1, 1)$, the sum is $4 + 5 = 9$ and no other policy leads to a higher sum. We say that $(c_1, c_2) = (1, 1)$ is a *joint maximum.*

Suppose each player is trying to hurt the other one. Each strives to minimize the other's payoff. If player 1 uses $c_1 = 1$, player 2, by using $c_2 = 2$, can limit player 1 to a payoff of 2. If player 1 uses $c_1 = 2$, player 2, by using $c_2 = 2$, can limit player 1 to a payoff of 3. If player 2 uses $c_2 = 1$, player 1's

Table 9-3 Payoffs in a one-period game: $[m_1(c_1, c_2), m_2(c_1, c_2)]$

	Player 2's policy c_2	
Player 1's policy c_1	1	2
1	4, 5	2, 6
2	5, 3	3, 2

Table 9-4 Payoffs for an ill-fare equilibrium point

	c_2	
c_1	1	2
1	−5, −4	−6, −2
2	−3, −5	−2, −3

best rejoinder (i.e., policy which minimizes player 2's payoff) is $c_1 = 2$. If player 2 uses $c_2 = 2$, player 1's best rejoinder is $c_1 = 2$. Therefore, each component of $(c_1, c_2) = (2, 2)$ hurts the opponent as much as possible. This is called an *ill-fare equilibrium point*. Observe that $(c_1, c_2) = (2, 2)$ is an equilibrium point for the following bimatrix game used in place of Table 9-3. The entry at (c_1, c_2) in Table 9-4 is $[-m_2(c_1, c_2), -m_1(c_1, c_2)]$, where $m_1(c_1, c_2)$ and $m_2(c_1, c_2)$ are taken from Table 9-3.

Player 1 prefers $(c_1, c_2) = (2, 1)$ to $(c_1, c_2) = (1, 1)$ because the payoff would be 5 instead of 4. However, player 2's payoff would drop from 5 to 3. Similarly, player 2 prefers $(c_1, c_2) = (1, 2)$ to (1, 1) because the payoff would be 6 instead of 5. However, player 1's payoff would drop from 4 to 2. Therefore, $(c_1, c_2) = (1, 1)$ has the property that neither player's welfare can be improved without injuring the other player. We say that $(c_1, c_2) = (1, 1)$ is *Pareto optimal* (called *efficient* or *admissible* in some literatures). Also, (1, 2) and (2, 1) are Pareto optimal.

Suppose each player tries to maximize the amount by which his or her payoff is higher than the opponent's payoff. Then we replace the entries in Table 9-3 with $[m_1(c_1, c_2) - m_2(c_1, c_2), m_2(c_1, c_2) - m_1(c_1, c_2)]$ as in Table 9-5. The entries in each cell then sum to zero. We call this criterion *maxmin the difference*. The policy $(c_1, c_2) = (2, 2)$ is an equilibrium point in Table 9-5. That is, $m_1(\cdot, 2) - m_2(\cdot, 2)$ is maximized at $c_1 = 2$, and $m_2(2, \cdot) - m_1(2, \cdot)$ is maximized at $c_2 = 2$.

Suppose, instead, that each player tries to maximize the amount by which his or her payoff exceeds the average payoff. If $(c_1, c_2) = (1, 1)$, the payoffs are (4, 5); so the average is 4.5 and, with respect to this average, the

Table 9-5 Payoffs for the criterion: maxmin the difference

	Player 2's policy c_2	
Player 1's policy c_1	1	2
1	−1, 1	−4, 4
2	2, −2	1, −1

players' increments are $(-0.5, 0.5)$. If $(c_1, c_2) = (1, 2)$, the increments are $(-2, 2)$; if $(c_1, c_2) = (2, 1)$, the increments are $(1, -1)$; and if $(c_1, c_2) = (2, 2)$, the increments are $(0.5, -0.5)$. Multiplying these increments by 2 yields the entries in Table 9-5. Therefore, (c_1, c_2) is an equilibrium point for the criterion maxmin the difference if, and only if, it is an equilibrium point for the criterion *beat the average*. If the game had more than two players, beat the average would be well defined, whereas maxmin the difference would be ill defined. □

We observe in Example 9-6 that an EP causes the players to balance each other in the sense that no player's welfare can be unilaterally improved. Thus, if all players but one announce in advance that they will adhere to their portion of an EP, the remaining player cannot do better than to adhere to his or her portion of the EP. However, an EP does not necessarily protect a player from damage due to shifts in the other players' choices. In Example 9-6, for example, $(c_1, c_2) = (2, 1)$ is an EP and yields a payoff of 5 to player 1. If player 1 adheres to $c_1 = 2$ and player 2 shifts from $c_2 = 1$ to $c_2 = 2$, player 1's payoff drops from 5 to 3.

Definitions for Sequential Games

Now we define equilibrium points and Pareto optima of SGs. The other solution concepts in Example 9-6 were equilibrium points of transformations of the original game.

It is convenient to write the policy $\Pi = (\Pi_q; q \in \mathcal{Q})$ as the pair (Π_k, Π_{-k}) for a specific player $k \in \mathcal{Q}$. Then Π_{-k} represents the policies of all the players except player k. Recall the notation $v_s^q(\Pi, N)$, $v_s^q(\Pi)$, and $g_s^q(\Pi)$ [from (9-4*a*), (9-4*b*), and (9-4*c*)] for the expected present values during finite and infinite durations and expected average reward per period. Let $\mathcal{S}'$ be a subset of states, π_q be a subset of player q's policies, and $\pi = \times_{q \in \mathcal{Q}} \pi_q$.

Definition 9-10 A policy $\Pi^* = (\Pi_q^*; q \in \mathcal{Q})$ is an *N-period equilibrium point (EP) with respect to* initial states in $\mathcal{S}'$ and policies in π if

$$v_s^q(\Pi^*, N) \geq v_s^q[(\xi_q, \Pi_{-q}^*), N] \qquad \text{for all } s \in \mathcal{S}',\ \xi_q \in \pi_q, \text{ and } q \in \mathcal{Q} \quad (9\text{-}8)$$

A policy Π^* is a *discounted equilibrium point (EP) with respect to* $\mathcal{S}'$ and π if

$$v_s^q(\Pi^*) \geq v_s^q[(\xi_q, \Pi_{-q}^*)] \qquad \text{for all } s \in \mathcal{S}',\ \xi_q \in \pi_q, \text{ and } q \in \mathcal{Q} \quad (9\text{-}9)$$

A policy Π^* is an *average-reward equilibrium point (EP) with respect to* $\mathcal{S}'$ and π if

$$g_s^q(\Pi^*) \geq g_s^q[(\xi_q, \Pi_{-q}^*)] \qquad \text{for all } s \in \mathcal{S}',\ \xi_q \in \pi_q, \text{ and } q \in \mathcal{Q} \quad (9\text{-}10)$$

Usually the type of equilibrium point, $\mathcal{S}'$, and π are clear from the context; so we say merely *equilibrium point* and use the abbreviation EP.

There are several reasons why sometimes $\mathcal{S}'$ and π are proper subsets of $\mathcal{S}$

and the set of all policies, respectively. First, proper subsets may be necessary to ensure existence of the expectations $v_s^q(\Pi, N)$, $v_s^q(\Pi)$, and $g_s^q(\Pi)$ (and to ensure that they correspond to bona fide r.v.'s). Second, we may be able to prove existence of EPs only for proper subsets. Third, there may be an EP with special properties only for proper subsets. An example of the second reason occurs in Section 9-3 when existence of discounted EPs is proved at first with respect to $\boldsymbol{\pi}$ as the set of stationary policies. An example of the third reason occurs in Section 9-4, where we study EP analogs of the myopic optima in Chapter 3.

An EP of a sequential game is closely related to the idea of EP for static games. We use discounted EPs to illustrate the relationship. A collection of static games corresponds to an SG with $\beta_q < 1$ for each $q \in \mathcal{Q}$. Suppose $\boldsymbol{\pi} = \times_{q \in \mathcal{Q}} \pi_q$; let $W = \boldsymbol{\pi}$ and $W_q = \boldsymbol{\pi}_q$, $q \in \mathcal{Q}$. Fix $s \in \mathcal{S}$; for each $q \in \mathcal{Q}$ and $\Pi \in \boldsymbol{\pi}$, let $m_q(\Pi) = v_s^q(\Pi)$. Call this the sth *static game*. There is an sth static game for each $s \in \mathcal{S}$.

Theorem 9-1 A policy Π^* is a discounted EP with respect to $\mathcal{S}'$ and $\boldsymbol{\pi}$ if, and only if, Π^* is an EP with respect to $W = \boldsymbol{\pi}$ for every sth static game with $s \in \mathcal{S}'$.

PROOF Left as Exercise 9-3. □

It does *not* follow from Theorem 9-1 that the issue of existence of a discounted EP can be settled by the straightforward application of Nash's theorem (Proposition 9-1). Theorem 9-1 requires that the *same* policy Π^* be an EP with respect to W for *all* of the sth static games with $s \in \mathcal{S}'$. The optimum of a discounted infinite-horizon MDP can be construed as the solution of a vector-maximization problem. Similarly, a discounted EP can be regarded as the solution of a vector static EP problem. In both cases, the components of the vector correspond to possible initial states.

Exercise 9-4 asks you to specify static games whose EP corresponds to an N-period EP and other static games whose EP corresponds to an average-reward EP.

Some of the results in Section 9-7 concern Pareto optima of sequential games.

Definition 9-11 Π^* is *Pareto optimal with respect to* $\boldsymbol{\pi}$ if $\xi \in \boldsymbol{\pi}$, $q \in \mathcal{Q}$, and $s \in \mathcal{S}$ with $v_s^q(\Pi^*) < v_s^q(\xi)$ implies the existence of $j \in \mathcal{Q}$ and $u \in \mathcal{S}$ with $v_u^j(\Pi^*) > v_u^j(\xi)$.

The definition of Pareto optimality can be altered so that the criterion is expected present value during a finite duration or expected average reward per period. Following the three parts of Definition 9-10, the formalities should be obvious.

We abbreviate Pareto optimality PO. A policy Π^* is PO if no player can be made better off from some initial state without some player becoming worse off from some initial state. Either the players or the initial states (but not both) may be the same.

EXERCISES

9-2 Define each of the following solution concepts for a sequential game with the criterion of expected present value.

(*a*) Joint maximum.

(*b*) Ill-fare equilibrium point.

(*c*) Maxmin the difference (assuming $Q = 2$).

(*d*) Beat-the-average equilibrium point.

9-3 Prove Theorem 9-1.

9-4 State analogs of Theorem 9-1 for N-period EPs and average-reward EPs.

9-3* EXISTENCE OF EQUILIBRIUM POINTS

An equilibrium point has the property that the players' strategies balance each other. No player's expected payoff can be raised if all the other players adhere to their components of the strategy. The players implicitly hold each other to the (joint) strategy. Appending the idea of balance to a mathematical model usually leads to fixed-point theorems. Therefore, it is hardly surprising that numerous authors have used fixed-point theorems† to prove the existence of EP (equilibrium point) solutions to sequential game models.

In this section, we present sufficient conditions for the existence of discounted, average-reward, and N-period equilibrium points. For discounted EPs, the existence proof is a generalization of John Nash's use of the Brouwer fixed-point theorem in (static) game theory. An alternative proof could be given based on the contraction-mapping fixed-point theorem (cf. Section 5-4) and the Kakutani fixed-point theorem.

All existence proofs depend on the following observation. Suppose all the players but one are using stationary policies. Then the remaining player faces an MDP (Markov decision process) whose criterion is the expected present value of the rewards. Therefore, Theorem 4-3 asserts that the player can confine a search for an optimal Markov policy to the set of stationary policies. As a result, the material in Sections 4-4, 4-5, and 5-2 can be invoked in existence proofs.

This section considers only *finite* SGs, that is, games with finitely many players, states, and actions. Here we introduce the notation which specifies the MDP faced by one player when all the other players use stationary policies. From Section 9-2, recall that a policy is written $\Pi = (\Pi_q;\ q \in \mathscr{Q})$, with player q's policy being the component $\Pi_q = (\pi_{1q}, \pi_{2q}, \ldots)$, in which π_{nq} specifies player q's decision rule in the nth period. The following definition of a single-stage decision rule permits randomized rules. As Section 9-2 illustrates, even in static games there may not exist an EP unless randomized rules are permitted.

Definition 9-12 A *single-stage decision rule for player q* is a collection $\delta_q = \{\delta_q(s)\colon s \in \mathcal{S}\}$ such that $\delta_q(s)$ is a probability distribution on A_s^q for each $s \in \mathcal{S}$.

† The theorems invoked include the contraction-mapping, Brouwer, and Kakutani fixed-point theorems and their corollaries. These may be found in many books on topology such as N. Dunford and J. T. Schwartz, *Linear Operators, Part I*, Interscience Publishers, New York (1958).

A *single-stage decision rule* δ is a collection $\{\delta_q: q \in \mathscr{Q}\}$ such that δ_q is a single-stage decision rule for player q, for each $q \in \mathscr{Q}$. If there is a single-stage decision rule δ_q such that $\pi_{nq}(H_n) = \delta_q(s_n)$ for all H_n and n, then Π_q is a *stationary policy for player q*. The policy $\Pi = (\Pi_q;\ q \in \mathscr{Q})$ is a *stationary policy* if, for each $q \in \mathscr{Q}$, Π_q is a stationary policy for player q. The symbols $\mathbf{Z}_q$ and $\mathbf{Z}$ denote the set of stationary policies for player q and the set of stationary policies, respectively.

Comments

(i) From the definitions,

$$\mathbf{Z} = \underset{q \in \mathscr{Q}}{\times} \mathbf{Z}_q \tag{9-11}$$

(ii) Recall that Z denotes the set of stationary policies in an MDP (Section 4-3).

(iii) Since the SG is finite, that is, $\#\mathscr{C} < \infty$, there are only finitely many elements in the sets of players, states, and actions.

(iv) A single-stage decision rule is a collection of probability distributions, one for each player in each state. The set A_s^q is finite; so for $k \in A_s^q$ we interpret $\delta_q(s)_k$ as the probability that player q will choose action k when the state is s.

A single-stage decision rule corresponds to each stationary policy, and conversely. Therefore, we also use $\mathbf{Z}$ and $\mathbf{Z}_q$ to denote the set of single-stage decision rules and the set of single-stage decision rules for player q, respectively. It will be clear from each context whether $\mathbf{Z}$ ($\mathbf{Z}_q$) represents single-stage decision rules (for player q) or stationary policies (for player q).

Let $\nu(s, q) = \# A_s^q$; $\nu(s, q) < \infty$ for each s and q in a finite SG. The set of all probability distributions on A_s^q is the set of all solutions of the system

$$D_{sk}^q \geq 0 \qquad k = 1, \ldots, \nu(s, q) \qquad \sum_{k=1}^{m(s,q)} D_{sk}^q = 1 \tag{9-12}$$

Let $\mathbf{Z}_{sq}$ denote the set of all solutions to (9-12); then define

$$\mathbf{Z}_q = \underset{s \in \mathcal{S}}{\times} \mathbf{Z}_{sq} \qquad \text{and} \qquad \mathbf{Z} = \underset{q \in \mathscr{Q}}{\times} \mathbf{Z}_q = \underset{q \in \mathscr{Q}}{\times} \ \underset{s \in \mathcal{S}}{\times} \mathbf{Z}_{sq} \tag{9-13}$$

Let $\boldsymbol{\delta} = (\delta_q;\ q \in \mathscr{Q}) \in \mathbf{Z}$. Then for each $s \in \mathcal{S}$ and $q \in \mathscr{Q}$, $\delta_q(s)$ is specified by exactly one solution to (9-12). Let $\{D_{sk}^q: k \in A_s^q,\ q \in \mathscr{Q},\ s \in \mathcal{S}\}$ correspond to $\boldsymbol{\delta}$. Recall that we use Q to denote the number of players, i.e., the size of $\mathscr{Q}$.

Let

$$b_{sj}(\boldsymbol{\delta}) = \sum_{k_1=1}^{\nu(s,1)} \cdots \sum_{k_Q=1}^{\nu(s,Q)} p_{sj}(k_1, \ldots, k_Q) \prod_{q \in \mathscr{Q}} D_{sk_q}^q \tag{9-14}$$

which is the expected value of the transition probability induced by $\boldsymbol{\delta}$. Let $B_{\boldsymbol{\delta}}$ represent the matrix of these transition probabilities and let $\boldsymbol{\delta}^\infty$ indicate the stationary policy in which player q uses the policy δ_q^∞, for each $q \in \mathscr{Q}$.

Let $\boldsymbol{\rho}^q(\boldsymbol{\delta})$ be the vector whose sth component is

$$\rho_s^q(\boldsymbol{\delta}) = \sum_{k_1=1}^{v(s,1)} \cdots \sum_{k_Q=1}^{v(s,Q)} \prod_{i \in \mathscr{Q}} D_{sk}^i \, r_q(s, k_1, \ldots, k_Q) \tag{9-15}$$

This quantity is player q's expected single-stage reward induced by policy $\boldsymbol{\delta}^\infty$ whenever the state is s. Let $v_s^q(\boldsymbol{\delta})$ be player q's expected present value induced by policy $\boldsymbol{\delta}^\infty$ from the initial state s; let $\mathbf{v}^q(\boldsymbol{\delta})$ denote the vector whose sth component is $v_s^q(\boldsymbol{\delta})$. Then $\beta_q < 1$ and Theorem 4-7 imply†

$$\mathbf{v}^q(\boldsymbol{\delta}) = \sum_{i=0}^{\infty} (\beta_q B_{\boldsymbol{\delta}})^i \boldsymbol{\rho}^q(\boldsymbol{\delta}) = (I - \beta_q B_{\boldsymbol{\delta}})^{-1} \boldsymbol{\rho}^q(\boldsymbol{\delta}) \tag{9-16}$$

(where $B_{\boldsymbol{\delta}}^0 = I$).

Existence of Discounted EPs

The main result (Theorem 9-2 below) asserts that a finite SG necessarily has a discounted EP among the class of stationary policies. The proof has two parts. First, we construct a mapping τ and use Brouwer's fixed-point theorem to prove that τ has a nonempty set of fixed points. Second, we show that the fixed points of τ are necessarily EPs and conversely.

Definition 9-13 A function f with domain and range $\mathscr{D}$ has *fixed point* d if $d \in \mathscr{D}$ and

$$f(d) = d$$

We abbreviate fixed point with FP. Brouwer's theorem presents sufficient conditions for the existence of an FP in $\mathscr{D}$.

Proposition 9-3: Brouwer's fixed-point theorem Let f be a function with domain and range $\mathscr{D}$. If $\mathscr{D}$ is a closed, bounded, and convex subset of $\mathbb{R}^m (m < \infty)$,‡ and if f is continuous on $\mathscr{D}$, then f has an FP in $\mathscr{D}$.

PROOF See Dunford and Schwartz (1958). □

Theorem 9-2 If a finite SG has $\beta_q < 1$ for all $q \in \mathscr{Q}$, there exists a discounted EP with respect to $\mathcal{S}$ and $\mathbf{Z}$.

PROOF Step (i) constructs a mapping τ on $\mathbf{Z}$ and uses Brouwer's theorem to establish that τ has an FP in $\mathbf{Z}$. Step (ii) proves that $\boldsymbol{\delta} \in \mathbf{Z}$ is an FP of τ if, and only if, $\boldsymbol{\delta}^\infty$ is an EP.

(i) In order to use Brouwer's theorem, $\mathbf{Z}$ must be a closed, bounded, and

† See Exercise 9-10 for an alternative proof based on Theorem 5-7.

‡ Brouwer's theorem requires only that $\mathscr{D}$ be a compact and convex set, but closed, bounded, and convex subsets of $\mathbb{R}^m$ with $m < \infty$ are sufficient for our purposes.

convex subset of $\mathbb{R}^m$ for some $m < \infty$. If $\mathbf{Z}_q$ has the required properties for each $q \in \mathcal{Q}$, then $\mathbf{Z} = \times_{q \in \mathcal{Q}} \mathbf{Z}_q$ has them too. The SG is finite so $v(s, q) = \# A_s^q < \infty$ for each s and q. Using (9-12) and (9-13), it is simple to verify that $\mathbf{Z}_{sq}$ is a closed, bounded, and convex subset of $\mathbb{R}^{v(s, q)}$ (Exercise 9-5 asks you to do this). Let

$$v \triangleq \sum_{s \in \mathcal{S}} \sum_{q \in \mathcal{Q}} v(s, q)$$

Then $\mathbf{Z} \subset \mathbb{R}^v$, $v < \infty$, and $\mathbf{Z}$ is closed, bounded, and convex.

Let $\boldsymbol{\delta}_{sk}^q$ denote the modification of $\boldsymbol{\delta}$ in which player q takes action $k \in A_s^q$ with probability 1 if the state is s; that is, $D_{sk}^q = 1$ and $D_{sj}^q = 0$ if $j \neq k$. Then

$$\rho_s^q(\boldsymbol{\delta}_{sk}^q) + \beta_q \sum_{j \in \mathcal{S}} b_{sj}(\boldsymbol{\delta}_{sk}^q) v_j^q(\boldsymbol{\delta})$$

is player q's expected discounted reward from the initial state s if all the other players adhere to their portions of $\boldsymbol{\delta}^\infty$ while player q uses $\boldsymbol{\delta}_{sk}^q$ in the first period and, thereafter, uses δ_q^∞. Let

$$\phi_{sk}^q(\boldsymbol{\delta}) = [\rho_s^q(\boldsymbol{\delta}_{sk}^q) + \beta_q \sum_{j \in \mathcal{S}} b_{sj}(\boldsymbol{\delta}_{sk}^q) v_j(\boldsymbol{\delta}) - v_s^q(\boldsymbol{\delta})]^+ \tag{9-17}$$

indicate the increase in player q's expected discounted reward, if any, from deferring δ_q^∞ for one period during which $\boldsymbol{\delta}_{sk}^q$ is used. Define a function $\tau: \mathbf{Z} \to \mathbf{Z}$ with†

$$\tau(\boldsymbol{\delta})[s, k, q] = \frac{D_{sk}^q + \phi_{sk}^q(\boldsymbol{\delta})}{1 + \sum_{i=1}^{m(s, q)} \phi_{si}^q(\boldsymbol{\delta})} \qquad k \in A_s^q \qquad s \in \mathcal{S} \qquad q \in \mathcal{Q} \tag{9-18}$$

Exercise 9-6 asks you to verify that $\tau(\boldsymbol{\delta}) \in \mathbf{Z}$ and that $\tau(\cdot)$ is continuous on $\mathbf{Z}$. Therefore, Brouwer's theorem implies existence of an FP of τ.

(ii) $\{\text{EP}\} \subset \{\text{FP}\}$: Suppose that $\boldsymbol{\delta}^\infty$ is an EP and all players except q use their components of $\boldsymbol{\delta}^\infty$. Then player q faces an MDP whose transition probabilities and rewards are specified by

$$\{b_{sj}(\boldsymbol{\delta}_{sk}^q): k \in A_s^q, s \in \mathcal{S}\} \qquad \text{and} \qquad \{\rho_s^q(\boldsymbol{\delta}_{sk}^q): k \in A_s^q, s \in \mathcal{S}\}$$

respectively. The definition of an EP (Definition 9-8) implies that δ_q^∞ is an optimal policy for this MDP. Then Theorem 4-9 implies $\phi_{sk}^q(\boldsymbol{\delta}) = 0$ for all s, k, and q. Therefore, $\tau(\boldsymbol{\delta}) = \boldsymbol{\delta}$ from (9-18); so $\boldsymbol{\delta}$ is an FP.

$\{\text{EP}\} \supset \{\text{FP}\}$: Suppose δ is an FP so $\tau(\boldsymbol{\delta}) = \boldsymbol{\delta}$. We shall prove $\phi_{sk}^q(\boldsymbol{\delta}) = 0$ for all s, k, and q so, by Theorem 4-9, $\boldsymbol{\delta}^\infty$ is an EP. From (9-18),

$$D_{sk}^q = \frac{D_{sk}^q + \phi_{sk}^q(\boldsymbol{\delta})}{1 + \sum_{i=1}^{m(s, q)} \phi_{si}^q(\boldsymbol{\delta})} \qquad k \in A_s^q \qquad s \in \mathcal{S} \qquad q \in \mathcal{Q}$$

so

$$\phi_{sk}^q = D_{sk}^q \sum_{i=1}^{m(s, q)} \phi_{si}^q \qquad k \in A_s^q \qquad s \in \mathcal{S} \qquad q \in \mathcal{Q} \tag{9-19}$$

† Since the elements of Z are called $\boldsymbol{\delta}$, the argument of $\tau(\cdot)$ is $\boldsymbol{\delta}$. Since $\tau(\boldsymbol{\delta})$ is a single-stage decision rule, for each s, k, and q it must specify the conditional probability that player q takes action k when the state is s. Therefore, we use the notation $\tau(\boldsymbol{\delta})[s, k, q]$.

where we suppress the notational dependence of ϕ_{si}^q on $\boldsymbol{\delta}$. From (9-19), if $D_{sk}^q = 0$, then $\phi_{sk}^q = 0$. Suppose, for some q and s,

$$\sum_{j=1}^{m(s, q)} \phi_{sj}^q > 0$$

Then

$$D_{sk}^q = \frac{\phi_{sk}^q}{\sum_{j=1}^{m(s, q)} \phi_{sj}^q} \qquad k \in A_s^q$$

so $D_{sk}^q > 0$ implies $\phi_{sk}^q > 0$. Also $D_{sk}^q > 0$ for some $k \in A_s^q$ from (9-12). But we shall prove that $\phi_{sk}^q = 0$ for some k such that $D_{sk}^q > 0$. From (9-14), (9-15), and (9-16),

$$\begin{aligned} v_s^q(\boldsymbol{\delta}) &= \sum_{k=1}^{m(s, q)} D_{sk}^q \, v_{sk}^q(\boldsymbol{\delta}) \\ &\geq \min \left\{ v_{sk}^q(\boldsymbol{\delta}) \colon k \in A_s^q \text{ and } D_{sk}^q > 0 \right\} \end{aligned}$$

Let j be a value of k at which min $\{\dots\}$ is attained so $v_{sj}^q(\boldsymbol{\delta}) - v_s^q(\boldsymbol{\delta}) \leq 0$ while $D_{sj}^q > 0$. Therefore, $\phi_{sj}^q(\boldsymbol{\delta}) = 0$ while $D_{sj}^q > 0$. Hence, $\phi_{sj}^q(\boldsymbol{\delta}) = 0$ for all s, j, and q so $\boldsymbol{\delta}^\infty$ is an EP. □

Under the assumptions of Theorem 9-2, if $\boldsymbol{\delta}^\infty$ is an EP, no player has an incentive to move to another stationary policy if the other players adhere to their portions of $\boldsymbol{\delta}^\infty$. The following result shows that no player can obtain an improvement by choosing a Markov policy (Definition 4-10) that is not *stationary*, i.e., outside **Z**.

Corollary 9-2 If an SG is finite and $\beta_q < 1$ for each $q \in \mathcal{Q}$, then there is a stationary policy which is a discounted EP with respect to $\mathcal{S}$ and the set of all Markov policies.

PROOF From Theorem 9-2, there exists a stationary policy $\boldsymbol{\delta}^\infty$ which is an EP with respect to $\mathcal{S}$ and **Z**. For any player q, if all the players but q use $\boldsymbol{\delta}_{-q}^\infty$, their portion of $\boldsymbol{\delta}^\infty$, then player q faces an MDP which satisfies the conditions of Theorem 4-3. That result asserts that an MDP has an optimal stationary policy if it has an optimal Markov policy. Therefore, within the class of Markov policies player q cannot improve upon δ_q^∞ as a rejoinder to $\boldsymbol{\delta}_{-q}^\infty$. This is true of all players $q \in \mathcal{Q}$. □

Existence of an Average-Reward Equilibrium Point

Recall that $g_s^q(\boldsymbol{\Pi})$ denotes player q's long-run average reward per period if the initial state is s and the players use the policy $\boldsymbol{\Pi}$. From Definition 9-10, $\boldsymbol{\Pi}^* = (\boldsymbol{\Pi}_q^*\,;\, q \in \mathcal{Q})$ is an average-reward equilibrium point with respect to initial states in $\mathcal{S}'$ and policies in $\boldsymbol{\pi}$ if

$$g_s^q(\boldsymbol{\Pi}^*) \geq g_s^q[(\xi_q, \boldsymbol{\Pi}_{-q}^*)] \qquad \text{for all } s \in \mathcal{S}', \quad \xi \in \pi, \quad \text{and} \quad q \in \mathcal{Q} \qquad (9\text{-}20)$$

The search for $\mathbf{\Pi}^*$ to satisfy (9-20) is at least as hard as the search for an optimal policy in an MDP with the criterion of average return. From Section 4-6, the MDP problem is simpler if every stationary policy induces a Markov chain with only one communicating class of states. We make the corresponding assumption below, although the existence of an equilibrium point can be proved under weaker conditions.†

Unichain Assumption Every single-stage decision rule $\boldsymbol{\delta}$ has a transition matrix $B_{\boldsymbol{\delta}}$ that induces a Markov chain with one communicating class of states and a (possibly empty) set of transient states.

The unichain assumption and finitely many states implies (cf. Section 7-6 in Volume I) existence of a stationary distribution which is the same regardless of the initial state. That is, for each $\boldsymbol{\delta}$ there is a probability vector $\mathbf{c}(\boldsymbol{\delta})$ with components $c_s^{\boldsymbol{\delta}}$, $s \in \mathcal{S}$, such that

$$\mathbf{c}(\boldsymbol{\delta}) \geq \mathbf{0} \qquad \mathbf{c}(\boldsymbol{\delta}) \cdot \mathbf{e} = 1 \qquad \text{and} \qquad \mathbf{c}(\boldsymbol{\delta}) = \mathbf{c}(\boldsymbol{\delta}) B_{\boldsymbol{\delta}} \tag{9-21}$$

where $\mathbf{e}$ denotes the column vector whose components are all 1. Moreover, if $a_n = \boldsymbol{\delta}(s_n)$ for all n, then regardless of the initial state

$$c_s^{\boldsymbol{\delta}} = \lim_{N \to \infty} \frac{1}{N} \sum_{n=1}^{N} P\{s_n = s\} \qquad s \in \mathcal{S}$$

so $c_s^{\boldsymbol{\delta}}$ is the long-run average probability of being in state s. Therefore, player q's long-run average reward per period is the same number $g^q(\boldsymbol{\delta}) = \mathbf{c}(\boldsymbol{\delta}) \cdot \boldsymbol{\rho}^q(\boldsymbol{\delta})$ regardless of the initial state (where the sth component of $\boldsymbol{\rho}^q(\boldsymbol{\delta})$ is defined by (9-15) and is player q's expected single-stage reward in period n if $s_n = s$ and $\mathbf{a}_n = \boldsymbol{\delta}(s)$. The importance of the unichain assumption in the following proof is that it simplifies the verification that τ is continuous on $\mathbf{Z}$. *Some* assumption is needed (see Exercise 9-7).

Theorem 9-3 Suppose a finite SG satisfies the unichain assumption. Then there exists an average-reward EP with respect to $\mathcal{S}$ and $\mathbf{Z}$.

PROOF‡ We show that the average reward EP in a SG is equivalent to an EP in a static game; then we invoke Nash's existence theorem, Proposition 9-2.

A static game (cf. Definition 9-8) consists of a nonempty set $\mathcal{Q}$, a nonempty set W_q for each $q \in \mathcal{Q}$, and a real-valued function $m_q(\cdot)$ on

$$W = \underset{i \in \mathcal{Q}}{\times} W_i$$

for each $q \in \mathcal{Q}$.

† See Federgruen (1980).

‡ This proof seems to be new, but it is in the same spirit as remarks on two-person zero-sum SGs made by Dr. Alan Hoffman to one of the authors circa 1973.

For the SG, the set of players $\mathscr{Q}$ is assumed to be nonempty, and let W_q denote player q's alternative randomized stationary policies so

$$\mathbf{Z} = \times_{i \in \mathscr{Q}} W_i$$

For each $\boldsymbol{\delta} \in \mathbf{Z}$, the unichain assumption implies that $m_q(\boldsymbol{\delta}) = \mathbf{c}(\boldsymbol{\delta}) \cdot \boldsymbol{\rho}^q(\boldsymbol{\delta})$ is well defined (and is the long-run average reward per period).

Proposition 9-2 asserts that a static game with finitely many players has an EP with respect to W if, for each q, W_q consists of all the probability distributions on a finite set, say Y_q. In the SG, let Y_q denote the set of player q's unrandomized stationary policies. Then this theorem follows from the observation (whose verification is left to Exercise 9-7) that the set of convex combinations of unrandomized stationary policies coincides with the set of randomized stationary policies. □

Existence of an N-Period Equilibrium Point

From Section 9-1, recall the notation

$$V^q(\mathbf{\Pi}, N) = \sum_{n=1}^{N} \beta_q^{n-1} r_q(s_n, \mathbf{a}_n) + \beta_q^N L_q(s_{N+1}) \tag{9-22a}$$

and

$$v_s^q(\mathbf{\Pi}, N) = E[V^q(\mathbf{\Pi}, N) \mid s_1 = s] \tag{9-22b}$$

which makes explicit the dependence on the players' policy $\mathbf{\Pi} = (\Pi_q;\ q \in \mathscr{Q})$. From Definition 9-8, a policy $\mathbf{\Pi}^*$ is an N-period EP with respect to initial states in $\mathcal{S}'$ and policies in $\boldsymbol{\pi} = \times_{q \in \mathscr{Q}} \boldsymbol{\pi}_q$ if

$$v_s^q(\mathbf{\Pi}^*, N) \geq v_s^q[(\xi_q, \mathbf{\Pi}^*_{-q}), N] \qquad s \in \mathcal{S}',\ \ \xi_q \in \pi_q, \text{ and } q \in \mathscr{Q} \tag{9-23}$$

Recall the notation in Definitions 9-8 and 9-9 of a static game and an EP of a static game. From Exercise 9-4, $\mathbf{\Pi}^*$ is an N-period EP with respect to $\mathcal{S}'$ and $\boldsymbol{\pi}$ if, and only if, it is an EP with respect to $W = \boldsymbol{\pi}$ for every static game, called the sth *static game*, obtained by fixing $s \in \mathcal{S}'$ and letting $w_q = \boldsymbol{\pi}_q$ and $m_q(\mathbf{\Pi}) = v_s^q(\mathbf{\Pi}, N)$, $q \in \mathscr{Q}$.

Theorem 9-4 For every $N \in I_+$, a finite SG has an N-period EP with respect to $\mathcal{S}$ and the set of all policies.

Proof Fix $N \in I_+$. We shall construct a policy which is simultaneously an EP for every sth static game, $s \in \mathcal{S}$. Fix $s \in \mathcal{S}$. A policy for player q is a sequence $\Pi_q = (\pi_{1q}, \pi_{2q}, \ldots, \pi_{Nq})$ with $\pi_{nq}(H_n)$ being a probability distribution on $A_{s_n}^q$ (where s_n is the last element of history H_n). For fixed n, a finite SG can have only finitely many sequences $H_n = (s_1, a_1, \ldots, s_{n-1}, a_{n-1}, s_n)$, hence only finitely many sequences $(H_1, \ldots, H_N)$, hence only finitely many unrandomized sequences $\Pi_q = (\pi_{1q}, \ldots, \pi_{Nq})$. Label player q's unrandomized policies $\Pi_q^1, \ldots, \Pi_q^{d(q)}$ with $d(q) < \infty$. Exercise 9-12 asks you to prove that any Π_q is a convex combination of these unrandomized policies. Therefore, the set of all policies for player q is a closed, bounded, and convex polyhedron. For each q, player q's payoff in the sth static game is the expected payoff. Therefore, Proposition 9-2 implies existence of an EP $\mathbf{\Pi}^*(s)$ in the sth static game.

Continue to keep $s \in \mathcal{S}$ fixed. Since $\mathbf{\Pi}^*(s)$ is a bona fide policy, it stipulates a randomized action $\mathbf{a}_n$ for each possible history H_n, for $n = 1, \ldots, N$. The possible histories include all possible initial states, but the payoffs $v_s^q(\mathbf{\Pi}, N)$ in the sth static game depend only on randomized actions in response to histories whose initial state is s. Let $\Pi^*(s)_j$ denote the portion of $\Pi^*(s)$ which stipulates randomized actions $\mathbf{a}_1, \ldots, \mathbf{a}_N$ for histories with initial state $j \in \mathcal{S}$. Therefore, in $\mathbf{\Pi}^*(s)$, if $j \neq s$, we may alter $\Pi^*(s)_j$ and the altered $\mathbf{\Pi}^*(s)$ will remain an EP in the sth static game.

Now let s vary in $\mathcal{S}$ and construct a policy with components $\Pi^*(s)_s$, $s \in \mathcal{S}$. For each $s \in \mathcal{S}$, if the initial state is s, the randomized actions are part of an EP in the sth static game. From the argument above, such a constructed policy is an EP in every sth static game, $s \in \mathcal{S}$, hence is an N-period EP. □

Recall the discussion on page 443 which observes that our definition of "policy" admits non-Markov decision rules.

Definition 9-14 A *Markov policy* $\mathbf{\Pi} = (\Pi_q;\, q \in \mathcal{Q})$ in an N-period SG is a policy in which π_{nq}, in player q's policy $\Pi_q = (\pi_{1q}, \ldots, \pi_{Nq})$, for each n, depends on H_n only through its last element s_n.

It follows from the definition that a Markov policy is a sequence $(\boldsymbol{\delta}^1, \boldsymbol{\delta}^2, \ldots, \boldsymbol{\delta}^N)$ in which $\boldsymbol{\delta}^n$ depends only on the state when n periods remain until the game ends.

The construction in the proof of Theorem 9-4 does not yield any insight into the structure of an N-period EP. The following proof is essentially an alternative proof of Theorem 9-4 which shows that there is an N-period EP which is a Markov policy. Thus there is an EP $\mathbf{\Pi}^*$ in which player q's component Π_q^* consists of a sequence $(\delta_q^1, \ldots, \delta_q^N)$ where, for each n, $\delta_q^n(s)$ is a probability distribution on A_s^q. We interpret $\delta_q^n(s)_k$ as the probability that player q will take action k in period n if $s_n = s$.

Theorem 9-5 For every $N \in I_+$, a finite SG has a Markov policy which is an N-period EP with respect to $\mathcal{S}$ and the set of all policies.

PROOF Let V be the set of all real-valued functions on $S \times \mathcal{Q}$ and let

$$h(s, a, q, v) = r_q(s, a) + \beta_q \sum_{j \in \mathcal{S}} p_{sj}(a) v(j, q) \tag{9-24}$$

$$(s, a) \in \mathscr{C},\ q \in \mathcal{Q},\ \text{and}\ v \in V$$

To initiate an inductive proof, let $N = 1$. Then Theorem 9-4 asserts† the existence of a one-period EP $\boldsymbol{\delta}^1$ with respect to $\mathcal{S}$ and the set of all policies. Trivially, $\boldsymbol{\delta}^1$ is a Markov policy.

Suppose the theorem is valid for $N - 1$. From the comment below Definition 9-14, there is an $(N - 1)$-period EP $(\boldsymbol{\delta}^{N-1}, \boldsymbol{\delta}^{N-2}, \ldots, \boldsymbol{\delta}^2, \boldsymbol{\delta}^1)$ in which $\boldsymbol{\delta}^n$ is the decision rule which determines the randomized action when n periods remain until the game ends. Also, $\boldsymbol{\delta}^n$ depends only on the state at the beginning of the period, that is, s_{N-n+1}.

Let $v(j, q) = v_j^q[(\boldsymbol{\delta}^{N-1}, \ldots, \boldsymbol{\delta}^1), N]$, which is player q's expected return in the N-1-period game, from initial state j, if the players use the $(N - 1)$-period EP $(\boldsymbol{\delta}^{N-1}, \ldots, \boldsymbol{\delta}^1)$. Then $h(s, a, q, v)$ in (9-24) is player q's expected return in the N-period game in which the initial state is s, the players take action a, and then use the $(N - 1)$-period EP $(\boldsymbol{\delta}^{N-1}, \ldots, \boldsymbol{\delta}^1)$. Fix $s \in \mathcal{S}$ and let $m_q(a) = h(s, a, q, v)$. The SG is finite; so there are only finitely many $a \in \times_{q \in \mathcal{Q}} A_s^q$. Consider the static game in which each player q's payoff is the expected value of $m_q(\cdot)$ and in which player q may randomize on A_s^q. From Proposition 9-2, this static game has an EP $\boldsymbol{\delta}^N(s)$. Let $\boldsymbol{\delta}^N = [\boldsymbol{\delta}^N(s); s \in \mathcal{S}]$. We shall prove that $(\boldsymbol{\delta}^N, \boldsymbol{\delta}^{N-1}, \ldots, \boldsymbol{\delta}^1)$ is an N-period EP.

We write $V_s^q(\boldsymbol{\Pi}, N)$ for the r.v. $V^q(\boldsymbol{\Pi}, N)$ in (9-22a) if $s_1 = s$ and $E_{\boldsymbol{\Pi}}$ for an expectation with respect to probabilities induced by policy $\boldsymbol{\Pi}$. Let $\mathbf{z}^{N-1}$ denote $(\boldsymbol{\delta}^{N-1}, \ldots, \boldsymbol{\delta}^1)$ and $\mathbf{z}^N = (\boldsymbol{\delta}^N, \mathbf{z}^{N-1})$; let $v(j, q) = v_j^q(\mathbf{z}^{N-1}, N - 1)$. Suppose all the players but q use policy $\mathbf{z}_{-q}^N = (\boldsymbol{\delta}_{-q}^N, \mathbf{z}_{-q}^{N-1})$ and player q uses an N-period policy ξ_q. Then player q's expected return from initial state s is

$$\begin{aligned} &v_s^q[(\xi_q, \mathbf{z}_{-q}^N), N] \\ &\quad = E_{(\xi_q, \mathbf{z}_q^N)}\left\{r_q(s, \mathbf{a}) + \beta_q \sum_{j \in \mathcal{S}} p_{sj}(\mathbf{a}) V_j^q[(\xi_q, \mathbf{z}_{-q}^N), N - 1]\right\} \\ &\quad = E_{(\xi_q, \mathbf{z}_{-q}^N)}\left\{r_q(s, \mathbf{a}) + \beta_q \sum_{j \in \mathcal{S}} p_{sj}(\mathbf{a}) v_j^q[(\xi_q, \mathbf{z}_{-q}^{N-1}), N - 1]\right\} \\ &\quad \le E_{(\xi_q, \mathbf{z}_{-q}^N)}\left[r_q(s, \mathbf{a}) + \beta_q \sum_{j \in \mathcal{S}} p_{sj}(\mathbf{a}) v_j^q(\mathbf{z}^{N-1}, N - 1)\right] \qquad (9\text{-}25) \\ &\quad = E_{(\xi_q, \mathbf{z}_{-q}^N)}[h(s, \mathbf{a}, q, v)] \le E_{\boldsymbol{\delta}^N}[h(s, \mathbf{a}, q, v)] \qquad (9\text{-}26) \end{aligned}$$

where $\mathbf{a}$ is the random action with N periods remaining. The inequality in (9-25) is due to $\mathbf{z}^{N-1}$ being an $(N - 1)$-period EP. The inequality in (9-26) is caused by the EP property of $\boldsymbol{\delta}^N$ and $\mathbf{z}^N = (\boldsymbol{\delta}^N, \mathbf{z}^{N-1})$. Therefore, $\mathbf{z}^N = (\boldsymbol{\delta}^N, \ldots, \boldsymbol{\delta}^1)$ is an N-period EP. □

EXERCISES

9-5 (a) Using notation in part (i) of the proof of Theorem 9-2, prove that $\mathbf{Z}_{sq}$ is a closed, bounded, and convex subset of $\mathbb{R}^{\nu(s, q)}$.

† Alternatively, for $N = 1$ appeal directly to Proposition 9-2 in the proof of Theorem 9-4 with $N = 1$.

(*b*) Use the result in (*a*) to prove that **Z** is a closed, bounded, and convex subset of $\mathbb{R}^v$ where

$$v \triangleq \sum_{s \in \mathcal{S}} \sum_{q \in \mathcal{Q}} v(s, q)$$

9-6 (*a*) Using notation in part (i) of the proof of Theorem 9-2, prove $\boldsymbol{\delta} \in \mathbf{Z}$ implies $\tau(\boldsymbol{\delta}) \in \mathbf{Z}$.

(*b*) Prove that $\tau(\cdot)$ is continuous on **Z**. [Hint: Use (9-14) through (9-18) to prove that $\phi^q_{sk}(\cdot)$, hence τ, is continuous.]

9-7 In a finite SG, let Y_q be the set of player q's unrandomized stationary policies. Prove that the convex hull of Y_q is the set of randomized stationary policies.

9-8 Let $P_1, P_2, \ldots, P_m$ be a finite number of stochastic matrices of the same size. Suppose that P_1 induces a Markov chain with one communicating class (and, perhaps, some transient states). Let $h_1, h_2, \ldots, h_m$ be nonnegative numbers which sum to 1 and $h_1 > 0$. Prove the following properties of

$$M = \sum_{i=1}^{m} h_i P_i \tag{9-27}$$

(*a*) If states k and j communicate in P_1, they communicate in M.

(*b*) M induces a Markov chain with one communicating class of states (and, perhaps, some transient states). (Hint for both parts:

$$M^n = \sum_{i=1}^{m} h_i^n P_i^n + \cdots = h_1^n P_1^n + \cdots)$$

9-9 In Exercise 9-8, suppose $P_1, P_2, \ldots, P_m$ each induce a Markov chain with one communicating class (and, perhaps, some transient states); the classes may differ from one matrix to the next. Prove the following properties of M in (9-27) without requiring $h_1 > 0$.

(*a*) M induces a Markov chain with one communicating class of states (and, perhaps, some transient states).

(*b*) Let H be the set of all $\mathbf{h} = (h_1, \ldots, h_m) \geq \mathbf{0}$ such that $\sum_{i=1}^{m} h_i = 1$. Let $M(\mathbf{h})$ make the dependence on $\mathbf{h}$ explicit. From (*a*), for each $\mathbf{h} \in H$, there exists a unique solution $\mathbf{c}(\mathbf{h})$ to $\mathbf{c}(\mathbf{h}) \geq \mathbf{0}$, $\mathbf{c}(\mathbf{h}) \cdot \mathbf{e} = 1$, and $\mathbf{c}(h) = \mathbf{c}(\mathbf{h})M(\mathbf{h})$. Then $\mathbf{c}(\cdot)$ is continuous on H. [Hint: When $N(x)$ is a nonsingular matrix with finitely many rows for all $x \in X$ and $N(\cdot)$ is continuous on X, $[N(\cdot)]^{-1}$ is continuous on X.]

9-10 This exercise† is along the lines of Section 5-4. Suppose $\mathcal{S}$ and $\mathcal{Q}$ are countable and A^q_s is countable for each s and q. Let V be the set of all bounded real-valued functions on $\mathcal{S} \times \mathcal{Q}$ and let

$$d(u, v) = \sup \{|u(s, q) - v(s, q)| : s \in \mathcal{S}, q \in \mathcal{Q}\} \qquad u \in V \qquad v \in V$$

(*a*) Prove that V with $d(\cdot, \cdot)$ is a complete metric space.

(*b*) Let $h(s, \mathbf{a}, q, v) = r_q(s, \mathbf{a}) + \beta_q \sum_{j \in \mathcal{S}} p^{\mathbf{a}}_{sj} v(j, q)$ for $(s, \mathbf{a}) \in \mathscr{C}$, $q \in \mathcal{Q}$, and $v \in V$. Assume that there exists $\beta < 1$ and $u < \infty$ such that $\beta_q \leq \beta$ for all $q \in \mathcal{Q}$ and $|r_q(s, \mathbf{a})| \leq u$ for all $q \in \mathcal{Q}$ and $(s, \mathbf{a}) \in \mathscr{C}$. State analogs of the contraction and boundedness assumptions in Section 5-4, and prove that $h(\cdot, \cdot, \cdot, \cdot)$ satisfies these assumptions.

(*c*) For each single-stage decision rule δ and $v \in V$, let $H_\delta v$ be the mapping on $\mathcal{S} \times \mathcal{Q}$ which assigns to (s, q) the value $h[s, \boldsymbol{\delta}(s), q, v]$. Prove for each $\boldsymbol{\delta}$ that $H_{\boldsymbol{\delta}}$ has a unique fixed point $v_{\boldsymbol{\delta}} \in V$, and

$$d(v_\delta, v) \leq \frac{d(H_{\boldsymbol{\delta}} v, v)}{1 - \beta} \qquad v \in V$$

(*d*) For each single-stage decision rule $\boldsymbol{\delta}$ and $v \in V$, let $L_{\boldsymbol{\delta}} v$ be the following mapping on $\mathcal{S} \times \mathcal{Q}$:

$$[L_{\boldsymbol{\delta}} v](s, q) = \sup \{h[s, a_q, \boldsymbol{\delta}_{-q}(s), q, v] : a_q \in A^q_s\} \qquad (s, q) \in \mathcal{S} \times \mathcal{Q}$$

Prove that $L_{\boldsymbol{\delta}}$ is a contraction mapping on V.

(*e*) Prove that h satisfies an analog of the monotonicity assumption in Section 5-4. Let $F_{\boldsymbol{\delta}}$ be the

† Exercises 9-10 and 9-11 ask you to provide the details in a proof in Denardo (1967).

unique FP in V of $L_{\boldsymbol{\delta}}$, and define $f_{\boldsymbol{\delta}}$ on $\mathcal{S} \times \mathcal{Q}$ as

$$f_{\boldsymbol{\delta}}(s, q) = \sup \{v_{(\gamma, \boldsymbol{\delta}_{-q})}(s, q): \gamma \in \Delta_q\}$$

where Δ_q is player q's set of single-stage decision rules, and $v_{\boldsymbol{\delta}}$ is the unique FP of $H_{\boldsymbol{\delta}}$. Prove that $f_{\boldsymbol{\delta}} = F_{\boldsymbol{\delta}}$ for each single-stage decision rule $\boldsymbol{\delta}$.

9-11 (Continuation) The steps in Exercise 9-10 do *not* prove existence of an EP. For that result, $v_{\boldsymbol{\delta}} = f_{\boldsymbol{\delta}}$ for some $\boldsymbol{\delta}$ is needed; then $\boldsymbol{\delta}^\infty$ is an EP. This last step has been proved† using the Kakutani fixed-point theorem. However, the problem is simpler if, as we assume now, there are only two players, rewards are zero-sum, that is, $r_1(s, \mathbf{a}) + r_2(s, \mathbf{a}) = 0$ for all $(s, \mathbf{a}) \in \mathscr{C}$, and $\#\mathscr{C} < \infty$. Define h as in Exercise 9-10; here $h(s, \mathbf{a}, 1, v) = -h(s, \mathbf{a}, 2, v)$ for all $(s, a) \in \mathscr{C}$ and $v \in V$; so in the remainder of the exercise, let $h(\cdot, \cdot, \cdot) = h(\cdot, \cdot, 1, \cdot)$. From the minimax theorem for matrix games, for each $v \in V$, define a mapping Gv on $\mathcal{S}$ via

$$[Gv](s) = \sup \{\inf \{h(s, \delta_1, \delta_2, v): \delta_2 \in \Delta_2\}: \delta_1 \in \Delta_1\}$$

$$= \inf \{\sup \{h(s, \delta_1, \delta_2, v): \delta_1 \in \Delta_1\}: \delta_2 \in \Delta_2\} \qquad s \in \mathcal{S}$$

For each $\boldsymbol{\delta} = (\delta_1, \delta_2)$ and $v \in V$, define a mapping $H_{\boldsymbol{\delta}} v$ on $\mathcal{S}$ with $[H_{\boldsymbol{\delta}} v](s) = h[s, \boldsymbol{\delta}(s), v]$.

(*a*) Explain why an FP of G is an EP.

(*b*) Prove that $H_{\boldsymbol{\delta}}$ is a contraction mapping; hence it has a unique FP $v_{\boldsymbol{\delta}}$.

(*c*) Prove that G is a contraction mapping; hence it has a unique FP. Hint: For each δ_1 and v let

$$[\mathscr{H}_{\delta_1} v](s) = \inf \{[H_{(\delta_1, \delta_2)} v](s): \delta_2 \in \Delta_2\}$$

and prove that $\mathscr{H}_{\delta_1}$ is a contraction mapping. Note that

$$[Gv](s) = \sup \{[\mathscr{H}_{\delta_1} v](s): \delta_1 \in \Delta_1\}$$

and use Theorem 5-8.

(*d*) Let $v^0 \in V$ be arbitrary and for each $i \in I_+$ let

$$v^i = Gv^{i-1} = G^i v^0$$

where $G^1 = G$ and $G^{i+1} = GG^i$. From (c), $\lim_{i \to \infty} v^i$ exists and is the FP of G. Fix i and explain (in detail), how to compute v^i starting from v^0.

9-12 Fix $N \in I_+$. In the proof of Theorem 9-4, it is claimed that any Π_q is a convex combination of the unrandomized Π_q's. Prove it.

9-13 Prove that $m_q(\cdot)$, defined in the proof of Theorem 9-5, is continuous on W (for each N, s, and q).

9-14 Theorems 9-4 and 9-5 are alternative existence proofs for N-period EPs. Construct a third proof by invoking Theorem 9-2. (Hint: For N fixed, construct an infinite-horizon game with state space $\mathcal{S} \times \{1, \ldots, N+1\}$.)

9-15 This exercise concerns the unichain assumption in Theorem 9-3. There is an alternative proof of that theorem which depends on a mapping $\tau: \mathbf{Z} \to \mathbf{Z}$ which is similar to the mapping τ in the proof of Theorem 9-2. For each $\boldsymbol{\delta} \in \mathbf{Z}$, the unichain assumption and Theorem 4-13 [cf. (4-111) through (4-116) in Section 4-6] imply for each q that there is a unique number $g^q(\boldsymbol{\delta})$ and vector $\mathbf{w}^q(\boldsymbol{\delta})$ with components $w_s^q(\boldsymbol{\delta})$ such that

$$g^q(\boldsymbol{\delta}) = \mathbf{c}(\boldsymbol{\delta}) \cdot \boldsymbol{\rho}^q(\boldsymbol{\delta}) \tag{9-28}$$

$$\mathbf{c}(\boldsymbol{\delta}) \cdot \mathbf{w}^q(\boldsymbol{\delta}) = 0 \tag{9-29}$$

$$\mathbf{e} \cdot g^q(\boldsymbol{\delta}) + \mathbf{w}^q(\boldsymbol{\delta}) = \boldsymbol{\rho}^q(\boldsymbol{\delta}) + B_\delta \mathbf{w}^q(\boldsymbol{\delta}) \tag{9-30}$$

† A. M. Fink: "Equilibrium in a Stochastic *n*-Person Game," *J. Sci. Hiroshima Univ. Ser. A-I* **28**: 89–93 (1964).

where $\mathbf{c}(\boldsymbol{\delta})$ is the stationary distribution which uniquely solves (9-21). Let

$$\phi_{sk}^q(\boldsymbol{\delta}) = \left[\rho_s^q(\boldsymbol{\delta}_{sk}^q) + \sum_{j \in \mathcal{S}} b_{sj}(\boldsymbol{\delta}_{sk}^q) w_j^q(\boldsymbol{\delta}) - g^q(\boldsymbol{\delta}) - w_s^q(\boldsymbol{\delta}) \right]^+ \tag{9-31}$$

which is similar to the test quantity (4-136) in the policy-improvement algorithm for the average-return criterion under the unichain assumption. Let $\{D_{sk}^q\}$ correspond to $\boldsymbol{\delta}$. Then define $\tau\colon \mathbf{Z} \to \mathbf{Z}$ with

$$\tau(\boldsymbol{\delta})(s, k, q) = \frac{D_{sk}^q + \phi_{sk}^q(\boldsymbol{\delta})}{1 + \sum_{i=1}^{v(s,q)} \phi_{si}^q(\boldsymbol{\delta})} \qquad k \in A_s^q \qquad s \in \mathcal{S} \qquad q \in \mathcal{Q}$$

which is the same as (9-18) except that $\phi_{sk}^q(\boldsymbol{\delta})$ is specified by (9-31) instead of (9-17).

Construct an example, with at most three states and two players, where τ is not continuous on Z. (Hint: Such an example cannot satisfy the unichain assumption.)

9-4 MYOPIC EQUILIBRIUM POINTS

Section 9-3 establishes that every finite SG, for every initial state, has an N-period EP (equilibrium point) for every N. Also, it has a discounted EP, and under certain conditions, it has an average-reward EP. However, the computation of an EP can be a formidable task.† This section presents sufficient conditions for the computation of an EP in an SG to be replaced by the computation of an EP in a static game. The latter task is much easier than the former.

The results in this section are analogous to the simplification for MDPs in Chapter 3. We find in Chapter 3 that myopic optima facilitate the qualitative analysis of optimal policies. In an example at the end of this section and throughout Section 9-5, we find that myopic EPs simplify the analysis of qualitative properties of EPs of SGs.

From Section 3-3, an MDP is said to have a *myopic optimum* if its data can be used easily to specify a single-period optimization problem with the following property: *ad infinitum* repetition of a solution to the single-period problem comprises an optimal MDP solution. Similarly, an SG is said to have a *myopic EP* if its data can be used easily to specify a static game with the following property: *ad infinitum* repetition of an EP of the static game comprises an EP for the SG.

The following assumptions are similar to Assumptions I through IV on pages 84 and 85 in Section 3-3. It is convenient to define

$$A^q = \bigcup_{s \in \mathcal{S}} A_s^q \qquad q \in \mathcal{Q} \qquad A = \times_{q \in \mathcal{Q}} A^q$$

and

$$S(\mathbf{a}) = \{s\colon (s, \mathbf{a}) \in \mathscr{C}\} \qquad \mathbf{a} \in A$$

Hence, $S(\mathbf{a})$ is the set of states from which the multiplayer action $\mathbf{a}$ is feasible.

† It follows from Exercise 9-11 that successive approximations can be used in the two-person zero-sum case. However, no finite algorithm can be devised for some SGs because an SG with rational data can have a unique irrational EP [J. F. Nash, and L. S. Shapley, "A Simple Three Person Poker Game," in *Contributions to the Theory of Games I*, Annals of Mathematical Studies, 24; edited by H. W. Kuhn and A. W. Tucker, Princeton University Press, Princeton, N.J. (1950), pp. 105–116].

Recall Definition 9-4, which introduces the notation $L_q(s)$ for player q's salvage value if the ultimate state is s. Finally, in notation similar to that in Section 9-2, we write $\mathbf{d} = (d_q, \mathbf{d}_{-q}) \in A$, where d_q is player q's action and $\mathbf{d}_{-q}$ denotes the actions of all the players except player q.

Assumption I

$$r_q(s, \mathbf{a}) = K_q(\mathbf{a}) + L_q(s) \qquad (s, \mathbf{a}) \in \mathscr{C} \qquad q \in \mathscr{Q} \tag{9-32}$$

Assumption II The transition function satisfies

$$p(J \mid s, \mathbf{a}) = p(J \mid \mathbf{a}), \qquad (s, \mathbf{a}) \in \mathscr{C} \qquad \text{so } s_{n+1} \sim \xi(a_n),\ n \in I_+ \tag{9-33}$$

Let

$$\gamma_q(\mathbf{a}) = K_q(\mathbf{a}) + \beta_q E\{L_q[\xi(\mathbf{a})]\} \qquad \mathbf{a} \in A \tag{9-34}$$

and let Γ denote the following static game among the players in $\mathscr{Q}$. Player q has available the set of moves A^q and $\gamma_q(a)$ is player q's payoff when the players choose $a \in A$. First, suppose Γ has an EP in pure (i.e., unrandomized) strategies.

Assumption III There exists $a^* \in A$ such that

$$\gamma_q(\mathbf{a}^*) \geq \gamma_q(k, \mathbf{a}^*_{-q}) \qquad k \in A^q \qquad q \in \mathscr{Q} \tag{9-35}$$

Assumption IV†

$$P\{\xi(\mathbf{a}^*) \in S(\mathbf{a}^*)\} = 1 \tag{9-36}$$

The MDP version of these assumptions is discussed in Section 3-3. In particular, the transition function $p(\cdot \mid s, \mathbf{a})$, which ordinarily depends on both s and $\mathbf{a}$, is assumed in (9-33) to depend only on $\mathbf{a}$. As a result, the $n + 1$st state s_{n+1}, which ordinarily depends on both the nth state s_n and the nth multiplayer action $\mathbf{a}_n$, is assumed in (9-33) to depend only on $\mathbf{a}_n$. Therefore, s_{n+1} has the same probability distribution as a random variable $\xi(\mathbf{a}_n)$ whose distribution, from (9-33), is

$$p(J \mid \mathbf{a}) = P\{\xi(\mathbf{a}) \in J\}$$

Following Corollary 9-6 below, we relax Assumption III, which requires Γ to have an EP in pure (i.e., unrandomized) strategies.

Let $\boldsymbol{\alpha}$ be a single-stage decision rule which specifies $\boldsymbol{\alpha}(s) = \mathbf{a}^*$ (with probability 1) if $s \in S(\mathbf{a}^*)[\Leftrightarrow \mathbf{a}^* \in \times_{q \in \mathscr{Q}} A_s^q]$ and specifies an arbitrary element of $\times_{q \in \mathscr{Q}} A_s^q$ if $s \notin S(\mathbf{a}^*)$. Let $\boldsymbol{\alpha}^N$ denote the N-period policy where $\mathbf{a}_n = \alpha(s_n)$ for $n = 1, \ldots, N$.

Theorem 9-6 Assumptions I through IV imply:
(*a*) $\boldsymbol{\alpha}^\infty$ is a discounted EP with respect to $S(\mathbf{a}^*)$ and the set of all policies;

† Let $(\Omega, \mathscr{B}, P)$ be the probability space here. By (9-36), we mean $P\{\xi(\mathbf{a}^*) \in H\} = 1$ for all $H \in \mathscr{B}$ with $S(\mathbf{a}^*) \subset H$.

(*b*) $\boldsymbol{\alpha}^N$ is an N-period EP, for every $N \in I_+$, with respect to $S(\mathbf{a}^*)$ and the set of all policies.

Proof As in the proof of Theorem 3-2, the substitution of (9-32) and (9-33) in

$$V^q(N) = \sum_{n=1}^{N} \beta_q^{n-1} r_q(s_n, \mathbf{a}_n) + \beta_q^N L_q(s_{N+1}) \tag{9-37}$$

yields

$$V^q(N) \sim L_q(s_1) + \sum_{n=1}^{N} \beta_q^{n-1}\{K_q(\mathbf{a}_n) + \beta_q L_q[\xi(\mathbf{a}_n)]\}$$

or

$$E[V^q(N)] = L_q(s_1) + E\left[\sum_{n=1}^{N} \beta_q^{n-1} \gamma_q(\mathbf{a}_n)\right] \tag{9-38}$$

where $\gamma_q(\cdot)$ is defined by (9-34).

Suppose $s_1 = s \in \mathcal{S}(\mathbf{a}^*)$ and all the players except player q use the single-stage decision rule $\boldsymbol{\alpha}_{-q}^N$ in periods $1, \ldots, N$. Let E_Π denote an expectation evaluated with probabilities induced by policy $\boldsymbol{\Pi}$. If player q uses the N-period policy ξ, then q's expected present value is

$$\begin{aligned} v_s^q[(\xi, \boldsymbol{\alpha}_{-q}^N), N] &= L_q(s) + E_{(\xi, \boldsymbol{\alpha}_{-q}^N)}\left[\sum_{n=1}^{N} \beta_q^{n-1} \gamma_q(\mathbf{a}_n)\right] \\ &= L_q(s) + E_\xi\left[\sum_{n=1}^{N} \beta_q^{n-1} \gamma_q(a_n^q, \mathbf{a}_{-q}^*)\right] \\ &\le L_q(s) + \gamma_q(\mathbf{a}^*) \sum_{n=1}^{N} \beta_q^{n-1} \end{aligned} \tag{9-39}$$

with the inequality due to (9-35), the static EP property of $\mathbf{a}^*$. Player q attains the right side of (9-39) by using the qth component of $\mathbf{a}^*$ in every period. This conclusion is valid for all $q \in \mathcal{Q}$, which completes the proof of (*b*).

For (*a*), let $N \to \infty$ above. The EP claim is valid with respect to all policies, rather than only the Markov policies, for the following reason. From (9-35), if all the players but q are using α_{-q}^∞ and player q uses any policy ξ, then

$$\begin{aligned} v_s^q(\xi, \alpha_{-q}^\infty) &= L_q(s) + E_{(\xi, \alpha_{-q}^\infty)}\left[\sum_{n=1}^{\infty} \beta_q^{n-1} \gamma_q(a_n^q, a_{-q}^*)\right] \\ &\le L_q(s) + \frac{\gamma_q(a^*)}{1-\beta} \end{aligned} \tag{9-40}$$

Player q attains the right side of (9-40) by using the qth component of α^∞.

□

Corollary 9-6 Suppose there is $u < \infty$ such that

$$|L_q(s)| \leq u \qquad s \in \mathcal{S} \qquad q \in \mathcal{Q} \tag{9-41}$$

and Assumptions I through IV are satisfied [with $\beta_q = 1$ in (9-34)]. Then α^∞ is an average-reward EP with respect to $S(a^*)$ and all policies.

PROOF Let $\beta_q = 1$ in (9-37). Then

$$\sum_{n=1}^{N} r_q(s_n, a_n) = V^q(N) - L_q(s_{N+1})$$

so

$$\begin{aligned} G^q &= \liminf_{N \to \infty} \frac{1}{N} \sum_{n=1}^{N} r_q(s_n, a_n) \\ &= \liminf_{N \to \infty} \frac{V^q(N)}{N} - \liminf_{N \to \infty} \frac{L_q(s_{N+1})}{N} \end{aligned} \tag{9-42}$$

From Theorem 9-6, if $s_1 \in \mathcal{S}(\mathbf{a}^*)$, then for each fixed N, each player's component of $\mathbf{a}_n = \boldsymbol{\alpha}(s_n)$, $n = 1, \ldots, N$, is an optimal rejoinder if all the other players do likewise; that is,

$$\gamma_q(\mathbf{a}^*) = \frac{v_s^q(\boldsymbol{\alpha}^\infty, N)}{N} \geq \frac{v_s^q[\xi_q, \boldsymbol{\alpha}_{-q}^\infty), N]}{N} \qquad s \in \mathcal{S}(\mathbf{a}^*) \tag{9-43}$$

for all policies ξ_q available to player q. Also, (9-41) implies $L_q(s_{N+1})/N \to 0$ for all policies. From (9-42) and (9-43), taking expected values in (9-42) and letting $N \to \infty$,

$$g_s^q(\boldsymbol{\alpha}^\infty) = \gamma_q(\mathbf{a}^*) \geq g_s^q[(\xi_q, \boldsymbol{\alpha}_{-q}^\infty)] \qquad s \in \mathcal{S}(\mathbf{a}^*)$$

for all ξ_q. Hence, $\boldsymbol{\alpha}^\infty$ satisfies (9-9), the definition of an average-reward EP. $\square$

Randomized Policies

Recall from Section 9-3, where $\# A_s^q < \infty$, that $\mathbf{Z}_{sq}$ denotes the set of all probability distributions on A_s^q. If A_s^q is not denumerable, technical difficulties require a modification of the definition of $\mathbf{Z}_{sq}$. For the remainder of this section, we avoid such modifications by considering only finite SGs. Now suppose Γ lacks a pure EP so Assumption III is not valid. Since the SG is finite, it follows that $A = \times_{q \in \mathcal{Q}} \bigcup_{s \in \mathcal{S}} A_s^q$ is a finite set so there necessarily exists at least a randomized EP (due to Proposition 9-2).

Formally,

$$\mathbf{Z}^q = \bigcup_{s \in \mathcal{S}} \mathbf{Z}_{sq} \qquad q \in \mathcal{Q} \qquad \text{and} \qquad \times_{q \in \mathcal{Q}} \mathbf{Z}^q$$

correspond to the sets A^q and A defined on page 465 prior to Assumption I. In fact, $\mathbf{Z}^q$ is contained in the set of all probability distributions on A^q:

$$\mathbf{Z}^q \subset \left\{(z_{iq};\ i \in A^q): z_{iq} \geq 0 \qquad \text{for all } i \in A^q \text{ and } \sum_{i \in A^q} z_{iq} = 1\right\}$$

We write $\mathbf{z}_q \in \mathbf{Z}^q$ for an element of $\mathbf{Z}^q$. Then $\times_{q \in \mathcal{Q}} \mathbf{Z}^q$ is the set of all $\mathbf{z} = (\mathbf{z}_q;\ q \in \mathcal{Q})$. If the players in Γ use the randomized strategy $\mathbf{z}$, player q's expected return is

$$\rho^q(\mathbf{z}) = \sum_{i_1 \in A^1} \cdots \sum_{i_Q \in A^Q} \prod_{j \in \mathcal{Q}} z_{i_j j} \gamma_q(i_1, \ldots, i_Q)$$

which extends the notation (9-15). An EP $\mathbf{z}$ of Γ, which necessarily exists, satisfies

$$\rho^q(\mathbf{z}) \geq \rho^q(\xi, \mathbf{z}_{-q}) \qquad \xi \in \mathbf{Z}^q \qquad q \in \mathcal{Q} \tag{9-44}$$

For a particular EP $\mathbf{z}$ of Γ, let A' denote the elements of A which are given positive probability:

$$A' = \{(k_1, \ldots, k_Q): (k_1, \ldots, k_Q) \in A \text{ and } z_{k_q q} > 0 \text{ for all } q\}$$

Let S' denote the set of states from which every element of A' is feasible:

$$S' = \left\{s: s \in \mathcal{S} \text{ and } a \in \mathop{\times}_{q \in \mathcal{Q}} A_s^q \qquad \text{for all } a \in A'\right\} = \bigcap_{a \in A'} S(a)$$

The SG is assumed to be finite; so let $p_{sj}(\mathbf{a})$ denote the transition probability $p(\{j\} \mid s, \mathbf{a})$. Assumption II is equivalent to the existence of numbers $p_j(\mathbf{a})$, $j \in \mathcal{S}$, $\mathbf{a} \in A$, such that

$$p_j(\mathbf{a}) = p_{sj}(\mathbf{a}) \qquad (s, \mathbf{a}) \in \mathscr{C}, j \in \mathcal{S}$$

In place of Assumption IV, there is

Assumption IV′

$$\sum_{j \in \mathcal{S}'} p_j(\mathbf{a}) = 1 \qquad \text{for all } a \in A' \tag{9-45}$$

The elements of A' are the actions which are given positive probability by $\mathbf{z}$. Then (9-45) asserts that, with probability 1, the subsequent state will be in S'. But S' is exactly the set of states from which all the elements of A' are feasible.

Let λ be a single-stage decision rule which uses $\mathbf{z}$ if $s \in S'$ and is arbitrary if $s \notin S'$. Let λ^N denote the N-period policy which uses the single-stage decision rule λ in periods $1, \ldots, N$.

Theorem 9-7 Suppose a finite SG satisfies Assumptions I, II, and IV′.

(*a*) λ^∞ is a discounted EP and an average-reward EP with respect to S' and all policies.

(*b*) For all $N \in I_+$, λ^N is an N-period EP with respect to S' and all policies.

PROOF Left as Exercise 9-16. □

Computation

Suppose an SG satisfies Assumptions I through IV or IV′. It follows from Theorems 9-6 and 9-7 that an EP solution of the SG is equivalent to finding an EP of the static game Γ specified between (9-34) and (9-35). Suppose that the SG is finite so that each player in Γ is randomizing over only finitely many actions; that is, A^q is finite for each q. The Bibliographic Guide at the end of this chapter includes references to several algorithms which may be used to compute or approximate an EP of Γ.

A Useful Special Case

Instances of the following structure arise later in this section and in Sections 9-5 and 9-6. It is convenient to write $\mathbf{a} = (a_1, a_2, \ldots, a_Q)$ instead of $(a^1, a^2, \ldots, a^Q)$ in stating and proving the following result.

Theorem 9-8 Suppose $Q \geq 2$ and for all $q \in \mathscr{Q}$ that $\gamma_q(\mathbf{a})$ in (9-34) takes the form

$$\gamma_q(\mathbf{a}) = \alpha_q + \frac{\omega_q a_q}{\sum_{i \in \mathscr{Q}} a_i} - \xi_q a_q$$

with $\omega_q > 0$ and $\xi_q \geq 0$. Let $\eta_q = \xi_q/\omega_q$ and suppose $\sum_{i \in \mathscr{Q}} \eta_i \neq 0$. Let $\sigma = (Q-1)/\sum_{i \in \mathscr{Q}} \eta_i$ and $\mathbf{a}^* = (a_q^*; q \in \mathscr{Q})$, where

$$a_q^* = \sigma - \eta_q \sigma^2 \qquad q \in \mathscr{Q}$$

If $\eta_q \leq \sigma^{-1}$ for each q, $\eta_q < \sigma^{-1}$ for some q, and $a^* \in A$, then $\mathbf{a}^*$ is an EP of Γ [the static game specified below (9-34)]. If also $\eta_q = \eta$ for all $q \in \mathscr{Q}$, then

$$a_q^* = \frac{Q-1}{Q^2 \eta} \qquad q \in \mathscr{Q}$$

Proof The assumed structure of $\gamma_q(\cdot)$ yields

$$\frac{\partial \gamma_q(\mathbf{a})}{\partial a_q} = \omega_q \sum_{i \neq q} \frac{a_i}{(\sum_{i \in \mathscr{Q}} a_i)^2} - \xi_q$$

$$\frac{\partial^2 \gamma_q(\mathbf{a})}{\partial a_q^2} = \frac{-2\omega_q \sum_{i \neq q} a_i}{(\sum_{i \in \mathscr{Q}} a_i)^3}$$

The assumptions imply $a_i^* \geq 0$ for each i; so the second derivative is nonpositive at $\mathbf{a} = \mathbf{a}^*$. Setting the first derivative equal to zero yields

$$\sum_{i \in \mathscr{Q}} a_i = a_q + \eta_q \left(\sum_{i \in \mathscr{Q}} a_i \right)^2 \tag{9-46}$$

Sum both sides over $q \in \mathscr{Q}$ to obtain

$$Q \sum_{i \in \mathscr{Q}} a_i = \sum_{q \in \mathscr{Q}} a_q + \left(\sum_{q \in \mathscr{Q}} \eta_q \right) \left(\sum_{i \in \mathscr{Q}} a_i \right)^2$$

Dividing both sides by $\sum_{i \in Q} a_i \ (\neq 0)$ leads to

$$\sum_{i \in \mathcal{Q}} a_i = \frac{Q-1}{\sum_{i \in \mathcal{Q}} \eta_i} = \sigma$$

which, upon substitution in (9-46), yields $a_q = \sigma - \eta_q \sigma^2$. If $\eta_q = \eta$ for all q, then $\sigma = (Q-1)/(Q\eta)$ so

$$a_q = \sigma - \eta_q \sigma^2 = \frac{Q-1}{Q\eta}\left(1 - \frac{Q-1}{Q}\right)$$

$$= \frac{Q-1}{Q^2 \eta}$$

□

"Guns or Butter" Example

Interactions among nations and among biological species are sometimes discussed in terms of trade-offs between consumption and investment. The idea is that a nation (species) is more likely to survive in the future if it invests more and consumes less. However, one of the purposes of survival is to be able to consume; so one would rather not forgo consumption. These considerations are implicit in the following model. It can be regarded as a sequential game generalization of the resource-management and capital-accumulation model in Section 8-4.

Suppose there are two players whose wealths at the beginning of period n are (s_n^1, s_n^2). Each period n, player q decides how much to invest in "guns," g_n^q, and how much to consume as "butter," b_n^q. A player cannot use more than the available wealth; so $g_n^q + b_n^q \leq s_n^q$ and

$$A^q_{(s^1,\, s^2)} = \{(g, b)\colon 0 \leq g,\ 0 \leq b, \text{ and } g + b \leq s^q\}$$

for each q and $0 \leq s^q$.

Let $(X_1^1, X_1^2), (X_2^1, X_2^2), (X_3^1, X_3^2), \ldots$ be independent and identically distributed random vectors. We assume

$$s_{n+1}^q = \begin{cases} X_n^q\left(s_n^q - b_n^q - g_n^q + h_q \dfrac{g_n^q}{g_n^1 + g_n^2}\right) & \text{if } g_n^q > 0 \\ X_n^q(s_n^q - b_n^q) & \text{if } g_n^q = 0 \end{cases} \tag{9-47}$$

where $h_q > 0$. In (9-47), $s_n^q - b_n^q - g_n^q$ is the residual wealth after consumption and investment, and $h_q g_n^q/(g_n^1 + g_n^2)$ is proportional to player q's fraction of the total investment. Hence, s_{n+1}^q is a random multiple of the sum of these two amounts. The term $h_q g_n^q/(g_n^1 + g_n^2)$ has the property that the effectiveness of player q's investment in "guns" diminishes as the other player's investment in "guns" increases.

Suppose that player q's reward in period n is the amount of "butter" that q consumes, b_n^q. Then

$$r_q\{(s^1, s^2), [(g^1, b^1), (g^2, b^2)]\} = b^q \qquad q = 1, 2 \tag{9-48}$$

This model satisfies Assumption I because, in (9-48), $r_q(s, \mathbf{a}) = K_q(\mathbf{a}) + L_q(s)$ with $L_q(\cdot) \equiv 0$ and $K_q(\mathbf{a}) = K_q[(g^1, b^1), (g^2, b^2)] = b^q$. However, Assumption II is violated by (9-47).

Assumption II stipulates that s_{n+1} depends only on $\mathbf{a}_n$. Therefore, we transform the definition of player q's action from (g^q, b^q) to $(s^q - b^q, g^q)$ and replace A_s^q in (9-44) with its equivalent for the new action (d^q, g^q), where $d^q = s^q - b^q$:

$$A^q_{(s^1,\ s^2)} = \{(d, g)\colon 0 \le g \le d \le s^q\}$$

Recall the notation

$$A^q = \bigcup_{s \in \mathcal{S}} A_s^q$$

Here, $\mathcal{S} = \mathbb{R}_+^2$ so

$$A^q = \{(d, g)\colon 0 \le g \le d\} \qquad q = 1, 2 \tag{9-49}$$

Also, $S(\mathbf{a})$ denotes $\{s\colon (s, a) \in \mathscr{C}\}$ so (9-48) yields

$$S[(d^1, g^1), (d^2, g^2)] = \{(s^1, s^2)\colon 0 \le g^q \le d^q \le s^q,\ q = 1, 2\} \tag{9-50}$$

Let $\mathbf{a}^q$ now denote the new action (d^q, g^q), let $\mathbf{a} = (a^1, a^2)$, and let $\mathbf{s} = (s^1, s^2)$. Then (9-48) becomes

$$r_q(\mathbf{s}, \mathbf{a}) = s^q - d^q$$

because $s^q - d^q = s^q - (s^q - b^q) = b^q$. In this form, $r_q(\cdot, \cdot)$ satisfies Assumption I with $K_q(\mathbf{a}) = -d^q$ and $L_q(\mathbf{s}) = s^q$. Also, (9-47) becomes

$$s_{n+1}^q = \begin{cases} X_n^q\left(d_n^q - g_n^q + h_q \dfrac{g_n^q}{g_n^1 + g_n^2}\right) & \text{if } g_n^q > 0 \\ X_n^q d_n^q & \text{if } g_n^q = 0 \end{cases} \tag{9-51}$$

Assumption II is satisfied because the right side of (9-51) involves only elements of $\mathbf{a}_n$.

For Assumption III, we must first specify the static game Γ whose payoffs $\gamma_q(\cdot)$ are defined by (9-34). Let $\mu_q = E(X_1^q)$. *For the remainder if this section, we write q as a subscript rather than a superscript.* Here, if $g_q > 0$,

$$\begin{aligned} \gamma_q(\mathbf{a}) &= K_q(\mathbf{a}) + \beta_q E\{L_q[\xi(\mathbf{a})]\} \\ &= -d_q + \beta_q E\left\{X_{1q}\left(d_q - g_q + h_q \frac{g_q}{g_1 + g_2}\right)\right\} \end{aligned}$$

so

$$\gamma_q(\mathbf{a}) = -d_q + \beta_q \mu_q\left(d_q - g_q + h_q \frac{g_q}{g_1 + g_2}\right) \tag{9-52}$$

Player q's available actions in Γ are A^q given by (9-49). Suppose $0 \le \beta_q \mu_q \le 1$,

$q = 1, 2$. Exercise 9-17 asks you to prove that Γ has the following EP:

$$d_1 = g_1 = \frac{1/h_2}{(1/h_1 + 1/h_2)^2}$$
$$d_2 = g_2 = \frac{1/h_1}{(1/h_1 + 1/h_2)^2} \tag{9-53}$$

The assumption $\beta_q \mu_q \le 1$ restricts $E(X_{1q})$ to at most $1/\beta_q$. Hence, if $\beta_q \mu_q \le 1$ for $q = 1, 2$, Assumption III is satisfied by $\mathbf{a}^*$ specified by (9-53).

Assumption IV is $P\{\xi(\mathbf{a}^*) \in \mathcal{S}(\mathbf{a}^*)\} = 1$. With (9-49), (9-50), and (9-52), this condition is

$$P\left\{X_{1q} \frac{h_q g_q}{g_1 + g_2} \ge g_q,\ q = 1, 2\right\} = 1 \tag{9-54}$$

Since g_1 and g_2 are specified by (9-53),

$$g_1 + g_2 = \frac{1}{1/h_1 + 1/h_2}$$

Hence, (9-54) becomes

$$P\left\{X_{1q} \ge \frac{1/h_1 + 1/h_2}{h_q} \qquad q = 1, 2\right\} = 1 \tag{9-55}$$

Therefore, if $(s_1^1, s_1^2) \ge (g^1, g^2)$ and $\beta_q \mu_q \le 1$ for $q = 1, 2$, so (9-55) is satisfied, then $(d_n^q, g_n^q) = (d_q, g_q)$ specified by (9-53) for all $n = 2, 3, \ldots, g_1^q = g_q$ and $b_1^q = s_1^q - g_q$ comprise an EP for the SG. Using (9-38), the resulting expected present value in the infinite-duration case is

$$s_1^1 - \frac{1/h_2}{(1/h_1 + 1/h_2)^2} \frac{1 - \beta_1 \mu_1 h_1(1/h_1 + 1/h_2)}{1 - \beta_1} \tag{9-56}$$

for player 1, with a similar expression for player 2. Notice that neither the actions (9-53) nor their consequent expected return (9-56) depend on the joint probability distribution of (X_1^1, X_1^2) except through μ_1 and μ_2, that is, the expected values of the marginal distributions.

A *symmetric game* is one in which each player has the same opportunities and rewards. In the present game, symmetry means $\beta \triangleq \beta_1 = \beta_2$, $\mu \triangleq \mu_1 = \mu_2$, and $h_1 = h_2$; then $J \triangleq J_1 = J_2$. If $h_1 = h_2 = h$, then (9-53) becomes

$$d^1 = g^1 = d^2 = g^2 = \frac{h}{4}$$

and (9-55) is reduced to

$$P\{X_1^q \ge 1/2,\ q = 1, 2\} = 1 \tag{9-57}$$

The entire restriction on (X_1^1, X_1^2) is $\mu \le 1/\beta$ and (9-57).

EXERCISES

9-16 Prove Theorem 9-7.

9-17 In (9-52), suppose $0 \leq \beta_q \mu_q \leq 1$ for $q = 1, 2$. Verify that (9-53) specifies an EP for Γ with payoffs given by (9-52) and actions given by (9-49). (Hint: Theorem 9-8.)

9-18 Suppose the guns and butter model is modified so H_n^q replaces h_q in (9-47) and we assume $(X_n^1, H_n^1, X_n^2, H_n^2)$, $n = 1, 2, \ldots$ are independent and identically distributed random vectors. Let $h_q = E(X_1^q H_1^q)/E(X_1^q)$ and let (9-52) define d^q and g^q, $q = 1, 2$. Prove that this modified model has the same myopic EP as the original model if (9-55) is replaced by

$$P\left\{X_1^q H_1^q \geq \frac{1}{h_1} + \frac{1}{h_2}, q = 1, 2\right\} = 1$$

9-19 The model in this section is *stationary*, i.e., time-invariant. Suppose, instead, that $\mathcal{S}_n$ is the set of states in period n, A_{sn}^q is player q's set of alternative actions in period n if

$$s_n = s \qquad p_n(J \mid s, \mathbf{a}) = P\{s_{n+1} \in J \mid s_n = s, a_n = \mathbf{a}\}$$

for $J \subset \mathcal{S}_{n+1}$ and $(s, \mathbf{a}) \in \mathscr{C}_n = \{(s, \mathbf{a}) : \mathbf{a} \in \times_{q \in \mathscr{Q}} A_{sn}^q \text{ and } s \in \mathcal{S}_n\}$, $r_{qn}(s, \mathbf{a})$ is player q's reward in period n if $(s_n, \mathbf{a}_n) = (s, \mathbf{a}) \in \mathscr{C}_n$, and one monetary unit at the start of $n + 1$ is worth β_{qn} at the start of period n. State and prove a version of Theorem 9-6 for this nonstationary model. Hint: Let

$$\gamma_{qn}(\mathbf{a}) = K_{qn}(\mathbf{a}) + \beta_{qn} E\{L_{q,n+1}[\xi_n(\mathbf{a})]\}$$

9-20 A more general version of the model in Theorem 9-8 is

$$\gamma_q(\mathbf{a}) = \alpha_q + \frac{\omega_q a_q}{(\sum_{i \in \mathscr{Q}} a_i)^m} - \xi_q a_q$$

with $m \in I_+$ and $m < Q$. Suppose also that $\eta = \xi_q/\omega_q$ is the same for all q. Show that

$$a_q^* = \left(\frac{Q - m}{Q^{m+1}}\right)^{1/m} \qquad \text{for all } q$$

is an EP of the static game Γ.

9-21 Prove that the conclusions of Theorem 9-8 are valid for the following model (which is different from the one in Theorem 9-8):

$$\gamma_q(\mathbf{a}) = \alpha_q - \frac{\omega_q \sum_{i \neq q} a_i}{\sum_{i \in \mathscr{Q}} a_i} - \xi_q a$$

9-5* COMPETITIVE ADVERTISING DECISIONS

Section 3-4 presents an MDP model of a firm's advertising decisions. That model does not include the competitive reaction of competing firms, a feature which is added in this section. It is important to use an SG model of advertising decisions instead of an MDP model if one's competitors react quickly and their decisions greatly affect one's own demand.

This section analyzes a duopoly model, i.e., a model of two competing firms. The principal conclusions are valid in a similar model with more than two firms, but the exposition would be more cumbersome. Duopoly models are sometimes useful to model competitive decisions in industries with more than two firms but where one firm is dominant. Then the two firms in the model are the dominant firm and a pseudo-"firm" which aggregates all the firms except the dominant one. Some examples of industries where the leading firm is indeed dominant are IBM in computers and General Mills in breakfast cereals.

The notation in the following model generally consists of appending q superscripts and subscripts to the notation in Section 3-4. The model describes two interacting firms; so $q \in \mathcal{Q} = \{1, 2\}$. We assume that the effect of firm q's advertising on its "goodwill" depreciates at a rate $1 - \theta_q$ per time period, $0 \leq \theta_q \leq 1$. Let z_n^q denote firm q's advertising expenditure in period n; so $z_n^q \theta_q^j$ is the impact of z_n^q on goodwill in period $n + j$. Firm q's goodwill is the aggregate impact of its advertising expenditures; so a_n^q, its goodwill in period n, is

$$a_n^q = z_1^q \theta_q^{n-1} + z_2^q \theta_q^{n-2} + \cdots + z_{n-1}^q \theta_q + z_n$$

so

$$a_n^q = \sum_{k=0}^{n-1} z_{n-k}^q \theta_q^k = z_n^q + \theta_q a_{n-1}^q = z_n^q + s_n^q \tag{9-58}$$

where s_n^q denotes $\theta_q a_{n-1}^q$. Let $\mathbf{a}_n = (a_n^1, a_n^2)$ and $\mathbf{s}_n = (s_n^1, s_n^2)$.

Let D_n^q be firm q's demand in period n, measured in physical units, and let $\mathbf{D}_n = (D_n^1, D_n^2)$. We assume for each n that the distribution of $\mathbf{D}_n$ depends only on $\mathbf{a}_n$; that is, demand depends only on current goodwill. Let

$$\mu_q(\mathbf{a}) = E(D_1^q \mid \mathbf{a}_1 = \mathbf{a}) \qquad \mathbf{a} \geq (0, 0) \tag{9-59}$$

where we emphasize that the distribution of D_1^q may depend on *both* firms' goodwills.

Let r_q be firm q's gross profit per unit of demand, not including advertising expenditures; we assume $r_q > 0$. Then $r_q D_n^q - z_n^q$ is the gross profit in period n; so

$$V^q = \sum_{n=1}^{\infty} \beta_q^{n-1}(r_q D_n^q - z_n^q) \tag{9-60}$$

is firm q's sum of discounted profits; we assume $\beta_q < 1$, $q = 1, 2$.

Let $s_1^q = \theta_q a_0^q$ denote the initial goodwill so $a_1^q = z_1^q + s_1^q$; then (9-58) is valid for all $n \in I_+$. Substitution of $z_n^q = a_n^q - \theta_q a_{n-1}^q$ in (9-60) yields

$$V^q = \theta_q a_0^q + \sum_{n=1}^{\infty} \beta_q^{n-1}[r_q D_n^q - (1 - \beta_q \theta_q) a_n^q]$$

Let

$$\gamma_q^*(\mathbf{a}) = r_q \mu_q(\mathbf{a}) - (1 - \beta_q \theta_q) a^q \qquad \mathbf{a} = (a^1, a^2) \geq (0, 0) \tag{9-61}$$

Let E_Π denote an expectation with respect to probabilities induced by the firms' use of policy $\mathbf{\Pi} = (\Pi_1, \Pi_2)$. Then

$$v_{\mathbf{s}}^q(\mathbf{\Pi}) = s^q + E_\Pi \left[\sum_{n=1}^{\infty} \beta_q^{n-1} \gamma_q^*(\mathbf{a}_n) \mid \mathbf{s}_1 = \mathbf{s} \right] \tag{9-62}$$

where $\mathbf{s} = (s^1, s^2)$.

The constraint on $\mathbf{a}_n$ is

$$\mathbf{0} \le \mathbf{z}_n = \mathbf{a}_n - \mathbf{s}_n = (a_n^1, a_n^2) - (\theta_1 a_{n-1}^1, \theta_2 a_{n-1}^2) \tag{9-63}$$

or $\theta_q a_{n-1}^q \le a_n^q$ for each q. Note that the values taken by the demands $\mathbf{D}_1, \mathbf{D}_2, \ldots$ do not affect the feasibility of $\mathbf{a}_1, \mathbf{a}_2, \ldots$.

Suppose there exists $\mathbf{a}^* = (a_1^*, a_2^*)$ such that

$$\gamma_1^*(\mathbf{a}^*) \ge \gamma_1^*(a^1, a_2^*) \qquad \text{and} \qquad \gamma_2^*(\mathbf{a}^*) \ge \gamma_2^*(a_1^*, a^2) \qquad a^1 \ge 0, a^2 \ge 0 \tag{9-64}$$

Then the model satisfies Assumptions I through IV on page 466 in Section 9-4 (Exercise 9-22 asks you to verify this claim); so the following result is a consequence of Theorem 9-6.

Proposition 9-4 If (9-64) is valid, $\mathbf{a}_n = \mathbf{a}^*$ for all n is a discounted EP† with respect to $\{\mathbf{s}: \mathbf{s} \le \mathbf{a}^*\}$ and the set of all policies.

Proposition 9-4 states that, if $\mathbf{s}_1 \le \mathbf{a}^*$, then $\mathbf{a}_n = \mathbf{a}^*$ for all n comprises an EP. From (9-63), the consequent advertising expenditures are

$$z_1^q = a_q^* - s_1^q \qquad \text{and} \qquad z_n^q = a_q^*(1 - \theta_q) \qquad \text{for } n > 1$$

In order to analyze the static game with payoffs in (9-61), it is convenient to alter the notation and write $\mathbf{a} = (a_1, a_2)$ instead of (a^1, a^2). Observe that $\mathbf{a}^*$ is an EP of the static game with payoffs in (9-61) if, and only if, it is an EP for the static game with payoffs $\gamma_q(\cdot) = \gamma_q^*(\cdot)/r_q$:

$$\gamma_q(\mathbf{a}) = \mu_q(\mathbf{a}) - h_q a_q \qquad h_q = \frac{1 - \beta_q \theta_q}{r_q} \tag{9-65}$$

If $\gamma_q(\cdot)$ is differentiable at $(a_1, a_2) > (0, 0)$, a necessary condition for an interior EP is

$$\frac{\partial \mu_q(a_1, a_2)}{\partial a_q} - h_q = 0 \qquad q = 1, 2 \tag{9-66}$$

Example 9-7 Let $w_q(\mathbf{a}) = a_q/(a_1 + a_2)$, which is firm q's fraction of the total goodwill. Suppose D_n^q for each n and q has a marginal distribution which is a uniform distribution on the interval $[g_q, g_q + 2J_q w_q(a)]$. Then $\mu_q(\mathbf{a}) = g_q + J_q w_q(\mathbf{a})$; so (9-66) is

$$\frac{J_1 a_2}{(a_1 + a_2)^2} - h_1 = \frac{J_2 a_1}{(a_1 + a_2)^2} - h_2 = 0 \tag{9-67}$$

Let $H_q = h_q/J_q$. From (9-67),

$$a_1 = H_2(a_1 + a_2)^2 \qquad \text{and} \qquad a_2 = H_1(a_1 + a_2)^2$$

† The rule $\mathbf{a}_n = \mathbf{a}^*$ is not a policy because $\mathbf{a}_n = \mathbf{a}^*$ is not feasible if $\mathbf{s}_n \nleq \mathbf{a}^*$. Let $\mathbf{a}_n$ be arbitrary, but feasible, if $\mathbf{s}_n \nleq \mathbf{a}^*$. Of course, if $\mathbf{s}_1 \le \mathbf{a}^*$, the rule $\mathbf{a}_n = \mathbf{a}^*$ causes $\mathbf{s}_n \le \mathbf{a}^*$ for all n.

so $a_1 + a_2 = (H_1 + H_2)^{-1}$, whose substitution in (9-67) yields

$$a_1^* = \frac{H_2}{(H_1 + H_2)^2} \qquad \text{and} \qquad a_2^* = \frac{H_1}{(H_1 + H_2)^2} \tag{9-68}$$

In the symmetric duopoly case, $H_1 = H_2 = H$; so (9-68) becomes $a_1^* = a_2^* = (4H)^{-1}$. □

EXERCISES

9-22 Verify that the duopoly advertising model satisfies Assumptions I through IV on page 466 if (9-64) is valid.

9-23 Suppose there are Q firms instead of only two, and $Q \geq 2$. In place of (9-64), if

$$\gamma_q(\mathbf{a}^*) \geq \gamma_q(a_q, \mathbf{a}^*_{-q}) \qquad a_q \geq 0 \qquad q = 1, \ldots, Q$$

then Proposition 9-4 is still valid. Work out the details of Example 9-7 if $w_q(\mathbf{a}) = a_q / \sum_{q=1}^{Q} a_q$ and D_n^q has a uniform marginal distribution on $[g_q, g_q + 2J_q w_q(\mathbf{a})]$ for each n and q. (Hint: Theorem 9-8.)

9-6* DYNAMIC OLIGOPOLY

Sequential games constitute a reasonable framework in which to analyze many of the economic phenomena associated with imperfect competition. The preceding section presents a duopoly model which is a natural generalization of Section 3-4. This section contains a multifirm generalization of Sections 3-1 and 3-2, i.e., a dynamic oligopoly model in which firms hold inventories from one period to the next. A firm's decisions each period are the amount to produce and the price at which it is willing to sell its goods.

Let $\mathcal{Q}$ be a set of firms and let s_n^q be firm q's inventory level at the beginning of period n. We assume that production is sufficiently rapid relative to a period's length that the quantity produced can be used to satisfy demand in the same period. Let z_n^q and D_n^q be firm q's production quantity and demand in period n, respectively. We assume $P\{D_n^q \geq 0\} = 1$ for all n and q. Suppose that excess demand is backlogged so $s_{n+1}^q = s_n^q + z_n^q - D_n^q$. Let $y_n^q = s_n^q + z_n^q$ and let $\mathbf{s}_n$, $\mathbf{z}_n$, $\mathbf{y}_n$, and $\mathbf{D}_n$ denote the vectors whose qth components are s_n^q, z_n^q, y_n^q, and D_n^q, respectively. Then

$$\mathbf{s}_{n+1} = \mathbf{s}_n + \mathbf{z}_n - \mathbf{D}_n = \mathbf{y}_n - \mathbf{D}_n$$

Let ρ_n^q be the price announced by firm q in period n and let $\boldsymbol{\rho}_n$ be the vector whose qth component is ρ_n^q. Let $\mathbf{a}_n^q = (y_n^q, \rho_n^q)$ and let $\mathbf{a}_n$ be the vector whose qth component is a_n^q. We assume that the distribution of $\mathbf{D}_n$, given $\mathbf{a}_n = \mathbf{a} = (\boldsymbol{\rho}, \mathbf{y})$, is conditionally independent of the history $\mathbf{s}_1, \mathbf{a}_1, \mathbf{D}_1, \ldots, \mathbf{s}_{n-1}, \mathbf{a}_{n-1}, \mathbf{D}_{n-1}, \mathbf{s}_n$, and depends only on $\boldsymbol{\rho}$ and, possibly, on $\mathbf{y}$. Let

$$\mu_q(\mathbf{a}) = E(D_n^q \mid \mathbf{a}_n = \mathbf{a})$$

Thus the distribution of each firm's demand may be affected by the prices and quantities set by competing firms as well as by its own price and quantity.

Suppose that each firm has production costs and inventory-related costs. We assume that the production cost is proportional to the amount produced and let $c_q \cdot z_n^q$ be firm q's production cost in period n. Also, if the duration N is finite, suppose that firm q's salvage value of s_{N+1}^q is $-c_q s_{N+1}^q$.

Let $g_q(\mathbf{y}, \boldsymbol{\rho}, \mathbf{d})$ denote firm q's revenue minus its inventory-related costs in period n if $\mathbf{a}_n = (\mathbf{y}, \boldsymbol{\rho})$ and $\mathbf{D}_n = \mathbf{d}$. This representation encompasses many cases.

Example 9-8 Let θ_n^q be the "raw" demand faced by firm q in period n. Suppose $\boldsymbol{\theta}_n$, the vector with qth component θ_n^q, depends only on the price vector $\boldsymbol{\rho}_n$. Let the amount sold be

$$D_n^q = \begin{cases} y_n^q & \text{if } y_n^q \le \theta_n^q \\ \theta_n^q + \min\left\{m_q \sum_{j \in \mathcal{Q}} (\theta_n^j - y_n^j)^+, \quad y_n^q - \theta_n^q\right\} & \text{if } y_n^q > \theta_n^q \end{cases} \tag{9-69}$$

where $m_q \ge 0$ and $\sum_{q \in \mathcal{Q}} m_q \le 1$. Thus the general back-ordering assumption includes the case in which excess demand is lost. The simplest such case would have $m_q = 0$ for all $q \in \mathcal{Q}$. Suppose that consumers in this industry are well informed and will pay only the lowest price set by any of the firms. Then $g_q(\mathbf{y}, \boldsymbol{\rho}, \mathbf{d})$ might take the form

$$g_q(\mathbf{y}, \boldsymbol{\rho}, \mathbf{d}) = d_q \cdot \min\{\rho^j : j \in \mathcal{Q}\} - h_q \cdot (y^q - d^q)^+ - b_q \cdot (d^q - y^q)^+ \tag{9-70}$$

where h_q and b_q are respective unit costs of inventory and shortage. In this example, other firms' actions affect firm q via demand and via the price that traffic will bear. □

Let β_q denote firm q's discount factor. Exercise 9-24 asks you to verify that the oligopoly model satisfies Assumptions I and II on page 466 in Section 9-4 with (9-34) taking the form

$$\gamma_q(\mathbf{a}) = E[g_q(\mathbf{a}, \mathbf{D}_1) - \beta_q c_q D_1^q \mid \mathbf{a}_1 = \mathbf{a}] - c_q(1 - \beta_q) y^q \tag{9-71}$$

where y^q is the qth component of $\mathbf{y}$ in $\mathbf{a} = (\mathbf{y}, \boldsymbol{\rho})$.

Let the constraints on $\mathbf{a}_n^q = (y_n^q, \rho_n^q)$ be

$$0 \le z_n^q = y_n^q - s_n^q \qquad \text{so} \qquad s_n^q \le y_n^q \qquad \text{and} \qquad 0 \le \rho_n^q$$

That is, production quantities and prices must both be nonnegative. With back ordering in the model, $\mathcal{S} = \mathbb{R}^Q$; so

$$A_s^q = [s^q, \infty) \times [0, \infty) \qquad A^q = \bigcup_{s \in \mathcal{S}} A_s^q = \mathbb{R}_+^2$$
$$\text{and} \qquad A = \times_{q \in \mathcal{Q}} A^q = \mathbb{R}^{2Q} \tag{9-72}$$

where s^q is the qth component of $\mathbf{s}$. As in the general case in Section 9-3, let Γ denote the following static game among the firms in $\mathcal{Q}$: firm q's payoff is $\gamma_q(a)$ and its set of alternative moves is A^q.

Exercise 9-24 asks you to verify that, if Γ has an EP (equilibrium point) in pure strategies, the model satisfies Assumption IV in Section 9-3. The following result is then an immediate corollary of Theorem 9-6.

Theorem 9-9 If Γ has an unrandomized EP $\mathbf{a}^* = (\mathbf{y}^*, \boldsymbol{\rho}^*)$, let α be a single-stage decision rule which specifies $\alpha(\mathbf{s}) = \mathbf{a}^*$ (with probability 1) if $\mathbf{s} \leq \mathbf{y}^*$ and is arbitrary but feasible if $\mathbf{s} \not\leq \mathbf{y}^*$. Suppose $\beta_q < 1$ for all $q \in \mathcal{Q}$.
(*a*) α^∞ is a discounted EP with respect to $(-\infty, \mathbf{y}^*]$ and the set of all policies.
(*b*) α^N is an N-period EP, for every $N \in I_+$, with respect to $(-\infty, \mathbf{y}^*]$ and the set of all policies.

An industry which has the structure of the oligopoly model would have the following properties if Γ has an unrandomized EP and $\mathbf{s}_1 \leq \mathbf{y}^*$:

(i) Each firm's price would be time-invariant. However, different firms may have different prices (the theorem does not assert that all the components of $\boldsymbol{\rho}^*$ are the same).
(ii) Each firm's maximum inventory would be time-invariant.

Suppose that the model remains myopic but no longer time-invariant (see Exercise 9-19). Then (i) and (ii) would no longer be valid. The myopic structure itself would vanish if the model included bankruptcy and other financial details.

We have assumed $P\{\mathbf{D}_n \geq \mathbf{0}\} = 1$ for all n. Therefore, from (9-69), if $\mathbf{y}_n = \mathbf{y}^*$,

$$\mathbf{s}_{n+1} = \mathbf{y}_n - \mathbf{D}_n = \mathbf{y}^* - \mathbf{D}_n \leq \mathbf{y}^*$$

so $\mathbf{y}_{n+1} = \mathbf{y}^*$ is again feasible (with probability 1). As a result, the requirement in Theorem 9-9 that Γ has an unrandomized EP is stronger than necessary.

Corollary 9-9 Suppose Γ has an EP $\mathbf{z}$, possibly randomized, which assigns $P\{\mathbf{y} = \mathbf{y}^*\} = 1$ for some $\mathbf{y}^* \in \mathbb{R}^{\mathcal{Q}}$. Let $\boldsymbol{\lambda}$ denote a single-stage decision rule which specifies $\boldsymbol{\lambda}(\mathbf{s}) = \mathbf{y}^*$ if $\mathbf{s} \leq \mathbf{y}^*$ and is arbitrary if $\mathbf{s} \not\leq \mathbf{y}^*$. Suppose $\beta_q < 1$ for all $q \in \mathcal{Q}$. Then (*a*) and (*b*) of Theorem 9-9 are valid with $\boldsymbol{\lambda}^\infty$ and $\boldsymbol{\lambda}^N$ in place of $\boldsymbol{\alpha}^\infty$ and $\boldsymbol{\alpha}^N$.

Linear Inventory and Back-Order Costs

The rest of this section presents some of the cases in which the condition of the corollary is satisfied. Suppose

$$g_q(\mathbf{y}, \boldsymbol{\rho}, \mathbf{d}) = \rho^d d^q - h \cdot (y^q - d^q)^+ - b \cdot (d^q - y^q)^+ \tag{9-73}$$

where, unlike (9-70), the unit costs of inventory and back orders are assumed the same for all firms (that is, h and b are not subscripted with q). Similarly, let $\beta_q = \beta$

and $c_q = c$ for all q. These assumptions are made merely to simplify the notation, but they are reasonable in many industries.

Suppose for each q that the marginal distribution function $F_q(\cdot)$ of D_1 depends on $\mathbf{a}_1 = (y_1, \rho_1)$ only via the price vector $\boldsymbol{\rho}_1$. Let

$$F_q(x \mid \boldsymbol{\rho}) = P\{D_1^q \leq x \mid \boldsymbol{\rho}_1 = \boldsymbol{\rho}\}$$

so that the expected value of D_1^q can be written

$$\mu_q(\boldsymbol{\rho}) = E[D_1^q \mid \mathbf{a}_n = (\mathbf{y}, \boldsymbol{\rho})]$$

With these simplifications, the substitution of (9-73) in (9-71) yields

$$\gamma_q(\mathbf{a}) = (\rho^q - \beta c)\mu_q(\boldsymbol{\rho}) - c(1 - \beta)y^q - E[h(y^q - D_1^q)^+ + b(D_1^q - y^q)^+ \mid \boldsymbol{\rho}_1 = \boldsymbol{\rho}] \quad (9\text{-}74')$$

Since we shall analyze several versions of Γ with payoffs as in (9-74′), it clarifies the exposition to write D_q for D_1^q, y_q for y^q, and ρ_q for ρ^q. Then (9-74′) becomes

$$\gamma_q(\mathbf{a}) = (\rho_q - \beta c)\mu_q(\boldsymbol{\rho}) - c(1 - \beta)y_q - E[h(y_q - D_q)^+ + b(D_q - y_q)^+ \mid \boldsymbol{\rho}_1 = \boldsymbol{\rho}] \quad (9\text{-}74)$$

We assume for each q that $F_q(\cdot \mid \boldsymbol{\rho})$ has a density function $\phi_q(\cdot \mid \boldsymbol{\rho})$. If Γ has an unrandomized EP $\mathbf{a}^* = (\mathbf{y}^*, \boldsymbol{\rho}^*)$ then

$$0 = \left.\frac{\partial \gamma_q(\mathbf{a})}{\partial y_q}\right|_{\mathbf{a}=\mathbf{a}^*} \quad (9\text{-}75)$$

is necessary. From (9-74),

$$\left.\frac{\partial \gamma_q(\mathbf{a})}{\partial y_q}\right|_{\boldsymbol{\rho}=\boldsymbol{\rho}^*} = -c(1 - \beta) - hF_q(y_q \mid \boldsymbol{\rho}^*) + b[1 - F_q(y_q \mid \boldsymbol{\rho}^*)]$$

Therefore, (9-75) is equivalent to

$$F_q(y_q \mid \boldsymbol{\rho}^*) = \frac{b - c(1 - \beta)}{b + h}$$

Let

$$F_q^{-1}(r \mid \boldsymbol{\rho}) = \sup\{x\colon F_q(x \mid \boldsymbol{\rho}) \leq r\}$$

and

$$v = \frac{b - c(1 - \beta)}{b + h} \quad (9\text{-}76)$$

where $b > c(1 - \beta)$ is assumed and satisfied in practice. Now (9-75) is equivalent to

$$y_q = F_q^{-1}(v \mid \boldsymbol{\rho}^*) \quad (9\text{-}77)$$

This equation has exactly the same form as (3-25), which specifies the optimal base stock level for a firm in a perfectly competitive market (or a monopolist whose price is not a decision variable).

From (9-77), for each j we may substitute

$$y_j = F_j^{-1}(v \mid \boldsymbol{\rho}) \tag{9-78}$$

in (9-74) to obtain $\gamma_q(\cdot)$ as a function only of $\boldsymbol{\rho}$. Let $M_q(\boldsymbol{\rho})$ denote $\gamma_q(\mathbf{a}) = \gamma_q(\mathbf{y}, \boldsymbol{\rho})$ when each component of $\mathbf{y}$ is replaced by $J_q(\boldsymbol{\rho}) \triangleq F^{-1}(v \mid \boldsymbol{\rho})$:

$$\begin{aligned} M_q(\boldsymbol{\rho}) &= (\rho_q - \beta c)\mu_q(\boldsymbol{\rho}) - c(1-\beta)F_q^{-1}(v \mid \boldsymbol{\rho}) \\ &\quad - h\int_0^{J_q(\boldsymbol{\rho})} [J_q(\boldsymbol{\rho}) - x]\phi_q(x \mid \boldsymbol{\rho})\,dx - b\int_{J_q(\boldsymbol{\rho})}^{\infty} [x - J_q(\boldsymbol{\rho})]\phi_q(x \mid \boldsymbol{\rho})\,dx \end{aligned} \tag{9-79}$$

Then a necessary condition for an EP is

$$0 = \left.\frac{\partial M_q(\boldsymbol{\rho})}{\partial \rho_q}\right|_{\boldsymbol{\rho} = \boldsymbol{\rho}^*} \tag{9-80}$$

if $M_q(\cdot)$ is concave on $\mathbb{R}_+^Q$ and if there is any $\boldsymbol{\rho}^*$ which satisfies (9-80).

In order to check (9-80), let

$$f_q(\boldsymbol{\rho}) = \frac{\partial J_q(\boldsymbol{\rho})}{\partial \rho_q} \qquad \mu_q'(\boldsymbol{\rho}) = \frac{\partial \mu_q(\boldsymbol{\rho})}{\partial \rho_q} \qquad \text{and} \qquad \phi_q'(x \mid \boldsymbol{\rho}) = \frac{\partial \phi_q(x \mid \boldsymbol{\rho})}{\partial \rho_q}$$

which are assumed to exist. Then Leibnitz' rule (see page 518) and (9-78) yield

$$\begin{aligned} \frac{\partial M_q(\boldsymbol{\rho})}{\partial \rho_q} &= \mu_q(\boldsymbol{\rho}) + (\rho_q - \beta c)\mu_q'(\boldsymbol{\rho}) - c(1-\beta)f_q(\boldsymbol{\rho}) \\ &\quad - h\int_0^{J_q(\boldsymbol{\rho})} [J_q(\boldsymbol{\rho}) - x]\phi_q'(x \mid \boldsymbol{\rho})\,dx - hf_q(\boldsymbol{\rho})F_q[J_q(\boldsymbol{\rho}) \mid \boldsymbol{\rho}] \\ &\quad - b\int_{J_q(\boldsymbol{\rho})}^{\infty} [x - J_q(\boldsymbol{\rho})]\phi_q'(x \mid \boldsymbol{\rho})\,dx + bf_q(\boldsymbol{\rho})\{1 - F_q[J_q(\boldsymbol{\rho}) \mid \boldsymbol{\rho}]\} \end{aligned} \tag{9-81}$$

Now $0 = \partial M_q(\boldsymbol{\rho})/\partial \rho_q$ and $F_q[J_q(\boldsymbol{\rho}) \mid \boldsymbol{\rho}] = F_q[F_q^{-1}(v \mid \boldsymbol{\rho}) \mid \boldsymbol{\rho}] = v$ yield

$$\begin{aligned} \mu_q(\boldsymbol{\rho}) + (\rho_q - \beta c)\mu_q'(\boldsymbol{\rho}) &= b\int_{J_q(\boldsymbol{\rho})}^{\infty} [x - J_q(\boldsymbol{\rho})]\phi_q'(x \mid \boldsymbol{\rho})\,dx \\ &\quad + h\int_0^{J_q(\boldsymbol{\rho})} [J_q(\boldsymbol{\rho}) - x]\phi_q'(x \mid \boldsymbol{\rho})\,dx \end{aligned} \tag{9-82}$$

Direct verification of the concavity of $M_q(\cdot)$ via its Hessian matrix seems too painful without a specific assumption concerning F_q, hence f_q, μ_q, μ_q', ϕ_q, and ϕ_q'. Nevertheless, observe that neither (9-77) nor (9-80) depends on the joint distribution of all the firms' demands except through the marginal distributions. In order to simplify the analysis, the following cases have $\phi_q'(x \mid \boldsymbol{\rho})$ constant with respect to x.

Uniform Marginal Distributions

Example 9-9 Suppose that there are two firms which we label $q = 1$ and $q = -1$ for convenience of exposition. We assume that the marginal distribution of D_q, firm q's demand, is uniform on the interval $[0, w\rho_{-q}/\{1 + (\rho_1 + \rho_{-1})^2\}]$. These marginal distributions include the following cases:

(i) For each q, $D_q = w\rho_{-q}U/[1 + (\rho_1 + \rho_{-1})^2]$, where U is uniformly distributed on $[0, w]$.

(ii) For each q, $D_q = \rho_{-q}U$, where U is uniformly distributed on $[0, w/\{1 + (\rho_{-1} + \rho_1)^2\}]$.

(iii) D_1 and D_{-1} are independent r.v.'s with the uniform distributions specified above.

Uniform marginal distributions yield

$$\mu_q = \frac{w\rho_{-q}/2}{1 + (\rho_1 + \rho_{-1})^2}$$

$$\mu_q' = \frac{-w\rho_{-q}(\rho_1 + \rho_{-1})}{[1 + (\rho_1 + \rho_{-1})^2]^2} \le 0 \qquad (\rho_1, \rho_{-1} \ge 0)$$

$$\frac{\partial \rho_q \mu_q}{\partial \rho_q} = \mu_q + \rho_q \mu_q' = \frac{(w-2)\rho_1\rho_{-1}(\rho_1 + \rho_{-1}) + w\rho_{-q}(1 + \rho_{-q}^2)}{2[1 + (\rho_1 + \rho_{-1})^2]^2}$$

These relationships show that:

(*a*) The firm's average demand decreases as its price rises.

(*b*) For certain values of w (for example, $w = 1$), firm q's expected revenue is unimodal (quasiconcave) in ρ_q.

(*c*) $D_q \to 0$ as $\rho_{-q} \to 0$, and $D_q \to w\rho_{-q}/(1 + \rho_{-q}^2)$ as $\rho_q \to 0$ (both with probability 1).

In order to use (9-77) and (9-81) to obtain $\mathbf{y}^*$ and $\boldsymbol{\rho}^*$, we need $F_q^{-1}(\cdot \mid \boldsymbol{\rho})$, $J_q(\cdot)$, and $\phi_q'(\cdot)$. By assumption,

$$\phi_q(x \mid \boldsymbol{\rho}) = \frac{1 + (\rho_1 + \rho_{-1})^2}{w\rho_{-q}} \qquad 0 \le x \le \frac{w\rho_{-q}}{1 + (\rho_1 + \rho_{-1})^2}$$

so

$$\phi_q'(x \mid \boldsymbol{\rho}) = \frac{2(\rho_1 + \rho_{-1})}{w\rho_{-q}}$$

$$F_q^{-1}(u \mid \boldsymbol{\rho}) = \frac{w\rho_{-q}u}{1 + (\rho_1 + \rho_{-1})^2} \qquad 0 < u < 1$$

$$J_q(\boldsymbol{\rho}) = F_q^{-1}(v \mid \boldsymbol{\rho}) = \frac{w\rho_{-q}v}{1 + (\rho_1 + \rho_{-1})^2} \tag{9-83}$$

It is convenient to define

$$\xi = \frac{-2\beta c + b(1-v)^2 + hv^2}{2}$$

and $$\theta_q = 2\rho_{-q}\xi - \rho_{-q}^2 - 1 \tag{9-84}$$

Substitution in (9-82) yields

$$\mu_q - (\rho_q - \beta c)(\rho_1 + \rho_{-1}) \frac{\mu_q/2}{1 + (\rho_1 + \rho_{-1})^2}$$

$$= \frac{2(\rho_1 + \rho_{-1})}{w\rho_{-q}} \left\{ b \int_{J_q(\mathbf{p})}^{\mu_q/2} [x - J_q(\mathbf{p})]\, dx + h \int_0^{J_q(\mathbf{p})} [J_q(\mathbf{p}) - x]\, dx \right\}$$

$$= 4\mu_q^2(\rho_1 + \rho_{-1}) \frac{b(1-v)^2 + hv^2}{w\rho_{-q}}$$

Therefore,

$$1 + (\rho_1 + \rho_{-1})^2 = 2(\rho_1 + \rho_{-1})^2(\rho_q + \xi)$$

$$\rho_q^2 + 2\xi\rho_q + \theta_q = 0$$

and $$\rho_q = -\xi + \sqrt{\xi^2 - \theta_q} \tag{9-85}$$

if $\xi^2 \geq \theta_q$, that is, if $\xi^2 - 2\rho_{-q}\xi + \rho_{-q}^2 + 1 \geq 0$. We shall have to verify

$$\rho_{-q}^2 - 2\xi\rho_{-q} + \xi^2 + 1 \geq 0 \tag{9-86}$$

when $\mathbf{p} = \mathbf{p}^*$.

The substitution of (9-84) in (9-85) yields

$$(\rho_q + \xi)^2 = \xi^2 + 1 + \rho_{-q}^2 - 2\xi\rho_{-q}$$

so $$2\xi(\rho_1 + \rho_{-1}) - 1 = \rho_{-q}^2 - \rho_q^2 \tag{9-87}$$

This is valid for $q = 1$ and $q = -1$; so $\rho_1 = \rho_{-1}$ reduces (9-87) to

$$\rho_1^* = \rho_{-1}^* = (4\xi)^{-1} \tag{9-88}$$

The substitution of (9-88) in (9-83) yields the optimal "order-up-to" quantity

$$y_1^* = y_2^* = \frac{wv\xi}{1 + 4\xi^2}$$

In order to verify (9-86), substitute (9-87) to obtain

$$\rho_{-q}^2 - 2\xi\rho_{-q} + \xi^2 + 1 = (16\xi^2)^{-1} - 1/2 + \xi^2 + 1$$

$$= (16\xi^2)^{-1} + \xi^2 + 1/2 \geq 0 \qquad \square$$

EXERCISES

9-24 (*a*) Verify that the oligopoly model satisfies Assumptions I and II in Section 9-4, with (9-34) taking the form (9-72).

(*b*) Suppose that the static game Γ [described below (9-72)] has an unrandomized EP **a*** = (**y***, **ρ***). Verify that the model satisfies Assumption IV in Section 9-4.

9-25 Consider the nonsymmetric version of Example 9-9 in which ξ_1 and ξ_{-1} may differ. Prove that EP prices satisfy

$$\frac{\rho_{-1}^*}{\rho_1^*} = \frac{\xi_1 - \xi_2 + [(\xi_1 - \xi_2)^2 - 4(\xi_2 - \xi_1 - 1)]^{1/2}}{2}$$

9-7* ORDINAL SEQUENTIAL GAMES

Section 9-3 presents sufficient conditions for an SG (sequential game) to possess an EP (equilibrium point). The proofs there depend on the following observation: if all players but one use Markov policies, the remaining player faces an MDP (Markov decision process). That fact permits us to invoke existence theorems for MDP optima, but it does not lead to the straightforward application of MDP algorithms to the computation of SG EPs.

This section has two purposes. First, we show that SGs with discounted payoffs can be embedded in the ordinal framework of Sections 4-3, 4-4, and 6-3. The second purpose is to justify a version of the policy-improvement algorithm for EPs of discounted SGs.

Let $\mathcal{S}$ be a nonempty and countable or finite set of *states*, $\mathcal{Q}$ a nonempty set of *players*, A_s^q a nonempty set of *actions* for player q in states s,

$$A_s = \times_{q \in \mathcal{Q}} \mathcal{A}_s^q$$

$$\mathcal{C} = \{s, a): a \in A_s, s \in \mathcal{S}\} \qquad \Delta = \times_{s \in \mathcal{S}} A_s \qquad \text{and} \qquad Y = \times_{n=1}^{\infty} \Delta \tag{9-89}$$

Elements of Δ are multiplayer *single-stage decision rules*, and elements of Y are multiplayer *Markov policies*.

Let W be the set of all probability distributions on $\mathcal{S}$ so $\mathbf{w} = [w(j), j \in \mathcal{S}] \in W \Leftrightarrow \mathbf{w} \geq \mathbf{0}$ and $\sum_{j \in \mathcal{S}} w(j) = 1$. For each $\mathbf{w} \in W$, $\delta \in \Delta$, and $j \in \mathcal{S}$ let

$$M(\mathbf{w}, \delta)(j) = \sum_{i \in \mathcal{S}} w(i) p_{ij}^{\delta(i)}$$

denote the conditional probability of being in state j next period if $\mathbf{w}$ is the distribution of the state this period and the players use rule δ. Let $\mathbf{w}_n$ denote the probability distribution of the state in period n and suppose that the multiplayer Markov policy $(\delta_1, \delta_2, \ldots)$ is used. Then $\mathbf{w}_{n+1} = M(\mathbf{w}_n, \delta_n)$.

A posterity is a feasible sequence $(\mathbf{w}_1, \delta_1, \mathbf{w}_2, \delta_2, \ldots)$ of successive distributions and multiplayer decision rules. The set $\Phi_{\mathbf{w}}$ of all posterities with the initial distribution $\mathbf{w}$ is

$$\Phi_{\mathbf{w}} = \{\mathbf{w}_1, \delta_1, \mathbf{w}_2, \delta_2, \ldots): \mathbf{w}_1 = \mathbf{w}, \delta_n \in \Delta, \text{ and } \mathbf{w}_{n+1} = M(\mathbf{w}_n, \delta_n) \text{ for all } n\} \tag{9-90}$$

Definitions (9-89) and (9-90) are formally identical to (6-77) and (6-78), which specify an ordinal framework for multiple-criteria MDPs. In Section 6-3, $\mathcal{Q}$ denotes a set of criteria; here, the elements of $\mathcal{Q}$ are players. With that difference in interpretation, the following results are immediately applicable to SGs: Proposition 6-1, Lemma 6-6, Theorems 6-9 and 6-10, and Corollary 6-10. The following exposition is based partly on those results and uses the same definitions and notation. Therefore, the reader should review the portion of Section 6-3 labeled "Stationary Policies" on pages 286 to 289.

Definition 9-15 A *coalition* is a nonempty subset of players. The set Ω of *latent coalitions* is a nonempty collection of coalitions.

For each $s \in \mathcal{S}$ and $\omega \in \Omega$, let $\theta_{\mathbf{w}}^{\omega} \subset \Phi_{\mathbf{w}} \times \Phi_{\mathbf{w}}$ indicate the preferences of coalition ω among posterities when $\mathbf{w}$ is the distribution of the initial state. We interpret $(\tau, \tau') \in \theta_{\mathbf{w}}^{\omega}$ as "coalition ω regards posterity τ as being at least as desirable as posterity τ' if $\mathbf{w}_1 = \mathbf{w}$."

As in Section 9-2, let Π_q label the qth player's portion of $\Pi \in Y$, where

$$Y = \mathop{\times}_{n=1}^{\infty} \mathop{\times}_{s \in \mathcal{S}} \mathop{\times}_{q \in \mathcal{Q}} A_s^q$$

The portion of Π due to all the other players is labeled Π_{-q}, and we sometimes write $\Pi = (\Pi_q, \Pi_{-q})$ instead of $\Pi = (\delta_1, \delta_2, \ldots)$, where the qth component of $a_n = \delta_n(s_n)$ is player q's action in period n. For each $s \in \mathcal{S}$ and $\Pi = (\delta_1, \delta_2, \ldots) \in Y$, let $\tau_{\mathbf{w}}(\Pi)$ denote the posterity generated by the Markov policy Π if $\mathbf{w}$ is the distribution of the initial state:

$$\tau_{\mathbf{w}}(\Pi) = (\mathbf{w}, \delta_1, M(\mathbf{w}, \delta_1), \delta_2, M[M(\mathbf{w}, \delta_1), \delta_2], \delta_3, \ldots)$$

Definition 9-16 Let Π and ξ be Markov policies. The symbols $\gtrsim_e$ and $\gtrsim_p$ mean:

$\Pi \gtrsim_p \xi \Leftrightarrow$ either $[\tau_{\mathbf{w}}(\Pi), \tau_{\mathbf{w}}(\xi)] \in \theta_{\mathbf{w}}^{\omega}$ for all $w \in W$ and $\omega \in \Omega$, or there are $\mathbf{w}$ and $\mathbf{y}$ in W and ω and u in Ω such that

$$[\tau_{\mathbf{w}}(\Pi), \tau_{\mathbf{w}}(\xi)] \notin \theta_{\mathbf{w}}^{\omega} \qquad \text{and} \qquad [\tau_{\mathbf{y}}(\xi), \tau_{\mathbf{y}}(\Pi)] \notin \theta_{\mathbf{y}}^{u} \tag{9-91}$$

If $\mathcal{Q} \subset \Omega$, then

$$\Pi \gtrsim_e \xi \Leftrightarrow [\tau_{\mathbf{w}}(\Pi), \tau_{\mathbf{w}}(\xi_q, \Pi_{-q})] \in \theta_{\mathbf{w}}^{q} \qquad \text{for all } \mathbf{w} \in W \text{ and } q \in \mathcal{Q} \tag{9-92}$$

We repeat Definition 6-3.

Definition 9-17 If D is a nonempty set and $B \subset D \times D$, then $b \in D$ is *B-maximal* if $(b, c) \in B$ for all $c \in D$.

These definitions lead to specifications of the core and an EP in terms of $\gtrsim_p$ and $\gtrsim_e$, respectively.

Definition 9-18 A Markov policy Π is an *equilibrium point* (EP) $\Leftrightarrow \Pi$ is $\gtrsim_e$-optimal. A Markov policy Π is in the *core* $\Leftrightarrow \Pi$ is $\gtrsim_p$-optimal. If $\Omega = \mathcal{Q}$, policies in the core are called *Pareto optima* (PO).

The core consists of those latent coalitions which are "undominated," in game-theory parlance. We refer the reader to Aubin (1979) and Shubik (1982) for discussions of the core in game theory and economics.

Neither $\succsim_e$ nor $\succsim_p$ is necessarily transitive, and as we observe in Section 6-3, intransitivity prevents the straightforward use of the arguments in Sections 4-3 and 4-4.

Example 9-10 Consider the following static bimatrix game (i.e., two-player noncooperative game) in which each player has two alternative actions:

	Player 2's action	
Player 1's action	1	2
1	0, 1	3, 0
2	2, 2	0, 0

The entries in each cell are the rewards garnered by players 1 and 2, respectively. This is the same array of numbers as in Example 6-9. Let (i, j) denote the policy in which player 1 takes action i and player 2 takes action j. Then $(1, 1) \succsim_p (1, 2)$ because player 2's reward is 1 at (1, 1) but only 0 at (1, 2). Also, $(1, 2) \succsim_p (2, 1)$ because player 1's reward is 3 at (1, 2) but only 2 at (2, 1). However, $(1, 1) \not\succsim_p (2, 1)$ because both players' rewards are lower at (1, 1) than at (2, 1). Therefore, $\succsim_p$ is *not* transitive. □

Example 9-11 Consider the following bimatrix game:

	Player 2's action		
Player 1's action	1	2	3
1	0, 0	0, −1	0, 1
2	−1, 0	0, 0	0, −1
3	1, 0	−1, 0	0, 0

Using the same notation as in Example 9-10, $(1, 1) \succsim_e (2, 2)$ because player 1 is no better off at (2, 1) than at (1, 1), and player 2 is no better off at (1, 2) than at (1, 1). Also, $(2, 2) \succsim_e (3, 3)$ because player 1 is no better off at (3, 2) than at (2, 2) and player 2 is no better off at (2, 3) than at (2, 2). However, $(1, 1) |_e (3, 3)$ because player 1 is better off at (3, 1) than at (1, 1) and player 2 is better off at (1, 3) than at (1, 1). Therefore, $\succsim_e$ is *not* transitive. □

Partial Resolution of Intransitivity

We repeat Definition 6-6.

Definition 9-19 Let D be a nonempty set and $B \subset D \times D$. An *inconsistency cycle connects x and y under B* if there is a finite sequence $x_1, \ldots, x_n$ such that $x_1 = x_n = x$, $x_k = y$ for some $1 < k < n$, $(x_i, x_{i+1}) \in B$ for all $i < n$, and $(x_{i+1}, x_i) \notin B$ for some i. The *completion of a binary relation* (B, D), written (B', D), is

$$B' = B \cup \{(x, y) : (x, y) \in D \times D, (x, y) \notin B, \text{ and } (y, x) \notin B\}$$

The *transitive completion of a binary relation* (B, D), written (B_c, D), is

$$B_c = B' \cup \{(x, y) : \text{there is an inconsistency cycle which connects } x \text{ and } y \text{ under } B'\}$$

See Example 6-10 for the completion and the transitive completion of $\gtrsim_p$ in the bimatrix game of Example 9-10.

Let $(\geq_c, Y)$ denote the transitive completion of $(\gtrsim_p, Y)$. The following restatement of Corollary 6-10 uses terminology in Definitions 6-7 and 6-8.

Theorem 9-10 If $\mathcal{S}$ is at most countably infinite, W is reachable, and $\{\theta_{\mathbf{w}}^{\omega}: \mathbf{w} \in W \text{ and } \omega \in \Omega\}$ has consistent choice and is continuous, then $\gtrsim_p$ and $\geq_c$ have the following properties:

$$T_\delta \Pi \geq_c \Pi \Rightarrow \delta^\infty \geq_c \Pi \tag{9-93}$$

$$\text{If } \Pi \in Y \text{ is } \gtrsim_p\text{-maximal, then } \Pi \text{ is } \geq_c\text{-maximal and there is a } \geq_c\text{-maximal } \delta^\infty \tag{9-94}$$

$$\text{If } \Pi \in Y \text{ is } \gtrsim_p\text{-maximal, then } \Pi \geq_c T_\delta \Pi \text{ for all } \delta \in \Delta \tag{9-95}$$

$$\text{If } \#\mathcal{C} < \infty, \text{ then there is a } \geq_c\text{-maximal } \delta^\infty \tag{9-96}$$

$$\text{For every } \gamma \in \Delta, \text{ either } \gamma^\infty \text{ is } \geq_c\text{-maximal or there is another } \delta \in \Delta \text{ such that } \delta^\infty \geq_c \gamma^\infty \tag{9-97}$$

The interpretations of Theorem 9-10 and Corollary 6-10 are similar. From (9-93), if it is worth (in the sense of the transitive completion) delaying a policy, it is worth delaying it forever. According to (9-94), if the core is nonempty, there is a stationary policy in the transitive completion of the core. If a policy is in the core, (9-95) asserts that it is not worth (in the sense of the transitive completion) delaying the use of the policy. From (9-96), the transitive completion of the core of a finite SG is nonempty and contains a stationary policy. Then (9-97) asserts that if a stationary policy γ^∞ is not in the transitive completion of the core, there is some other stationary policy which is at least as good as γ^∞ ("as good" in the sense of the transitive completion).

Recall that $\gtrsim_e$ is the binary relation underlying an EP. Let $(\geq, Y)$ denote the transitive completion of $(\gtrsim_e, Y)$. Exercise 9-26 asks you to prove the following result.

Theorem 9-11 If $\mathcal{S}$ is at most countably infinite, $\mathcal{Q} \subset \Omega$, and $\{\theta^q_{\mathbf{w}}: \mathbf{w} \in W \text{ and } q \in \mathcal{Q}\}$ is continuous and has consistent choice, then (9-93) through (9-97) are valid with $\gtrsim_e$ and $\geq$ in place of $\gtrsim_p$ and $\geq_c$, respectively.

EXERCISES

9-26 Prove Theorem 9-11.

9-27 Exercise 4-30 observes that several results in Sections 4-3 and 4-4 are valid for deterministic MDPs without restricting $\mathcal{S}$ to be a countable (or finite) set. Verify that Theorems 9-10 and 9-11 are valid for deterministic SGs if we merely assume that $\mathcal{S}$ is a nonempty set and A_s is nonempty for each $s \in \mathcal{S}$.

9-28 (Continuation) In a deterministic SG, suppose that the state-to-state mapping M is nonstationary. That is, for each t, let $\mathcal{S}_t$, $A^q_s(t)$, and M_t depend on $t = 1, 2, \ldots$ and suppose $M_t(s, a) \in \mathcal{S}_{t+1}$ for all $s \in \mathcal{S}_t$ and $a \in \times_{q \in \mathcal{Q}} A^q_s(t)$ with $s_{t+1} = M_t(s_t, a_t)$. Say that $s \in \mathcal{S}_{t+1}$ is *reachable* if there exists $v \in \mathcal{S}_t$ and $a \in \times_{q \in \mathcal{Q}} A^q_s(t)$ such that $s = M_t(v, a)$; and $\mathcal{S}_{t+1}$ is reachable if all $s \in \mathcal{S}_{t+1}$ are reachable. (*a*) Prove that Theorem 9-11 remains valid. (*b*) Prove that Theorem 9-10 remains valid if $\mathcal{S}_t$ is reachable for all $t = 2, 3, \ldots$.

BIBLIOGRAPHIC GUIDE

Modern game theory and its applications were given an enormous impetus by von Neumann and Morgenstern (1944), which cites earlier contributions. Two-person zero-sum stochastic games were introduced in the elegant paper by Shapley (1953). Shapley's paper prompted a series of papers concerned with the existence of minimax solutions and algorithms for their computation: Beniest (1963), Bewley and Kohlberg (1976*a*, 1976*b*, 1978). Blackwell and Ferguson (1968), Charnes and Schroeder (1967), Denardo (1967), Eaves (1977), Everett (1957), Gillette (1957), Hoffman and Karp (1966), Liggett and Lippman (1968), Mertens and Neyman (1981, 1982), Monash (1982), Pollatschek and Avi-Itzhak (1969), Rao, Chandrasekaran, and Nair (1973), and Zachrisson (1964).

Nash (1950, 1951) used the Brouwer and Kakutani fixed-point theorems to prove existence of equilibrium points in noncooperative (static) games. The subsequent research on noncooperative stochastic games includes Federgruen (1978, 1980), Fink (1964), Friedman (1977), Hordijk, Vrieze, and Wanrooij (1976), Kirman and Sobel (1974), Parthasarathy (1973), Rogers (1969), Shubik and Whitt (1973), Sobel (1971, 1981), Takahashi (1964), van der Wal (1977), van der Wal and Wessels (1977), and Whitt (1980).

The preceding references concern a sequential game model in which the players are identifiable at "time zero," players receive real-valued rewards each period, and each player optimizes some function of the sequence of rewards (present value or average reward per period). For examples of alternative approaches to sequential game models, see Rosenthal (1979), Sanghvi (1978), Sanghvi and Sobel (1976), Shofield (1980), and Sobel (1980).

Differential games are continuous-time sequential games. See Friedman (1971, 1974) and Kuhn and Szëgo (1971) for an exposition of the continuous-time approach. Algorithms for EP solutions of static games are presented in Garcia, Lemke, and Luethi (1973), Rosenmüller (1971), Scarf (1973), and Wilson (1971).

For outstanding introductions and surveys of game theory we recommend von Neumann and Morgenstern (1944), Luce and Raiffa (1957), Aubin (1979), and Shubik (1982). Parthasarathy and Stern (1977) survey the stochastic game literature up to 1976 and provide many references besides the ones mentioned above and cited below.

References

Aubin, J.-P.: *Mathematical Methods of Game and Economic Theory*, North-Holland Publishing Co., Amsterdam (1979).

Başar, T., and G. J. Olsder: *Dynamic Noncooperative Game Theory*, Academic Press, New York (1982).

Beniest, W.: "Jeux Stochastiques Totalement Cooperatifs Arbitres," *Cahiers du Centre d'Etudes de Recherche Operationnelle* **5**: 124–138 (1963).

Bewley, T., and E. Kohlberg: "The Asymptotic Theory of Stochastic Games," *Math. Oper. Res.* **1**: 197–208 (1976*a*).

———: "The Asymptotic Solution of a Recursive Equation Arising in Stochastic Games," *Math. Oper. Res.* **1**: 321–336 (1976*b*).

———: "On Stochastic Games with Stationary Optimal Strategies," *Math. Oper. Res.* **3**: 104–126 (1978).

Blackwell, D., and T. S. Ferguson: "The Big Match," *Ann. Math. Stat.* **39**: 159–163 (1968).

Charnes, A., and R. G. Schroeder: "On Some Stochastic Antisubmarine Games," *Nav. Res. Logistics Quart.* **14**: 291–311 (1967).

Denardo, E. V.: "Contraction Mappings in the Theory Underlying Dynamic Programming," *SIAM Rev.* **9**: 165–177 (1967).

Eaves, B. C.: "Complementary Pivot Theory and Markovian Decision Chains," in *Fixed Points: Algorithms and Applications*, edited by S. Karamardian, Academic Press, New York (1977), pp. 59–85.

Everett, H.: "Recursive Games," in *Contributions to the Theory of Games*, vol. III, edited by M. Dresher, A. W. Tucker, and P. Wolfe, Princeton University Press, Princeton, N.J. (1957), pp. 47–78.

Federgruen, A.: "On N-Person Stochastic Games with Denumerable State Space," *Adv. Appl. Probability* **10**: 452–471 (1978).

———: "Successive Approximation Methods in Undiscounted Stochastic Games," *Oper. Res.* **28**: 794–809 (1980).

Fink, A. M.: "Equilibrium in a Stochastic n-Person Game," *J. Sci. Hiroshima Univ. Ser. A-I* **28**: 89–93 (1964).

Friedman, A.: *Differential Games*, Wiley-Interscience, New York (1971).

———: *Differential Games*, American Mathematics Society, Providence, R.I. (1974).

Friedman, J. W.: *Oligopoly and the Theory of Games*, North-Holland, Amsterdam (1977).

Garcia, C. B., C. E. Lemke, and H. Luethi: "Simplicial Approximation of an Equilibrium Point for Non-Cooperative N-Person Games," *Mathematical Programming*, edited by T. C. Hu and S. M. Robinson, Academic Press, New York (1973), pp. 227–260.

Gillette, D.: "Stochastic Games with Zero Stop Probabilities," in *Contributions to the Theory of Games*, vol. III, edited by M. Dresher, A. W. Tucker, and P. Wolfe, Princeton University Press, Princeton, N.J. (1957), pp. 179–188.

Groenewegen, L. P. J., and J. Wessels: "On the Relation between Optimality and Saddle-Conservation in Markov Games," in *Dynamische Optimierung* edited by M. Schäl, Bonner Mathematische Schriften, no. 98, Bonn (1977), pp. 19–31.

———: "On Equilibrium Strategies in Non-cooperative Dynamic Games," in *Game Theory and Related Topics* edited by O. Moeschlin and D. Pallaschke, North-Holland, Amsterdam (1979), pp. 47–57.

Himmelberg, C. J., T. Parthasarathy, T. E. S. Raghavan, and F. S. van Vleck: "Existence of p-Equilibrium and Optimal Stationary Strategies in Stochastic Games," *Proc. Am. Math. Soc.* **60**: 245–251 (1976).

Hoffman, A. J., and R. M. Karp: "On Nonterminating Stochastic Games," *Manage. Sci.* **12**: 359–370 (1966).

Hordijk, A. O. Vrieze, and G. Wanrooij: "Semi-Markov Strategies in Stochastic Games," Report BW68/76, Mathematical Centre, Amsterdam (1976).

Kirman, A. P., and M. J. Sobel: "Dynamic Oligopoly with Inventories," *Econometrica* **42**: 279–287 (1974).

Kuhn, H. W., and G. P. Szëgo, eds.: *Differential Games and Related Topics*, North-Holland, Amsterdam (1971).

Kushner, H. J., and S. G. Chamberlain: "Finite State Stochastic Games: Existence Theorems and Computational Procedures," *IEEE Trans. Auto. Control* **Ac-14**: 248–255 (1969).

Liggett, T. M., and S. A. Lippman: "Stochastic Games with Perfect Information and Time Average Pay-off," *SIAM Rev.* **11**: 604–607 (1969).

Luce, R. D., and H. Raiffa: *Games and Decisions*, Wiley, New York (1957).

Mertens, J. F., and A. Neyman: "Stochastic Games," *Int. J. Game Theory* **10**: 53–66 (1981).

———: "Stochastic Games have a Value," *Proc. Natl. Acad. Sci. USA* **79**: 2145–2146 (1982).

Monash, C. A.: "Stochastic Games II: The Minimax Theorem," Cowles Foundation D.P. No. 624, Yale University (1982).

Nash, J.: "Equilibrium Points in n-Person Games," *Proc. Nat. Acad. Sci. USA* **36**: 48–49 (1950).

———: "Non-cooperative Games," *Ann. Math.* **54**: 286–295 (1951).

Parthasarathy, T.: "Discounted, Positive, and Noncooperative Stochastic Games," *Int. J. Game Theory* **2**: 25–37 (1973).

——— and M. Stern: "Markov Games—A Survey," *Differential Games and Control Theory*, edited by E. Roxin, P-T. Liu, and R. Sternberg, Marcel Dekker, New York (1977), pp. 1–46.

——— and T. E. S. Raghavan: "An Orderfield Property for Stochastic Games When One Player Controls Transition Probabilities," *J. Optimization Theory Appl.* **33**: 375–392 (1981).

Pollatschek, M. A., and B. Avi-Itzhak: "Algorithms for Stochastic Games," *Manage. Sci.* **15**: 399–415 (1969).

Rao, S. S., R. Chandrasekaran, and K. P. K. Nair: "Algorithms for Discounted Stochastic Games," *J. Optimization Theory Appl.* **11**: 627–637 (1973).

Rogers, P. D.: "Nonzero-sum Stochastic Games," Report ORC 69-8, Operations Research Center, University of California, Berkeley (1969).

Rosenmüller, J.: "On a Generalization of the Lemke-Howson Algorithm to Non-cooperative n-Person Games, "*SIAM J. Appl. Math.* **21**: 73–79 (1971).

Rosenthal, R. W.: "Sequences of Games with Varying Opponents," *Econometrica* **47**: 1353–1366 (1979).

Sanghvi, A. P.: "Sequential Games as Stochastic Processes," *Stochastic Processes and Their Applications* **6**: 323–336 (1978).

——— and M. J. Sobel: "Bayesian Games as Stochastic Processes," *Int. J. Game Theory* **5**: 1–22 (1976).

Scarf, H.: *The Computation of Economic Equilibria* (in collaboration with T. Hansen) Yale University Press, New Haven, Conn. (1973).

Shapley, L.: "Stochastic Games," *Proc. Nat. Acad. Sci. USA* **39**: 1095–1100 (1953).

Shofield, N.: "The Theory of Dynamic Games," Discussion Paper 141, Department of Economics, University of Essex, Essex, U.K. (1980).

Shubik, M.: *Game Theory in the Social Sciences*, Vol. 1, MIT Press, Cambridge, Mass. (1982).

——— and W. Whitt: "Fiat Money in an Economy with One Nondurable Good and No Credit (A Noncooperative Sequential Game" in *Topics in Differential Games*, edited by A. Blaquiere, North-Holland, Amsterdam (1973), pp. 401–448.

Sobel, M. J.: "Noncooperative Stochastic Games," *Ann. Math. Stat.* **42**: 1930–1935 (1971).

———: "Ordinal Sequential Games," *Économies et Sociétés* **XIV**: 1571–1582 (1980).

———: "Myopic Solutions of Markov Decision Processes and Stochastic Games," *Oper. Res.* **29**: 995–1009 (1981).

Takahashi, M.: "Equilibrium Points of Stochastic Non-Cooperative n-Person Games," *J. Sci. Hiroshima Univ. Ser. A-I* **28**: 95–99 (1964).

van der Wal, J.: "Successive Approximations for Average Reward Markov Games," Memo. -COSOR 77-10, Department of Technology, Eindhoven University of Technology, Eindhoven, Netherlands (1977).

———: "Discounted Markov Games: Generalized Policy Iteration Method," *J. Optimization Theory Appl.* **25**: 125–138 (1978).

——— and J. Wessels: "Successive Approximation Methods for Markov Games," in *Markov Decision Theory*, edited by H. C. Tijms and J. Wessels, MC-Tract 93, Mathematical Centre, Amsterdam (1977), pp. 39–56.

von Neumann, J., and O. Morgenstern: *The Theory of Games and Economic Behavior*, Princeton University Press, Princeton, N.J. (1944).

Whitt, W.: "Representation and Approximation of Noncooperative Sequential Games," *SIAM J. Control Optimization* **18**: 33–48 (1980).

Wilson, R.: "Computing Equilibria of N-Person Games," *SIAM J. Appl. Math.* **21**: 80–87 (1971).

Zachrisson, L. E.: "Markov Games," in *Advances in Game Theory* edited by M. Dresher, L. Shapley, and A. W. Tucker, Princeton University Press, Princeton, N.J. (1964), pp. 211–253.

APPENDIX

A

BACKGROUND MATERIAL

The factual information necessary to understand the material in this book usually is presented in an undergraduate calculus sequence and a calculus-based course on probability. Not all courses cover and emphasize the particular facts utilized in this book; so this appendix is designed to be a ready reference. Most of the material on probability theory in Section A-1 should be familiar. The concepts of conditional probability and conditional expectation are used repeatedly in studying stochastic processes, and you should become comfortable with them. The simple, but crucial, properties of the exponential distribution given in Section A-2 are not brought out in some probability courses and may be new to you.

Section A-3 contains a review of facts from advanced calculus, such as the definition of lim sup, conditions for interchanging the order of integration, and "little oh" notation.

A-1 RUDIMENTS OF PROBABILITY THEORY

The basic concepts of probability theory include sample space, event, probability measure (or, simply, probability), random variable, and distribution function.

Sample Spaces and Probability Measures

A *sample space*, Ω say, is the set of all possible outcomes of a potential experiment. An event is a subset of the sample space; i.e., it is a collection of outcomes. If A is an event and the outcome of the experiment is the member ω of Ω, then A occurs if $\omega \in A$. Events are the objects to which probabilities are assigned.

In any application, we are free to choose Ω and the events in any way we please. Our choice should be related to the phenomenon we are modeling. In discussing the time to failure of a piece of equipment, for example, a convenient choice for Ω might be $[0, \infty)$; if we knew that the item always lasted at least a but never more than b, we might choose $\Omega = [a, b]$. It would not be a good idea to choose $\Omega = (-b, a/2)$. The events have some more natural structure to them. We would like to ensure that Ω is an event because it is a "sure thing." To continue with our example, if A_1 is the event "the time to a failure is no more than 1," that is,

$$\Omega = [0, \infty) \qquad \text{and} \qquad A_1 = \{\omega\colon \omega \le 1\}$$

then the set

$$A_1^c = \{\omega\colon \omega > 1\}$$

represents the time to failure being greater than 1 (the *complement* of A_1). We would like to have A_1^c be an event, so that both the occurrence and the nonoccurrence of something substantial are events.

If A_1 and A_2 both are events, then their union $A_1 \cup A_2$ is the set $\{\omega\colon \omega \in A_1$ and/or $\omega \in A_2\}$. With A_1 as above and $A_2 = \{\omega\colon \frac{1}{2} \le \omega \le 2\}$, $A_1 \cup A_2 = \{\omega\colon 0 \le \omega \le 2\}$; that is, $A_1 \cup A_2$ is the set of outcomes such that the time to failure is between 0 and 2. We would like $A_1 \cup A_2$ to be an event too. Continuing in this way, we are led to want $A_1 \cup A_2 \cup A_3 \cdots$ to be an event when each A_i is an event.

Thus, a *family of events*, $\mathscr{F}$ say, is constructed such that

$$\Omega \in \mathscr{F} \tag{A-1}$$

$$A \in \mathscr{F} \Leftrightarrow A^c \in \mathscr{F} \qquad \text{where} \qquad A^c \triangleq \Omega - A \tag{A-2}$$

and $\quad$ if $A_1, A_2, \ldots$ are in $\mathscr{F}$, then so is their union $\quad$ (A-3)

The technical name for a family of sets satisfying (A-1) through (A-3) is a *σ-field*.

We write $\bigcup_{i=1}^{n} A_i$ for the union of $A_1, A_2, \ldots, A_n$. With these axioms, we can readily conclude that (1) the empty set, $\varnothing$ say, is in $\mathscr{F}$; (2) if A_1 and A_2 are in $\mathscr{F}$, then so is $A_1 \cap A_2$ (the intersection, consisting of points common to A_1 and A_2); and (3) the union of a finite number of sets is in $\mathscr{F}$.

Now let us look at the axioms that define a probability measure. Before we start, it is appropriate to mention that there are ways other than the axiomatic method to define a probability measure, and more than one set of axioms has been proposed by the proponents of the axiomatic method. The axiomatic method is the most common way to define probability, and the axioms presented here are almost universally accepted.

Definition A-1 Given a sample space Ω and a family of events $\mathscr{F}$, a *probability measure* (or *probability*) is a function, $P\{\cdot\}$ say, which assigns a real number to each event A in $\mathscr{F}$ such that

(*a*) For any $A \in \mathscr{F}, 0 \leq P\{A\} \leq 1$
(*b*) $P\{\Omega\} = 1$
(*c*) For any infinite sequence of disjoint events $A_1, A_2, \ldots$ (that is, $A_i \cap A_j = \varnothing$ whenever $i \neq j$,

$$P\left\{\bigcup_{i=1}^{\infty} A_i\right\} = \sum_{i=1}^{\infty} P\{A_i\}$$

From this definition it is readily established that (1) $P\{\varnothing\} = 0$; (2) if $A_1, \ldots, A_n$ is a finite sequence of disjoint events, then $P\{\bigcup_{i=1}^{n} A_n\} = \sum_{i=1}^{n} P\{A_i\}$; (3) $P\{A^c\} = 1 - P\{A\}$; (4) if $A_1 \subset A_2$ (A_1 is a subset of A_2), then $P\{A_2\} \geq P\{A_1\}$; and (5) for any two events A_1 and A_2, $P\{A_1 \cup A_2\} = P\{A_1\} + P\{A_2\} - P\{A_1 \cap A_2\}$.

In every probability model, there are a sample space Ω, a family of events $\mathscr{F}$, and a probability measure $P\{\cdot\}$, which are chosen in that order. The triple $\{\Omega, \mathscr{F}, P\}$ is called a *probability space*. The sample space and probability measure arise naturally in applications, but usually the family $\mathscr{F}$ is not mentioned. When Ω is countable, $\mathscr{F}$ is typically the set of all subsets, and it need not be mentioned explicitly. When Ω is uncountable, $\mathscr{F}$ can be very complicated. Suppose $\Omega = [0, \infty)$ and we want to assign probabilities to all intervals of the form $(a, b]$. Then typically $\mathscr{F}$ is chosen to be the smallest σ-field that includes all such intervals; the family of all subsets of $[0, \infty)$ is not chosen because that leads to technical difficulties. Thus, there are subsets of Ω whose probability is not defined; these sets are very hard to find and are of no practical importance. The important thing to remember is that the foundation of every probability model is a probability space, and only sets in $\mathscr{F}$ have probability.

When Ω is uncountable, usually there is no way to construct $\mathscr{F}$ without including some nonempty sets that will be assigned probability zero if we wish to include all those sets which we think deserve positive probability. Often one finds interesting statements that are true except when one of these events of probability zero occurs, in which case we say the statement is true "with probability 1," abbreviated w.p.1. This is a fine point that is mentioned to explain why "w.p.1" is written after some statements.

Random Variables and Distribution Functions

A random variable is, as its name suggests, a variable whose value is determined by a random mechanism. In mathematics generally, things whose values are determined by some mechanism are functions, so a random variable is a function. A real-valued random variable is a real-valued function, $X(\cdot)$ say, defined on Ω such that if the outcome $\omega \in \Omega$ occurs, the random variable assumes the value $X(\omega)$.

To fix these ideas firmly, let us examine a very simple random variable. A single flip of a coin can result in either a head or a tail showing. Denote these outcomes by H and T, respectively, so the sample space for a single flip of the coin is $\Omega = \{\text{H, T}\}$. If a dollar is won when the coin lands heads and lost when the coin lands tails, then the payoff from a flip of the coin is a random variable. This random variable is described by

$$X(\omega) = \begin{cases} 1 & \text{if } \omega = \text{H} \\ -1 & \text{if } \omega = \text{T} \end{cases}$$

Some functions from Ω to the real numbers cannot be random variables, but any function we are likely to choose will be a random variable.† We frequently abbreviate random variable as r.v.

An r.v. that achieves a countable number of values is called *discrete*; all other r.v.'s are called *continuous.*

Just as the function $h(\cdot) = g[f(\cdot)]$ denotes that function which assumes the value $g[f(t)]$ at the point t, the random variable $Y(\cdot) = g[X(\cdot)]$ is the r.v. which assumes the value $g[X(\omega)]$ when outcome ω occurs. We typically drop the arguments when denoting an r.v., writing X for $X(\cdot)$ and $Y = g(X)$ for $Y(\cdot) = g[X(\cdot)]$. The ω's are used occasionally for emphasis.

The probability measure for events is employed to attach probabilities to values (or *realizations*) of r.v.'s in this manner. For a given set of real numbers B, let $A = \{\omega: X(\omega) \in B\}$. Then

$$P\{X \in B\} = P\{A\}$$

In order for the left-side to be defined, we must have $A \in \mathscr{F}$. This is where restrictions on $X(\cdot)$ appear.

Associated with every random variable X is a unique function, $F(\cdot)$ say, called its *distribution function*, which is abbreviated by d.f. This function is defined by

$$F(x) = P\{X \leq x\} \qquad -\infty < x < \infty$$

If‡ $F(\infty) = 1$, then $F(\cdot)$ is called *honest* (or *proper*); otherwise, it is *defective*, and $1 - F(\infty)$ is called the *defect*. The interval $[a, b]$ is the *support* of $F(\cdot)$ if it is the smallest interval such that $F(x) = 0$ for all $x < a$ and $F(x) = 1$ for all $x > b$. The d.f. contains all the probabilistic information about the random variable. For example,

$$\begin{aligned} P\{a < X \leq b\} &= P\{X \leq b\} - P\{X \leq a\} \\ &= F(b) - F(a) \qquad a < b \end{aligned}$$

When X is discrete and takes the values $x_1 < x_2 < x_3 < \cdots$, say, then

$$P\{X = x_i\} = F(x_i) - F(x_{i-1})$$

† For those readers familiar with measure theory, a random variable is a measurable function. In this book, only measurable functions are allowed.

‡ A distribution function $F(\cdot)$ is a bounded monotone function, so $\lim_{x\to\infty} F(x)$ necessarily exists. We write this limit as $F(\infty)$ although $F(\cdot)$ is defined on $(-\infty, \infty)$ which does not include $\pm\infty$. Similarly, we write $F(\infty, x_2, \ldots, x_n)$ for $\lim_{x\to\infty} F(x, x_2, \ldots, x_n)$.

When X is continuous, often $F(\cdot)$ has a derivative; this derivative is called the *(probability) density function* of X, and is abbreviated p.d.f. Let $f(\cdot)$ be the derivative of $F(\cdot)$. When it is notationally inconvenient to denote the p.d.f. explicitly, we write $\text{Pd}\{X = x\}$ for the density function of the random variable X evaluated at x. By definition,

$$f(x) = \lim_{\Delta x \to 0} \frac{F(x + \Delta x) - F(x)}{\Delta x}$$

so that for small† positive values of Δx,

$$f(x)\Delta x = F(x + \Delta x) - F(x) + o(\Delta x)$$

Since

$$F(x + \Delta x) - F(x) = P\{x < X \leq x + \Delta x\}$$

we have

$$f(x)\, \Delta x \approx P\{x < X \leq x + \Delta x\}$$

where "$\approx$" means equal except for terms of order Δx or smaller [i.e., except for $o(\Delta x)$]. Notice that $f(x)$ by itself is not the probability on the right, and $f(x)\,\Delta x$ is not the probability that $X = x$. In fact, the latter probability is zero because

$$P\{X = x\} = \lim_{\Delta x \to 0} [F(x + \Delta x) - F(x - \Delta x)] = F(x) - F(x) = 0$$

whenever $F(\cdot)$ is continuous at x. In this case, $X \neq x$ is an example of a statement that holds w.p.1.

These ideas can be extended to cover more than one random variable at a time. Let

$$F(x_1, x_2, \ldots, x_n) = P\{X_1 \leq x_1, X_2 \leq x_2, \ldots, X_n \leq x_n\}$$

for any nonempty set of random variables $X_1, X_2, \ldots, X_n$ and real numbers $x_1, x_2, \ldots, x_n$. Then $F(\cdot, \cdot, \ldots, \cdot)$ is the *joint distribution function* of the random variables $X_1, X_2, \ldots, X_n$. If $P\{X_1 < \infty\} = 1$, then

$$F(\infty, x_2, \ldots, x_n) = P\{X_2 \leq x_2, \ldots, X_n \leq x_n\}$$

and similar statements hold when various other x's are set at $+\infty$. When all the x's but one, x_i say, are set at $+\infty$, the resulting function of x_i alone is called the *marginal distribution function* of X_i. The derivative of the marginal d.f. (when it exists) is called the *marginal density function.* The function (assuming it exists)

$$f(x_1, \ldots, x_n) = \frac{\partial^n}{\partial x_1 \cdots \partial x_n} F(x_1, \ldots, x_n)$$

is called the *joint density function* of $X_1, \ldots, X_n$. The marginal density function of X_i is obtained from the joint density function by integrating the joint density function, with respect to x_j for each $j \neq i$, from $-\infty$ to $+\infty$. For example, let

† The notation $o(x)$ stands for any function of the real variable x, say $g(x)$, such that $\lim_{x \to 0} [g(x)/x] = 0$. The beginning of Section A-3 contains a discussion of this notation.

$f_1(\cdot)$ be the marginal density of X_1. Then

$$f_1(x_1) = \int_{-\infty}^{\infty} \cdots \int_{-\infty}^{\infty} f(x_1, x_2, \ldots, x_n)\, dx_2 \cdots dx_n$$

A very important concept is *stochastic independence*, commonly referred to simply as *independence*.

Definition A-2 The random variables X_1 and X_2 with joint distribution function $F(\cdot, \cdot)$ are said to be *independent* if for all x_1 and x_2,

$$F(x_1, x_2) = F(x_1, \infty)F(\infty, x_2)$$

That is, a pair of r.v.'s is independent if their joint distribution function can be expressed as the product of the two marginal distributions. This definition is extended in the obvious way to any finite number of random variables. Observe that when X_1 and X_2 are independent, the joint probability density function satisfies

$$f(x_1, x_2) = f_1(x_1)f_2(x_2)$$

for all x_1 and x_2.

An infinite collection of random variables is independent if every finite set of at least two of them consists of independent random variables. An important consideration when there are three or more random variables is *mutual independence*. It may happen that X_1 and X_2 and X_2 and X_3 are two pairs of independent random variables, but X_1 and X_3 are not independent (in which case we say that they are *dependent*). If every subset of a collection of (at least two) random variables (i.e., all possible pairs, triplets, and so forth) contains only independent random variables, the collection consists of mutually independent random variables.

It is also useful to define independent events. The events A and B are independent if

$$P\{A \cap B\} = P\{A\}P\{B\}$$

When $X_1, X_2, \ldots, X_n$ are mutually independent and have the same d.f., we say they are *independent and identically distributed*, which is abbreviated i.i.d.

Expectations

The intuitive idea behind the expected value of a random variable is that it is a weighted sum of its potential values, where the weight associated with each value is the probability of obtaining that value. The symbol $E(X)$ is commonly used to denote the expected value of the random variable X.

Definition 4-3 The *expected value* of an r.v. with distribution function $F(\cdot)$ is

$$E(X) = \int_{-\infty}^{\infty} x \, dF(x)$$

provided the integral† is absolutely convergent.

Other common names for $E(X)$ are *mean* and *first moment*.

When $F(\cdot)$ has a density function $f(\cdot)$, $E(X)$ is

$$E(X) = \int_{-\infty}^{\infty} xf(x) \, dx$$

When X is discrete, we obtain

$$E(X) = \sum_{i=-\infty}^{\infty} x_i P\{X = x_i\}$$

which conforms exactly to our intuitive notion of $E(X)$.

When X can assume both positive and negative values,

$$\int_{-\infty}^{\infty} x \, dF(x) = \int_{-\infty}^{0} x \, dF(x) + \int_{0}^{\infty} x \, dF(x)$$

and the integral on the left may not exist because both integrals on the right may diverge, the first to $-\infty$ and the second to $+\infty$. In this case, we say that $E(X)$ does not exist. However, when $X \geq 0$, the contribution of the first integral on the right is zero, so $E(X)$ either exists or is represented by an integral that diverges to $+\infty$. In the latter situation, we say $E(X) = +\infty$. Similar remarks apply when $X \leq 0$.

When X is a random variable and $g(\cdot)$ is a reasonable‡ function, $Y = g(X)$ also is a random variable and may have an expected value. Suppose that X and Y are discrete. Then $Y = y_i$ whenever $X \in S_i = \{x_j : g(x_j) = y_i\}$, and so

$$E(Y) \triangleq \sum_{i=-\infty}^{\infty} y_i P\{Y = y_i\} = \sum_{i=-\infty}^{\infty} y_i P\{X \in S_i\} = \sum_{j=-\infty}^{\infty} g(x_j) P\{X = x_j\}$$

This analysis can be extended to arbitrary random variables, and the formula

$$E(Y) = E[g(X)] = \int_{-\infty}^{\infty} g(x) \, dF(x) \tag{A-4}$$

is always valid when Y is a bona fide random variable. Let us look at some examples of this result (assume the integrals exist):

(*a*) When $g(X) = X^n$, we obtain

$$E(X^n) = \int_{-\infty}^{\infty} x^n \, dF(x)$$

This is called the *n*th *moment* of X.

† See Section A-3 if you are not familiar with this notation for an integral.

‡ For those readers familiar with measure theory, reasonable means measurable.

(*b*) When $g(X) = cX$ for some constant c,

$$E(cX) = c\int_{-\infty}^{\infty} x\, dF(x) = cE(X)$$

(*c*) When $g(X) = [X - E(X)]^2 = Y$,

$$E(Y) = \int_{-\infty}^{\infty} [x - E(X)]^2\, dF(x) = E(X^2) - [E(X)]^2$$

and $E(Y)$ is called the *variance of* X, denoted by Var (X) [often the notation $V(X)$ or σ_X^2 is used].

Since the integrand is nonnegative, Var $(X) \geq 0$. The positive square root of Var (X) is called the *standard deviation* of X and is typically denoted by σ_X. Using (*b*) and (*c*), we deduce that

$$\text{Var}\,(cX) = E(c^2X^2) - [cE(X)]^2 = c^2\,\text{Var}\,(X)$$

for any constant c.

In the next four examples we assume that $F(\cdot, \cdot)$ has a p.d.f. $f(\cdot, \cdot)$; this simplifies the notation. The conclusions obtained are true for any d.f.

(*d*) When $g(X_1, X_2) = X_1 + X_2$,

$$\begin{aligned} E(X_1 + X_2) &= \int_{-\infty}^{\infty}\int_{-\infty}^{\infty} (x_1 + x_2) f(x_1, x_2)\, dx_1\, dx_2 \\ &= \int_{-\infty}^{\infty} x_1 \int_{-\infty}^{\infty} f(x_1, x_2)\, dx_2\, dx_1 \\ &\quad + \int_{-\infty}^{\infty} x_2 \int_{-\infty}^{\infty} f(x_1, x_2)\, dx_1\, dx_2 \\ &= \int_{-\infty}^{\infty} x_1 f_1(x_1)\, dx_1 + \int_{-\infty}^{\infty} x_2 f_2(x_2)\, dx_2 \\ &= E(X_1) + E(X_2) \end{aligned}$$

Combining (*b*) and (*d*) and using mathematical induction, we obtain

$$E\left(\sum_{i=1}^{n} c_i X_i\right) = \sum_{i=1}^{n} c_i E(X_i)$$

for any finite and positive integer n.

(*e*) When $g(X_1, X_2) = X_1X_2$ and X_1 and X_2 are independent,

$$E(X_1X_2) = \int_{-\infty}^{\infty}\int_{-\infty}^{\infty} x_1x_2 f(x_1, x_2)\, dx_1\, dx_2$$

$$= \int_{-\infty}^{\infty} \int_{-\infty}^{\infty} x_1 f_1(x_1) x_2 f_2(x_2)\, dx_1\, dx_2$$

$$= E(X_1)E(X_2)$$

where $f_1(\cdot)$ and $f_2(\cdot)$ are the p.d.f's of X_1 and X_2, respectively.

(*f*) When $g(X_1, X_2) = [X_1 - E(X_1)][X_2 - E(X_2)] = Y, E(Y)$ is called the *covariance* of X_1 and X_2, denoted by Cov (X_1, X_2) (sometimes the notation $\sigma_{x_1;x_2}$ is used).

From (*b*),

$$\begin{aligned} \text{Cov}\,(X_1, X_2) &= \int_{-\infty}^{\infty} \int_{-\infty}^{\infty} [x_1 x_2 - x_1 E(X_2) - x_2 E(X_1) - E(X_1)E(X_2)] \\ &\quad \times f(x_1, x_2)\, dx_1\, dx_2 \\ &= E(X_1 X_2) - 2E(X_1)E(X_2) + E(X_1)E(X_2) \\ &= E(X_1 X_2) - E(X_1)E(X_2) = \text{Cov}\,(X_2, X_1) \end{aligned}$$

When X_1 and X_2 are independent, (*e*) yields

$$\text{Cov}\,(X_1, X_2) = E(X_1)E(X_2) - E(X_1)E(X_2) = 0$$

However, it is not necessarily true that X_1 and X_2 are independent if Cov $(X_1, X_2) = 0$.

(*g*) When $g(X_1, X_2) = [c_1 X_1 + c_2 X_2 - E(c_1 X_1 + c_2 X_2)]^2$, where c_1 and c_2 are constants, use (*b*), (*c*), (*d*), and (*e*) to obtain

$$\begin{aligned} E[g(x)] &= \text{Var}\,(c_1 X_1 + c_2 X_2) \\ &= E(c_1[X_1 - E(X_1)])^2 + E(c_2[X_2 - E(X_2)])^2 \\ &\quad + 2c_1 c_2 E[X_1 - E(X_1)]E[X_2 - E(X_2)] \\ &= c_1^2 \text{ Var}\,(X_1) + c_2^2 \text{ Var}\,(X_2) + 2c_1 c_2 \text{ Cov}\,(X_1, X_2) \end{aligned}$$

When X_1 and X_2 are independent, Cov $(X_1, X_2) = 0$, so

$$\text{Var}\,(c_1 X_1 + c_2 X_2) = c_1^2 \text{ Var}\,(X_1) + c_2^2 \text{ Var}\,(X_2)$$

(*h*) When $g(X) = e^{-sX}$ and $X \geq 0$, we obtain

$$E(e^{-sX}) = \int_0^{\infty} e^{-sx}\, dF(x)$$

This is the *Laplace-Stieltjes transform* (abbreviated LST) of $F(\cdot)$. When $F(\cdot)$ has p.d.f. $f(\cdot)$,

$$E(e^{-sX}) = \int_0^{\infty} e^{-sx} f(x)\, dx$$

which is the *Laplace transform*† of $f(\cdot)$.

† We also call $E(e^{-sX})$ the Laplace (or Laplace-Stieltjes) transform of X.

(*i*) Let $v = E(X)$ and $\sigma^2 = \text{Var}(X)$, and choose $Y = (X - v)^2$. For any number $z > 0$,

$$\sigma^2 = E(Y) \geq \int_{y=z^2}^{\infty} y\, dF(y) \geq z^2 \int_{y=z^2}^{\infty} dF(y) = z^2 P\{Y \geq z^2\}$$

hence

$$P\{|X - v| \geq z\} \leq \frac{\sigma^2}{z^2}$$

This is called *Chebychev's inequality.*

Conditional Probability

The basic idea of conditional probability can be easily illustrated by the roll of a fair die. If X is the number of pips showing after a roll, then $P\{X = i\} = \frac{1}{6}$ for $i = 1, 2, \ldots, 6$. But if we are told that an even number of pips is showing, we would naturally say that $P\{X = i\} = \frac{1}{3}$ when i is 2, 4, or 6 and $P\{X = i\} = 0$ otherwise. What has happened is that the set of possible outcomes has been reduced and the assignments of probabilities have been adjusted accordingly. We now want to do this in a formal way so that conditional probability calculations can be made rigorously and systematically.

Definition A-4 For any two events A and B with $P\{B\} > 0$, the *conditional probability* of A given B is denoted by $P\{A \mid B\}$ and is defined by

$$P\{A \mid B\} = \frac{P\{A \cap B\}}{P\{B\}}$$

When $P\{B\} = 0$, $P\{A \mid B\}$ is not defined. It is often convenient to adopt the *convention* that $P\{A \mid B\} = 0$ when $P\{B\} = 0$.

The first consequence of this definition is a rule for finding $P\{A \cap B\}$. In many applications, we naturally have $P\{B\}$ and $P\{A \mid B\}$. Then

$$P\{A \cap B\} = P\{A \mid B\}P\{B\}$$

When A and B are independent, $P\{A \cap B\} = P\{A\}P\{B\}$, and so

$$P\{A \mid B\} = P\{A\}$$

This is the intuitive idea behind the use of the word independence.

If $A_1, A_2, \ldots$ are disjoint events with $\bigcup_{i=1}^{\infty} A_i = \Omega$, it is easily shown that

$$P\{B\} = \sum_{i=1}^{\infty} P\{B \cap A_i\}$$

Since $P\{B \cap A_i\} = P\{B \mid A_i\}P\{A_i\}$,

$$P\{B\} = \sum_{i=1}^{\infty} P\{B \mid A_i\}P\{A_i\}$$

which is called the *theorem of total probability*. Substituting this formula for $P\{B\}$ into the definition of $P\{A \mid B\}$, we obtain *Bayes' formula:*

$$P\{A_j \mid B\} = \frac{P\{A_j \cap B\}}{P\{B\}}$$
$$= \frac{P\{B \mid A_j\}P\{A_j\}}{\sum_{i=1}^{\infty} P\{B \mid A_i\}P\{A_i\}}$$

This notion of conditional probability can be expressed in terms of random variables and distribution functions, and we often do so. Let X and Y be random variables with joint distribution function $F(\cdot, \cdot)$, and let the marginal d.f.'s of X and Y be $F_1(\cdot)$ and $F_2(\cdot)$, respectively. From the definition of conditional probability,

$$P\{X \le x \mid Y \le y\} = \frac{P\{X \le x, Y \le y\}}{P\{Y \le y\}}$$

Denoting the left-side by $F_1(x \mid Y \le y)$, which is called the *conditional d.f.* of X. we obtain

$$F_1(x \mid Y \le y) = \frac{F(x, y)}{F_2(y)}$$

When the appropriate derivatives exist, we can write the *conditional density function*

$$f_1(x \mid Y \le y) = \frac{\partial}{\partial x} \frac{F(x, y)}{F_2(y)}$$

When X and Y are discrete r.v.'s and $P\{Y = y\} > 0$,

$$P\{X \le x \mid Y = y\} = \frac{P\{X \le x, Y = y\}}{P\{Y = y\}}$$

is a direct consequence of Definition A-4. When Y is not discrete, $P\{Y = y\} > 0$ is the exception, so we have to offer a different definition. If the theorem of total probability *were* valid when Y is continuous with distribution function $G(\cdot)$, it would state

$$P\{X \le x\} = \int_{-\infty}^{\infty} P\{X \le x \mid Y = y\}\, dG(y)$$

We *define* $P\{X \le x \mid Y = y\}$ as *any* solution (there is always one, but there may be many) of this equation. We usually use $P\{X \le x \mid Y = y\}$ to obtain $P\{X \le x\}$ from the definition of the conditional probability, so nonuniqueness is not a major issue. The important fact is that it is permissible to condition on the value of an r.v., say $Y = y$, when $P\{Y = y\} = 0$.

The preceding notions extend directly from two to n random variables. In particular, if X, Y, and Z are random variables on the same probability space then

$$P\{X \leq x \mid Y \leq y, Z \leq z\} = \frac{P\{X \leq x, Y \leq y, Z \leq z\}}{P\{Y \leq y, Z \leq z\}}$$

We say that X and Y are *conditionally independent given* Z (or X is *conditionally independent of* Y *given* Z) if

$$P\{X \leq x, Y \leq y \mid Z \leq z\} = P\{X \leq x \mid Z \leq z\}P\{Y \leq y \mid Z \leq z\}$$

is valid for all values of x, y, and z. This extension of Definition A-2 is equivalent to the validity of

$$P\{X \leq x \mid Y \leq y, Z \leq z\} = P\{X \leq x \mid Z \leq z\}$$

for all values of x, y, and z. We usually encounter conditional independence in models where Y or Z (or both) are vector-valued random variables.

Conditional Expectation

Conditional probabilities can be viewed as establishing a new sample space B and new probabilities for events. Random variables defined over the original sample space Ω remain random variables when conditioned on the event B, but have a smaller domain of definition. They may still have expectations, and our intuitive idea of what an expectation represents remains the same. To be both intuitive and precise, we use discrete random variables to introduce the conditional expectation. The ideas and results are applicable to continuous random variables if we replace sums by integrals.

For any event B and discrete random variable X, the *conditional expectation of* X *given* B is defined by

$$E(X \mid B) = \sum_{i=-\infty}^{\infty} iP\{X = i \mid B\}$$

and it is a number. If we take the event B to be $\{Y = b\}$ for some discrete random variable Y, we obtain

$$E(X \mid Y = b) = \sum_{i=-\infty}^{\infty} iP\{X = i \mid Y = b\}$$

Different values of b may produce different values of $E(X \mid Y = b)$. We usually write the *function* $E(X \mid Y = \cdot)$ as $E(X \mid Y)$ and express it as "the expected value of X given Y." Since $E(X \mid Y)$ is a function whose value is determined by the value taken on by Y, *it is a random variable*.

One of the main reasons we study conditional expectations is that they frequently provide a convenient way to obtain an unconditional (i.e., "ordinary") expectation. Since $E(X \mid Y)$ is a random variable, form its expectation,

$E[E(X \mid Y)]$; then

$$E[E(X \mid Y)] = \sum_{b=-\infty}^{\infty} E(X \mid Y = b)P\{Y = b\}$$

$$= \sum_{b=-\infty}^{\infty} \sum_{i=-\infty}^{\infty} iP\{X = i \mid Y = b\}P\{Y = b\}$$

$$= \sum_{i=-\infty}^{\infty} \sum_{b=-\infty}^{\infty} P\{X = i \mid Y = b\}P\{Y = b\}$$

$$= \sum_{i=-\infty}^{\infty} iP\{X = i\} = E(X)$$

This procedure for obtaining $E(X)$ is employed frequently in the study of stochastic processes. To illustrate its use, let X be the number of pips showing on a fairly rolled die, and define Y to be 1 if X is even and 0 if X is odd. Then

$$E(X \mid Y = 1) = 4 \qquad E(X \mid Y = 0) = 3$$

$$P\{Y = 1\} = P\{Y = 0\} = \tfrac{1}{2}$$

and so

$$E(X) = E[E(X \mid Y)] = \frac{4 + 3}{2}$$

Since $E(X \mid Y)$ is a random variable, it has all the endowments of random variables we have studied previously, such as a variance, a d.f., and a Laplace transform.

Sums of Random Variables

Let X_1 and X_2 be random variables, and define the random variable S by $S = X_1 + X_2$. When X_1 and X_2 are discrete, it is easy to see that

$$P\{S = k\} = \sum_{i=-\infty}^{\infty} P\{X_2 = k - i \mid X_1 = i\}P\{X_1 = i\}$$

When X_1 and X_2 have distribution functions $F_1(\cdot)$ and $F_2(\cdot)$, respectively, then the d.f. of S, say $G(\cdot)$, is given by

$$G(s) = \int_{-\infty}^{\infty} F_2(s - x \mid X_1 = x)\, dF_1(x)$$

If either X_1 or X_2 possesses a density function, then so does S.

When X_1 and X_2 are independent and nonnegative, then the above formulas reduce to

$$P\{S = k\} = \sum_{i=0}^{k} P\{X_2 = k - i\}P\{X_1 = i\}$$

and

$$G(s) = \int_0^s F_2(s - x)\, dF_1(x) = \int_0^s F_1(x)\, dF_2(s - x)$$

The expressions on the right-side of the last two equations are called *convolutions.*

Once we know how to find the distribution of the sum of two random variables, in principle, we can find the distribution of a sum of any finite number of them. We can find the distribution of $X_1 + X_2$, then of $X_3 + (X_1 + X_2)$, then of $X_4 + (X_1 + X_2 + X_3)$, etc.

Now we consider sums of a large number of random variables and the famous limit theorems for them. To this end, define $S_0 = 0$ and

$$S_n = X_1 + X_2 + \cdots + X_n \qquad n \in I_+$$

If each X_i has mean μ, then

$$E(S_n) = n\mu \qquad n \in I$$

Furthermore, if the X_i's are independent and have the common variance σ^2, then

$$\text{Var}\,(S_n) = n\sigma^2$$

and

$$\text{Var}\left(\frac{S_n}{n}\right) = \text{Var}\left(\frac{X_1}{n} + \frac{X_2}{n} + \cdots + \frac{X_n}{n}\right)$$

$$= \frac{n\sigma^2}{n^2} = \frac{\sigma^2}{n}$$

Thus, $E(S_n/n) = \mu$, which is a constant, and Var $(S_n/n) \to 0$ as $n \to \infty$, which suggests that S_n/n, which is a random variable, approaches a constant as $n \to \infty$. This is a typical interpretation of "statistical regularity" or a "law of averages," but it is not a precise mathematical statement. The precise statements are the laws of large numbers given now.

Theorem A-1: Weak law of large numbers Let $X_1, X_2, \ldots$ be mutually independent random variables, and let $v_k = E(X_k)$ and $\sigma_k^2 = \text{Var}\,(X_k)$ for each k. Set $S_n = X_1 + \cdots + X_n$ so that $E(S_n) \triangleq m_n = \sum_1^n v_k$ and Var $(S_n) \triangleq s_n^2 = \sum_1^n \sigma_k^2$. If $s_n^2/n \to 0$, then for any $\epsilon > 0$,

$$\lim_{n \to \infty} P\left\{\frac{|S_n - m_n|}{n} < \epsilon\right\} = 1$$

This version of the weak law of large numbers requires that the random variables possess a variance. This restriction is not necessary when $X_1, X_2, \ldots$ have the same distribution.

The weak law states that when all the v_k's are the same, v say, given any $\epsilon > 0$, there exists a $\delta > 0$ (possibly depending on ϵ) and an N (possibly depending on ϵ and δ) such that for any $n > N$,

$$P\left\{\left|\frac{S_n}{n} - v\right| < \epsilon\right\} > 1 - \delta$$

That is, for all sufficiently large n, S_n/n is close to v with probability close to 1. What the weak law does *not* say is that if $|S_n/n - v|$ is small for one large value of n, it will remain small for all larger values of n. To draw that conclusion, we need to take the limit inside the probability statement, and that stronger result is the strong law of large numbers.

Theorem A-2: Strong law of large numbers With the notation of Theorem A-1, assume that either

(*a*) $\sum_{k=1}^{\infty} \frac{\sigma_k^2}{k^2} < \infty$

or

(*b*) $X_1, X_2, \ldots$ are i.i.d. with $E(X_1) < \infty$

Then

$$\lim_{n\to\infty} \frac{S_n - m_n}{n} = 1 \qquad \text{(w.p.1)} \tag{A-5}$$

Equivalently, for every pair $\epsilon > 0$ and $\delta > 0$, there is an N such that for every $r \in I$,

$$P\left\{\frac{|S_N - m_N|}{N} < \epsilon, \ldots, \frac{|S_{N+r} - m_{N+r}|}{N+r} < \epsilon\right\} > 1 - \delta$$

In case (*b*), $m_n = n\mu$ and $\lim_{n\to\infty} (S_n/n) = \mu$. Note that the left side of (A-5) is an r.v. and the right side is a constant.

The strong law justifies the interpretation of probability measure as the long-run frequency of occurrence. To see this, consider many independent replications of an experiment performed under identical circumstances, and let ω_n be the outcome of the nth experiment. For each $n = 1, 2, \ldots$ and any event A, define the random variable X_n by

$$X_n = \begin{cases} 1 & \text{if } \omega_n \in A \\ 0 & \text{if } \omega_n \notin A \end{cases}$$

Hence $E(X_n) = P\{A\}$. The strong law implies

$$\frac{1}{n} \sum_{i=1}^{n} X_j \to P\{A\} \qquad \text{(w.p.1)}$$

and the left side is the relative frequency of occurrences of A.

The weak and strong laws both state that the difference between S_n and its expectation is small compared to n, when n is large. In absolute terms, S_n may be (in fact, usually is) far from m_n. To determine how far S_n is from m_n, divide $|S_n - m_n|$ by a function of n that goes to infinity more slowly than n does.

Theorem A-3: Law of the iterated logarithm When $X_1, X_2, \ldots$ are i.i.d. with mean 0 and variance σ^2,

$$\limsup_{n\to\infty} \frac{|S_n|}{\sigma\sqrt{2n \ln(\ln n)}} = 1 \qquad \text{(w.p.1)}$$

This theorem states that, for $\lambda > 1$,

$$|S_n| > \lambda\sigma\sqrt{2n\ln(\ln n)}$$

holds w.p.1. for only finitely many values of n, and for $0 < \lambda < 1$, it holds for infinitely many values of n.

The next limit theorem is perhaps the most widely known, used, and abused theorem in probability theory. The simple version given here suffices for our purposes.

Theorem A-4: Central-limit theorem When $X_1, X_2, \ldots$ are i.i.d. with mean v and variance σ^2, then for every fixed z,

$$\lim_{n\to\infty} P\left\{\frac{S_n - nv}{\sigma\sqrt{n}} < z\right\} = \Phi(z)$$

where $\Phi(z)$ is the standard normal d.f. given by

$$\Phi(x) = \frac{1}{\sqrt{2\pi}}\int_{-\infty}^{x} e^{-u^2/2}\,du$$

This theorem asserts that if one adds a large number of i.i.d. random variables and *normalizes* (subtracts from S_n its mean nv and divides by the standard deviation $\sigma\sqrt{n}$) so the resulting r.v. has mean 0 and variance 1, then this latter r.v. has a d.f. that is close (pointwise) to the distribution function $\Phi(\cdot)$. An important feature of the central-limit theorem to remember is that the conclusion is typically known to be true only when the summands are mutually independent.

A-2 EXPONENTIAL FAMILY OF DISTRIBUTIONS

The exponential distribution is very important in applied probability models. This distribution often fits data on times to complete a telephone conversation, times between failures of complex equipment, and times between job arrivals at a computer. It is also useful as a building block. By combining exponential distributions a wide variety of nonexponential distributions can be obtained.

Exponential Distribution

Definition A-5 The distribution function $F(\cdot)$ is *exponential* with parameter $\alpha > 0$ if

$$F(x) = \begin{cases} 1 - e^{-\alpha x} & x \geq 0 \\ 0 & x < 0 \end{cases}$$

An r.v. with an exponential d.f. is called *exponential*.

The following properties of the exponential d.f. follow immediately from the definition.

Proposition A-1 If X is exponential with parameter α, then

(*a*) X has a density function $f(\cdot)$ given by

$$f(x) = \begin{cases} \alpha e^{-\alpha x} & x \geq 0 \\ 0 & x < 0 \end{cases}$$

(*b*) $F^c(x) \triangleq P\{X > x\} = \begin{cases} 1 & x < 0 \\ e^{-\alpha x} & x \geq 0 \end{cases}$

(*c*) $E(X^n) = \alpha \int_0^\infty x^n e^{-\alpha x}\, dx = \dfrac{n!}{\alpha^n} \quad n \in I; \quad$ in particular, $\quad E(X) = 1/\alpha$

(*d*) $\text{Var}(X) = \dfrac{1}{\alpha^2}$

(*e*) $E(e^{-sx}) = \alpha \int_0^\infty e^{-sx} e^{-\alpha x}\, dx = \dfrac{\alpha}{s + \alpha}$

PROOF This is left to you. □

Observe that (*c*) and (*d*) imply that the ratio of the standard deviation to the mean is 1.

The importance of the exponential distribution is due to its unique memoryless property. If X is exponentially distributed with parameter α, then

$$P\{X > t + s \mid X > s\} = \frac{e^{-\alpha(t+s)}}{e^{-\alpha s}}$$

$$= e^{-\alpha t} = P\{X > t\} \qquad t, s \geq 0 \tag{A-6}$$

Suppose X represents the time to failure of some device. Equation (A-6) asserts that if the device has lasted until time s, the distribution of the remaining life is independent of s; in particular, it is the same as the time to failure of a new device. In other words, when failure times are exponentially distributed, a used item is as good as new. We call the property expressed by (A-6) the *memoryless property* of the exponential distribution.

Here is another version of the memoryless property. Suppose $F(\cdot)$ is the d.f. of a nonnegative random variable X and $F(\cdot)$ has a density $f(\cdot)$. The function

$$h(t) \triangleq \frac{f(t)}{F^c(t)} \qquad t \geq 0$$

is the *hazard rate* (*or failure rate*) of X [of $F(\cdot)$]. If X represents the time to failure of a device, then the probability that the device fails during $(t, t + \Delta t]$ is $h(t)\,\Delta t + o(\Delta t)$. When $F(t) = 1 - e^{-\alpha t}$,

$$h(t) = \frac{\alpha e^{-\alpha t}}{e^{-\alpha t}} = \alpha \qquad t \geq 0 \tag{A-7}$$

i.e., the exponential distribution has a constant hazard rate. In other words, no matter how long the device has been working, the probability that it will fail during an interval of length Δt is $\alpha \, \Delta t + o(\Delta t)$.

The exponential distribution is singled out for so much attention because it is the only d.f. with a density that is memoryless.

Proposition A-2 Let X be a nonnegative r.v. with distribution function $F(\cdot)$ possessing a density $f(\cdot)$, and let $E(X) = 1/\alpha$, $0 < \alpha < \infty$. Then the following three statements are equivalent:

(*a*) X has the exponential distribution

$$F(t) = 1 - e^{-\alpha t}$$

(*b*) X has the *memoryless property*

$$P\{X > t + s \mid X > s\} = P\{X > t\} \qquad \text{for all } t, s > 0$$

(*c*) X has the *constant-hazard-rate property*

$$\frac{f(t)}{F^c(t)} = \text{const} \qquad \text{for all } t \geq 0$$

PROOF We showed above that $(a) \Rightarrow (b)$. We now show $(b) \Rightarrow (c) \Rightarrow (a)$. To show $(b) \Rightarrow (c)$, observe that (b) implies

$$\begin{aligned} F(t + s) - F(t) &= P\{t < X \leq t + s\} = P\{X \leq t + s \mid X > t\}P\{X > t\} \\ &= P\{X \leq s\}P\{X > t\} = F(s)F^c(t) \end{aligned}$$

Thus

$$\begin{aligned} f(t) &= \lim_{s \downarrow 0} \frac{F(t + s) - F(t)}{s} \\ &= \lim_{s \downarrow 0} \frac{F(s)}{s} F^c(t) = f(0)F^c(t) \end{aligned}$$

and hence $f(t)/F^c(t)$ equals the constant $f(0)$. To show (c) implies (a), let the constant in (c) be β. Since $f(t) = -dF^c(t)/dt$, (c) implies

$$\beta = \frac{-1}{F^c(t)} \frac{dF^c(t)}{dt} = -\frac{d}{dt}[\ln F^c(t)]$$

Hence

$$\int d[\ln F^c(t)] = -\beta \int dt + \ln \gamma$$

where $\ln \gamma$ is the constant of integration. Thus,

$$F^c(t) = \gamma e^{-\beta t} \qquad t \geq 0$$

Since $1 = F^c(0) = \gamma, f(t) = \beta e^{-\beta t}$. Since $E(X) = 1/\alpha$,

$$\frac{1}{\alpha} = \int_0^\infty t\beta e^{-\beta t}\, dt = \frac{1}{\beta}$$

and the demonstration is complete. □

The condition that $F(\cdot)$ have a density is crucial for the "only if" part of Proposition A-2. When X is a discrete r.v., the geometric distribution is memoryless (see Exercises A-1 through A-3).

The next proposition is used in Chapter 5.

Proposition A-3 Let $X_1, X_2, \ldots, X_n$ be independent r.v.'s with $P\{X_i \le t\} = 1 - e^{-\alpha_i t}$, $i = 1, 2, \ldots, n$. Let $M = \min\{X_1, X_2, \ldots, X_n\}$. Then

$$P\{M \le t\} = 1 - e^{-(\alpha_1 + \cdots + \alpha_n)t} \qquad t \ge 0$$

PROOF Observe that $\{M > t\}$ occurs if, and only if, each $X_i > t$. Using the independence assumption yields

$$P\{M > t\} = \prod_{i=1}^{n} P\{X_i > t\} = \prod_{i=1}^{n} e^{-\alpha_i t} = e^{-(\alpha_1 + \cdots + \alpha_n)t}$$

and the desired result follows immediately. □

Gamma Distribution

Let X_1 and X_2 be i.i.d. exponential r.v.'s with mean $1/\lambda$. The random variable $S_2 = X_1 + X_2$ has density

$$f_2(t) = \int_0^t \lambda e^{-\lambda(t-x)} \lambda e^{-\lambda x}\, dx = \lambda^2 t e^{-\lambda t} \qquad t \ge 0$$

When $X_1, X_2, \ldots, X_k$ are i.i.d. exponential r.v.'s with mean $1/\lambda$, a simple induction will show that the random variable $S_k = X_1 + \cdots + X_k$ has density

$$f_k(t) = \frac{\lambda^k t^{k-1} e^{-\lambda t}}{(k-1)!} \qquad k \in I_+ \tag{A-8}$$

We call $f_k(t)$ the *gamma density with shape parameter k and scale parameter λ*. There is no reason to restrict k to being an integer. Suppose we want a density function of the form

$$f_\kappa(t) = ct^{\kappa-1}e^{-\lambda t} \qquad t \ge 0$$

where $\kappa > 0$, $\lambda > 0$, and c is chosen so that $f_\kappa(\cdot)$ integrates to 1. Setting $x = \lambda t$ yields

$$1 = c\int_0^\infty t^{\kappa-1}e^{-\lambda t}\, dt = \frac{c}{\lambda^{\kappa-1}}\int_0^\infty x^{\kappa-1}e^{-x}\, dx$$

The last integral is the gamma function

$$\Gamma(\kappa) = \int_0^\infty x^{\kappa-1}e^{-x}\,dx$$

Integration by parts establishes

$$\Gamma(\kappa) = (\kappa - 1)\Gamma(\kappa - 1)$$

and $\Gamma(k) = (k-1)!$ for $k \in I_+$.

Thus

$$f_\kappa(t) = \frac{\lambda(\lambda t)^{\kappa-1}e^{-\lambda t}}{\Gamma(\kappa)} \qquad t > 0 \tag{A-9}$$

is a bona fide density function for any $\kappa > 0$; it is called the *gamma density with shape parameter* κ *and scale parameter* λ.

The reason why κ is called the shape parameter and λ is called the scale parameter can be seen in Figures A-1 and A-2. For fixed λ, changes in κ lead to qualitative changes in the shape of the density, while for fixed κ, changes in λ lead primarily to changes in the scale of the function.

The properties of the gamma density that we use are given in this proposition.

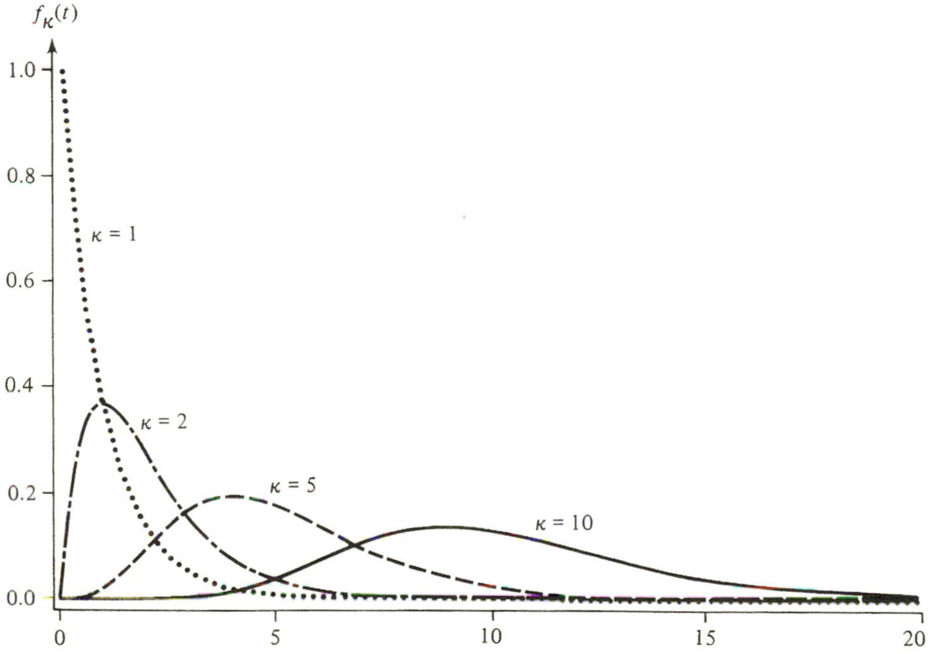

Figure A-1 $f_\kappa(t)$ with $\lambda = 1$ and $\kappa = 1, 2, 5, 10$.

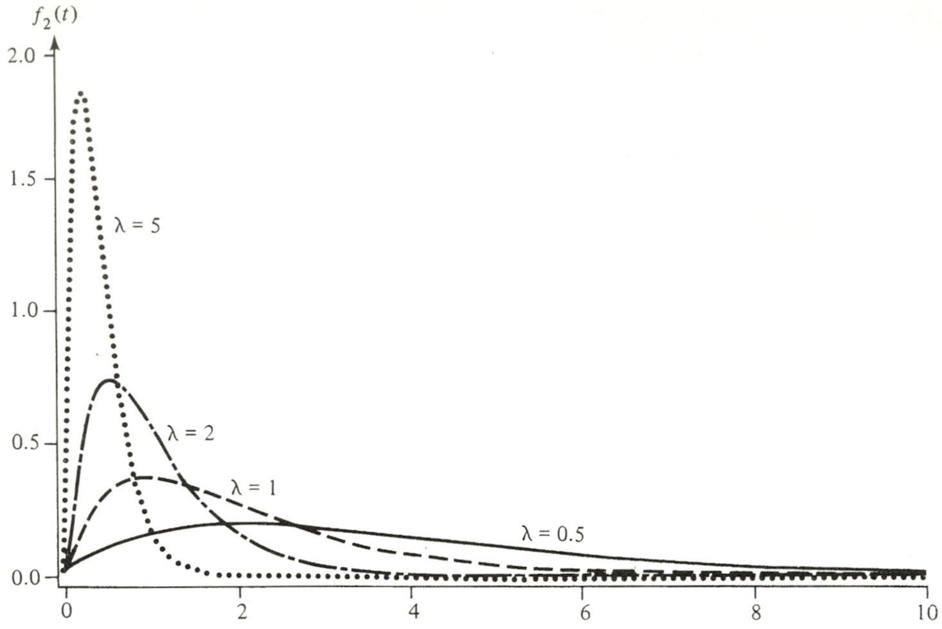

Figure A-2 $f_\kappa(t)$ with $\kappa = 2$ and $\lambda = \frac{1}{2}$, 1, 2, 5.

Proposition A-4 Let X be an r.v. whose density function is given by (A-9). Then

(i) $E(X) = \dfrac{\kappa}{\lambda}$

(ii) $\text{Var}\,(X) = \dfrac{\kappa}{\lambda^2}$

(iii) $E(e^{-sX}) = \dfrac{\lambda^\kappa}{(s + \lambda)^\kappa}$

PROOF This is left as Exercise A-4. □

Observe that $+\sqrt{\text{Var}\,(X)}/E(X) = 1/\sqrt{\kappa}$, which is less than 1 when $\kappa > 1$ and larger than 1 when $\kappa < 1$. Thus, large values of κ lead to distributions that are "more regular" than the exponential, and small values of κ correspond to distributions that are "more irregular" than the exponential.

By putting $\kappa = k \in I_+$ and $\lambda = k\alpha$ in (A-8), a family of distributions with mean $1/\alpha$ and variance $(k\alpha^2)^{-1}$ is achieved. This special form of the gamma density is called the *Erlang* density and denoted by E_k. The Erlang density frequently is used in queueing models to describe r.v.'s where the data indicate that the exponential d.f. is not appropriate because the ratio of the standard deviation to the mean is much smaller than 1. Since the Erlang distribution represents the sum of i.i.d. exponential r.v.'s, some properties of the exponential d.f. can be used in the subsequent analysis of models with Erlang distributions (see Example 8-5 in Volume I).

Hyperexponential Distribution

Let X_1 and X_2 be independent exponential r.v.'s with $E(X_i) = 1/\alpha_i$. Define the random variable Y by

$$Y = \begin{cases} X_1 & \text{with probability } p \\ X_2 & \text{with probability } 1 - p \end{cases}$$

The density function of Y is

$$f_2(t) = p\alpha_1 e^{-\alpha_1 t} + (1 - p)\alpha_2 e^{-\alpha_2 t} \qquad t \geq 0 \tag{A-10}$$

This is called the *hyperexponential density of order* 2 and denoted by H_2. A physical model that leads to the H_2 density for failure times (say) occurs when there are two manufacturing plants and plant i produces objects which have exponentially distributed failure times with mean $1/\alpha_i$. Plant 1 produces $100p$ percent of the total objects. When the objects are prepared for sale to the public, the outputs from both plants are mixed. The failure time of a randomly chosen object is then given by the random variable Y defined above.

The density in (A-10) has the obvious generalization

$$f_k(t) = \sum_{i=1}^{k} p_i \alpha_i e^{-\alpha_i t} \qquad t \geq 0 \tag{A-11}$$

where $p_i \geq 0$, $\sum_{i=1}^{k} p_i = 1$, and each $\alpha_i > 0$. This is the hyperexponential density of order k and is denoted by H_k. The salient features of the H_k density are requested in Exercise A-5.

EXERCISES

A-1 A nonnegative discrete random variable N is memoryless if $P\{N = m + n \mid N > m\} = P\{N = n\}$ for all $m, n \in I$. The hazard rate at n is $P\{N = n\}/P\{N \geq n\}$. Suppose $P\{N = n\} = (1 - \pi)\pi^n$, $n \in I$, for some π between 0 and 1. Is the failure rate a constant? Is the r.v. memoryless?

A-2 (Continuation) Show that $(1 - \pi)\pi^{n-1}$, $n \in I_+$, is the unique discrete memoryless distribution.

A-3 (Continuation) Prove that when $P\{N = 0\} = 0$, the memoryless property holds if, and only if, the constant-hazard-rate property holds.

A-4 Prove Proposition A-4.

A-5 Let the random variable Y have an H_k density. Find $E(Y)$, $\text{Var}(Y)$, $P\{Y \leq y\}$ and $E(e^{-sY})$.

A-3 FACTS FROM MATHEMATICAL ANALYSIS

In the study of stochastic models frequently we use limiting operations. This section contains statements that are utilized to justify various formal operations, such as interchanging the order of integration and taking the derivative of an infinite sum of functions.

Big-oh, Little-oh Notation

The function $f(\cdot)$ is $o[g(\cdot)]$ (read "f is little-oh of g") as $x \to a$ if $\lim_{x \to a}[f(x)/g(x)] = 0$. When $g(x) = x$, we write $f(\cdot)$ is $o(\cdot)$ and say "$f(\cdot)$ is little-oh of x." The most common circumstance is for $a = 0$ and $g(x) = x$; then $f(\cdot)$ is $o(\cdot)$ means $\lim_{x \to 0}[f(x)/x] = 0$. The purpose of this notation is to describe an essential property of a function without specifying the function in detail. For example, let X be an r.v. with density function $f(\cdot)$. We may write

$$P\{x < X < x + \Delta x\} = f(x)\,\Delta x + o(\Delta x)$$

where Δx is small. The most common example of an $o(\cdot)$ function that we encounter is $x^{1+\delta}$ for some $\delta > 0$, with $x \to 0$.

There are two other important special cases. Choosing $g(x) \equiv 1$ gives $f(\cdot)$ is $o(1)$ if $\lim_{x \to a} f(x) = 0$. Choosing $g(x) = 1/x$ gives $f(\cdot)$ is $o(1/x)$ if $\lim_{x \to a} xf(x) = 0$.

It follows directly from the definition that the sum of two (and hence of any *finite* number) of little-oh functions is a little-oh function, and multiplying a little-oh function by a constant yields another little-oh function.

For $a < \infty$, we write $f(\cdot)$ is $O[g(\cdot)]$ (read "f is big-oh of g") if $f(x)/g(x)$ is bounded for all x in a neighborhood of a. For $a = \infty$, $f(\cdot)$ is $O[g(\cdot)]$ if $f(x)/g(x)$ is bounded for all sufficiently large x. In particular, $f(\cdot)$ is $O(x)$ if $\lim_{x \to \infty}[f(x)/x]$ is finite. Big-oh notation is not used in Volume II.

Convergence Concepts

The *supremum* (abbreviated *sup*) of a set is its least upper bound, and the *infinum* (abbreviated *inf*) of a set is its greatest lower bound. The sup and inf need not be members of the set; if they are members of the set, they are called the *maximum* and *minimum*, respectively. For example, the set $(0, 1]$ has an inf of 0 and a maximum of 1.

Let $f(x)$ be a function that is defined for some values of x near the point a. Let $\delta > 0$ be chosen arbitrarily and define the functions

$$\phi_a(\delta) \triangleq \sup\,\{f(x)\colon 0 < |x - a| < \delta\}$$

and

$$\psi_a(\delta) \triangleq \inf\,\{f(x)\colon 0 < |x - a| < \delta\}$$

Observe that these are monotone in δ, so they have limits (possibly infinite) as $\delta \downarrow 0$. The *limit superior* (abbreviated *lim sup*) and the *limit inferior* (abbreviated *lim inf*) of $f(\cdot)$ at point a are defined by

$$\limsup_{x \to a} f(x) \triangleq \lim_{\delta \to 0+} \phi_a(\delta)$$

and

$$\liminf_{x \to a} f(x) \triangleq \lim_{\delta \to 0+} \psi_a(\delta)$$

respectively.

The limit superior is clearly at least as large as the limit inferior. When equality holds, their common value is $\lim_{x \to a} f(x)$; when equality does not hold, the limit does not exist. A useful way to prove $\lim_{x \to a} f(x) = L$ is to establish

$$\liminf_{x \to a} f(x) \geq L \geq \limsup_{x \to a} f(x)$$

A sequence of functions $\{f_n(\cdot)\}$, each defined on a set A, *converges pointwise* to the function $f(\cdot)$ if $\lim_{n \to \infty} f_n(x) = f(x)$ for each $x \in A$. Sometimes a stronger mode of convergence of functions is required. The sequence $\{f_n(\cdot)\}$ *converges uniformly* on A to $f(\cdot)$ if for any $\epsilon > 0$ there exists a number $N = N(\epsilon)$ (which does not depend on x) such that $n > N$ implies $|f_n(x) - f(x)| < \epsilon$ for all $x \in A$. Alternatively, if $\sup \{|f_n(x) - f(x)| : x \in A\} < \epsilon$ for all $n > N$, then $f_n(\cdot)$ converges to $f(\cdot)$ uniformly.

To establish that a sum of functions converges uniformly, one may use this proposition.

Proposition A-5 If $|f_n(x)| \leq a_n$ for all $x \in A$ and $\sum_{n=0}^{\infty} a_n < \infty$, then $\sum_{n=1}^{\infty} f_n(x)$ converges uniformly (and absolutely) on A.

Uniform convergence is connected to integration and differentiation of infinite sums.

Proposition A-6 If $f_n(\cdot)$ is integrable on $[a, b]$ for each $n \in I_+$ and $f_n(\cdot)$ converges uniformly to $f(\cdot)$ on $[a, b]$, then $f(\cdot)$ is integrable on $[a, b]$ and

$$\lim_{n \to \infty} \int_a^b f_n(x)\, dx = \int_a^b \lim_{n \to \infty} f_n(x)\, dx = \int_a^b f(x)\, dx$$

In particular, choosing $f_n(x) = \sum_{i=1}^{n} g_i(x)$ yields

$$\int_a^b \sum_{i=1}^{\infty} g_i(x)\, dx = \sum_{i=1}^{\infty} \int_a^b g_i(x)\, dx \tag{A-12}$$

Proposition A-7 If (*a*) $f_n(\cdot)$ is differentiable on $[a, b]$ for each $n \in I_+$; (*b*) for some $x_0 \in [a, b]$, $\lim_{n \to \infty} f_n(x_0)$ exists; and (*c*) the first derivatives of $f_n(\cdot)$ converge uniformly on $[a, b]$, then $f_n(\cdot)$ converges uniformly on $[a, b]$ to some differentiable function $f(\cdot)$ and

$$\frac{d}{dx} f(x) = \lim_{n \to \infty} \frac{d}{dx} f_n(x)$$

Choosing $f_n(x) = \sum_{i=1}^{n} g_i(x)$ yields

$$\frac{d}{dx} \sum_{i=0}^{\infty} g_i(x) = \sum_{i=0}^{\infty} \frac{d}{dx} g_i(x)$$

whenever $f_n(\cdot)$ satisfies Proposition A-7.

Integration Theorems

The purpose of uniform convergence in Proposition A-6 is to guarantee that $f(\cdot)$ is integrable. If Proposition A-6 were relied on to justify bringing limits inside integrals, considerable effort would be expended in establishing uniform convergence. We can avoid these tasks by using a notion of integral that encompasses a larger class of functions. The Lebesgue integral is the appropriate integral for this task. Those readers familiar with Lebesgue integration can interpret the integrals in this book in that sense. *This book does not require a knowledge of Lebesgue integration.*

The integrals used in this book are of the *Riemann-Stieltjes* type. This is their definition.

Definition A-6: Riemann-Stieltjes integral Let $f(\cdot)$ and $g(\cdot)$ be defined and bounded on a closed interval $[a, b]$. Then $f(\cdot)$ is *Riemann-Stieltjes-integrable with respect to $g(\cdot)$ on* $[a, b]$, with *Riemann-Stieltjes integral* $\mathfrak{J}$, if for any $\epsilon > 0$ there exists $\delta > 0$ such that whenever

(*a*) $$a = a_0 < a_1 < \cdots < a_n = b$$

with $d \triangleq \max\{a_i - a_{i-1};\ i = 1, 2, \ldots, n\} < \delta$ and (*b*) $a_{i-1} \le x_i \le a_i$, $i = 1, 2, \ldots, n$, hold then

(*c*) $$\left| \sum_{i=1}^{n} f(x_i)[g(a_i) - g(a_{i-1})] - \mathfrak{J} \right| < \epsilon$$

Letting $\Delta g_i = g(a_i) - g(a_{i-1})$, we write (*c*) in limit notation as

$$\mathfrak{J} \triangleq \int_a^b f(x)\, dg(x) = \lim_{d \to 0} \sum_{i=1}^{n} f(x_i)\, \Delta g_i$$

Riemann-Stieltjes integrals where a or b is infinite are obtained by taking limits as $a \to -\infty$ or $b \to +\infty$. The following are the properties of Riemann-Stieltjes integrals we employ.

Proposition A-8 Riemann-Stieltjes integrals possess these properties:

(*a*) If $\int_a^b f(x)\, dg(x)$ exists and $g(\cdot)$ has derivative $g'(\cdot)$, then

$$\int_a^b f(x)\, dg(x) = \int_a^b f(x) g'(x)\, dx$$

(*b*) If $f(\cdot)$ is integrable with respect to $g(\cdot)$ on $[a, b]$, then $g(\cdot)$ is integrable with respect to $f(\cdot)$ on $[a, b]$, and the *integration-by-parts formula*

$$\int_a^b f(x)\, dg(x) + \int_a^b g(x)\, df(x) = f(b)g(b) - f(a)g(a)$$

is valid.

(*c*) If, on the interval $[a, b]$, one of the functions $f(\cdot)$ and $g(\cdot)$ is continuous and the other is monotonic, then $\int_a^b f(x)\,dg(x)$ exists.

(*d*) If $f(\cdot)$ is continuous and $g(\cdot)$ is monotonic on $[a, b]$, then there exists a point y, $a \le y \le b$, such that

$$\int_a^b f(x)\,dg(x) = f(y)[g(b) - g(a)]$$

This is called the *first mean-value theorem for Riemann-Stieltjes integrals.*

The problem with using Riemann-Stieltjes integrals is that $\int_a^b f(x)\,dg(x)$ is not defined if $f(\cdot)$ and $g(\cdot)$ both have a jump at the same point. To circumvent this problem, we adopt the following convention throughout this book.

Convention If $f(\cdot)$ and $g(\cdot)$ both have a jump at y, $a < y < b$, then

$$\int_a^b f(x)\,dg(x) = \int_a^{y-} f(x)\,dg(x) + f(y)[g(y^+) - g(y^-)] + \int_{y+}^b f(x)\,dg(x) \quad \text{(A-13)}$$

The notations $\int_a^{y-} f(x)\,dg(x)$ and $\int_{y+}^b f(x)\,dg(x)$ mean that the contribution at y is excluded. More precisely, in part (*c*) of Proposition A-8, we take $g(a_n) = \lim_{u \uparrow y} g(u)$ in the former integral and $g(a_0) = \lim_{u \downarrow y} g(u)$ in the latter integral. The term $g(y^+) - g(y^-)$ is the value of the jump at y.

The convention is extended to a countable number of jumps by iteration. With this convention, if $f(\cdot)$ and $g(\cdot)$ are step functions with common jump points, the integral becomes a sum. This means that we do not have to distinguish sums from integrals. More important, if $g(\cdot)$ is nonnegative and nondecreasing (e.g., if it is a d.f.), then the integral in (A-13) agrees with the Lebesgue integral, and the powerful theorems of that theory apply.

Proposition A-9: Monotone convergence theorem Assume (*a*) $f_n(\cdot) \ge 0$ for each $n \in I_+$, (*b*) $f_n(x) \uparrow f(x)$ pointwise as $n \to \infty$, and (*c*) $g(\cdot) \ge 0$ and nondecreasing. Then

$$\lim_{n \to \infty} \int_a^b f_n(x)\,dg(x) = \int_a^b f(x)\,dg(x)$$

whenever $0 \le a \le b$.

Choosing $f_n(x) = \sum_{i=1}^n g_i(x)$ and $g(x) = x$ establishes (A-12) for nonnegative functions.

Here is another theorem of this type.

Proposition A-10: Dominated convergence theorem Suppose that for each† $x \in [a, b]$, $\lim_{t \to s} f(t, x) = f(s, x)$ and $|f(t, x)| \le h(x)$ for each x and some

† This theorem remains true for the intervals $(-\infty, b]$, $[a, \infty)$, and $(-\infty, \infty)$.

function $h(\cdot)$. Then if $\int_a^b h(x)\,dx$ exists and $g(\cdot) \geq 0$ and is nondecreasing, then as $t \to s$,

$$\int_a^b f(t, x)\,dg(x) \to \int_a^b f(s, x)\,dg(x)$$

In particular, if $\partial f(t, x)/\partial t$ exists at $t = s$ and is bounded by some function $h(x)$ which is integrable, then

$$\frac{d}{dt}\int_a^b f(t, x)\,dx = \int_a^b \frac{\partial f(t, x)}{\partial t}\,dx$$

When the roles of $f(\cdot, \cdot)$ and $g(\cdot)$ are interchanged, this is the corresponding theorem.

Proposition A-11: Helley-Bray theorem Let $g(\cdot)$ be continuous and $F_n(\cdot)$ be a d.f. for each $n \in I$. If $F_n(\cdot) \to F(\cdot)$ pointwise and $F(\cdot)$ is a d.f., then

$$\lim_{n\to\infty} \int_a^b g(x)\,dF_n(x) = \int_a^b g(x)\,dF(x)$$

whenever $-\infty < a < b < \infty$. If $g(\cdot)$ is bounded, then

$$\lim_{n\to\infty} \int_{-\infty}^{\infty} g(x)\,dF_n(x) = \int_{-\infty}^{\infty} g(x)\,dF(x)$$

The interchange of the order of integration is justified by the next proposition.

Proposition A-12: Fubini's theorem If $f(x, y) \geq 0$ for all (x, y) in its domain and $g_1(\cdot)$ and $g_2(\cdot)$ are nonnegative and nondecreasing, then

$$\int_{-\infty}^{\infty}\left[\int_{-\infty}^{\infty} f(x, y)\,dg_2(y)\right] dg_1(x) = \int_{-\infty}^{\infty}\left[\int_{-\infty}^{\infty} f(x, y)\,dg_1(x)\right] dg_2(y) \qquad \text{(A-14)}$$

If $f(x, y) < 0$ for some (x, y), (A-14) holds if the integrals are absolutely convergent.

In many applications of Fubini's theorem, $g_1(\cdot)$ and $g_2(\cdot)$ are d.f.'s; then (A-14) holds whenever $f(\cdot, \cdot)$ is bounded.

This is the formula for differentiating a definite integral.

Proposition A-13: Leibnitz' rule Let

$$F(x) = \int_{a(x)}^{b(x)} f(x, y)\,dy$$

If $f(x, y)$ has a continuous derivative with respect to x in the region

$c \le x \le d$, if $a(x) \le y \le b(x)$, and if $a(\cdot)$ and $b(\cdot)$ are differentiable, then

$$\frac{d}{dx} F(x) = \int_{a(x)}^{b(x)} \frac{\partial f(x, y)}{\partial x} \, dy + f[x, b(x)] \frac{db(x)}{dx} - f[x, a(x)] \frac{da(x)}{dx}$$

whenever $c \le x \le d$.

In particular,

$$\frac{d}{dx} \int_a^b f(x, y) \, dy = \int_a^b \frac{\partial f(x, y)}{\partial x} \, dy$$

This equation is valid when $b = \infty$ or $a = -\infty$ provided the right-side is finite.

Taylor's Series

A Taylor's series expansion of a function is often useful in obtaining asymptotic behavior.

Proposition A-14: Taylor's theorem Let $f(y)$ be a function possessing $n + 1$ continuous derivatives for all y between a and x inclusive. Then

$$f(x) = f(a) + f^{(1)}(a)(x - a) + \frac{f^{(2)}(a)(x - a)^2}{2!} + \cdots + \frac{f^{(n)}(x)(x - a)^n}{n!} + R_n(x, a) \quad \text{(A-15)}$$

where

$$R_n(x, a) = \int_a^x \frac{(x - y)^n}{n!} f^{(n+1)}(y) \, dy$$

and

$$f^{(n)}(y) \triangleq \frac{d^n}{dy^n} f(y)$$

The mean-value theorem for integrals may be employed to obtain Lagrange's form of the remainder:

$$R_n(x, a) = \frac{f^{(n+1)}(t)(x - a)^{n+1}}{(n + 1)!} \quad \text{(A-16)}$$

for some t between a and x.

When $a = 0$, (A-15) is called a *Maclaurin series.*

The next proposition is a form of Taylor's theorem for a real-valued function of m variables.

Proposition A-15 Let $f(\cdot)$ be a twice differentiable function on $\mathbb{R}^m$. Then for all $\mathbf{x}$ and $\mathbf{y}$ in $\mathbb{R}^m$ there is a number θ such that $0 \leq \theta \leq 1$ and

$$f(\mathbf{y}) = f(\mathbf{x}) + \sum_{i=1}^{m} (y_i - x_i) \frac{\partial f(\mathbf{x})}{\partial x_i} + \frac{1}{2} \sum_{i=1}^{m} \sum_{j=1}^{m} (y_i - x_i)(y_j - x_j) \frac{\partial^2 f[\theta \mathbf{x} + (1 - \theta)\mathbf{y}]}{\partial x_i \partial y_j}$$

BIBLIOGRAPHIC GUIDE

References

There are many excellent books on probability theory. Included among those that do not use measure theory are the following:

Clark, A. Bruce, and Ralph L. Disney: *Probability and Random Processes for Engineers and Scientists*, Wiley, New York (1970).

Feller, William: *An Introduction to Probability Theory and Its Applications*, vol. 1, 3d ed., Wiley, New York (1968).

Neuts, Marcel F.: *Probability*, Allyn and Bacon, Boston (1973).

Parzen, Emanuel: *Modern Probability Theory and Its Applications*, Wiley, New York (1960).

Ross, Sheldon M.: *A First Course in Probability*, Macmillan, New York (1976).

Among the probability books that use measure theory are:

Breiman, Leo: *Probability*, Addison-Wesley, Reading, Mass. (1968).

Chung, Kai Lai: *A Course in Probability Theory*, 2d ed., Academic, New York (1974).

Feller, William: *An Introduction to Probability Theory and Its Applications*, vol. 2, 2d ed., Wiley, New York (1971).

Loève, Michel: *Probability Theory*, 3d ed., Van Nostrand, Princeton, N.J. (1963).

Books on mathematical analysis that do not treat measure theory include:

Apostle, Tom M.: *Mathematical Analysis*, 2d ed., Addison-Wesley, Reading, Mass. (1974).

Bartle, Robert G.: *Elements of Real Analysis*, 2d ed., Wiley, New York (1976).

Olmsted, John M. H.: *Advanced Calculus*, Appleton-Century-Crofts, New York (1961).

Rudin, Walter: *Principles of Mathematical Analysis*, 2d ed., McGraw-Hill, New York (1964).

Books on measure theory and integration include:

Bartle, Robert G.: *The Elements of Integration*, Wiley, New York (1966).

Halmos, Paul R.: *Measure Theory*, Van Nostrand, Princeton, N.J. (1950).

Royden, H. L.: *Real Analysis*, Macmillan, New York (1963).

APPENDIX B

CONVEXITY

Convex sets and convex functions are encountered frequently in the analysis of operations research models. This appendix presents some of the basic definitions and properties of convex sets and functions. Much of this material relates convexity to optimization.

B-1 DEFINITIONS

Definition B-1 $X \subset \mathbb{R}^n$ is said to be a *convex set* if $\lambda \mathbf{x} + (1 - \lambda)\mathbf{y} \in X$ whenever $\mathbf{x} \in X$, $\mathbf{y} \in X$, and $0 < \lambda < 1$.

The set† of points $\{\lambda \mathbf{x} + (1 - \lambda)\mathbf{y}: 0 \leq \lambda \leq 1\}$ is the line segment joining $\mathbf{x}$ and $\mathbf{y}$. So a set is convex if it contains as subsets all line segments joining pairs of its points.

Example B-1 For all numbers b, the following half-lines are convex sets: $(-\infty, b)$, $(-\infty, b]$, $[b, \infty)$, and (b, ∞). □

† The notion of convexity applies to more general vector spaces than $\mathbb{R}^n$, but $\mathbb{R}^n$ is sufficient in this book.

Example B-2 For all numbers a and b with $a \le b$, the following intervals are convex sets: (a, b), $[a, b)$, $(a, b]$, and $[a, b]$. The statement remains true in $\mathbb{R}^n$ if $\mathbf{a} \le \mathbf{b}$ means that $a_i \le b_i$ for all $i = 1, \ldots, n$ where $\mathbf{a} = (a_i)$ and $\mathbf{b} = (b_i)$. □

Example B-3 For all $\mathbf{a} \in \mathbb{R}^n$ and $b \in \mathbb{R}$, the set $\{\mathbf{x}: \mathbf{x} \in \mathbb{R}^n, \mathbf{a} \cdot \mathbf{x} \le b\}$ is convex. □

Example B-4 The perimeter of a circle is not convex (if the radius is positive). □

Proposition B-1 Let K be a nonempty index set and $X_k \subset \mathbb{R}^n$ be a convex set for each $k \in K$. Let

$$W = \bigcap_{k \in K} X_k$$

Then W is a convex set.

Proof Pick $\mathbf{x}$ and $\mathbf{y}$ in W and $\lambda \in (0, 1)$. Then $\lambda\mathbf{x} + (1 - \lambda)\mathbf{y} \in X_k$ for each k because X_k is a convex and $\mathbf{x}$ and $\mathbf{y}$ in W implies $\mathbf{x}$ and $\mathbf{y}$ in X_k. Therefore, $\lambda\mathbf{x} + (1 - \lambda)\mathbf{y} \in W$. □

But unions of convex sets generally are not convex.

Exercise B-1 asks you to prove that the two-point definition of a convex set is equivalent to the following definition based on any finite number of points.

Definition B-1′ $X \subset \mathbb{R}^n$ is a *convex set* if $m \in I_+$, $\mathbf{x}_1 \in X, \mathbf{x}_2 \in X, \ldots, \mathbf{x}_m \in X$, and $\lambda_1 \ge 0, \lambda_2 \ge 0, \ldots, \lambda_m \ge 0$ with $\sum_{i=1}^m \lambda_i = 1$ implies $\sum_{i=1}^m \lambda_i \mathbf{x}_i \in X$.

The *convex hull* of a set is the intersection of all the convex sets which contain it as a subset. If X is the intersection of finitely many half-spaces, it is a convex set and called a *convex polyhedron* (or *convex polygon* or *convex polytope*). The sum $\sum_{i=1}^m \lambda_i \mathbf{x}_i$ in Definition B-1′ is called a *convex combination.* Exercise B-2 asks you to show that the convex hull of a set X is the set of all convex combinations of points from X. Hence, a trivial consequence of Definition B-1′ is that X is a convex set if, and only if, it coincides with its own convex hull. Also, X is a closed convex polyhedron if, and only if, it is the convex hull of a finite set of points.

Example B-5 For all $m \times n$ matrices A and vectors $\mathbf{b} \in \mathbb{R}^m$, the set $\{\mathbf{x}: \mathbf{x} \in \mathbb{R}^n, A\mathbf{x} \le \mathbf{b}\}$ is a convex set. □

Example B-6 For all $a \le b$, $(-\infty, a) \cup (b, \infty)$ is not a convex subset of $\mathbb{R}$. □

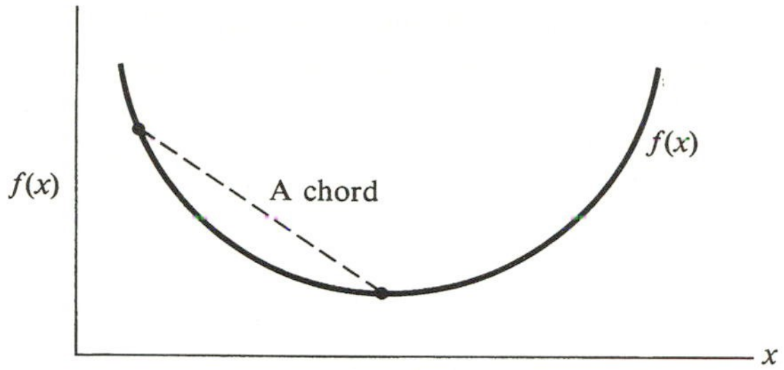

Figure B-1 A convex function does not lie above any chord connecting two of its points.

Definition B-2 Let f be a real-valued function defined on a convex subset X of $\mathbb{R}^n$. Then f is a *convex function* if for every $\mathbf{x} \in X$, $\mathbf{y} \in X$, and $\lambda \in [0, 1]$,

$$f[\lambda \mathbf{x} + (1 - \lambda)\mathbf{y}] \leq \lambda f(\mathbf{x}) + (1 - \lambda) f(\mathbf{y}) \tag{B-1}$$

The set of points $\{\lambda f(\mathbf{x}) + (1 - \lambda) f(\mathbf{y}) \colon 0 \leq \lambda \leq 1\}$ is the line segment that connects $f(\mathbf{x})$ with $f(\mathbf{y})$; and $\{f[\lambda \mathbf{x} + (1 - \lambda)\mathbf{y}] \colon 0 \leq \lambda \leq 1\}$ are the values taken by $f(\cdot)$ on the line segment that connects $\mathbf{x}$ with $\mathbf{y}$. So $f(\cdot)$ is convex if the line segment that connects any two points on the graph of $f(\cdot)$ is never below $f(\cdot)$ itself between the two end points. Figure B-1 illustrates the idea. In other words, linear interpolation never underestimates the actual value of a convex function. The multipoint inequality

$$f\left(\sum_{i=1}^{m} \lambda_i \mathbf{x}_i\right) \leq \sum_{i=1}^{m} \lambda_i f(\mathbf{x}_i) \tag{B-2}$$

where $\lambda_i \geq 0$ for all i, $\sum_{i=1}^{m} \lambda_i = 1$, and $\mathbf{x}_i \in X$ for all i, is implied by (B-1) and called Jensen's inequality (Exercise B-3 asks you to prove the inequality).†

Our definition of a convex function precludes function values of $+\infty$ and $-\infty$. It is straightforward to develop the theory with $+\infty$ values permitted, but some proofs become slightly more elaborate and various propositions become restricted to the subset of the domain where the function is finite. We choose to suppress these nuisances and allow only real-valued functions.

A linear function $f(\cdot)$ satisfies (B-1) with an equality everywhere on its domain. The other extreme is a function that satisfies (B-1) with the inequality strict for all $\lambda \in (0, 1)$ and $\mathbf{x} \neq \mathbf{y}$. Such a function is *strictly convex*.

We say that a function $f(\cdot)$ is *concave* (*strictly concave*) if $-f(\cdot)$ is convex (strictly convex). The notions of convexity of functions and convexity of sets are intimately related via Proposition B-8 later in this appendix. However, there is no concept of concave *sets* that corresponds to concave functions.

Example B-7 x^2 is convex and strictly convex on $\mathbb{R}$ and $-x^2$ is concave and strictly concave on $\mathbb{R}$. □

† Actually, Jensen's inequality concerns more general sums and limits of such sums. The general result is that, if $f(\cdot)$ is a convex function on $\mathbb{R}$ and X is an r.v. with $E(X) < \infty$, then $f[E(X)] \leq E[f(X)]$.

Example B-8 Let $\delta(\cdot)$ denote the *Heavyside function:* $\delta(x) = 0$ if $x \le 0$ and $\delta(x) = 1$ if $x > 0$. In this example, we restrict the domain of $\delta(\cdot)$ to $\mathbb{R}_+$. If $H \ge 0$, then $H\delta(x) + cx$ is concave on $\mathbb{R}_+$. If $H \le 0$, then $H\delta(x) + cx$ is convex on $\mathbb{R}_+$. In neither case is the function continuous if $H \ne 0$. However, the function is continuous on the interior of its domain. This is true in general. □

Often we take sums of convex functions and, more generally, expected values. Fortunately, the resulting functions are again convex. Exercise B-4 asks you to prove the next result.

Proposition B-2 Let K be a nonempty index set, let X be a convex set, and for each $k \in K$ let $f_k(\cdot)$ be a convex function on X and let $p_k \ge 0$. Then

$$\sum_{k \in K} p_k f_k(\mathbf{x}) \qquad \mathbf{x} \in X$$

is a convex function on any convex subset of X, where the sum takes finite values.

Example B-9 Let $f(\cdot)$ be a convex function on $\mathbb{R}$ and D a random variable for which $g(x) = E[f(x - D)]$ exists and is finite for all $x \in \mathbb{R}$. Then $g(\cdot)$ is a convex function on $\mathbb{R}$.† □

EXERCISES

B-1 Prove the equivalence of Definitions B-1 and B-1′.

B-2 Prove that the convex hull of a set is the set of all convex combinations of points in the set.

B-3: Jensen's inequality Let f be a real-valued function on a convex set $X \subset \mathbb{R}^n$. Prove that f is convex on X if, and only if,

$$f\left(\sum_{j=1}^{r} \lambda_j \mathbf{x}_j\right) \le \sum_{j=1}^{r} \lambda_j f(\mathbf{x}_j)$$

for every $r \in I_+$, $\mathbf{x}_1$ through $\mathbf{x}_r$ in X, and $\lambda_1 \ge 0, \ldots, \lambda_r \ge 0$ with $\sum_{j=1}^{r} \lambda_j = 1$.

B-4 Prove Proposition B-2.

B-2 MINIMIZATION AND DIRECTIONAL DERIVATIVES

A point $\mathbf{x}$ in a convex set X is called an *extreme point* of X if it does not lie on any line segment connecting two other points of X, that is, if there do not exist $\mathbf{y}$ and $\mathbf{z}$ in X and $0 < \lambda < 1$ such that $\mathbf{x} = \lambda\mathbf{y} + (1 - \lambda)\mathbf{z}$. If X is an open set, it cannot contain extreme points. To fix the idea geometrically, the extreme points

† Convexity of $g(\cdot)$ follows from (*a*) our Definition A-6 of the Riemann-Stieltjes integral and (*b*) the fact that if a sequence $f_1(\cdot), f_2(\cdot), \ldots$ of convex functions on $\mathbb{R}$ converges pointwise on $\mathbb{R}$ to $f(\cdot)$ with $|f(x)| < \infty$ for all x, then $f(\cdot)$ is convex on $\mathbb{R}$.

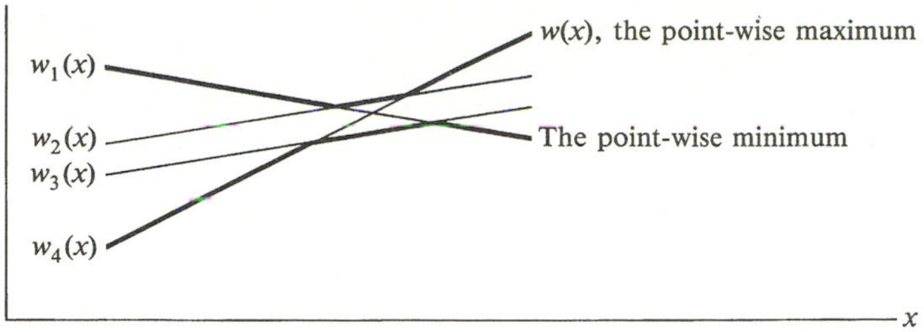

Figure B-2 The point-wise maximum (minimum) of convex (concave) functions is a convex (concave) function.

of a square are its four corner points. Part (*c*) of the next proposition relates maximization of convex functions to extreme points.

Proposition B-3 Let f be a convex function on a convex set X and let K index a collection of convex functions $\{w_k(\cdot)\}$ on X. Let

$$g = \inf\{f(\mathbf{x}): \mathbf{x} \in X\} \qquad h = \sup\{f(\mathbf{x}): \mathbf{x} \in X\}$$

$$X^* = \{\mathbf{x}: \mathbf{x} \in X, g = f(\mathbf{x})\}$$

$$W(\mathbf{x}) = \sup\{w_k(\mathbf{x}): k \in K\} \qquad \mathbf{x} \in X$$

Then:

(*a*) X^* is a convex set.
(*b*) $W(\cdot)$ is a convex function on the convex set $\{\mathbf{x}: W(\mathbf{x}) < \infty, \mathbf{x} \in X\}$.
(*c*) If X is closed and bounded, $h < \infty$, and the supremum is attained, then there is an extreme point $\mathbf{x}$ of X with $h = f(\mathbf{x})$.

If the w_k are concave and "sup" is replaced by "inf," then $W(\cdot)$ is concave. Exercise B-5 asks you to prove parts (*a*) and (*b*).†

Example B-10 Let $X = \mathbb{R}$ and $w_k(x) = a_k + b_k x$. A line is both convex and concave; so (*b*) implies that the pointwise maximum is convex and the pointwise minimum is concave. See Figure B-2 for an illustration. □

The following result is used many times in this book.

Proposition B-4 Let X be a nonempty set with $A_{\mathbf{x}}$ a nonempty set for each $\mathbf{x} \in X$. Let $C = \{(\mathbf{x}, \mathbf{y}): \mathbf{y} \in A_{\mathbf{x}}, \mathbf{x} \in X\}$, let J be a real-valued function on C, and define

$$f(\mathbf{x}) = \inf\{J(\mathbf{x}, \mathbf{y}): \mathbf{y} \in A_{\mathbf{x}}\} \qquad \mathbf{x} \in X \tag{B-3}$$

If C is a convex set and J is a convex function on C, then f is a convex function on any convex subset of $X^* = \{\mathbf{x}: \mathbf{x} \in X, f(\mathbf{x}) > -\infty\}$.

† For a proof of part (*c*), see pages 91–92 in G. Hadley: *Nonlinear and Dynamic Programming*, Addison Wesley, Reading, Mass. (1964).

PROOF Pick $\mathbf{x}_1$ and $\mathbf{x}_2$ in X^* so $f(\mathbf{x}_1) > -\infty < f(\mathbf{x}_2)$. Then for all $\gamma > 0$ there are $\mathbf{y}_1$ and $\mathbf{y}_2$ with $(\mathbf{x}_i, \mathbf{y}_i) \in C$, $i = 1$ and 2, such that $f(\mathbf{x}_i) + \gamma > J(\mathbf{x}_i, \mathbf{y}_i)$. Pick $\lambda \in (0, 1)$ and let $(\mathbf{x}, \mathbf{y}) = \lambda(\mathbf{x}_1, \mathbf{y}_1) + (1 - \lambda)(\mathbf{x}_2, \mathbf{y}_2)$, which is in C because C is convex. Now

$$\begin{aligned}\lambda f(\mathbf{x}_1) + (1 - \lambda) f(\mathbf{x}_2) &\geq \lambda J(\mathbf{x}_1, \mathbf{y}_1) + (1 - \lambda) J(\mathbf{x}_2, \mathbf{y}_2) - \gamma \\ &\geq J(\mathbf{x}, \mathbf{y}) - \gamma \geq f(\mathbf{x}) - \gamma\end{aligned}$$

with the second inequality due to convexity of J on C. Letting $\gamma \to 0$ yields convexity of f. □

Example B-11 In some inventory and production models, $X = \mathbb{R}$, $A_x = [x, \infty)$ for $x \in \mathbb{R}$, and $C = \{(x, y): y \geq x, x \in \mathbb{R}\}$; so C and X are convex sets. Therefore, $f(\cdot)$ is convex on $\mathbb{R}$ if $J(\cdot, \cdot)$ is convex on C. □

You know now that nonnegative sums of convex functions are convex and that minimization preserves convexity. But you lack analytical tests of convexity. These are particularly necessary for convex functions on sets in $\mathbb{R}^n$ with $n > 1$.

Shortly we shall present necessary and sufficient conditions for a differentiable function to be convex. The basic ideas, however, depend on the monotonicity of functions related to directional derivatives rather than on differentiability. Let f be a real-valued function on $X \subset \mathbb{R}^n$ and define *directional differences*

$$\begin{aligned} d_{\mathbf{x}}^{+}(h, \mathbf{r}) &= \frac{f(\mathbf{x} + h\mathbf{r}) - f(x)}{h} \\ d_{\mathbf{x}}^{-}(h, \mathbf{r}) &= \frac{f(\mathbf{x}) - f(\mathbf{x} - h\mathbf{r})}{h} \end{aligned} \tag{B-4}$$

where $\mathbf{x} \in X$, $0 < h$ is a number, and $\mathbf{r} \in \mathbb{R}^n$ such that $\mathbf{x} + h\mathbf{r} \in X$ and $\mathbf{x} - h\mathbf{r} \in X$.

Proposition B-5 Let f be a convex function on a convex set $X \subset \mathbb{R}^n$, $0 < g \leq h$, and $0 < m \leq s$. Then

$$d_{\mathbf{x}}^{-}(h, \mathbf{r}) \leq d_{\mathbf{x}}^{-}(g, \mathbf{r}) \leq d_{\mathbf{x}}^{+}(m, \mathbf{r}) \leq d_{\mathbf{x}}^{+}(s, \mathbf{r})$$

at each $\mathbf{x} \in X$ and $\mathbf{r}$ for which $\mathbf{x} - h\mathbf{r} \in X$ and $\mathbf{x} + s\mathbf{r} \in X$.

PROOF Take $\mathbf{x}$ and $\mathbf{y}$ in X and $\lambda \in (0, 1]$. Since f is convex,

$$0 \leq \lambda f(\mathbf{y}) + (1 - \lambda) f(\mathbf{x}) - f[\lambda \mathbf{y} + (1 - \lambda)\mathbf{x}]$$

or

$$\lambda[f(\mathbf{x}) - f(\mathbf{y})] \leq f(\mathbf{x}) - f[\lambda \mathbf{y} + (1 - \lambda)\mathbf{x}]$$

Let $h\mathbf{r} = \mathbf{x} - \mathbf{y}$ and $0 < g \leq h$ so $\lambda = g/h \in (0, 1]$. Then the preceding inequality asserts

$$\frac{g}{h}[f(\mathbf{x}) - f(\mathbf{x} - h\mathbf{r})] \leq f(\mathbf{x}) - f(\mathbf{x} - g\mathbf{r})$$

so
$$\frac{f(\mathbf{x}) - f(\mathbf{x} - h\mathbf{r})}{h} \leq \frac{f(\mathbf{x}) - f(\mathbf{x} - g\mathbf{r})}{g}$$

which is $d_{\mathbf{x}}^{-}(h, \mathbf{r}) \leq d_{\mathbf{x}}^{-}(g, \mathbf{r})$. The other inequalities are obtained similarly. □

Suppose X in Proposition B-5 is an open set so for each $\mathbf{x} \in X$ and $\mathbf{r} \in \mathbb{R}^n$ there is $\lambda > 0$ such that $\mathbf{x} + \lambda\mathbf{r} \in X$ and $\mathbf{x} - \lambda\mathbf{r} \in X$. Then the proposition implies that $d_{\mathbf{x}}^{-}(\cdot, \mathbf{r})$ and $d_{\mathbf{x}}^{+}(\cdot, \mathbf{r})$ are bounded monotone functions on $(0, \lambda)$ so their limits at 0 must exist. Let

$$\theta^{-}(\mathbf{x}, \mathbf{r}) = \lim_{h \downarrow 0} d_{\mathbf{x}}^{-}(h, \mathbf{r}) \qquad \theta^{+}(\mathbf{x}, \mathbf{r}) = \lim_{h \downarrow 0} d_{\mathbf{x}}^{+}(h, \mathbf{r}) \tag{B-5}$$

The proposition implies $\theta^{-}(\mathbf{x}, \mathbf{r}) \leq \theta^{+}(\mathbf{x}, \mathbf{r})$. These are called the left- and right-hand directional derivatives, respectively. *The left- and right-hand directional derivatives of a convex function exist everywhere except (possibly) on the boundary of its domain. Also, the right-hand one is always at least as great as the left-hand one.* As you might imagine, $f(\cdot)$ is differentiable at $\mathbf{x}$ if $\theta^{-}(\mathbf{x}, \mathbf{r}) = \theta^{+}(\mathbf{x}, \mathbf{r})$ for all directions $\mathbf{r}$. An important motivation for a treatment of convexity without differentiability is that piecewise linear functions are used frequently in practice but they are not differentiable at their breakpoints. See Figure B-3.

In most of the book we assume that integrals exist wherever we need them. The same is not true of differentiability. Suppose $f(\cdot)$ is a convex function on $\mathbb{R}$ and D is an r.v. for which $w(x) = E[f(x - D)]$ exists, $x \in \mathbb{R}$. Let $\theta_w^{+}(x)$ and $\theta_f^{+}(x)$ denote $\theta^{+}(x, 1)$ for w and f, respectively. Then

$$\theta_w^{+}(x) = E[\theta_f^{+}(x - D)]$$

i.e., differentiation under the integral, is justified by the monotonicity in h of $[f(x + h) - f(x)]/h$ and the monotone convergence theorem (Proposition A-9).

Several of the most useful properties of convex functions stem from the following corollary of Proposition B-5.

Proposition B-6 Let f be a convex function on a convex set $X \subset \mathbb{R}^n$. Then

$$f(\mathbf{y}) - f(\mathbf{x}) \geq d_{\mathbf{x}}^{+}(m, \mathbf{y} - \mathbf{x}) \geq d_{\mathbf{x}}^{-}(g, \mathbf{y} - \mathbf{x}) \geq f(\mathbf{x}) - f(\mathbf{y}) \tag{B-6}$$

for all $\mathbf{x}$ and $\mathbf{y}$ in X, $m \in (0, 1]$, and $g \in (0, 1]$ for which $\mathbf{x} - g \cdot (\mathbf{y} - \mathbf{x}) \in X$.

PROOF In Proposition B-5, take $h = s = 1$ and $\mathbf{r} = \mathbf{y} - \mathbf{x}$. □

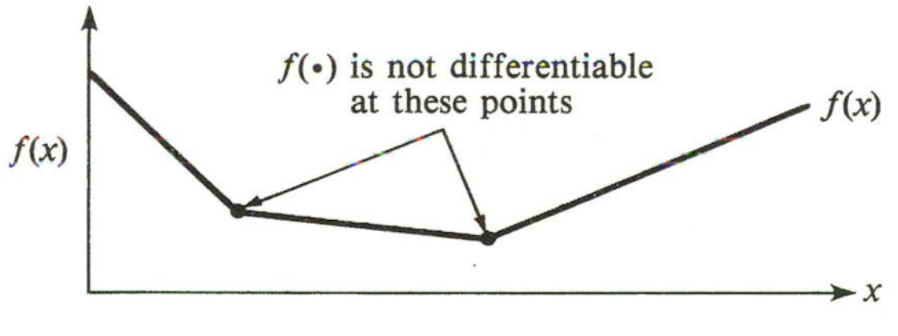

Figure B-3 A piece-wise linear convex function.

Proposition B-5 implies that a continuously differentiable convex function has zero derivatives at a point if, and only if, it is a global minimum. In other words, convex functions do not have inflection points. The basic idea has nothing to do with differentiability. First, we observe that Proposition B-6 implies

$$f(\mathbf{y}) - f(\mathbf{x}) \geq \theta^+(\mathbf{x}, \mathbf{y} - \mathbf{x}) \geq \theta^-(\mathbf{x}, \mathbf{y} - \mathbf{x}) \tag{B-7}$$

for all $\mathbf{x}$ and $\mathbf{y}$ in X. Let $V(\mathbf{x})$ denote the *set of feasible directions at* $\mathbf{x}$, that is,

$$V(\mathbf{x}) \triangleq \{\mathbf{r}: \mathbf{r} \in \mathbb{R}^n,\ \mathbf{x} + h\mathbf{r} \in X \text{ for some number } h > 0\},\ \mathbf{x} \in X$$

Proposition B-7 Let f be a convex function on a convex set $X \subset \mathbb{R}^n$ and let

$$g = \inf \{f(\mathbf{x}): \mathbf{x} \in X\}$$

For any $\mathbf{x}^* \in X$, $g = f(\mathbf{x}^*)$ if, and only if,

$$\theta^+(\mathbf{x}^*, \mathbf{r}) \geq 0 \qquad \text{for all } \mathbf{r} \in V(\mathbf{x}^*) \tag{B-8a}$$

If $\mathbf{x}^*$ is not on the boundary of X, then $g = f(\mathbf{x}^*)$ if, and only if,

$$\theta^-(\mathbf{x}^*, \mathbf{r}) \leq 0 \leq \theta^+(\mathbf{x}^*, \mathbf{r}) \qquad \text{for all } \mathbf{r} \in \mathbb{R}^n \tag{B-8b}$$

PROOF Suppose $g = f(\mathbf{x}^*)$. For all $\mathbf{r} \in V(\mathbf{x}^*)$ there is $h^* > 0$ with $\mathbf{x}^* + h\mathbf{r} \in X$ for all $h \in (0, h^*)$ (because X is a convex set). Now $g = f(\mathbf{x}^*) \leq f(\mathbf{y})$ for all $\mathbf{y} \in X$ so $f(\mathbf{x}^*) \leq f(\mathbf{x}^* + h\mathbf{r})$ and $d^+_{\mathbf{x}^*}(h, \mathbf{r}) \geq 0$ for all $h \in (0, h^*)$. Taking the limit as $h \downarrow 0$ yields (B-8a). On the other hand, suppose (B-8a) is true. Pick any $\mathbf{y} \in X$. Then $\mathbf{y} - \mathbf{x}^* \in V(\mathbf{x}^*)$ and

$$f(\mathbf{y}) - f(\mathbf{x}^*) = d^+_{\mathbf{x}^*}(1, \mathbf{y} - \mathbf{x}^*) \geq \theta^+(\mathbf{x}^*, \mathbf{y} - \mathbf{x}^*) \geq 0$$

with the first inequality due to Proposition B-6 and the second due to (B-8a).

If $\mathbf{x}^*$ is not on the boundary of X, then $V(\mathbf{x}^*) = \mathbb{R}^n$; so $\mathbf{r} \in V(\mathbf{x}^*)$ and $-\mathbf{r} \in V(\mathbf{x}^*)$ for every $\mathbf{r} \in \mathbb{R}^n$. But $d^+_{\mathbf{x}^*}(h, -\mathbf{r}) = -d^-_{\mathbf{x}^*}(h, \mathbf{r})$; so $\theta^+(\mathbf{x}^*, -\mathbf{r}) \geq 0$ if, and only if, $\theta^-(\mathbf{x}^*, \mathbf{r}) \leq 0$. □

Note that the hypothesis of (B-8b) is necessarily true if g is attained on an open set X. The importance of convex (and concave) functions to optimization theory rests on this proposition. It asserts that a point is a local optimal if, and only if, it is a global optimum.

EXERCISES

B-5 Prove parts (*a*) and (*b*) of Proposition B-3.

B-6 Prove part (*c*) of Proposition B-3 under the simplifying assumption that X is a polyhedron.

B-7 Suppose f is convex on $\mathbb{R}$ and D is an r.v. Let

$$v(y) = E[f(y - D)]$$

and suppose $|v(y)| < \infty$, $y \in \mathbb{R}$.

(*a*) Suppose D has a density function $\phi(\cdot)$ so

$$P\{D \leq a\} = \int_{-\infty}^{a} \phi(u)\, du$$

for all $a \in \mathbb{R}$. Carefully prove that $v(\cdot)$ is convex on $\mathbb{R}$.

(*b*) Suppose D is neither a discrete r.v. nor one with a density function. Carefully prove that $v(\cdot)$ is convex on $\mathbb{R}$.

B-8 Let f be a convex function on an open convex set $X \subset \mathbb{R}^n$. Then f is continuous on X. Give a counterexample if X is not an open set.

B-9 (Continuation) Prove the assertion, *in detail*, for $n = 1$.

B-10 Let $h = f \cdot g$ be the composition of two functions g: $\mathbb{R}^n \to \mathbb{R}^m$ and f: $\mathbb{R}^m \to \mathbb{R}$. Verify the four entries in the following table that stipulate conditions under which h is convex or concave.

	g	
f	Concave	Convex
Concave	h is concave if f is nondecreasing	h is concave if f is nonincreasing
Convex	h is convex if f is nonincreasing	h is convex if f is nondecreasing

B-3 NECESSARY AND SUFFICIENT CONDITIONS

This section focuses on conditions under which functions are convex. First we relate convex sets to convex functions. The connection between convex sets and convex functions is via epigraphs.

Definition B-3 Let f be a real-valued function on $X \subset \mathbb{R}^n$. The *epigraph* and *hypograph* of $f(\cdot)$ are

$$\text{epi } f = \{(\mathbf{x}, a)\colon \mathbf{x} \in X,\ a \in \mathbb{R}, f(\mathbf{x}) \leq a\} \subset \mathbb{R}^{n+1}$$

$$\text{hypo } f = \{(\mathbf{x}, a)\colon \mathbf{x} \in X,\ a \in \mathbb{R}, f(\mathbf{x}) \geq a\} \subset \mathbb{R}^{n+1}$$

Notice that epi $f \subset \mathbb{R}^{n+1}$ when $X \subset \mathbb{R}^n$. The epigraph of f consists of the graph of f (that is, $\{[\mathbf{x}, f(\mathbf{x})]\colon x \in X\}$) and the area above the graph. Figure B-4 shows the epigraph of a convex function on the interval $[0, 1]$.

Proposition B-8 Let $X \subset \mathbb{R}^n$ be convex and let $f(\cdot)$ be a real-valued function on X. Then $f(\cdot)$ is a convex (concave) function if, and only if, epi f (hypo f) is a convex set.

Definition B-4 Let f be a real-valued function on $X \subset \mathbb{R}^n$ and $b \geq \inf\{f(\mathbf{x})\colon \mathbf{x} \in X\}$. The *level sets at b* are $\{\mathbf{x}\colon \mathbf{x} \in X, f(\mathbf{x}) \leq b\}$ and $\{\mathbf{x}\colon \mathbf{x} \in X, f(\mathbf{x}) < b\}$.

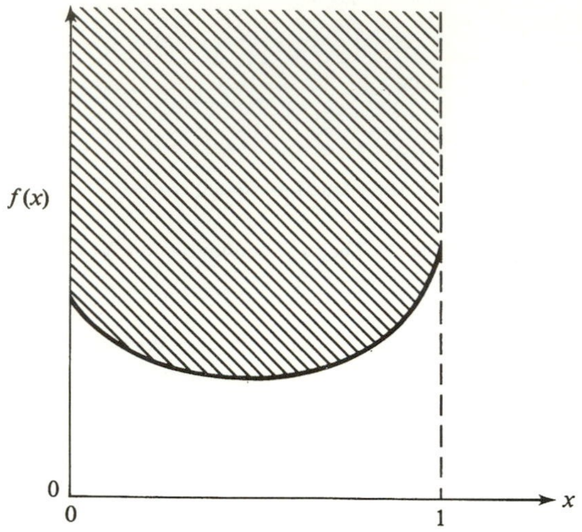

Figure B-4 The epigraph of f is the graph of f and the shaded area above the graph.

You might wonder if Proposition B-8 can be strengthened by replacing "epi f" with "level sets." However, Exercise B-11 asks you to given an example of a nonconvex function which has convex level sets.

Convex functions on $\mathbb{R}^n$ and $\mathbb{R}$ are connected by the next result, which you are asked to prove in Exercise B-13.

Proposition B-9 Let f be a real-valued function on a convex set $X \subset \mathbb{R}^n$. Then f is convex if, and only if, $\phi(\cdot, \mathbf{x}, \mathbf{y})$ is convex on $[0, 1]$ for all $\mathbf{x}$ and $\mathbf{y}$ in X, where

$$\phi(\lambda, \mathbf{x}, \mathbf{y}) = f[\lambda\mathbf{x} + (1 - \lambda)\mathbf{y}]$$

The remainder of this section develops second-order properties of convex functions. The second-order properties turn out to be implied by the first-order properties asserted in Proposition B-6.

Proposition B-10 Let f be a convex function on a convex set X. Then

$$d_{\mathbf{x}}^{+}(m, \mathbf{y} - \mathbf{x}) + d_{\mathbf{y}}^{+}(s, \mathbf{x} - \mathbf{y}) \leq 0 \tag{B-9}$$

for all $\mathbf{x}$ and $\mathbf{y}$ in X and m and s in $(0, 1]$.

PROOF From (B-6) in Proposition B-6,

$$f(\mathbf{y}) - f(\mathbf{x}) - d_{\mathbf{x}}^{+}(m, \mathbf{y} - \mathbf{x}) \geq 0$$
$$f(\mathbf{x}) - f(\mathbf{y}) - d_{\mathbf{y}}^{+}(s, \mathbf{x} - \mathbf{y}) \geq 0$$

whose sum yields the proposition. □

Also from (B-6), with the same proof, either $d_{\mathbf{x}}^{+}(m, \mathbf{y} - \mathbf{x})$ or $d_{\mathbf{y}}^{+}(s, \mathbf{x} - \mathbf{y})$ or both could have been replaced in Proposition B-10 by $d_{\mathbf{x}}^{-}(m, \mathbf{y} - \mathbf{x})$ and $d_{\mathbf{y}}^{-}(s, \mathbf{x} - \mathbf{y})$. We use this fact in the next result.

Proposition B-11 Let f be a convex function on a convex set $X \subset \mathbb{R}^n$. Let $\mathbf{x} \in X$, $h \in (0, 1]$, and $\mathbf{r} \in \mathbb{R}^n$ such that $\mathbf{x} + \mathbf{r}(1 + h) \in X$. Then

$$f(\mathbf{x} + \mathbf{r}h) - f(\mathbf{x}) \leq f(\mathbf{x} + \mathbf{r}h + \mathbf{r}) - f(\mathbf{x} + \mathbf{r}) \qquad \text{(B-10)}$$

PROOF We could have replaced (B-9) by

$$d_{\mathbf{x}}^{+}(m, \mathbf{y} - \mathbf{x}) + d_{\mathbf{y}}^{-}(s, \mathbf{x} - \mathbf{y}) \leq 0$$

For $\mathbf{x} \in X$ and $\mathbf{r} \in \mathbb{R}^n$ let $\mathbf{y} = \mathbf{x} + \mathbf{r}$ and $m = s = h \in (0, 1)$. Then

$$0 \geq d_{\mathbf{x}}^{+}(h, \mathbf{r}) + d_{\mathbf{x}+\mathbf{r}}^{-}(h, -\mathbf{r})$$

which yields (B-10). □

It follows from (B-10) that the directional differences $d_{\mathbf{x}}^{+}$ and $d_{\mathbf{x}}^{-}$ defined in (B-4) have a version of monotonicity in $\mathbf{x}$. Suppose $X \subset \mathbb{R}^2$. Then these differences are nondecreasing as you move "northeasterly." If $n = 1$ and f is differentiable, take $r > 0$ and let $h \downarrow 0$ to obtain $f'(x) \leq f'(x + r)$ so $f''(x) \geq 0$ if f is twice differentiable.

EXERCISES

B-11 Can Proposition B-8 be strengthened? Find a nonconvex function $f(\cdot)$ defined on a convex set $X \subset \mathbb{R}^n$ for which $\{\mathbf{x}: \mathbf{x} \in X, f(\mathbf{x}) \leq r\}$ is a convex set for each $r \in \mathbb{R}$.

B-12 Prove Proposition B-8.

B-13 Prove Proposition B-9.

B-14 Let f be convex on X, $\mathbf{x} \in X$, and let $\mathbf{u}_1, \ldots, \mathbf{u}_k$ be in $V(\mathbf{x})$ and $\lambda_1 \geq 0, \ldots, \lambda_k \geq 0$. Prove that

$$\theta^{+}\left(\mathbf{x}, \sum_{j=1}^{k} \lambda_j \mathbf{u}_j\right) \leq \sum_{j=1}^{k} \lambda_j \theta^{+}(\mathbf{x}, \mathbf{u}_j)$$

Hint: First prove that $\theta^{+}(\mathbf{x}, \cdot)$ is positively homogeneous, that is, $\mathbf{u} \in V(\mathbf{x})$ and $\lambda > 0$ implies $\theta^{+}(\mathbf{x}, \lambda\mathbf{u}) = \lambda\theta^{+}(\mathbf{x}, \mathbf{u})$. Then prove that $\theta^{+}(\mathbf{x}, \cdot)$ is subadditive, that is, $\mathbf{u}_1$ and $\mathbf{u}_2$ in $V(\mathbf{x})$ imply $\theta^{+}(\mathbf{x}, \mathbf{u}_1 + \mathbf{u}_2) \leq \theta^{+}(\mathbf{x}, \mathbf{u}_1) + \theta^{+}(\mathbf{x}, \mathbf{u}_2)$.

B-15 Propositions B-2 through B-11 remain true, of course, for strictly convex functions. In fact, some of the claims can be strengthened, by replacement of weak inequalities with strict inequalities. Review the propositions and, for each, decide whether or not strict convexity leads to a stronger statement. If so, state it and prove it. If not, offer a counterexample. Hint: Proposition B-5 is nontrivial.

B-16 Prove the following assertion or give a counterexample. Let D be a discrete r.v., g a function on $\mathbb{R}^2$, and suppose

$$G(y) = E[g(y, D)]$$

exists for all $y \in \mathbb{R}$. Then G is convex on $\mathbb{R}$ if, and only if, $g(\cdot, \mathrm{b})$ is convex on $\mathbb{R}$ for all b in the sample space of D.

B-17 A real-valued function on a convex set is said to be *quasiconvex* if its level sets (cf. Definition B-4) are convex sets.

(*a*) Prove that a convex function is quasiconvex.

(*b*) Let f be quasiconvex on $\mathbb{R}^n$ and consider $d^+_{\mathbf{x}+g\mathbf{r}}(h, \mathbf{r}) = [f(\mathbf{x} + g\mathbf{r} + h\mathbf{r}) - f(\mathbf{x} + g\mathbf{r})]/h$ for any fixed $\mathbf{x}$ and $\mathbf{r}$ in $\mathbb{R}^n$ and $h > 0$. Prove that there exists m, possibly $-\infty$ or $+\infty$, such that $d^+_{\mathbf{x}+g\mathbf{r}}(h, \mathbf{r})$ is nonpositive for $g \in (-\infty, m]$ and nonnegative for $g \in [m, \infty)$.

B-18 Let f be quasiconvex on $\mathbb{R}$ and b and c numbers with $b \leq c$. Suppose x^* minimizes f on $\mathbb{R}$. Show that

$$\text{Minimum } \{f(x): b \leq x \leq c\} = \begin{cases} f(b) & \text{if } x^* \leq b \\ f(x^*) & \text{if } b \leq x^* \leq c \\ f(c). & \text{if } c \leq x^* \end{cases}$$

B-19 Let $g, g_1, g_2, \ldots$ be convex functions on an open convex subset X of $\mathbb{R}$ such that g_n is nonincreasing (nondecreasing) in n and $g_n(x) \to g(x)$ as $n \to \infty$, $x \in X$. Let g_n^-, g^-, g_n^+, and g^+ denote derivatives from the left and from the right.

(*a*) Prove

$$g^-(x) \leq \liminf_{n\to\infty} g_n^-(x) \leq \limsup_{n\to\infty} g_n^+(x) \leq g^+(x) \qquad x \in X$$

(*b*) Present an example where $g^-(x) < \liminf_{n\to\infty} g_n^-(x)$ and another where $g^+(x) > \limsup_{n\to\infty} g_n^+(x)$.

(*c*) State and prove an extension of (*a*) which concerns directional derivatives of a monotonely convergent sequence of convex functions on a convex subset $X \subset \mathbb{R}^n$.

B-20 Let $C = \{(x, y): x \in \mathbb{R}, x \leq y \leq u(x)\}$ where $x \leq u(x) < \infty$ for each x. Prove that C is a convex set if, and only if, there exists $m \in \mathbb{R}_+$ such that $u(x) = x + m$ for all $x \in \mathbb{R}$.

B-21 A real-valued function $h(\cdot)$ on $X \subset \mathbb{R}^n$ is *lower semicontinuous* on X if $h(\mathbf{x}) \leq \liminf_{\mathbf{y}\to\mathbf{x}} h(\mathbf{y})$ for all $\mathbf{x} \in X$. If $-h$ is lower semicontinuous, h is *upper semicontinuous.* Suppose X is a convex set and h is a convex function on X. For each of (*a*) and (*b*), either prove the assertion or present a counter-example.

(*a*) h is lower semicontinuous on X.

(*b*) h is upper semicontinuous on X.

B-22 Let f be a convex function on an open convex set X. Prove that f is continuous on X. (Hint: Less difficult if you first show that f is bounded above on some neighborhood of any given point.)

B-4 DIFFERENTIABLE CONVEX FUNCTIONS

It is rather surprising that properties (B-6) and (B-9) imply convexity as well as being implied by it if $f(\cdot)$ is differentiable. We introduce some definitions to state these results.

Recall that a real-valued function $g(\cdot)$ on $\mathbb{R}^n$ is said to be $o(\mathbf{h})$ if $g(\mathbf{h})/\|\mathbf{h}\| \to 0$ as $\mathbf{h} \to \mathbf{0} \in \mathbb{R}^n$ [where $\mathbf{h} = (h_i) \in \mathbb{R}^n$ and $\|\mathbf{h}\| = (\sum_{i=1}^n h_i^2)^{1/2}$]. Then $g(\cdot)$ is said to be *differentiable* at $\mathbf{x} \in X$ if there is a vector $\boldsymbol{\nabla} \in \mathbb{R}^n$ such that $g(\mathbf{x} + \mathbf{h}) = g(\mathbf{x}) + \boldsymbol{\nabla} \cdot \mathbf{h} + o(\mathbf{h})$ where $\boldsymbol{\nabla} \cdot \mathbf{h}$ denotes the "inner product" $\sum_{i=1}^n \nabla_i h_i$. The vector $\boldsymbol{\nabla}$ turns out to be unique (for g at $\mathbf{x}$); it is called the *gradient* and written as $\boldsymbol{\nabla}\mathbf{g}(\mathbf{x})$, and its ith coordinate is $\partial g(\mathbf{x})/\partial x_i$. If $g(\cdot)$ is differentiable at every $\mathbf{x} \in X$, then $g(\cdot)$ is said to be differentiable on† X.

The following result sharpens Proposition B-6.

† More generally, without assuming differentiability, the set $\partial f(\mathbf{x})$ of *subgradients* of $f(\cdot)$ at $\mathbf{x}$ is the set of all $\mathbf{y}$ such that $f(\mathbf{z}) \geq f(\mathbf{x}) + (\mathbf{z} - \mathbf{x}) \cdot \mathbf{y}$ for all $\mathbf{z} \in X$.

Proposition B-12 Suppose f is a real-valued function defined and differentiable on an open convex set $X \subset \mathbb{R}^n$. Then f is convex if, and only if,

$$f(\mathbf{y}) \geq f(\mathbf{x}) + \nabla \mathbf{f}(\mathbf{x}) \cdot (\mathbf{y} - \mathbf{x}) \tag{B-11}$$

for all $\mathbf{x} \in X$ and $\mathbf{y} \in X$.

PROOF First suppose f is convex and differentiable. From (B-6) in Proposition B-6 and (B-5) [or (B-7)],

$$f(\mathbf{y}) - f(\mathbf{x}) \geq d_{\mathbf{x}}^{+}(\mathbf{h}, \mathbf{y} - \mathbf{x}) \geq \lim_{h \downarrow 0} d_{\mathbf{x}}^{+}(h, \mathbf{y} - \mathbf{x}) = \theta^{+}(\mathbf{x}, \mathbf{y} - \mathbf{x}) \tag{B-12}$$

From definitions (B-4) of directional differences and of differentiability,

$$\begin{aligned} d_{\mathbf{x}}^{+}(h, \mathbf{y} - \mathbf{x}) &= \frac{f[\mathbf{x} + h(\mathbf{y} - \mathbf{x})] - f(\mathbf{x})}{h} \\ &= \frac{\nabla f(\mathbf{x}) \cdot (\mathbf{y} - \mathbf{x})h}{h} + o[h(\mathbf{y} - \mathbf{x})] \\ &= \frac{\nabla f(\mathbf{x}) \cdot (\mathbf{y} - \mathbf{x})}{h} + o(h) \end{aligned}$$

so the limit in (B-12) has the value needed to establish (B-11).

Suppose (B-11) is valid on X. Choose $\mathbf{x}$ and $\mathbf{y}$ in X and λ in [0, 1] and use (B-11) twice:

$$f(\mathbf{y}) - f[\lambda \mathbf{x} + (1 - \lambda)y] \geq \nabla f[\lambda \mathbf{x} + (1 - \lambda)\mathbf{y}] \cdot (\mathbf{x} - \mathbf{y})(-\lambda)$$

$$f(\mathbf{x}) - f[\lambda \mathbf{x} + (1 - \lambda)\mathbf{y}] \geq \nabla f[\lambda \mathbf{x} + (1 - \lambda)\mathbf{y}] \cdot (\mathbf{x} - \mathbf{y})(1 - \lambda)$$

Multiply the first inequality by $1 - \lambda$, the second by λ, and sum the products to obtain

$$\lambda f(\mathbf{x}) + (1 - \lambda) f(\mathbf{y}) - f[\lambda \mathbf{x} + (1 - \lambda)\mathbf{y}] \geq 0$$

so f is convex on X. □

Why is differentiability needed in Proposition B-12? Suppose X is an open convex set in $\mathbb{R}^n$; then the directional derivative $\theta^{+}(\mathbf{x}, \mathbf{r})$ exists for all $\mathbf{x} \in X$ and $\mathbf{r} \in \mathbb{R}^n$. Is it true that $f(\cdot)$ is convex on X if, and only if,

$$f(\mathbf{y}) \geq f(\mathbf{x}) + \theta^{+}(\mathbf{x}, \mathbf{y} - \mathbf{x})$$

for all $\mathbf{x}$ and $\mathbf{y}$ in X? Our proof of Proposition B-12 depended, below (B-12), on

$$\theta^{+}(\mathbf{x}, \mathbf{y} - \mathbf{x}) = \sum_{i=1}^{n} (y_i - x_i)\theta^{+}(\mathbf{x}, \mathbf{e}_i)$$

where $\mathbf{e}_i$ is the ith unit vector in $\mathbb{R}^n$. This fact appeared in the proof as

$$d_{\mathbf{x}}^{+}(h, \mathbf{y} - \mathbf{x}) = \frac{\nabla f(\mathbf{x}) \cdot (\mathbf{y} - \mathbf{x})}{h} + o(h)$$

But, without differentiability, is it true that

$$\theta^+(\mathbf{x}, \mathbf{r}) = \sum_{i=1}^{n} r_i \theta^+(\mathbf{x}, \mathbf{e}_i)$$

where $\mathbf{r} = r_i = \sum_{i=1}^{n} r_i \mathbf{e}_i$, for all $\mathbf{x} \in X$ and $\mathbf{r} \in V(\mathbf{x})$ [where $V(\mathbf{x})$, defined between (B-7) and (B-8), is the set of feasible directions at $\mathbf{x}$]? The answer, in general, is "no." In general, it is true only that

$$\theta^+\left(\mathbf{x}, \sum_{j=1}^{k} \lambda_j \mathbf{u}_j\right) \le \sum_{j=1}^{k} \lambda_j \theta^+(\mathbf{x}, \mathbf{u}_j)$$

for $\mathbf{u}_j \in V(\mathbf{x}) \subset \mathbb{R}^n$ and $\lambda_j \ge 0, j = 1, \ldots, k$ (Exercise B-14).

The following result sharpens Proposition B-10.

Proposition B-13 Suppose f is a real-valued function defined and differentiable on an open convex set $X \subset \mathbb{R}^n$. Then f is convex if, and only if, for all $\mathbf{x}$ and $\mathbf{y}$ in X,

$$[\nabla f(\mathbf{y}) - \nabla f(\mathbf{x})] \cdot (\mathbf{y} - \mathbf{x}) \ge 0 \tag{B-13}$$

PROOF If f is convex, Proposition B-7 yields

$$f(\mathbf{y}) - f(\mathbf{x}) - \nabla f(\mathbf{x}) \cdot (\mathbf{y} - \mathbf{x}) \ge 0$$

$$f(\mathbf{x}) - f(\mathbf{y}) - \nabla f(\mathbf{y}) \cdot (\mathbf{x} - \mathbf{y}) \ge 0$$

whose sum is (B-13). Now suppose (B-13) is valid. Because f is differentiable on X, from the mean-value theorem of calculus, there is a point $\mathbf{z}$ on the line segment connecting any two points $\mathbf{x}$ and $\mathbf{y}$ in X such that

$$f(\mathbf{y}) = f(\mathbf{x}) + \nabla f(\mathbf{z}) \cdot (\mathbf{y} - \mathbf{x})$$

There exists $\lambda \in [0, 1]$ with $\mathbf{z} = \lambda \mathbf{x} + (1 - \lambda)\mathbf{y}$ and $\mathbf{z} \in X$ because X is a convex set. Therefore, (B-13) yields

$$\{\nabla f[\lambda \mathbf{x} + (1 - \lambda)\mathbf{y}] - \nabla f(\mathbf{x})\} \cdot (\mathbf{y} - \mathbf{x})(1 - \lambda) \ge 0$$

so

$$\begin{aligned} f(\mathbf{y}) - f(\mathbf{x}) &= \nabla f(\mathbf{z}) \cdot (\mathbf{y} - \mathbf{x}) \\ &= \nabla f[\lambda \mathbf{x} + (1 - \lambda]\mathbf{y}) \cdot (\mathbf{y} - \mathbf{x}) \ge \nabla f(\mathbf{x}) \cdot (\mathbf{y} - \mathbf{x}) \end{aligned}$$

Hence, Proposition B-12 implies convexity of f. □

EXERCISE

B-23 Let f be a convex function on a closed bounded convex set $X \subset \mathbb{R}^n$ with nonempty interior. Suppose there is a number $u < \infty$ such that $|\partial f(\mathbf{x})/\partial x_j| \le u, j = 1, \ldots, n$, for all $\mathbf{x} \in X$. Show that there is a convex function g on $\mathbb{R}^n$ such that $g(\mathbf{x}) = f(\mathbf{x})$, $\mathbf{x} \in X$. Such a function is called an *extension* of f from X to $\mathbb{R}^n$. (Hint: Less difficult if you use Proposition B-12.)

B-5 TWICE DIFFERENTIABLE CONVEX FUNCTIONS

In (B-13), suppose $\mathbf{y} = \mathbf{x} + \mathbf{e}$, where $\mathbf{e}$ is the n-vector whose components are all unity. Then (B-13) asserts

$$\sum_{i=1}^{n} \left[\frac{\partial f(\mathbf{x} + \mathbf{e})}{\partial x_i} - \frac{\partial f(\mathbf{x})}{\partial x_i} \right] \geq 0$$

which is *not* the same as each component of $\nabla f(\cdot)$ being a nondecreasing function on $X \subset \mathbb{R}^n$. Indeed, from (B-13) with $\mathbf{y} = h\mathbf{e}_i$ where $h \in \mathbb{R}$ and $\mathbf{e}_i \in \mathbb{R}^n$ is the ith unit vector,

$$\left[\frac{\partial f(\mathbf{x} + h\mathbf{e}_i)}{\partial x_i} - \frac{\partial f(x)}{\partial x_i} \right] h \geq 0$$

so $\partial^2 f(\mathbf{x})/\partial x_i^2 \geq 0$ if f is twice differentiable. Recall that a differentiable function f defined on an open set $X \subset \mathbb{R}^n$ is said to be *twice differentiable* at $\mathbf{x} \in X$ if there is an $n \times n$ matrix $H = (h_{ij})$ of real numbers such that†

$$f(\mathbf{x} + \mathbf{r}) = f(\mathbf{x}) + \nabla f(\mathbf{x}) \cdot \mathbf{r} + \frac{\mathbf{r}'H\mathbf{r}}{2} + o(\| \mathbf{r} \|^2)$$

where $\mathbf{r} \in \mathbb{R}^n$ such that $\mathbf{x} + \mathbf{r} \in X$ and $\mathbf{r}'H\mathbf{r}$ denotes the quadratic form $\sum_{i=1}^n \sum_{j=1}^n r_i r_j h_{ij}$. The matrix H for f at $\mathbf{x}$ turns out to be unique, it is called the *Hessian* matrix, we write it as $H_f(\mathbf{x})$, and its (i, j)th component is $\partial^2 f(\mathbf{x})/\partial x_i \partial x_j$. We say that f is *twice differentiable* on X if f is twice differentiable at every $\mathbf{x} \in X$.

The argument above based on (B-13) shows that the Hessians of convex functions possess nonnegative diagonals. Are the off-diagonal elements constrained too?

Proposition B-14 Let f be a real-valued twice differentiable function defined on an open convex set $X \subset \mathbb{R}^n$. Then f is convex if, and only if,

$$\mathbf{y}'H_f(\mathbf{x})\mathbf{y} \geq 0 \qquad \mathbf{x} \in X, \mathbf{y} \in \mathbb{R}^n \tag{B-14}$$

PROOF First, suppose f is convex and pick $\mathbf{x} \in X$ and $\mathbf{y} \in \mathbb{R}^n$. Then there exists $\lambda^* > 0$ for which $\mathbf{x} + \lambda\mathbf{y} \in X$ (because X is an open set) for all $\lambda \in (0, \lambda^*]$. From Proposition B-12 and twice differentiability,

$$\begin{aligned} 0 &\leq f(\mathbf{x} + \lambda\mathbf{y}) - f(\mathbf{x}) - \nabla f(\mathbf{x}) \cdot \mathbf{y}\lambda \\ &= \mathbf{y}'H_f(\mathbf{x})\mathbf{y}\lambda^2 + o(\lambda^2) \end{aligned}$$

Divide by λ^2 and let $\lambda \to 0$ to obtain (B-14).

Now suppose (B-14) is valid and pick $\mathbf{x}$ and $\mathbf{y}$ in X. The twice differentiability of f and Taylor's theorem imply there is a point $\mathbf{z}$ on the line segment

† We write $\mathbf{r}'$ for the transpose of the vector $\mathbf{r}$.

connecting **x** and **y** such that

$$f(\mathbf{y}) = f(\mathbf{x}) + \nabla f(\mathbf{x}) \cdot (\mathbf{y} - \mathbf{x}) + \frac{(\mathbf{y} - \mathbf{x})H_f(\mathbf{z})(\mathbf{y} - \mathbf{x})}{2}$$

But (B-14) implies nonnegativity of the third term; so Proposition B-12 implies convexity of f. □

See Stoer and Witzgall (1970), pp. 152–153, for a more direct proof of Proposition B-14.

EXERCISES

B-24 Let f be a twice differentiable function defined on an open convex set $X \subset \mathbb{R}^n$. Prove that f is strictly convex on X if $H_f(\mathbf{x})$ is positive definite for all $\mathbf{x} \in X$.

B-25 Let f be a continuously differentiable convex function on $\mathbb{R}^n$. Prove that **x** minimizes f on $\mathbb{R}^n$ if, and only if, $\nabla f(\mathbf{x}) = \mathbf{0}$.

B-6 CONVEX QUADRATIC FORMS

Definition B-5 The symmetric $n \times n$ matrix $Y = (y_{ij})$ is

$$\left.\begin{array}{l} \textit{Positive definite} \\ \textit{Positive semidefinite} \\ \textit{Negative definite} \\ \textit{Negative semidefinite} \end{array}\right\} \quad \text{if } \mathbf{a}'Y\mathbf{a} \left\{\begin{array}{c} > \\ \geq \\ < \\ \leq \end{array}\right\} 0$$

for all $\mathbf{a} \neq \mathbf{0}$.

The symmetry assumption in the definition is without loss of generality for the following reason. Let Y be an $n \times n$ matrix which is not necessarily symmetric. Then $\frac{1}{2}(Y' + Y)$ is a symmetric matrix and

$$\mathbf{a}'[\tfrac{1}{2}(Y' + Y)]\mathbf{a} = \mathbf{a}'Y\mathbf{a}$$

Proposition B-14 asserts that twice differentiable functions are convex if, and only if, their Hessians are positive semidefinite. The conjecture that this statement remains true when "convex" and "semidefinite" are replaced by "strictly convex" and "definite" is false (Exercise B-15)!

Positive semidefinite and positive definite matrices are encountered frequently in the optimization of convex functions. Here are three fundamental properties some of whose parts are proved in many books on matrix algebra.†

† For example, see P. Lancaster, *Theory of Matrices*, Academic Press, New York (1969). For a clear and concise presentation, see pp. 9–11 in G. Zoutendijk, *Methods of Feasible Directions*, Elsevier Publishing Co., Amsterdam (1960).

Proposition B-15 For every matrix Y, the matrix YY' is symmetric and positive semidefinite.

Proposition B-16 If Y is positive definite, then Y is nonsingular and Y^{-1} is positive definite.

Proposition B-17 If Y is an $m \times m$ positive definite matrix and Z is an $m \times q$ matrix with rank q ($q \leq m$), then $Z'YZ$ is positive definite.

Let Y be a symmetric $n \times n$ matrix, $\mathbf{w}$ an $n \times 1$ vector, and

$$f(\mathbf{x}) = \mathbf{x}'Y\mathbf{x} + 2\mathbf{w}'\mathbf{x} \qquad \mathbf{x} \in \mathbb{R}^n$$

If Y is positive (negative) definite, there is a simple formula for the value of $\mathbf{x}$ which minimizes (maximizes) $f(\cdot)$.

Proposition B-18 If Y is positive definite (negative definite), the global minimum (maximum) of f is achieved at

$$\mathbf{x} = -Y^{-1}\mathbf{w} \tag{B-15}$$

PROOF By definition,

$$f(\mathbf{x}) = \sum_{i=1}^{n} \sum_{j=1}^{n} x_i y_{ij} x_j + 2 \sum_{i=1}^{n} w_i x_i$$

so

$$\frac{\partial f(\mathbf{x})}{\partial x_i} = 2x_i y_{ii} + \sum_{j \neq i} x_j(y_{ij} + y_{ji}) + 2w_i$$

$$= 2 \sum_{j=1}^{n} x_j y_{ij} + 2w_i$$

with the second equality due to the symmetry of Y. Therefore,

$$\nabla f(\mathbf{x}) = 2Y\mathbf{x} + 2\mathbf{w} \tag{B-16}$$

By Exercise B-24, f is strictly convex on $\mathbb{R}^n$ and, by Exercise B-25, $\mathbf{x}$ minimizes f only if $\nabla f(\mathbf{x}) = \mathbf{0}$. By Proposition B-16 and (B-16), $\nabla f(\mathbf{x}) = \mathbf{0}$ if, and only if, (B-15) is satisfied. Uniqueness follows from Exercise B-26.

If Y is negative definite, $-Y$ is positive definite and f is maximized if, and only if, $-f$ is minimized. But

$$-f(\mathbf{x}) = \mathbf{x}'(-Y)\mathbf{x} - 2\mathbf{w}'\mathbf{x}$$

which is strictly convex and, by the part already proved, minimized uniquely at

$$\mathbf{x} = -(-Y)^{-1}(-\mathbf{w}) = -Y^{-1}\mathbf{w}$$

which is (B-15). □

EXERCISES

B-26 Let f be a strictly convex function on the convex set X. Prove that there is at most one point $\mathbf{x} \in X$ such that

$$f(\mathbf{x}) = \inf \{f(\mathbf{y}) : \mathbf{y} \in X\}$$

B-27 (This exercise uses concepts and terminology from Chapter 4.) A set $X \subset \mathbb{R}^n$ is a *cone* if $\lambda > 0$ and $\mathbf{x} \in X$ imply $\lambda \mathbf{x} \in X$. A *convex cone* is a cone which is a convex set. If X is a cone and f is a real-valued function on X, then f is *homogeneous of degree m* if $f(\lambda \mathbf{x}) = \lambda^m f(\mathbf{x})$ for all $\lambda > 0$ and $\mathbf{x} \in X$. In the notation of Chapter 4, let

$$f(s) = \sup \{f(s, a) + \beta E(f[M(s, a, \xi_{sa})]) : a \in A_s\} \qquad s \in \mathcal{S} \tag{B-17}$$

be an MDP written as a dynamic program (assume that the expectation is well defined) in which $\xi_{s,a}$ is a real r.v. whose distribution depends only on s and a. Let $C = \{(s, a) : a \in A_s, s \in \mathcal{S}\}$ and let $f(s, \lambda)$ denote $f(s)$ in (B-17) if $\xi_{s,a}$ is replaced by $\lambda \xi_{s,a}$. Prove the following† proposition: If C is a cone, $\mathcal{S} \subset \mathbb{R}$, and $r(\cdot, \cdot)$ and $M(\cdot, \cdot, \cdot)$ are homogeneous of degree 1, then $f(\cdot, \cdot)$ is homogeneous of degree 1, that is, $f(s, \lambda) = \lambda f(s/\lambda, 1)$, $\lambda > 0$. You may be surprised to learn that we have found this arcane result to be useful.

BIBLIOGRAPHIC GUIDE

The general theory of convex sets, convex functions, and their applications has developed throughout this century. The book Rockafellar (1970) is generally accepted as the most comprehensive treatment of convex sets and convex functions at this time. We also like the less detailed expositions in Fenchel (1953), Mangasarian (1969), and Stoer and Witzgall (1970). See Schaible and Ziemba (1981) for some directions of current research.

References

Fenchel, W.: *Convex Cones, Sets, and Functions*, mimeographed lecture notes, Princeton University, Princeton, N.J. (1953).

Mangasarian, O. L.: *Nonlinear Programming*, McGraw-Hill, New York (1969).

Rockafellar, R. T.: *Convex Analysis*, Princeton University Press, Princeton, N.J. (1970).

Schaible, S., and W. T. Ziemba, eds.: *Generalized Concavity in Optimization and Economics*, Academic Press, New York (1981).

Stoer, J., and C. Witzgall: *Convexity and Optimization in Finite Dimensions, I*, Springer-Verlag, Berlin and New York (1970).

† The proposition was first proved in unpublished notes by A. F. Veinott, Jr.

INDEXES

NAME INDEX

SUBJECT INDEX